Extracellular Matrix in Development and Disease

Extracellular Matrix in Development and Disease

Special Issue Editor

Julia Thom Oxford

MDPI • Basel • Beijing • Wuhan • Barcelona • Belgrade

Special Issue Editor
Julia Thom Oxford
Boise State University
USA

Editorial Office
MDPI
St. Alban-Anlage 66
4052 Basel, Switzerland

This is a reprint of articles from the Special Issue published online in the open access journal *International Journal of Molecular Sciences* (ISSN 1422-0067) from 2017 to 2019 (available at: https://www.mdpi.com/journal/ijms/special_issues/extracellular_matrix)

For citation purposes, cite each article independently as indicated on the article page online and as indicated below:

LastName, A.A.; LastName, B.B.; LastName, C.C. Article Title. *Journal Name* **Year**, *Article Number*, Page Range.

ISBN 978-3-03897-570-0 (Pbk)
ISBN 978-3-03897-571-7 (PDF)

Cover image courtesy of Roxanne Stone and Stephanie Frahs; scanning electron micrograph acquired by Makenna Hardy, Boise State University.
The cover image depicts a decellularized cartilage ECM scaffold seeded with chondrocytes.

Contents

About the Special Issue Editor

Julia Thom Oxford, Ph.D., is a Distinguished Professor and the Stueckle Endowed Chair in the Department of Biological Sciences. She received a Doctor of Philosophy degree in Biochemistry and Biophysics in 1986 from Washington State University and completed a postdoctoral fellowship at Shriners Hospital for Children focusing on the role of extracellular matrix in skeletal development. She began working at Boise State University in 2000 and has taught courses on the Extracellular Matrix and Developmental Biology since 2001. Her current research projects focus on rare diseases, molecular mechanisms of cellular mechanoreception in bone and cartilage, the regulation of cell signaling by Col11a1 during craniofacial development, the extracellular matrix as a key factor in cancer progression, the effects of simulated microgravity on articular cartilage, the induction of early stages of osteoarthritis, biomaterials, 3-D cell culture model systems, and alternative splicing of mRNA during embryonic development. She is the Program Director of the NIH Center of Biomedical Research Excellence in Matrix Biology.

International Journal of
Molecular Sciences

Editorial

Extracellular Matrix in Development and Disease

Julia Thom Oxford [1,2,3,4,*], Jonathon C. Reeck [1,2] and Makenna J. Hardy [1,3]

1   Center of Biomedical Research Excellence in Matrix Biology, Boise State University, Boise, ID 83725-1511,
    USA; jonathonreeck@boisestate.edu (J.C.R.); makennahardy@u.boisestate.edu (M.J.H.)
2   Biomolecular Research Center, Boise State University, Boise, ID 83725-1511, USA
3   Biomolecular Sciences Graduate Program, Boise State University, Boise, ID 83725, USA
4   Department of Biological Science, Boise State University, Boise, ID 83725-1515, USA
*   Correspondence: joxford@boisestate.edu; Tel.: +1-208-426-2395

Received: 19 December 2018; Accepted: 4 January 2019; Published: 8 January 2019

Keywords: extracellular matrix; hereditary diseases; reproduction; cancer; muscle; tissue engineering;
integrins

1. Introduction

The evolution of multicellular metazoan organisms was marked by the inclusion of an extracellular
matrix (ECM), a multicomponent, proteinaceous network between cells that contributes to the spatial
arrangement of cells and the resulting tissue organization. The development of an ECM that provides
support in larger organisms may have represented an advantage in the face of selection pressure for
the evolution of the ECM.

The study of the ECM, or material outside the cells of tissues, began with the cell theory of life in
the 1850s [1]. Advances in our understanding of the composition of the ECM occurred prior to 1975
with the application of new techniques in chemical, physical, and biological research [2]. These new
techniques provided information about the molecular structure of collagen and the chemical nature of
the crosslinks that stabilize the collagen higher-ordered structure. Analysis of biosynthetic pathways
has revealed that ECM components are synthesized and secreted by conserved mechanisms throughout
metazoan organisms. Additionally, a comparison between organisms has revealed that ECM proteins
are oligomeric, commonly forming trimers and pentamers of protein monomers that often possess
repeated motifs that predate the evolutionary origin of animals. With the advance of molecular
biology and cell biology fields, new collagens and noncollagenous constituents were discovered and
characterized to give our current view of the ECM as a complex and dynamic biomaterial. Today, we
recognize that the ECM plays an essential role in development and disease, supporting cell survival,
differentiation, and tissue organization.

The ECM has been the focus of research for over a century. Scientists strive to understand the
processes involved in biosynthesis and turnover, and many important discoveries have helped to pave
the way to the application of extracellular matrices in ways that translate to patient care. Despite these
efforts, there are still many unanswered questions that remain. This book deals with molecular and
cellular aspects of the role of ECM in development and disease. Cells exist in three-dimensional
scaffolding that holds together the millions of cells that make up our blood vessels, organs, skin, and
all tissues of the body. The matrix serves as a reservoir of signaling molecules as well. In bacterial
cultures, biofilms form as an ECM and play essential roles in disease and drug resistance. Topics such
as matrix structure and function, cell attachment, and cell surface proteins mediating cell–matrix
interactions, synthesis, regulation, composition, structure, assembly, remodeling, and function of the
matrix are covered. A common thread uniting the topics is the essential nature that the matrix plays
in normal development and pathophysiology. Providing new knowledge will lead us to improved

diagnostics, preventions to disease progression, and therapeutic strategies for repair and regeneration of tissues.

2. Extracellular Matrix and Hereditary Diseases

The link between ECM molecules and phenotypic characteristics of specific hereditary diseases is reviewed by Arseni and colleagues [3], with an emphasis on the most prevalent of all ECM proteins, collagens, and mutations that result in bone fragility, muscle weakness, and skin defects. The authors review the dynamic state of collagen molecules (chains), domains, and molecular interaction, highlighting the dynamic nature of the ECM and the complexity of understanding how mutations contribute to hereditary diseases.

Extending the focus on collagens and point mutations that lead to inherited diseases, primarily affecting the cartilage and skeletal systems such as Stickler syndrome and congenital spondyloepiphyseal dysplasia, Chakkalakal and colleagues [4] reported new findings on mutations of collagen type II that impact protein secretion, the unfolded protein response, and ultimately, cell survival. This research reinforces the findings that collagen mutations near the C-terminus contribute to disease by causing retention in the ER and inducing the ER-stress response.

Another prevalent ECM molecule is hyaluronan. Stridh and colleagues [5] provided novel findings on the role of hyaluronan produced by renomedullary interstitial cells in kidney function. Hyaluronan is vital to the balance of fluid electrolytes. Through this knowledge, they discovered a rapid hyaluronan turnover mechanism alternative to regulating synthesis. They build on previous work showing that hyaluronan levels change corresponding to body hydration levels. This new work adds that hormone-regulated hyaluronidase activity decreases hyaluronan.

3. Extracellular Matrix and Reproduction

Hyaluronan plays an essential role during embryonic development. Eva Nagyova [6] reviews the biological role of hyaluronan-rich oocyte-cumulus ECM in female reproduction. In her review, she focuses on key molecules that play an important role in the formation of the cumulus ECM, generated by the oocyte-cumulus complex. One of these key molecules, inter-alpha-trypsin inhibitor protein forms covalent interactions with hyaluronan to create the major structural component of the cumulus ECM.

During implantation and placentation, plasminogen activator inhibitor type 1 (PAI-1) is responsible for inhibiting ECM degradation, thereby causing inhibition of trophoblast invasion. An increase of PAI-1 in the blood is associated with an increased risk for infertility and pregnancy complications. PAI-1 levels are increased in patients with recurrent pregnancy losses, preeclampsia, intrauterine growth restriction, gestational diabetes mellitus in the previous pregnancy, endometriosis and polycystic ovary syndrome. Ye and colleagues [7] provide an overview of the current knowledge of the role of PAI-1 in reproductive diseases. Based on this review, PAI-1 may represent a promising biomarker for reproductive diseases.

4. Extracellular Matrix and Cancer

Cancer is defined as a disease of uncontrolled cell proliferation and dysregulation of the microenvironment, which included the ECM. Walker et al. [8] reviewed the role of the ECM in cancer development and progression from a biochemical as well as biophysical perspective while Wang et al. [9] reviewed the link between increased risk of thrombosis in cancer patients. Also related to cancer is the recent article by Del Ben et al. [10], which focuses on the role of the ECM molecule vitronectin in nonalcoholic steatohepatitis (NASH), a leading cause of liver cirrhosis and hepatocellular carcinoma. The results of their studies indicate that a fragment of vitronectin may serve as an effective biomarker for noninvasive diagnosis of NASH. Rijal et al. [11] and Song et al. [12] address novel three-dimensional ECM scaffolds with application to cancer research. Rijal et al. [11] reported the generation of breast-specific ECM that can be used to form a hydrogel porous scaffold for studying human breast cancer. Song et al. [12] further reviewed the three-dimensional scaffolds that may accurately reflect the complexity of native tissues and cancer microenvironments for future studies.

5. Extracellular Matrix and Muscle

Mast cells have been implicated in the development of cardiac fibrosis. The controversial role that mast cells may play in promoting a profibrotic environment in cardiac muscle is reviewed by Levick and Widiapradja [13]. The role of the members of the thrombospondin family is reviewed by Chistiakov et al. [14], highlighting the potential for the therapeutic treatment of cardiovascular disease by targeting cardiac thrombospondin-mediated signaling in cardiovascular disease. Rohm and colleagues [15] present their current findings that will impact patients with pulmonary hypertension, right ventricular dysfunction and vascular remodeling. In their studies, they investigated the potential of elevated levels of a novel isoform of tenascin-C as a diagnostic serum biomarker.

McRae et al. [16] provide novel information about the dysregulated ECM in Duchenne muscular dystrophy and the effect of glucocorticoids on the versican-rich transitional matrix in both hindlimb and diaphragm muscle of a mouse model system. Versican may serve as a therapeutic target to promote myoblast fusion during muscle development and regeneration in patients with Duchenne muscular dystrophy.

Using a zebrafish model for muscle development, Dancevic et al. [17] demonstrated that ADAMTS5 plays an essential role in somite differentiation due to a new and novel role associated with the sonic hedgehog (shh) signaling pathway, in addition to its role in ECM remodeling. Their results suggest that a loss of ADAMTS5 causes changes in shh signaling which has been associated with musculoskeletal diseases.

6. Extracellular Matrix and Tissue Engineering Applications

Alvarèz Fallas et al. [18] and Uricuiolo and De Coppi [19] addressed the challenge of muscle regeneration. Alvarèz Fallas et al. [18] addressed the application of decellularized skeletal muscle tissue that preserves the tissue-specific ECM to support tissue engineering and regenerative medicine using both ex vivo and in vivo models. Their work shows that decellularized diaphragm is a suitable scaffold for skeletal muscle tissue engineering and regeneration. Uricuiolo and De Coppi [19] review the literature on the repair of volumetric muscle loss and the promising aspects of applying decellularized tissues as natural scaffolds for therapeutic outcomes.

7. Integrins

Finally, LaFoya and colleagues [20] review the many diverse roles for members of the integrin family. Integrins are traditionally thought of as receptors for the molecules of the ECM, however, in addition to this important role, they also play roles in cell–cell interactions, as growth factors and hormone receptors, and as interaction sites for viruses and bacteria. The many non-ECM integrin ligands are actively being characterized and may find many interesting uses in biotechnology. For example, it has been shown that RGD peptides attached to liposomes or viral particles increase tissue specificity.

8. Concluding Remarks

We thank all the authors who have generously contributed their articles to this book. By disseminating and sharing our research, we ensure that advances in the field can be made. It is by dissemination of our work that the next innovation will arise.

Funding: This project was funded by the National Institutes of Health National Institute for General Medical Sciences IDeA Program grant number P20GM109095.

Conflicts of Interest: The authors declare no conflicts of interest.

Abbreviations

ECM	Extracellular matrix
ER	Endoplasmic reticulum
PAI-1	Plasminogen activator inhibitor type 1
NASH	nonalcoholic steatohepatitis
ADAMTS5	A disintegrin and metalloproteinase with thrombospondin motifs 5
Shh	Sonic hedgehog
RGD	Arginine-glycine-aspartic acid

References

1. Turner, W. The Cell Theory, Past and Present. *J. Anat. Physiol.* **1890**, *24*, 253–287. [PubMed]
2. Piez, K.A. History of extracellular matrix: A personal view. *Matrix Biol.* **1997**, *16*, 85–92. [CrossRef]
3. Arseni, L.; Lombardi, A.; Orioli, D. From Structure to Phenotype: Impact of Collagen Alterations on Human Health. *Int. J. Mol. Sci.* **2018**, *19*, 1407. [CrossRef] [PubMed]
4. Chakkalakal, S.A.; Heilig, J.; Baumann, U.; Paulsson, M.; Zaucke, F. Impact of Arginine to Cysteine Mutations in Collagen II on Protein Secretion and Cell Survival. *Int. J. Mol. Sci.* **2018**, *19*, 541. [CrossRef] [PubMed]
5. Stridh, S.; Palm, F.; Takahashi, T.; Ikegami-Kawai, M.; Friederich-Persson, M.; Hansell, P. Hyaluronan Production by Renomedullary Interstitial Cells: Influence of Endothelin, Angiotensin II and Vasopressin. *Int. J. Mol. Sci.* **2017**, *18*, 2701. [CrossRef] [PubMed]
6. Nagyova, E. The Biological Role of Hyaluronan-Rich Oocyte-Cumulus Extracellular Matrix in Female Reproduction. *Int. J. Mol. Sci.* **2018**, *19*, 283. [CrossRef] [PubMed]
7. Ye, Y.; Vattai, A.; Zhang, X.; Zhu, J.; Thaler, C.J.; Mahner, S.; Jeschke, U.; von Schönfeldt, V. Role of Plasminogen Activator Inhibitor Type 1 in Pathologies of Female Reproductive Diseases. *Int. J. Mol. Sci.* **2017**, *18*, 1651. [CrossRef] [PubMed]
8. Walker, C.; Mojares, E.; Del Río Hernández, A. Role of Extracellular Matrix in Development and Cancer Progression. *Int. J. Mol. Sci.* **2018**, *19*, 3028. [CrossRef] [PubMed]
9. Wang, S.; Li, Z.; Xu, R. Human Cancer and Platelet Interaction, a Potential Therapeutic Target. *Int. J. Mol. Sci.* **2018**, *19*, 1246. [CrossRef] [PubMed]
10. Del Ben, M.; Overi, D.; Polimeni, L.; Carpino, G.; Labbadia, G.; Baratta, F.; Pastori, D.; Noce, V.; Gaudio, E.; Angelico, F.; et al. Overexpression of the Vitronectin V10 Subunit in Patients with Nonalcoholic Steatohepatitis: Implications for Noninvasive Diagnosis of NASH. *Int. J. Mol. Sci.* **2018**, *19*, 603. [CrossRef] [PubMed]
11. Rijal, G.; Wang, J.; Yu, I.; Gang, D.R.; Chen, R.K.; Li, W. Porcine Breast Extracellular Matrix Hydrogel for Spatial Tissue Culture. *Int. J. Mol. Sci.* **2018**, *19*, 2912. [CrossRef] [PubMed]
12. Song, K.; Wang, Z.; Liu, R.; Chen, G.; Liu, L. Microfabrication-Based Three-Dimensional (3-D) Extracellular Matrix Microenvironments for Cancer and Other Diseases. *Int. J. Mol. Sci.* **2018**, *19*, 935. [CrossRef] [PubMed]
13. Levick, S.P.; Widiapradja, A. Mast Cells: Key Contributors to Cardiac Fibrosis. *Int. J. Mol. Sci.* **2018**, *19*, 231. [CrossRef] [PubMed]
14. Chistiakov, D.A.; Melnichenko, A.A.; Myasoedova, V.A.; Grechko, A.V.; Orekhov, A.N. Thrombospondins: A Role in Cardiovascular Disease. *Int. J. Mol. Sci.* **2017**, *18*, 1540. [CrossRef] [PubMed]
15. Rohm, I.; Grün, K.; Müller, L.M.; Kretzschmar, D.; Fritzenwanger, M.; Yilmaz, A.; Lauten, A.; Jung, C.; Schulze, P.C.; Berndt, A.; et al. Increased Serum Levels of Fetal Tenascin-C Variants in Patients with Pulmonary Hypertension: Novel Biomarkers Reflecting Vascular Remodeling and Right Ventricular Dysfunction? *Int. J. Mol. Sci.* **2017**, *18*, 2371. [CrossRef] [PubMed]
16. McRae, N.; Forgan, L.; McNeill, B.; Addinsall, A.; McCulloch, D.; Van der Poel, C.; Stupka, N. Glucocorticoids Improve Myogenic Differentiation In Vitro by Suppressing the Synthesis of Versican, a Transitional Matrix Protein Overexpressed in Dystrophic Skeletal Muscles. *Int. J. Mol. Sci.* **2017**, *18*, 2629. [CrossRef] [PubMed]
17. Dancevic, C.M.; Gibert, Y.; Berger, J.; Smith, A.D.; Liongue, C.; Stupka, N.; Ward, A.C.; McCulloch, D.R. The ADAMTS5 Metzincin Regulates Zebrafish Somite Differentiation. *Int. J. Mol. Sci.* **2018**, *19*, 766. [CrossRef] [PubMed]

18. Alvarèz Fallas, M.E.; Piccoli, M.; Franzin, C.; Sgrò, A.; Dedja, A.; Urbani, L.; Bertin, E.; Trevisan, C.; Gamba, P.; Burns, A.J.; et al. Decellularized Diaphragmatic Muscle Drives a Constructive Angiogenic Response In Vivo. *Int. J. Mol. Sci.* **2018**, *19*, 1319. [CrossRef]
19. Urciuolo, A.; De Coppi, P. Decellularized Tissue for Muscle Regeneration. *Int. J. Mol. Sci.* **2018**, *19*, 2392. [CrossRef] [PubMed]
20. LaFoya, B.; Munroe, J.A.; Miyamoto, A.; Detweiler, M.A.; Crow, J.J.; Gazdik, T.; Albig, A.R. Beyond the Matrix: The Many Non-ECM Ligands for Integrins. *Int. J. Mol. Sci.* **2018**, *19*, 449. [CrossRef] [PubMed]

International Journal of
Molecular Sciences

Review

From Structure to Phenotype: Impact of Collagen Alterations on Human Health

Lavinia Arseni [1,†], Anita Lombardi [2,†] and Donata Orioli [2,*]

[1]   Department of Molecular Genetics, German Cancer Research Center (DKFZ), 69120 Heidelberg, Germany; l.arseni@dkfz-heidelberg.de
[2]   Istituto di Genetica Molecolare, Consiglio Nazionale delle Ricerche, 27100 Pavia, Italy; anita.lombardi@igm.cnr.it
*   Correspondence: orioli@igm.cnr.it; Tel.: +49-0382-546330.
†   These authors contributed equally to this work.

Received: 31 March 2018; Accepted: 4 May 2018; Published: 8 May 2018

Abstract: The extracellular matrix (ECM) is a highly dynamic and heterogeneous structure that plays multiple roles in living organisms. Its integrity and homeostasis are crucial for normal tissue development and organ physiology. Loss or alteration of ECM components turns towards a disease outcome. In this review, we provide a general overview of ECM components with a special focus on collagens, the most abundant and diverse ECM molecules. We discuss the different functions of the ECM including its impact on cell proliferation, migration and differentiation by highlighting the relevance of the bidirectional cross-talk between the matrix and surrounding cells. By systematically reviewing all the hereditary disorders associated to altered collagen structure or resulting in excessive collagen degradation, we point to the functional relevance of the collagen and therefore of the ECM elements for human health. Moreover, the large overlapping spectrum of clinical features of the collagen-related disorders makes in some cases the patient clinical diagnosis very difficult. A better understanding of ECM complexity and molecular mechanisms regulating the expression and functions of the various ECM elements will be fundamental to fully recognize the different clinical entities.

Keywords: collagen; extracellular matrix; skin defects; bone fragility; muscle weakness

---

1. Introduction

The extracellular matrix (ECM) is a non-cellular complex network that provides a structural scaffold to the surrounding cells and, at the same time, a deposit of cytokines and growth factors capable of influencing cell behaviour.

The main ECM components include collagens, proteoglycans and glycoproteins [1]. In addition, many proteins such as growth factors, cytokines and proteolytic enzymes are associated to the ECM. Through all these components the ECM provides environmental information to the cells, which in turn respond by adapting their behaviour and adjusting proliferation, migration and differentiation. The ECM is highly dynamic and its remodelling has to be tightly regulated in order to maintain tissue homeostasis. Deregulation of the ECM structure or composition contributes to the onset of a variety of pathological conditions characterized by a wide range of tissue alterations, further straightening the functional relevance of the ECM. After a general overview of the ECM main features and functions we will focus on the most abundant elements of the ECM, the collagens. We will discuss the different collagen types, their synthesis and structural organization as well as the relevance of properly assembled collagen fibres by discussing their impact on human health.

2. The ECM: Molecular and Structural Diversity

The ECM is a highly dynamic and heterogeneous structure. Each tissue has an ECM with a unique composition generated in early embryonic stages and then maintained and remodelled throughout the entire life. The great complexity of ECM structure and functions makes the research in this field very challenging. A significant input derives from the definition of the "matrisome," a list of proteins contributing to the ECM in different organisms and tissues, which has been predicted by integrating the information derived from experimental knowledge, genome comparative studies and bioinformatics tools [2]. This plastic and evolving list of ECM proteins, comprising between 1% and 1.5% of the mammalian proteome, requires further investigations and biochemical characterization [3].

2.1. Chemical Composition and Mechanical Properties

Hundreds of ECM components are known, many of which are capable of binding to other ECM components on specific sites, thus making the matrix a highly intricate structure (Figure 1).

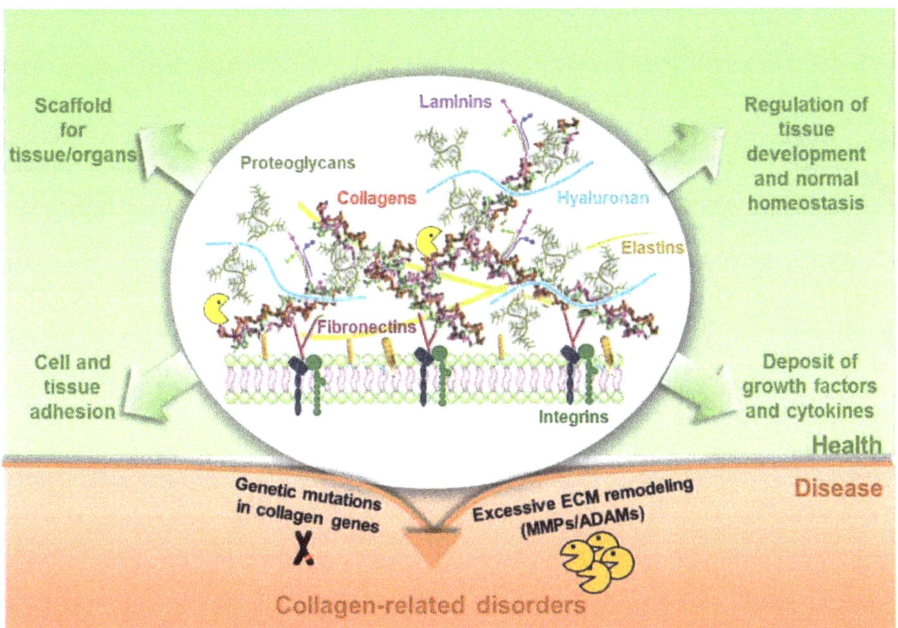

Figure 1. Schematic representation of the complex meshwork of proteins forming the extracellular matrix (ECM). The main ECM components, namely collagens, proteoglycans, hyaluronan, fibronectin, laminin and elastin, as well as the integrin ECM receptors, are depicted. The ECM provides mechanical support and anchoring for cells and tissues but it also acts as a reservoir of growth factors and cytokines and regulator of normal tissue development and homeostasis. Alterations in any of these functions result in a pathological status characterized by various tissue abnormalities.

Although the molecular composition can vary widely, the ECM main components are: proteoglycans, hyaluronan, adhesive glycoproteins (fibronectins and laminins) and fibrous proteins (collagens and elastin). Proteoglycans are constituted by large carbohydrates (generically referred as glycosaminoglycans, GAGs) attached to a protein core. The anionic polysaccharides GAGs allow the sequestration of water and other cations such as calcium [3]. Several types of protein core exist and various types of GAGs can bind to each other in a wide variety of combinations. Proteoglycans promote cell-ECM adhesion but also bind to secreted proteins and growth factors in the ECM. Many components of this category have space-filling and lubrication functions. A special type of GAG is hyaluronan (HA), the only GAG element of the ECM that lacks a protein core. It is a linear polysaccharide composed of the repeating disaccharide units *N*-acetyl-D-glucosamine and D-glucuronate [4]. It is particularly abundant in human tissues, including joints, eyes, umbilical cord, synovial fluids, skeletal tissues, hurt and lung. In spite of its simple structure, HA plays a role in several biological processes, including cell signalling, inflammation, wound healing and cell development [5] and contributes to maintain tissue homeostasis by interacting with other ECM proteins or proteoglycans [6]. Thanks to its hygroscopic features, it acts as space filler among cells and contributes to the maintenance of tissue hydration. Its biocompatibility makes HA suitable for tissue engineering and clinical applications, where it can be used as a diagnostic marker [7].

In mammals, around 200 glycoproteins provide interactions with other ECM components thus allowing ECM formation and assembly. They share multiple repeating domains and motifs typical of the ECM constituents and promote cell adhesion, cell signalling and binding to growth factors [8]. The best-studied ECM glycoproteins are fibronectins and laminins. Fibronectins are proteins encoded by a single gene through multiple alternative splicing. The dimerization of two fibronectin monomers results in the final molecule, which contains repeating units named type I, II and III. Type I and II are tight together by disulphide bonds, the third one has seven-stranded β-barrel composition. All these modules are organized in order to contain binding sites for a variety of other molecules, such as heparin sulphate proteoglycans (HSPGs), integrins and collagens [9]. Fibronectins can be soluble or associated to fibrils thus acting as bridging factors among different ECM components and anchoring cells to matrix fibres. Laminins are large cross-shaped ECM proteins composed of α, β and γ chains. In particular, twelve genes encode 5α, 4β and 3γ chains, which can differentially combine to generate many types of laminins. Among these, laminin G domain-like (LG) of the laminin α2 chain is composed by a β sandwich structure and contains a calcium-binding site, surrounded by a large number of epitopes involved in the interaction with cellular receptors and extracellular ligands ([10,11] and references therein).

Fibrous proteins include collagens and elastin. Collagens are the most abundant proteins of the ECM and a detailed description of their structure and functions is found below (Section 4). Elastin is a key element of the ECM that provides elasticity and flexibility to different tissues including large arteries, ligaments, tendon, lung, skin and cartilage. It is synthesized and secreted as tropoelastin, a soluble precursor implicated in the formation of elastic fibres through the interaction with the N-terminal domains of fibrillins 1 and 2. Tropoelastin contains several hydrophobic domains (consisting in proline, glycine, valine and alanine) that are responsible for the extensibility properties of the protein. The tensile strength provided by collagen fibrils is therefore counterbalanced by the extensibility of elastic fibres, which, conversely to collagen, can undergo progressive stretching and relaxation cycles. Dynamic tissues are thus able to sustain mechanical stress without being permanently affected but reverting the tissues back to their original shape, a property known as viscoelasticity [12]. The ECM tensile strength is determined by the dynamic activities of lysil oxidase (LOX) and lysil hydroxylase, two enzymes involved in regulating the cross-linking between collagens and elastin [13].

2.2. ECM-Bound Growth and Secreted Factors

Although not structural components of the matrix, many growth factors can bind to elements of the ECM and therefore are categorized as ECM constituents. Indeed, it was shown that vascular endothelial growth factor (VEGF) binds to the type III modules of fibronectin and this interaction depends on the heparin-binding residues of fibronectin [14]. VEGF and fibroblast growth factor (FGF) can bind to HSPGs from where they are cleaved off as soluble ligands by the heparanase enzyme. Hepatocyte growth factor (HGF) binds to the 70 kDa N-terminal and the 40 kDa C-terminal fragments of fibronectin [15], whereas platelet-derived growth factor (PDGF) binds the type III and the variable domains of fibronectin [16]. Differently, transforming growth factor-beta (TGFβ) binds to the latent transforming growth factor beta binding protein (LTBPs), which in turn binds to fibrillins and fibronectin-rich matrices [14,15,17,18]. Also, fibrillin-containing microfibrils of the ECM regulate the availability and activity of bone morphogenetic proteins (BMPs) and growth and differentiation factor-5 (GDF-5), cytokines of the TGFβ family [19]. Overall, the different growth factors associated to the ECM could be released locally and become available for the interaction with their canonical receptors. Thus, the ECM serves as a storage for growth factors and chemokines, whose interactions with the matrix control their half-life, local concentration and biological activity.

3. ECM Functions

For many years, the ECM has been defined as a static structure whose unique function was to provide support and shape to cells and tissues. Although this passive role is definitively fundamental for organism organization and maintenance, it is now clear that the ECM is much more than simple scaffolding. Synthesized and organized by the cells, the matrix itself can actively regulate cell behaviour [20]. Indeed, the ECM provides a substrate over which cells can adhere and migrate by sensing the ECM constituents. In turn, cells will secrete new elements that result in ECM remodelling. Therefore, through the coordinated action of its main molecular constituents, the ECM can influence cell proliferation, adhesion and migration as well as differentiation and cell death [21].

3.1. Structural Roles of ECM

The ECM plays important structural roles during development and in particular during the formation of the skeleton. In tissues with mechanical functions such as cartilage, bone and tendons, the ECM is the major component that confers structural properties.

Thanks to its biologically diverse array of macromolecules, the ECM provides a robust and dynamic scaffold capable to evolve during the normal physiological development but also to face insults that could disrupt tissue homeostasis. ECM reactions to these conditions are mediated by its physical, biochemical and biomechanical properties. Depending on its chemical composition, its topography and dimensionality, the ECM exerts different stimuli on the cells, which sense these forces and in turn respond to them. Cell migration, which is critical for proper normal embryonic development, is one of the best examples. Both motile (like the immune cells) and non-motile cells (such as adult epithelial cells) sense the composition and density of the surroundings and respond by migrating towards or moving away from the source.

Tissue elasticity also depends on the ECM chemical composition, which defines soft or stiff matrices. Human tumours for example are surrounded by a stiff matrix with high collagen concentrations and the ECM rigidity might represent an optimal growing milieu for some surrounding cells that are therefore attracted towards this source [22]. This phenomenon, known as desmoplasia, is usually associated with malignant tumours and it is present in many solid tumours. Tissue stiffness can drive malignant transformation via integrin-mediated mechanisms [23] and such fibrotic "stiff" lesions are associated with a poor prognosis [24].

ECM protein receptors, including integrin, syndecan, discoidin domain receptors (DDRs) and proteoglycans, can act synergistically to anchor the cells to the matrix and favour the reciprocal matrix organization. These adhesion dynamics are particularly important to maintain the right balance between self-renewal and differentiation of stem cells [25]. In this respect, it has been shown that the genomic loss of integrin β1 encoding gene in the basal cells of mouse mammary epithelium affects stem cell regeneration and results in irregular branching ducts due to developmental defects of the mammary gland [26].

3.2. Signalling Modulation

As previously mentioned, the ECM can sequester several growth factors that are not structural components of the matrix per se but become active elements of the ECM. Chemokines, cytokines and growth factors, such as VEGFs, Wnts and FGFs, can be retained by the ECM and therefore create a "reservoir" of signalling molecules. By retaining these factors, the ECM may preserve the ligand source in proximity of the receiving cells and prevent their diffusion to the extracellular space. In addition, through the adhesion with ECM components, the ECM can modulate the ligand-receptors interaction and control the formation of morphgen gradients, whose concentration regulates developmental processes [21]. One example of ECM molecule directly implicated in the regulation of morphogens gradients is represented by the HSPGs, which binds morphogens but also many cell surface co-receptors, thus acting as a linking platform that mediates the interactions of morphogens with the other ECM components [27].

The ECM also contributes to ligand maturation. One representative example is the TGFβ proteins, which indirectly connect to fibrillins and fibronectins [18] and are stored in the ECM in their inactive form until proteolitically activated by matrix metalloproteinases (MMP) or by mechanical forces. Furthermore, the ECM can trigger signalling events, as shown by the biologically active fragments derived from the proteolytic cleavage of collagens, proteoglycans, elastin and laminins ([28] and references therein). Matrikines were first described by Maquart and colleagues in 1999 [29] and then named matricryptins one year later by Davis and colleagues with the following definition "biologically active sites that are not exposed in the mature, secreted form of ECM molecules but which become exposed after structural or conformational alterations" [30]. Among them, endostatin derived from collagen XVIII [31,32] is the most extensively studied matricryptin. The ectodomains of membrane collagens XIII, XVII, XXIII and XXV [33] are matricryptins involved in cell adhesion, migration or proliferation [34]. Matrikines from collagens IV are involved in angiogenesis [35] and synapse formation [36]. The ectodomains of syndecans 1–4 are also matricryptins, whereas fragments derived from hyaluronan degradation regulate inflammation and wound healing ([28] and references therein). Therefore, these ECM-fragments may act by regulating cell proliferation, cell death, cell differentiation and angiogenesis [35]. Also, ECM receptors play a role in signal transduction, in particular the collagen receptors DDRs with their intracellular tyrosine kinase activity and the integrin proteins capable of transmitting chemical signals into the cells [37]. Upon ligand binding, receptors get activated and trigger intracellular signalling events that through the involvement of Rho, Rock and the pathway of MAP kinases modulate cellular survival, proliferation and differentiation [21].

3.3. ECM in Development

The ECM provides several different contributions to the developmental events where it plays a dual role, both as functional as well as structural supporting element. A well-known example is provided by the morphogens, soluble factors contributing to define the patterning of surrounding cells during embryonic development. Morphogens are produced in restricted areas of the embryo from which they diffuse and, thanks to the presence of the ECM, generate gradients of signalling molecules that influence cell migration, adhesion and contractility by the activation of intracellular signalling pathways. Meanwhile, the ECM behaves as a structural element by defining the roads for cell migration, delimiting differentiating tissues and maintaining the shape of developing organs. In this context, the ECM assumes architectural roles, such as insulation of tissue to avoid nonspecific adhesion between tissues or, conversely, mediating adhesion between different tissue layers. The opposite functions of insulating or gluing embryonic tissues together show the flexibility of the ECM, which can select one or the other or even synchronize both events according to environmental stimuli. An example of coexistence is observed during muscle differentiation, where a sticky ECM is required to bind the extremities of the cells, whereas a slippery matrix coats the lateral sides [38]. A slippery ECM is defined as an intact basement membrane (composed by laminin, integrin and glycoproteins) that allows tissues to freely slide on each other. GAGs with their negative charges and the resulting chain-chain repulsions, are the main cause of tissue slippery. Conversely, the removal of laminin induces a fragmentation of the basement membrane that, together with the presence of cell adhesion molecules (CAM), results in tissue sticking.

ECM plasticity is also fundamental during branching morphogenesis of several vertebrate organs such as lung, kidney and mammary gland but also in skeletal development. During osteogenesis, skeletal progenitor cells undergo several morphological changes to ultimately give rise to the adult bone [39]. In mature bone, the ECM is the result of active and opposing remodelling events exerted by the osteoclasts and osteoblasts, which degrade and deposit the bone matrix, respectively. An imbalance between degradation and deposition leads to alteration of bone density and disease.

3.4. Cell Migration

The ability of cells to move is central for embryo development as well as maintenance of multicellular organisms. Cell movement strictly depends on the balance between adherence to and release from the ECM in a dynamic fashion. The adhesive properties of the cells are mainly regulated by integrins, which play both structural and signalling roles. Integrins can sense the physical state of the matrix, by interacting with specific ECM molecules, such as collagens and laminins and activate downstream intracellular signalling cascades involving the focal adhesion kinase (FAK) signalling pathway, the mitogen-activated protein (MAP) kinases and the Rho family GTPases [40,41]. Cells tend to migrate along oriented fibrils in a non-random movement. The removal of specific ECM components at a specific time, such as MMP-dependent proteolysis, can instead reorganize the ECM structure and therefore alter or promote the migration process. Thus, the ECM is not only a substrate but it plays dynamic and opposing roles in regulating cell migration. On one side, the basement membrane with its dense fibrillar protein network acts as a barrier to migrating cells, on the other, the ECM promotes cell movement by exposing chemotactic factors that can attract or repulse cells. The ECM remodelling contributes to the formation of organized pathways along which the cells can migrate in an oriented way. Collective cell migration along oriented patterns is an essential aspect of wound healing, a multi-steps process in which the skin repairs itself after injury. During this process, the ECM regulates the interactions between epidermal, dermal and bone marrow cells, it influences cell proliferation and orchestrates the deposition of new connective tissue and the migration of keratinocytes to the wound site.

3.5. ECM Remodelling

To fulfil its activities, the ECM requires a constant and regulated remodelling whose precise orchestration is crucial for tissue homeostasis and developmental processes characterized by transient and dynamic signalling events. ECM remodelling implies changes in ECM composition (novel synthesis or degradation of specific ECM components) or ECM architecture (modification of the macromolecule organization). Several enzymes are involved in ECM remodelling, below is provided a short description of the most known and characterized enzymes.

Matrix metalloproteinases (MMPs) are a large family of enzymes that participate in the degradation of all major ECM components, including those of the basement membrane. They are zinc-dependent endopeptidases initially secreted in the extracellular environment as inactive zymogens with a pro-peptide domain that needs to be removed to allow enzyme activation. The MMPs can be either soluble or membrane-bound and present a substrate-specificity. At least 24 MMPs proteins have been so far identified and, based on their structural organization and substrate specificity, MMPs can be classified into: collagenases, gelatinases, stromelysins, matrilysins and membrane-type I [42]. Under physiological conditions MMP activities are tightly regulated but they may increase during pathological events.

Adamlysins, also called ADAMs (a disintegrin and metalloproteinases) and ADAMTS (ADAMs with a thrombospondin motif), are ECM proteinases involved in cell phenotype regulation, adhesion and migration. The ADAMs include both transmembrane and secreted proteins, whereas the ADAMTSs only contain secreted proteins. 21 ADAMs and 19 ADAMTS are known. They share several structural domains including the metalloproteinase as well as the disintegrin domain, the latest being involved in the binding to integrins. ADAMs are involved in cytokines processing and growth factor receptor shedding [43] while ADAMTS play a role in degradation of ECM components, particularly collagens and proteoglycans [44].

Meprins are membrane-bound or secreted metalloproteinases capable to cleave ECM molecules including the collagen type IV and fibronectins. In addition, they are involved in the synthesis of mature collagen molecules and in the activation of other metalloproteinases including MMPs and ADAMs [45].

The right balance between ECM degradation and deposition has to be guaranteed for the correct tissue integrity. It is therefore evident the relevance of ECM proteinase inhibitors. The tissue inhibitors of metalloproteinases (TIMPs) represent a small family composed of only four members with the function of reversibly inhibiting the activity of MMPs and ADAMs. TIMPs present two distinct domains, one at the N- and one at the C-terminal region of the protein, which are responsible for the binding and inhibition of MMP activity, respectively. Although each TIMP molecule is active against various MMPs, they all show some substrate preferences [46].

Other enzymes may be involved in ECM remodelling. Various proteolytic enzymes, including serine proteinases, cathepsins, heparanases, sulphatases and hyaluronidases, were shown to target ECM proteins. Plasmin and elastase are serine proteinases, the first degrades fibrin, fibronectin and laminin [47], whereas the latter degrades fibronectin and elastin [48]. Cathepsins are lysosomal proteases, which are appointed to the degradation of intracellular or endocytosed proteins. Under specific circumstances cathepsins can be secreted in the extracellular environment where they contribute to the ECM protein degradation. Heparanases and sulphatases cleave heparin sulphate [49] and remove its 6-*O*-sulphate residues, respectively. They affect the ability of heparin sulphate to bind several growth factors such as VEGF, PDGF and FGF, thus altering the downstream signalling events [50]. Hyaluronidases are a family of enzymes capable of degrading HA [51].

4. Collagens

Collagens are the major insoluble fibrous proteins in humans and other vertebrates, accounting for about a quarter of their total protein mass. So far 28 different types of collagens have been identified in vertebrates. They assemble to adopt a triple-helix conformation that gives rise to long thin fibrils or two-dimensional reticulum or even associate with other ECM elements. The different types of collagens and their structure are crucial to provide mechanical stability, elasticity and strength to tissues and organs.

4.1. Collagen Synthesis and Organization

Fibrillar collagens are the most abundant collagens in humans and they are synthetized as long precursors, known as procollagens, which contain a large polypeptide extension at both the N- and C-terminal ends. The C-propeptide has an essential role inside the rough endoplasmic (ER) reticulum where it initiates the assembly of three coiled subunits (α chains) one around the other and along a central axis in order to generate right-handed triple-helix. In vertebrates, over 40 genes encode collagen α chains, which are differentially combined to form 28 different collagen types. Despite the different structural organization, all collagen types share the triple-helix structure. An essential element for the assembly of the three α chains is the proline-rich tripeptide Gly-X-Y repetition, which characterizes all collagens. In the triple helix, glycine residues are localised in the central part, thus allowing a close packing of the molecule [52,53]. Proline and hydroxyproline residues usually occupy the X and Y positions of the tripeptide. Moreover, hydroxylation of prolines and lysines in the middle region of the chains allows the formation of intra-molecular hydrogen bonds that stabilize the entire complex. The extent of lysine hydroxylation varies between tissues and collagen types. Some of the hydroxylysines are further modified by glycosylation with galactose and glucose [54]. Notably, the short N- and C-terminal portion of the chains, which do not assemble in the triple-helix, are required for the extracellular secretion of the polypeptide and the formation of collagen fibrils. The N- and C-propeptides (telopeptides) are subsequently removed by procollagen aminoproteinases and procollagen carboxyproteinases, respectively, giving rise to tropocollagen units [55,56]. Finally, adjacent tropocollagens are bound together through the formation of intermolecular interactions that involve lysine and hydroxylysine residues, thus providing the tensile strength of collagen fibrils. Finally, fibrils assemble into fibres of larger diameter [52]. Details on the synthesis of the other collagen types are included (Section 4.2) and shown in Figure 2.

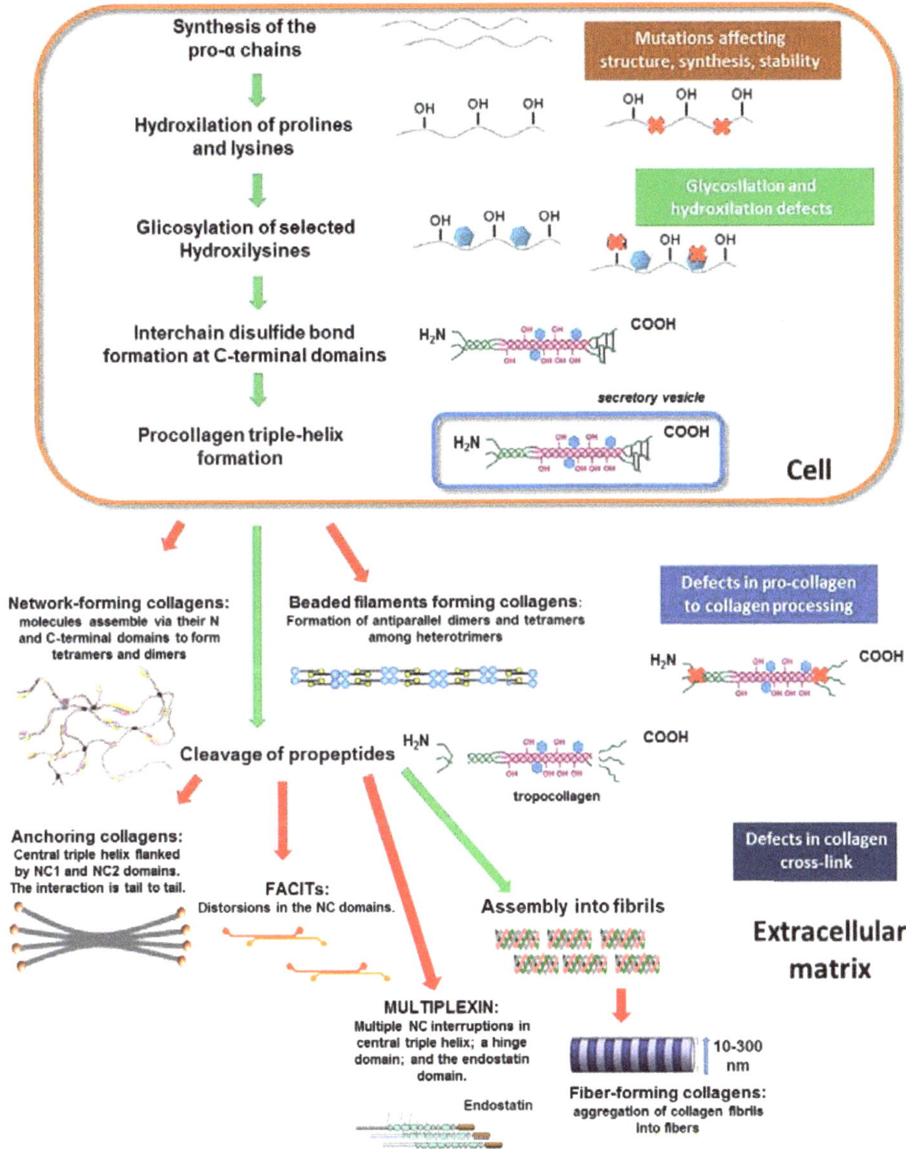

Figure 2. Schematic representation of collagen biosynthesis steps. The main biosynthesis steps of various collagen types are indicated. Green arrows highlight the contiguous processing steps whereas red arrows indicate the final step of collagen assembly into the different structural conformations. Coloured boxes on the right indicate potential alterations occurring during collagen processing at different steps, thus causing abnormalities in the structure and/or assembly. Abbreviations: NC, non-collagenous domain.

4.2. Nomenclature and Classification

Collagens can be grouped based on their structure, function and tissue distribution. They are designated by Roman numerals according to the order of their discovery (I-XXVIII) [53]. They are formed by three identical chains (homotrimers) or by two/three different chains (heterotrimers).

The most abundant collagen of the human body, the interstitial type I collagen, is made by two identical α1 and one α2 chain, which shows high sequence homology with α1 [57]. In most other cases, including the collagen type II, they are homotrimers made by three identical α1 chains. The length of the triple helical region differs among the various collagens. The tripeptide repetition is the predominant motif in fibril collagens whereas it is much shorter and frequently interrupted by non-triple helical domains in other collagen types (such as the non-fibrillar collagens). Non-collagenous (NC) regions may also have structural function as shown by the transmembrane collagens [58].

Following a classification based on collagen function and composition, we can distinguish:

Fiber-forming collagens: they are characterized by a fibrillar shape and a rope structure. Under electron microscopy they show characteristic banding pattern. Fibril collagens assemble to form fibres whose diameter ranges from 12 to >500 nm and the length varies depending on the tissue and developmental stage. They are stabilized by non-reducible covalent crosslinks among specific triple-helix domains and telopeptides [59]. They include the most abundant collagens of the organisms, such as the interstitial collagens (types I, II and III) and the collagen types V and XI, whose main functions consist in providing structural support, balance of pulling forces and enabling cell movement.

FACITs (fibril-associated collagens with interrupted triple helices): they contain short collagenous regions with interruptions in the triple helix intercalated by four NC regions. These molecules are mostly heterotrimers and carry a glycosaminoglycan side chain. They include collagen types IX, XII and XIV, which associate with various collagen fibrils.

Network-forming collagens: they are non-fibrillar collagens that aggregate linearly or laterally to form open networks. They are longer than classical collagens and can give rise to different kinds of networks depending on the collagen type. In particular, collagen type IV, the main component of epithelial basement membranes as well as vascular basal lamina, is irregularly assembled. Differently, collagen types VIII and X form regular hexagonal networks. The function of these network-forming collagens varies and likely depends on their structural organization [53,60].

Transmembrane collagens: they are expressed in many different tissues and cells. This group of collagens plays an important role in epithelial and neural cell adhesion as well as in epithelial–mesenchymal interaction during morphogenesis. They are characterized by the presence of several triple helical regions in the extracellular C-terminal domain interspersed by NC stretches. Next to the extracellular portion of the protein, there is a conserved coiled-coil domain essential for the trimerization of transmembrane collagens. Collagen types XIII and XVII are included in this group [61].

MULTIPLEXINs (multiple triple-helix domains and interruptions): collagen types XV and XVIII consist of several collagen domains with NC interruptions in the triple helixes, which are able to form oligomeric assemblies. They are found in some basement membranes covalently linked to glycosaminoglycan chains. The NC1 domain of collagen types XV and XVIII includes a peptide (endostatin) that following proteolytic cleavage is released in the extracellular environment. Several studies show the anti-angiogenic properties of endostatin as inhibitor of endothelial cell migration and tumour growth [62,63].

Anchoring fibrils: collagen type VII is the major component of the anchoring fibrils, whose function is to secure the adhesion of the epidermal and dermal layers. It consists of a central collagenous triple-helical domain flanked by NC1 and NC2 domains. The NC2 domain is proteolytically cleaved while NC1 is preserved to anchor other ECM elements, including collagens and laminins. The anchoring filaments are assembled in an antiparallel manner, tail to tail with some C-terminal overlap [62,64].

Beaded-filament-forming collagen: collagen type VI is the archetypal beaded filament-forming collagen. It is widely expressed and holds up tissue integrity. Collagen VI monomers are made up of short triple helical domains, which aggregate linearly to form beaded filaments or laterally through their globular domains, thus creating 3D networks. For this reason collagen type VI can also be included among the network-forming collagens [60]. The N and C non-collagenous regions of the monomers are preserved and antiparallel dimers and tetramers are assembled intracellularly [53,62].

Notably, collagen-like triple helical domains are found in several other proteins that do not have structural function and therefore are not considered as real collagens [65].

4.3. Collagen Degradation

Collagens have a great structural stability, resulting in high resistance against degradation by bacterial collagenases and other peptidases. Nevertheless, under physiological conditions most connective tissues undergo to a persistent turnover and continuous remodelling. Collagen degradation is a multi-step process that relies first on the activity of extracellular proteases to break down the ECM collagen fibrils and subsequently on the cellular uptake and intracellular lysosomal degradation of fragmented fibrils [66]. The extracellular fragmentation of collagens is mainly mediated by proteinases such as the MMPs (collagenases and stromelysin), cysteine cathepsins and serine proteinases (plasmin). MMPs can target a wide range of ECM proteins, not only collagens. They act at neutral pH and recognize specific cleavage sites on the target molecules [67]. Cathepsins are lysosomal proteases active at acidic pH, which can be active both intracellularly and upon secretion. Cathepsin S was shown to target and degrade collagens [68]. An indirect collagen digestion can be achieved in response to plasminogen activation to plasmin. The pro-MMP-2 enzyme is activated by plasmin into MMP-2, also known as gelatinase A. Upon activation, MMP-2 can degrade several collagen types, fibronectin, elastin as well as gelatin, the denatured form of collagen [69]. The extracellular collagen fragments are then recruited through phagocytosis from the neighbouring cells, mainly fibroblasts and macrophages, which send them to degradation via the lysosomal pathway [70]. The relationship between the extracellular and intracellular pathways is complex and not fully understood. It has been shown that in macrophages uPARAP/Endo180 acts as collagen internalization receptor after the interaction with pro-uPA (pro-urokinase plasminogen activator) and uPAR (urokinase plasminogen activator receptor [71]) proteins. Moreover, collagen internalization requires the expression of specific integrins and cytokines, including TGFβ and interleukin 1α. Finally, lysosomes fuse together to generate large structures containing collagen and ECM fragments that undergo enzymatic digestion by cysteine cathepsins [72,73].

5. Collagen Alterations in Pathological Events

The tight regulation of ECM synthesis and remodelling is fundamental for human health, as attested by the high number of hereditary disorders caused by mutations in genes encoding structural elements of the ECM or proteases implicated in the remodelling process. Alterations of ECM remodelling can also influence the course and progression of several other pathological conditions, including fibrosis, skin disorders and cancer [74]. The excessive ECM production and the concomitant loss of degradation leading to fibrosis will not be discussed in this review, which is focused on the human genetic disorders associated to an altered collagen structure (Table 1) or resulting in excessive degradation of specific collagen elements (Table 2). A general overview of the clinical features associated to these collagen-related disorders is also provided (Tables 1 and 2).

Since collagens are present throughout the entire body, alterations impairing the quality or quantity of collagen structures can affect any tissue or organ. Since each collagen is generally expressed in several different tissues and it is tightly associated with other ECM elements, alterations result in widely overlapping features that make the diagnosis difficult. Nevertheless, we thereafter propose a tentative classification of the collagen-related disorders according to the major clinical features and affected organs listed in Tables 1 and 2.

Skeletal and cartilage abnormalities: collagen type I is the major ECM component secreted by osteoblasts during bone development. Therefore, alterations in this molecule can give rise to the osteogenesis imperfecta (OI) characterized by bone fragility, as well as the Caffey disease with its infantile episodes of excessive new bone formation (hyperostosis). Also, mutations in *COL2A1* gene result in skeletal abnormalities including the incomplete bone ossification in patients with achondrogenesis type II before birth or the short stature (dwarfism) of patients with Kniest dysplasia. Dwarfism can also be caused by alterations of collagen type IX or X (multiple epiphyseal dysplasia or the Schmid-type metaphyseal chondrodysplasia, respectively) or by mutations in *COL11A1*, *COL11A2* or *COL27A1* genes. Moreover, collagen types II, IX and XI are implicated in the formation and maintenance of the cartilage, thus they are relevant for the joint health and long bone development. Mutations in *COL2A1* gene may result in cartilage alterations characterized by progressive degeneration at the joints (patients with osteoarthritis with mild chondrodysplasia), by hypercellular cartilage with large chondrocytes (Torrance type of platyspondylic lethal skeletal dysplasia) or by a translucent and abnormal gelatinous texture (achondrogenesis type II). Additionally, the presence of fibrous cartilage can be found associated with skeleton defects in patients with fibrochondrogenesis-1 or multiple epiphyseal dysplasia, due to mutations in *COL11A* or *COL9A* gene, respectively.

Skin alterations: some of the collagen-related disorders present severe skin alterations. The dystrophic forms of epidermolysis bullosa (EB) with mutations in *COL7A1* or *COL17A1* gene is one of the major forms of EB where patients present a fragile skin, which can shed at the slightest touch. In milder cases blistering may affect the hands, feet, knees and elbows but in severe cases blistering may lead to vision loss, disfigurement and strictures of the gastrointestinal tract. Skin defects are also the main features of patients with the Ehlers-Danlos syndrome (EDS), a genetically heterogeneous disorder with more than nineteen causative genes. Mutations in *COL5A1* or *COL5A2* are responsible for the classical form of EDS. Patients present a soft, velvety skin that is highly stretchy (skin hyperextensibility) and fragile. Affected individuals tend to bruise easily and in some cases, they show atrophic scars. Skin alterations are also found in patients with Bethlem myopathy-1 caused by mutations in *COL6A* genes. Even though muscle dystrophy is the main clinical feature in this group of patients, they present follicular hyperkeratosis on the arms and legs, soft, velvety skin on the hand palms and feet soles, abnormal wound healing that leads to shallow scars.

Finally, alterations of *MMP1* gene expression have been associated to disorders with skin defects: overexpression of *MMP1* in primary dermal fibroblasts of patients with trichothiodystrophy is responsible for collagen type I degradation and altered wound healing features whereas a functional single nucleotide polymorphism in *MMP1* promoter is associated with increased collagen type VII degradation and high severity of recessive dystrophic EB.

Hearing loss and visual defects: many of the collagen-related disorders are characterized by sensorineural hearing loss. In particular, all the disorders due to alterations of collagen type XI (see Table 1), the Alport syndrome and the X-linked deafness-6 due to alterations of collagen type IV and the OI and EDS disorders by collagen type I defects may reveal sensorineural hearing loss. Notably, some of these disorders are also associated to visual problems that may include myopia, cataract and in some cases (Stickler syndrome) retinal detachment. More severe vision defects can be observed in patients with the Knobloch syndrome-1 due to alterations of the type XVIII collagen. These patients are affected by high myopia, cataract, dislocated lens, vitreoretinal degeneration and retinal detachment. Finally, alterations of type IV collagen can also result in visual defects as observed in patients with retinal arterial tortuosity, Axenfeld-Rieger anomaly and Small vessel disease of the brain.

Muscle weakness: collagen types VI certainly plays a relevant role in skeletal muscle maintenance and regeneration. Alterations of its major elements, the $\alpha1$ and $\alpha2$ chains, result in Bethlem myopathy-1, Ullrich congenital muscular dystrophy-1 or the autosomal recessive myosclerosis, all characterized by progressive muscle weakness (hypotonia) and joint stiffness (contractures) with different degree of severity. In the most severe cases weakness of respiratory muscles are reported. Differently, alterations affecting the $\alpha3$ chain of collagen type VI are found in rare cases with dystonia 27. These patients reveal dystonic action and postural tremor mainly involving the face, neck, bulbar muscles and upper limbs. A severe generalized hypotonia leading to exercise intolerance, feeding difficulties and respiratory insufficiency are present in patients with congenital myasthenic syndrome type 19 due to mutations in *COL13A1* gene. Also, patients with EDS reveal weak muscle tone and hypermobile joints, which can delay the development of motor skills such as sitting, standing and walking. Two cases with OI due to mutations in *SPARC* gene present underdeveloped muscles of the lower extremities, muscle hypotonia and gross motor developmental delay, whereas a single family with congenital fibrosis of extraocular muscles-5 caused by mutations in *COL25A1* gene showed ophthalmoplegia of the extraocular muscles.

Small vessel anomalies and kidney disease: the type IV collagen is the major constituent of the basement membranes. It is a non-fibrillar collagen made of three distinct heterotrimers generated by the products (α chains) of 6 distinct genes. Mutations in *COL4A1* and *COL4A2* result in thickness and damaged vascular basement membranes that affect the straightness of the vessels in patients with susceptibility to intracerebral haemorrhage, hereditary angiopathy or small vessel disease of the brain. Also, the retinal arterial tortuosity derives from mutations in the *COL4A1* gene. Differently, alterations impairing the α3, α4 or α5 chains of collagen type IV result in defects of the glomerular basement membrane that affect kidney functionality. This is observed in patients with Alport syndrome who experience high levels of haematuria and proteinuria due to the progressive loss of kidney activity. Persistent and recurrent haematuria is also observed in the benign form of familial haematuria due to mutations in *COL4A3* or *COL4A4* genes.

Most of the collagen-related disorders affect early childhood health but mild situations can also occur during adulthood. Indeed, the severity of clinical features in patients affected by *Stickler syndrome* varies among individuals and mild cases with late onset have also been reported. Conversely, when collagen alterations result in severe deformations, the survival of the entire organism is compromised as shown by *PPIB* mutations in *type IX osteogenesis imperfecta*. Patients affected by such severe form of OI die during gestation or shortly after birth. It is worthwhile considering that the severity of the disorder relies on several different factors where tissue distribution and function of the affected collagen play leading roles. The OI with its various degree of severity is the result of structural alterations of collagen type I, the most abundant collagen in humans. In particular, mutations inactivating one of *COL1A1* alleles and resulting in reduced levels of an otherwise normal type I collagen are usually responsible for the mild forms of OI whereas dominant negative mutations in *COL1A1* or *COL1A2* genes account for the most severe forms. Notably, among the most common mutations responsible for the severe form of OI are those involving the substitutions of the glycine amino acid in the G-X-Y repeats, essential for the formation of the triple helix. This strongly points to the notion that structural alterations are more detrimental to human health than collagen impoverishment. Similarly, mutations in *COL2A1* gene result in several rare autosomal dominant clinical entities that share skeletal dysplasia, short stature and sensorial defects. The wide range of clinical manifestations were not fully understood but a recent study on over 700 cases (harbouring 415 different mutations) revealed that one-third of the mutations affect the glycine amino acid in the G-X-Y repeats and give rise to the severe *achondrogenesis type II* disease, which is typically identified in utero and may result in embryo death. In contrast, mutations resulting in a premature stop codon or the p.Arg275Cys substitution are responsible for the less severe cases affected by *Stickler syndrome* or *Czech dysplasia* [75]. In summary, the type of mutation, the tissue distribution and the function of the affected collagens all impact on the clinical spectrum of collagen-related disorders.

Table 1. Hereditary disorders resulting from collagen structural alterations [a].

Disorder	Collagen Type	Genetic Alteration [b]	Major Clinical Features [b]
Ehlers-Danlos syndrome (EDS)	**COL1** **COL3** **COL5**	Mutations in more than a dozen genes have been found to cause EDS (#130000). The classical type 1 and 2 (#130010) result from mutations in either the *COL5A1* (*120215) or *COL5A2* (*120190) gene. Other genes involved are *COL1A1* (*120150), *COL1A2* (*120160), *COL3A1* (*120180). Mutation in *COL1A1* or *COL1A2* lead to the deletion of exon 6 of the mRNA coding the α1 (EDS VIIA, #130060) or the α2 chain (EDS VIIB, #617821) of type I collagen, respectively. Inheritance is autosomal dominant.	EDS is the name associated with at least nine phenotypically characterized clinical entities, which result from different types of mutations in distinct collagen genes and several collagen processing genes. These disorders are biochemically and clinically distinct although they all manifest structural weakness in connective tissue as a result of defects in the structure and function of collagens [76]. Although all types of EDS affect joints and skin, additional features vary by type. Severity ranges from mild to severe. Joint hypermobility occurs with most forms of EDS. Infants with hypermobile joints often have weak muscle tone, which can delay the development of motor skills such as sitting, standing and walking. The loose joints are unstable and prone to dislocation and chronic pain. Many EDS patients have soft, velvety skin that is highly stretchy (skin hyperextensibility) and fragile. Affected individuals tend to bruise easily and in some cases, they show atrophic scars. People with the classical form of EDS experience wounds that split open with little bleeding and leave scars that widen over time to create characteristic "cigarette paper" scars.
		Mutations in the *COL3A1* have been identified in the vascular type of EDS (#130050).	The vascular type can involve serious and potentially life-threatening complications due to unpredictable tearing of blood vessels. This rupture can cause internal bleeding, stroke and shock. The EDS vascular type is also associated with an increased risk of organ rupture, including tearing of the intestine or the uterus (womb) during pregnancy.
		Mutations in *procollagen-lysine, 2-oxoglutarate 5-dioxygenase 1 gene* (*PLOD1*, *153454) and *FK506 binding protein 14* (*FKBP14*, *614505) are responsible for the EDS kyphoscoliotic type I and II (EDSKMH, #225400; #614557). Inheritance is autosomal dominant in both cases. PLOD1 catalyses the hydroxylation of lysyl residues in collagen-like peptides, which are critical for the stability of intermolecular crosslinks. FKBP14 acts at the level of the protein folding in the ER, including components of the ECM (COL1, COL3, COL6 and fibronectin).	In addition to the classical symptoms of EDS, patients with EDSKMH I and II are characterised also by progressive kyphoscoliosis with muscle hypotonia from birth, joint laxity, gross motor delay, severe skin hyperelasticity, easy bruising, fragility of sclerae, myopathy and hearing loss [77].
		The Musculocontractural Type I and type II form of EDS (EDSMC1, #601776; EDSMC2, #615539) are caused by recessive loss-of-function mutations in the *carbohydrate sulfotransferase 14* (*CHST14*, *608429) and in the *dermatan sulphate epimerase* (*DSE*, *605942) genes, respectively. The genes encode enzymes involved in the dermatan sulphate (DS) bio-synthesis that is involved in the assembly of collagen fibril. Mutations in both genes lead to the intracellular retention of COL1 and COL3 and a reduced deposition of collagen types I, III, V and VI in the ECM.	EDSMC1 and 2 share most of the clinical features, even though the majority of cases (31) refer to the EDSMC1 and only three cases are reported for the EDSMC2 type [78]. The two syndromes are characterised by progressive kyphoscoliosis, adducted thumbs in infancy or clenched fists and talipes equinovarus, hands with atypically shallow palmar creases and tapering fingers, joint hypermobility, clubfoot, arachnodactyly, elastic skin and poor wound healing. Craniofacial features include brachycephaly, large fontanel, hypertelorism, downslanting palpebral fissures, microcorneae, strabismus, prominent nasolabial folds, short philtrum, thin upper lip, small mouth, high palate and microretrognathia. EDSMC neonates show distal arthrogryposis and muscular hypotonia [78,79].

Table 1. *Cont.*

Disorder	Collagen Type	Genetic Alteration [b]	Major Clinical Features [b]
		The Spondylodysplastic Type 1 (also known as progeroid form of EDS) and 2 forms of EDS (EDSSPD1, #130070; EDSSPD2, #615349) are caused by mutations in the *β-1,4-Galactosyltransferase 7* gene (*B4GALT7*, *604327) and *β-1,3-Galactosyltransferase 6* (*B3GALT6*, *615291), respectively. The genes encode enzymes involved in the production and proper folding of collagen in connective tissue. The Spondylocheiro dysplastic form of EDS (SCD-EDS) results from mutations in the membrane-bound zinc transporter *SLC39A13* (*608735) and has a reliable clinical overlap with EDSSPD1-2. Mutations in *SLC39A13* result in increased Zn^{2+} content inside the endoplasmic reticulum, which inhibits the proper collagen crosslinking and the stability of the collagen triple helix. EDSSPD1 is an autosomal dominant disease whereas EDSSPD2 and SCD-EDS have an autosomal recessive inheritance.	Patients with EDSSPD1-2 showed short stature, muscle hypotonia, radioulnar synostosis and mild to severe intellectual disability (ID). In addition, they present facial dysmorphism, hyperextensible skin, joint hypermobility (JHM), single transverse palmar crease, severe hypermetropia, limb bowing and osteopenia [80].
		The Dermatosparaxis type of EDS (EDSDERMS, #225410) results from mutations in disintegrin and metalloproteinase with thrombospondin motifs (*ADAMTS2*,*604539), the gene encoding the procollagen peptidase that cleavages the N-propeptide of the fibrillar procollagens types I-III and V	The EDSDERMS is characterized by skin that sags and wrinkles. Extra (redundant) folds of skin may be present as affected children get older [81].
Osteogenesis Imperfecta (OI)	COL1	Mutations in the *COL1A1* (*120150) and *COL1A2* (*120160) genes are responsible for more than 90% of all cases of OI (#166200). Most of the mutations causative of OI type I affect *COL1A1* gene and result in reduced levels of COL1, whereas those responsible for most of OI types II (#166210), III (#259420) and IV (#166220) cases occur in *COL1A1* or *COL1A2* genes and impair COL1 structure. The inheritance is autosomal dominant.	At least four biochemically and clinically distinguishable forms of OI have been identified associated to defects in COL1. These are named as OI type I (mild), type II (perinatal lethal), type III (deforming) and type IV (mild deforming). A defect in COL1 structure weakens connective tissues, particularly bones. All four forms of OI present reduced levels of COL1 and brittle bones that break easily. Multiple fractures result in bone deformities. Additional symptoms may include blue sclera, short height, loose joints, hearing loss, breathing and teeth problems, cervical artery dissection and aortic dissection [82,83].
		Mutations in *Cartilage associated protein CRTAP*, *605497), *Prolyl 3-hydroxylase* (*P3H1*, *610339) and *Peptidyl-prolyl isomerase B* (*PPIB*, *123841) genes results in OI type VII (#610682), VIII (#610915) and IX (#259440), respectively. *CRTAP* encodes a cartilage-associated protein whereas *P3H1* an enzyme belonging to the collagen propyl hydroxylase family. *PPIB* encodes for a cyclophilins (Cyps) protein that catalyses the *cis–trans* isomerisation of peptide bonds. All these proteins are required for proper collagen synthesis, assembly and secretion. In these cases, the inheritance is autosomal recessive.	In addition to the four forms of OI previously described, eleven additional phenotypically related disorders in the OI family exist, all associated with bone fragility and low bone mass. Among the OI associated to collagen alterations, the type VII (mutations in *CRTAP*) is sometimes considered a lethal form with multiple fractures, long bone deformities, scoliosis and short stature [84]. The type VIII form of OI (mutations in *P3H1*) includes severe growth defects, skeletal demineralization, scoliosis, round face and proptosis [85]. The type IX (mutations in *PPIB*) is a very severe form of OI. Embryos dye during pregnancy or few months after birth. Radiographs and an autopsy showed the presence of shortened, bowed and fractured long bones without evident rhizomelia [86].

Table 1. *Cont.*

Disorder	Collagen Type	Genetic Alteration [b]	Major Clinical Features [b]
		The *Serpin family H member 1* (*SERPINH1*, *600943) gene encodes a collagen-binding protein that has chaperone activity in the endoplasmic reticulum. Mutations in *SERPINH1* cause the type X OI (#613848), whose inheritance is autosomal recessive.	Types X OI is a severe deforming form of the disorder characterized by aberrant collagen crosslinking, folding and chaperoning [87].
		Absence of *FK506 binding protein 10* (*FKBP10*, *607063) in recessive type XI OI (#610968) leads to reduced collagen cross-linking and deposition. *FKBP10* encodes a chaperone that contributes to type I procollagen folding. Mutations in this gene affect its secretion.	Clinical hallmarks of OI type XI are congenital contractures. All the others clinical data on the 29 patients with OI type XI (mutations in *FKBP10*) are limited and heterogeneous regarding the age of onset, the number of fractures, the type of affected bones and the severity of the disorder [88].
		The recessive form of OI type XIII (OI13, #614856) is caused by mutations in the *BMP1* (*112264) gene, which is involved in the processing of the C-propeptides of procollagens types I-III and the proteolytic activation of the enzyme lysyl oxidase, necessary for the formation of covalent cross-links in collagen and elastic fibres.	The OI type XIII is characterized by normal teeth, faint blue sclerae, severe growth deficiency, borderline osteoporosis and an average of 10–15 fractures a year affecting both the upper and lower limbs and with severe bone deformity.
		The *Secreted protein acidic and cysteine rich* (*SPARC*,*182120) gene encodes a glycoprotein that binds to COL1 and other ECM proteins. Mutations in this gene are responsible for the type XVII OI (#616507) and seem to result in the over-modification of collagen during triple-helical formation. The inheritance is autosomal recessive.	Two clinical cases have been reported: the first is a girl from North Africa with low bone mineral density (BMD), scoliosis, short stature, mild joint hyperlaxity, weak underdeveloped muscles of the lower extremities, bowing of both humeri and speech delay. The second patient is an Indian girl, who had a left hip dislocation at the age of 10 weeks, muscle hypotonia and gross motor developmental delay. Other features are decreased calf muscle mass, joint hyperlaxity and soft skin [89].
Caffey disease, also called infantile cortical hyperostosis	COL1	The *COL1A1* (*120150) variant c.3040C>T (p.Arg836Cys) in exon 41 is the pathogenic variant currently identified in all individuals with Caffey disease (#114000). Inheritance is autosomal dominant but not all people who inherit the mutation develop signs and symptoms. The amino acid change leads to COL1 fibrils that are variable in size and shape.	Caffey disease is characterised by excessive new bone formation (hyperostosis) in early infants. Affected bones may double or triple in width and include jawbone, scapulae, clavicles and the shafts (diaphyses) of long bones in arms and leg. Affected babies are frequently feverish and irritable. They show swelling of joints, pain and redness of affected areas. Usually, there is spontaneous resolution of the inflammatory signs within few months or years. Rare cases of recurrence have also been described [90].
Alpha-2-Deficient Collagen Disease	COL1		In 1974 Meigel and co-authors [91] described a 10-year-old son of consanguineous parents, with an apparently 'new' connective tissue disorder. The clinical and radiologic abnormalities were reminiscent of both Marfan syndrome and osteogenesis imperfecta. Study of cultured fibroblasts showed a complete failure of synthesis of α-2 chains of collagen.
Spondyloepiphyseal dysplasia congenita (SED)	COL2	SED congenita (# 183900) is caused by heterozygous mutation in *COL2A1* gene (*120140) on chromosome 12q13. The inheritance is autosomal dominant.	SED congenita is a chondrodysplasia characterized by short spine, barrel-shaped chest, abnormal epiphyses and flattened vertebral bodies. Skeletal features are manifested at birth and evolve with time. Other features include myopia and/or retinal degeneration with retinal detachment and cleft palate [92].

Table 1. *Cont.*

Disorder	Collagen Type	Genetic Alteration [b]	Major Clinical Features [b]
Stanescu type of spondyloepiphyseal dysplasia (SEDSTN)	COL2	SEDSTN (#616583) is caused by heterozygous mutation in *COL2A1* gene (*120140) on chromosome 12q13. The inheritance is autosomal dominant.	Spondyloepiphyseal dysplasia with accumulation of glycoprotein in chondrocytes has been designated the "Stanescu type". Clinical hallmarks include progressive joint contracture with premature degenerative joint disease, particularly in the knee, hip and finger joints and swollen interphalangeal joints of the hands. The affected individuals are not short, despite the presence of a short trunk. Radiologically, spondylar and epiphyseal abnormalities are quite conspicuous. Other clinical characteristics are generalized platyspondyly, hypoplastic pelvis, epiphyseal flattening with metaphyseal splaying of the long bones and enlarged phalangeal epimetaphyses of the hands [93,94].
Multiple epiphyseal dysplasia with myopia and conductive deafness (EDMMD)	COL2	EDMMD (#132450) is caused by heterozygous mutation in *COL2A1* gene (*120140) on chromosome 12q13. The inheritance is autosomal dominant.	EDMMD is characterized by epiphyseal dysplasia associated with progressive myopia, retinal thinning, crenated cataracts, conductive deafness, joint pain, deformity, waddling gait and short stature. In 1978 Beighton and colleagues [95] described an Afrikaner family in South Africa in which the mother, two sons and one daughter had a syndrome of multiple epiphyseal dysplasia, myopia and conductive deafness. The patients had short stature, brachydactyly, genu valgus deformity and dysplasia of the epiphyses. The epiphyses around the knee joint were flattened, the femoral necks were widened and the vertebral bodies were mildly reduced in height and were concave on their upper and lower surfaces.
Achondrogenesis type II (ACG2)	COL2	ACG2 (#200610) is caused by mutations in *COL2A1* gene (*120140) on chromosome 12q13. The inheritance is autosomal dominant but somatic and germline mosaicism have also been reported [96].	ACG2 is characterized by severe micromelic dwarfism with small chest and prominent abdomen. Other clinical features include incomplete bone ossification and disorganization of the costochondral junction. The cartilage appears as abnormal gelatinous texture and translucent [75].
Czech dysplasia	COL2	Czech dysplasia (#609162) is caused by heterozygous mutations in *COL2A1* gene (*120140) on chromosome 12q13. The inheritance is autosomal dominant.	Czech dysplasia is a skeletal dysplasia characterized by early and progressive onset, shortening of the third and fourth toes caused by metatarsal hypoplasia [97]. Affected individuals have a normal stature but usually complain of severe joint pain before adolescence. Clinical signs are restricted mobility in the lower limb joints and kyphoscoliosis. Skeletal radiographs reveal signs of pseudorheumatoid. Narrow joint spaces and flattened epiphyses platyspondyly with irregular endplates and elongated vertebrae can be observed in the most severe cases. Only five affected families from the Czech Republic have been so far reported [98].
Legg-Calve-Perthes disease (LCPD)	COL2	LCPD (#150600) is caused by heterozygous mutation in the *COL2A1* (*120140) gene on chromosome 12q13. The inheritance is autosomal dominant.	LCPD is a form of avascular necrosis of the femoral head (ANFH; #608805) that affects hip development in growing children. It is due to loss of circulation in the femoral head. Radiology does not permit an early diagnosis that depends on the phase of disease progression through ischemia, revascularization, fracture and collapse, repair and remodelling of the bone. LCPD affects more often boys who are usually shorter than their peers [99].

Table 1. *Cont.*

Disorder	Collagen Type	Genetic Alteration [b]	Major Clinical Features [b]
Osteoarthritis with mild chondrodysplasia (OSCDP)	COL2	OSCDP (#604864) is caused by heterozygous mutation in *COL2A1* gene (*120140) on chromosome 12q13. The inheritance is autosomal dominant.	OSCDP is a common disease that produces joint pain and stiffness together with radiologic evidence of progressive degeneration of joint cartilage. Several cases have been reported, included family members over various generations [95,100,101]. Major features are primary osteoarthritis associated with mild chondrodysplasia. Over the years the range of motion becomes limited. In about 60% of affected persons, abnormalities of the vertebral bodies consistent with mild chondrodysplasia have been described, including platyspondyly, irregular end plates, herniations within the vertebral bodies (Schmorl nodes) and anterior wedging. Other minor changes include iliac exostoses.
Torrance type of platyspondylic lethal skeletal dysplasia (PLSD-T)	COL2	PLSD-T (#151210) can be caused by heterozygous mutation in *COL2A1* gene (*120140) on chromosome 12q13. The disease is transmitted in an autosomal dominant manner. All the patients analysed so far have mutations in the C-propeptide domain of COL2A1, which lead to altered biosynthesis. The phenotype could result from a combination of diminished collagen fibril formation, toxic effects through the accumulation of unfolded collagen chains inside the chondrocytes and/or alteration of a putative signalling function of the C-propeptide.	PLSD-T is a rare skeletal dysplasia characterized by platyspondyly, brachydactyly and metaphyseal changes. Radiology reveals decreased ossification of the skull base, short thin ribs, hypoplastic pelvis with wide sacrosciatic notches and flat acetabular roof, short tubular long bones with ragged metaphyses and bowing of the radius. Histologically, the growth plate appeared relatively normal. The resting cartilage appeared hypercellular with large chondrocytes [102,103]. Though generally lethal in the perinatal period, a few long-term survivors with PLSD-T have been reported [104]. Some patients also present shortening of long bones, degenerative changes in the proximal femora, limited elbow extension, midface hypoplasia, myopia, deafness and mental retardation [105].
Strudwick type of spondyloepimeta-physe dysplasia (SEMD)	COL2	SEMD (#184250) is an autosomal dominant disorder caused by heterozygous mutation in *COL2A1* gene (*120140) on chromosome 12q13.	SEMD clinical features include severe dwarfism, marked pectus carinatum and scoliosis. Cleft palate and retinal detachment are frequently associated. Distinctive radiographic feature is irregular sclerotic changes, described as "dappled" in the metaphyses of the long bones that are caused by alternating zones of osteopenia and osteosclerosis [106].
Spondyloperipheral dysplasia	COL2	Spondyloperipheral dysplasia (#271700) is autosomal dominant disorder caused by heterozygous mutation in *COL2A1* gene (*120140) on chromosome 12q13.	The disorder is a skeletal dysplasia with platyspondyly and brachydactyly E-like changes (short meta-carpals and metatarsals, short distal phalanges in the hand and feet) [107].
Stickler syndrome (STL)	COL2, COL9 COL11	Pathogenic variants in one of six genes (*COL2A1, COL11A1, COL11A2, COL9A1, COL9A2* and *COL9A3*) can be associated with Stickler syndrome. STL is inherited in autosomal dominant manner when mutated in *COL2A1, COL11A1* or *COL11A2*, in autosomal recessive manner when mutated in *COL9A1, COL9A2, or COL9A3*.	STL is a genetically heterogeneous connective tissue disorder characterized by myopia, cataract and retinal detachment, conductive and sensorineural hearing loss. Additional findings may include mid–facial underdevelopment and cleft palate, mild spondyloepiphyseal dysplasia and/or precocious arthritis. Variable phenotypic expression occurs within and among families. Interfamilial variability is partially explained by locus and allelic heterogeneity [108].
Stickler syndrome type I (STL1)		STL1 (#108300), also called the membranous vitreous type, is caused by heterozygous mutation in *COL2A1* gene (*120140) on chromosome 12q13.	STL1 patients usually display a congenital vitreous abnormality consisting of a vestigial gel in the retrolental space, bounded by a highly folded membrane. Most affected individuals are at high risk for retinal detachment. Systemic features typically seen in STL1 are premature osteoarthritis, cleft palate, hearing impairment and craniofacial abnormalities [109].

Table 1. Cont.

Disorder	Collagen Type	Genetic Alteration [b]	Major Clinical Features [b]
Stickler syndrome type II (STL2)		STL2 (#604841), sometimes called the beaded vitreous type, is caused by heterozygous mutation in *COL11A1* gene (*120280) on chromosome 1p21.	Patients affected by STL2 are myopic, rarely with paravascular lattice retinopathy. They frequently present cataract or are aphakic or pseudophakic. Retinal detachment, either mono- or bi-lateral may appear in the 3rd decade. Moreover, *COL11A1* mutations are associated by early-onset hearing loss [110].
Stickler syndrome type III (STL3)		STL3 (#184840) or "nonocular Stickler syndrome" has been recently reclassified as form of otospondylomegaepiphyseal dysplasia or Weissenbacher-Zweymuller syndrome (OSMEDA or WZS). It is caused by heterozygous mutations in *COL11A2* gene (*120290) on 6p21 chromosome.	Patients affected by STL3 have typical facial features, including midface hypoplasia combined with hearing impairment. No ocular abnormalities are reported. They present relatively short extremities with abnormally large knees and elbows but normal total body length. Diagnostic radiologic findings are enlarged epiphyses combined with moderate platyspondyly, mainly in the lower thoracic region [111].
Stickler syndrome type IV (STL4)		STL4 (#614134) is caused by homozygous mutation in *COL9A1* gene (*120210) on chromosome 6q13.	Individuals affected by STL4 have moderate-to-severe sensorineural hearing loss, moderate-to-high myopia with vitreoretinopathy, cataracts and epiphyseal dysplasia [112]. The vitreous abnormality resembles an aged vitreous rather than the typical membranous, beaded or non-fibrillar type.
Stickler syndrome type V (STL5)		STL5 (#614284) is caused by homozygous mutation in *COL9A2* gene (*120260) on chromosome 1p34.	One family with STL5 has been reported. Major clinical findings are high myopia, vitreoretinal degeneration, retinal detachment, hearing loss and short stature. None of the family members was known to have cleft palate and, although there was short stature in childhood, normal height was found in adults [108].
Stickler syndrome atypical		The atypical form of STL (#609508) with predominantly ocular findings is caused by mutation in *COL2A1* gene (*120140). The inheritance is autosomal dominant.	Patients display high myopia and retinal detachment. Systemic features of premature osteoarthritis, cleft palate, hearing impairment and craniofacial abnormalities are very mild or absent [113].
Familial avascular necrosis of the femoral head-1 (ANFH1)	COL2	ANFH1 (#608805) is an autosomal dominant disorder caused by heterozygous mutation in *COL2A1* gene (*120140) on chromosome 12q13.	ANFH1 is a debilitating disease affecting young adults between 35 and 55 years of age. The disorder is characterized by progressive pain in the groin, mechanical failure of the subchondral bone and degeneration of the hip joint. Nearly half of patients require hip replacement before 40 years of age [114].
Kniest dysplasia	COL2	Kniest dysplasia (#156550) is caused by mutations in *COL2A1* gene (*120140). The inheritance is autosomal dominant.	Patients have short stature, flat facial profile, high myopia, risk of retinal detachment, cleft palate, deafness, high risk of severe degenerative joint disease and odontoid hypoplasia leading to risk of atalantoaxial instability and paralysis. Other features include neonatal respiratory distress, infantile hypotonia, abnormal oval-shaped vertebra at birth and later platyspondyly, shortened, "dumbbellshaped" long bones, with splaying of the epiphyses and metaphyses [115].

Table 1. *Cont.*

Disorder	Collagen Type	Genetic Alteration [b]	Major Clinical Features [b]
Alport syndrome	COL4	Alport syndrome is a clinically and genetically heterogeneous nephropathy. Approximately 80% of cases are transmitted as an X-linked semi-dominant condition due to *COL4A5* mutations. 20% of cases are autosomal recessive due to mutation in either *COL4A3* or *COL4A4*. Same families with autosomal-dominant Alport syndrome have been reported, either caused by *COL4A3* or *COL4A4* mutations.	Alport syndrome is characterized by progressive nephritis associated with hearing loss and sometime ocular lesions. Patients experience progressive loss of kidney function. The majority of affected individuals have blood (haematuria) and high levels of proteins (proteinuria) in their urine, which indicate impaired kidney function. Many patients also develop hypertension and at end-stage renal disease. Ocular anomalies are frequent in Alport syndrome and they can precede proteinuria in 40% of patients. Anterior lenticonus, abnormal coloration of the retina, lens rupture, cataracts and corneal erosions can be found [116].
Alport syndrome autosomal dominant		The autosomal dominant form of Alport syndrome (#104200) is caused by heterozygous mutation in *COL4A3* gene (*120070).	Pregnancy of patients with Alport syndrome is very challenging and often complicated by deterioration of renal function, preeclampsia, severe placental dysfunction and sometime acute renal failure. Preterm delivery is frequent [117].
Alport syndrome X-LINKED (ATS)		ATS (#301050) is caused by mutations in *COL4A5* (*303630) gene. The inheritance is dominant.	ATS males are more severely affected than females. Men have a 90% chance of developing end-stage kidney disease by age 40. Patients with large deletions or nonsense mutations have significantly earlier onset than those with missense mutations. The majority (95.5%) of women with *COL4A5* mutations develop microscopic haematuria [118].
Leiomyomatosis, diffuse, with Alport syndrome (DL-ATS)		DL-ATS (#308940) is caused by large deletions involving *COL4A5* (*303630) and *COL4A6* (*303631) genes. Likely an X-linked semi-dominant inheritance.	DL-ATS reveals the Alport syndrome features associated with diffuse leiomyomatosis [119].
Alport syndrome autosomal recessive		This form of Alport syndrome (#203780) is caused by mutations in *COL4A3* (*120070) or *COL4A4* (*120131) gene.	Autosomal recessive Alport syndrome presents as gross proteinuria in childhood and progression to end-stage kidney disease often before the fourth decade [120].
Autosomal dominant mental retardation-34 (MRD34)	COL4	MRD34 (#616351) is caused by heterozygous mutation in *COL4A3BP* (*604677) gene on chromosome 5q13. The inheritance is autosomal dominant.	Patients with MRD34 present unremarkable perinatal history and delivery with a normal birth weight. Neonatal feeding difficulties may occur. Psychomotor development is delayed and speech skills limited. Auto-mutilation behaviour and anxiety are observed. Normal growth parameters and no evident dysmorphism are recorded in adults [121,122].
Retinal arterial tortuosity (RATOR)	COL4	RATOR (#180000) is caused by heterozygous mutation in *COL4A1* gene (*120130) on chromosome 13q34. The inheritance is autosomal dominant. One single family with approximately 20 familial cases has been reported so far.	RATOR is an uncommon condition characterized by marked tortuosity of second- and third-order retinal arteries with normal first-order arteries and venous system. Typically, the vascular tortuosity is predominantly located at the macular and peripapillary area and develops during childhood or early adulthood. Although the disease may be asymptomatic, most patients complain of variable degrees of transient vision loss due to retinal haemorrhage following physical exertion or minor trauma. Involvement of non-ocular vascular beds has not been demonstrated in most cases but occasionally other associated vascular abnormalities have been recorded, including malformations in the Kieselbach nasal septum, spinal cord vascular mass, telangiectasis of bulbar conjunctiva and internal carotid artery aneurysm [123].

Table 1. *Cont.*

Disorder	Collagen Type	Genetic Alteration [b]	Major Clinical Features [b]
Hereditary angiopathy with nephropathy, aneurysms and muscle cramps (HANAC)	COL4	HANAC (#611773) is caused by heterozygous mutation in *COL4A1* gene (*120130) on chromosome 13q34. The inheritance is autosomal dominant.	HANAC syndrome is characterized by angiopathy that affects several parts of the body. Patients present kidney alterations consisting of multiple renal cysts and sometimes haematuria. The brain is only mildly affected and intracranial aneurysms causing haemorrhagic stroke can occur. Leukoencephalopathy is found in about half of affected individuals whereas muscle cramps are experienced by most of patients in early childhood. In addition, patients may manifest eye problems, like arterial retinal tortuosity, cataract and abnormality called Axenfeld-Rieger anomaly [124].
Small vessel disease of the brain with or without ocular anomalies (BSVD)	COL4	BSVD (#607595) is caused by heterozygous mutation in *COL4A1* gene (*120130) on chromosome 13q34. The inheritance is autosomal dominant.	BSVD is characterized by a wide spectrum of symptoms of varying severity including porencephaly variably associated with eye defects (retinal arterial tortuosity, Axenfeld-Rieger anomaly, cataract) and systemic findings such as kidney involvement, muscle cramps, cerebral aneurysms, Raynaud phenomenon, cardiac arrhythmia and haemolytic anaemia. Stroke is often the first symptom and is usually caused by haemorrhagic rather than ischemic stroke. Patients also have leukoencephalopathy and may experience infantile hemiparesis, seizures and migraine headaches accompanied by visual auras [125].
Porencephaly	COL4	Porencephaly is an autosomal dominant disorder characterize by mutations in *COL4A1* (*120130) or *COL4A2* (*120090) genes on chromosome 13q34.	It is a neurological disorder characterized by fluid-filled cysts or cavities in the brain and is thought to result from disturbed vascular supply leading to cerebral degeneration. Affected individuals have delayed growth and development, hypotonia, spastic hemiplegia, seizures, migraine headaches, speech problems and intellectual disability with variable severity [126].
Porencephaly-1 (POREN1)		POREN1 (#175780) is caused by mutations in *COL4A1* gene.	POREN1 is more common. It is usually unilateral and results from destructive lesions.
Porencephaly-2 (POREN2)		POREN2 (#614483) is caused by mutations in *COL4A2* gene.	POREN2 is usually symmetrical and results from developmental malformation.
Schizencephaly	COL4	Some patients with schizencephaly (#269160) have mutations in *COL4A1* (*120130) gene.	Schizencephaly is a very rare cortical malformation that results in grey matter line clefts impacting one or both sides of the brain. Two types of schizencephaly have been described, depending on the size of the area involved and on the separation of the cleft lips. The clinical picture is mainly based on the presence of motor deficits and mental retardation but the severity of the symptoms varies depending on the size and location of the clefts and on the presence of associated cerebral malformations. Patients with type I are almost normal, they may have seizures or motor impairment. Type II is associated with mental retardation, seizures, hypotonia, spasticity, inability to walk or speak and blindness [127].
Susceptibility to intracerebral haemorrhage (ICH)	COL4	ICH (#614519) may be due to mutations in *COL4A2* (*120090) or *COL4A1* (*120130) genes on chromosome 13q34. The inheritance is autosomal dominant.	Few patients with adult-onset haemorrhagic stroke have been reported. The mutated vascular collagen diminishes the tensile strength of vessels and increases their fragility, which can lead to haemorrhage [128].

Table 1. *Cont.*

Disorder	Collagen Type	Genetic Alteration [b]	Major Clinical Features [b]
X-linked deafness-6 (DFNX6)	COL4	DFNX6 (#300914) is caused by mutation in *COL4A6* gene (*303631) on chromosome Xq22. One family has been reported so far.	The symptoms vary in male and female patients affected by this disorder. The severe bilateral sensorineural hearing loss apparent in infancy affects only males, who present bilateral malformation of the cochlea with incomplete separation from the internal auditory canal. Language skills in these patients are severely restricted. Female patients develop mild to moderate hearing impairment in the third/fourth decades of life and rarely hearing loss in the first decade of life [129].
Benign familial haematuria (BFH)	COL4	BFH (#141200) are caused by mutations in *COL4A3* (*120070) or *COL4A4* (*120131) gene, both of which map on chromosome 2q36. The inheritance is autosomal dominant.	BFH is characterized by the presence of persistent or recurrent haematuria, usually detected in childhood. Haematuria remains isolated and never results in end-stage renal disease. Diffuse attenuation of the glomerular basement membrane is usually considered the hallmark of the condition but it is not specific [130].
Bethlem myopathy-1 (BTHLM1)	COL6	BTHLM1 (#158810) is caused by mutations in *COL6A1* (*120220), *COL6A2* (*120240) or *COL6A3* (*120250) genes, giving rise to the altered or even lack of type VI collagen. Both recessive and dominant mutations have been reported.	The disease is characterized by progressive muscle weakness and joint stiffness (contractures). The features can appear at any age, in some cases before birth (decreased foetal movements) in other cases during infancy with joint laxity (loose joints) and hypotonia (weak muscle tone). Later, during childhood, patients develop contractures in their fingers, wrists, elbows and ankles. When adult, they may develop weakness in respiratory muscles, which result in breathing difficulty. The mild form may also reveal skin abnormalities, including follicular hyperkeratosis on the arms and legs; soft, velvety skin on the hand palms and feet soles; abnormal wound healing resulting in shallow scars [131].
Ullrich congenital muscular dystrophy-1 (UCMD1)	COL6	UCMD1 (#254090) is caused by mutations in *COL6A1* (*120220), *COL6A2* (*120240) or *COL6A3* (*120250) genes, giving rise to the altered or even lack of type VI collagen. The disease is transmitted in an autosomal recessive manner and only in rare cases in a dominant pattern.	Patients suffer from a severe muscle weakness beginning soon after birth. Some affected individuals are never able to walk and others can walk only with support. Several lose ambulation ability in adolescence. Progressive scoliosis and deterioration of respiratory function is a typical feature. Some patients need continuous mechanical ventilation to help them breathing. Affected individuals develop contractures in their neck, hips and knees, which further impair movement. There may be joint laxity in patient fingers, wrists, toes, ankles and other joints. As in BTHLM1, some people with UCMD1 have follicular hyperkeratosis [132].
Autosomal recessive myosclerosis	COL6	The autosomal recessive myosclerosis (#255600) has an autosomal recessive inheritance and is caused by mutations in *COL6A2* gene (*120240). One family has been reported so far.	The disorder is characterized by chronic inflammation of skeletal muscle with hyperplasia of the interstitial connective tissue. The clinical symptoms include slender muscles with "woody" consistency and restriction of movement of many joints because of muscle contractures. Muscles are thin and may result sclerotic on palpation. The few patients so far described showed difficulty in running and climbing stairs and had Achilles tendon contractures during early childhood. Skeletal muscle biopsies showed a myopathic pattern with fibrosis, proliferation of endomysial and perimysial connective tissue, variation of myofibre diameter. Increased serum creatine kinase was also found [133].
Dystonia 27 (DYT27)	COL6	DYT27 (#616411) is caused by compound heterozygous mutations in *COL6A3* gene (*120250) on chromosome 2q37. It is an autosomal recessive disorder.	Neurological disorder characterized by the onset of segmental isolated dystonia involving the face, neck, bulbar muscles and upper limbs in the first two decades of life. Few cases have been reported and the symptoms included dystonic action and postural tremor, writer's cramp, oromandibular and laryngeal dystonia [134].

Table 1. *Cont.*

Disorder	Collagen Type	Genetic Alteration [b]	Major Clinical Features [b]
The dystrophic forms of epidermolys s bullosa (DEB)	COL7 COL17	The autosomal dominant form of epidermolysis bullosa dystrophica (DDEB, #131750) is caused by heterozygous mutations in *COL7A1* gene (*120120) on chromosome 3p21. The autosomal recessive dystrophic form of epidermolysis bullosa (RDEB, #226600) and the RDEB localized variant (#226650) are caused by homozygous or compound heterozygous mutations in *COL17A1* gene (*113811).	Epidermolysis bullosa (EB) is a term referring to a family of disorders that are associated with excessive blistering in response to mechanical injury or trauma. Microscopic examination of the skin shows cleavage below the basement membrane within the papillary dermis. The signs and symptoms of this condition vary widely among affected individuals. In mild cases, blistering may primarily affect the hands, feet, knees and elbows. Severe cases involve widespread blistering leading to vision loss, disfigurement and other serious medical problems such as strictures of the gastrointestinal tract leading to poor nutrition. Patients show an increased risk of developing aggressive squamous cell carcinoma. Kids with EB are often defined "butterfly wing" children because of their extremely fragile skin, which can shed at the slightest touch. DEB is one of the major forms of EB. DDEB and RDEB are also known as Cockayne-Touraine disease and Hallopeau-Siemens disease, respectively [135]. Variations in severity are observed among the different forms of RDEB. Notably, a functional SNP in *MMP1* (*120353) promoter is associated with high severity in RDEB. Since COL7 is degraded by MMP1, an imbalance between COL7 synthesis and degradation could worsen the RDEB phenotype [136].
Nonsyndromic congenital nail disorder-8 (NDNC8)	COL7	NDNC8 (#607523) is caused by heterozygous mutations in *COL7A1* gene (*120120) on chromosome 3p21.1. The disorder is inherited in an autosomal dominant manner.	This form of isolated toenail dystrophy has been found in few Japanese families in which other members had the autosomal recessive dystrophic epidermolysis bullosa (RDEB, #226600) or the transient bullous dermolysis of the newborn (#131705), the features of which include dystrophic nails. The nail plates of the toes were buried in the nail bed and the free edge of the toenail was deformed and narrow [137].
Fuchs endothel al corneal dystrophy-1 (FECD1)	COL8	FECD1 (#136800) is caused by heterozygous mutations in *COL8A2* gene (*120252) on chromosome 1p34. It is an autosomal dominant disorder.	FECD is a progressive, bilateral condition leading to reduced vision quality due to dysfunction of the corneal endothelial cells, a thin layer of cells in the back of the cornea that regulates the amount of fluid inside the cornea. FECD occurs when the endothelial cells die and the cornea becomes swollen with too much fluid. Corneal endothelial cells continue to die over time, resulting in further vision problems. Ultrastructural features include loss and attenuation of endothelial cells with thickening and excrescences (guttae) of the underlying basement membrane that are the clinical hallmark of FECD and that worsen with disease progression. As the endothelial layer develops confluent guttae in the central cornea, the cornea becomes dehydrated and clear [138]. In the USA about 5% of the over 40 population is affected by FECD and some early-onset cases are due to *COL8A2* mutations.
Posterior polymorphous corneal dystrophy (PPCD2)	COL8	A single family with PPCD2 (#609140) caused by heterozygous missense mutation in *COL8A2* gene (*120252) has been described. Another family with one PPCD2 patient and few FECD cases, due to heterozygous missense mutation in *COL8A2*, has been described [139].	Father and daughter with PPCD2 have been reported. The patients show a bilateral penetrating keratoplasty at the age of twenties (daughter) and fifties (father). The authors suggested that the underlying pathogenesis of FECD and PPCD2 may be related to disturbance of the role of COL8 in influencing the terminal differentiation of the neural crest-derived corneal endothelial cell [140].

Table 1. *Cont.*

Disorder	Collagen Type	Genetic Alteration [b]	Major Clinical Features [b]
Multiple epiphyseal dysplasia (EDM)	COL9	There are two types of EDM, which can be distinguished by their pattern of inheritance, the dominant and recessive types. EDM caused by mutations affecting collagen structures have an autosomal dominant transmission. Mutations in *COL9A1*, *COL9A2* or *COL9A3* genes are found in less than 5% of individuals with dominant EDM.	EDM is a clinically and genetically heterogeneous skeletal disorder, which is characterized by joint pain and stiffness, mild short stature and degenerative joint features. Both cartilage and bone development are affected, mainly at the ends of the long bones in the arms and legs (epiphyses). It has been suggested that mutations in *COL9A1*, *COL9A2* or *COL9A3* genes may cause COL9 to accumulate inside the cell or interact abnormally with other cartilage components.
Multiple epiphyseal dysplasia-2 (EDM2)		EDM2 (#600204) is caused by heterozygous mutation in *COL9A2* gene (*120260) on chromosome 1p34.	EDM2 onset is usually in childhood, around 3–4 years of age and clinical variability is observed even within the same family [141].
Multiple epiphyseal dysplasia-3 (EDM3)		EDM3 (#600969) is caused by heterozygous mutation in *COL9A3* gene (*120270).	EDM3 patients show early-onset short stature, waddling gait and pain/stiffness in the knees. Few patients experience involvement of elbow, wrist or ankle [142].
Multiple epiphyseal dysplasia-6 (EDM6)		EDM6 (#614135) is caused by heterozygous mutation in *COL9A1* gene (*120210) on chromosome 6p13. One single family has been reported.	A 30-year-old proband was reported with knee pains and difficulty walking since 10 years of age. Radiographs showed early osteoarthritis of one knee, Schmorl nodes, endplate irregularities, anterior osteophytes in the thoracolumbar vertebrae and normal hips. The mother had the same mutation but she did not reveal any symptom before age 45 years [143]
Schmid-type metaphyseal chondrodysplasia (MCDS)	COL10	MCDS (#156500) is caused by heterozygous mutation in *COL10A1* (*120110) gene on chromosome 6q22. MCDS is transmitted as an autosomal dominant trait.	MCDS is a rare genetic disorder characterized by short stature, short arms and legs (short-limbed dwarfism) and bowing of the long bones. Radiographic features include widening and irregularity of the growth plates, especially in the distal and proximal femora. These defects give rise to unusual "waddling" walk (gait) [144].
Marshall syndrome (MRSHS)	COL11	MRSHS (#154780) is an autosomal dominant genetic disorder caused by mutations in *COL11A1* gene (*120280) on chromosome 1p21.	Patients have a distinctive flat midface with a flattened nasal bridge (saddle nose), nostrils that turn upward, widely spaced eyes, high myopia, cataracts and sensorineural hearing loss. Other symptoms include crossed eyes (esotropia), retinal detachment, glaucoma, protruding upper incisors (teeth) and a small or missing nasal bone [145].
Fibrochondrogenesis-1 (FBCG1)	COL11	FBCG1 (#228520) is a severe, autosomal recessive disorder caused by mutations in *COL11A1* gene (*120280) on chromosome 1p21.	FBCG1 and FBCG2 are short-limbed skeletal dysplasia frequently lethal. The disorder is named for the disorganized cartilage growth plate in which chondrocytes have a fibroblastic appearance and the presence of fibrous cartilage extracellular matrix. Patients are characterized by short stature (dwarfism) and skeletal abnormalities. Affected individuals have shortened long bones in the arms and legs that are unusually wide at the ends (described as dumbbell-shaped). Hands and feet are relatively normal. Vertebrae are flattened (platyspondyly) and have a characteristic pinched or pear shape that is noticeable on x-rays. Ribs are typically short and wide and have metaphyseal cupping at both ends. Affected infants have a very narrow chest, which prevents the lungs from developing normally. Most infants are stillborn or die shortly after birth from respiratory failure. Some affected individuals have lived into childhood. Affected individuals who survive the neonatal period have high myopia, mild to moderate hearing loss and severe skeletal dysplasia [146].

Table 1. *Cont.*

Disorder	Collagen Type	Genetic Alteration [b]	Major Clinical Features [b]
Fibrochondrogenesis-2 (FBCG2)		FBCG2 (#614524) can have an autosomal recessive or dominant inheritance due to mutations in *COL11A2* gene (*120290) on chromosome 6p21.3.	
Autosomal dominant deafness-13 (DFNA13)	COL11	DFNA13 (#601868) is an autosomal dominant disorder caused by heterozygous mutation in *COL11A2* gene (*120290) on chromosome 6p21.	A single family has been described, characterized by a dominant nonsyndromic postlingual hearing loss. The affected individuals experienced progressive hearing loss beginning in the second to fourth decades [147].
Otospondylo-megaepip dysplasia, autosomal dominant (OSMEDA)	COL11	The autosomal dominant OSMEDA (#184840), also known as Weissenbacher-Zweymuller syndrome (WZS), is caused by heterozygous mutation in *COL11A2* gene (*120290) on chromosome 6p21. The disorder has an autosomal dominant transmission.	OSMED is characterized by skeletal abnormalities, distinctive facial features and severe hearing loss. The term "otospondylomegaepiphyseal" refers to the parts of the body that are affected: ears (oto-), bones of the spine (spondylo-) and the ends (epiphyses) of long bones in the arms and legs. The disorder is characterized by sensorineural hearing loss, relatively short extremities with abnormally large knees and elbows (enlarged epiphyses), vertebral body anomalies and characteristic facies. The diagnostic radiologic findings are enlarged epiphyses combined with moderate platyspondyly, mainly in the lower thoracic region. No ocular abnormalities are reported. Patients have typical
Autosomal recessive (OSMEDB)		The autosomal recessive OSMEDB (#215150) is also caused by mutation in the *COL11A2* gene.	
Congenital myasthenic syndrome type 19 (CMS19)	COL13	CMS19 (#616720) is an autosomal recessive disorder resulting from mutations in *COL13A1* gene (*120350) on chromosome 10q22.	The congenital myasthenic syndromes (CMSs) are a heterogeneous group of inherited disorders resulting from impaired neuromuscular transmission and caused by mutations in genes involved in the formation or integrity of neuromuscular junctions (NMJs). CMS19 result in generalized muscle weakness, exercise intolerance and respiratory insufficiency. Patients present hypotonia, feeding difficulties and respiratory problems soon after birth. The severity of the weakness and disease course is variable [149].
Epithelial recurrent erosion dystrophy (ERED)	COL17	ERED (#122400) is caused by heterozygous mutation in *COL17A1* gene (*113811) on chromosome 10q24. The disorder is transmitted as an autosomal dominant trait.	ERED is characterized by bilaterally painful recurrent corneal erosions. Erosions often are precipitated by relatively minor trauma and are often difficult to treat, lasting for up to a week. Fortunately, the erosions become less frequent as patients age and may cease altogether by the fifth decade of life. The onset is in the first decade of life (even in the first year of life) often with some subepithelial haze or blebs while denser centrally located opacities develop with time. Small grey anterior stromal flecks associated with larger focal grey-white disc-shaped, circular or wreath-like lesions with central clarity, in the Bowman layer and immediately subjacent anterior stroma, varying from 0.2 to 1.5 mm in diameter, may be diagnostic of ERED [150].
Knobloch syndrome-1 (KNO1)	COL18	KNO1 (#267750) is a hereditary autosomal recessive disorder caused by mutations in *COL18A1* gene (*120328) on chromosome 21q22.3.	KNO1 is primarily characterized by severe vision problems and skull defects. Eye abnormalities include high myopia, cataracts, dislocated lens, vitreoretinal degeneration and retinal detachment. Skull defects range from occipital encephalocele to occult cutis aplasia [151].

Table 1. *Cont.*

Disorder	Collagen Type	Genetic Alteration [b]	Major Clinical Features [b]
Congenital fibrosis of extraocular muscles-5 (CFEOM5)	**COL25**	CFEOM5 (#616219) has an autosomal recessive inheritance and is caused by mutations in *COL25A1* gene (*610004) on chromosome 4q25. A single family had been reported so far.	CFEOM include several different inherited strabismus syndromes characterized by congenital restrictive ophthalmoplegia affecting extraocular muscles innervated by the oculomotor and/or trochlear nerves. CFEOM5 has been reported in a single family with 3 sibs showing a congenital cranial dysinnervation affecting the ocular muscles. The patients had variable abnormal ocular motility without other systemic defects. Two sibs showed congenital ptosis with levator palpebrae muscle dysinnervation of one or both orbits. The levator palpebrae muscle was normally innervated by cranial nerve III (oculomotor nerve). The third sib had no ptosis but showed bilateral Duane retraction syndrome, exotropic in the right eye and esotropic in the left [152].
Steel syndrome (STLS)	**COL27**	STLS (#615155) displays an autosomal recessive inheritance due to mutations in *COL27A1* gene (*608461) on 9q32 chromosome. Few cases have been reported who belong to the same family.	Patients affected by STLS present a characteristic facies, dislocated hips and radial heads, carpal coalition (fusion of carpal bones), short stature, scoliosis and cervical spine anomalies. The dislocated hips are resistant to surgical intervention [153].

[a] Only hereditary disorders resulting in structural collagen alterations are listed; [b] Information mainly based on OMIM, the Online Mendelian Index in Man at http://www.ncbi.nlm.nih.gov/entrez/query.fcgi?db=OMIM. Grey and white rows are used to distinguish the different disorders. Collagen types are in bold. Abbreviations: COL: collagen. The acronyms of the pathologies are all specified inside the Table.

Table 2. Hereditary disorders associated to reduced synthesis or excessive degradation of specific collagen types.

Disorder	Genetic Alteration	Link with the Disorder	Major Clinical Features [a]
Autosomal recessive dystrophic epidermolysis bullosa (RDEB)	A defect in collagenase *MMP1* (*120353) has been implicated in RDEB (#226600). An association between disease severity and specific SNP in *MMP1* gene (*120353) was found in three affected members of one family and in a cohort of 31 unrelated French RDEB patients [136]. The SNP results in increased transcript and active MMP1 protein levels.	COL7 is susceptible to degradation by the collagenase matrix metalloproteinases-1 (MMP1). An imbalance between COL7 synthesis and degradation could conceivably worsen the RDEB phenotype.	Patients with RDEB present generalized blisters at birth that result in extensive scarring and pseudosyndactyly. After birth, extensive blisters may affect the mucous membranes particularly the oral cavity, oesophagus and anal canal. Caused by chronic blood loss, inflammation, infection and poor nutrition, patients develop anaemia, failure to thrive, delayed puberty and osteoporosis. Patients usually do not survive more than 30 years due to severe renal complications or aggressive squamous cell carcinoma arising in the areas of repeated scarring [154].
Aneurysm, abdominal aortic (AAA)	Mapped loci for AAA (#100070) include *AAA1* (*100070) on chromosome 19q13, *AAA2* (*609782) on chromosome 4q31, *AAA3* (*611891) on chromosome 9p21 and AAA4 (*614375) on chromosome 12q13. Inheritance is autosomal dominant.	Several studies pointed to a role of MMPs in the end-stage of AAA. MMPs are enzymes capable of degrading connective tissue that may affect arterial walls by degrading collagens and other ECM components. Polymorphisms in *MMP2*, *MMP3*, *MMP9* and *MMP13* genes result in increased protein levels significantly associated to AAA risk.	AAA is characterized by chronic inflammation and ECM degradation of the aortic wall. The main symptoms of this condition are dysphasia, frontotemporal cerebral atrophy and frontotemporal dementia, speech disorder, memory impairment [155].

Table 2. *Cont.*

Disorder	Genetic Alteration	Link with the Disorder	Major Clinical Features [a]
Trichothiodystrophy I, photosensitive form (TTD1)	TTD (#601675) is caused by homozygous or compound heterozygous mutation in the *ERCC2/XPD* gene (*126340) on chromosome 19q13. The gene encodes a helicase subunit of the transcription/repair factor TFIIH. The inheritance is autosomal recessive.	A reduced expression of *COL6A1* (*120220), an abundant collagen of skin and connective tissue, has been shown in the skin of TTD patients with mutations in the *ERCC2/XPD* gene [156]. It has been shown that specific transcription deregulations in the cells of TTD patients with mutations in the *ERCC2/XPD* gene result in the overexpression of *MMP1* gene. This event leads to hyper-secretion of active MMP1 enzyme and degradation of collagen type I in the dermis of TTD patient skin [157].	TTD is characterized by hair abnormalities, physical and mental retardation, ichthyosis, signs of premature aging and cutaneous photosensitivity. The clinical spectrum of TTD varies widely from patients with only brittle, fragile hair to patients with the most severe neuroectodermal symptoms. TTD patients present sulphur-deficient brittle hair with a diagnostic alternating light and dark banding pattern (called 'tiger' tail banding) under polarizing microscopy. Common additional clinical features include collodion baby, characteristic facies, ocular abnormalities, short stature, decreased fertility and recurrent infections. TTD patients present a 20-fold higher mortality compared to the US general population [158,159].
Atopic dermatitis (ATOD)	ATOD (#603165) is caused by the presence of a specific SNP (rs4668761) in *COL29A1* gene (*611916), which encodes a novel epidermal collagen. The gene is on chromosome 3q22.1. Inheritance is autosomal dominant.	*COL29A1* shows a specific gene expression pattern with the highest transcript levels in skin, lung and gastrointestinal tract, which are the major sites of allergic disease manifestation. Lack of *COL29A1* expression in the outer layers of the epidermis of ATOD patients points to a role of collagen XXIX in epidermal integrity, whose breakdown is a clinical hallmark of AD [160].	ATOD is a chronic inflammatory skin disease characterized by intensely itchy skin lesions. The onset of disease is typically observed during the first two years of life [161]. The hallmarks of atopic dermatitis are a chronic relapsing form of skin inflammation, a disturbance of epidermal barrier function that culminates in dry skin and IgE-mediated sensitization to food and environmental allergens.
Bruck syndrome (BRKS)	BRKS is a very rare autosomal recessive syndrome. Two forms are found: BRKS1 (#259450) is caused by mutations in *FKBP10* (*607067) gene whereas BRKS2 (#609220) by mutations in *PLOD2* (*601865) gene.		BRKS is characterized by bone fragility associated with congenital joint contractures. Patients commonly show short stature, skull wormian bones and kyphoscoliosis. Most cases had normal teeth, white sclera, normal cognitive functions and normal hearing. A few cases had dysmorphic features including triangular face and brachycephaly [162].
Bruck syndrome 1 (BRKS1)	BRKS1 (#259450) is caused by homozygous mutations in *FKBP10* gene (*607063) on chromosome 17q21 resulting in FKBP65 loss of function. Inheritance is autosomal recessive.	Mutations in *FKBP10* result in delay of type 1 procollagen secretion, incomplete stabilization of collagen trimer, reduced hydroxylation of the telopeptide lysyl residue (involved in intermolecular collagen cross-linking).	BRKS1 patients have short stature, high incidence of joint contractures, frequent fractures and scoliosis.
Bruck syndrome 2 (BRKS2)	BRKS2 (#609220) is caused by homozygous mutation in *PLOD2* gene (*601865) on chromosome 3q24. Inheritance is autosomal recessive.	*PLOD2* encodes the telopeptide lysyl hydroxylase required for the triple-helical cross-linking of collagen molecules. Mutations in this gene affect the instalment and secretion of collagen fibres from osteoblasts [163].	No phenotypic differences between BRKS1 and BRKS2 have been reported.
Ehlers-Danlos syndrome (EDS) subtypes	The EDS subtypes are due to mutations in several genes, including *PLOD1, FKBP14, ADAMTS2, ZNF469* and *PRDM5*.		
EDS Kyphoscoliotic Type 1 (EDSKSCL1)	EDSKSCL1 (#225400) previously designated EDS6, is caused by homozygous or compound heterozygous mutation in the *PLOD1* (*153454) gene on chromosome 1p36. Inheritance is autosomal recessive.	*PLOD1* encodes a lysyl hydroxylase that catalyses the hydroxylation of lysine residues in X-lys-gly sequences of collagens and other proteins with collagen-like domains. This hydroxylation is essential for the stability of intermolecular collagen crosslinks.	EDSKSCL1 is characterized by skin fragility (easy bruising, friable skin, poor wound healing, widened atrophic scarring), scleral and ocular fragility/rupture, microcornea, facial dysmorphology. General features also include congenital muscle hypotonia, congenital or early onset kyphoscoliosis, joint hypermobility with subluxations or dislocations of shoulders, hips and knees [164].
EDS Kyphoscoliotic Type, 2 (EDSKSCL2)	EDSKSCL2 (#614557) is caused by homozygous or compound heterozygous mutations in *FKBP14* gene (*614505) on chromosome 7p15. Inheritance is autosomal recessive.	*FKBP14* is an ER-resident protein belonging to the family of FK506-binding peptidyl-prolyl *cis–trans* isomerases (PPIases). It catalyses the folding of COL3 and interacts with COL3, COL4 and COLX [165].	EDSKSCL2 is characterised by congenital hearing impairment (sensorineural, conductive, or mixed), follicular hyperkeratosis, muscle atrophy, bladder diverticula.
EDS dermatosparaxis Type (EDSDERMS)	EDSDERMS (#225410) is caused by mutation in *ADAMTS2* (*604539) gene on chromosome 5q35. Inheritance is autosomal recessive.	*ADAMTS2* encodes a procollagen protease that takes part to the processing of type I procollagen.	Dermatosparaxis means 'tearing of skin.' Patients present extreme skin laxity and fragility, easy bruising, extensive scar formation and joint laxity. Blue sclerae, micrognathia, umbilical hernia and postnatal growth retardation are reported [164].
Brittle Cornea Syndrome1 (BCS1)	BCS1 #229200) can be caused by homozygous mutation in the *ZNF469* gene (*612078) on chromosome 16q24. Inheritance is autosomal recessive.	*ZNF469* encodes a zinc-finger protein that likely acts as a transcription factor or extra-nuclear regulator factor for the synthesis or organization of collagen fibres.	BCS1 and BCS2 are associated with retinal microvascular abnormalities, keratoconus or keratoglobus, blue sclerae, extreme corneal thinning and a high risk of corneal rupture. Hyperelasticity of the skin without excessive fragility and hypermobility of the joints are other hallmarks of the disease [164].

Table 2. *Cont.*

Disorder	Genetic Alteration	Link with the Disorder	Major Clinical Features [a]
Brittle Cornea Syndrome2 (BCS2)	BCS2 (#614170) is caused by mutation in *PRDM5* gene (*614161) on chromosome 4q27. Inheritance is autosomal recessive.	PRDM5 seems to regulate the expression of proteins involved in extracellular matrix development and maintenance, including COL4A1 and COL11A1.	BCS2 features overlap with BCS1. Systemic abnormalities included increased skin laxity, pectus excavatum, scoliosis, congenital hip dislocation, recurrent shoulder dislocation, high-frequency hearing loss, high-arched palate and mitral valve prolapse [166].
CUTIS LAXA	Cutis laxa can be caused by mutations in either *PYCR1* (*179035) or *ALDH18A1* (#614438) gene.		Cutis laxa is a rare skin disorder characterized by wrinkled, redundant, inelastic and sagging skin due to defective synthesis of elastic fibres and other proteins of the ECM [167].
Cutis Laxa, autosomal recessive Type IIB (ARCL2B)	ARCL2B (#612940) is caused by homozygous or compound heterozygous mutation in the *PYCR1* gene (*179035) on chromosome 17q25.3. Inheritance is autosomal recessive.	*PYCR1* encodes the enzyme pyrroline-5-carboxylate reductase1, which catalyses the last step of proline synthesis. PYCR1 deficiency can affect the proper collagen formation.	ARCL2 is a more benign form of cutis laxa present at birth. Growth and developmental delay and skeletal anomalies are reported. Intellectual deficit and seizures have been reported in older patients [167]. Systemic manifestations are mild whereas pulmonary emphysema and cardiac anomalies are rare.
Cutis Laxa, autosomal recessive Type IIIB (ARCL3B)	ARCL3B (#614438) is caused by mutation in *PYCR1* gene (179035) on chromosome 17q25. Inheritance is autosomal recessive.		ARCL3B is a rare autosomal recessive disorder characterized by a progeria-like appearance with distinctive facial features, sparse hair, ophthalmologic abnormalities and intrauterine growth retardation [168].
Cutis Laxa, autosomal recessive, Type IIIA (ARCL3A)	ARCL3A (#219150) is caused by mutation in the *ALDH18A1* gene (*138250) on chromosome 10q24. Inheritance is autosomal recessive.	The protein encoded by *ALDH18A1* catalyses the reduction of glutamate to delta1-pyrroline-5-carboxylate, a critical step in the de novo biosynthesis of proline, ornithine and arginine.	ARCL3A is characterized by cutis laxa (a progeria-like appearance) and ophthalmologic abnormalities [169]. In some case, additional features have been described, including delayed development, intellectual disability, seizures and problems with movement that can worsen over time.
Cutis Laxa, autosomal dominant, Type III (ADCL3)	ADCL3 (#616603) is caused by mutation in *ALDH18A1* gene (*138250) on chromosome 10q24. Inheritance is autosomal dominant.		ADCL3 has a progeroid appearance characterized by thin skin with visible veins and wrinkles, ophthalmological abnormalities, clenched fingers, pre- and postnatal growth retardation and moderate intellectual disability. Patients also exhibit a combination of muscular hypotonia with brisk muscle reflexes [170].
Keratoconus-1 (KTCN1)	KTCN1 (#148300) is caused by heterozygous mutation in the *Visual system homeobox gene 1* (*VSX1*, *605020) gene on 20p11 chromosome. Inheritance is autosomal dominant.	*VSX1* encodes a homeoprotein that regulates the expression of the cone opsin genes early in development. Recent studies showed that the structural deformity of the cornea in KCTN patients may be due to reduced expression of collagens (*COL1A1* and *COL4A1*) and LOX family oxidases, as well as on the concomitant increased expression of *MMP9* [171].	KTCN1 is the most common corneal dystrophy. It is a bilateral, often asymmetrical, non-inflammatory progressive corneal ectasia that causes visual morbidity. In affected individuals, the cornea becomes progressively thin and conical in shape, resulting in myopia, irregular astigmatism and corneal scarring. It typically appears in the teenage years and then it progresses until the third and fourth decades. No specific treatment exists except corneal transplantation when visual acuity can no longer be corrected by contact lenses [172].

[a] Information mainly based on OMIM, the Online Mendelian Index in Man at http://www.ncbi.nlm.nih.gov/entrez/query.fcgi?db=OMIM. Grey and white rows are used to distinguish the different disorders. Causative genes are in bold. Abbreviations: the acronyms of the pathologies are all specified inside the Table.

6. Conclusions

The large number of genetic disorders associated to collagen alterations clearly strengthens the relevance of this wide group of proteins, which have been for long time considered inert elements with no other function than maintenance of tissue shape and architecture. The observation that mutations in collagen coding genes result in alterations of relevant developmental processes (skeletal and cartilage development) or defects of tissue homeostasis (skin, sensorineural, visual and muscle alterations) clearly demonstrate a regulatory role for this type of molecules. Therefore, not only the collagens but also the ECM with its broad number of elements and its wide complexity plays a role of primary importance among the mechanisms implicated in embryonic development, normal organ physiology and human health. Moreover, the wide and largely overlapping spectrum of clinical features of collagen, or even ECM-related disorders, makes in some instances the clinical diagnosis and patient management very difficult. A better understanding of the signalling events regulating the expression, functions and dynamic interplay of the various ECM elements will improve our knowledge on the pathogenesis of ECM-related disorders and, in parallel, will provide the tools for the identification of potential therapeutic targets.

Author Contributions: All authors equally contributed to the writing of the manuscript.

Acknowledgments: We are grateful to Associazione Italiana per la Ricerca sul Cancro (AIRC, IG 17710 to DO) for supporting our research activities; to Fondazione Umberto Veronesi and the German Cancer Research Center (DKFZ) for the fellowships to L.A. and progetto Bandiera Epigen for the fellowship to A.L.

Conflicts of Interest: The authors declare no conflicts of interest.

References

1. Järveläinen, H.; Sainio, A.; Koulu, M.; Wight, T.N.; Penttinen, R. Extracellular matrix molecules: Potential targets in pharmacotherapy. *Pharmacol Rev.* **2009**, *61*, 198–223. [CrossRef] [PubMed]
2. Naba, A.; Clauser, K.R.; Hoersch, S.; Liu, H.; Carr, S.A.; Hynes, R.O. The matrisome: In silico definition and in vivo characterization by proteomics of normal and tumor extracellular matrices. *Mol. Cell. Proteom.* **2012**, *11*, M111.014647. [CrossRef] [PubMed]
3. Hynes, R.O.; Naba, A. Overview of the matrisome-an inventory of extracellular matrix constituents and functions. *Cold Spring Harb. Perspect. Biol.* **2012**, *4*, a004903. [CrossRef] [PubMed]
4. Almond, A. Hyaluronan. *Cell. Mol. Life Sci.* **2007**, *64*, 1591–1596. [CrossRef] [PubMed]
5. Chen, W.Y.; Abatangelo, G. Functions of hyaluronan in wound repair. *Wound Repair Regen.* **1999**, *7*, 79–89. [CrossRef] [PubMed]
6. Dicker, K.T.; Gurski, L.A.; Pradhan-Bhatt, S.; Witt, R.L.; Farach-Carson, M.C.; Jia, X. Hyaluronan: A simple polysaccharide with diverse biological functions. *Acta Biomater.* **2014**, *10*, 1558–1570. [CrossRef] [PubMed]
7. Volpi, N.; Schiller, J.; Stern, R.; Soltés, L. Role, metabolism, chemical modifications and applications of hyaluronan. *Curr. Med. Chem.* **2009**, *16*, 1718–1745. [CrossRef] [PubMed]
8. Kim, S.H.; Turnbull, J.; Guimond, S. Extracellular matrix and cell signalling: The dynamic cooperation of integrin, proteoglycan and growth factor receptor. *J. Endocrinol.* **2011**, *209*, 139–151. [CrossRef] [PubMed]
9. Singh, P.; Carraher, C.; Schwarzbauer, J.E. Assembly of fibronectin extracellular matrix. *Annu. Rev. Cell Dev. Biol.* **2010**, *26*, 397–419. [CrossRef] [PubMed]
10. Aumailley, M. The laminin family. *Cell. Adhes. Migr.* **2013**, *7*, 48–55. [CrossRef] [PubMed]
11. Davies, J.A. *Extracellular Matrix*; John Wiley & Sons Ltd.: Chichester, UK, 2001.
12. Mithieux, S.M.; Weiss, A.S. Elastin. *Adv. Protein Chem.* **2005**, *70*, 437–461. [PubMed]
13. Frantz, C.; Stewart, K.M.; Weaver, V.M. The extracellular matrix at a glance. *J. Cell Sci.* **2010**, *123*, 4195–4200. [CrossRef] [PubMed]
14. Wijelath, E.S.; Rahman, S.; Namekata, M.; Murray, J.; Nishimura, T.; Mostafavi-Pour, Z.; Patel, Y.; Suda, Y.; Humphries, M.J.; Sobel, M. Heparin-II domain of fibronectin is a vascular endothelial growth factor-binding domain: Enhancement of VEGF biological activity by a singular growth factor/matrix protein synergism. *Circ. Res.* **2006**, *99*, 853–860. [CrossRef] [PubMed]

15. Rahman, S.; Patel, Y.; Murray, J.; Patel, K.V.; Sumathipala, R.; Sobel, M.; Wijelath, E.S. Novel hepatocyte growth factor (HGF) binding domains on fibronectin and vitronectin coordinate a distinct and amplified Met-integrin induced signalling pathway in endothelial cells. *Bmc Cell. Biol.* **2005**, *6*, 8. [CrossRef] [PubMed]

16. Lin, F.; Ren, X.D.; Pan, Z.; Macri, L.; Zong, W.X.; Tonnesen, M.G.; Rafailovich, M.; Bar-Sagi, D.; Clark, R.A. Fibronectin growth factor-binding domains are required for fibroblast survival. *J. Investig. Dermatol.* **2011**, *131*, 84–98. [CrossRef] [PubMed]

17. Ramirez, F.; Rifkin, D.B. Extracellular microfibrils: Contextual platforms for TGFβ and BMP signaling. *Curr. Opin. Cell Biol.* **2009**, *21*, 616–622. [CrossRef] [PubMed]

18. Munger, J.S.; Sheppard, D. Cross talk among TGF-β signaling pathways, integrins, and the extracellular matrix. *Cold Spring Harb. Perspect. Biol.* **2011**, *3*, a005017. [CrossRef] [PubMed]

19. Rahman, M.S.; Akhtar, N.; Jamil, H.M.; Banik, R.S.; Asaduzzaman, S.M. TGF-β/BMP signaling and other molecular events: Regulation of osteoblastogenesis and bone formation. *Bone Res.* **2015**, *3*, 15005. [CrossRef] [PubMed]

20. Geiger, B.; Yamada, K.M. Molecular architecture and function of matrix adhesions. *Cold Spring Harb. Perspect. Biol.* **2011**, *3*, a005033. [CrossRef] [PubMed]

21. Hynes, R.O. The extracellular matrix: Not just pretty fibrils. *Science* **2009**, *326*, 1216–1219. [CrossRef] [PubMed]

22. Klein, E.A.; Yin, L.; Kothapalli, D.; Castagnino, P.; Byfield, F.J.; Xu, T.; Levental, I.; Hawthorne, E.; Janmey, P.A.; Assoian, R.K. Cell-cycle control by physiological matrix elasticity and in vivo tissue stiffening. *Curr. Biol.* **2009**, *19*, 1511–1518. [CrossRef] [PubMed]

23. Paszek, M.J.; Zahir, N.; Johnson, K.R.; Lakins, J.N.; Rozenberg, G.I.; Gefen, A.; Reinhart-King, C.A.; Margulies, S.S.; Dembo, M.; Boettiger, D.; et al. Tensional homeostasis and the malignant phenotype. *Cancer Cell* **2005**, *8*, 241–254. [CrossRef] [PubMed]

24. Colpaert, C.; Vermeulen, P.; Van Marck, E.; Dirix, L. The presence of a fibrotic focus is an independent predictor of early metastasis in lymph node-negative breast cancer patients. *Am. J. Surg. Pathol.* **2001**, *25*, 1557–1558. [CrossRef] [PubMed]

25. Li, L.; Xie, T. Stem cell niche: Structure and function. *Annu. Rev. Cell Dev. Biol.* **2005**, *21*, 605–631. [CrossRef] [PubMed]

26. Taddei, I.; Deugnier, M.A.; Faraldo, M.M.; Petit, V.; Bouvard, D.; Medina, D.; Fässler, R.; Thiery, J.P.; Glukhova, M.A. Beta1 integrin deletion from the basal compartment of the mammary epithelium affects stem cells. *Nat. Cell Biol.* **2008**, *10*, 716–722. [CrossRef] [PubMed]

27. Yan, D.; Lin, X. Shaping morphogen gradients by proteoglycans. *Cold Spring Harb. Perspect. Biol.* **2009**, *1*, a002493. [CrossRef] [PubMed]

28. Ricard-Blum, S.; Salza, R. Matricryptins and matrikines: Biologically active fragments of the extracellular matrix. *Exp. Dermatol.* **2014**, *23*, 457–463. [CrossRef] [PubMed]

29. Maquart, F.X.; Siméon, A.; Pasco, S.; Monboisse, J.C. Regulation of cell activity by the extracellular matrix: The concept of matrikines. *J. Soc. Biol.* **1999**, *193*, 423–428. [CrossRef] [PubMed]

30. Davis, G.E.; Bayless, K.J.; Davis, M.J.; Meininger, G.A. Regulation of tissue injury responses by the exposure of matricryptic sites within extracellular matrix molecules. *Am. J. Pathol.* **2000**, *156*, 1489–1498. [CrossRef]

31. Ricard-Blum, S.; Ballut, L. Matricryptins derived from collagens and proteoglycans. *Front. Biosci.* **2011**, *16*, 674–697. [CrossRef]

32. Folkman, J. Antiangiogenesis in cancer therapy—Endostatin and its mechanisms of action. *Exp. Cell Res.* **2006**, *312*, 594–607. [CrossRef] [PubMed]

33. Ricard-Blum, S. The collagen family. *Cold Spring Harb. Perspect. Biol.* **2011**, *3*, a004978. [CrossRef] [PubMed]

34. Väisänen, M.R.; Väisänen, T.; Pihlajaniemi, T. The shed ectodomain of type XIII collagen affects cell behaviour in a matrix-dependent manner. *Biochem. J.* **2004**, *380*, 685–693. [CrossRef] [PubMed]

35. Rhodes, J.M.; Simons, M. The extracellular matrix and blood vessel formation: Not just a scaffold. *J. Cell. Mol. Med.* **2007**, *11*, 176–205. [CrossRef] [PubMed]

36. Su, J.; Stenbjorn, R.S.; Gorse, K.; Su, K.; Hauser, K.F.; Ricard-Blum, S.; Pihlajaniemi, T.; Fox, M.A. Target-derived matricryptins organize cerebellar synapse formation through α3β1 integrins. *Cell Rep.* **2012**, *2*, 223–230. [CrossRef] [PubMed]

37. Fu, H.L.; Valiathan, R.R.; Arkwright, R.; Sohail, A.; Mihai, C.; Kumarasiri, M.; Mahasenan, K.V.; Mobashery, S.; Huang, P.; Agarwal, G.; et al. Discoidin domain receptors: Unique receptor tyrosine kinases in collagen-mediated signaling. *J. Biol. Chem.* **2013**, *288*, 7430–7437. [CrossRef] [PubMed]

38. Brown, N.H. Extracellular matrix in development: Insights from mechanisms conserved between invertebrates and vertebrates. *Cold Spring Harb. Perspect. Biol.* **2011**, *3*, a005082. [CrossRef] [PubMed]

39. Aiken, A.; Khokha, R. Unraveling metalloproteinase function in skeletal biology and disease using genetically altered mice. *Biochim. Biophys. Acta Mol. Cell. Res.* **2010**, *1803*, 121–132. [CrossRef] [PubMed]

40. Zaidel-Bar, R.; Ballestrem, C.; Kam, Z.; Geiger, B. Early molecular events in the assembly of matrix adhesions at the leading edge of migrating cells. *J. Cell Sci.* **2003**, *116*, 4605–4613. [CrossRef] [PubMed]

41. Huveneers, S.; Danen, E.H. Adhesion signaling- crosstalk between integrins, Src and Rho. *J. Cell Sci.* **2009**, *122*, 1059–1069. [CrossRef] [PubMed]

42. Sbardella, D.; Fasciglione, G.F.; Gioia, M.; Ciaccio, C.; Tundo, G.R.; Marini, S.; Coletta, M. Human matrix metalloproteinases: An ubiquitarian class of enzymes involved in several pathological processes. *Mol. Asp. Med.* **2012**, *33*, 119–208. [CrossRef] [PubMed]

43. Murphy, G. The adams: Signalling scissors in the tumour microenvironment. *Nat. Rev. Cancer* **2008**, *8*, 929–941. [CrossRef] [PubMed]

44. Apte, S.S. A disintegrin-like and metalloprotease (reprolysin-type) with thrombospondin type 1 motif (ADAMTS) superfamily: Functions and mechanisms. *J. Biol. Chem.* **2009**, *284*, 31493–31497. [CrossRef] [PubMed]

45. Bertenshaw, G.P.; Norcum, M.T.; Bond, J.S. Structure of homo- and hetero-oligomeric meprin metalloproteases. Dimers, tetramers, and high molecular mass multimers. *J. Biol. Chem.* **2003**, *278*, 2522–2532. [CrossRef] [PubMed]

46. Khokha, R.; Murthy, A.; Weiss, A. Metalloproteinases and their natural inhibitors in inflammation and immunity. *Nat. Rev. Immunol.* **2013**, *13*, 649–665. [CrossRef] [PubMed]

47. Smith, H.W.; Marshall, C.J. Regulation of cell signalling by uPAR. *Nat. Rev. Mol. Cell Biol.* **2010**, *11*, 23–36. [CrossRef] [PubMed]

48. Bonnefoy, A.; Legrand, C. Proteolysis of subendothelial adhesive glycoproteins (fibronectin, thrombospondin, and von Willebrand factor) by plasmin, leukocyte cathepsin G, and elastase. *Thromb. Res.* **2000**, *98*, 323–332. [CrossRef]

49. Ilan, N.; Elkin, M.; Vlodavsky, I. Regulation, function and clinical significance of heparanase in cancer metastasis and angiogenesis. *Int. J. Biochem. Cell Biol.* **2006**, *38*, 2018–2039. [CrossRef] [PubMed]

50. Uchimura, K.; Morimoto-Tomita, M.; Bistrup, A.; Li, J.; Lyon, M.; Gallagher, J.; Werb, Z.; Rosen, S.D. HSulf-2, an extracellular endoglucosamine-6-sulfatase, selectively mobilizes heparin-bound growth factors and chemokines: Effects on VEGF, FGF-1, and SDF-1. *BMC Biochem.* **2006**, *7*, 2. [CrossRef] [PubMed]

51. Bastow, E.R.; Byers, S.; Golub, S.B.; Clarkin, C.E.; Pitsillides, A.A.; Fosang, A.J. Hyaluronan synthesis and degradation in cartilage and bone. *Cell. Mol. Life Sci.* **2008**, *65*, 395–413. [CrossRef] [PubMed]

52. Shoulders, M.D.; Raines, R.T. Modulating collagen triple-helix stability with 4-chloro, 4-fluoro, and 4-methylprolines. *Adv. Exp. Med. Biol.* **2009**, *611*, 251–252. [PubMed]

53. Kadler, K.E.; Baldock, C.; Bella, J.; Boot-Handford, R.P. Collagens at a glance. *J. Cell Sci.* **2007**, *120*, 1955–1958. [CrossRef] [PubMed]

54. Karsdal, M.A. *Biochemistry of Collagens, Laminins and Elastin: Structure, Function and Biomarkers*; Elsevier: New York, NY, USA, 2016.

55. Lodish, H.; Berk, A.; Zipursky, S.L.; Matsudaira, P.; Baltimore, D.; Darnell, J. Collagen: The Fibrous Proteins of the Matrix. In *Molecular Cell Biology*, 4th ed.; W. H. Freeman: New York, NY, USA, 2000; Section 22.3.

56. Canty, E.G.; Kadler, K.E. Procollagen trafficking, processing and fibrillogenesis. *J. Cell Sci.* **2005**, *118*, 1341–1353. [CrossRef] [PubMed]

57. Sharma, U.; Carrique, L.; Vadon-Le Goff, S.; Mariano, N.; Georges, R.N.; Delolme, F.; Koivunen, P.; Myllyharju, J.; Moali, C.; Aghajari, N.; et al. Structural basis of homo- and heterotrimerization of collagen I. *Nat. Commun.* **2017**, *8*, 14671. [CrossRef] [PubMed]

58. Thiagarajan, G.; Li, Y.; Mohs, A.; Strafaci, C.; Popiel, M.; Baum, J.; Brodsky, B. Common interruptions in the repeating tripeptide sequence of non-fibrillar collagens: Sequence analysis and structural studies on triple-helix peptide models. *J. Mol. Biol.* **2008**, *376*, 736–748. [CrossRef] [PubMed]

59. Eyre, D.R.; Paz, M.A.; Gallop, P.M. Cross-linking in collagen and elastin. *Annu. Rev. Biochem.* **1984**, *53*, 717–748. [CrossRef] [PubMed]

60. Knupp, C.; Squire, J.M. Molecular packing in network-forming collagens. *Adv. Protein Chem.* **2005**, *70*, 375–403. [PubMed]

61. Franzke, C.W.; Tasanen, K.; Schumann, H.; Bruckner-Tuderman, L. Collagenous transmembrane proteins: Collagen XVII as a prototype. *Matrix Biol.* **2003**, *22*, 299–309. [CrossRef]

62. Fratzl, P. *Collagen: Structure and Mechanics*; Springer: New York, NY, USA, 2008.

63. Marneros, A.G.; Olsen, B.R. Physiological role of collagen XVIII and endostatin. *FASEB J.* **2005**, *19*, 716–728. [CrossRef] [PubMed]

64. Chen, M.; Marinkovich, M.P.; Veis, A.; Cai, X.; Rao, C.N.; O'Toole, E.A.; Woodley, D.T. Interactions of the amino-terminal noncollagenous (NC1) domain of type VII collagen with extracellular matrix components. A potential role in epidermal-dermal adherence in human skin. *J. Biol. Chem.* **1997**, *272*, 14516–14522. [CrossRef] [PubMed]

65. Van de Wetering, J.K.; van Golde, L.M.; Batenburg, J.J. Collectins: Players of the innate immune system. *Eur. J. Biochem.* **2004**, *271*, 1229–1249. [CrossRef] [PubMed]

66. Everts, V.; van der Zee, E.; Creemers, L.; Beertsen, W. Phagocytosis and intracellular digestion of collagen, its role in turnover and remodelling. *Histochem. J.* **1996**, *28*, 229–245. [CrossRef] [PubMed]

67. Bonnans, C.; Chou, J.; Werb, Z. Remodelling the extracellular matrix in development and disease. *Nat. Rev. Mol. Cell Biol.* **2014**, *15*, 786–801. [CrossRef] [PubMed]

68. Parks, W.C.; Mecham, R.P. *Extracellular Matrix Degradation*; Springer: Berlin, Germany; London, UK, 2011.

69. Monea, S.; Lehti, K.; Keski-Oja, J.; Mignatti, P. Plasmin activates pro-matrix metalloproteinase-2 with a membrane-type 1 matrix metalloproteinase-dependent mechanism. *J. Cell. Physiol.* **2002**, *192*, 160–170. [CrossRef] [PubMed]

70. Arora, P.D.; Wang, Y.; Bresnick, A.; Dawson, J.; Janmey, P.A.; McCulloch, C.A. Collagen remodeling by phagocytosis is determined by collagen substrate topology and calcium-dependent interactions of gelsolin with nonmuscle myosin IIA in cell adhesions. *Mol. Biol. Cell* **2013**, *24*, 734–747. [CrossRef] [PubMed]

71. Mohamed, M.M.; Sloane, B.F. Cysteine cathepsins: Multifunctional enzymes in cancer. *Nat. Rev. Cancer* **2006**, *6*, 764–775. [CrossRef] [PubMed]

72. Shingleton, W.D.; Hodges, D.J.; Brick, P.; Cawston, T.E. Collagenase: A key enzyme in collagen turnover. *Biochem. Cell Biol.* **1996**, *74*, 759–775. [CrossRef] [PubMed]

73. Madsen, D.H.; Ingvarsen, S.; Jürgensen, H.J.; Melander, M.C.; Kjøller, L.; Moyer, A.; Honoré, C.; Madsen, C.A.; Garred, P.; Burgdorf, S.; et al. The non-phagocytic route of collagen uptake: A distinct degradation pathway. *J. Biol. Chem.* **2011**, *286*, 26996–27010. [CrossRef] [PubMed]

74. Muschler, J.; Streuli, C.H. Cell-matrix interactions in mammary gland development and breast cancer. *Cold Spring Harb. Perspect. Biol.* **2010**, *2*, a003202. [CrossRef] [PubMed]

75. Barat-Houari, M.; Sarrabay, G.; Gatinois, V.; Fabre, A.; Dumont, B.; Genevieve, D.; Touitou, I. Mutation update for *COL2A1* gene variants associated with type II collagenopathies. *Hum. Mutat.* **2016**, *37*, 7–15. [CrossRef] [PubMed]

76. Malfait, F.; Wenstrup, R.; De Paepe, A. Ehlers-Danlos Syndrome, Classic Type. In *GeneReviews*® *[Internet]*; Adam, M.P., Ardinger, H.H., Pagon, R.A., Wallace, S.E., Bean, L.J.H., Stephens, K., Amemiya, A., Eds.; University of Washington: Seattle, WA, USA, 2007.

77. Yeowell, H.N.; Steinmann, B. Ehlers-Danlos Syndrome, Kyphoscoliotic Form. In *GeneReviews*® *[Internet]*; Adam, M.P., Ardinger, H.H., Pagon, R.A., Wallace, S.E., Bean, L.J.H., Stephens, K., Amemiya, A., Eds.; University of Washington: Seattle, WA, USA, 2000.

78. Müller, T.; Mizumoto, S.; Suresh, I.; Komatsu, Y.; Vodopiutz, J.; Dundar, M.; Straub, V.; Lingenhel, A.; Melmer, A.; Lechner, S.; et al. Loss of dermatan sulfate epimerase (DSE) function results in musculocontractural Ehlers-Danlos syndrome. *Hum. Mol. Genet.* **2013**, *22*, 3761–3772. [CrossRef] [PubMed]

79. Janecke, A.R.; Li, B.; Boehm, M.; Krabichler, B.; Rohrbach, M.; Müller, T.; Fuchs, I.; Golas, G.; Katagiri, Y.; Ziegler, S.G.; et al. The phenotype of the musculocontractural type of Ehlers-Danlos syndrome due to *CHST14* mutations. *Am. J. Med. Genet. A* **2016**, *170A*, 103–115. [CrossRef] [PubMed]

80. Ritelli, M.; Dordoni, C.; Cinquina, V.; Venturini, M.; Calzavara-Pinton, P.; Colombi, M. Expanding the clinical and mutational spectrum of *B4GALT7*-spondylodysplastic Ehlers-Danlos syndrome. *Orphanet J. Rare Dis.* **2017**, *12*, 153. [CrossRef] [PubMed]

81. Malfait, F.; De Coster, P.; Hausser, I.; van Essen, A.J.; Franck, P.; Colige, A.; Nusgens, B.; Martens, L.; De Paepe, A. The natural history, including orofacial features of three patients with Ehlers-Danlos syndrome, dermatosparaxis type (EDS type VIIC). *Am. J. Med. Genet. A* **2004**, *131*, 18–28. [CrossRef] [PubMed]

82. Van Dijk, F.S.; Sillence, D.O. Osteogenesis imperfecta: Clinical diagnosis, nomenclature and severity assessment. *Am. J. Med. Genet. A* **2014**, *164A*, 1470–1481. [CrossRef] [PubMed]

83. Marini, J.C.; Forlino, A.; Bächinger, H.P.; Bishop, N.J.; Byers, P.H.; Paepe, A.; Fassier, F.; Fratzl-Zelman, N.; Kozloff, K.M.; Krakow, D.; et al. Osteogenesis imperfecta. *Nat. Rev. Dis. Primers* **2017**, *3*, 17052. [CrossRef] [PubMed]

84. Morello, R.; Bertin, T.K.; Chen, Y.; Hicks, J.; Tonachini, L.; Monticone, M.; Castagnola, P.; Rauch, F.; Glorieux, F.H.; Vranka, J.; et al. CRTAP is required for prolyl 3- hydroxylation and mutations cause recessive osteogenesis imperfecta. *Cell* **2006**, *127*, 291–304. [CrossRef] [PubMed]

85. Fratzl-Zelman, N.; Barnes, A.M.; Weis, M.; Carter, E.; Hefferan, T.E.; Perino, G.; Chang, W.; Smith, P.A.; Roschger, P.; Klaushofer, K.; et al. Non-lethal type VIII osteogenesis imperfecta has elevated bone matrix mineralization. *J. Clin. Endocrinol. Metab.* **2016**, *101*, 3516–3525. [CrossRef] [PubMed]

86. Van Dijk, F.S.; Nesbitt, I.M.; Zwikstra, E.H.; Nikkels, P.G.; Piersma, S.R.; Fratantoni, S.A.; Jimenez, C.R.; Huizer, M.; Morsman, A.C.; Cobben, J.M.; et al. Ppib mutations cause severe osteogenesis imperfecta. *Am. J. Hum. Genet.* **2009**, *85*, 521–527. [CrossRef] [PubMed]

87. Shapiro, R.J.; Byers, P.H.; Glorieux, F.H.; Sponsellor, P.D. *Osteogenesis Imperfecta: A Translational Approach to Brittle Bone Disease*; Academic Press: Cambridge, MA, USA, 2014; ISBN 978-0-12-397165-4.

88. Kelley, B.P.; Malfait, F.; Bonafe, L.; Baldridge, D.; Homan, E.; Symoens, S.; Willaert, A.; Elcioglu, N.; Van Maldergem, L.; Verellen-Dumoulin, C.; et al. Mutations in FKBP10 cause recessive osteogenesis imperfecta and bruck syndrome. *J. Bone Miner. Res.* **2011**, *26*, 666–672. [CrossRef] [PubMed]

89. Mendoza-Londono, R.; Fahiminiya, S.; Majewski, J.; Tétreault, M.; Nadaf, J.; Kannu, P.; Sochett, E.; Howard, A.; Stimec, J.; Dupuis, L.; et al. Recessive osteogenesis imperfecta caused by missense mutations in SPARC. *Am. J. Hum. Genet.* **2015**, *96*, 979–985. [CrossRef] [PubMed]

90. Kamoun-Goldrat, A.; le Merrer, M. Infantile cortical hyperostosis (*Caffey disease*): A review. *J. Oral Maxillofac. Surg.* **2008**, *66*, 2145–2150. [CrossRef] [PubMed]

91. Meigel, W.N.; Müller, P.K.; Pontz, B.F.; Sörensen, N.; Spranger, J. A constitutional disorder of connective tissue suggesting a defect in collagen biosynthesis. *Klin. Wochenschr.* **1974**, *52*, 906–912. [CrossRef] [PubMed]

92. Anderson, I.J.; Goldberg, R.B.; Marion, R.W.; Upholt, W.B.; Tsipouras, P. Spondyloepiphyseal dysplasia congenita: Genetic linkage to type II collagen (*COL2A1*). *Am. J. Hum. Genet.* **1990**, *46*, 896–901. [PubMed]

93. Nishimura, G.; Saitoh, Y.; Okuzumi, S.; Imaizumi, K.; Hayasaka, K.; Hashimoto, M. Spondyloepiphyseal dysplasia with accumulation of glycoprotein in the chondrocytes: Spondyloepiphyseal dysplasia, stanescu type. *Skelet. Radiol.* **1998**, *27*, 188–194. [CrossRef]

94. Hammarsjö, A.; Nordgren, A.; Lagerstedt-Robinson, K.; Malmgren, H.; Nilsson, D.; Wedrén, S.; Nordenskjöld, M.; Nishimura, G.; Grigelioniene, G. Pathogenenic variant in the *COL2A1* gene is associated with Spondyloepiphyseal dysplasia type Stanescu. *Am. J. Med. Genet. A* **2016**, *170A*, 266–269. [CrossRef] [PubMed]

95. Beighton, P.; Goldberg, L.; Hof, J.O. Dominant inheritance of multiple epiphyseal dysplasia, myopia and deafness. *Clin. Genet.* **1978**, *14*, 173–177. [CrossRef] [PubMed]

96. Comstock, J.M.; Putnam, A.R.; Sangle, N.; Lowichik, A.; Rose, N.C.; Opitz, J.M. Recurrence of achondrogenesis type 2 in sibs: Additional evidence for germline mosaicism. *Am. J. Med. Genet. A* **2010**, *152A*, 1822–1824. [CrossRef] [PubMed]

97. Kozlowski, K.; Marik, I.; Marikova, O.; Zemkova, D.; Kuklik, M. Czech dysplasia metatarsal type. *Am. J. Med. Genet. A* **2004**, *129A*, 87–91. [CrossRef] [PubMed]

98. Hoornaert, K.P.; Marik, I.; Kozlowski, K.; Cole, T.; Le Merrer, M.; Leroy, J.G.; Coucke, P.J.; Sillence, D.; Mortier, G.R. Czech dysplasia metatarsal type: Another type II collagen disorder. *Eur. J. Hum. Genet.* **2007**, *15*, 1269–1275. [CrossRef] [PubMed]

99. Chen, W.M.; Liu, Y.F.; Lin, M.W.; Chen, I.C.; Lin, P.Y.; Lin, G.L.; Jou, Y.S.; Lin, Y.T.; Fann, C.S.; Wu, J.Y.; et al. Autosomal dominant avascular necrosis of femoral head in two Taiwanese pedigrees and linkage to chromosome 12q13. *Am. J. Hum. Genet.* **2004**, *75*, 310–317. [CrossRef] [PubMed]

100. Knowlton, R.G.; Katzenstein, P.L.; Moskowitz, R.W.; Weaver, E.J.; Malemud, C.J.; Pathria, M.N.; Jimenez, S.A.; Prockop, D.J. Genetic linkage of a polymorphism in the type II procollagen gene (*COL2A1*) to primary osteoarthritis associated with mild chondrodysplasia. *N. Engl. J. Med.* **1990**, *322*, 526–530. [CrossRef] [PubMed]

101. Learmonth, I.D.; Christy, G.; Beighton, P. Namaqualand hip dysplasia. Orthopedic implications. *Clin. Orthop. Relat. Res.* **1987**, *218*, 142–147. [CrossRef]

102. Nishimura, A.L.; Mitne-Neto, M.; Silva, H.C.; Richieri-Costa, A.; Middleton, S.; Cascio, D.; Kok, F.; Oliveira, J.R.; Gillingwater, T.; Webb, J.; et al. A mutation in the vesicle-trafficking protein vapb causes late-onset spinal muscular atrophy and amyotrophic lateral sclerosis. *Am. J. Hum. Genet.* **2004**, *75*, 822–831. [CrossRef] [PubMed]

103. Zankl, A.; Neumann, L.; Ignatius, J.; Nikkels, P.; Schrander-Stumpel, C.; Mortier, G.; Omran, H.; Wright, M.; Hilbert, K.; Bonafé, L.; et al. Dominant negative mutations in the C-propeptide of *COL2A1* cause platyspondylic lethal skeletal dysplasia, torrance type, and define a novel subfamily within the type 2 collagenopathies. *Am. J. Med. Genet. A* **2005**, *133A*, 61–67. [CrossRef] [PubMed]

104. Neumann, L.; Kunze, J.; Uhl, M.; Stöver, B.; Zabel, B.; Spranger, J. Survival to adulthood and dominant inheritance of platyspondylic skeletal dysplasia, Torrance-Luton type. *Pediatr. Radiol.* **2003**, *33*, 786–790. [CrossRef] [PubMed]

105. Zankl, A.; Scheffer, H.; Schinzel, A. Ectodermal dysplasia with tetramelic deficiencies and no mutation in p63: Odontotrichomelic syndrome or a new entity? *Am. J. Med. Genet. A* **2004**, *127A*, 74–80. [CrossRef] [PubMed]

106. Anderson, C.E.; Sillence, D.O.; Lachman, R.S.; Toomey, K.; Bull, M.; Dorst, J.; Rimoin, D.L. Spondylometepiphyseal dysplasia, strudwick type. *Am. J. Med. Genet.* **1982**, *13*, 243–256. [CrossRef] [PubMed]

107. Tiller, G.E.; Polumbo, P.A.; Weis, M.A.; Bogaert, R.; Lachman, R.S.; Cohn, D.H.; Rimoin, D.L.; Eyre, D.R. Dominant mutations in the type II collagen gene, *COL2A1*, produce spondyloepimetaphyseal dysplasia, strudwick type. *Nat. Genet.* **1995**, *11*, 87–89. [CrossRef] [PubMed]

108. Baker, S.; Booth, C.; Fillman, C.; Shapiro, M.; Blair, M.P.; Hyland, J.C.; Ala-Kokko, L. A loss of function mutation in the *COL9A2* gene causes autosomal recessive Stickler syndrome. *Am. J. Med. Genet. A* **2011**, *155A*, 1668–1672. [CrossRef] [PubMed]

109. Faber, J.; Winterpacht, A.; Zabel, B.; Gnoinski, W.; Schinzel, A.; Steinmann, B.; Superti-Furga, A. Clinical variability of stickler syndrome with a *COL2A1* haploinsufficiency mutation: Implications for genetic counselling. *J. Med. Genet.* **2000**, *37*, 318–320. [CrossRef] [PubMed]

110. Majava, M.; Hoornaert, K.P.; Bartholdi, D.; Bouma, M.C.; Bouman, K.; Carrera, M.; Devriendt, K.; Hurst, J.; Kitsos, G.; Niedrist, D.; et al. A report on 10 new patients with heterozygous mutations in the *COL11A1* gene and a review of genotype-phenotype correlations in type XI collagenopathies. *Am. J. Med. Genet. A* **2007**, *143A*, 258–264. [CrossRef] [PubMed]

111. Vikkula, M.; Mariman, E.C.; Lui, V.C.; Zhidkova, N.I.; Tiller, G.E.; Goldring, M.B.; van Beersum, S.E.; de Waal Malefijt, M.C.; van den Hoogen, F.H.; Ropers, H.H. Autosomal dominant and recessive osteochondrodysplasias associated with the *COL11A2* locus. *Cell* **1995**, *80*, 431–437. [CrossRef]

112. Nikopoulos, K.; Schrauwen, I.; Simon, M.; Collin, R.W.; Veckeneer, M.; Keymolen, K.; Van Camp, G.; Cremers, F.P.; van den Born, L.I. Autosomal recessive Stickler syndrome in two families is caused by mutations in the *COL9A1* gene. *Investig. Ophthalmol. Vis. Sci.* **2011**, *52*, 4774–4779. [CrossRef] [PubMed]

113. McAlinden, A.; Majava, M.; Bishop, P.N.; Perveen, R.; Black, G.C.; Pierpont, M.E.; Ala-Kokko, L.; Männikkö, M. Missense and nonsense mutations in the alternatively-spliced exon 2 of *COL2A1* cause the ocular variant of stickler syndrome. *Hum. Mutat.* **2008**, *29*, 83–90. [CrossRef] [PubMed]

114. Malizos, K.N.; Karantanas, A.H.; Varitimidis, S.E.; Dailiana, Z.H.; Bargiotas, K.; Maris, T. Osteonecrosis of the femoral head: Etiology, imaging and treatment. *Eur. J. Radiol.* **2007**, *63*, 16–28. [CrossRef] [PubMed]

115. Gilbert-Barnes, E.; Langer, L.O.; Opitz, J.M.; Laxova, R.; Sotelo-Arila, C. Kniest dysplasia: Radiologic, histopathological, and scanning electronmicroscopic findings. *Am. J. Med. Genet.* **1996**, *63*, 34–45. [CrossRef]

116. Kruegel, J.; Rubel, D.; Gross, O. Alport syndrome—Insights from basic and clinical research. *Nat. Rev. Nephrol.* **2013**, *9*, 170–178. [CrossRef] [PubMed]

117. Crovetto, F.; Moroni, G.; Zaina, B.; Acaia, B.; Ossola, M.W.; Fedele, L. Pregnancy in women with Alport syndrome. *Int. Urol. Nephrol.* **2013**, *45*, 1223–1227. [CrossRef] [PubMed]

118. Bekheirnia, M.R.; Reed, B.; Gregory, M.C.; McFann, K.; Shamshirsaz, A.A.; Masoumi, A.; Schrier, R.W. Genotype-phenotype correlation in X-linked Alport syndrome. *J. Am. Soc. Nephrol.* **2010**, *21*, 876–883. [CrossRef] [PubMed]

119. Anker, M.C.; Arnemann, J.; Neumann, K.; Ahrens, P.; Schmidt, H.; König, R. Alport syndrome with diffuse leiomyomatosis. *Am. J. Med. Genet. A* **2003**, *119A*, 381–385. [CrossRef] [PubMed]

120. Longo, I.; Porcedda, P.; Mari, F.; Giachino, D.; Meloni, I.; Deplano, C.; Brusco, A.; Bosio, M.; Massella, L.; Lavoratti, G.; et al. *COL4A3/COL4A4* mutations: From familial hematuria to autosomal-dominant or recessive Alport syndrome. *Kidney Int.* **2002**, *61*, 1947–1956. [CrossRef] [PubMed]

121. De Ligt, J.; Willemsen, M.H.; van Bon, B.W.; Kleefstra, T.; Yntema, H.G.; Kroes, T.; Vulto-van Silfhout, A.T.; Koolen, D.A.; de Vries, P.; Gilissen, C.; et al. Diagnostic exome sequencing in persons with severe intellectual disability. *N. Engl. J. Med.* **2012**, *367*, 1921–1929. [CrossRef] [PubMed]

122. Hamdan, F.F.; Srour, M.; Capo-Chichi, J.M.; Daoud, H.; Nassif, C.; Patry, L.; Massicotte, C.; Ambalavanan, A.; Spiegelman, D.; Diallo, O.; et al. De novo mutations in moderate or severe intellectual disability. *PLoS Genet.* **2014**, *10*, e1004772. [CrossRef] [PubMed]

123. Zenteno, J.C.; Crespí, J.; Buentello-Volante, B.; Buil, J.A.; Bassaganyas, F.; Vela-Segarra, J.I.; Diaz-Cascajosa, J.; Marieges, M.T. Next generation sequencing uncovers a missense mutation in col4a1 as the cause of familial retinal arteriolar tortuosity. *Graefes Arch. Clin. Exp. Ophthalmol.* **2014**, *252*, 1789–1794. [CrossRef] [PubMed]

124. Plaisier, E.; Chen, Z.; Gekeler, F.; Benhassine, S.; Dahan, K.; Marro, B.; Alamowitch, S.; Paques, M.; Ronco, P. Novel *COL4A1* mutations associated with HANAC syndrome: A role for the triple helical CB3[IV] domain. *Am. J. Med. Genet. A* **2010**, *152A*, 2550–2555. [CrossRef] [PubMed]

125. Lanfranconi, S.; Markus, H.S. *COL4A1* mutations as a monogenic cause of cerebral small vessel disease: A systematic review. *Stroke* **2010**, *41*, e513–e518. [CrossRef] [PubMed]

126. Yoneda, Y.; Haginoya, K.; Kato, M.; Osaka, H.; Yokochi, K.; Arai, H.; Kakita, A.; Yamamoto, T.; Otsuki, Y.; Shimizu, S.; et al. Phenotypic spectrum of *COL4A1* mutations: Porencephaly to schizencephaly. *Ann. Neurol.* **2013**, *73*, 48–57. [CrossRef] [PubMed]

127. Granata, T.; Freri, E.; Caccia, C.; Setola, V.; Taroni, F.; Battaglia, G. Schizencephaly: Clinical spectrum, epilepsy, and pathogenesis. *J. Child Neurol.* **2005**, *20*, 313–318. [CrossRef] [PubMed]

128. Carpenter, A.M.; Singh, I.P.; Gandhi, C.D.; Prestigiacomo, C.J. Genetic risk factors for spontaneous intracerebral haemorrhage. *Nat. Rev. Neurol.* **2016**, *12*, 40–49. [CrossRef] [PubMed]

129. Rost, S.; Bach, E.; Neuner, C.; Nanda, I.; Dysek, S.; Bittner, R.E.; Keller, A.; Bartsch, O.; Mlynski, R.; Haaf, T.; et al. Novel form of X-linked nonsyndromic hearing loss with cochlear malformation caused by a mutation in the type IV collagen gene COL4A6. *Eur. J. Hum. Genet.* **2014**, *22*, 208–215. [CrossRef] [PubMed]

130. Badenas, C.; Praga, M.; Tazón, B.; Heidet, L.; Arrondel, C.; Armengol, A.; Andrés, A.; Morales, E.; Camacho, J.A.; Lens, X.; et al. Mutations in the *COL4A4* and *COL4A3* genes cause familial benign hematuria. *J. Am. Soc. Nephrol.* **2002**, *13*, 1248–1254. [PubMed]

131. Bertini, E.; Pepe, G. Collagen type VI and related disorders: Bethlem myopathy and Ullrich scleroatonic muscular dystrophy. *Eur. J. Paediatr. Neurol* **2002**, *6*, 193–198. [CrossRef] [PubMed]

132. Yonekawa, T.; Nishino, I. Ullrich congenital muscular dystrophy: Clinicopathological features, natural history and pathomechanism(s). *J. Neurol. Neurosurg. Psychiatry* **2015**, *86*, 280–287. [CrossRef] [PubMed]

133. Merlini, L.; Martoni, E.; Grumati, P.; Sabatelli, P.; Squarzoni, S.; Urciuolo, A.; Ferlini, A.; Gualandi, F.; Bonaldo, P. Autosomal recessive myosclerosis myopathy is a collagen VI disorder. *Neurology* **2008**, *71*, 1245–1253. [CrossRef] [PubMed]

134. Zech, M.; Lam, D.D.; Francescatto, L.; Schormair, B.; Salminen, A.V.; Jochim, A.; Wieland, T.; Lichtner, P.; Peters, A.; Gieger, C.; et al. Recessive mutations in the α3 (VI) collagen gene *COL6A3* cause early-onset isolated dystonia. *Am. J. Hum. Genet.* **2015**, *96*, 883–893. [CrossRef] [PubMed]

135. Rashidghamat, E.; McGrath, J.A. Novel and emerging therapies in the treatment of recessive dystrophic epidermolysis bullosa. *Intractable Rare Dis Res.* **2017**, *6*, 6–20. [CrossRef] [PubMed]

136. Titeux, M.; Pendaries, V.; Tonasso, L.; Décha, A.; Bodemer, C.; Hovnanian, A. A frequent functional SNP in the MMP1 promoter is associated with higher disease severity in recessive dystrophic epidermolysis bullosa. *Hum. Mutat.* **2008**, *29*, 267–276. [CrossRef] [PubMed]

137. Sato-Matsumura, K.C.; Yasukawa, K.; Tomita, Y.; Shimizu, H. Toenail dystrophy with *COL7A1* glycine substitution mutations segregates as an autosomal dominant trait in 2 families with dystrophic epidermolysis bullosa. *Arch. Dermatol.* **2002**, *138*, 269–271. [CrossRef] [PubMed]

138. Baratz, K.H.; Tosakulwong, N.; Ryu, E.; Brown, W.L.; Branham, K.; Chen, W.; Tran, K.D.; Schmid-Kubista, K.E.; Heckenlively, J.R.; Swaroop, A.; et al. E2-2 protein and Fuchs's corneal dystrophy. *N. Engl. J. Med.* **2010**, *363*, 1016–1024. [CrossRef] [PubMed]

139. Gottsch, J.D.; Sundin, O.H.; Liu, S.H.; Jun, A.S.; Broman, K.W.; Stark, W.J.; Vito, E.C.; Narang, A.K.; Thompson, J.M.; Magovern, M. Inheritance of a novel *COL8A2* mutation defines a distinct early-onset subtype of fuchs corneal dystrophy. *Investig. Ophthalmol. Vis. Sci.* **2005**, *46*, 1934–1939. [CrossRef] [PubMed]

140. Biswas, S.; Munier, F.L.; Yardley, J.; Hart-Holden, N.; Perveen, R.; Cousin, P.; Sutphin, J.E.; Noble, B.; Batterbury, M.; Kielty, C.; et al. Missense mutations in *COL8A2*, the gene encoding the α2 chain of type VIII collagen, cause two forms of corneal endothelial dystrophy. *Hum. Mol. Genet.* **2001**, *10*, 2415–2423. [CrossRef] [PubMed]

141. Jackson, G.C.; Marcus-Soekarman, D.; Stolte-Dijkstra, I.; Verrips, A.; Taylor, J.A.; Briggs, M.D. Type IX collagen gene mutations can result in multiple epiphyseal dysplasia that is associated with osteochondritis dissecans and a mild myopathy. *Am. J. Med. Genet. A* **2010**, *152A*, 863–869. [CrossRef] [PubMed]

142. Muragaki, Y.; Mundlos, S.; Upton, J.; Olsen, B.R. Altered growth and branching patterns in synpolydactyly caused by mutations in HOXD13. *Science* **1996**, *272*, 548–551. [CrossRef] [PubMed]

143. Czarny-Ratajczak, M.; Lohiniva, J.; Rogala, P.; Kozlowski, K.; Perälä, M.; Carter, L.; Spector, T.D.; Kolodziej, L.; Seppänen, U.; Glazar, R.; et al. A mutation in *COL9A1* causes multiple epiphyseal dysplasia: Further evidence for locus heterogeneity. *Am. J. Hum. Genet.* **2001**, *69*, 969–980. [CrossRef] [PubMed]

144. Mäkitie, O.; Susic, M.; Ward, L.; Barclay, C.; Glorieux, F.H.; Cole, W.G. Schmid type of metaphyseal chondrodysplasia and *COL10A1* mutations—findings in 10 patients. *Am. J. Med. Genet. A* **2005**, *137A*, 241–248. [CrossRef] [PubMed]

145. Çalışkan, E.; Açıkgöz, G.; Yeniay, Y.; Özmen, İ.; Gamsızkan, M.; Akar, A. A case of Marshall's syndrome and review of the literature. *Int. J. Dermatol.* **2015**, *54*, e217–e221. [CrossRef] [PubMed]

146. Tompson, S.W.; Faqeih, E.A.; Ala-Kokko, L.; Hecht, J.T.; Miki, R.; Funari, T.; Funari, V.A.; Nevarez, L.; Krakow, D.; Cohn, D.H. Dominant and recessive forms of fibrochondrogenesis resulting from mutations at a second locus, *COL11A2*. *Am. J. Med. Genet. A* **2012**, *158A*, 309–314. [CrossRef] [PubMed]

147. Brown, M.R.; Tomek, M.S.; Van Laer, L.; Smith, S.; Kenyon, J.B.; Van Camp, G.; Smith, R.J. A novel locus for autosomal dominant nonsyndromic hearing loss, DFNA13, maps to chromosome 6p. *Am. J. Hum. Genet.* **1997**, *61*, 924–927. [CrossRef] [PubMed]

148. Van Steensel, M.A.; Buma, P.; de Waal Malefijt, M.C.; van den Hoogen, F.H.; Brunner, H.G. Oto-spondylo-megaepiphyseal dysplasia (OSMED): Clinical description of three patients homozygous for a missense mutation in the *COL11A2* gene. *Am. J. Med. Genet.* **1997**, *70*, 315–323. [CrossRef]

149. Logan, C.V.; Cossins, J.; Rodríguez Cruz, P.M.; Parry, D.A.; Maxwell, S.; Martínez-Martínez, P.; Riepsaame, J.; Abdelhamed, Z.A.; Lake, A.V.; Moran, M.; et al. Congenital myasthenic syndrome type 19 is caused by mutations in *COL13A1*, encoding the atypical non-fibrillar collagen type XIII α1 chain. *Am. J. Hum. Genet.* **2015**, *97*, 878–885. [CrossRef] [PubMed]

150. Oliver, V.F.; van Bysterveldt, K.A.; Cadzow, M.; Steger, B.; Romano, V.; Markie, D.; Hewitt, A.W.; Mackey, D.A.; Willoughby, C.E.; Sherwin, T.; et al. A *COL17A1* splice-altering mutation is prevalent in inherited recurrent corneal erosions. *Ophthalmology* **2016**, *123*, 709–722. [CrossRef] [PubMed]

151. Aldahmesh, M.A.; Khan, A.O.; Mohamed, J.; Alkuraya, F.S. Novel recessive BFSP2 and PITX3 mutations: Insights into mutational mechanisms from consanguineous populations. *Genet. Med.* **2011**, *13*, 978–981. [CrossRef] [PubMed]

152. Shinwari, J.M.; Khan, A.; Awad, S.; Shinwari, Z.; Alaiya, A.; Alanazi, M.; Tahir, A.; Poizat, C.; Al Tassan, N. Recessive mutations in *COL25A1* are a cause of congenital cranial dysinnervation disorder. *Am. J. Hum. Genet.* **2015**, *96*, 147–152. [CrossRef] [PubMed]

153. Flynn, J.M.; Ramirez, N.; Betz, R.; Mulcahey, M.J.; Pino, F.; Herrera-Soto, J.A.; Carlo, S.; Cornier, A.S. Steel syndrome: Dislocated hips and radial heads, carpal coalition, scoliosis, short stature, and characteristic facial features. *J. Pediatr. Orthop.* **2010**, *30*, 282–288. [CrossRef] [PubMed]

154. Intong, L.R.; Murrell, D.F. Inherited epidermolysis bullosa: New diagnostic criteria and classification. *Clin. Dermatol.* **2012**, *30*, 70–77. [CrossRef] [PubMed]

155. Saratzis, A.; Sarafidis, P.; Melas, N.; Khaira, H. Comparison of the impact of open and endovascular abdominal aortic aneurysm repair on renal function. *J. Vasc. Surg.* **2014**, *60*, 597–603. [CrossRef] [PubMed]

156. Orioli, D.; Compe, E.; Nardo, T.; Mura, M.; Giraudon, C.; Botta, E.; Arrigoni, L.; Peverali, F.A.; Egly, J.M.; Stefanini, M. *XPD* mutations in trichothiodystrophy hamper collagen VI expression and reveal a role of TFIIH in transcription derepression. *Hum. Mol. Genet.* **2013**, *22*, 1061–1073. [CrossRef] [PubMed]

157. Arseni, L.; Lanzafame, M.; Compe, E.; Fortugno, P.; Afonso-Barroso, A.; Peverali, F.A.; Lehmann, A.R.; Zambruno, G.; Egly, J.M.; Stefanini, M.; et al. TFIIH-dependent MMP-1 overexpression in trichothiodystrophy leads to extracellular matrix alterations in patient skin. *Proc. Natl. Acad. Sci. USA* **2015**, *112*, 1499–1504. [CrossRef] [PubMed]

158. Stefanini, M.; Botta, E.; Lanzafame, M.; Orioli, D. Trichothiodystrophy: From basic mechanisms to clinical implications. *DNA Repair* **2010**, *9*, 2–10. [CrossRef] [PubMed]

159. Faghri, S.; Tamura, D.; Kraemer, K.H.; Digiovanna, J.J. Trichothiodystrophy: A systematic review of 112 published cases characterises a wide spectrum of clinical manifestations. *J. Med. Genet.* **2008**, *45*, 609–621. [CrossRef] [PubMed]

160. Söderhäll, C.; Marenholz, I.; Kerscher, T.; Rüschendorf, F.; Esparza-Gordillo, J.; Worm, M.; Gruber, C.; Mayr, G.; Albrecht, M.; Rohde, K.; et al. Variants in a novel epidermal collagen gene (*COL29A1*) are associated with atopic dermatitis. *PLoS Biol.* **2007**, *5*, e242. [CrossRef] [PubMed]

161. Kay, J.; Gawkrodger, D.J.; Mortimer, M.J.; Jaron, A.G. The prevalence of childhood atopic eczema in a general population. *J. Am. Acad. Dermatol.* **1994**, *30*, 35–39. [CrossRef]

162. Moravej, H.; Karamifar, H.; Karamizadeh, Z.; Amirhakimi, G.; Atashi, S.; Nasirabadi, S. Bruck syndrome—A rare syndrome of bone fragility and joint contracture and novel homozygous FKBP10 mutation. *Endokrynol. Polska* **2015**, *66*, 170–174. [CrossRef] [PubMed]

163. Ha-Vinh, R.; Alanay, Y.; Bank, R.A.; Campos-Xavier, A.B.; Zankl, A.; Superti-Furga, A.; Bonafé, L. Phenotypic and molecular characterization of Bruck syndrome (osteogenesis imperfecta with contractures of the large joints) caused by a recessive mutation in PLOD2. *Am. J. Med. Genet. A* **2004**, *131*, 115–120. [CrossRef] [PubMed]

164. Malfait, F.; Francomano, C.; Byers, P.; Belmont, J.; Berglund, B.; Black, J.; Bloom, L.; Bowen, J.M.; Brady, A.F.; Burrows, N.P.; et al. The 2017 international classification of the Ehlers-Danlos syndromes. *Am. J. Med. Genet. C Semin. Med. Genet.* **2017**, *175*, 8–26. [CrossRef] [PubMed]

165. Baumann, M.; Giunta, C.; Krabichler, B.; Rüschendorf, F.; Zoppi, N.; Colombi, M.; Bittner, R.E.; Quijano-Roy, S.; Muntoni, F.; Cirak, S.; et al. Mutations in FKBP14 cause a variant of Ehlers-Danlos syndrome with progressive kyphoscoliosis, myopathy, and hearing loss. *Am. J. Hum. Genet.* **2012**, *90*, 201–216. [CrossRef] [PubMed]

166. Avgitidou, G.; Siebelmann, S.; Bachmann, B.; Kohlhase, J.; Heindl, L.M.; Cursiefen, C. Brittle cornea syndrome: Case report with novel mutation in the *PRDM5* gene and review of the literature. *Case Rep. Ophthalmol. Med.* **2015**, *2015*, 637084. [CrossRef] [PubMed]

167. Morava, E.; Guillard, M.; Lefeber, D.J.; Wevers, R.A. Autosomal recessive cutis laxa syndrome revisited. *Eur. J. Hum. Genet.* **2009**, *17*, 1099–1110. [CrossRef] [PubMed]

168. Lin, D.S.; Yeung, C.Y.; Liu, H.L.; Ho, C.S.; Shu, C.H.; Chuang, C.K.; Huang, Y.W.; Wu, T.Y.; Huang, Z.D.; Jian, Y.R.; et al. A novel mutation in PYCR1 causes an autosomal recessive cutis laxa with premature aging features in a family. *Am. J. Med. Genet. A* **2011**, *155A*, 1285–1289. [CrossRef] [PubMed]

169. Kivuva, E.C.; Parker, M.J.; Cohen, M.C.; Wagner, B.E.; Sobey, G. De barsy syndrome: A review of the phenotype. *Clin. Dysmorphol.* **2008**, *17*, 99–107. [CrossRef] [PubMed]

170. Fischer-Zirnsak, B.; Escande-Beillard, N.; Ganesh, J.; Tan, Y.X.; Al Bughaili, M.; Lin, A.E.; Sahai, I.; Bahena, P.; Reichert, S.L.; Loh, A.; et al. Recurrent de novo mutations affecting residue Arg138 of Pyrroline-5-carboxylate synthase cause a progeroid form of Autosomal-Dominant cutis laxa. *Am. J. Hum. Genet.* **2015**, *97*, 483–492. [CrossRef] [PubMed]

171. Shetty, R.; Sathyanarayanamoorthy, A.; Ramachandra, R.A.; Arora, V.; Ghosh, A.; Srivatsa, P.R.; Pahuja, N.; Nuijts, R.M.; Sinha-Roy, A.; Mohan, R.R. Attenuation of lysyl oxidase and collagen gene expression in keratoconus patient corneal epithelium corresponds to disease severity. *Mol. Vis.* **2015**, *21*, 12–25. [PubMed]

172. Mas Tur, V.; MacGregor, C.; Jayaswal, R.; O'Brart, D.; Maycock, N. A review of keratoconus: Diagnosis, pathophysiology, and genetics. *Surv. Ophthalmol.* **2017**, *62*, 770–783. [CrossRef] [PubMed]

International Journal of
Molecular Sciences

Article

Impact of Arginine to Cysteine Mutations in Collagen II on Protein Secretion and Cell Survival

Salin A. Chakkalakal [1],[†] , Juliane Heilig [2],[3],[†] , Ulrich Baumann [4] , Mats Paulsson [2],[3],[5],[6] and Frank Zaucke [2],[3],[7],*

1   Center for Research in FOP and Related Disorders, Perelman School of Medicine,
    University of Pennsylvania, Philadelphia, PA 19104, USA; salin.chakkalakal@gmail.com
2   Center for Biochemistry, Medical Faculty, University of Cologne, 50931 Cologne, Germany;
    juliane.heilig@uni-koeln.de (J.H.); mats.paulsson@uni-koeln.de (M.P.)
3   Cologne Center for Musculoskeletal Biomechanics (CCMB), 50931 Cologne, Germany
4   Institute of Biochemistry, University of Cologne, 50931 Cologne, Germany; ubaumann@uni-koeln.de
5   Cluster of Excellence Cellular Stress Responses in Aging-Associated Diseases (CECAD),
    University of Cologne, 50931 Cologne, Germany
6   Center for Molecular Medicine Cologne (CMMC), University of Cologne, 50931 Cologne, Germany
7   Dr. Rolf M. Schwiete Research Unit for Osteoarthritis, Orthopedic University Hospital Friedrichsheim,
    60528 Frankfurt/Main, Germany
*   Correspondence: frank.zaucke@friedrichsheim.de; Tel.: +49-69-6705-372; Fax: +49-69-6705-280
†   These authors contributed equally to this work.

Received: 31 December 2017; Accepted: 6 February 2018; Published: 11 February 2018

Abstract: Inherited point mutations in collagen II in humans affecting mainly cartilage are broadly classified as chondrodysplasias. Most mutations occur in the glycine (Gly) of the Gly-X-Y repeats leading to destabilization of the triple helix. Arginine to cysteine substitutions that occur at either the X or Y position within the Gly-X-Y cause different phenotypes like Stickler syndrome and congenital spondyloepiphyseal dysplasia (SEDC). We investigated the consequences of arginine to cysteine substitutions (X or Y position within the Gly-X-Y) towards the N and C terminus of the triple helix. Protein expression and its secretion trafficking were analyzed. Substitutions R75C, R134C and R704C did not alter the thermal stability with respect to wild type; R740C and R789C proteins displayed significantly reduced melting temperatures (T_m) affecting thermal stability. Additionally, R740C and R789C were susceptible to proteases; in cell culture, R789C protein was further cleaved by matrix metalloproteinases (MMPs) resulting in expression of only a truncated fragment affecting its secretion and intracellular retention. Retention of misfolded R740C and R789C proteins triggered an ER stress response leading to apoptosis of the expressing cells. Arginine to cysteine mutations towards the C-terminus of the triple helix had a deleterious effect, whereas mutations towards the N-terminus of the triple helix (R75C and R134C) and R704C had less impact.

Keywords: collagen II; chondrodysplasia; mutation; unfolded protein response; triple helix

1. Introduction

Collagens are the major proteins in the connective tissues, and collagen II is the major collagen type in cartilage. However, it is also expressed in the vitreous of the eye and is detected during early embryogenesis [1]. It is also expressed in nonchondrogenic tissues including notochord and several parts of the developing eye [2]. Structurally, collagen II consists of three α chains, which are wound around each other into a right-handed triple helix, which consists of repeats of three amino acids (Gly-X-Y repeats). In various heritable connective tissue disorders, glycine substitutions in the collagen chains cause structural changes that result in reduced thermal stability [3–5].

Cells synthesize collagen II α chains as longer precursors called procollagens. Simultaneously, the growing polypeptide chains are co-translationally transported into the rough endoplasmic reticulum (ER) where they undergo a series of post-translational modifications [6]. In addition to post-translational modifications, the ER performs a stringent quality control for unfolded molecules, and these are either retained or degraded. The correct folding and assembly of proteins within the endoplasmic reticulum (ER) are prerequisites for subsequent transport from this organelle to the Golgi apparatus [7]. Only recently, the pathway of collagen secretion from the Golgi complex to the plasma membrane in large cargo vesicles has been elucidated [8]. The recognition and retention of unassembled or misfolded proteins requires an interaction with molecular chaperones within the ER [9]. One classic example of this process occurs during the biosynthesis of procollagen. Incompletely folded intermediates or misfolded products are recognized by chaperones that prevent their secretion and eventually lead to intracellular accumulation [10]. Procollagens are secreted into the extracellular matrix (ECM) through the Golgi apparatus. The collagen II structure is susceptible to mutations as it involves this complex process of collagen synthesis, folding and assembly, leading to disease states that are collectively termed chondrodysplasias and comprise a wide range of well-characterized clinical phenotypes [11–13]. Most chondrodysplasias are due to point mutations in the gene encoding collagen II. Glycin replacement within the Gly-X-Y repeats accounts for 34% of all mutations [14]. These changes result in abnormal conformation and destabilization of the triple helix [15]. Only a few non-glycine missense mutations have been reported, and among these, the arginine to cysteine substitutions predominate [16].

Point mutations leading to a change from arginine to cysteine are interesting since they produce a broader spectrum of unusual phenotypes with either normal or short stature, but never lethal conditions, i.e., congenital spondyloepiphyseal dysplasia (SEDC), Stickler syndrome, Czech dysplasia metatarsal type and osteoarthritis-associated SED. Different amino acid substitutions in the X position of Gly-X-Y repeats have been shown to cause variable phenotypes in Stickler syndrome [17], while it was speculated that substitutions in the Y position might lead to SEDC.

In the present study, we analyzed the impact of arginine to cysteine mutations at the protein level and their effect on intracellular trafficking and secretion. A panel of mutations towards the N- (R75C, R134C) and C- (R704C, R740C and R789C) terminus of the triple helix was selected. These mutations cause a spectrum of different clinical phenotypes in humans including Czech dysplasia metatarsal type (R75C) [18]), Stickler syndrome (R704C) [19,20] and spondyloepiphyseal dysplasia congenita (R789C) [21]. All of these mutations are located in the triple helical region and in the X or Y position of the Gly-X-Y repeats. In addition to these mutants, two artificial mutations R134C and R740C, which are not naturally identified, but lying in the X position of a Gly-X-Y repeat in the triple helix, were also studied.

2. Results

2.1. Collagen II Mutations towards the C Terminus Affect Secretion and Increase Susceptibility to Proteases

Collagen II variants were detected in supernatants and cells lysates of transiently-transfected 293 Epstein–Barr nuclear antigen (EBNA) and HT1080 cells (Figure 1A). All variants were expressed in both cell lines, but R740C and R789C were not as efficiently secreted as the other proteins and found in higher amounts in the cell lysate, indicating an intracellular retention. In addition, the R789C protein migrated much faster on SDS PAGE gels, suggesting a proteolytic processing already in the cells (Figure 1A). All secreted proteins were purified using affinity chromatography on a nickel column and detectable with specific antibodies directed against collagen II and the myc epitope, respectively (Figure 1B). Purified collagen II variants were digested with trypsin to evaluate the formation of a correctly-aligned and stable triple helix. WT, R75C, R134C and R704C collagens were trypsin resistant, while R740C and R789C collagens were completely degraded (Figure 1C). This implies that the secreted R740C and R789C proteins are not forming stable triple helical structures.

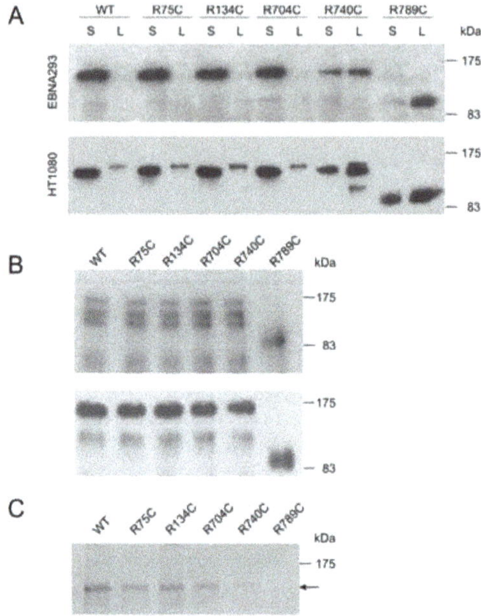

Figure 1. Recombinant expression of collagen II variants. (**A**) 293 Epstein–Barr nuclear antigen (EBNA) (upper panel) and HT1080 cells (lower panel) were transfected with collagen II constructs (WT, R75C, R134C, R704C, R740C and R789C) and the expression and secretion was analyzed using immunoblots. Supernatants (S) and cell lysates (L) were harvested three days post-transfection and separated by SDS PAGE (8% gel) under reducing conditions. Collagen II was detected with an antibody directed against the myc epitope. (**B**) Coomassie blue stained SDS-PAGE gel and immunoblot of purified collagen II variants using an antibody directed against collagen II (upper panel) and the myc epitope (lower panel). (**C**) Collagen II variants were treated with trypsin at 25 °C for 2 min followed by SDS-PAGE. The collagens were visualized by silver staining. WT, as well as R75C, R134C and R704C collagens were resistant to trypsin and displayed a single band of the size of the triple helical domain without the N- and C-propeptides (indicated by an arrow), whereas R740C and R789C collagens were completely degraded.

2.2. C-Terminal Mutations (R740C and R789C) Lead to Thermal Instability of the Triple Helix

Circular dichroism (CD) spectroscopy of WT, R75C, R134C and R704C proteins displayed typical collagen CD spectra with a positive ellipticity at 222 nm, indicating a correctly folded triple helical structure (Figure 2A). In contrast, the collagen II mutants R740C and R789C (yellow and orange line) showed a negative ellipticity at 222 nm, indicating a decreased triple helical content in these proteins and confirming the trypsin sensitivity.

Melting temperatures (T_m) of purified collagens were determined by incremental heating with simultaneous measurement of the molar ellipticity at 222 nm. Melting curves of all purified collagen II proteins are depicted in Figure 2B using a curve fit model assuming that 100% of the molecules were folded initially. The curves for WT, R75C, R134C and R704C collagens show similar profiles, while R740C and R789C proteins showed a decreased T_m value indicating a decreased thermal instability as compared to other collagens. The calculated melting temperatures are summarized in Table 1.

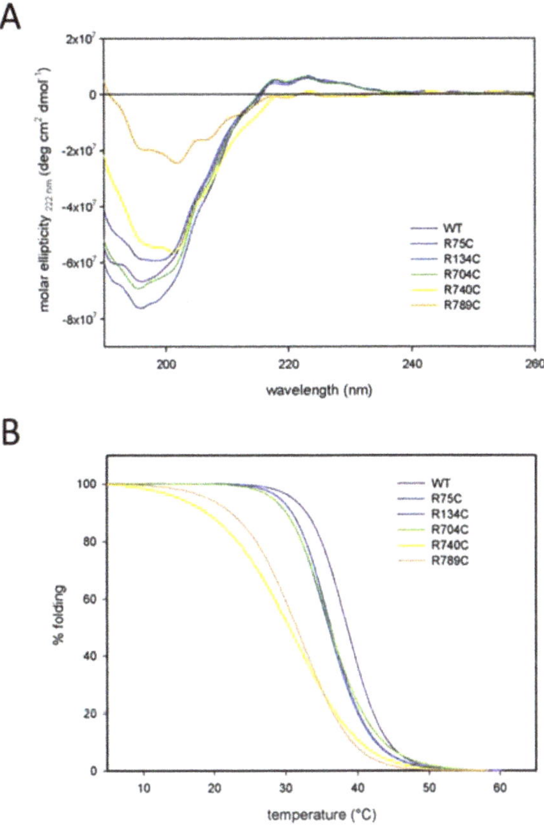

Figure 2. CD spectra and melting curves of purified collagen II variants. (**A**) Spectra were recorded using collagens at a concentration of 60 µg/mL after dialysis against 100 mM acetic acid. The structure of the mutant collagen II proteins R740C and R789C was altered as compared to other collagen II variants, and the shape of the spectra already indicated a decrease in triple helical structure. (**B**) Melting curves of collagen II proteins (60 µg/mL) were recorded after dialysis into 100 mM acetic acid at 222 nm with a 1 °C/min temperature gradient from 10–55 °C. To calculate the percentage of folding, a curve fit model was used assuming that initially 100% of the collagen molecules were folded. The R740C and R789C proteins displayed a decreased thermal stability when compared to all other collagen II variants.

Table 1. Melting temperatures (Tm) of the purified collagen II variants.

Protein	Melting Temperature (T_m)
WT	38.6 °C
R75C	36.2 °C
R134C	36.1 °C
R704C	36.1 °C
R740C	30.2 °C
R789C	31.5 °C

2.3. Inhibition of MMP Activity Prevents Cleavage of R789C

To investigate if the shift in R789C protein mobility is due to MMP cleavage, cells expressing wild type and R789C collagen were incubated for two days with the MMP inhibitor GM6001. In the presence of GM6001, significant amounts of R789C proteins remained uncleaved, in contrast to what was seen in the absence of the inhibitor (Figure 3). This suggests that the mutation R789C increases the susceptibility and/or accessibility for MMPs to cleave collagen II. Interestingly, the resulting uncleaved R789C protein was now detected in the supernatant in large amounts indicating that inhibition of cleavage results in increased secretion. Even though some uncleaved protein was secreted, a large proportion of both uncleaved and cleaved collagen was still retained intracellularly.

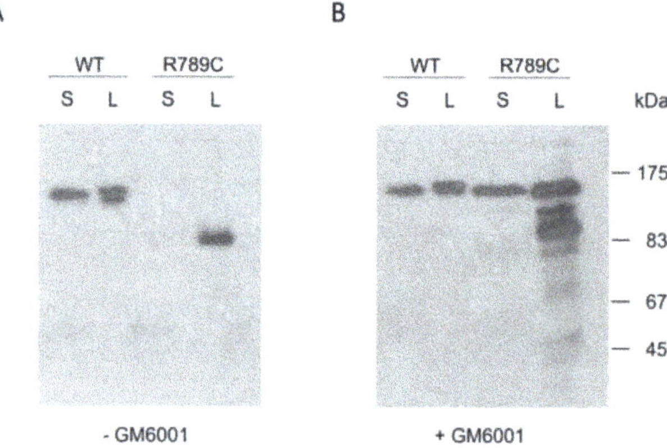

Figure 3. Analysis of proteolytic processing using the MMP inhibitor GM6001. Cells expressing wildtype (WT) and R789C collagens were cultured for two days in the absence (**A**) or presence (**B**) of the MMP inhibitor GM6001 (25 µM). Supernatants (S) and cell lysates (L) were harvested and analyzed by immunoblot with an antibody direct against the myc epitope. After treatment with GM6001, the R789C collagen was partially protected from degradation, indicating that the reduced mass and shift in mobility are caused by a proteolytic cleavage by MMPs.

2.4. Effect of Mutations on Intracellular Collagen Trafficking in Transfected HT1080 Cells

Collagen II trafficking was visualized by co-staining with antibodies directed against PDI and the 58K protein as marker for the ER and the Golgi apparatus, respectively, in transfected HT1080 cells. WT, R75C, R134C and R704C collagen II proteins co-localized mainly with the perinuclear Golgi apparatus, demonstrating that the intracellular trafficking of these proteins from ER to Golgi apparatus is not affected by the mutation (Figure 4). In contrast, the mutant R740C and R789C were mainly detected in the ER, which is typically spread throughout the whole cell. Overlapping staining for collagen II and PDI in cells expressing R740C and R789C proteins suggests that the protein is retained in the ER compartment due to the mutation (Figure 4).

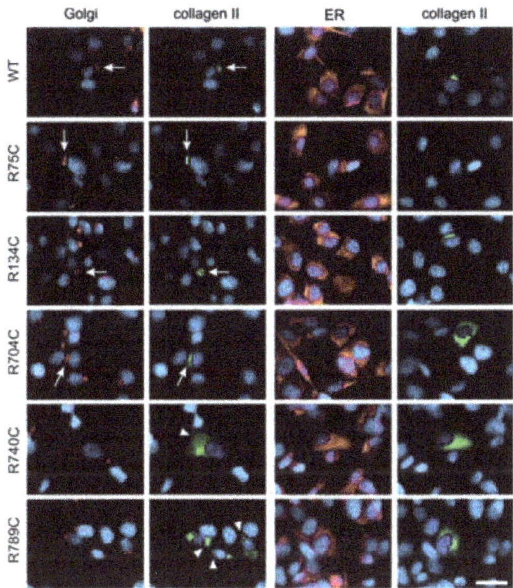

Figure 4. Immunofluorescence microscopy of transfected HT1080 cells showing the intracellular localization of wildtype and mutant collagen II variants. HT1080 cells transfected with collagen II constructs were analyzed three days post-transfection with antibodies directed against the myc epitope of collagen II (green) and either against PDI as a marker for ER or 58K as a marker for the Golgi apparatus (both in red). Colocalization of collagen with 58K in the Golgi compartment is seen in cells expressing WT, R75C, R134C and R704C collagen (arrows). In cells expressing R740C and R789C collagens, significant amounts of protein colocalize with PDI in the ER detected outside the Golgi apparatus (arrowheads). Scale bar, 25 μm.

2.5. C-Terminal Mutations Induce Activation of the Unfolded Protein Response Due to Accumulation of Misfolded Collagen II Proteins

Intracellular retention of misfolded proteins in the ER leads to activation of a complex signal transduction pathway called the unfolded protein response (UPR) [22,23]. UPR results in the induction of ER stress response genes by two potent transcription factors, activating transcription factor-6 and XBP-1 (X-box DNA binding protein-1). Active XBP-1 is generated by excision of a 26-nucleotide sequence from the XBP-1 transcript by IRE1 endonuclease in response to accumulation of misfolded proteins in the ER [22,23]. XBP-1 splicing was investigated by RT-PCR from RNA isolated from HT1080 cells using specific primers giving rise to a 248-bp band in the case of unspliced XBP-1 and to a 222-bp band where XBP-1 is spliced. Figure 5A shows the PCR products resolved on a 2.5% agarose gel. In cells expressing WT, R75C, R134C and R704C collagen, only the unspliced form of XBP-1 was detected, while in cells expressing R740C and R789C collagen, the spliced variant (XBP-S) could also be detected. This suggests that the intracellular retention of misfolded proteins has triggered an ER stress response.

Expression and Association of Chaperones with Mutant Collagen II Proteins

Activation of XBP-1 results in upregulation of ER-resident molecular chaperones. To investigate whether the splicing of XBP-1 associated with accumulation of R740C and R789C mutant collagen II proteins results in increased expression and association of BiP in transfected HT1080 cells, cells were stained for collagen II and BiP. BiP expression was observed in cells expressing WT collagen, and collagen II staining was observed around the Golgi compartment in the transfected cells (Figure 5B) with only little co-staining detected between BiP and collagen II. In cells expressing R740C and R789C collagens, significant co-staining of BiP and collagen II proteins was observed, and the intensity of BiP staining was also increased in these cells, compared to the cells transfected with wildtype collagen II.

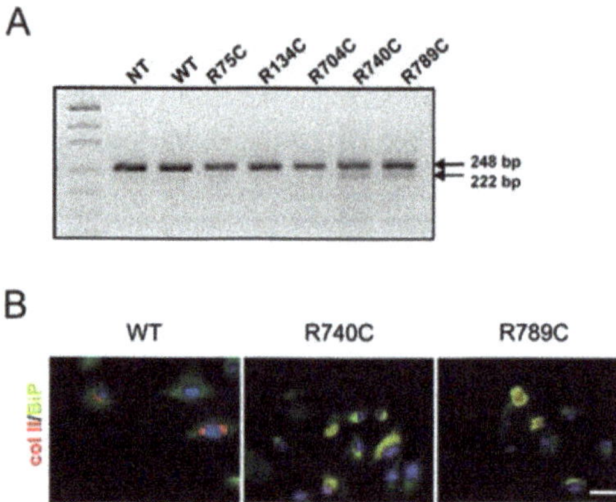

Figure 5. Detection of stress-induced XBP-1 splicing and BiP expression in transfected HT1080 cells. (**A**) mRNA isolated from HT1080 cells was used for RT-PCR with primers specific for XBP-1. The PCR products were separated on 2.5% agarose gels. XBP-1 mRNA (248 bp) could be detected in all cells. Splicing of XBP-1 due to ER stress gives rise to a 222-bp fragment that was seen only in cells transfected with R740C and R789C constructs. (**B**) HT1080 cells transfected with collagen II constructs were analyzed three days post-transfection by staining with antibodies directed against collagen (red) and BiP (green). BiP expression was observed in all the cells. However, increased BiP expression and colocalization with collagen II resulting in strong yellow signals were seen only in cells expressing R740C and R789C collagens. Scale bar, 25 μm.

2.6. Intracellular Accumulation of Misfolded Proteins Induces Apoptosis in Cells Expressing Mutant Collagens

Retention of large amounts of misfolded proteins affects cell viability. We therefore analyzed if HT1080 cells expressing mutant collagens undergo apoptosis as a consequence of ER stress. The executioner caspase caspase-3 has to be activated by other proteases like caspase-8 and -9. Caspase-3 then cleaves a large number of cellular proteins, and their degradation finally disrupts cellular homeostasis and causes cell death.

Three days after transfection, R740C- and R789C-expressing HT1080 cells were positive for activated caspase-3 detected by immunostaining using a FITC-conjugated anti-active caspase-3 antibody. In non-transfected cells or cells expressing WT, R75C, R134C and R704C collagens, no active caspase-3 staining could be detected (Figure 6 and Supplemental Figure S1).

Thus, cells expressing R740C and R789C collagens become apoptotic, most likely due to an ER stress response triggered by irreversible accumulation of misfolded proteins. Apoptosis is accompanied by cleavage and fragmentation of nuclear DNA. We therefore analyzed if DNA fragmentation could be detected in HT1080 cells expressing R740C and R789C collagens. Nick-labeled DNA was not observed in non-transfected cells and cells transfected with wild type and R704C constructs, whereas nick-labeled DNA was observed in cells expressing R740C and R789C (Figure 6).

Apoptosis was quantified in HT1080 cells four days post-transfection by comet assays (single cell gel electrophoresis) (Figure 7). Nuclear DNA fragmentation leads to DNA tailing that can be visualized by a comet-shaped appearance. Such comets tails were observed in cells transfected with R740C and R789C constructs (Figure 7A). From each genotype (20 cells per genotype), the mean tail length was determined using the CometScore software (Figure 7B). Apoptosis in cells expressing

R740C and R789C collagens was indicated by a significant increase in tail length corresponding to DNA fragmentation.

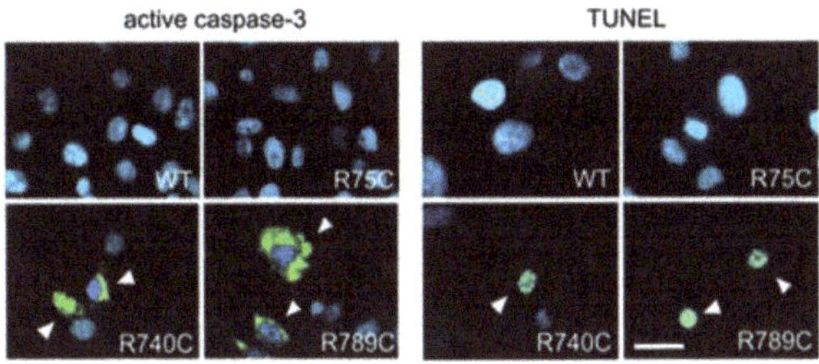

Figure 6. Detection of apoptosis in HT1080 cells transfected with collagen II constructs. HT1080 cells transfected with collagen II constructs were analyzed three days post-transfection for the presence of active caspase-3 (green) as a marker for apoptosis. Active caspase-3 was only found in cells expressing R740C and R789C collagens (left panel, indicated by arrow heads). Four days after transfection, HT1080 cells transfected with R740C and R789C constructs were positive for TUNEL staining (green, right panel, indicated by arrow heads). Nick labeling was neither seen in non-transfected cells, nor in cells transfected with other collagen constructs (Supplemental Figure S1). Scale bar, 25 μm.

A

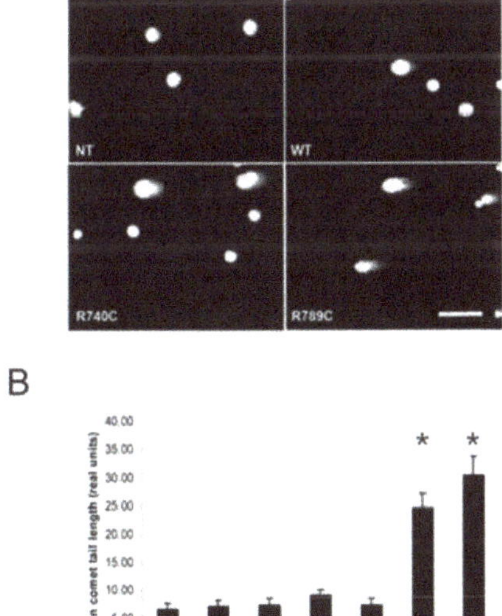

B

Figure 7. Single cell gel electrophoresis (comet assay) of HT1080 cells transfected with collagen II variants. (**A**) Nuclei of non-transfected (NT) and WT collagen-expressing cells were round when visualized after electrophoresis. In contrast, a comet-shaped appearance of fragmented nuclei was seen in cells transfected with R740C and R789C, indicating ongoing DNA fragmentation. Scale bar, 25 μm. (**B**) The mean tail length was evaluated for these comets (20 cells for each genotype) using the CometScore software. A significant increase in tail length was observed in cells transfected with R740C and R789C collagen constructs, while cells transfected with all other constructs behaved similar to non-transfected controls. Asterisks indicate significant difference compared to and WT collagen-expressing cells with $p < 0.05$.

3. Discussion

Point mutations in fibrillar collagens cause a number of abnormalities in connective tissues, leading for example to brittle bone disease, osteoarthritis and osteochondrodysplasias [24,25]. In the present study, a set of *COL2A1* mutations leading to substitution of an arginine to cysteine residue in the triple helix were studied. The mutations were selected based on their localization within the triple helix and position with the Gly-X-Y repeats. Interestingly, the selected mutations cause a rather heterogeneous disease spectrum in humans including Czech dysplasia metatarsal type (R75C), Stickler syndrome (R704C) and spondyloepiphyseal dysplasia congenital (R789C). We included two artificial mutations (R134C and R740C) and analyzed their effects on intracellular protein trafficking, secretion and cell survival.

We were able to express all variants in both 293 EBNA and HT1080 cells. WT, R75C, R134C and R704C proteins were mainly detected in the cell culture supernatant indicating a normal secretion. The variant R740C showed a retarded secretion with similar protein amounts in the supernatant and cell lysate, respectively. R789C collagen was not only retained intracellularly, but also processed, resulting in a prominent band at around 90–100 kDa. This cleavage took place already in the intracellular compartment. Altered secretion and moderate intracellular retention of R789C collagen was reported earlier when this construct was expressed in SW-1353 and HT1080 cells [26,27]. However, the processing was observed for the first time and is in contrast to earlier reports [27,28]. This apparent difference might be explained by expression levels depending on the vectors used and it is likely that a cell starts to degrade the protein when a certain amount has been accumulated. Even though two variants were not properly secreted, we were able to purify sufficient amounts of all constructs to perform biochemical analysis. CD spectroscopy with collagen-specific spectra indicated correct folding of the variants WT, R75C, R134C and R704C. The melting temperature of wild type collagen II with 38.6 °C was 2.4 °C lower than for collagen extracted from bovine nasal cartilage [29]. This difference in the absolute melting temperature might be caused by inefficient hydroxylation by 293 EBNA cells [30,31]. The R75C, R134C and R704C proteins had a 2.5 °C lower melting temperature than the wild type protein, suggesting slight structural differences even though these changes were not pronounced enough to loose trypsin resistance. In contrast and in agreement with earlier studies [26,27], R740C and R789C collagens had a significantly reduced melting temperature, and incubation with trypsin led to complete degradation, confirming an unstable triple helix in these variants. Such an instability could well contribute to the disease mechanism in human patients and has been reported to be involved for mutations in *COL2A1* causing Kniest dysplasia [32], in *COL3A1* causing Ehlers-Danlos syndrome type IV [33] and in COL17A1 causing junctional epidermolysis bullosa [34].

The cleavage of partially-misfolded R789C collagen might be due to an increased accessibility for proteases at regions around the site of mutation. MMP-1, MMP-8 and MMP-13 [35,36] are able to initiate the intrahelical cleavage of triple helical collagen at neutral pH. These collagenases cleave the collagens types I, II and III specifically at a single site (Gly_{775}–Leu/Ile_{776}) within each α chain of the triple helical collagen molecule [37]. The exact cleavage position in the R789C protein is yet to be identified, but it is attractive to speculate that the mutation in close proximity of the well-known MMP cleavage site facilitates cleavage due to local unfolding [38–40]. The fact that pretreatment of the cells with the MMP inhibitor GM6001 abrogated the processing further supports the notion that MMPs are responsible for the cleavage. Interestingly, the inhibition of cleavage leads to an increased secretion of the now fully-intact protein into the supernatant. This is in good agreement with earlier studies in which mutating the MMP cleavage site resulted in an increased secretion of R789C collagen [27,28,41]. Increased collagen cleavage by MMPs at this position has indeed been shown in patients with specific forms of osteoarthritis, as well as in transgenic mice with osteoarthritis [42–44]. This might explain why specific collagen mutations predispose for cartilage degeneration.

The fibroblastic cell line HT1080 is commonly used to study protein expression, trafficking and secretion of collagen II [45,46]. Immunoblotting already suggested that a considerable amount of R740C and R789C protein was present in cell lysates owing to intracellular retention. Co-staining with compartment-specific antibodies revealed that WT, R75C, R134C and R704C proteins were found mainly in the Golgi compartment, indicating that these proteins are being transported efficiently to the extracellular space. In contrast, R740C and R789C proteins showed only weak co-localization with the 58k Golgi marker and were instead detected in other cellular compartments. Co-staining with PDI, an ER-resident chaperone, suggests that mutant and misfolded R740C and R789C proteins are retained in the ER [47,48]. A similar intracellular retention of the misfolded proteins in the ER due to point mutations was reported for other cartilage proteins such as COMP, collagen IX and matrilin-3 in patients and animal models [49,50].

The retention of misfolded protein in the ER leads to the activation of a complex signal transduction pathway called the unfolded protein response (UPR) [22,23]. The fact that a targeted induction of ER stress alone is able to induce cartilage pathology [51] underlines the crucial role of this process in disease initiation. The splicing of XBP-1 mRNA is a hallmark of the mammalian UPR, and we indeed detected this unconventional splicing event in cells expressing R740C and R789C collagen. A similar observation was made in response to misfolded protein accumulation due to a mutation in the NC1 domain of collagen X [52]. XBP-1 splicing in turn leads to an upregulation of ER-resident molecular chaperones like BiP [53]. We also detected an upregulation and colocalization of BiP with R740C and R789C procollagen chains, and similar findings were reported for collagen chains harboring mutations in type I collagen from patients with osteogenesis imperfecta [54].

Prolonged accumulation of misfolded proteins in the ER without degradation can lead to programmed cell death [55,56]. Indeed, the large amount of rounded and dead cells observed in cultures expressing R740C and R789C collagens may indicate that the cells are undergoing apoptosis. We were able to detect an increase of active caspase-3 and TUNEL staining in cells transfected with R740C and R789C constructs, confirming earlier observations [57]. An increased apoptosis may also explain the reduced number of chondrocytes observed in both patients and animal models [37,58,59]. In addition to intracellular disturbances described above, changes in the ECM might also contribute to the phenotype. It has been shown earlier for mutations in COMP that both intra- and extra-cellular pathways are involved in the pathogenesis of pseudoachondroplasia [60] and that disruption of the ECM structure may cause phenotypes even in the absence of impaired secretion [61]. The structure of the ECM could be affected by reduced amounts, the complete absence or the presence of mutated collagen II. The amount and quality of extracellular collagen II depends on the specific mutation, and this might explain why different mutations cause such a spectrum of disease phenotypes.

In our study, mutations at the N-terminus (R75C, R134C) of the triple helix resulted in less pronounced effects on all parameters investigated compared to mutations at the C-terminus. Similar findings were reported for mutations in collagen I in which glycine substitutions towards the C-terminus of the collagen I chains are clinically more severe than those towards the N-terminus [62]. This might be due to the fact that triple helix formation and propagation initiates at the C-terminus, and mutations at this structurally-important site will interfere with protein stability [63–65]. Our findings suggest that the phenotype of patients harboring the C-terminal mutation R789C in the collagen II chain might be caused by accumulation of the mutated collagen II protein in the ER, leading to ER-stress and apoptosis of chondrocytes. However, this seems not to be a universal disease mechanism common for all collagen mutations. According to our study, even the rather C-terminal R704C and also the N-terminal R75C mutation neither induce ER stress nor apoptosis and might cause a disease predominantly via extracellular interference. In addition, it is tempting to speculate that other factors such as genetic modifiers or the presence of additional mutations in other gene loci yet to be identified contribute to a chondrodysplasia phenotype. Similar observations were made in patients with multiple epiphyseal dysplasia (MED) [66].

4. Materials and Methods

4.1. Collagen II cDNA, Site-Directed Mutagenesis and Removal of Endogenous Signal Peptide

Human collagen II cDNA of 4.5 kb including the poly-adenylation signal was kindly provided by Fibrogen Europe. The eukaryotic expression vector pCEP-Pu containing a BM-40 signal peptide, puromycin resistance [67] and an N-terminal sequence coding for the his$_6$-myc tag [68] was used for expression of collagen II variants in the mammalian cells. Site-directed mutagenesis was carried out using the XL Quick Change mutagenesis kit (Stratagene, La Jolla, CA, USA). Mutations were introduced into collagen II cDNA resulting in exchange of R75C, R134C, R704C, R740C and R789C. The primer pairs that were used for sited-directed mutagenesis are represented in Table 2. Nucleotide change leading to mutation is indicated in bold. The mutated constructs were inserted into the episomal expression vector pCEP-Pu- his$_6$-myc tag (N-terminal) in-frame with the sequence of the signal peptide of BM-40 as described earlier [66].

Table 2. Primer pairs used for site directed mutagenesis.

Primer	Sequence
75F	5′-GGTCCTCAGGGTGCTT**G**TGGTTTCCCAGG-3′
75R	5′-CCTGGGAAACCAC**A**AGCACCCTGAGGACC-3′
134F	5′-CTGGTGAAAGAGGA**T**GCACTGGCCCTGCTG-3′
134R	5′-CCAGCAGGGCCAGTGC**A**TCCTCTTTCACC-3′
704F	5′-GGAGCTGCTGGGTGC**G**TTGGACCCCC-3′
704R	5′-GGGGGTCCAACGC**A**CCCAGCAGCTCC-3′
740F	5′-CCCCCTGGCTGCG**C**TGGTGAACCCGG-3′
740R	5′-GGGTTCACCAGC**G**CAGCCAGGGGGG-3′
789F	5′-GGTCTGCCTGGGCAA**T**GTGGTGAGAGAGGATTCC-3′
789R	5′-GGAATCCTCTCTCACCAC**A**TTGCCCAGGCAGACC-3′

Numbers 75, 134, 704, 740 and 789 stand for the amino acid residue numbered from the start of the triple helix where the mutation of interest is located. Nucleotides marked in bold indicate the nucleotide change to introduce the desired mutation.

4.2. Cell Culture and Transfection

Human embryonic kidney-derived 293 EBNA cells (Invitrogen) and human fibrosarcoma-derived HT1080 (CCL-121 from ATTC) cells were cultured in DMEM-F12 medium containing 200 U/mL penicillin, 200 µg/mL streptomycin, 20 mM L-glutamine and 10% FBS (Biochrom, Berlin, Germany), from now on referred to as standard medium, at 37 °C in a humified incubator with a 5% CO_2 atmosphere. One hundred micrograms per milliliter of ascorbate were added to the cell culture medium during expression of recombinant collagens. Cells were transfected using FuGene6 reagent (Roche, Munich, Germany) following manufacturers instruction.

4.3. Purification of Collagen II Proteins

After 24 h post-transfection of 293 EBNA cells with collagen II constructs, cells were selected using puromycin, and the supernatants were harvested from stably-transfected cells and the secreted proteins purified on a nickel column by ion exchange chromatography, as previously described in [68].

4.4. CD Spectroscopy and Melting Curves

CD spectroscopy was performed with purified collagen II proteins, which were dialyzed in 100 mM acetic acid, and the concentration was adjusted to 60 µg/mL. Two hundred microliters of collagen solution were used, and a spectrum between 190 nm and 280 nm was recorded using a Jasco J-715 Polarimeter at 4 °C. Melting curves were determined at 222 nm in the temperature range of 10–55 °C at increments of 1 °C/min.

4.5. SDS-PAGE and Immunoblotting

SDS polyacrylamide gel electrophoresis was performed using the buffer system of Laemmli. Protein samples were mixed with an equal volume of 2× SDS sample buffer (50 mM Tris-HCl, pH 6.8, 2% (*w/v*) SDS, 20% glycerol and 0.025% (*w/v*) bromophenol blue), whereas cells were lysed and resuspended in 1× SDS sample buffer. Before loading, samples were reduced and denatured by adding 5% β-mercaptoethanol and boiled at 95 °C for 5 min. Equal amounts of supernatants and cell lysates obtained 72 h after transfection were resolved on 8% (*w/v*) SDS-PAGE gels and were analyzed by immunoblotting. After SDS-PAGE, proteins were transferred to nitrocellulose and incubated for 1 h in TBS in the presence of 5% (*w/v*) milk protein. Rabbit anti-myc antibodies or goat anti-collagen II antibodies were used at a dilution of 1:1500, and the bound primary antibodies were detected using a 1:2000 dilution of horseradish peroxidase-conjugated rabbit anti-goat or swine anti-rabbit IgG (Dako Corp., Glostrup, Denmark). Blots were developed with a self-made enhanced chemiluminescence (ECL) reaction.

4.6. Trypsin Digestion

Protease treatment of purified collagen II variant proteins (3 micrograms) was carried out in a 50-μL reaction volume containing digestion buffer (50 mM Tris-HCl, pH 7.4, containing 150 mM NaCl and 10 mM EDTA). Trypsin was added to the samples at a concentration of 100 μg/mL and was incubated for 2 min at 25 °C. The digestion was stopped by addition of soybean trypsin inhibitor (Sigma, Munich, Germany) at a final concentration of 5 μg/mL. SDS-PAGE sample buffer containing 5% β-mercaptoethanol was added, and the samples were resolved on 8% SDS-PAGE. The resolved proteins were visualized by silver staining.

4.7. Inhibition of Matrix Metalloproteinase Cleavage

HT1080 cells were transfected with collagen II constructs, and three days after transfection, the supernatants were collected. Fresh culture medium was supplemented with 25 μM GM6001 (1 mM stock in DMSO) and 100 μg/mL of ascorbate. Supernatants and cell lysates were harvested 48 h after addition of the inhibitor. Proteins were resolved by SDS-PAGE on 8% gels and collagen II, and after blotting, collagen II and fragments thereof were detected using the anti-myc antibody.

4.8. Reverse Transcription-PCR for Analysis of Spliced XBP1 Transcripts

RNA was extracted using the TRIZOL reagent (Invitrogen, Carlsbad, CA, USA) according to the manufacturer's instruction from cells 72 h post-transfection. To avoid ER stress induced by glucose starvation, fresh medium with serum was added 2 h prior to RNA extraction. One hundred nanograms of total RNA/reaction were used for cDNA synthesis using reverse transcriptase and random hexamers (Roche Diagnostics, Basel, Switzerland). XBP1 transcripts were analyzed by PCR of the cDNA using primers corresponding to the spliced and unspliced forms of the XBP1 transcripts (5′-GGAGTTAAGACAGCGCTTGG-3′, bp 401–420) and antisense (5′-ACTGGGTCCAAGTTGTCCAG-3′, bp 648–629) primers spanning the XBP1 RNA processing sequence (GenBank Accession Number AB076383) [37]. The PCR products corresponding to unspliced and spliced *XBP1* (248 and 222 bp, respectively) were obtained after 35 cycles using a primer annealing temperature of 60 °C. The products were resolved on 2.5% (*w/v*) agarose gels and visualized under ultraviolet light.

4.9. Immunofluorescence Staining of HT1080 Cells

Transfected and non-transfected cells were grown on glass cover slips in 24-well dishes. After 3 days of transfection, the cells were fixed with 2% paraformaldehyde and permeabilized using 0.2% Triton X-100 in PBS. Cells were blocked with 10% normal goat serum in PBS after three washes with PBS. Primary antibodies were added at a dilution of 1:1000 and incubated for

60 min, followed by washings to remove unbound antibodies. Bound primary antibodies were detected with a secondary fluorescent-labelled antibody for a further 60 min. Antibodies directed against protein disulfide isomerase (PDI) (Biomol, Hamburg, Germany) and the 58K protein Sigma (Munich, Germany) were used as markers for the ER and the Golgi apparatus, respectively. Mouse anti-myc antibodies were used to detect the myc epitope of recombinantly-expressed collagen II. Rabbit anti-BiP was used for the detection of BiP. Secondary anti-rabbit and secondary anti-mouse Cy3- and Alexa488-conjugated antibodies were from Molecular Probes (Leiden, The Netherlands). Nuclei were stained with bisbenzimide (Sigma, 0.1 μg/mL). The slides were finally mounted in DAKO fluorescent mounting medium and examined under an Axiophot fluorescence microscope (Zeiss, Oberkochen, Germany).

4.10. Nick Labelling

TUNEL staining was performed to detect apoptosis using the DeadEnd™ Fluorometric TUNEL System (Promega, Madison, WI, USA) following a slightly modified manufacturer's protocol. Cells were fixed by 25-min immersion in 4% paraformaldehyde solution four days post-transfection. After two washes in PBS for 5 min cells were permeabilized with 0.2% Triton X-100 solution in PBS for 5 min. After rinsing cells in PBS, 100 μL of equilibration buffer were added for 5–10 min at room temperature. Cells were then incubated with 50 μL of reaction buffer (rTdT incubation buffer) for 60 min inside a dark humidified chamber at 37 °C. The tailing reaction was terminated by adding 2× saline sodium citrate solution (SSC) for 15 min. Nuclei were stained with bisbenzimide (0.1 μg/mL in PBS) for 5 min. After washing with deionized water for 5 min at room temperature, cover slips were mounted on histo-slides with DAKO mounting medium.

4.11. Comet Assay or Single Cell Gel Electrophoresis

Four days after transfection, cells were trypsinized, counted and diluted to give approximately 5×10^4 cells/mL. Eighty microliters of the cell suspension were added to 400 μL of 0.5% low melting agarose maintained at 37 °C, and 90 μL of this suspension were pipetted onto a slide, which was precoated with 1.0% agarose. An additional 1% low melting agarose layer without cells was added after solidification of the layer. Alkaline lysis solution (2.5 M NaCl, 100 mM EDTA, 10 mM Tris-HCl, 1% Triton X-100, 10% DMSO, pH to 10.0) was added to the slides at 4 °C for 1 h in the dark to lyse cells. After lysis, the slides were placed in neutralizing solution (400 mM Tris-HCl, pH 7.5) and rinsed three times for 30 min to remove salts and detergents. Slides were then placed in a horizontal electrophoresis chamber with alkaline buffer (300 mM NaOH, 1 mM EDTA, pH > 13) and incubated for 20–60 min to allow unwinding of the DNA. Electrophoresis of the slides was run for 30 min at 0.6 V/cm. After electrophoresis, the slides were rinsed with neutralization buffer twice and stained with 100 μL ethidium bromide solution (10 mg/mL). Slides were scored immediately or alternatively dried in cold 100% ethanol before storage. DNA tailing due to fragmentation was visualized as comets using an inverted fluorescence microscope and evaluated using TriTek CometScore™ software (TriTek Corp, Sumerduck, VA, USA).

Supplementary Materials: Supplementary materials can be found at www.mdpi.com/1422-0067/19/2/541/s1.

Acknowledgments: This project was supported by grants from the Deutsche Forschungsgemeinschaft (SFB829/B11 to Frank Zaucke and Ulrich Baumann and the European Community's Seventh Framework Programme to F.Z. under Grant Agreement No. 602300 (SYBIL). Salin A. Chakkalakal obtained a fellowship and funding from the International Graduate School in Genetics and Functional Genomics at the University of Cologne, Germany.

Author Contributions: Salin A. Chakkalakal, Mats Paulsson and Frank Zaucke conceived of and designed the experiments. Salin A. Chakkalakal and Juliane Heilig performed the experiments. Salin A. Chakkalakal, Juliane Heilig, Ulrich Baumann, Mats Paulsson and Frank Zaucke analyzed the data. Salin A. Chakkalakal, Juliane Heilig and Frank Zaucke wrote the paper. All authors have approved the submitted version.

Conflicts of Interest: The authors declare no conflict of interest.

References

1. Myllyharju, J.; Kivirikko, K.I. Collagens, modifying enzymes and their mutations in humans, flies and worms. *Trends Genet.* **2004**, *20*, 33–43. [CrossRef] [PubMed]
2. Brewton, R.G.; Mayne, R. Mammalian vitreous humor contains networks of hyaluronan molecules: Electron microscopic analysis using the hyaluronan-binding region (G1) of aggrecan and link protein. *Exp. Cell Res.* **1992**, *198*, 237–249. [CrossRef]
3. Kivirikko, K.I. Collagens and their abnormalities in a wide spectrum of diseases. *Ann. Med.* **1993**, *25*, 113–126. [CrossRef] [PubMed]
4. Dalgleish, R. The human type I collagen mutation database. *Nucleic Acids Res.* **1997**, *25*, 181–187. [CrossRef] [PubMed]
5. Dalgleish, R. The human collagen mutation database 1998. *Nucleic Acids Res.* **1998**, *26*, 253–255. [CrossRef] [PubMed]
6. Bulleid, N.J.; John, D.C.; Kadler, K.E. Recombinant expression systems for the production of collagen. *Biochem. Soc. Trans.* **2000**, *28*, 350–353. [CrossRef] [PubMed]
7. Canty, E.G.; Kadler, K.E. Procollagen trafficking, processing and fibrillogenesis. *J. Cell Sci.* **2005**, *118*, 1341–1353. [CrossRef] [PubMed]
8. Malhotra, V.; Erlmann, P. The pathway of collagen secretion. *Annu. Rev. Cell Dev. Biol.* **2015**, *31*, 109–124. [CrossRef] [PubMed]
9. Gregersen, N.; Bolund, L.; Bross, P. Protein misfolding, aggregation, and degradation in disease. *Mol. Biotechnol.* **2005**, *31*, 141–150. [CrossRef]
10. Walmsley, A.R.; Batten, M.R.; Lad, U.; Bulleid, N.J. Intracellular retention of procollagen within the endoplasmic reticulum is mediated by prolyl 4-hydroxylase. *J. Biol. Chem.* **1999**, *274*, 14884–14892. [CrossRef] [PubMed]
11. Mundlos, S.; Olsen, B.R. Heritable diseases of the skeleton. Part II: Molecular insights into skeletal development-matrix components and their homeostasis. *FASEB J.* **1997**, *11*, 227–233. [CrossRef] [PubMed]
12. Myllyharju, J.; Kivirikko, K.I. Collagens and collagen-related diseases. *Ann. Med.* **2001**, *33*, 7–21. [CrossRef] [PubMed]
13. Vikkula, M.; Metsaranta, M.; Ala-Kokko, L. Type II collagen mutations in rare and common cartilage diseases. *Ann. Med.* **1994**, *26*, 107–114. [PubMed]
14. Barat-Houari, M.; Sarrabay, G.; Gatinois, V.; Fabre, A.; Dumont, B.; Genevieve, D.; Touitou, I. Mutation Update for COL2A1 Gene Variants Associated with Type II Collagenopathies. *Hum. Mutat.* **2016**, *37*, 7–15. [CrossRef] [PubMed]
15. Hoornaert, K.P.; Vereecke, I.; Dewinter, C.; Rosenberg, T.; Beemer, F.A.; Leroy, J.G.; Bendix, L.; Björck, E.; Bonduelle, M.; Boute, O.; et al. Stickler syndrome caused by COL2A1 mutations: Genotype-phenotype correlation in a series of 100 patients. *Eur. J. Hum. Genet.* **2010**, *18*, 872–880. [CrossRef] [PubMed]
16. Hoornaert, K.P.; Dewinter, C.; Vereecke, I.; Beemer, F.A.; Courtens, W.; Fryer, A.; Fryssira, H.; Lees, M.; Müllner-Eidenböck, A.; Rimoin, D.L.; et al. The phenotypic spectrum in patients with arginine to cysteine mutations in the COL2A1 gene. *J. Med. Genet.* **2006**, *43*, 406–413. [CrossRef] [PubMed]
17. Richards, A.J.; Baguley, D.M.; Yates, J.R.; Lane, C.; Nicol, M.; Harper, P.S.; Scott, J.D.; Snead, M.P. Variation in the vitreous phenotype of stickler syndrome can be caused by different amino acid substitutions in the x position of the type II collagen gly-x-y triple helix. *Am. J. Hum. Genet.* **2000**, *67*, 1083–1094. [PubMed]
18. Williams, C.J.; Considine, E.L.; Knowlton, R.G.; Reginato, A.; Neumann, G.; Harrison, D.; Buxton, P.; Jimenez, S.; Prockop, D.J. Spondyloepiphyseal dysplasia and precocious osteoarthritis in a family with an $Arg_{75} \rightarrow cys$ mutation in the procollagen type Ii gene (COL_2A_1). *Hum. Genet.* **1993**, *92*, 499–505. [CrossRef] [PubMed]
19. Ballo, R.; Beighton, P.H.; Ramesar, R.S. Stickler-like syndrome due to a dominant negative mutation in the COL2A1 gene. *Am. J. Med. Genet. A* **1998**, *80*, 6–11. [CrossRef]
20. Richards, A.J.; McNinch, A.; Martin, H.; Oakhill, K.; Rai, H.; Waller, S.; Treacy, B.; Whittaker, J.; Meredith, S.; Poulson, A.; et al. Stickler syndrome and the vitreous phenotype: Mutations in COL2A1 and COL11A1. *Hum. Mutat.* **2010**, *31*, E1461–E1471. [CrossRef] [PubMed]

21. Chan, D.; Taylor, T.K.; Cole, W.G. Characterization of an arginine 789 to cysteine substitution in alpha 1 (ii) collagen chains of a patient with spondyloepiphyseal dysplasia. *J. Biol. Chem.* **1993**, *268*, 15238–15245. [PubMed]

22. Rutkowski, D.T.; Kaufman, R.J. A trip to the ER: Coping with stress. *Trends Cell Biol.* **2004**, *14*, 20–28. [CrossRef] [PubMed]

23. Zhang, K.; Kaufman, R.J. Signaling the unfolded protein response from the endoplasmic reticulum. *J. Biol. Chem.* **2004**, *279*, 25935–25938. [CrossRef] [PubMed]

24. Bruckner-Tuderman, L.; Bruckner, P. Genetic diseases of the extracellular matrix: More than just connective tissue disorders. *J. Mol. Med.* **1998**, *76*, 226–237. [CrossRef] [PubMed]

25. Bonafe, L.; Cormier-Daire, V.; Hall, C.; Lachman, R.; Mortier, G.; Mundlos, S.; Nishimura, G.; Sangiorgi, L.; Savarirayan, R.; Sillence, D.; et al. Nosology and classification of genetic skeletal disorders: 2015 Revision. *Am. J. Med. Genet. A* **2015**, *167A*, 2869–2892. [CrossRef] [PubMed]

26. Ito, H.; Rucker, E.; Steplewski, A.; McAdams, E.; Brittingham, R.J.; Alabyeva, T.; Fertala, A. Guilty by association: Some collagen II mutants alter the formation of ECM as a result of atypical interaction with fibronectin. *J. Mol. Biol.* **2005**, *352*, 382–395. [CrossRef] [PubMed]

27. Steplewski, A.; Ito, H.; Rucker, E.; Brittingham, R.J.; Alabyeva, T.; Gandhi, M.; Ko, F.K.; Birk, D.E.; Jimenez, S.A.; Fertala, A. Position of single amino acid substitutions in the collagen triple helix determines their effect on structure of collagen fibrils. *J. Struct. Biol.* **2004**, *148*, 326–337. [CrossRef] [PubMed]

28. Steplewski, A.; Majsterek, I.; McAdams, E.; Rucker, E.; Brittingham, R.J.; Ito, H.; Hirai, K.; Adachi, E.; Jimenez, S.A.; Fertala, A. Thermostability gradient in the collagen triple helix reveals its multi-domain structure. *J. Mol. Biol.* **2004**, *338*, 989–998. [CrossRef] [PubMed]

29. Arnold, W.V.; Fertala, A.; Sieron, A.L.; Hattori, H.; Mechling, D.; Bächinger, H.P.; Prockop, D.J. Recombinant procollagen II: Deletion of D period segments identifies sequences that are required for helix stabilization and generates a temperature-sensitive N-proteinase cleavage site. *J. Biol. Chem.* **1998**, *273*, 31822–31828. [CrossRef] [PubMed]

30. Mizuno, K.; Hayashi, T.; Bächinger, H.P. Hydroxylation-induced stabilization of the collagen triple helix. Further characterization of peptides with 4(R)-hydroxyproline in the XAA position. *J. Biol. Chem.* **2003**, *278*, 32373–32379. [CrossRef] [PubMed]

31. Wagner, K.; Pöschl, E.; Turnay, J.; Baik, J.; Pihlajaniemi, T.; Frischholz, S.; von der Mark, K. Coexpression of alpha and beta subunits of prolyl 4-hydroxylase stabilizes the triple helix of recombinant human type X collagen. *Biochem. J.* **2000**, *352*, 907–911. [CrossRef] [PubMed]

32. Weis, M.A.; Wilkin, D.J.; Kim, H.J.; Wilcox, W.R.; Lachman, R.S.; Rimoin, D.L.; Cohn, D.H.; Eyre, D.R. Structurally abnormal type II collagen in a severe form of Kniest dysplasia caused by an exon 24 skipping mutation. *J. Biol. Chem.* **1998**, *273*, 4761–4768. [CrossRef] [PubMed]

33. Superti-Furga, A.; Steinmann, B.; Ramirez, F.; Byers, P.H. Molecular defects of type III procollagen in Ehlers-Danlos syndrome type IV. *Hum. Genet.* **1989**, *82*, 104–108. [CrossRef] [PubMed]

34. Väisänen, L.; Has, C.; Franzke, C.; Hurskainen, T.; Tuomi, M.L.; Bruckner-Tuderman, L.; Tasanen, K. Molecular mechanisms of junctional epidermolysis bullosa: Col 15 domain mutations decrease the thermal stability of collagen XVII. *J. Investig. Dermatol.* **2005**, *125*, 1112–1118. [CrossRef] [PubMed]

35. Freije, J.M.; Diez-Itza, I.; Balbin, M.; Sanchez, L.M.; Blasco, R.; Tolivia, J.; Lopez-Otin, C. Molecular cloning and expression of collagenase-3, a novel human matrix metalloproteinase produced by breast carcinomas. *J. Biol. Chem.* **1994**, *269*, 16766–16773. [PubMed]

36. Knauper, V.; Lopez-Otin, C.; Smith, B.; Knight, G.; Murphy, G. Biochemical characterization of human collagenase-3. *J. Biol. Chem.* **1996**, *271*, 1544–1550. [CrossRef] [PubMed]

37. Billinghurst, R.C.; Dahlberg, L.; Ionescu, M.; Reiner, A.; Bourne, R.; Rorabeck, C.; Mitchell, P.; Hambor, J.; Diekmann, O.; Tschesche, H.; et al. Enhanced cleavage of type ii collagen by collagenases in osteoarthritic articular cartilage. *J. Clin. Investig.* **1997**, *99*, 1534–1545. [CrossRef] [PubMed]

38. Bruckner, P.; Prockop, D.J. Proteolytic enzymes as probes for the triple-helical conformation of procollagen. *Anal Biochem.* **1981**, *110*, 360–368. [CrossRef]

39. Cabral, W.A.; Chernoff, E.J.; Marini, J.C. G76e substitution in type I collagen is the first nonlethal glutamic acid substitution in the alpha1(i) chain and alters folding of the N-terminal end of the helix. *Mol. Genet. Metab.* **2001**, *72*, 326–335. [CrossRef] [PubMed]

40. Galicka, A.; Wolczynski, S.; Gindzienski, A. Studies on type I collagen in skin fibroblasts cultured from twins with lethal osteogenesis imperfecta. *Acta Biochim. Pol.* **2003**, *50*, 481–488. [PubMed]

41. Majsterek, I.; McAdams, E.; Adachi, E.; Dhume, S.T.; Fertala, A. Prospects and limitations of the rational engineering of fibrillar collagens. *Protein Sci.* **2003**, *12*, 2063–2072. [CrossRef] [PubMed]

42. Salminen, H.J.; Saamanen, A.M.; Vankemmelbeke, M.N.; Auho, P.K.; Perala, M.P.; Vuorio, E.I. Differential expression patterns of matrix metalloproteinases and their inhibitors during development of osteoarthritis in a transgenic mouse model. *Ann. Rheum. Dis.* **2002**, *61*, 591–597. [CrossRef] [PubMed]

43. Tchetina, E.V.; Squires, G.; Poole, A.R. Increased type II collagen degradation and very early focal cartilage degeneration is associated with upregulation of chondrocyte differentiation related genes in early human articular cartilage lesions. *J. Rheumatol.* **2005**, *32*, 876–886. [PubMed]

44. Wu, W.; Billinghurst, R.C.; Pidoux, I.; Antoniou, J.; Zukor, D.; Tanzer, M.; Poole, A.R. Sites of collagenase cleavage and denaturation of type II collagen in aging and osteoarthritic articular cartilage and their relationship to the distribution of matrix metalloproteinase 1 and matrix metalloproteinase 13. *Arthritis Rheum.* **2002**, *46*, 2087–2094. [CrossRef] [PubMed]

45. Fertala, A.; Sieron, A.L.; Ganguly, A.; Li, S.W.; Ala-Kokko, L.; Anumula, K.R.; Prockop, D.J. Synthesis of recombinant human procollagen II in a stably transfected tumour cell line (ht1080). *Biochem. J.* **1994**, *298 Pt 1*, 31–37. [CrossRef] [PubMed]

46. Steplewski, A.; Brittingham, R.; Jimenez, S.A.; Fertala, A. Single amino acid substitutions in the c-terminus of collagen II alter its affinity for collagen ix. *Biochem. Biophys. Res. Commun.* **2005**, *335*, 749–755. [CrossRef] [PubMed]

47. John, D.C.; Grant, M.E.; Bulleid, N.J. Cell-free synthesis and assembly of prolyl 4-hydroxylase: The role of the beta-subunit (PDI) in preventing misfolding and aggregation of the alpha-subunit. *EMBO J.* **1993**, *12*, 1587–1595. [PubMed]

48. Kivirikko, K.I.; Myllyharju, J. Prolyl 4-hydroxylases and their protein disulfide isomerase subunit. *Matrix Biol.* **1998**, *16*, 357–368. [CrossRef]

49. Hecht, J.T.; Makitie, O.; Hayes, E.; Haynes, R.; Susic, M.; Montufar-Solis, D.; Duke, P.J.; Cole, W.G. Chondrocyte cell death and intracellular distribution of COMP and type IX collagen in the pseudoachondroplasia growth plate. *J. Orthop. Res.* **2004**, *22*, 759–767. [CrossRef] [PubMed]

50. Vranka, J.; Mokashi, A.; Keene, D.R.; Tufa, S.; Corson, G.; Sussman, M.; Horton, W.A.; Maddox, K.; Sakai, L.; Bachinger, H.P. Selective intracellular retention of extracellular matrix proteins and chaperones associated with pseudoachondroplasia. *Matrix Biol.* **2001**, *20*, 439–450. [CrossRef]

51. Rajpar, M.H.; McDermott, B.; Kung, L.; Eardley, R.; Knowles, L.; Heeran, M.; Thornton, D.J.; Wilson, R.; Bateman, J.F.; Poulsom, R.; et al. Targeted induction of endoplasmic reticulum stress induces cartilage pathology. *PLoS Genet.* **2009**, *5*, e1000691. [CrossRef] [PubMed]

52. Wilson, R.; Freddi, S.; Chan, D.; Cheah, K.S.; Bateman, J.F. Misfolding of collagen x chains harboring schmid metaphyseal chondrodysplasia mutations results in aberrant disulfide bond formation, intracellular retention, and activation of the unfolded protein response. *J. Biol. Chem.* **2005**, *280*, 15544–15552. [CrossRef] [PubMed]

53. Oyadomari, S.; Mori, M. Roles of chop/gadd153 in endoplasmic reticulum stress. *Cell Death Differ.* **2004**, *11*, 381–389. [CrossRef] [PubMed]

54. Chessler, S.D.; Byers, P.H. BiP binds type I procollagen pro alpha chains with mutations in the carboxyl-terminal propeptide synthesized by cells from patients with osteogenesis imperfecta. *J. Biol. Chem.* **1993**, *268*, 18226–18233. [PubMed]

55. Breckenridge, D.G.; Germain, M.; Mathai, J.P.; Nguyen, M.; Shore, G.C. Regulation of apoptosis by endoplasmic reticulum pathways. *Oncogene* **2003**, *22*, 8608–8618. [CrossRef] [PubMed]

56. Rao, R.V.; Ellerby, H.M.; Bredesen, D.E. Coupling endoplasmic reticulum stress to the cell death program. *Cell Death Differ.* **2004**, *11*, 372–380. [CrossRef] [PubMed]

57. Hintze, V.; Steplewski, A.; Ito, H.; Jensen, D.A.; Rodeck, U.; Fertala, A. Cells expressing partially unfolded R789C/p.R989C type II procollagen mutant associated with spondyloepiphyseal dysplasia undergo apoptosis. *Hum. Mutat.* **2008**, *29*, 841–851. [CrossRef] [PubMed]

58. Gaiser, K.G.; Maddox, B.K.; Bann, J.G.; Boswell, B.A.; Keene, D.R.; Garofalo, S.; Horton, W.A. Y-position collagen II mutation disrupts cartilage formation and skeletal development in a transgenic mouse model of spondyloepiphyseal dysplasia. *J. Bone Miner. Res.* **2002**, *17*, 39–47. [CrossRef] [PubMed]

59. Hecht, J.T.; Hayes, E.; Haynes, R.; Cole, W.G. Comp mutations, chondrocyte function and cartilage matrix. *Matrix Biol.* **2005**, *23*, 525–533. [CrossRef] [PubMed]

60. Dinser, R.; Zaucke, F.; Kreppel, F.; Hultenby, K.; Kochanek, S.; Paulsson, M.; Maurer, P. Pseudoachondroplasia is caused through both intra- and extracellular pathogenic pathways. *J. Clin. Investig.* **2002**, *110*, 505–513. [CrossRef] [PubMed]

61. Schmitz, M.; Becker, A.; Schmitz, A.; Weirich, C.; Paulsson, M.; Zaucke, F.; Dinser, R. Disruption of extracellular matrix structure may cause pseudoachondroplasia phenotypes in the absence of impaired cartilage oligomeric matrix protein secretion. *J. Biol. Chem.* **2006**, *281*, 32587–35295. [CrossRef] [PubMed]

62. Bateman, J.F.; Moeller, I.; Hannagan, M.; Chan, D.; Cole, W.G. Characterization of three osteogenesis imperfecta collagen alpha 1(i) glycine to serine mutations demonstrating a position-dependent gradient of phenotypic severity. *Biochem. J.* **1992**, *288*, 131–135. [CrossRef] [PubMed]

63. Jakkula, E.; Mäkitie, O.; Czarny-Ratajczak, M.; Jackson, G.C.; Damignani, R.; Susic, M.; Briggs, M.D.; Cole, W.G.; Ala-Kokko, L. Mutations in the known genes are not the major cause of MED; distinctive phenotypic entities among patients with no identified mutations. *Eur. J. Hum. Genet.* **2005**, *13*, 292–301. [CrossRef] [PubMed]

64. Bonadio, J.; Byers, P.H. Subtle structural alterations in the chains of type I procollagen produce osteogenesis imperfecta type II. *Nature* **1985**, *316*, 363–366. [CrossRef] [PubMed]

65. Byers, P.H. Brittle bones-fragile molecules: Disorders of collagen gene structure and expression. *Trends Genet.* **1990**, *6*, 293–300. [CrossRef]

66. Kuivaniemi, H.; Tromp, G.; Prockop, D.J. Mutations in collagen genes: Causes of rare and some common diseases in humans. *FASEB J.* **1991**, *5*, 2052–2060. [CrossRef] [PubMed]

67. Kohfeldt, E.; Maurer, P.; Vannahme, C.; Timpl, R. Properties of the extracellular calcium binding module of the proteoglycan testican. *FEBS Lett.* **1997**, *414*, 557–561. [CrossRef]

68. Wuttke, M.; Muller, S.; Nitsche, D.P.; Paulsson, M.; Hanisch, F.G.; Maurer, P. Structural characterization of human recombinant and bone-derived bone sialoprotein. Functional implications for cell attachment and hydroxyapatite binding. *J. Biol. Chem.* **2001**, *276*, 36839–36848. [CrossRef] [PubMed]

International Journal of
Molecular Sciences

Article

Hyaluronan Production by Renomedullary Interstitial Cells: Influence of Endothelin, Angiotensin II and Vasopressin

Sara Stridh [1,2] , Fredrik Palm [1], Tomoko Takahashi [3], Mayumi Ikegami-Kawai [3], Malou Friederich-Persson [1] and Peter Hansell [1,*]

[1]   Department of Medical Cell Biology, Uppsala University, Biomedical Center, SE-75123 Uppsala, Sweden; strs@rkh.se (S.S.); fredrik.palm@mcb.uu.se (F.P.); malou.friederich@mcb.uu.se (M.F.-P.)
[2]   Department of Health Sciences, Red Cross University College, SE-14152 Stockholm, Sweden
[3]   Faculty of Pharmaceutical Sciences, Hoshi University, Tokyo 142-8501, Japan; tomoko-takahashi@mue.biglobe.ne.jp (T.T.); m-kawai@hoshi.ac.jp (M.I.-K.)
*   Correspondence: peter.hansell@mcb.uu.se; Tel.: +46-184-714-130

Received: 24 October 2017; Accepted: 10 December 2017; Published: 13 December 2017

Abstract: The content of hyaluronan (HA) in the interstitium of the renal medulla changes in relation to body hydration status. We investigated if hormones of central importance for body fluid homeostasis affect HA production by renomedullary interstitial cells in culture (RMICs). Simultaneous treatment with vasopressin and angiotensin II (Ang II) reduced HA by 69%. No change occurred in the mRNA expressions of hyaluronan synthase 2 (HAS2) or hyaluronidases (Hyals), while Hyal activity in the supernatant increased by 67% and CD44 expression reduced by 42%. The autocoid endothelin (ET-1) at low concentrations (10^{-10} and 10^{-8} M) increased HA 3-fold. On the contrary, at a high concentration (10^{-6} M) ET-1 reduced HA by 47%. The ET-A receptor antagonist BQ123 not only reversed the reducing effect of high ET-1 on HA, but elevated it to the same level as low concentration ET-1, suggesting separate regulating roles for ET-A and ET-B receptors. This was corroborated by the addition of ET-B receptor antagonist BQ788 to low concentration ET-1, which abolished the HA increase. HAS2 and Hyal2 mRNA did not alter, while Hyal1 mRNA was increased at all ET-1 concentrations tested. Hyal activity was elevated the most by high ET-1 concentration, and blockade of ET-A receptors by BQ123 prevented about 30% of this response. The present study demonstrates an important regulatory influence of hormones involved in body fluid balance on HA handling by RMICs, thereby supporting the concept of a dynamic involvement of interstitial HA in renal fluid handling.

Keywords: hyaluronan; kidney; interstitium; medulla; endothelin; vasopressin; angiotensin II

1. Introduction

Hyaluronan (HA) is a negatively charged interstitial glycosaminoglycan with large water attracting ability [1]. In the kidney, HA is predominantly found in the inner medulla during normal physiological conditions [2–5]. This site is responsible for the fine tuning of fluid electrolyte balance primarily under the influence of hormones such as angiotensin II (Ang II), aldosterone and vasopressin. The differential intrarenal distribution of HA is important for urine concentration and dilution [6]. We have previously demonstrated that during acute hydration medullary HA increases and, in contrast, HA content decreases during water deprivation [2,3]. It has been suggested that apart from the changes occurring in vasopressin-regulated aquaporins, the physicochemical properties of the interstitial matrix changes in relation to hydration status, which influences primarily water permeability [6]. Ginetzinsky [7] demonstrated that the HA-degrading enzyme hyaluronidase (Hyal) is excreted in the urine in larger amounts during dehydration and drops to virtually zero during hydration. It was suggested that the interstitial matrix was altered by Hyal activity, thereby changing the level of diuresis. Under pathological conditions of altered kidney function, HA is inappropriately regulated [6]. During renal ischemia-reperfusion injury or tubulointerstitial inflammation, renal cortical levels of HA increase, while medullary levels are largely unaltered, which may contribute to the pathophysiology of the disease process [4,8,9]. Indeed, suppression of HA accumulation during renal ischemia-reperfusion improves renal function, suggesting a protecting effect against ischemic insults [10]. During diabetes, both cortical and medullary levels are elevated [6], which may contribute to the phenotype due to the pro-inflammatory and water-attracting properties of HA.

We have previously demonstrated that a major contributor of interstitial HA in renal medulla is the renomedullary interstitial cell (RMIC) [11,12]. These cells express receptors for hormones and autocoids known to be involved in the regulation of fluid and electrolyte balance [13]. These receptors include Ang II AT1, vasopressin V1a, endothelin (ET)-A and -B, and bradykinin B2. RMICs in culture produce less HA during hyperosmotic conditions than during iso- and hypo-osmotic conditions [11,12], supporting the in vivo observations that HA content decreases during dehydration, while it increases during hydration [2,3].

Despite the observations from our and other laboratories that renal HA content changes during physiological and pathophysiological conditions, the mechanisms regulating HA content under these conditions are unclear. Thus, the aim of the present study was to determine the effects of hormones and an autocoid involved in normal regulation of body fluid balance on HA turnover by RMICs in culture. Furthermore, the importance of Hyal activity and of the HA scavenging receptor CD44 expression [14,15] for HA turnover were also investigated.

2. Results

2.1. Hyaluronan (HA) in Supernatant

The HA content in the supernatant of RMICs grown under iso-osmotic conditions was 0.29 ± 0.11 ng HA/ng cell protein (Figure 1). Changing growth media to hypo-osmotic conditions resulted in a more than 4-fold elevation of the HA content ($p < 0.05$). A similar increase occurred after treatment with the Hyal inhibitor L-ascorbic acid 6-hexadecanoate (Asc-P) (more than 3-fold elevation, $p < 0.05$), while the HA synthesis inhibitor 4-methylumbelliferone (4-MU) reduced the HA content by 52% ($p < 0.05$).

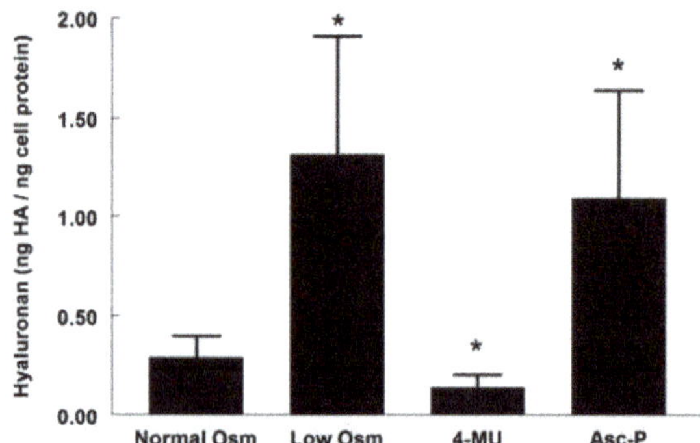

Figure 1. Hyaluronan (HA) content in supernatants of cultured renomedullary interstitial cells (RMICs) during control conditions (normal osmolality) and after 24 h exposure to hypo-osmotic media conditions (200 mOsm/kg H_2O), a hyaluronidase inhibitor (Asc-P) or the HA synthesis inhibitor 4-MU.* $p < 0.05$ vs. corresponding value of control cells (normal osmolality).

Neither Ang II nor vasopressin alone reduced HA in the supernatant significantly (-16% and -58%, respectively, ns). However, when the combination of Ang II and vasopressin was used, HA was reduced by 69% ($p < 0.05$) (Figure 2).

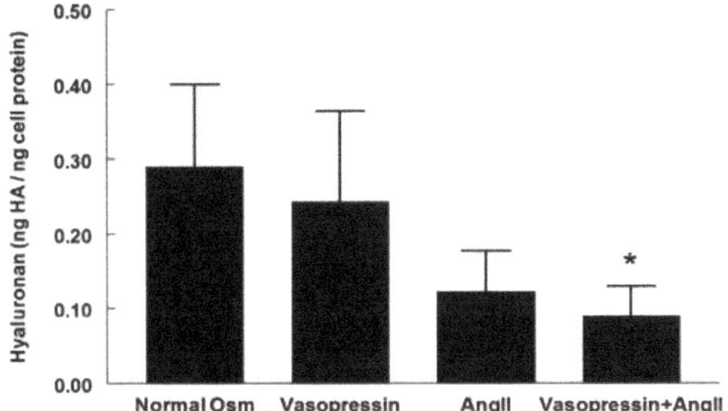

Figure 2. Hyaluronan (HA) content in the supernatant of cultured RMICs during control conditions (normal osmolality) and after 24 h exposure to angiotensin II (Ang II, 10^{-6} M), vasopressin (10^{-6} M), and a combination of Ang II and vasopressin. * $p < 0.05$ vs. corresponding value of control cells (normal osmolality).

Endothelin-1 (ET-1) at 10^{-10} M and 10^{-8} M increased supernatant HA more than 3-fold ($p < 0.05$), while, on the contrary, a high concentration of ET-1 (10^{-6} M) reduced HA by 47% ($p < 0.05$, Figure 3) as compared with untreated control cells. The ET-A receptor antagonist BQ123 not only reversed the reducing effect of the high concentration of ET-1 on HA, but elevated it to the same level as low concentration ET-1 (10^{-10} M), suggesting an important mechanism involving the still active ET-B receptor. This was corroborated by the addition of ET-B receptor antagonist BQ788 to low concentration ET-1, which abolished the increase in HA.

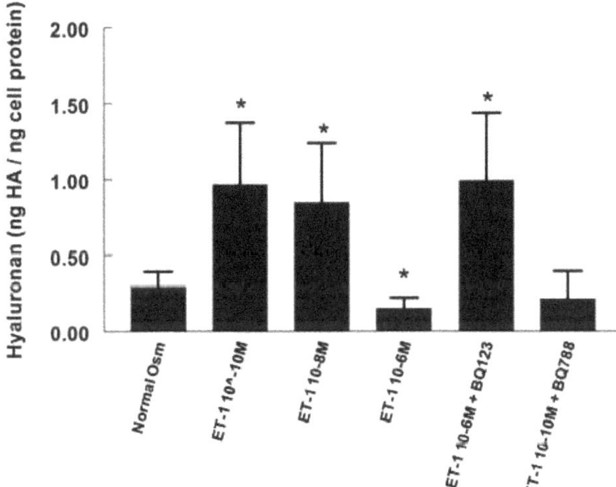

Figure 3. Hyaluronan (HA) content in the supernatant of cultured RMICs during control conditions and after 24 h exposure to endothelin (ET-1), with or without the ET-A receptor antagonist BQ123 or the ET-B receptor antagonist BQ788. * $p < 0.05$ vs. corresponding value of control cells (normal osmolality).

2.2. Hyaluronidase (Hyal) Activity in Supernatant

The Hyal activity in the supernatant of cells grown in low-osmolality media was 44% lower ($p < 0.05$) compared to during iso-osmotic conditions (Figure 4). Neither Ang II nor vasopressin alone altered the activity. However, the combination of Ang II and vasopressin increased the Hyal activity by 67% ($p < 0.05$). ET-1 increased the supernatant activity at 10^{-8} M and 10^{-6} M. At 10^{-8} M the activity increased by 54%, while the high concentration (10^{-6} M) elevated the Hyal activity by 137%. When the ET-A receptor antagonist BQ123 was added, the elevation in activity after the high concentration of ET-1 (10^{-6} M) was reduced and thus similar to that of the lower concentrations used (55%).

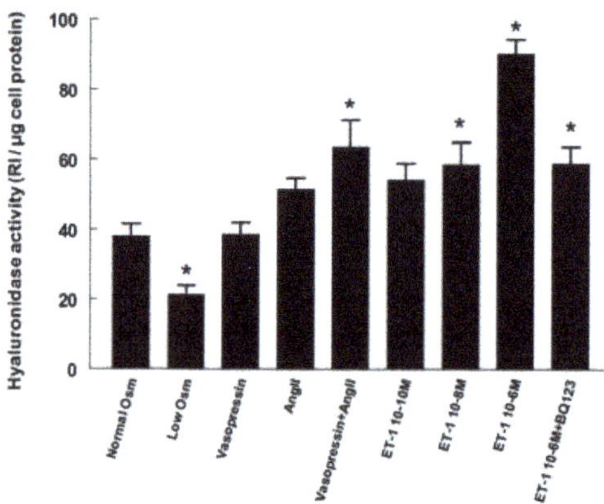

Figure 4. Hyaluronidase (Hyal) activity in supernatants of cultured RMICs during different treatments. Values are related to the amount of total cell protein in each culture dish. RI, relative intensity. * $p < 0.05$ vs. control cells (normal osmolality).

2.3. Hyaluronan Synthase (HAS) and Hyaluronidase mRNA in Renomedullary Interstitial Cells in Culture (RMICS)

HAS2 and Hyal2 mRNA expression did not change in most of the treatment groups (Figures 5 and 6), however, HAS2 was increased in ET-1 groups of low and medium concentration, and BQ788 together with low concentration ET-1 increased expression of both HAS2 and Hyal2.

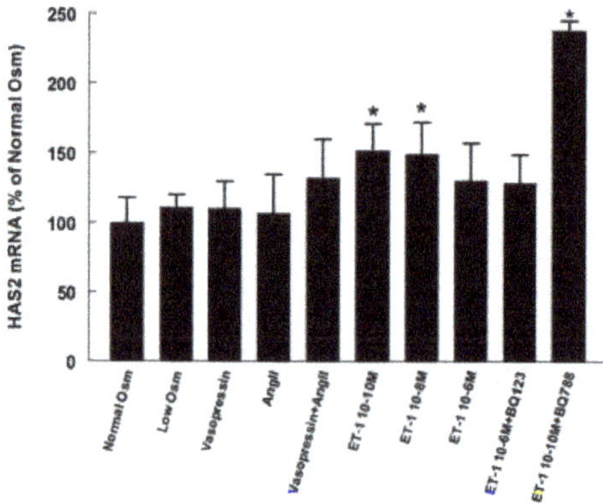

Figure 5. mRNA expressions of hyaluronan synthase 2 (HAS2) in RMICs during different treatments. All values are in relation to cells grown at normal osmolality = 100%. * $p < 0.05$ vs. control cells (normal osmolality).

In RMICs grown in hypo-osmolar conditions the Hyal1 mRNA was reduced by 35% as compared to iso-osmolar conditions ($p < 0.05$, Figure 7), while Ang II and vasopressin, alone or in combination, did not change the expression. All concentrations of ET-1 elevated the Hyal1 mRNA expression similarly. This elevation was not affected by the ET-A receptor blocker BQ123, but was abolished by the ET-B receptor blocker BQ788.

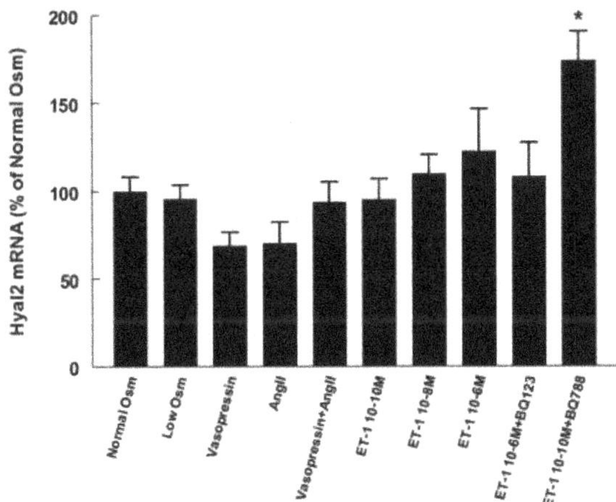

Figure 6. mRNA expressions of hyaluronidase 2 (Hyal 2) in RMICs during different treatments. All values are in relation to cells grown at normal osmolality = 100%. *$p < 0.05$ vs. control cells (normal osmolality).

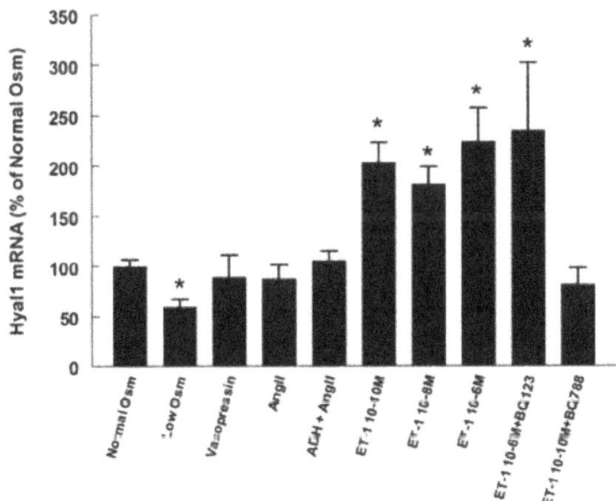

Figure 7. mRNA expressions of hyaluronidase 1 (Hyal 1) in RMICs during different treatments. All values are in relation to cells grown at normal osmolality = 100%. *$p < 0.05$ vs. control cells (normal osmolality).

2.4. CD44 on RMIC Surface

The CD44 surface expression on RMICs grown in hypo-osmolar conditions was 68% lower than that on cells grown at iso-osmolar conditions ($p < 0.05$, Figure 8). No change in expression occurred by Ang II or vasopressin separately, but when combined the CD44 expression was reduced by 42% ($p < 0.05$). The lowest concentration of ET-1 (10^{-10} M) increased CD44 expression more than 4-fold ($p < 0.05$). The intermediate concentration of ET-1 (10^{-8} M) also increased the CD44 expression, but to a lesser extent. Low concentration of ET-1 in combination with ET-B receptor blocker BQ788 normalized CD44 expression. The higher concentration of ET-1 together with the ET-A receptor blocker BQ123 tended to increase CD44 expression, suggesting an effect over ET-B receptors.

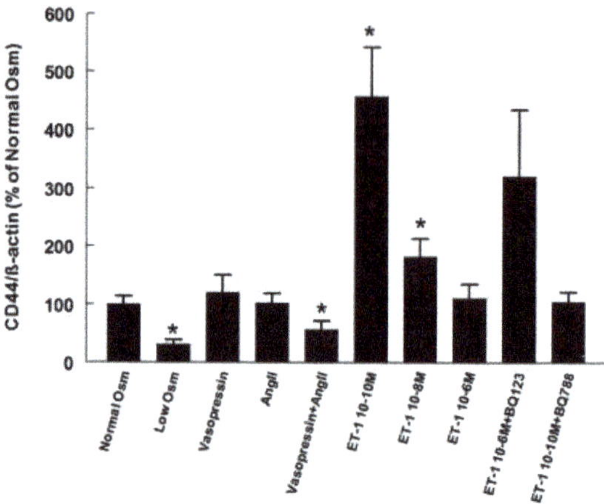

Figure 8. Expression of the scavenging receptor CD44 on the cell surface of RMICs during different treatments. All values are in relation to cells grown at normal osmolality = 100%. * $p < 0.05$ vs. control cells (normal osmolality).

3. Discussion

The present study demonstrates an important regulatory role of hormones involved in body fluid balance on HA turnover by cultured RMICs. Furthermore, the results demonstrate an important mechanism over altered Hyal activity to regulate HA turnover, which provides a rapid way to change HA as opposed to primarily regulating synthesis. Our previous studies have suggested an important role for HA in renomedullary water handling. During acute hydration the rat medullary interstitial HA content increases, while the opposite occurs during water deprivation [2,3]. The elevation in medullary interstitial HA content during excess water intake will antagonize medullary water reabsorption by changing the interstitial matrix properties, thereby resulting in reduced fluid conductance. The opposite occurs during water deprivation in conjunction with increased vasopressin-regulated aquaporins. Our present findings widen our understanding on how the composition of the renomedullary interstitial matrix changes in response to hydration status and also sets focus on hyaluronidases in order to achieve rapid responses.

Changes in overall HA content are due to changes in HA synthesis and/or HA degradation. The present study suggests that regulation of the degradation pathway is of major importance for the changes seen in supernatant HA both after a hypo-osmolar challenge as well as after hormonal action. The mRNA expression of HAS2 and Hyal2 in RMICs did not change after hypo-osmotic challenge, Ang II, or vasopressin, suggesting a change in Hyal activity. Our previous in vivo data in rats show no changes in the mRNA levels of HAS or Hyals after 2 h hydration when medullary HA is elevated [16], again suggesting a change in the activity. In the case of ET-1, mRNA levels of Hyal1 were elevated, while hypo-osmolality reduced the Hyal1 expression, which strengthens the proposition that the degradation pathway is an important way to change HA. It is of interest in this context that lack of hyaluronidases exacerbates renal post-ischemic injury, inflammation, and fibrosis, pointing to an ongoing defense mechanism [17].

The mechanism(s) underlying Hyal activation in the present study are not known but it is generally accepted that mRNA levels may not reflect activity or protein levels. Changing activity would be a more rapid way to increase catabolic function as opposed to primarily producing more of the enzyme *per se*. In a study by Albeiroti and colleagues [18] on platelets, it was demonstrated that Hyal2 is activated by being transported from intracellular sites to the surface of the cell, thereby achieving catabolic activity. This activation in an embryonic kidney cell line (as opposed to that in platelets) requires coexistence with CD44 expression on the plasma membrane [19]. In the present study on RMICs, we demonstrate that CD44 surface expression decreases during hypo-osmolar conditions of the media and lowers Hyal activity, which results in elevated amounts of HA in the media, thus suggesting reduced catabolic and internalization processes. The same line of reasoning regarding effects of Hyal activity and CD44 expression cannot be performed after treatment with Ang II, vasopressin, or ET-1, thus suggesting another pathway of Hyal activation which may include intracellular calcium and IP3 signaling [13].

In a previous in vitro study [11] in cultured rat RMICs, we found that the HA binding receptor CD44 is downregulated under hypo-osmotic conditions (mimicking in vivo hydration), while it is upregulated under hyperosmotic conditions (mimicking water deprivation). This was corroborated by the present study, which showed reduced CD44 expression when reducing growth media osmolality. Once HA binds to CD44 it can be internalized and degraded [14,20,21]. Furthermore, in an in vitro study [22] it was demonstrated that low ionic strength inhibits HA hydrolysis catalyzed by Hyal. This would imply that during hypo-osmotic conditions when low ionic strength applies both the cellular uptake of HA over CD44 and the breakdown of HA by Hyal, is reduced. The acute hydration-induced elevation in HA in vivo (within 2 h) [3] suggests an important role over inhibition of Hyals and not primarily increased HAS expression or activity.

ET-1 resulted in a biphasic response on supernatant HA levels which was related to agonist concentration. Such a concentration dependent biphasic response has previously been described for ET-1 on the effect on vascular smooth muscle (i.e., low concentration results in dilation and high concentration results in contraction) [23,24]. In the present study, low concentration ET-1 increased HA in the supernatant, while the high concentration reduced HA. When the high concentration was combined with the ET-A receptor antagonist BQ123, HA was returned to a level comparable with low concentration ET-1. It can be hypothesized that low concentration ET-1 primarily affects ET-B receptors, which increase HA by increasing nitric oxide (NO) production [25–27]. This was corroborated by the abolished increase in HA by the addition of the ET-B receptor antagonist BQ788 to low concentration ET-1. Endothelins are also known to enhance the release of prostaglandins by stimulation of ET-B receptors located on vascular endothelial cells [28,29], which also elevate HA production [27,30,31]. The high concentration of ET-1 may primarily affect ET-A receptors, which have been shown to increase CD44 expression with a BQ123-sensitive mechanism [32]. We have previously demonstrated an inverse relationship between elevated levels of surface CD44 on RMICs and supernatant HA, suggesting increased internalization [11] and thereby reduced levels of HA in the supernatant. When the ET-A receptor antagonist BQ-123 was included in the present study HA was elevated, not only back to control levels, but to the levels corresponding to the low concentration of ET-1, suggesting an action on the still active and intact ET-B receptor. When the ET-B receptor antagonist BQ788 was included, the HA amount no longer differed from the control. The observed changes in HA in the present study fit well with the demonstrated effects of ET-A vs. ET-B receptor activation on medullary fluid handling. ET-B activation reduces fluid reabsorption over NO, while ET-A activation increases fluid reabsorption [33]. This fit well with our previous finding that an intact NO-system is required for the hydration-induced medullary HA-elevation to occur [34].

Ang II and vasopressin in combination reduced the HA content in the supernatant of cultured RMICs and we have previously shown that vasopressin infusion in vivo reduces papillary HA [34], presumably over the V_1-receptor, since the selective V_2-receptor agonist desmopressin failed to produce such a response [3], although this may be a concentration issue. The finding of V_{1a} receptors but not of the V_2 subtype on RMICs [13] corroborates our suggestion. Vasopressin stimulates the activity of Hyal in the rat renal papilla. The activation of these enzymes is associated with a decrease in the content of HA [35]. In homozygous Brattleboro rats lacking vasopressin, the urine osmolality and Hyal activity of renal papillary tissue were closely related after vasopressin treatment [36]. In support of this notion, we have demonstrated that outer medullary HA content is increased in Brattleboro rats, thus inferring reduced breakdown [3]. It has also been demonstrated that antisera against rat kidney Hyal blocks the hydro-osmotic effect of vasopressin [37]. Furthermore, in a study by Ivanova et al. [38], the mRNA expressions of Hyal1 and Hyal2 were increased in the medullary tissue after treatment with a vasopressin analogue. In the present study, vasopressin alone did not reduce HA with statistical significance, only when combined with Ang II. We have, however, in a previous study found a reduction of HA after vasopressin treatment [34]. The discrepancy in results is not clear. The effector mechanism underlying the reduction of HA by Ang II treatment (in combination with vasopressin) seems to be, at least partly, due to increased Hyal activity. This would fit with previous data on neonatal angiotensin converting enzyme (ACE) inhibition showing reduced Hyal1 mRNA expression in the renal medulla and reduced urine Hyal activity early in the newborn rat [39]. It is also noteworthy that these two hormones are simultaneously elevated during dehydration.

In the present study on RMICs, it is clear that the HA content is reduced after high concentration ET-1 (ET-A receptor mediated) and the combination of Ang II and vasopressin. Furthermore, it is well known that both Ang II and vasopressin levels in plasma are elevated during dehydration/antidiuresis (i.e., when medullary HA levels in vivo are reduced). However, previous studies have shown that ET-1, through its action over the ET-A receptor, and Ang II, over the AT_1 receptor, increase proliferation and extracellular matrix production (ECM) by RMICs in culture [40,41]. What could be the underlying cause for this apparent contradiction? The standard index for ECM production (i.e., ^{35}S-methionine/cysteine incorporation) is not a measure of HA production, since HA does not incorporate methionine/cysteine, being that it is a sugar compound. This index is more correct for estimating collagen-related production, and the true relationship between different matrix components has not been demonstrated in parallel with giving the ECM-index. It has, however, been shown that laminin production by RMICs increases after Ang II [41]. A reduced HA content in the medullary interstitium provides favorable conditions for an increase in the permeability of glycosaminoglycan structures adjacent to the cell surface. Since RMICs seem to provide structural support for the renal medulla, it could be speculated that the increase in ECM production, like laminin, would maintain structural integrity when HA levels are reduced in parallel to increase in vivo interstitial water permeability. However, a reduced HA content in parallel with elevation of collagen in a tissue would, during pathological conditions, provide for fibrosis, since the HA reduction leads to reduced viscoelasticity and hydration.

CD44 is the main cell surface receptor for HA [15]. Besides providing a signal response to HA, it participates in HA endocytosis as a scavenger receptor [42]. In the present study, RMIC surface expression of CD44 was reduced by low media osmolality when supernatant HA content was increased, which could provide a pathway for regulation via reduced internalization and degradation. However, in the two other situations when CD44 expression was altered (Ang II + vasopressin and low concentration ET-1, respectively), supernatant HA content changed in an opposite direction, which does not enable a causal relationship. The underlying mechanism to the reduced CD44 expression after Ang II + vasopressin and elevated expression after low concentration of ET-1 is unclear and warrants further investigation.

The Hyal inhibitor used in the present study (L-ascorbic Acid 6-hexadecanoate, Asc-P) is a documented potent inhibitor of different Hyal activities [43]. The importance of Hyal activity for regulating HA turnover in RMICs is clear. When inhibiting Hyal activity during iso-osmotic conditions, supernatant HA increases to similar levels as RMICs grown under hypo-osmotic conditions. As stated above, this would fit well with the suggestion that the elevation in HA during hypo-osmotic conditions occurs through a reduced activity of intracellular Hyals.

Upon continuous hydration in the rat, the medullary HA-levels increase and peak after about 2 h and return to control levels after about 4 h of hydration [2]. Interestingly, peak urine flow rate was also observed after 2 h of hydration. It could be speculated that medullary HA in the rat is primarily important for the acute and rapid exclusion of water during excessive intake.

Dwyer et al. [44] found that in obese rabbits the medullary HA is selectively elevated. It was suggested that a possible distension could occur in the renal medulla with consequences for interstitial hydrostatic pressure and reabsorption. Such a constitutively elevated amount of HA could therefore result in a reduced reabsorptive capacity leading to hypohydration. Whether this can partly explain the higher incidence of hypohydration in obese US adults remains to be established [45].

The concentrations of drugs used in the present study are generally high. However, this was done to achieve maximum response from the systems observed. It is, furthermore, worth noting that the various peptide receptors on RMICs are spatially close and overlapping [46] and complex interactions may therefore occur between components of the intracellular chain of reactions which are transmitted as the hormonal signal. How this affects the results of the present study is unknown.

In conclusion, the present study demonstrates an important regulatory influence of hormones involved in body fluid balance on HA handling by RMICs thereby supporting the concept of a dynamic involvement of interstitial HA on renal fluid handling.

4. Materials and Methods

4.1. Animals

All animal procedures were approved by the local animal ethics committee at Uppsala University (Approval code C290/11, approval date 25 November, 2011). Male Sprague-Dawley rats weighing 80–90 g (Charles River, Sulzfeld, Germany) were used.

4.2. Culture of RMICs

RMICs were isolated from kidneys of young Sprague-Dawley rats as previously described [47,48]. For the first weeks, the culture is primarily epithelial, but after that it consists of a homogenous non-epithelial population. The cells were used for experiments at passage 7 and later. The described isolation and culturing methodology produce cells containing the characteristic lipid vacuoles.

4.3. Experimental Protocol—In Vitro

RMICs were plated at a density of about 5×10^4 cells/cm^2 for 48 h in a 1:1 mixture of two media: Roswell Park Memorial Institute (RPMI) 1640 culture medium and Dulbecco's Modified Eagle's Medium (DMEM) culture medium conditioned by 3T3 mouse fibroblasts, the mixture containing a total of 15% fetal bovine serum, as previously described [47]. The cells were then treated for 24 h with different compounds described below. Following 24 h treatment, the supernatant was collected and analyzed for HA content and Hyal activity. Cells were harvested and analyzed for CD44 expression and gene expression of HAS and Hyals. The amount of protein was determined using a routine method (DC Protein Assay, Bio-Rad Laboratories, Hercules, CA, USA). Ang II (Bachem, Bubendorf, Switzerland), vasopressin (Sigma-Aldrich, St. Louis, MO, USA), or a combination of the two were given in concentrations of 10^{-6} M. ET-1 (Sigma-Aldrich) was administrated to provide final concentrations of 10^{-10} M, 10^{-8} M, or 10^{-6} M. The highest ET-1 concentration (10^{-6} M) was also given together with the selective ET-A receptor antagonist BQ123 (10^{-6} M) (Sigma-Aldrich). The lowest ET-1 concentration (10^{-10} M) was also given together with the selective ET-B receptor antagonist BQ788 (10^{-6} M) (Sigma-Aldrich). The Hyal inhibitor L-ascorbic acid 6-hexadecanoate (Asc-P) was purchased from Sigma-Aldrich and administered in a final concentration of 10^{-7} M. The HA synthesis inhibitor 4-methylumbelliferone (4-MU; Sigma-Aldrich) was used in a concentration of 10^{-6} M. Growth media osmolality was reduced to 200 mOsm/kg H$_2$O by 2:3 dilution with distilled water.

4.4. Analysis of HA

HA content in supernatants from RMICs in culture was measured using a commercially available enzyme-linked immunosorbent assay (Echelon Biosciences Inc., Salt Lake City, UT, USA) by following the enclosed instructions and was then related to the amount of protein.

4.5. Hyaluronidase Activity in Supernatants

HA from rooster comb and all reagents for the polymerization of electrophoretic gels were obtained from Wako Pure Chemical Industries (Osaka, Japan), Alcian blue 8GX from Fluka Chemical (Buchs, Switzerland), and Actinase E from Kaken Pharmaceutical (Tokyo, Japan). Supernatant Hyal activity was determined by quantitative zymography [49] with a slight modification because of its very low activity. Briefly, three volumes of supernatant were mixed with one volume of $2\times$ Laemmli sample buffer containing 8% sodium dodecyl sulfate (SDS) and no reducing reagent. A control rat serum used as a standard was diluted 200-fold with 0.15 M NaCl containing 0.1 mg/mL Bovine serum albumin (BSA) and mixed with an equivalent volume of Laemmli sample buffer containing 4% SDS and no reducing reagent. After incubation for 1 h at 37 °C, 32 µL of the supernatant mixture and 2–20 µL of control serum mixture were applied to 7% SDS-polyacrylamide gels containing 0.17 mg/mL HA. After electrophoretic run at 20 mA for approximately 90 min at 4 °C, gels were rinsed with 2.5% Triton X-100 for 80 min at room temperature and incubated with 0.1 M formate buffer (pH 3.5 and containing 0.03 M NaCl) for 24 h at 37 °C on an orbital shaker. Gels were then treated with 0.1 mg/mL Actinase E in 20 mM Tris-HCl buffer (pH 8.0) for 2 h at 37 °C. To visualize digestion of HA, gels were stained with 0.5% Alcian blue in 25% ethanol: 10% acetic acid. After destaining, gels were counterstained with Coomassie brilliant blue R-250. For the determination of Hyal activity, the stained gel was scanned on an Image Scanner (GE healthcare Japan, Tokyo, Japan) and scans were analyzed using Image J 1.42q software (National Institutes of Health, Rockville, MD, USA). The relative band intensity (RI) of supernatant Hyal activity was calculated from the ratio of the band intensity of Hyal activity from 0.05 µL of a control rat serum.

4.6. CD44 Analysis

Prior to CD44 analysis by western blot the surface proteins were isolated (Pierce® Cell surface protein isolation kit, Pierce Biotechnology, Rockford, IL, USA). Molecular weight separation was performed on 10% Tris-HCl gels with Tris/glycine/SDS buffer, the proteins transferred to nitrocellulose membranes, and CD44 detected with sheep anti-rat CD44 (0.1 µg/mL; R&D Systems, Minneapolis, MN, USA) and HRP-conjugated rabbit anti-sheep (1:5000; Kirkegaard and Perry Laboratories, Gaithersburg, MD, USA). Luminescent signal was captured on an enhanced chemiluminescence (ECL)-camera system (Kodak image station 2000; New Haven, CT, USA). β-actin was detected with mouse anti-rat β-actin antibody (1:20,000, Sigma-Aldrich, St Louis, MO, USA) and secondary horseradish peroxidase (HRP)-conjugated goat-anti mouse antibody (1:10,000; Kirkegaard and Perry Laboratories, Gaithersburg, MD, USA). CD44 western blot analysis of samples from isolated surface proteins was normalized to the β-actin expression.

4.7. Gene Expression Analysis

Total RNA was isolated from the cells (RNAquous®-4PCR, Ambion, Austin, TX, USA). cDNA was obtained from the RNA (iScript™cDNA Synthesis Kit, Bio Rad Laboratories, Hercules, CA, USA), and the following semi-quantitative real-time PCR was performed by LightCycler® FastStart DNA MasterPLUS SYBR Green I (Roche Diagnostics, Mannheim, Germany) in a Lightcycler system (Roche Diagnostics, Mannheim, Germany). PCR products were verified by agarose gel electrophoresis. The mRNA analyzed were Hyal1 and 2, and HAS2. All values were normalized for reference genes TATA-binding protein (TBP), β-actin (Actb), and glucose-6-phosphate dehydrogenase (G6PDH). Values were then expressed as normalized values for the means of the reference genes by using the formula: 2^Ct(reference genes) – Ct(gene of interest), where Ct is the cycle number and Ct for the reference genes is a mean of the cycle numbers for the reference genes, which did not differ much from each other. Primers were obtained from MWG Biotech (Ebersberg, Germany), with sequences presented in Table 1. Primers were evaluated in terms of efficiency (Table 2), melt curves (Figure 9), and verified product size.

Table 1. Primer sequences for *HAS2*, *Hyal1*, and *Hyal2*.

Gene	Gene Accession	Forward Primer	Reverse Primer
G6PDH	NM_017006.2	GTCATGCAGAACCACCTCCT	ACATACTGGCCAAGGACCAC
AktB	NM_031144.3	GCCCTGGCTCCTAGCACC	CCACCAATCCACACAGAGTACTTG
TBP	NM_001004198.1	ACCCTTCACCAATGACTCCTATG	ATGATGACTGCAGCAAATCGC
HAS2	NM_013153.1	GTGACTGCACCAGTTCCGCTAA	CATGTCTAATGGGACTGCACACAAG
Hyal1	NM_207616.1	TCGGACCCTTTATCCTGAAC	TTCTTACACCACTCTCCACTC
Hyal2	NM_172040.2	CGTTACGTCAAGGCAGTCAG	AGGTACACGGAGGGAAAGAG

Table 2. Primer efficiency for *HAS2*, *Hyal1*, and *Hyal2*.

Gene	Ct Average NOsm ($n = 8$)	Ct Average LOsm ($n = 8$)	Primer Efficiency (%)
G6PDH	25.2 ± 0.2	25.5 ± 0.2	96.6
AKTB	24.4 ± 0.2	22.5 ± 0.2	98.3
TBP	17.1 ± 0.1	17.0 ± 0.2	120.1
HAS2	27.3 ± 0.2	29.0 ± 0.3	104.0
Hyal 1	27.3 ± 0.4	26.7 ± 0.4	124.5
Hyal 2	23.6 ± 0.2	24.3 ± 0.2	125.1

NOsm, normal osmolality; LOsm, low osmolality.

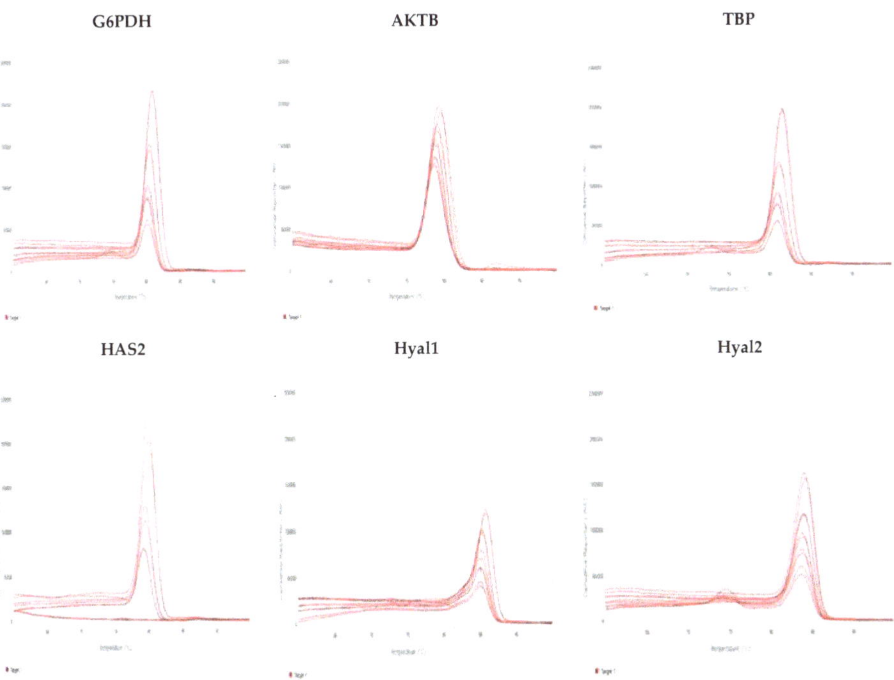

Figure 9. Melt curves for the primers used.

Int. J. Mol. Sci. **2017**, *18*, 2701

4.8. Statistical Analysis

Data are given as mean values ± SEM. The comparison between groups was evaluated with one-way ANOVA followed by Fisher's Least Significant Difference (LSD) post-hoc test. Parameters presented as percentage in graphs were statistically evaluated using the original values. A p-value of <0.05 was considered statistically significant.

Acknowledgments: The skillful technical assistance of Angelica Fasching is gratefully acknowledged. Financial support was provided by grants from the Swedish Science Council-Medicine which included funds for covering the costs to publish in open access.

Author Contributions: Sara Stridh and Peter Hansell conceived and designed the experiments; Sara Stridh, Mayumi Ikegami-Kawai, and Malou Friederich-Persson performed the experiments and analyzed samples; Sara Stridh, Peter Hansell, and Tomoko Takahashi analyzed the data; Fredrik Palm contributed reagents; Sara Stridh, Peter Hansell, and Fredrik Palm wrote the paper. All authors have edited the manuscript up to and including its final version.

Conflicts of Interest: The authors declare no conflict of interest.

Abbreviations

HA	Hyaluronan
HAS	Hyaluronan synthase
Hyal	Hyaluronidase
Ang II	Angiotensin II
RMICs	Renomedullary interstitial cells
ET-1	Endothelin-1

References

1. Laurent, T.C.; Fraser, J.R.E. Hyaluronan. *FASEB J.* **1992**, *6*, 2397–2404. [PubMed]
2. Göransson, V.; Johnsson, C.; Nylander, O.; Hansell, P. Renomedullary and intestinal hyaluronan content during body water excess: A comparative study in rats and gerbils. *J. Physiol.* **2002**, *542*, 315–322. [CrossRef] [PubMed]
3. Hansell, P.; Göransson, V.; Odlind, C.; Gerdin, B.; Hällgren, R. Hyaluronan content in the kidney in different states of body hydration. *Kidney Int.* **2000**, *58*, 2061–2068. [CrossRef] [PubMed]
4. Johnsson, C.; Tufveson, G.; Wahlberg, J.; Hällgren, R. Experimentally-induced warm renal ischemia induces cortical accumulation of hyaluronan in the kidney. *Kidney Int.* **1996**, *50*, 1224–1229. [CrossRef] [PubMed]
5. Wells, A.F.; Larsson, E.; Tengblad, A.; Fellstrom, B.; Tufveson, G.; Klareskog, L.; Laurent, T.C. The localization of hyaluronan in normal and rejected human kidneys. *Transplantation* **1990**, *50*, 240–243. [CrossRef] [PubMed]
6. Stridh, S.; Palm, F.; Hansell, P. Renal interstitial hyaluronan: Functional aspects during normal and pathological conditions. *Am. J. Physiol. Regul. Physiol.* **2012**, *302*, R1235–R1249. [CrossRef] [PubMed]
7. Ginetzinsky, A.G. Role of hyaluronidase in the re-absorption of water in renal tubules: The mechanism of action of the antidiuretic hormone. *Nature* **1958**, *182*, 1218–1219. [CrossRef] [PubMed]
8. Göransson, V.; Johnsson, C.; Jacobson, A.; Heldin, P.; Hallgren, R.; Hansell, P. Renal hyaluronan accumulation and hyaluronan synthase expression after ischaemia-reperfusion injury in the rat. *Nephrol. Dial. Transplant.* **2004**, *19*, 823–830. [CrossRef] [PubMed]
9. Nilsson, A.B.M.; Johnsson, C.; Friberg, P.; Hansell, P. Renal cortical accumulation of hyaluronan in adult rats neonatally exposed to ACE inhibition. *Acta Physiol. Scand.* **2001**, *173*, 343–350. [CrossRef] [PubMed]
10. Colombaro, V.; Declèves, A.E.; Jadot, I.; Voisin, V.; Giordano, L.; Habsch, I.; Nonclercq, D.; Flamion, B.; Caron, N. Inhibition of hyaluronan is protective against renal ischaemia-reperfusion injury. *Nephrol. Dial. Transplant.* **2013**, *28*, 2484–2493. [CrossRef] [PubMed]
11. Göransson, V.; Hansell, P.; Moss, S.; Alcorn, D.; Johnsson, C.; Hällgren, R.; Maric, C. Renomedullary interstitial cells in culture; the osmolality and oxygen tension influence the extracellular amounts of hyaluronan and cellular surface expression of CD44. *Matrix Biol.* **2001**, *20*, 129–136. [CrossRef]

12. Hansell, P.; Maric, C.; Alcorn, D.; Göransson, V.; Johnsson, C.; Hällgren, R. Renomedullary interstitial cells regulate hyaluronan turnover depending on growth media osmolality suggesting a role in renal water handling. *Acta Physiol. Scand.* **1999**, *165*, 115–116. [CrossRef] [PubMed]

13. Zhuo, J.L. Renomedullary interstitial cells: A target for endocrine and paracrine actions of vasoactive peptides in the renal medulla. *Clin. Exp. Pharmacol. Physiol.* **2000**, *27*, 465–473. [CrossRef] [PubMed]

14. Culty, M.; Nguyen, H.A.; Underhill, C.B. The hyaluronan receptor (CD44) participates in the uptake and degradation of hyaluronan. *J. Cell Biol.* **1992**, *116*, 1055–1062. [CrossRef] [PubMed]

15. Aruffo, A.; Stamenkovic, I.; Melnick, M.; Underhill, C.B.; Seed, B. CD44 is the principal cell surface receptor for hyaluronate. *Cell* **1990**, *61*, 1303–1313. [CrossRef]

16. Rügheimer, L.; Olerud, J.; Johnsson, C.; Takahashi, T.; Shimizu, K.; Hansell, P. Hyaluronan synthases and hyaluronidases in the kidney during changes in hydration status. *Matrix Biol.* **2009**, *28*, 390–395. [CrossRef] [PubMed]

17. Colombaro, V.; Jadot, I.; Declèves, A.E.; Voisin, V.; Giordano, L.; Habsch, I.; Malaisse, J.; Flamion, B.; Caron, N. Lack of hyaluronidases exacerbates renal post-ischemic injury, inflammation, and fibrosis. *Kidney Int.* **2015**, *88*, 61–71. [CrossRef] [PubMed]

18. Albeiroti, S.; Ayasoufi, K.; Hill, D.R.; Shen, B.; de la Motte, C.A. Platelet hyaluronidase-2: An enzyme that translocates to the surface upon activation to function in extracellular matrix degradation. *Blood* **2015**, *125*, 1460–1469. [CrossRef] [PubMed]

19. Harada, H.; Takahashi, M. CD44-dependent intracellular and extracellular catabolism of hyaluronic acid by hyaluronidase-1 and -2. *J. Biol. Chem.* **2007**, *282*, 5597–5607. [CrossRef] [PubMed]

20. Formby, B.; Stern, R. Lactate-sensitive response elements in genes involved in hyaluronan catabolism. *Biochem. Biophys. Res. Commun.* **2003**, *305*, 203–208. [CrossRef]

21. Hua, Q.; Knudson, C.B.; Knudson, W. Internalization of hyaluronan by chondrocytes occurs via receptor-mediated endocytosis. *J. Cell Sci.* **1993**, *106*, 365–375. [PubMed]

22. Asteriou, T.; Vincent, J.C.; Tranchepain, F.; Deschrevel, B. Inhibition of hyaluronan hydrolysis catalysed by hyaluronidase at high substrate concentration and low ionic strength. *Matrix Biol.* **2006**, *25*, 166–174. [CrossRef] [PubMed]

23. Edwards, R.M.; Pullen, M.; Nambi, P. Activation of endothelin ETB receptors increases glomerular cGMP via an L-arginine-dependent pathway. *Am. J. Physiol. Ren. Physiol.* **1992**, *263*, F1020–F1025.

24. Harris, P.J.; Zhuo, J.; Mendelsohn, F.A.; Skinner, S.L. Haemodynamic and renal tubular effects of low doses of endothelin in anaesthetized rats. *J. Physiol.* **1991**, *433*, 25–39. [CrossRef] [PubMed]

25. Chenevier-Gobeaux, C.; Morin-Robinet, S.; Lemarechal, H.; Poiraudeau, S.; Ekindjian, J.C.; Borderie, D. Effects of pro- and anti-inflammatory cytokines and nitric oxide donors on hyaluronic acid synthesis by synovial cells from patients with rheumatoid arthritis. *Clin. Sci.* **2004**, *107*, 291–296. [CrossRef] [PubMed]

26. Deliu, E.; Brailoiu, G.C.; Mallilankaraman, K.; Wang, H.; Madesh, M.; Undieh, A.S.; Koch, W.J.; Brailoiu, E. Intracellular endothelin type B receptor-driven Ca^{2+} signal elicits nitric oxide production in endothelial cells. *J. Biol. Chem.* **2012**, *287*, 41023–41031. [CrossRef] [PubMed]

27. Rügheimer, L.; Johnsson, C.; Hansell, P. Nitric Oxide and Prostaglandins Influence the Renomedullary Hyaluronan Content. In *Hyaluronan—Its Structure, Metabolism, Biological Activities and Therapeutic Applications*; Balazs, E.A., Hascall, V.C., Eds.; Winmar Enterprises: Edgewater, NJ, USA, 2005; Volume II, pp. 773–776, ISBN 0-9771359-0.

28. De Nucci, G.; Thomas, R.; D'Orleans-Juste, P.; Antunes, E.; Walder, C.; Warner, T.D.; Vane, J.R. Pressor effects of circulating endothelin are limited by its removal in the pulmonary circulation and by the release of prostacyclin and endothelium-derived relaxing factor. *Proc. Natl. Acad. Sci. USA* **1988**, *85*, 9797–9800. [CrossRef] [PubMed]

29. Warner, T.D.; Mitchell, J.A.; de Nucci, G.; Vane, J.R. Endothelin-1 and endothelin-3 release EDRF from isolated perfused arterial vessels of the rat and rabbit. *J. Cardiovasc. Pharmacol.* **1989**, *13*, S85–S88. [CrossRef] [PubMed]

30. Honda, A.; Sekiguchi, Y.; Mori, Y. Prostaglandin E2 stimulates cyclic AMP-mediated hyaluronan synthesis in rabbit pericardial mesothelial cells. *Biochem. J.* **1993**, *292*, 497–502. [CrossRef] [PubMed]

31. Mahadevan, P.; Larkins, R.G.; Fraser, J.R.; Fosang, A.J.; Dunlop, M.E. Increased hyaluronan production in the glomeruli from diabetic rats: A link between glucose-induced prostaglandin production and reduced sulphated proteoglycan. *Diabetologia* **1995**, *38*, 298–305. [CrossRef] [PubMed]

32. Tanaka, Y.; Makiyama, Y.; Mitsui, Y. Endothelin-1 is involved in the growth promotion of vascular smooth muscle cells by hyaluronic acid. *Int. J. Cardiol.* **2000**, *76*, 39–47. [CrossRef]

33. Hyndman, K.A.; Pollock, J.S. Nitric oxide and the A and B of endothelin of sodium homeostasis. *Curr. Opin. Nephrol. Hypertens.* **2013**, *22*, 26–31. [CrossRef] [PubMed]

34. Rügheimer, L.; Johnsson, C.; Maric, C.; Hansell, P. Hormonal regulation of renomedullary hyaluronan. *Acta Physiol. (Oxford)* **2008**, *193*, 191–198. [CrossRef] [PubMed]

35. Ivanova, L.N.; Goryunova, T.E. Mechanism of the renal hyaluronate hydrolases activation in response to ADH. In *Proceedings of the 28th International Congress of Physiological Sciences, Budapest, Hungary, 13–19 July 1980, Kidney and Body Fluids*; Takacs, L., Ed.; Elsevier: Amsterdam, The Netherlands, 1981; pp. 587–591, ISBN 9781483153803.

36. Ivanova, L.N.; Goryunova, T.E.; Nikiforovskaya, L.F.; Tishchenko, N.I. Hyaluronate hydrolase activity and glycosaminoglycans in the Brattleboro rat kidney. *Ann. N. Y. Acad. Sci.* **1982**, *394*, 503–508. [CrossRef] [PubMed]

37. Law, R.O.; Rowen, D. The influence of hyaluronidase on urinary and renal medullary composition following antidiuretic stimulus in the rat. *J. Physiol.* **1981**, *311*, 341–354. [CrossRef] [PubMed]

38. Ivanova, L.N.; Babina, A.V.; Baturina, G.; Katkova, L.E. The effect of vasopressin on the expression of genes of key enzymes of the interstitial hyaluronan turnover and concentration ability in WAG rat kidneys. *Russ. J. Genet. Appl. Res.* **2017**, *7*, 249–257. [CrossRef]

39. Stridh, S.; Kerjaschki, D.; Chen, Y.; Rügheimer, L.; Astrand, A.B.; Johnsson, C.; Friberg, P.; Olerud, J.; Palm, F.; Takahashi, T.; et al. Angiotensin converting enzyme inhibition blocks interstitial hyaluronan dissipation in the neonatal rat kidney via hyaluronan synthase 2 and hyaluronidase 1. *Matrix Biol.* **2011**, *30*, 62–69. [CrossRef] [PubMed]

40. Maric, C.; Aldred, G.P.; Antoine, A.M.; Eitle, E.; Dean, R.G.; Williams, D.A.; Harris, P.J.; Alcorn, D. Actions of endothelin-1 on cultured rat renomedullary interstitial cells are modulated by nitric oxide. *Clin. Exp. Pharmacol. Physiol.* **1999**, *26*, 392–398. [CrossRef] [PubMed]

41. Maric, C.; Zheng, W.; Walther, T. Interactions between Angiotensin II and Atrial Natriuretic Peptide in Renomedullary Interstitial Cells: The Role of Neutral Endopeptidase. *Nephron Physiol.* **2006**, *103*, 149–156. [CrossRef] [PubMed]

42. Lee, J.Y.; Spicer, A.P. Hyaluronan: A multifunctional, megaDalton, stealth molecule. *Curr. Opin. Cell Biol.* **2000**, *12*, 581–586. [CrossRef]

43. Botzki, A.; Rigden, D.J.; Braun, S.; Nukui, M.; Salmen, S.; Hoechstetter, J.; Bernhardt, G.; Dove, S.; Jedrzejas, M.J.; Buschauer, A. L-Ascorbic acid 6-hexadecanoate, a potent hyaluronidase inhibitor. X-ray structure and molecular modeling of enzyme-inhibitor complexes. *J. Biol. Chem.* **2004**, *279*, 45990–45997. [CrossRef] [PubMed]

44. Dwyer, T.M.; Banks, S.A.; Alonso-Galicia, M.; Cockrell, K.; Carroll, J.F.; Bigler, S.A.; Hall, J.E. Distribution of renal medullary hyaluronan in lean and obese rabbits. *Kidney Int.* **2000**, *58*, 721–729. [CrossRef] [PubMed]

45. Rosinger, A.Y.; Lawman, H.G.; Akinbami, L.J.; Ogden, C.L. The role of obesity in the relation between total water intake and urine osmolality in US adults, 2009–2012. *Am. J. Clin. Nutr.* **2016**, *104*, 1554–1561. [CrossRef] [PubMed]

46. Zhuo, J.; Dean, R.; Maric, C.; Aldred, P.G.; Harris, P.; Alcorn, D.; Mendelsohn, F.A. Localization and interactions of vasoactive peptide receptors in renomedullary interstitial cells of the kidney. *Kidney Int. Suppl.* **1998**, *67*, S22–S28. [CrossRef] [PubMed]

47. Fontoura, B.M.A.; Nussenzveig, D.R.; Pelton, K.M.; Maack, T. Atrial natriuretic factor receptors in cultured renomedullary interstitial cells. *Am. J. Physiol. Cell Physiol.* **1990**, *258*, C692–C699.

48. Maric, C.; Aldered, G.P.; Antoine, A.M.; Dean, R.G.; Eitle, E.; Mendelsohn, F.A.O.; Williams, D.A.; Harris, P.J.; Alcorn, D. Effects of angiotensin II on cultured rat renomedullary interstitial cells are mediated by AT1A receptors. *Am. J. Physiol. Ren. Physiol.* **1996**, *271*, F1020–F1028.
49. Ikegami-Kawai, M.; Okuda, R.; Nemoto, T.; Inada, N.; Takahashi, T. Enhanced activity of serum and urinary hyaluronidases in streptozotocin-induced diabetic Wistar and GK rats. *Glycobiology* **2004**, *14*, 65–72. [CrossRef] [PubMed]

International Journal of
Molecular Sciences

Review

The Biological Role of Hyaluronan-Rich Oocyte-Cumulus Extracellular Matrix in Female Reproduction

Eva Nagyova

Institute of Animal Physiology and Genetics, Academy of Sciences of the Czech Republic, 27721 Libechov, Czech Republic; nagyova@iapg.cas.cz; Tel.: +420-315-639564

Received: 8 December 2017; Accepted: 16 January 2018; Published: 18 January 2018

Abstract: Fertilization of the mammalian oocyte requires interactions between spermatozoa and expanded cumulus extracellular matrix (ECM) that surrounds the oocyte. This review focuses on key molecules that play an important role in the formation of the cumulus ECM, generated by the oocyte-cumulus complex. In particular, the specific inhibitors (AG1478, lapatinib, indomethacin and MG132) and progesterone receptor antagonist (RU486) exerting their effects through the remodeling of the ECM of the cumulus cells surrounding the oocyte have been described. After gonadotropin stimulus, cumulus cells expand and form hyaluronan (HA)-rich cumulus ECM. In pigs, the proper structure of the cumulus ECM depends on the interaction between HA and serum-derived proteins of the inter-alpha-trypsin inhibitor (IαI) protein family. We have demonstrated the synthesis of HA by cumulus cells, and the presence of the IαI, tumor necrosis factor-alpha-induced protein 6 and pentraxin 3 in expanding oocyte-cumulus complexes (OCC). We have evaluated the covalent linkage of heavy chains of IαI proteins to HA, as the principal component of the expanded HA-rich cumulus ECM, in porcine OCC cultured in medium with specific inhibitors: AG1478 and lapatinib (both inhibitors of epidermal growth factor receptor tyrosine kinase activity); MG132 (a specific proteasomal inhibitor), indomethacin (cyclooxygenase inhibitor); and progesterone receptor antagonist (RU486). We have found that both RU486 and indomethacin does not disrupt the formation of the covalent linkage between the heavy chains of IαI to HA in the expanded OCC. In contrast, the inhibitors AG1478 and lapatinib prevent gonadotropin-induced cumulus expansion. Finally, the formation of oocyte-cumulus ECM relying on the covalent transfer of heavy chains of IαI molecules to HA has been inhibited in the presence of MG132.

Keywords: extracellular matrix; hyaluronan; inter-alpha-trypsin inhibitor; tumor necrosis factor-alpha-induced protein 6; pentraxin 3; oocyte-cumulus complexes

---

1. Extracellular Matrix in General

The extracellular matrix (ECM) is an important structure that is present in all tissues. The ECM interacts with cells to regulate a wide range of functions, including adhesion, proliferation, apoptosis and differentiation. The ECM can also locally release growth factors, such as epidermal growth factor (EGF), fibroblast growth factor and other signaling molecules such as transforming growth factor (TGFβ) and amphiregulin [1]. Naba et al. [2] defined ECM proteins of the mammalian matrisome by analysis of the human and mouse genome. It comprises 1–1.5% of the mammalian proteome. There are almost 300 proteins, including 43 collagen subunits, 36 proteoglycans (e.g., aggrecan, versican, perlecan and decorin) and ~200 complex glycoproteins (e.g., laminins, elastin, fibronectins, thrombospondins, tenascins or nidogen). Moreover, there are large numbers of ECM-modifying enzymes, ECM-binding growth factors, and other ECM-associated proteins [3]. Proteoglycans are important structural macromolecules in tissues. They consist of a core protein with attached glycosaminoglycan side chains. There are six types of glycosaminoglycans: chondroitin sulfate, dermatan sulfate, heparin sulfate, heparin, keratin sulfate and hyaluronan (HA). Hyaluronan is the only glycosaminoglycan synthesized at the cell membrane and it is present in a protein-free form [4]. The size of HA depends on the relative activity of HA-synthesizing and degrading enzymes. There are three hyaluronan synthase isoforms (HAS1, 2 and 3) [5]. The expression of the three HAS isoforms is regulated by various stimuli, suggesting different functions of the three proteins [6]. In contrast, hyaluronidases 1–4 degrade HA into several fragments with size-dependent functions [7]. Dysregulation of ECM structure causes tissue malfunction such as inflammation, infertility and cancer. The importance of the function of the ECM is demonstrated by the embryonic lethality caused by mutations in genes that encode components of the ECM [8,9].

2. Cumulus–Oophorus Extracellular Matrix in Ovarian Follicles: Characterization of Essential Components

It has been shown that the complex of heavy chains of inter-alpha-trypsin-inhibitor (IαI) to HA is the principal structural component of the cumulus ECM in ovarian follicles in mice [10,11] and pigs [12]. Genetic deletion of specific ECM proteins such as bikunin (light chain of IαI) and tumor necrosis factor alpha-induced protein 6 (*Tnfaip6*-null mice were unable to transfer heavy chains from IαI to HA) exhibit infertility in female mice [11,13]; see Table 1. Suzuki et al. [14] identified the full repertoire of the IαI deficiency-related genes from bikunin-knockout female mice. They suggested that proteins of the IαI family have additional effects on reproductive biology by modulating the expression of a large number of cellular genes.

Table 1. Gonadotropin-induced matrix components essential for the formation and stability of the HA-rich oocyte-cumulus ECM.

ECM Component	Description	Tissue Location (Oocyte-Cumulus Complexes-OCC)	Species	References
HA	Hyaluronan	OCC	Mice	[15–18]
		OCC	Pigs	[19]
IαI	Inter-alpha-trypsin inhibitor (also called inter-alpha-trypsin inhibitor (ITI))	OCC	Mice	[10,18,20,21]
		OCC	Pigs	[12,22]
TNFAIP6	Tumor necrosis factor alpha-induced protein 6 (also called Tumor necrosis factor stimulated gene-6 (TSG-6))	OCC	Mice	[13,18,23,24]
		Ovary	Rats	[25]
		OCC	Pigs	[26,27]
		Ovarian follicles	Equine	[28]
PTX3	Pentraxin 3	OCC	Mice	[29–33]
		OCC	Pigs	[34]

2.1. IαI Family Proteins

It has been demonstrated that mice lacking intact IαI family proteins fail to form a stable cumulus ECM, and the naked ovulated oocytes are not fertilized in vivo [11]. Importantly, HA-rich oocyte-cumulus ECM does not form in IαI immune-depleted serum, while it does in the presence of purified IαI molecules [20]. The proper structure of the cumulus ECM depends on the interaction between HA and serum-derived proteins of the IαI family [11,12,21]. The IαI family proteins are synthesized and assembled in the liver and secreted into the blood at high concentrations (0.15–0.5 mg/mL of plasma) [35]. These IαI molecules consist of a small protein, named bikunin or light chain, with a chondroitin sulfate moiety that contains one or two evolutionarily related proteins, named heavy chains (HC1, HC2, and HC3). The first member, IαI, carries two heavy chains, HC1 and HC2, while the next two members, pre-α-inhibitor (PαI) and inter-α-like inhibitor (IαLI) have one heavy chain, i.e., HC, HC3 and HC2, respectively [36]. Positive bands of about 220, 130, and 120 kDa were detected in porcine serum [12] that likely correspond to IαI (bikunin plus HC1 and HC2), PαI (bikunin plus HC3), and IαLI (bikunin plus HC2), respectively [37,38]. In addition, porcine follicular fluids, collected at different stages of folliculogenesis were analyzed for the presence of IαI proteins [12]. Importantly, the levels of IαI molecules in porcine follicular fluid did not change in PMSG-primed follicles or in 8 h hCG-stimulated follicles, while a detectable increase in concentration was observed at 24 h post-hCG injection. In pigs, there is no apparent barrier to the transfer of IαI family molecules from the blood to the follicles, and the LH/hCG stimulation only facilitates their diffusion. Inside the follicle, the heavy chains are transferred from the glycosaminoglycan of IαI-related molecules to HA through a transesterification process [21]. To determine, whether HCs of serum-derived IαI-related molecules are covalently linked to HA in porcine expanded OCC in vivo, the experiments with OCC isolated from the antral follicles of pigs treated with PMSG (unexpanded OCC) or PMSG followed by hCG for 24 h (expanded OCC) were performed [12]. The authors found that expanded OCC contained positive bands of about 220, 130, and 120 kDa that likely correspond to IαI, PαI and IαLI, respectively [12,37,38]. After digestion of the expanded complexes with hyaluronidase, two additional immunopositive bands of about 85 kDa and 95 kDa were detected in the matrix extracts, the former likely corresponding to the relative molecular mass of single HC1 and HC2, and the latter to that of a single HC3 [36,39]. In addition, IαI proteins were detected in porcine OCC expanded in vitro after their culture in FSH- and serum-supplemented medium [12]. Interestingly, it has been shown that HCs from each IαI-related molecule identified in the serum are transferred and covalently linked to HA during the cumulus expansion of OCC. Analysis of the cumulus matrix and cell extracts clearly confirmed that the immunoreactivity was associated with the HA-rich cumulus ECM [12]. In mice and pigs [12,15,21], it has been confirmed that the covalent linkage of the heavy chains of IαI to HA is critical for cross-linking HA strands and stabilizing expanded HA-rich cumulus ECM.

2.2. TNFAIP6

It has been shown that the covalent transfer of heavy chains (HCs) of IαI proteins to HA does not occur in the OCC of *Tnfaip6*-null mice, indicating that TNFAIP6 is actively involved in this process [13]. TNFAIP6 is an inflammation-associated protein with the ability to bind HA, IαI, and other ligands and participate in the cumulus ECM formation and remodeling [40–42]. TNFAIP6 is produced by cumulus and granulosa cells after an ovulatory stimulus in mice, rats and pigs [13,18,24–28]. Previously, in pigs, four bands were detected with the antibody specific for TNFAIP6 in total and matrix protein extracts from OCC expanded in vivo. The major positive band had an apparent molecular weight of 35 kDa that also correlated well with the size of the free TNFAIP6 protein in mouse OCC. A doublet at ~120 kDa (HC-TNFAIP6 complex in mouse and pig) was also immunoreactive with the anti-TNFAIP6 antibody. In addition, the TNFAIP6 protein was detected in porcine OCC expanded in vitro after their culture in FSH- and serum-supplemented medium for 24 and 42 h [26]. It has been demonstrated that TNFAIP6, which binds to HA and interacts with heavy chains of IαI proteins, is another protein essential for the formation and stability of expanded HA-rich oocyte- cumulus ECM in mice and pigs [18,23,26]. Together with the high sequence similarity found among human, murine, rat, equine and porcine TNFAIP6 [23,25,27,28] and the expression of this protein and/or the respective gene in ovarian follicles of all of the examined species [18,23,25–28,43], it strongly supports the concept that TNFAIP6-mediated covalent binding of HCs (of IαI proteins) onto HA is a mechanism that mammalian OCC have in common [22].

2.3. PTX3

Experiments performed with *Ptx3* knockout mice [29] have shown that complexes from *Ptx3*−/−mice have defective cumulus matrix organization. Interestingly, hormone stimulation of *Ptx3*−/−OCC in vitro showed that while cumulus cells synthesized HA at a normal rate they were unable to organize this polymer in the cumulus ECM [30]. PTX3 is essential protein for organizing the HA polymer in the cumulus ECM in mice [30–32]. PTX3 protein plays role in cumulus ECM assembly, where HCs transferred from IαI to HA by the catalytic activity of TNFAIP6 bind distinct protomers of multimeric PTX3 [33]. In pigs, *PTX3* transcripts were significantly increased in OCC 24 h after in vivo hCG or in vitro FSH/LH stimulation [34]. Western blot analysis with PTX3 antibody revealed that cumulus ECM extracts from both in vivo hCG-stimulated pigs and in vitro FSH/LH-stimulated OCC cultured in medium supplemented either with follicular fluid or porcine serum, contain high levels of PTX3 protein. The localization of PTX3 protein in the porcine OCC was confirmed by immunostaining [34]. The mouse data concerning the integrity of HA-rich oocyte-cumulus ECM [30,31] together with porcine data [34] demonstrated the importance of PTX3 protein in the ovarian follicles.

3. Effect of Specific Inhibitors (AG1478, Lapatinib, Indomethacin and MG132) and Progesterone Receptor Antagonist (RU486) on the Formation of HA-Rich Cumulus Extracellular Matrix

3.1. Inhibition of EGFR Signaling Pathway (with AG1478) Affects Meiotic Maturation, Cumulus Expansion and Hyaluronan and Progesterone Synthesis

Several observations support the finding that EGF-like growth factors, i.e., amphiregulin and epiregulin, produced by granulosa cells and cumulus cells play a major role in triggering oocyte maturation and the cumulus expansion of OCC in mice [44]. Epidermal growth factor (EGF) is a poor inducer of porcine cumulus expansion in vitro [45]. Nevertheless, FSH pre-treatment strongly enhances EGF response within 3 h, as evidenced by a high increase in HA production and cumulus expansion after sequential exposure to FSH and EGF [46]. FSH itself does not affect epidermal growth factor receptor (EGFR) concentration or the tyrosine phosphorylation of EGFR, but it enhances the EGF-induced tyrosine phosphorylation of EGFR, indicating that the FSH signaling pathways may stimulate or modulate specific EGFR–regulating proteins. FSH also rapidly induces porcine OCC to express EGF-like growth factors [47] and TACE/ADAM17, a protease that cleaves and activates the EGF transmembrane precursors [48]. It has been shown [45] that AG1478, the inhibitor of EGFR tyrosine kinase activity, reduces 50% of the synthesis and 90% of the HA retained in the cumulus ECM, and prevents the expansion of porcine OCC stimulated with FSH for 24 h in vitro culture. Furthermore, although EGF does not stimulate progesterone production by porcine OCC and granulosa cells, the pre-treatment of both cell types with inhibitor AG1478, significantly reduces the stimulatory effect of FSH on progesterone production. This result is in agreement with the previous finding showing that AG1478 reduced the FSH-induced expression of the steroidogenic enzyme P450 side chain cleavage, *Cyp11a1*, in rat granulosa cells [49]. Importantly, the addition of AG1478 to the culture medium, irrespective of the stimulation, inhibited nuclear maturation in pigs [45,47]. Similarly, Ashkenazi et al. [50] observed that after the local administration of AG1478 inhibitor into the rat ovary, the ratio of entrapped immature oocytes (in germinal vesicle stage) in the inhibitor-treated ovaries was 5-fold higher than in the contralateral untreated ovaries. Thus, results in pigs [45] showing that EGFR activation by EGF-like growth factors produced under the FSH stimulus is involved in initiating the ovulatory events in porcine OCC are consistent with the results obtained in mice and rats [44,50]. However, it is possible that FSH *trans*-activates EGFR via mechanisms independent of EGF shedding [48]. Finally, it is important to note that the FSH-induced synthesis of both HA and progesterone is reduced but not abolished by AG1478, indicating that other signaling pathways elicited by FSH are operating in parallel [45].

3.2. Inhibition of EGFR Tyrosine Kinase (with Lapatinib) Affects Meiotic Maturation, Cumulus Expansion, and Expression of Cumulus-Associated Transcripts

In the ovarian follicles, EGFR mediates the ovulatory response to LH and the sustained activity of EGFR is an absolute requirement for LH-induced oocyte maturation and cumulus expansion [51]. However, abnormally elevated EGFR kinase activity can lead to various pathological states, including cancer. The human epidermal growth factor receptor (HER) family consists of four closely related transmembrane receptors: HER1 (human epidermal growth factor receptor 1, EGFR), HER2/c-Erb-B2, HER3/Erb-B3 and HER4/Erb-B4. These members of the type I receptor tyrosine kinase family are frequently implicated in cancer [52,53]. HER family-related downstream signaling plays a crucial role in cell proliferation, survival, migration and differentiation [54,55]. Recently, it has been investigated the effect of lapatinib on processes essential for ovulation, such as oocyte meiotic maturation and cumulus expansion, since it has been demonstrated that using of biological agents for treating cancer in women increases the probability that some women will conceive while taking the inhibitor (lapatinib) of growth factor signaling [56]. Lapatinib (GW572016, Tykerb/Tyverb; GlaxoSmithKline) is an orally active small molecule that reversibly and selectively inhibits the tyrosine kinase domain of both EGFR and HER2 [57] by binding to the ATP-binding site of the kinase, and preventing autophosphorylation or the rapid development of resistance to monotherapies [58]. It has been found that lapatinib, through the EGFR signaling pathway, is able to inhibit oocyte maturation in pigs [59]. In addition, lapatinib is able to reduce the expression of cumulus expansion -related transcripts (*TNFAIP6*, *PTGS2*), HA synthesis, cumulus expansion and progesterone secretion by porcine OCC cultured in FSH/LH supplemented medium [59]. This is in good agreement with the previous study showing the reduction of FSH-induced synthesis of both HA and progesterone by AG1478, another inhibitor of EGFR tyrosine kinase activity [45].

3.3. Addition of Progesterone Receptor Antagonist (RU486) to Culture Medium Affects Meiotic Maturation; It Does Not Affect Formation of Cumulus Extracellular Matrix Relying on the Covalent Transfer of Heavy Chains of IαI Molecules to Hyaluronan

Progesterone is an ovarian steroid hormone that regulates key aspects of female reproduction and acts through the progesterone receptor (PR). The progesterone receptor is a member of the nuclear receptor superfamily and functions as a ligand-activated transcription factor. The functional roles of PR in the ovary have been investigated with genetically modified mouse models and PR antagonist (RU486). It has been demonstrated that PR-knockout mice do not ovulate and are infertile [60,61]. Interestingly, despite the failure of ovulation in PR-null mice, cumulus expansion proceeds normally. However, the addition of RU486 to the culture medium with porcine OCC significantly decreases FSH/LH-induced resumption of oocyte meiosis (~74%; $p < 0.05$) and progression of oocyte maturation to the MII stage (~44%; $p < 0.05$) [62]. Gonadotropins stimulate cumulus expansion as well as HA synthesis by porcine OCC during in vitro maturation [19]. The addition of RU486 did not change FSH/LH-stimulated total HA synthesis; however, the retained amount of HA within the complexes was significantly reduced ($p < 0.05$). The amount of HA retained in cumulus ECM was approximately 60% of the amount retained within the cumulus ECM of the OCC cultured with FSH/LH alone [62]. However, the immunodetection of HABP, TNFAIP6, and PTX3 proteins in FSH/LH-stimulated OCC treated with RU486 confirmed the spatial localization of cumulus-associated components [62]. Furthermore, western blot analysis detected the heavy chains of IαI proteins in the matrix extracts of FSH/LH stimulated-OCC, treated with RU486 [62]. Shimada et al. found [63] that porcine OCC cultured in vitro in the presence of FSH/LH and RU486 had little developmental competence to proceed to the blastocyst stage. Moreover, RU486 significantly impaired blastocyst development in mice [64] and in cows [65]. Also, the administration of RU486 by intraperitoneal injection to gonadotropin-primed mice reduced the number of ovulated oocytes [66] and in mouse follicles cultured with hCG/EGF, in the presence of RU486, the MII rate was significantly lower (62%) [67]. Surprisingly the addition of RU486 to the culture medium significantly increased progesterone production by porcine OCC compared to FSH/LH alone [62]. To summarize, in pigs, the inhibition of PR with RU486 does not affect HA synthesis, the formation of cumulus ECM and covalent linkage between HA and heavy chains of IαI, but it appears that progesterone may be critical for maintaining an optimal microenvironment for oocyte maturation and fertilization [62].

3.4. Addition of General COX Inhibitor (Indomethacin) to Culture Medium does not Affect Meiotic Maturation, nor Formation of Cumulus Extracellular Matrix

In mammalian species, the preovulatory surge in gonadotropins upregulates the follicular expression of cyclooxygenase-2 (COX-2), which elevates the levels of prostaglandins [68,69]. Mice genetically deficient in *Cox-2* exhibited ovulatory failure [70,71]. In addition, *Cox-2* and prostaglandin E2 receptor subtype *Ep2* null mice were infertile [72]. The administration of either general

COX inhibitors (indomethacin) or inhibitors selective for COX-2 (NS-398, celecoxib) reduced ovulation rates in rodents, domestic animals, and monkeys [73–75]. In porcine OCC cultured in FSH/LH supplemented medium for 44 h, neither the resumption of meiosis (~87%) nor progression of oocyte maturation to MII (~72%) was affected by indomethacin [62]. Concomitantly, the total HA synthesis and retained amount of HA within the complexes was similar to those of OCC stimulated with FH/LH alone. In addition, the covalent binding between heavy chains of IαI molecules to HA in the cumulus ECM extracts, as well as cumulus ECM-related proteins (HABP, TNFAIP6, and PTX3) were detected in FSH/LH-stimulated OCC treated with indomethacin [62]. In agreement, indomethacin did not block HA synthesis induced by FSH in Graafian follicles in mice [76]. Moreover, Western blot analysis confirmed that in cumulus cells of *Cox-2* and *Ep2* null mice, the TNFAIP6 protein remained covalently associated with the heavy chains of IαI molecules. This is clear evidence that the ovaries of *Cox-2* null mice maintain the capacity to produce *Has2* mRNA in response to an ovulatory dose of hCG as well as the ability to form expanded cumulus ECM [72]. Matsumoto et al. [77] showed that the ovulatory process, but not follicular growth, oocyte maturation or fertilization, was primarily affected in adult *Cox-2* or *Ep2*-deficient mice. Eppig [76] suggested that PGE2 might play a role in the indirect stimulation of cumulus expansion by LH. Interestingly, Ben-Ami et al. [78] have shown that LH may mediate its effects on COX-2 expression in cumulus cells via the induction of the EGF-related factors amphiregulin, epiregulin, and betacelulin produced in human granulosa cells. Similarly, in mice, these factors bind EGF receptor in cumulus cells and induce the *Cox-2* message [79]. Also Hsieh et al. [80] have demonstrated that in mice, the LH-induction of *Cox-2/Ptgs2* expression is dependent on the activation of EGF receptor signaling in cumulus and mural granulosa cells. Moreover, it has been shown that lapatinib, the inhibitor of EGFR tyrosine kinase activity, reduces the expression of *COX-2* mRNA in porcine OCC cultured in vitro [59]. To summarize, in pigs, the inhibition of COX-2 by indomethacin does not affect FSH/LH-stimulated HA synthesis and the formation of the covalent linkage between heavy chains of IαI to HA nor progesterone production by cultured OCC [62].

3.5. Inhibition of Proteasomal Proteolysis (with MG132) Strongly Affects Meiotic Maturation and Formation of Cumulus Extracellular Matrix

Protein turnover mediated by the ubiquitin-proteasome pathway plays an essential role in cell physiology and pathology. Ubiquitin is a small chaperone protein that forms covalently linked isopeptide chains on protein substrates to mark them for degradation by the 26S proteasome. The 26S proteasome is a multicatalytic protease complex that specifically recognizes and hydrolyzes proteins tagged with multiubiquitin chains. The subunits of the 26S proteasome comprise approximately 1% of the total proteome in mammalian cells; the ubiquitin-proteasome pathway serves as the main substrate-specific cellular protein degradation pathway [81–83]. MG132 is a cell-permeable peptide aldehyde that inhibits the chymotrypsin-like activity of the 20S proteasomal core [84,85]. The ubiquitin–proteasome pathway modulates mouse oocyte meiotic maturation and fertilization via regulation of the MAPK-cascade and cyclin B1 degradation [86,87]. In addition, the role of the 26S proteasome in the regulation of oocyte meiosis has been described in mammals, specifically rats, mice and pigs [83,88–90]. In pigs, the addition of MG132 to the gonadotropin-supplemented medium prevented cumulus expansion and significantly reduced HA synthesis by the cumulus cells. Moreover, the covalent binding between HA and heavy chains of IαI was not detected in the MG132-treated porcine OCC [89]. The formation of expanded HA-rich cumulus ECM depends on HA association with specific hyaluronan-binding proteins [41], such as IαI [12,20], TNFAIP6 [23–27] and PTX3 [30,34], which all have been detected in mice and pigs. The TNFAIP6-HC complex is likely a catalyst in the transfer of heavy chains (of IαI) onto HA [91]. This mediator reacts rapidly with any HA, leading to the formation of heavy chain-HA and release of the TNFAIP6 catalyst. While the mRNA expression of *HAS2* and *TNFAIP6* in the gonadotropin-stimulated OCC was increased in pigs [27], mice [17,23], and rats [25], in the presence of MG132, the expression of *HAS2* and *TNFAIP6* was markedly suppressed in porcine OCC. In addition no signal of HA was observed by immunostaining in porcine OCC [92]. Tsafriri et al. [93] used a broad-spectrum metalloprotease inhibitor, GM6001, to find whether proteolytic activity was involved in the action of LH on the resumption of meiosis in rats. His conclusion was that this inhibitor prevented the LH-induced resumption of meiosis. In agreement, the inhibition of proteasomal proteolysis with MG132 arrested 90% of porcine oocytes in the germinal vesicle stage. Moreover, MG132 blocked the degradation of F-actin-rich transzonal projections (TZPs) interconnecting cumulus cells with the oocyte and cumulus expansion in pigs [89]. The resumption of oocyte meiosis was accompanied by the disappearance of the zona pellucida-spanning and actin microfilament-rich TZPs, and an alteration of gap junction communication [16,89,94]. Since the maintenance of TZPs supports an oocyte meiotic block and porcine OCC treated with MG132 remain unexpanded, it has been suggested that proteasomal proteolysis participates in the process of the resumption of meiosis. The terminal differentiation of cumulus oophorus within the ovarian follicle plays a crucial role in the ability of the oocyte to resume meiosis and reach full developmental competence [95–97]. Progesterone has been shown to enhance the activity of proteolytic enzymes important for the rupture of the follicular wall at ovulation [98]. Gonadotropins induce PR expression in cumulus cells concomitantly with an increase in progesterone secretion by porcine OCC [99,100]. The involvement of the proteasome in the turnover of StAR has been described [101–103] with subsequent influence on progesterone synthesis [102]. The transfer of cholesterol across the mitochondrial membranes is promoted by StAR [104]. In MG132-treated porcine OCC the progesterone levels were reduced [92]. In contrast, Tajima et al. [102] found a significant elevation in progesterone synthesis in MG132-treated rat granulosa cells. This discrepancy can be explained by the differences in the cell culture regimen. The relation between progesterone and proteolytic enzyme activity during ovulation in the gonadotropin-treated immature rat ovary was studied by Iwamasa et al. [98]. Their results suggested that progesterone played an indispensable role during the first 4 h of the ovulatory process by regulating proteolytic enzyme activities. Our results showed that the ability of gonadotropin-stimulated porcine cumulus cells to produce progesterone to a level comparable with control OCC was not restored when MG132 was present for 20 h of the culture, but it was restored (50%) when MG132 was only present for 3 h. In summary, the specific proteasomal inhibitor MG132 prevents the gonadotropin-induced resumption of meiosis and subsequent cumulus expansion. In addition, it protects TZPs against breakdown, affects the terminal differentiation of cumulus cells, markedly reduces the expression of *HAS2* and *TNFAIP6* and prevents the formation of a covalent linkage between HA and the heavy chains of IαI [89,92].

4. Conclusions

Ovulation is controlled through multiple inputs including endocrine hormones, immune and metabolic signals, as well as intra-follicular paracrine factors from the theca, mural and cumulus granulosa cells and the oocyte itself. The ovulatory mediators exert their effects through remodeling of the cumulus ECM that surrounds the oocyte. The proper structure of the cumulus ECM, which is essential for ovulation, transport of the OCC to the oviduct and fertilization depends on the interaction between HA and HA-associated ECM proteins. HA cross-linking within the cumulus ECM represents an important new mechanism in the regulation of the ovulatory process in mammalian follicles. We suggest that the structural changes in cumulus ECM affect signaling pathways and consequently the resumption of meiosis. In addition, it is interesting to note that the synthesis of ECM molecules is controlled by specific growth factors and the life of ECM molecules is determined by proteases. Thus, growth factors signaling pathways and the control of ECM turnover by proteases become possible targets for new therapies.

Acknowledgments: This work was supported by EXCELENCE CZ.02.1.01/0.0/0.0/15_003/0000460 OP RDE; Institutional Research Concept RV067985904. I did not receive any funds for covering the costs to publish the manuscript in open access.

Author Contributions: I appreciate help of all people who have been involved in this research.

Conflicts of Interest: The author declares no conflict of interest.

Abbreviations

ECM	Extracellular matrix
HA	Hyaluronan
IαI	Inter-alpha-trypsin inhibitor
TNFAIP6	Tumor necrosis factor alpha-induced protein 6
PTX3	Pentraxin 3
OCC	Oocyte-cumulus complexes
COX2/PTGS2	Cyclooxygenase/prostaglandin endoperoxide synthase
EGF	Epidermal growth factor
EGFR	Epidermal growth factor receptor
GVBD	Germinal vesicle breakdown
M I	Metaphase I
M II	Metaphase II
hCG	Human chorion gonadotropin
PMSG	Pregnant mare gonadotropin
TFGβ	Transforming growth factor beta

References

1. Theocharis, A.D.; Skandalis, S.S.; Gialeli, C.; Karamanos, N.K. Extracellular matrix structure. *Adv. Drug Deliv. Rev.* **2016**, *97*, 4–27. [CrossRef] [PubMed]
2. Naba, A.; Clauser, K.R.; Hoersch, S.; Liu, H.; Carr, S.A.; Hynes, R.O. The matrisome: In silico definition and in vivo characterization by proteomics of normal and tumor extracellular matrices. *Mol. Cell. Proteomics* **2012**, *11*. [CrossRef] [PubMed]
3. Hynes, R.O.; Naba, A. Overview of the matrisome—An inventory of extracellular matrix constituents and functions. *Cold Spring Harb. Perspect. Biol.* **2012**, *4*. [CrossRef] [PubMed]
4. Theocharis, A.D.; Skandalis, S.S.; Tzanakakis, G.N.; Karamanos, N.K. Proteoglycans in health and disease: Novel roles for proteoglycans in malignancy and their pharmacological targeting. *FEBS J.* **2010**, *277*, 3904–3923. [CrossRef] [PubMed]
5. Itano, N.; Sawai, T.; Yoshida, M.; Lenas, P.; Yamada, Y.; Imagawa, M.; Shinomura, T.; Hamaguchi, M.; Yoshida, Y.; Ohnuki, Y.; et al. Three isoforms of mammalian hyaluronan synthases have distinct enzymatic properties. *J. Biol. Chem.* **1999**, *274*, 25085–25092. [CrossRef] [PubMed]
6. Jacobson, A.; Brinck, J.; Briskin, M.J.; Spicer, A.P.; Heldin, P. Expression of human hyaluronan synthases in response to external stimuli. *Biochem. J.* **2000**, *348 Pt 1*, 29–35. [CrossRef] [PubMed]
7. Stern, R.; Jedrzejas, M.J. Hyaluronidases: Their genomics, structures, and mechanisms of action. *Chem. Rev.* **2006**, *106*, 818–839. [CrossRef] [PubMed]
8. Jarvelainen, H.; Sainio, A.; Koulu, M.; Wight, T.N.; Pentinen, R. Extracellular matrix molecules: Potential targets in pharmacotherapy. *Pharmacol. Rev.* **2009**, *61*, 198–223. [CrossRef] [PubMed]
9. Bateman, J.F.; Boot-Handford, R.P.; Lamande, S.R. Genetic diseases of connective tissues: Cellular and extracellular effects of ECM mutations. *Nat. Rev. Gen.* **2009**, *10*, 173–183. [CrossRef] [PubMed]
10. Chen, L.; Mao, S.J.; McLean, L.R.; Powers, R.W.; Larsen, W.J. Proteins of the inter-alpha-trypsin inhibitor family stabilize the cumulus extracellular matrix through their direct binding with hyaluronic acid. *J. Biol. Chem.* **1994**, *269*, 28282–28287. [PubMed]
11. Zhuo, L.; Yoneda, M.; Zhao, M.; Yingsung, W.; Yoshida, N.; Kitagawa, Y.; Kawamura, K.; Suzuki, T.; Kimata, K. Defect in SHAP-hyaluronan complex causes severe female infertility. A study by inactivation of the bikunin gene in mice. *J. Biol. Chem.* **2001**, *276*, 7693–7696. [CrossRef] [PubMed]
12. Nagyova, E.; Camaioni, A.; Prochazka, R.; Salustri, A. Covalent transfer of heavy chains of inter-alpha-trypsin inhibitor family proteins to hyaluronan in in vivo and in vitro expanded porcine oocyte-cumulus complexes. *Biol. Reprod.* **2004**, *71*, 1838–1843. [CrossRef] [PubMed]
13. Fulop, C.; Szanto, S.; Mukhopadhyay, D.; Bardos, T.; Kamath, R.V.; Rugg, M.S.; Day, A.J.; Salustri, A.; Hascall, V.C.; Glant, T.T.; et al. Impaired cumulus mucification and female sterility in tumor necrosis factor-induced protein 6 deficient mice. *Development* **2003**, *130*, 2253–2261. [CrossRef] [PubMed]

14. Suzuki, M.; Kobayashi, H.; Tanaka, Y.; Kanayama, N.; Terao, T. Reproductive failure in mice lacking inter-alpha-trypsin inhibitor (ITI)-ITI target genes in mouse ovary identified by microarray analysis. *J. Endocrinol.* **2004**, *183*, 29–38. [CrossRef] [PubMed]

15. Salustri, A.; Yanagishita, M.; Hascall, V.C. Synthesis and accumulation of hyaluronic acid and proteoglycans in the mouse cell-oocyte complex during follicle-stimulating hormone-induced mucification. *J. Biol. Chem.* **1989**, *264*, 13840–13847. [PubMed]

16. Chen, L.; Wert, S.E.; Hendrix, E.M.; Russell, P.T.; Cannon, M.; Larsen, W.J. Hyaluronic acid synthesis and gap junction endocytosis are necessary for normal expansion of the cumulus mass. *Mol. Reprod. Dev.* **1990**, *26*, 236–247. [CrossRef] [PubMed]

17. Fulop, C.; Salustri, A.; Hascall, V.C. Coding sequence of a hyaluronan synthase homologue expressed during expansion of the mouse cumulus-oocyte complex. *Arch. Biochem. Biophys.* **1997**, *337*, 261–266. [CrossRef] [PubMed]

18. Carrette, O.; Nemade, R.V.; Day, A.J.; Brickner, A.; Larsen, W.J. TSG-6 is concentrated in the extracellular matrix of mouse cumulus oocyte complexes through hyaluronan and inter-alpha-inhibitor binding. *Biol. Reprod.* **2001**, *65*, 301–308. [CrossRef] [PubMed]

19. Nagyová, E.; Procházka, R.; Vanderhyden, B.C. Ooocytectomy does not influence synthesis of hyaluronic acid by pig cumulus cells: Retention of hyaluronic acid after insulin-like growth factor-I treatment in serum free-medium. *Biol. Reprod.* **1999**, *61*, 569–574. [CrossRef] [PubMed]

20. Chen, L.; Mao, S.J.; Larsen, W.J. Identification of a factor in foetal bovine serum that stabilizes the cumulus extracellular matrix. A role for a member of the inter-alpha-trypsin inhibitor family. *J. Biol. Chem.* **1992**, *267*, 12380–12386. [PubMed]

21. Chen, L.; Zhang, H.; Powers, R.W.; Russell, P.T.; Larsen, W.J. Covalent linkage between proteins of the inter-a-inhibitor family and hyaluronic acid is mediated by a factor produced by granulosa cells. *J. Biol. Chem.* **1996**, *271*, 19409–19414. [CrossRef] [PubMed]

22. Nagyova, E. Organization of the expanded cumulus-extracellular matrix in preovulatory follicles: A role for inter-alpha-trypsin inhibitor. *Endocr. Regul.* **2015**, *49*, 37–45. [CrossRef] [PubMed]

23. Fulop, C.; Kamath, R.V.; Li, Y.; Otto, J.M.; Salustri, A.; Olsen, B.R.; Glant, T.T.; Hascall, V.C. Coding sequence, exon-intron structure and chromosomal localization of murine TNF-stimulated gene 6 that is specifically expressed by expanding cumulus cell-oocyte complexes. *Gene* **1997**, *202*, 95–102. [CrossRef]

24. Mukhopadhyay, D.; Hascall, V.C.; Day, A.J.; Salustri, A.; Fulop, C. Two distinct populations of tumor necrosis factor-stimulated gene-6 protein in the extracellular matrix of expanded mouse cumulus cell-oocyte complexes. *Arch. Biochem. Biophys.* **2001**, *394*, 173–181. [CrossRef] [PubMed]

25. Yoshioka, S.; Ochsner, S.; Russell, D.L.; Ujioka, T.; Fujii, S.; Richards, J.S.; Espey, L.L. Expression of tumor necrosis factor -stimulated gene-6 in the rat ovary in response to an ovulatory dose of gonadotropin. *Endocrinology* **2000**, *141*, 4114–4119. [CrossRef] [PubMed]

26. Nagyova, E.; Camaioni, A.; Prochazka, R.; Day, A.J.; Salustri, A. Synthesis of tumor necrosis factor alpha-induced protein 6 in porcine preovulatory follicles: A study with A38 antibody. *Biol. Reprod.* **2008**, *78*, 903–909. [CrossRef] [PubMed]

27. Nagyova, E.; Nemcova, L.; Prochazka, R. Expression of tumor necrosis factor alpha-induced protein 6 messenger RNA in porcine preovulatory ovarian follicles. *J. Reprod. Dev.* **2009**, *55*, 231–235. [CrossRef] [PubMed]

28. Sayasith, K.; Dore, M.; Sirois, J. Molecular characterization of tumor necrosis alpha-induced protein 6 and its human choret alionic gonadotropin-dependent induction in theca and mural granulosa cells of equine preovulatory follicles. *Reproduction* **2007**, *133*, 135–145. [CrossRef] [PubMed]

29. Varani, S.; Elvin, J.A.; Yan, C.; De Mayo, J.; DeMayo, F.J.; Horton, H.F.; Byrne, M.C.; Matzuk, M.M. Knockout of pentraxin 3, a downstream target of growth differentiation factor-9, causes female subfertility. *Mol. Endocrinol.* **2002**, *16*, 1154–1167. [CrossRef] [PubMed]

30. Salustri, A.; Garlanda, C.; Hirsch, E.; De Acetis, M.; Taccagno, A.; Bottazzi, B.; Doni, A.; Bastone, A.; Mantovani, G.; Beck Peccoz, P.; et al. PTX3 plays a key role in the organization of the cumulus oophorus extracellular matrix and in in vivo fertilization. *Development* **2004**, *131*, 1577–1586. [CrossRef] [PubMed]

31. Scarchilli, L.; Camaioni, A.; Bottazzi, B.; Negri, V.; Doni, A.; Deban, L.; Bastone, A.; Salvatori, G.; Mantovani, A.; Siracusa, G.; et al. PTX3 interacts with inter-alpha-trypsin inhibitor: Implications for hyaluronan organization and cumulus oophorus expansion. *J. Biol. Chem.* **2007**, *282*, 30161–30170. [CrossRef] [PubMed]

32. Ievoli, E.; Lindstedt, R.; Inforzato, A.; Camaioni, A.; Palone, F.; Day, A.J.; Mantovani, A.; Salvatori, G.; Salustri, A. Implications of the oligomeric state of the N-terminal PTX3 domain in cumulus matrix assembly. *Matrix Biol.* **2011**, *30*, 330–337. [CrossRef] [PubMed]

33. Inforzato, A.; Rivieccio, V.; Morreale, A.P.; Bastone, A.; Salustri, A.; Scarchilli, L.; Verdoliva, A.; Vincenti, S.; Gallo, G.; Chiapparino, C.; et al. Structural characterization of PTX3 disulfide bond network and its multimeric status in cumulus matrix organization. *J. Biol. Chem.* **2008**, *283*, 10147–10161. [CrossRef] [PubMed]

34. Nagyova, E.; Kalous, J.; Nemcova, L. Increased expression of pentraxin 3 after in vivo and in vitro stimulation with gonadotropins in porcine oocyte-cumulus complexes and mural granulosa cells. *Domest. Anim. Endocrinol.* **2016**, *56*, 29–35. [CrossRef] [PubMed]

35. Mizon, C.; Balduyck, M.; Albani, D.; Michalski, C.; Burnouf, T.; Mizon, J. Development of an enzyme-linked immunosorbent assay for human plasma inter-alpha-trypsin inhibitor (ITI) using specific antibodies against each of the H1 and H2 heavy chains. *Immunol. Methods* **1996**, *190*, 61–70. [CrossRef]

36. Salier, J.P.; Rouet, P.; Raguenez, G.; Daveau, M. The inter-a-inhibitor family: From structure to regulation. *Biochem. J.* **1996**, *315*, 1–9. [CrossRef] [PubMed]

37. Rouet, P.; Daveau, M.; Salier, J.P. Electrophoretic pattern of the inter-alpha-inhibitor family proteins in human serum characterized by chain-specific antibodies. *Biol. Chem. Hoppe-Seyler* **1992**, *373*, 1019–1024. [CrossRef] [PubMed]

38. Carrette, O.; Mizon, C.; Sautiere, P.; Sesboue, R.; Mizon, J. Purification and characterization of pig inter-a-inhibitor and its constitutive heavy chains. *Biochem. Biophys. Acta* **1997**, *1338*, 21–30. [CrossRef]

39. Flahaut, C.; Capon, C.; Balduyck, M.; Ricart, G.; Sautiere, P.; Mizon, J. Glycosylation pattern of human inter-alpha-inhibitor heavy chains. *Biochem. J.* **1998**, *333 Pt 3*, 749–756. [CrossRef] [PubMed]

40. Milner, C.M.; Day, A.J. TSG-6: A multifunctional protein associated with inflammation. *J. Cell Sci.* **2003**, *116*, 1863–1873. [CrossRef] [PubMed]

41. Day, A.J.; de la Motte, C.A. Hyaluronan cross-linking: A protective mechanism in inflammation? *Trends Immunol.* **2005**, *28*, 637–643. [CrossRef] [PubMed]

42. Milner, C.M.; Higman, V.A.; Day, A.J. TSG-6: A pluripotent inflammatory mediator? *Biochem. Soc. Trans.* **2006**, *34*, 446–450. [CrossRef] [PubMed]

43. Jessen, T.E.; Odum, L. Role of tumour necrosis factor stimulated gene 6 (TSG-6) in the coupling of inter-alpha-trypsin inhibitor to hyaluronan in human follicular fluid. *Reproduction* **2003**, *125*, 27–31. [CrossRef] [PubMed]

44. Conti, M.; Hsieh, M.; Park, J.Y.; Su, Y.Q. Role of the epidermal growth factor network in ovarian follicles. *Mol. Endocrinol.* **2006**, *20*, 715–723. [CrossRef] [PubMed]

45. Nagyova, E.; Camaioni, A.; Scsukova, S.; Mlynarcikova, A.; Prochazka, R.; Nemcova, L.; Salustri, A. Activation of Cumulus Cell SMAD2/3 and Epidermal Growth Factor Receptor Pathways Are Involved in Porcine Oocyte-Cumulus Cell Expansion and Steroidogenesis. *Mol. Reprod. Dev.* **2011**, *78*, 391–402. [CrossRef] [PubMed]

46. Prochazka, R.; Kalab, P.; Nagyova, E. Epidermal growth factor-receptor tyrosine kinase activity regulates expansion of porcine oocyte-cumulus cell complexes in vitro. *Biol. Reprod.* **2003**, *68*, 797–803. [CrossRef] [PubMed]

47. Procházka, R.; Petlach, M.; Nagyová, E.; Nemcová, L. Effect of epidermal growth factor-like peptides on pig cumulus cell expansion, oocyte maturation, and acquisition of developmental competence in vitro: Comparison with gonadotropins. *Reproduction* **2011**, *141*, 425–435. [CrossRef] [PubMed]

48. Yamashita, Y.; Kawashima, I.; Yanai, Y.; Nishibori, M.; Richards, J.S.; Shimada, M. Hormone-induced expression of tumor necrosis factor alpha converting enzyme/A disintegrin and metaloprotease-17 impacts porcine cumulus cell oocyte complex expansion and meiotic maturation via ligand activation of the epidermal growth factor receptor. *Endocrinology* **2007**, *148*, 6164–6175. [CrossRef] [PubMed]

49. Wayne, C.M.; Fan, H.Y.; Cheng, X.; Richards, J.S. Follicle-stimulating hormone induces multiple signaling cascades: Evidence that activation of Rous sarcoma oncogene, RAS, and the epidermal growth factor receptor are critical for granulosa cell differentiation. *Mol. Endocrinol.* **2007**, *21*, 1940–1957. [CrossRef] [PubMed]

50. Ashkenazi, H.; Cao, X.; Motola, S.; Popliker, M.; Conti, M.; Tsafriri, A. Epidermal growth factor family members: Endogenous mediators of the ovulatory response. *Endocrinology* **2005**, *46*, 77–84. [CrossRef] [PubMed]

51. Reizel, Y.; Elbaz, Y.; Dekel, N. Sustained activity of the EGF receptor is an absolute requisite for LH-induced oocyte maturation and cumulus expansion. *Mol. Endocrinol.* **2010**, *24*, 402–411. [CrossRef] [PubMed]

52. Srinivasan, R.; Benton, E.; McCormick, F.; Thomas, H.; Gullick, W.J. Expression of the c-erbB-3/HER-3 and c-erbB-4/HER-4 growth factor receptors and their ligands, neuregulin-1 alpha, neuregulin-1 beta, and betacellulin, in normal endometrium and endometrial cancer. *Clin. Cancer Res.* **1999**, *5*, 2877–2883. [PubMed]

53. Santin, A.D.; Bellone, S.; Gokden, M.; Palmieri, M.; Dunn, D.; Agha, J. Overexpression of HER-2/neu in uterine serous papillary cancer. *Clin. Cancer Res.* **2002**, *8*, 1271–1279. [PubMed]

54. Yarden, Y.; Sliwkowski, M.X. Untangling the ErbB signaling network. *Nat. Rev. Mol. Cell Biol.* **2001**, *2*, 127–137. [CrossRef] [PubMed]

55. Baselga, J.; Arteaga, C.L. Critical update and emerging trends in epidermal growth factor receptor targeting in cancer. *J. Clin. Oncol.* **2005**, *23*, 2445–2459. [CrossRef] [PubMed]

56. Kelly, H.; Graham, M.; Humes, E. Delivery of a healthy baby after first-trimester maternal exposure to lapatinib. *Clin. Breast Cancer* **2006**, *7*, 339–341. [CrossRef] [PubMed]

57. Rusnak, D.W.; Lackey, K.; Affleck, K.; Wood, E.R.; Alligood, K.J.; Rhodes, N. The effects of the novel, reversible epidermal growth factor receptor/ErbB-2 tyrosine kinase inhibitor, GW2016, on the growth of human normal and tumor-derived cell lines in vitro and in vivo. *Mol. Cancer Ther.* **2001**, *1*, 85–94. [PubMed]

58. Eccles, S. The epidermal growth factor receptor /Erb-B/HER family in normal and malignant breast biology. *Int. J. Dev. Biol.* **2011**, *55*, 685–696. [CrossRef] [PubMed]

59. Nagyova, E.; Nemcova, L.; Mlynarcikova, A.; Scsukova, S.; Kalous, J. Lapatinib inhibits meiotic maturation of porcine oocyte-cumulus complexes cultured in vitro in gonadotropin -supplemented medium. *Fertil. Steril.* **2013**, *99*, 1739–1748. [CrossRef] [PubMed]

60. Lydon, J.P.; DeMayo, F.J.; Funk, C.R.; Mani, S.K.; Hughes, A.R.; Montgomery, C.A., Jr.; Shyamala, G.; Conneely, O.M.; O'Malley, B.W. Mice lacking progesterone receptor exhibit pleiotropic reproductive abnormalities. *Genes Dev.* **1995**, *9*, 2266–2278. [CrossRef] [PubMed]

61. Faivre, E.J.; Daniel, A.R.; Hillard, C.H.J.; Lange, C.A. Progesterone receptor rapid signaling mediates serine 345 phosphorylation and tethering to specificity protein 1 transcription factors. *Mol. Endocrinol.* **2008**, *22*, 823–837. [CrossRef] [PubMed]

62. Nagyova, E.; Scsukova, S.; Kalous, J.; Mlynarcikova, A. Effect of RU486 and indomethacin on meiotic maturation, formation of extracellular matrix, and progesterone production by porcine oocyte-cumulus complexes. *Domest. Anim. Endocrinol.* **2014**, *48*, 7–14. [CrossRef] [PubMed]

63. Shimada, M.; Nishibori, M.; Yamashita, Y.; Ito, J.; Mori, T.; Richards, J.S. Down-regulated expression of A disintegrin and metalloproteinase with thrombospondin-like repeats-1 by progesterone receptor antagonist is associated with impaired expansion of porcine cumulus-oocyte complexes. *Endocrinology* **2004**, *145*, 4603–4614. [CrossRef] [PubMed]

64. Roh, S.I.; Batten, B.E.; Friedman, C.I.; Kim, M.H. The effects of progesterone antagonist RU486 on mouse oocyte maturation, ovulation, fertilization, and cleavage. *Am. J. Obstet. Gynecol.* **1988**, *159*, 1584–1589. [CrossRef]

65. Aparicio, I.M.; Garcia-Herreros, M.; O'Shea, L.C.; Hensey, C.; Lonergan, P.; Fair, T. Expression, regulation, and function of progesterone receptors in bovine cumulus oocyte complexes during in vitro maturation. *Biol. Reprod.* **2011**, *84*, 910–921. [CrossRef] [PubMed]

66. Shao, R.; Markström, E.; Friberg, P.A.; Johansson, M.; Billig, H. Expression of progesterone receptor (PR) A and B isoforms in mouse granulosa cells: Stage-dependent PR-mediated regulation of apoptosis and cell proliferation. *Biol. Reprod.* **2003**, *68*, 914–921. [CrossRef] [PubMed]

67. Romero, S.; Smitz, J. Epiregulin can effectively mature isolated cumulus-oocyte complexes, but fails as a substitute for the hCG/epidermal growth factor stimulus on cultured follicles. *Reproduction* **2009**, *137*, 997–1005. [CrossRef] [PubMed]

68. Sirois, J. Induction of prostaglandin endoperoxide synthase-2 by human chorionic gonadotropin in bovine preovulatory follicles in vivo. *Endocrinology* **1994**, *135*, 841–848. [CrossRef] [PubMed]

69. Nuttinck, F.; Reinaud, P.; Tricoire, H.; Vigneron, C.; Peynot, N.; Mialot, J.P.; Mermillod, P.; Charpigny, G. Cyclooxygenase-2 is expressed by cumulus cells during oocyte maturation in cattle. *Mol. Reprod. Dev.* **2002**, *61*, 93–101. [CrossRef] [PubMed]

70. Davis, B.J.; Lennard, D.E.; Lee, C.A.; Tiano, H.F.; Morham, S.G.; Wetsel, W.C.; Langenbach, R. Anovulation in cyclooxygenase-2-deficient mice is restored by prostaglandin E(2) and interleukin-1 beta. *Endocrinology* **1999**, *140*, 2685–2695. [CrossRef] [PubMed]

71. Lim, H.; Paria, B.C.; Das, S.K.; Dinchuc, J.E.; Langebach, R.; Trzaskos, J.M.; Dey, S.K. Multiple female reproductive failures in cyclooxygenase 2- deficient mice. *Cell* **1997**, *91*, 197–208. [CrossRef]

72. Ochsner, S.A.; Russell, D.L.; Day, A.J.; Breyer, R.M.; Richards, J.S. Decreased expression of tumor necrosis factor-alpha-stimulated gene 6 in cumulus cells of the cyclooxygenase-2 and EP2 null mice. *Endocrinology* **2003**, *144*, 1008–1019. [CrossRef] [PubMed]

73. Janson, P.O.; Brannstrom, M.; Holmes, P.V.; Sogn, J. Studies on the mechanism of ovulation using the model of the isolated ovary. *Ann. N. Y. Acad. Sci.* **1988**, *541*, 22–29. [CrossRef] [PubMed]

74. Peters, M.W.; Pursley, J.R.; Smith, G.W. Inhibition of intrafollicular PGE2 synthesis and ovulation following ultrasound-mediated intrafollicular injection of the selective cyclooxygenase-2 inhibitor NS-398 in cattle. *J. Anim. Sci.* **2004**, *82*, 1656–1662. [CrossRef] [PubMed]

75. Duffy, D.M.; VandeVoort, C.A. Maturation and fertilization of non-human primate oocytes are compromised by oral administration of a COX-2 inhibitor. *Fertil. Steril.* **2011**, *95*, 1256–1260. [CrossRef] [PubMed]

76. Eppig, J.J. Prostaglandin E_2 stimulates cumulus expansion and hyaluronic acid synthesis by cumuli oophori isolated from mice. *Biol. Reprod.* **1981**, *25*, 191–195. [CrossRef] [PubMed]

77. Matsumoto, H.; Ma, W.G.; Smalley, W.; Trzaskos, J.; Breyer, R.M.; Dey, S.K. Diversification cyclooxygenase-2-derived prostaglandins in ovulation and implantation. *Biol. Rerpod.* **2001**, *64*, 1557–1565. [CrossRef]

78. Ben-Ami, I.; Freimann, S.; Armon, L.; Dantes, A.; Strassburger, D.; Friedler, S.; Raziel, A.; Seger, R.; Ron-El, R.; Amsterdam, A. PGE2 up-regulates EGF-like growth factor biosynthesis in human granulosa cells: New insights into the coordination between PGE2 and LH in ovulation. *Mol. Hum. Reprod.* **2006**, *12*, 593–599. [CrossRef] [PubMed]

79. Park, J.Y.; Su, Y.Q.; Ariga, M.; Law, E.; Jin, S.L.; Conti, M. EGF-like growth factors as mediators of LH action in the ovulatory follicle. *Science* **2004**, *303*, 682–684. [CrossRef] [PubMed]

80. Hsieh, M.; Lee, D.; Panigone, S.; Horner, K.; Chen, R.; Theologis, A.; Lee, D.C.; Threadgill, D.W.; Conti, M. Luteinizing hormone-dependent activation of the epidermal growth factor network is essential for ovulation. *Mol. Cell. Biol.* **2007**, *27*, 1914–1924. [CrossRef] [PubMed]

81. Goldberg, A.L.; Stein, R.; Adams, J. New insights into proteasome function: From archaebacteria to drug development. *Chem. Biol.* **1995**, *2*, 503–508. [CrossRef]

82. Coux, O.; Tanaka, K.; Goldberg, A.L. Structure and functions of the 20S and 26S proteasomes. *Annu. Rev. Biochem.* **1996**, *65*, 801–847. [CrossRef] [PubMed]

83. Josefsberg, L.B.; Galiani, D.; Dantes, A.; Amsterdam, A.; Dekel, N. The proteasome is involved in the first metaphase-to-anaphase transition of meiosis in rat oocytes. *Biol. Reprod.* **2000**, *62*, 1270–1277. [CrossRef] [PubMed]

84. Lee, D.H.; Goldberg, A.L. Proteasome inhibitors: Valuable new tools for cell biologists. *Trends Cell Biol.* **1998**, *8*, 397–403. [CrossRef]

85. Rock, K.L.; Gramm, C.; Rothstein, L.; Clark, K.; Stein, R., Dick, L.; Hwang, D.; Goldberg, A.L. Inhibitors of the proteasome block the degradation of most cell proteins and the generetaion of peptides presented on MHC class I molecules. *Cell* **1994**, *78*, 767–771. [CrossRef]

86. Huo, L.J.; Fan, H.Y.; Liang, C.G.; Yu, L.Z.; Zhong, Z.S.; Chen, D.Y.; Sun, Q.Y. Regulation of ubiquitin-proteasome pathway on pig oocyte meiotic maturation and fertilization. *Biol. Reprod.* **2004**, *71*, 853–862. [CrossRef] [PubMed]

87. Huo, L.J.; Fan, H.Y.; Zhong, Z.S.; Chen, D.Y.; Schatten, H.; Sun, Q.Y. Ubiquitin-proteasome pathway modulates mouse oocyte meiotic maturation and fertilization via regulation of MAPK cascade and cyclin B1 degradation. *Mech. Dev.* **2004**, *121*, 1275–1287. [CrossRef] [PubMed]

88. Chmelikova, E.; Sedmikova, M.; Rajmon, R.; Petr, J.; Svestkova, D.; Jilek, F. Effect of proteasome inhibitor MG132 on in vitro maturation of pig oocytes. *Zygote* **2004**, *12*, 157–162. [CrossRef] [PubMed]

89. Yi, Y.J.; Nagyova, E.; Manandhar, G.; Prochazka, R.; Sutovsky, M.; Park, C.S.; Sutovsky, P. Proteolytic activity of the 26S proteasome is required for the meiotic resumption, germinal vesicle breakdown, and cumulus expansion of porcine cumulus-oocyte complexes matured in vitro. *Biol. Reprod.* **2008**, *78*, 115–126. [CrossRef] [PubMed]

90. Mailhes, J.B.; Hilliard, C.; Lowery, M.; London, S.N. MG-132, an inhibitor of proteasomes and calpains, induced inhibition of oocyte maturation and aneuploidy in mouse oocytes. *Cell Chromosom.* **2002**, *1*, 2–7. [CrossRef]

91. Rugg, M.S.; Willis, A.C.; Mukhopadhyay, D.; Hascall, V.C.; Fries, E.; Fülöp, C.; Milner, C.M.; Day, A.J. Characterization of complexes formed between TSG-6 and inter-α-inhibitor that act as intermediates in the covalent transfer of heavy chains onto hyaluronan. *J. Biol. Chem.* **2005**, *280*, 25674–25686. [CrossRef] [PubMed]

92. Nagyova, E.; Scsukova, S.; Nemcova, L.; Mlynarcikova, A.; Yi, Y.-J.; Sutovsky, M.; Sutovsky, P. Inhibition of proteasomal proteolysis affects expression of extracellular matrix components and steroidogenesis in porcine oocyte-cumulus complexes. *Domest. Anim. Endocrinol.* **2012**, *42*, 50–62. [CrossRef] [PubMed]

93. Tsafriri, A.; Cao, X.; Ashkenazi, H.; Motola, S.; Popliker, M.; Pomerantz, S.H. Resumption of oocyte meiosis in mammals: On models, meiosis activating sterols, steroids and EGF-like factors. *Mol. Cell. Endocrinol.* **2005**, *234*, 37–45. [CrossRef] [PubMed]

94. Sutovsky, P.; Flechon, J.E.; Flechon, B.; Motlik, J.; Peynot, N.; Chesne, P.; Heyman, Y. Dynamic changes of gap junctions and cytoskeleton during in vitro culture of cattle oocyte cumulus complexes. *Biol. Reprod.* **1993**, *49*, 1277–1287. [CrossRef] [PubMed]

95. Luciano, A.M.; Lodde, V.; Beretta, M.S.; Colleoni, S.; Lauria, A.; Modina, S. Developmental capability of denuded bovine oocyte in co-culture system with intact cumulus-oocyte complexes; role of cumulus cells, cyclic adenosine-3′,5′-monophosphate, and glutathione. *Mol. Reprod. Dev.* **2005**, *71*, 389–397. [CrossRef] [PubMed]

96. Gutnisky, C.; Dalvit, G.C.; Pintos, L.N.; Thompson, J.G.; Beconi, M.T.; Cetica, P.D. Influence of hyaluronic acid synthesis and cumulus mucification on bovine oocyte in vitro maturation, fertilization and embryo development. *Reprod. Fertil. Dev.* **2007**, *19*, 488–497. [CrossRef] [PubMed]

97. Nuttinck, F.; Marquant-LeGuienne, B.; Clement, L.; Reinaud, P.; Charpigny, G.; Grimard, B. Expression of genes involved in prostaglandin E2 and progesterone production in bovine cumulu-oocyte complex during in vitro maturation and fertilization. *Reproduction* **2008**, *135*, 593–603. [CrossRef] [PubMed]

98. Iwamasa, J.; Shibata, S.; Tanaka, N.; Matsuura, K.; Okamura, H. The relationship between ovarian progesterone and proteolytic enzyme activity during ovulation in the gonadotropin-treated immature rat. *Biol. Reprod.* **1992**, *46*, 309–313. [CrossRef] [PubMed]

99. Jezova, M.; Scsukova, S.; Nagyova, E.; Vranova, J.; Prochazka, R.; Kolena, J. Effect of intraovarian factors on porcine follicular cells: Cumulus expansion, granulosa and cumulus cell progesterone production. *Anim. Reprod. Sci.* **2001**, *65*, 115–126. [CrossRef]

100. Shimada, M.; Terada, T. FSH and LH induce progesterone production and progesterone receptor synthesis in cumulus cells, a requirement for meiotic resumption in porcine oocytes. *Mol. Hum. Reprod.* **2002**, *8*, 612–618. [CrossRef] [PubMed]

101. Granot, Z.; Melamed-Book, N.; Bahat, A.; Orly, J. Turnover of STAR protein: Roles for the proteasome and mitochondrial proteases. *Mol. Cell. Endocrinol.* **2007**, *265–266*, 51–58. [CrossRef] [PubMed]

102. Tajima, K.; Babich, S.; Yoshida, Y.; Dantes, A.; Strauss, J.F., 3rd; Amsterdam, A. The proteasome inhibitor MG132 promotes accumulation of the steroidogenic acute regulatory protein (star) and steroidogenesis. *FEBS Lett.* **2001**, *490*, 59–64. [CrossRef]
103. Ziolkowska, A.; Tortorella, C.; Nussdorfer, G.G.; Rucinski, M.; Majchrzak, M.; Malendowicz, L.K. Accumulation of steroidogenic acute regulatory protein mRNA, and decrease in the secretory and proliferative activity of rat adrenocortical cells in the presence of proteasome inhibitors. *Int. J. Mol. Med.* **2006**, *17*, 865–868. [CrossRef] [PubMed]
104. Clark, B.J.; Wells, J.; King, S.R.; Stocco, D.M. The purification, cloning, and expression of a novel luteinizing hormone-induced mitochondrial protein in MA-10 mouse Leydig tumor cells. Characterization of the steroidogenic acute regulatory protein (StAR). *J. Biol. Chem.* **1994**, *269*, 28314–28322. [PubMed]

International Journal of
Molecular Sciences

Review

Role of Plasminogen Activator Inhibitor Type 1 in Pathologies of Female Reproductive Diseases

Yao Ye, Aurelia Vattai, Xi Zhang, Junyan Zhu, Christian J. Thaler, Sven Mahner, Udo Jeschke * and Viktoria von Schönfeldt

Department of Gynaecology and Obstetrics, Ludwig-Maximilians University of Munich, Campus Großhadern: Marchioninistr. 15, 81377 Munich and Campus Innenstadt: Maistr. 11, 80337 Munich, Germany; Yao.Ye@med.uni-muenchen.de (Y.Y.); Aurelia.Vattai@med.uni-muenchen.de (A.V.); Xi.Zhang@med.uni-muenchen.de (X.Z.); Junyan.Zhu@med.uni-muenchen.de (J.Z.); Thaler@med.uni-muenchen.de (C.J.T.); Sven.Mahner@med.uni-muenchen.de (S.M.); Viktoria.Schoenfeldt@med.uni-muenchen.de (V.v.S.)
* Correspondence: Udo.Jeschke@med.uni-muenchen.de; Tel.: +49-89-4400-74531

Received: 6 July 2017; Accepted: 27 July 2017; Published: 29 July 2017

Abstract: Normal pregnancy is a state of hypercoagulability with diminishing fibrinolytic activity, which is mainly caused by an increase of plasminogen activator inhibitor type 1 (PAI-1). PAI-1 is the main inhibitor of plasminogen activators, including tissue-type plasminogen activator (tPA) and urokinase-type plasminogen activator (uPA). In human placentas, PAI-1 is expressed in extravillous interstitial trophoblasts and vascular trophoblasts. During implantation and placentation, PAI-1 is responsible for inhibiting extra cellular matrix (ECM) degradation, thereby causing an inhibition of trophoblasts invasion. In the present study, we have reviewed the literature of various reproductive diseases where PAI-1 plays a role. PAI-1 levels are increased in patients with recurrent pregnancy losses (RPL), preeclampsia, intrauterine growth restriction (IUGR), gestational diabetes mellitus (GDM) in the previous pregnancy, endometriosis and polycystic ovary syndrome (PCOS). In general, an increased expression of PAI-1 in the blood is associated with an increased risk for infertility and a worse pregnancy outcome. GDM and PCOS are related to the genetic role of the 4G/5G polymorphism of *PAI-1*. This review provides an overview of the current knowledge of the role of PAI-1 in reproductive diseases. PAI-1 represents a promising monitoring biomarker for reproductive diseases and may be a treatment target in the near future.

Keywords: plasminogen activator inhibitor type 1; trophoblast invasion; recurrent pregnancy losses; preeclampsia; intrauterine growth restriction; gestational diabetes mellitus; endometriosis; polycystic ovary syndrome

---

1. Introduction

The fibrinolytic system plays a role in several physiological and pathophysiological processes, such as hemostatic balance, tissue remodeling, tumor invasion, angiogenesis and reproduction [1]. Normal pregnancy is a state of hypercoagulability with remarkable changes in all aspects of hemostasis—an increase of clotting factors and coagulability and a decrease of anticoagulants and fibrinolytic activity, thereby influencing placental function during pregnancy and meeting delivery's hemostatic challenge [2]. Fibrinolytic system is depressed during pregnancy, and this change partly explains the higher incidence of thromboembolic complications such as recurrent pregnancy losses, preeclampsia and intrauterine growth restriction [2]. The diminishing fibrinolytic activity is mainly caused by a continuous increase of the major inhibitor of the fibrinolytic system: plasminogen activator inhibitor type 1 (PAI-1) [2]. PAI-1 is responsible for approximately 60% of the PA-inhibitory activity in the plasma [3] and is the key inhibitor of fibrinolysis compared with PAI-2 and PAI-3 during pregnancy [4].

PAI-1 gene deficiency shows a transient impaired placentation in mice [5], while, in humans, *PAI-1* gene deficiency is associated with abnormal bleeding after a trauma or surgery [6,7]. Transgenic mice that overexpress PAI-1 exhibit thrombotic occlusion [8]. Former studies in humans suggest that increased PAI-1 levels are found to be crucial mediators of vascular disease, fibrosis, tumor metastasis, diabetes, and reproductive diseases [9–12]. PAI-1 acts as a major inhibitor of fibrinolysis, its overexpression leads to fibrin accumulation and insufficient placentation. In this review, we focus on the complex roles of PAI-1 in normal placentation and reproductive diseases, including recurrent pregnancy losses, preeclampsia, intrauterine growth restriction, endometriosis and polycystic ovary syndrome.

2. Fibrinolytic System and PAI-1 (Plasminogen Activator Inhibitor Type 1)

The prime fibrinolytic protease of the fibrinolytic system is plasminogen, which can be activated by urokinase-type plasminogen activator (uPA) and the tissue-type plasminogen activator (tPA) [1,8]. Plasminogen can then be converted into plasmin, and eventually cleaves fibrin into cross-linked fibrin degradation products (Figure 1) [8]. Plasminogen activator inhibitors include PAI-1, PAI-2, PAI-3, C1-esterase inhibitor and protease nexin (Figure 1) [8]. Plasmin inhibitors are α2-plasmin inhibitor (α2-PI), α2-macroglobulin (α2-MG) and protease nexin (Figure 1) [8]. Both uPA and tPA are serine proteases that cleave a single Arg-Val peptide bond to transfer plasminogen to plasmin; uPA functions mainly in pericellular proteolysis while tPA is involved in the circulation [13]. uPA plays an important role in a variety of physiological and pathological processes including tissue destruction, inflammatory reactions and invasion of trophoblasts [14] and cancer cells [15]. Both uPA and tPA consist of a single-chain form and a two-chain form [4,16]. During normal pregnancy, the levels of uPA, PAI-1, PAI-2 and α2-antiplasmin are increased and tPA levels are decreased [17].

PAI-1 is the primary inhibitor of tPA in the plasma during pregnancy [4]. It is a single-chain glycoprotein consisting of 379 or 381 amino acids (N-terminal heterogeneity) and belongs to the serine family of protease inhibitor, with a molecular weight of about 45 kDa. There are three different forms of PAI-1: active, inactive and substrate form. The active form can inhibit tPA or uPA by forming a 1:1 stoichiometric complex with each enzyme and the inactive form does not react with the target proteinase [18]. The conformational conversion from the active into the inactive form is completed by the P1-P1′ in a reactive center loop (RCL) of the serpin cleave, followed by the insertion of the RCL into the β-sheet A of the serpin [19]. *PAI-1* gene in humans is located on chromosome 7 (q21.3-q22), extends approximately 12.200 base pairs and consists of nine exons and eight introns [20]. *PAI-1* gene has several polymorphisms and the 4G allele of the 4G/5G polymorphism is related to high PAI-1 levels [21]. 4G polymorphism is located in the PAI-1 promotor, which is 675 bp upstream from the start site of transcription in the promoter region [21]. Circulating PAI-1 is mainly found in platelets, whilst a large range of cells can further express PAI-1, such as fibroblasts, smooth muscle cells, endothelial cells, hepatocytes, inflammatory cells and placental cells [22].

Both forms of tPA are inhibited by PAI-1, whereas PAI-2 inhibits mainly the two-chain form [23]. PAI-2 consists of two molecular forms: the low molecular weight (LMW) form with 43–48 kDa is intracellular and non-glycosylated, while the high molecular weight (HMW) form with 60 kDa is secreted and glycosylated [24]. PAI-2 is mainly expressed by placental trophoblasts and macrophages [23]. PAI-3, also known as protein C inhibitor (PCI), is prominently expressed in male reproductive organs and its low levels in seminal plasma are associated with infertility [25].

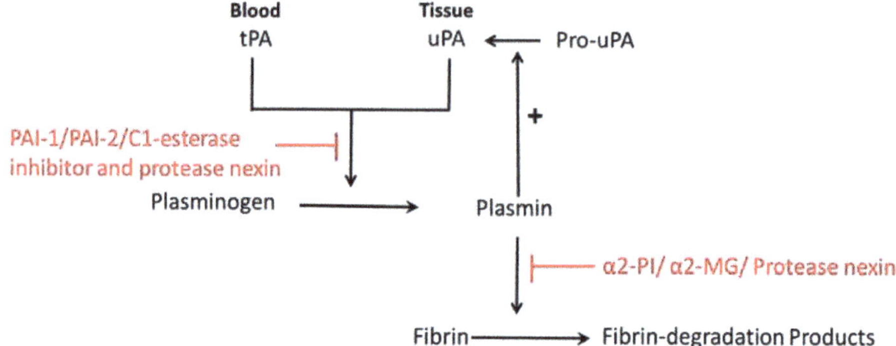

Figure 1. Schematic diagram of fibrinolysis: plasminogen is activated by plasminogen activator (tPA in blood or uPA in tissue), and then converted to plasmin. Then plasmin cleaves fibrin into fibrin-degradation products. Plasminogen activators inhibitors are PAI-1, PAI-2, C1-esterase inhibitor and protease nexin. Plasmin inhibitors are α2-plasmin inhibitor (α2-PI), α2-macroglobulin (α2-MG) and protease nexin. Pro-uPA can be converted to uPA, which is catalyzed by plasmin, the product of plasminogen.

3. Role of PAI-1 in the Female Reproduction System

During a healthy pregnancy, PAI-1 levels in the plasma gradually elevate during the second trimester of pregnancy and reach a maximum at 32–40 weeks of pregnancy. Within 5–8 weeks after delivery, PAI-1 levels fall again to the levels before the occurrence of pregnancy [2]. The concentration of PAI-1 in the plasma of healthy non-pregnant women varies (55 ± 17 ng/mL) [26]. The maximum of PAI-1 concentration during the last trimester of pregnancy is approximately 3–5 times higher than that of non-pregnant women [26]. These temporary changes of PAI-1 during pregnancy are accounted for hormonal influences [27]. Increases in PAI-2 levels are observed during pregnancy and during delivery [23]. Both PAI-1 and PAI-2 decrease quickly following the placenta separation from the uterus, but PAI-2 can still be found in the circulation up to eight weeks postpartum [2].

Trophoblasts can express PAI-1 and PAI-2 in vivo and in vitro [28]. Both PAI-1 and PAI-2 are localized in the cytoplasm of cytotrophoblasts and in the cytoplasm and plasma membrane of intermediate and syncytiotrophoblasts [28]. PAI-1 is localized in invading trophoblasts in the human placenta by immunostaining [29], especially in extravillous interstitial trophoblasts and vascular trophoblasts [30]. In the human placenta, PAI-1 protein and mRNA expression exist in most extravillous cytotrophoblast cells of the decidual layer, especially the chorionic villous tree and in cytotrophoblast cells of the chorionic plate, basal plate and intercotyledonary septae [31]. No expression of PAI-1 has been observed in the basal plate of endometrial stromal cells, chorionic plate connective tissue cells, septal endometrial stromal cells or villous core mesenchyme [31]. PAI-2 is the predominant PAI accumulated in villous syncytiotrophoblasts [29]. Both PAI-1 and PAI-2 mRNAs expressions are detected in cultured cytotrophoblasts isolated from both the first trimester and term placenta [29]. PAI-1 and PAI-2 are also expressed by uterine natural killer (uNK) cells [32].

3.1. PAI-1 Inhibits Trophoblast Invasion

Trophoblast invasion at the maternal-fetal interface is a key process during implantation and placentation, and during this process extravillous cytotrophoblasts (EVT) acquire invasive properties, which are able to invade and remodel maternal tissues (interstitial EVT) and uterine spiral artery (endovascular EVT) [33]. EVT can degrade extracellular matrix (ECM) to promote cell migration to the maternal side [33]. This process is precisely controlled by many factors expressed by maternal cells and trophoblasts (Figure 2) [33].

Trophoblasts and malignant tumors use the same biochemical mediators to facilitate invasion, including extracellular matrix degradation and immunosuppression of environmental conditions. PAI-1 can inhibit trophoblasts invasion while promoting tumor cell immigration [34,35]. PAI-1 is a biomarker for malignancies with poor prognosis because it facilitates tumor cell migration and invasion [35]. PAI-1/uPA/uPA receptor (uPAR)/low density lipoprotein receptor-related protein (LRP)/integrin complexes are initiating an "adhesion–detachment–re-adhesion" cycle to promote tumor cell migration [35,36]. Hyperinvasiveness in premalignant and malignant extravillous trophoblasts (JAR and JEG-3 choriocarcinoma cell lines) results from a downregulation of tissue inhibitors of metalloprotease (TIMP)-1 and *PAI-1* genes [34].

EVT invasion in early pregnancy occurs in a relatively low-oxygen (3%) environment, which is mediated by a general inhibition of the plasminogen activator system [37], as well as many adhesion molecules, growth factors, cytokines, interleukins, ECM components, and various placental hormones [14,33]. The anti-invasive action of EVT is caused by an upregulation of the tissue inhibitor of TIMP-1 and PAI-1 and a downregulation of uPA [14]. PAI-1 and PAI-2 are expressed in invading human extravillous trophoblast cells and they limit the depth of invasion [37,38]. PAI-1 is found to be absent in the placental bed of ectopic and molar pregnancies, suggesting that no expression of PAI-1 can contribute to an uncontrolled placental invasion [30]. Expression of PAI-1 is elevated in the process of EVT invasion treated by tumor necrosis factor α (TNF-α), and adding PAI-1-inactivating antibodies restores migration [39].

The limitation of EVT invasion is due to reduced ECM degradation (Figure 2), which requires the balance of promoting and restraining factors, such as metalloproteinases (MMPs) and tissue inhibitors of MMPs (TIMPs), uPA and PAI-1 [40]. PAI-1 restrains ECM degradation in three different ways. Firstly, PAI-1 can directly block uPA activity to constrain the proteolytic activity of plasmin [18]. Secondly, PAI-1 binds to uPA and causes its degradation via uPAR and low-density lipoprotein receptor-related protein (LRP) internalization [11]. The direct binding of occupied uPAR to LRP is essential for internalization and clearance of uPA-PAI-1-occupied uPAR [42]. The activity of PAI-1 depends on its interactions with LRP, which leads to the activation of the Janus kinase 2/signal transducer and activator of transcription protein (Jak/-Stat) signaling pathway [43]. Lastly, an increase in the *PAI-1* transcript and translation leads to the formation of keloids and fibrosis [44]. Renaud et al. (2005) suggested that macrophages participate in decreasing the depth of trophoblast invasion by secreting TNF-α, which couples TNFR and promotes EVT to release PAI-1 during placentation [45]. Huber et al. (2006) further reported that TNF-α stimulates PAI-1 level of HTR-8/SVneo cells by activation of the nuclear factor κ-light-chain-enhancer of activated B cells (NF-κB) pathway [46].

PAI-1 may also play a role in remodeling maternal uterine spiral arteries (Figure 2) [47]. PAI-1 mRNA positive cells in the maternal arteriole co-express cytokeratin, implying that PAI-1 may participate in the process of endovascular cytotrophoblast cells replacing cells of the arteriole wall [47]. Hypoinvasion and failed placental vascular remodeling are associated with reproductive diseases, such as recurrent pregnancy losses (RPL), preeclampsia and intrauterine growth restriction (IUGR) and even maternal as well as fetal death [48,49]. In terms of tumor cells, PAI-1 can both promote and inhibit tumor growth and angiogenesis. Low concentrations of PAI-1 can stimulate tumor angiogenesis while treatment of animals with high doses of PAI-1 suppresses angiogenesis and tumor growth [50].

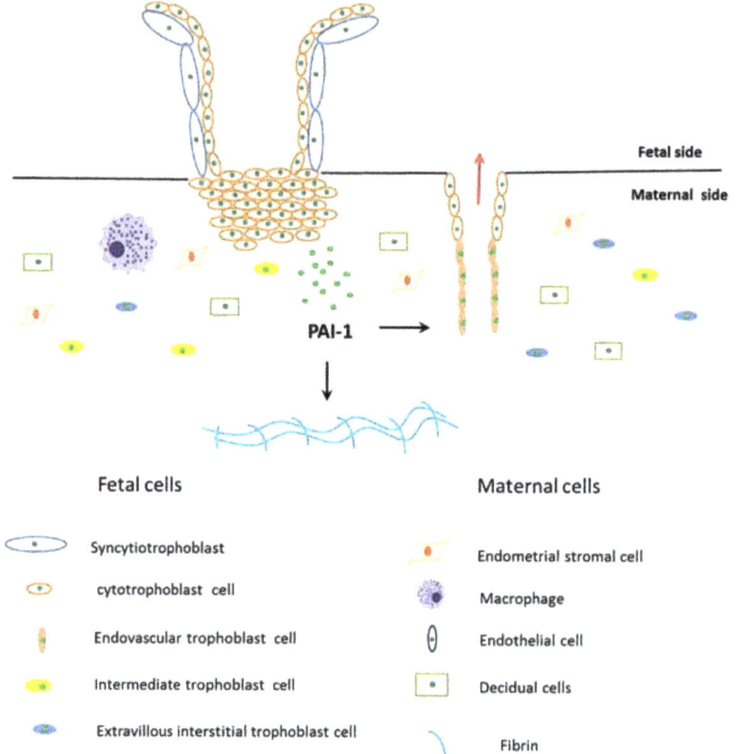

Figure 2. Schematic diagram of PAI-1's role during trophoblasts invasion. The fetal side of the human placenta mainly includes cytotrophoblasts and syncytiotrophoblasts, and sycytiotrophoblasts are differentiated and fused by cytotrophoblasts. The cytotrophoblasts invade into the maternal side and differentiate into extravillous interstitial trophoblasts, intermediate trophoblasts and endovascular trophoblasts. Among them, extravillous interstitial trophoblasts and endovascular trophoblasts express plasminogen activator inhibitor type 1 (PAI-1). Furthermore, cells from the maternal side take part in trophoblast invasion, such as endometrial stromal cells, decidual cells, macrophages and endothelial cells. Extravillous trophoblast invasion in early pregnancy is precisely controlled by many factors expressed by trophoblasts and maternal cells, where PAI-1 is the main anti-invasive factor. PAI-1 prevents trophoblast invasion by inhibiting extracellular matrix degradation, which leads to fibrin accumulation in the maternal side. PAI-1 may also play a role in remodeling maternal uterine spiral arteries.

3.2. Recurrent Pregnancy Losses

Recurrent pregnancy losses (RPL) affects approximately 1% of all couples worldwide [51], and its latest definition is two or more consecutive failed pregnancies as documented by ultrasonography or histopathologic examination before the 20th pregnancy week according to the Practice Committee of the American Society for Reproductive Medicine [52]. The identifiable reasons for RPL include genetic abnormalities, structural abnormalities, infection, endocrine abnormalities, immune dysfunction, and thrombophilic disorders [53]. However, there are still up to 40–50% of pregnancy losses that have unidentifiable causes [54]. PAI-1 plasma levels in RPL patients are increased in comparison to women with healthy pregnancies [55]. Among all the thrombophilic genes, functional *PAI-1-675*

4G/5G polymorphism is one of the most frequently analyzed PAI-1 genetic variants. However, the contribution of *PAI-1-675 4G/5G* to unexplained RPL has remained controversial.

Gris et al. (1997) analyzed a study with 500 women with unexplained RPL and demonstrated that increased PAI-1 levels are the most frequent hemostasis-related abnormality connected to unexplained RPL [56]. A recent meta-analysis by Li et al. (2015) including 22 studies with 4306 cases and 3076 controls showed that *PAI-1* 4G/5G polymorphism is associated with an increased RPL risk ($p = 0.0003$), especially in the Caucasian subgroup ($p < 0.001$) [57]. Khosravi et al. (2014) further found that a high prevalence of *PAI-1 -675* 4G/4G existed in RPL patients as well as in implantation failure (IF) patients ($p < 0.001$) [58].

Another group suggests that *PAI-1-675 4G/5G* alone is not responsible for RPL [59,60], and more thrombophilic gene mutations together can help predispose RPL risk. More than three gene mutations among the 10 thrombophilic gene mutations (*factor V* G1691A, *factor V* H1299R (R2), *factor V* Y1702C, *factor II pro-thrombin* G20210A, *factor XIII* V34L, *b-fibrinogen* -455G > A, *PAI-1* 4G/5G, *HPA1* a/b (L33P), *MTHFR* C677T, and *MTHFR* A1298C) are more prevalent in RPL patients [61,62]. Endothelial PAI-1 synthesis is induced by angiotensin II, which is generated by angiotensin I-converting enzyme (ACE) and this might be one of the main reasons of elevated PAI-1 concentrations in RPL [63].

In general, associations between *PAI-1* 4G/5G and unexplained RPL may have been masked by sample-size effects, ethnicity, enrollment criteria or combinations. In order to improve our understanding of the pathobiology of RPL, we need to identify not only novel genetic variants and the interaction, but also how genes, proteins and the environment contribute to RPL. Thus far, PAI-1 inhibiting fibrinolysis and fibrin accumulation are believed to be the principle reasons for RPL, but the understanding of the mechanism of PAI-1 in RPL still has to be further analyzed.

3.3. Preeclampsia

Preeclampsia affects about 2.5% to 3.0% of pregnant women [64] and it is a leading cause of perinatal morbidity and mortality both for the fetus and the mother [65]. Preeclampsia is a pregnancy disorder characterized by hypertension, proteinuria and placental abnormalities [66], which are generally manifested in the second to third trimester of pregnancy [67]. Although the mechanisms responsible for the pathogenesis of preeclampsia are poorly understood, there is an agreement that it is associated with hypoinvasion and failed conversion of maternal endometrial spiral arteries in the placenta [64], both of which are related to PAI-1 and PAI-2.

Maternal PAI-1 levels in the plasma are higher in patients with preeclampsia during the second trimester of pregnancy [67], and its mRNA levels are positively correlated with the severity of preeclampsia during 35–41 weeks of gestation [68]. The ratio of PAI-1 to PAI-2 is increased in women with early-onset preeclampsia (24–32 gestational weeks) in comparison to the control group, but not in late-onset preeclampsia (35–42 gestational weeks) [69]. Estelles et al. (1989) found that both antigenic and functional PAI-1 levels are increased while antigenic and functional levels of PAI-2 are decreased in the third trimester ($\geq$28 weeks) compared with healthy controls [70]. In contrast, in the first trimester (11–13 weeks), PAI-2 in the plasma is observed to be not differently expressed in preeclampsia [71]. PAI-1 antigen levels are positively correlated to proteinuria in women with preeclampsia [72]. Therefore, PAI-1 has been considered as a potentially useful predictor of preeclampsia. Although it is known that smoking may reduce the risk for preeclampsia [73], specifically chronic smoking upregulates PAI-1 levels [74].

It remains uncertain whether increased PAI levels are the primary mechanism leading to preeclampsia or a consequence of the associated endothelial and placental damage [75]. Cytotrophoblasts, which do not express Raf kinase inhibitor protein (RKIP), could be one of the reasons for impaired migration of cytotrophoblasts in preeclampsia, because locostatin (the inhibitor of RKIP) induces PAI-1 expression with the support of activation of NF-κB pathway and finally contributes to an inadequate trophoblast invasion [76].

PAI-1 expression in the plasma is increased following exposure to inflammatory cytokines, including interleukin 1β (IL-1β) [77], vascular endothelial growth factor (VEGF), epidermal growth factor (EGF) and fibroblast growth factor (FGF) or hypoxic conditions [78]. Hypoxia can directly stimulate PAI-1 mRNA and protein expression [35], and can also stimulate hypoxia-inducible transcription factors (HIF-1α and HIF-2α) to induce PAI-1 [79], both of which may also be the mechanisms of preeclampsia. When preeclampsia occurs in combination with increased levels of syncytial PAI-1, intervillous fibrin deposition and infarction may reduce the flow of nutrients from mother to fetus leading to IUGR [80].

Still, there is a dispute of the correlation between *PAI-1* polymorphism and preeclampsia. Gerhardt et al. (2005) discovered that women with *PAI-1* 5G/5G genotype are at risk for early onset of severe preeclampsia (17–35 gestational weeks) [81]. Morgan JA et al. (2013) found that *PAI-1-675* 4G/5G polymorphism is not associated with preeclampsia with a total of 5003 women involved in the meta-analysis [82].

3.4. Intrauterine Growth Restriction

Intrauterine growth restriction (IUGR) is defined as fetuses with pathological smallness caused by an underlying functional problem [83], occurring in 5–10% of all pregnancies [84]. IUGR is associated with increased risks of neonatal death or disability in the perinatal period [84,85] and predisposes the child to a lifelong increased risk for hypertension, cardiovascular disorders and renal disease [86]. Histopathological studies of IUGR placentas indicate abnormalities of the maternal spiral arterioles, dysregulated villous vasculogenesis, and abundant fibrin deposition, as well as oxidative stress and apoptosis in villous trophoblast [87].

PAI-1 is a potential marker of placental insufficiency and it is associated with fetal hypoxia and angiogenesis in IUGR [84]. PAI-1 levels in the umbilical cord blood are increased in patients with IUGR and it is associated with the plasma's angiogenic potency measured in vitro [84]. Cytotrophoblast cells isolated from the placenta of IUGR pregnancies express significantly higher levels of PAI-1, with a significant decrease in plasminogen activator activity, compared with trophoblast cells from normal pregnancy cultured in vitro [88]. This localized increased production of PAI-1 may play an important part in restricting endovascular trophoblast invasion in early pregnancy, increasing fibrin deposition and reducing uteroplacental blood flow in IUGR pregnancies [88].

PAI-1 placental levels are increased in pregnancies with IUGR and preeclampsia, but not in patients with isolated IUGR [89]. In contrast, PAI-2 expression is reduced in placentas of both IUGR women with and without preeclampsia compared with normal placentas [89]. In the first trimester, low PAI-2 is associated with a higher risk for the development of IUGR [90]. A significant decrease in PAI-2 in the plasma and amniotic fluid is observed in IUGR groups in comparison with normal pregnancies [91]. PAI-2 levels are correlated with fetal weight of IUGR pregnancies [92,93], indicating that PAI-2 is not only the marker of the quantity and quality of the placenta tissues but also a marker for fetal growth and development [17]. *PAI-1* polymorphism (4G/5G) has not been associated with an increased risk of IUGR in the case-control analysis [94,95].

3.5. Gestational Diabetes Mellitus

Gestational diabetes mellitus (GDM) is characterized by impaired glucose tolerance during pregnancy, with the prevalence of 0.6–20% of pregnancies [96]. Women with GDM can have complications including polyhydramnios, fetal macrosomia, preeclampsia, shoulder dystocia and increased risk for operative delivery. Postpartum, patients with GDM have a more than seven-fold increased risk of developing postpartum diabetes compared to women without GDM [97]. Insulin resistance and chronic subclinical inflammation are the two main pathways leading to GDM and possibly previous GDM [98]. Insulin resistance syndrome is associated with increased levels of leptin, TNF-α, tPA, PAI-1 and testosterone [99].

PAI-1 is partly secreted by adipose tissue, and can lead to an impairment of the fibrinolytic system [100]. Salmi et al. (2012) reported that PAI-1 levels are higher in the serum of women with GDM compared to healthy women during the early third trimester pregnancy [101]. Bugatto et al. (2017) stated that PAI-1 levels in maternal uterine blood do not change in women with GDM at the third trimester pregnancy compared to controls [102]. McManus et al. (2014) suggested that GDM women have lower concentrations of PAI-1 in comparison to the age- and weight-matched controls in maternal plasma, as well as in the umbilical artery and umbilical vein [103]. This study further demonstrated that GDM offspring also have decreased PAI-1 concentrations compared to controls [103].

PAI-1 levels in women with GDM are not consistent, but PAI-1 levels have been shown to be increased in women who had GDM during a previous pregnancy [98]. Women with previous GDM have elevated PAI-1 levels 0.25–4 years after delivery [98]. PAI-1 levels are correlated with the insulin sensitivity index (SI), and the PAI-1/SI ratio is increased in women with a previous GDM and impaired insulin [104].

PAI-1 expression is significantly correlated with a variety of adiposity features, including body mass index (BMI), total fat mass, waist circumference, visceral adipose tissue and subcutaneous adipose tissue, total cholesterol triglycerides, fasting plasma glucose and 2 h plasma glucose in the glucose tolerance test, insulin sensitivity as well as pancreatic beta-cell function [100,105,106]. Hyperglycemia inhibits the expression of uPA and PAI-1 by the induction of p38 mitogen-activated protein kinases (p38 MAPK) and peroxisome proliferator-activated receptor γ (PPAR-γ) stress signaling pathways, which are different from the PAI-1 levels induced during preeclampsia [107].

Leipold et al. (2006) reported that the 5G allele of the *PAI-1* gene is associated with normal glucose tolerance in pregnant women independent of maternal age or BMI and it is also related to low fasting glucose values in the oral glucose tolerance test [108]. Physical activities can lower cardiometabolic risk in women with previous GDM and PAI-1 levels, as well as the level of C-reactive protein, leptin and triglycerides [109].

Furthermore, Glueck et al. (2000) found that the polymorphism of *PAI-1* gene is not associated with eclampsia, abruptio placentae and intrauterine fetal death by stepwise logistic regression [110].

3.6. Endometriosis

Endometriosis is defined as the presence of endometrial-like tissue outside the uterus, causing dysmenorrhea, chronic pelvic pain, dyspareunia, and infertility [111]. Pathogenesis of endometriosis includes inflammation, angiogenesis, cytokine/chemokine expression and endocrine alterations such as estrogen receptors (ESRs) and progesterone receptors (PGRs) expression [111]. PAI-1 plays an important role in tumor cell migration and invasion [11], which is also the mechanism of endometriotic cells invasion [112].

Ovarian endometriotic tissues have higher antigenic levels of PAI-1 than normal endometrium [113]. Increased PAI-1 antigen levels in peritoneal fluid from patients with endometriosis contribute to an increase of peritoneal adhesions [114]. In vitro, isolated endometriotic stromal cells release a higher level of PAI-1 compared to the endometrial stromal and epithelial cells of women with endometriosis and controls [112].

There exists a dispute of the association between *PAI-1* gene polymorphism 4G/5G and endometriosis [115,116]. *PAI-1* 4G/5G polymorphism is associated with an increased risk for endometriosis-associated infertility [116]. However, Goodarzi et al. found that *PAI-1* genotype distribution is similar in patients with endometriosis and controls [115].

3.7. Polycystic Ovary Syndrome

Polycystic ovary syndrome (PCOS) is the most common endocrinopathy of women in the reproductive age, with a prevalence of up to 10% and symptoms include hyperandrogenism and/or hyperandrogenemia, oligo-ovulation and polycystic ovarian morphology [117]. PCOS is associated with infertility, RPL, type 2 diabetes mellitus and cardiovascular diseases [117]. Many studies in women with PCOS have reported that PAI-1 levels in the plasma are increased in both normal weight and overweight/obese women with PCOS compared with controls with matched body mass index (BMI) [100,118,119]. Serum PAI-1 activity is related to the BMI and homeostasis model assessment (HOMA) score [120], insulin levels and insulin sensitivity indices [121]. PAI-1 levels decline after treatment with sibutramine and metformin in normal weight and overweight women with PCOS [118,122].

Oligo-ovulation (defined as delayed menses >35 days or <8 spontaneous hemorrhagic episodes per year) is not completely understood yet [123]. uPA plays an essential role in the early growing follicles during cell proliferation and migration, and in the early corpus luteum (CL) formation related to ECM degradation and angiogenesis [124]. PAI-1 is localized in the granulosa and theca cells, indicating it possibly plays a role in human ovulation, but its role in PCOS needs to be further explained [125]. Increased PAI-1 expression in CL of rat and monkey at a later stage is correlated with a sharp decrease in progesterone production of CL [124]. The role of PAI-1 in ovulation needs to be studied further in humans. PAI-1 is a predominant independent risk factor for miscarriages in women with PCOS [126]. Elevated plasma PAI-1 levels are associated with an increased risk for both type 2 diabetes mellitus and cardiovascular diseases (i.e., coronary disease/ischemic stroke) [127–129].

PAI-1 4G/5G polymorphism is associated with susceptibility to PCOS in European, Turkish, and Asian populations [130,131].

4. Conclusions

PAI-1 is expressed in the placenta and maternal plasma of pregnant women. In the human placenta, PAI-1 is localized in invading trophoblasts, especially in extravillous trophoblasts. By inhibiting ECM degradation, PAI-1 plays a vital role in the prevention of trophoblast invasion in RPL, preeclampsia and IUGR. Increased expression of PAI-1 in the plasma is found in RPL, preeclampsia, IUGR, GDM in previous pregnancies, endometriosis and PCOS. Similar to tumor cells, PAI-1 promotes endometriotic cells invasion during endometriosis. In PCOS, PAI-1 may regulate ovulation but this hypothesis still requires further research. *PAI-1* 4G/5G polymorphism is associated with GDM and PCOS; however, contradictory results are found in patients with RPL, preeclampsia and endometriosis. PAI-1 expression can be reduced through physical exercise and pharmacological interventions, such as the oral intake of thiazolidinedines (troglitazone or pioglitazone) [100], statins [100] and metformin [118]. Altered PAI-1 levels have further been found in other pathologies. Studies on cardiovascular diseases have displayed a strong positive correlation between PAI-1 levels in the serum and cardiovascular risks for myocardial infarction (MI), recurrent MI, angina pectoris and atherosclerosis [9]. In addition, PAI-1 is a pivotal mediator of vascular diseases, cancer, asthma, insulin resistance and diabetes [100]. In the future, PAI-1 may function as a biological monitoring parameter and possibly represents a therapeutic approach in reproductive pathologies.

Acknowledgments: China Scholarship Council supported scholarships for Yao Ye, Xi Zhang and Junyan Zhu. The sponsors did not participate in the manuscript writing.

Author Contributions: Yao Ye and Aurelia Vattai performed the literature search, and wrote and revised the manuscript; Xi Zhang, Junyan Zhu, Christian J. Thaler, Sven Mahner, Udo Jeschke and Viktoria von Schönfeldt revised the manuscript.

Conflicts of Interest: The authors declare no conflict of interest.

Abbreviations

PAI	plasminogen activator inhibitor
tPA	tissue-type plasminogen activator
uPA	urokinase-type plasminogen activator
ECM	extra cellular matrix
α2-PI	α2-plasmin inhibitor
α2-MG	α2-macroglobulin

RCL	reactive center loop
LMW	low molecular weight
HMW	high molecular weight
PCI	protein C inhibitor
uNK	uterine natural killer
EVT	extravillous cytotrophoblast
uPAR	uPA receptor
LRP	low density lipoprotein receptor-related protein
TIMP	tissue inhibitors of metalloprotease
TNF-α	tumor necrosis factor α
MMP	metalloproteinase
Jak/-Stat	Janus kinase 2/signal transducer and activator of transcription protein
NF-κB	nuclear factor κ-light-chain-enhancer of activated B cells
ACE	angiotensin I-converting enzyme
IL-1β	interleukin 1β
VEGF	vascular endothelial growth factor
EGF	epidermal growth factor
FGF	fibroblast growth factor
HIF	hypoxia-inducible transcription factors
p38 MAPK	p38 mitogen-activated protein kinases
PPAR-γ	peroxisome proliferator-activated receptor γ
RKIP	raf kinase inhibitor protein
ESR	estrogen receptor
PGR	progesterone receptor
SI	sensitivity index
BMI	body mass index
HOMA	homeostasis model assessment
CL	corpus luteum
RPL	recurrent pregnancy losses
IF	implantation failure
IUGR	intrauterine growth restriction
GDM	gestational diabetes mellitus
PCOS	polycystic ovary syndrome
MI	myocardial infarction

References

1. Zorio, E.; Gilabert-Estelles, J.; Espana, F.; Ramon, L.A.; Cosin, R.; Estelles, A. Fibrinolysis: The key to new pathogenetic mechanisms. *Curr. Med. Chem.* **2008**, *15*, 923–929. [CrossRef] [PubMed]

2. Hellgren, M. Hemostasis during normal pregnancy and puerperium. *Semin. Thromb. Hemost.* **2003**, *29*, 125–130. [CrossRef] [PubMed]

3. Kluft, C.; Jie, A.F.; Sprengers, E.D.; Verheijen, J.H. Identification of a reversible inhibitor of plasminogen activators in blood plasma. *FEBS Lett.* **1985**, *190*, 315–318. [CrossRef]

4. Jorgensen, M.; Philips, M.; Thorsen, S.; Selmer, J.; Zeuthen, J. Plasminogen activator inhibitor-1 is the primary inhibitor of tissue-type plasminogen activator in pregnancy plasma. *Thromb. Haemost.* **1987**, *58*, 872–878. [PubMed]

5. Labied, S.; Blacher, S.; Carmeliet, P.; Noel, A.; Frankenne, F.; Foidart, J.M.; Munaut, C. Transient reduction of placental angiogenesis in PAI-1-deficient mice. *Physiol. Genom.* **2011**, *43*, 188–198. [CrossRef] [PubMed]

6. Fay, W.P.; Parker, A.C.; Condrey, L.R.; Shapiro, A.D. Human plasminogen activator inhibitor-1 (PAI-1) deficiency: Characterization of a large kindred with a null mutation in the PAI-1 gene. *Blood* **1997**, *90*, 204–208. [PubMed]

7. Heiman, M.; Gupta, S.; Shapiro, A.D. The obstetric, gynaecological and fertility implications of homozygous PAI-1 deficiency: Single-centre experience. *Haemophilia* **2014**, *20*, 407–412. [CrossRef] [PubMed]

8. Cesarman-Maus, G.; Hajjar, K.A. Molecular mechanisms of fibrinolysis. *Br. J. Haematol.* **2005**, *129*, 307–321. [CrossRef] [PubMed]

9. Yasar Yildiz, S.; Kuru, P.; Toksoy Oner, E.; Agirbasli, M. Functional stability of plasminogen activator inhibitor-1. *Sci. World J.* **2014**, *2014*, 858293. [CrossRef] [PubMed]

10. Ghosh, A.K.; Vaughan, D.E. PAI-1 in tissue fibrosis. *J. Cell Physiol.* **2012**, *227*, 493–507. [CrossRef] [PubMed]

11. Duffy, M.J.; McGowan, P.M.; Harbeck, N.; Thomssen, C.; Schmitt, M. uPA and PAI-1 as biomarkers in breast cancer: Validated for clinical use in level-of-evidence-1 studies. *Breast Cancer Res.* **2014**, *16*, 428. [CrossRef] [PubMed]

12. Lyon, C.J.; Hsueh, W.A. Effect of plasminogen activator inhibitor-1 in diabetes mellitus and cardiovascular disease. *Am. J. Med.* **2003**, *115*, 62–68. [CrossRef]

13. Lijnen, H.R.; Collen, D. Mechanisms of physiological fibrinolysis. *Baillieres Clin. Haematol.* **1995**, *8*, 277–290. [CrossRef]

14. Chakraborty, C.; Gleeson, L.M.; McKinnon, T.; Lala, P.K. Regulation of human trophoblast migration and invasiveness. *Can. J. Physiol. Pharmacol.* **2002**, *80*, 116–124. [CrossRef] [PubMed]

15. Blasi, F.; Vassalli, J.D.; Dano, K. Urokinase-type plasminogen activator: Proenzyme, receptor, and inhibitors. *J. Cell Biol.* **1987**, *104*, 801–804. [CrossRef] [PubMed]

16. Petersen, L.C.; Lund, L.R.; Nielsen, L.S.; Dano, K.; Skriver, L. One-chain urokinase-type plasminogen activator from human sarcoma cells is a proenzyme with little or no intrinsic activity. *J. Biol. Chem.* **1988**, *263*, 11189–11195. [PubMed]

17. Brenner, B. Haemostatic changes in pregnancy. *Thromb. Res.* **2004**, *114*, 409–414. [CrossRef] [PubMed]

18. Gils, A.; Declerck, P.J. The structural basis for the pathophysiological relevance of PAI-1 in cardiovascular diseases and the development of potential PAI-1 inhibitors. *Thromb. Haemost.* **2004**, *91*, 425–437. [CrossRef] [PubMed]

19. Andreasen, P.A.; Egelund, R.; Petersen, H.H. The plasminogen activation system in tumor growth, invasion, and metastasis. *Cell. Mol. Life Sci. CMLS* **2000**, *57*, 25–40. [CrossRef] [PubMed]

20. Gils, A. The pathophysiological relevance of PAI-1 in cardiovascular diseases and the development of monoclonal antibodies as PAI-1 inhibitors. *Verh K Acad Geneeskd Belg* **2006**, *68*, 179–198. [PubMed]

21. Eriksson, P.; Nilsson, L.; Karpe, F.; Hamsten, A. Very-low-density lipoprotein response element in the promoter region of the human plasminogen activator inhibitor-1 gene implicated in the impaired fibrinolysis of hypertriglyceridemia. *Arterioscler. Thromb. Vasc. Biol.* **1998**, *18*, 20–26. [CrossRef] [PubMed]

22. Saksela, O.; Rifkin, D.B. Cell-associated plasminogen activation: Regulation and physiological functions. *Annu. Rev. Cell Biol.* **1988**, *4*, 93–126. [CrossRef] [PubMed]

23. Astedt, B.; Lindoff, C.; Lecander, I. Significance of the plasminogen activator inhibitor of placental type (PAI-2) in pregnancy. *Semin. Thromb. Hemost.* **1998**, *24*, 431–435. [CrossRef] [PubMed]

24. Kruithof, E.K.; Baker, M.S.; Bunn, C.L. Biological and clinical aspects of plasminogen activator inhibitor type 2. *Blood* **1995**, *86*, 4007–4024. [PubMed]

25. Suzuki, K. The multi-functional serpin, protein C inhibitor: Beyond thrombosis and hemostasis. *J. Thromb. Haemost.* **2008**, *6*, 2017–2026. [CrossRef] [PubMed]

26. Kruithof, E.K.; Tran-Thang, C.; Gudinchet, A.; Hauert, J.; Nicoloso, G.; Genton, C.; Welti, H.; Bachmann, F. Fibrinolysis in pregnancy: A study of plasminogen activator inhibitors. *Blood* **1987**, *69*, 460–466. [CrossRef]

27. Stirling, Y.; Woolf, L.; North, W.R.; Seghatchian, M.J.; Meade, T.W. Haemostasis in normal pregnancy. *Thromb. Haemost.* **1984**, *52*, 176–182. [PubMed]

28. Hofmann, G.E.; Glatstein, I.; Schatz, F.; Heller, D.; Deligdisch, L. Immunohistochemical localization of urokinase-type plasminogen activator and the plasminogen activator inhibitors 1 and 2 in early human implantation sites. *Am. J. Obstet. Gynecol.* **1994**, *170*, 671–676. [CrossRef]

29. Feinberg, R.F.; Kao, L.C.; Haimowitz, J.E.; Queenan, J.T., Jr.; Wun, T.C.; Strauss, J.F., 3rd; Kliman, H.J. Plasminogen activator inhibitor types 1 and 2 in human trophoblasts. PAI-1 is an immunocytochemical marker of invading trophoblasts. *Lab. Investig. J. Tech. Methods Pathol.* **1989**, *61*, 20–26.

30. Floridon, C.; Nielsen, O.; Holund, B.; Sweep, F.; Sunde, L.; Thomsen, S.G.; Teisner, B. Does plasminogen activator inhibitor-1 (PAI-1) control trophoblast invasion? A study of fetal and maternal tissue in intrauterine, tubal and molar pregnancies. *Placenta* **2000**, *21*, 754–762. [CrossRef] [PubMed]

31. Hu, Z.Y.; Liu, Y.X.; Liu, K.; Byrne, S.; Ny, T.; Feng, Q.; Ockleford, C.D. Expression of tissue type and urokinase type plasminogen activators as well as plasminogen activator inhibitor type-1 and type-2 in human and rhesus monkey placenta. *J. Anat.* **1999**, *194*, 183–195. [CrossRef] [PubMed]

32. Naruse, K.; Lash, G.E.; Bulmer, J.N.; Innes, B.A.; Otun, H.A.; Searle, R.F.; Robson, S.C. The urokinase plasminogen activator (uPA) system in uterine natural killer cells in the placental bed during early pregnancy. *Placenta* **2009**, *30*, 398–404. [CrossRef] [PubMed]

33. Silva, J.F.; Serakides, R. Intrauterine trophoblast migration: A comparative view of humans and rodents. *Cell Adh. Migr.* **2016**, *10*, 88–110. [CrossRef] [PubMed]

34. Lala, P.K.; Lee, B.P.; Xu, G.; Chakraborty, C. Human placental trophoblast as an in vitro model for tumor progression. *Can. J. Physiol. Pharmacol.* **2002**, *80*, 142–149. [CrossRef] [PubMed]

35. Fitzpatrick, T.E.; Graham, C.H. Stimulation of plasminogen activator inhibitor-1 expression in immortalized human trophoblast cells cultured under low levels of oxygen. *Exp. Cell Res.* **1998**, *245*, 155–162. [CrossRef] [PubMed]

36. Crippa, M.P. Urokinase-type plasminogen activator. *Int. J. Biochem. Cell Biol.* **2007**, *39*, 690–694. [CrossRef] [PubMed]

37. Lash, G.E.; Otun, H.A.; Innes, B.A.; Bulmer, J.N.; Searle, R.F.; Robson, S.C. Low oxygen concentrations inhibit trophoblast cell invasion from early gestation placental explants via alterations in levels of the urokinase plasminogen activator system. *Biol. Reprod.* **2006**, *74*, 403–409. [CrossRef] [PubMed]

38. Xia, Y.; Wen, H.Y.; Kellems, R.E. Angiotensin II inhibits human trophoblast invasion through AT1 receptor activation. *J. Biol. Chem.* **2002**, *277*, 24601–24608. [CrossRef] [PubMed]

39. Bauer, S.; Pollheimer, J.; Hartmann, J.; Husslein, P.; Aplin, J.D.; Knofler, M. Tumor necrosis factor-α inhibits trophoblast migration through elevation of plasminogen activator inhibitor-1 in first-trimester villous explant cultures. *J. Clin. Endocrinol. Metab.* **2004**, *89*, 812–822. [CrossRef] [PubMed]

40. Estella, C.; Herrer, I.; Atkinson, S.P.; Quinonero, A.; Martinez, S.; Pellicer, A.; Simon, C. Inhibition of histone deacetylase activity in human endometrial stromal cells promotes extracellular matrix remodelling and limits embryo invasion. *PLoS ONE* **2012**, *7*, e30508. [CrossRef] [PubMed]

41. Degryse, B.; Sier, C.F.; Resnati, M.; Conese, M.; Blasi, F. PAI-1 inhibits urokinase-induced chemotaxis by internalizing the urokinase receptor. *FEBS Lett.* **2001**, *505*, 249–254. [CrossRef]

42. Czekay, R.P.; Kuemmel, T.A.; Orlando, R.A.; Farquhar, M.G. Direct binding of occupied urokinase receptor (uPAR) to LDL receptor-related protein is required for endocytosis of uPAR and regulation of cell surface urokinase activity. *Mol. Biol. Cell* **2001**, *12*, 1467–1479. [CrossRef] [PubMed]

43. Degryse, B.; Neels, J.G.; Czekay, R.P.; Aertgeerts, K.; Kamikubo, Y.; Loskutoff, D.J. The low density lipoprotein receptor-related protein is a motogenic receptor for plasminogen activator inhibitor-1. *J. Biol. Chem.* **2004**, *279*, 22595–22604. [CrossRef] [PubMed]

44. Zhang, Q.; Wu, Y.; Ann, D.K.; Messadi, D.V.; Tuan, T.L.; Kelly, A.P.; Bertolami, C.N.; Le, A.D. Mechanisms of hypoxic regulation of plasminogen activator inhibitor-1 gene expression in keloid fibroblasts. *J. Investig. Dermatol.* **2003**, *121*, 1005–1012. [CrossRef] [PubMed]

45. Renaud, S.J.; Postovit, L.M.; Macdonald-Goodfellow, S.K.; McDonald, G.T.; Caldwell, J.D.; Graham, C.H. Activated macrophages inhibit human cytotrophoblast invasiveness in vitro. *Biol. Reprod.* **2005**, *73*, 237–243. [CrossRef] [PubMed]

46. Huber, A.V.; Saleh, L.; Bauer, S.; Husslein, P.; Knofler, M. TNFα-mediated induction of PAI-1 restricts invasion of HTR-8/SVneo trophoblast cells. *Placenta* **2006**, *27*, 127–136. [CrossRef] [PubMed]

47. Feng, Q.; Liu, Y.; Liu, K.; Byrne, S.; Liu, G.; Wang, X.; Li, Z.; Ockleford, C.D. Expression of urokinase, plasminogen activator inhibitors and urokinase receptor in pregnant rhesus monkey uterus during early placentation. *Placenta* **2000**, *21*, 184–193. [CrossRef] [PubMed]

48. Soares, M.J.; Chakraborty, D.; Kubota, K.; Renaud, S.J.; Rumi, M.A. Adaptive mechanisms controlling uterine spiral artery remodeling during the establishment of pregnancy. *Int. J. Dev. Biol.* **2014**, *58*, 247–259. [CrossRef] [PubMed]

49. Kaufmann, P.; Black, S.; Huppertz, B. Endovascular trophoblast invasion: Implications for the pathogenesis of intrauterine growth retardation and preeclampsia. *Biol. Reprod.* **2003**, *69*, 1–7. [CrossRef] [PubMed]

50. McMahon, G.A.; Petitclerc, E.; Stefansson, S.; Smith, E.; Wong, M.K.; Westrick, R.J.; Ginsburg, D.; Brooks, P.C.; Lawrence, D.A. Plasminogen activator inhibitor-1 regulates tumor growth and angiogenesis. *J. Biol. Chem.* **2001**, *276*, 33964–33968. [CrossRef] [PubMed]

51. Stirrat, G.M. Recurrent miscarriage. *Lancet* **1990**, *336*, 673–675. [CrossRef]
52. Practice Committee of the American Society for Reproductive Medicine. Definitions of infertility and recurrent pregnancy loss: A committee opinion. *Fertil. Steril.* **2013**. [CrossRef]
53. Rai, R.; Regan, L. Recurrent miscarriage. *Lancet* **2006**, *368*, 601–611. [CrossRef]
54. Jaslow, C.R.; Carney, J.L.; Kutteh, W.H. Diagnostic factors identified in 1020 women with two versus three or more recurrent pregnancy losses. *Fertil. Steril.* **2010**, *93*, 1234–1243. [CrossRef] [PubMed]
55. Gris, J.C.; Neveu, S.; Mares, P.; Biron, C.; Hedon, B.; Schved, J.F. Plasma fibrinolytic activators and their inhibitors in women suffering from early recurrent abortion of unknown etiology. *J. Lab. Clin. Med.* **1993**, *122*, 606–615. [PubMed]
56. Gris, J.C.; Ripart-Neveu, S.; Maugard, C.; Tailland, M.L.; Brun, S.; Courtieu, C.; Biron, C.; Hoffet, M.; Hedon, B.; Mares, P. Respective evaluation of the prevalence of haemostasis abnormalities in unexplained primary early recurrent miscarriages. The Nimes Obstetricians and Haematologists (NOHA) Study. *Thromb. Haemost.* **1997**, *77*, 1096–1103. [PubMed]
57. Li, X.; Liu, Y.; Zhang, R.; Tan, J.; Chen, L.; Liu, Y. Meta-analysis of the association between plasminogen activator inhibitor-1 4G/5G polymorphism and recurrent pregnancy loss. *Med. Sci. Monit. Int. Med. J. Exp. Clin. Res.* **2015**, *21*, 1051–1056. [CrossRef] [PubMed]
58. Khosravi, F.; Zarei, S.; Ahmadvand, N.; Akbarzadeh-Pasha, Z.; Savadi, E.; Zarnani, A.H.; Sadeghi, M.R.; Jeddi-Tehrani, M. Association between plasminogen activator inhibitor 1 gene mutation and different subgroups of recurrent miscarriage and implantation failure. *J. Assist. Reprod. Genet.* **2014**, *31*, 121–124. [CrossRef] [PubMed]
59. Goodman, C.; Hur, J.; Goodman, C.S.; Jeyendran, R.S.; Coulam, C. Are polymorphisms in the ACE and PAI-1 genes associated with recurrent spontaneous miscarriages? *Am. J. Reprod. Immunol.* **2009**, *62*, 365–370. [CrossRef] [PubMed]
60. Su, M.T.; Lin, S.H.; Chen, Y.C.; Kuo, P.L. Genetic association studies of ACE and PAI-1 genes in women with recurrent pregnancy loss: A systematic review and meta-analysis. *Thromb. Haemost.* **2013**, *109*, 8–15. [CrossRef] [PubMed]
61. Coulam, C.B.; Jeyendran, R.S.; Fishel, L.A.; Roussev, R. Multiple thrombophilic gene mutations rather than specific gene mutations are risk factors for recurrent miscarriage. *Am. J. Reprod. Immunol.* **2006**, *55*, 360–368. [CrossRef] [PubMed]
62. Poursadegh Zonouzi, A.; Chaparzadeh, N.; Ghorbian, S.; Sadaghiani, M.M.; Farzadi, L.; Ghasemzadeh, A.; Kafshdooz, T.; Sakhinia, M.; Sakhinia, E. The association between thrombophilic gene mutations and recurrent pregnancy loss. *J. Assist. Reprod. Genet.* **2013**, *30*, 1353–1359. [CrossRef] [PubMed]
63. Buchholz, T.; Lohse, P.; Rogenhofer, N.; Kosian, E.; Pihusch, R.; Thaler, C.J. Polymorphisms in the ACE and PAI-1 genes are associated with recurrent spontaneous miscarriages. *Hum. Reprod.* **2003**, *18*, 2473–2477. [CrossRef] [PubMed]
64. Redman, C.W.; Sargent, I.L. Latest advances in understanding preeclampsia. *Science* **2005**, *308*, 1592–1594. [CrossRef] [PubMed]
65. Sibai, B.M.; Stella, C.L. Diagnosis and management of atypical preeclampsia-eclampsia. *Am. J. Obstet. Gynecol.* **2009**, *200*, 481. [CrossRef] [PubMed]
66. Irani, R.A.; Xia, Y. The functional role of the renin-angiotensin system in pregnancy and preeclampsia. *Placenta* **2008**, *29*, 763–771. [CrossRef] [PubMed]
67. Bodova, K.B.; Biringer, K.; Dokus, K.; Ivankova, J.; Stasko, J.; Danko, J. Fibronectin, plasminogen activator inhibitor type 1 (PAI-1) and uterine artery Doppler velocimetry as markers of preeclampsia. *Dis. Mark.* **2011**, *30*, 191–196. [CrossRef] [PubMed]
68. Purwosunu, Y.; Sekizawa, A.; Koide, K.; Farina, A.; Wibowo, N.; Wiknjosastro, G.H.; Okazaki, S.; Chiba, H.; Okai, T. Cell-free mRNA concentrations of plasminogen activator inhibitor-1 and tissue-type plasminogen activator are increased in the plasma of pregnant women with preeclampsia. *Clin. Chem.* **2007**, *53*, 399–404. [CrossRef] [PubMed]
69. Wikstrom, A.K.; Nash, P.; Eriksson, U.J.; Olovsson, M.H. Evidence of increased oxidative stress and a change in the plasminogen activator inhibitor (PAI)-1 to PAI-2 ratio in early-onset but not late-onset preeclampsia. *Am. J. Obstet. Gynecol.* **2009**, *201*, 597. [CrossRef] [PubMed]

70. Estelles, A.; Gilabert, J.; Aznar, J.; Loskutoff, D.J.; Schleef, R.R. Changes in the plasma levels of type 1 and type 2 plasminogen activator inhibitors in normal pregnancy and in patients with severe preeclampsia. *Blood* **1989**, *74*, 1332–1338. [PubMed]

71. Akolekar, R.; Cruz Jde, J.; Penco, J.M.; Zhou, Y.; Nicolaides, K.H. Maternal plasma plasminogen activator inhibitor-2 at 11 to 13 weeks of gestation in hypertensive disorders of pregnancy. *Hypertens. Pregnancy* **2011**, *30*, 194–202. [CrossRef] [PubMed]

72. Catarino, C.; Rebelo, I.; Belo, L.; Rocha, S.; Castro, E.B.; Patricio, B.; Quintanilha, A.; Santos-Silva, A. Relationship between maternal and cord blood hemostatic disturbances in preeclamptic pregnancies. *Thromb. Res.* **2008**, *123*, 219–224. [CrossRef] [PubMed]

73. Wei, J.; Liu, C.X.; Gong, T.T.; Wu, Q.J.; Wu, L. Cigarette smoking during pregnancy and preeclampsia risk: A systematic review and meta-analysis of prospective studies. *Oncotarget* **2015**, *6*, 43667–43678. [CrossRef] [PubMed]

74. Simpson, A.J.; Gray, R.S.; Moore, N.R.; Booth, N.A. The effects of chronic smoking on the fibrinolytic potential of plasma and platelets. *Br. J. Haematol.* **1997**, *97*, 208–213. [CrossRef] [PubMed]

75. Said, J.M.; Tsui, R.; Borg, A.J.; Higgins, J.R.; Moses, E.K.; Walker, S.P.; Monagle, P.T.; Brennecke, S.P. The PAI-1 4G/5G polymorphism is not associated with an increased risk of adverse pregnancy outcome in asymptomatic nulliparous women. *J. Thromb. Haemost.* **2012**, *10*, 881–886. [CrossRef] [PubMed]

76. Ciarmela, P.; Marzioni, D.; Islam, M.S.; Gray, P.C.; Terracciano, L.; Lorenzi, T.; Todros, T.; Petraglia, F.; Castellucci, M. Possible role of RKIP in cytotrophoblast migration: Immunohistochemical and in vitro studies. *J. Cell. Physiol.* **2012**, *227*, 1821–1828. [CrossRef] [PubMed]

77. Prutsch, N.; Fock, V.; Haslinger, P.; Haider, S.; Fiala, C.; Pollheimer, J.; Knofler, M. The role of interleukin-1β in human trophoblast motility. *Placenta* **2012**, *33*, 696–703. [CrossRef] [PubMed]

78. Anteby, E.Y.; Greenfield, C.; Natanson-Yaron, S.; Goldman-Wohl, D.; Hamani, Y.; Khudyak, V.; Ariel, I.; Yagel, S. Vascular endothelial growth factor, epidermal growth factor and fibroblast growth factor-4 and -10 stimulate trophoblast plasminogen activator system and metalloproteinase-9. *Mol. Hum. Reprod.* **2004**, *10*, 229–235. [CrossRef] [PubMed]

79. Meade, E.S.; Ma, Y.Y.; Guller, S. Role of hypoxia-inducible transcription factors 1α and 2α in the regulation of plasminogen activator inhibitor-1 expression in a human trophoblast cell line. *Placenta* **2007**, *28*, 1012–1019. [CrossRef] [PubMed]

80. Guller, S. Role of the syncytium in placenta-mediated complications of preeclampsia. *Thromb. Res.* **2009**, *124*, 389–392. [CrossRef] [PubMed]

81. Gerhardt, A.; Goecke, T.W.; Beckmann, M.W.; Wagner, K.J.; Tutschek, B.; Willers, R.; Bender, H.G.; Scharf, R.E.; Zotz, R.B. The G20210A prothrombin-gene mutation and the plasminogen activator inhibitor (PAI-1) 5G/5G genotype are associated with early onset of severe preeclampsia. *J. Thromb. Haemost.* **2005**, *3*, 686–691. [CrossRef] [PubMed]

82. Morgan, J.A.; Bombell, S.; McGuire, W. Association of plasminogen activator inhibitor-type 1 (-675 4G/5G) polymorphism with pre-eclampsia: Systematic review. *PLoS ONE* **2013**, *8*, e56907. [CrossRef] [PubMed]

83. Dall'Asta, A.; Brunelli, V.; Prefumo, F.; Frusca, T.; Lees, C.C. Early onset fetal growth restriction. *Matern. Health Neonatol. Perinat.* **2017**, *3*, 2. [CrossRef] [PubMed]

84. Seferovic, M.D.; Gupta, M.B. Increased Umbilical Cord PAI-1 Levels in Placental Insufficiency Are Associated with Fetal Hypoxia and Angiogenesis. *Dis. Markers* **2016**, *2016*, 7124186. [CrossRef] [PubMed]

85. Bernstein, I.M.; Horbar, J.D.; Badger, G.J.; Ohlsson, A.; Golan, A. Morbidity and mortality among very-low-birth-weight neonates with intrauterine growth restriction. The Vermont Oxford Network. *Am. J. Obstet. Gynecol.* **2000**, *182*, 198–206. [CrossRef]

86. Murphy, V.E.; Smith, R.; Giles, W.B.; Clifton, V.L. Endocrine regulation of human fetal growth: The role of the mother, placenta, and fetus. *Endocr. Rev.* **2006**, *27*, 141–169. [CrossRef] [PubMed]

87. Scifres, C.M.; Nelson, D.M. Intrauterine growth restriction, human placental development and trophoblast cell death. *J. Physiol.* **2009**, *587*, 3453–3458. [CrossRef] [PubMed]

88. Sheppard, B.L.; Bonnar, J. Uteroplacental hemostasis in intrauterine fetal growth retardation. *Semin. Thromb. Hemost.* **1999**, *25*, 443–446. [CrossRef] [PubMed]

89. Estelles, A.; Gilabert, J.; Keeton, M.; Eguchi, Y.; Aznar, J.; Grancha, S.; Espna, F.; Loskutoff, D.J.; Schleef, R.R. Altered expression of plasminogen activator inhibitor type 1 in placentas from pregnant women with preeclampsia and/or intrauterine fetal growth retardation. *Blood* **1994**, *84*, 143–150. [PubMed]

90. Coolman, M.; Timmermans, S.; de Groot, C.J.; Russcher, H.; Lindemans, J.; Hofman, A.; Geurts-Moespot, A.J.; Sweep, F.C.; Jaddoe, V.V.; Steegers, E.A. Angiogenic and fibrinolytic factors in blood during the first half of pregnancy and adverse pregnancy outcomes. *Obstet. Gynecol.* **2012**, *119*, 1190–1200. [CrossRef] [PubMed]

91. Gilabert, J.; Estelles, A.; Ayuso, M.J.; Espana, F.; Chirivella, M.; Grancha, S.; Mico, J.M.; Aznar, J. Evaluation of plasminogen activators and plasminogen activator inhibitors in plasma and amniotic fluid in pregnancies complicated with intrauterine fetal growth retardation. *Gynecol. Obstet. Investig.* **1994**, *38*, 157–162. [CrossRef]

92. De Boer, K.; ten Cate, J.W.; Sturk, A.; Borm, J.J.; Treffers, P.E. Enhanced thrombin generation in normal and hypertensive pregnancy. *Am. J. Obstet. Gynecol.* **1989**, *160*, 95–100. [CrossRef]

93. Estelles, A.; Gilabert, J.; Espana, F.; Aznar, J.; Galbis, M. Fibrinolytic parameters in normotensive pregnancy with intrauterine fetal growth retardation and in severe preeclampsia. *Am. J. Obstet. Gynecol.* **1991**, *165*, 138–142. [CrossRef]

94. Infante-Rivard, C.; Rivard, G.E.; Guiguet, M.; Gauthier, R. Thrombophilic polymorphisms and intrauterine growth restriction. *Epidemiology* **2005**, *16*, 281–287. [CrossRef] [PubMed]

95. Larciprete, G.; Rossi, F.; Deaibess, T.; Brienza, L.; Barbati, G.; Romanini, E.; Gioia, S.; Cirese, E. Double inherited thrombophilias and adverse pregnancy outcomes: Fashion or science? *J. Obstet. Gynaecol. Res.* **2010**, *36*, 996–1002. [CrossRef] [PubMed]

96. Miehle, K.; Stepan, H.; Fasshauer, M. Leptin, adiponectin and other adipokines in gestational diabetes mellitus and pre-eclampsia. *Clin. Endocrinol.* **2012**, *76*, 2–11. [CrossRef] [PubMed]

97. Moon, J.H.; Kwak, S.H.; Jang, H.C. Prevention of type 2 diabetes mellitus in women with previous gestational diabetes mellitus. *Korean J. Intern. Med.* **2017**, *32*, 26–41. [CrossRef] [PubMed]

98. Vrachnis, N.; Belitsos, P.; Sifakis, S.; Dafopoulos, K.; Siristatidis, C.; Pappa, K.I.; Iliodromiti, Z. Role of adipokines and other inflammatory mediators in gestational diabetes mellitus and previous gestational diabetes mellitus. *Int. J. Endocrinol.* **2012**, *2012*, 549748. [CrossRef] [PubMed]

99. Seely, E.W.; Solomon, C.G. Insulin resistance and its potential role in pregnancy-induced hypertension. *J. Clin. Endocrinol. Metab.* **2003**, *88*, 2393–2398. [CrossRef] [PubMed]

100. Cesari, M.; Pahor, M.; Incalzi, R.A. Plasminogen activator inhibitor-1 (PAI-1): A key factor linking fibrinolysis and age-related subclinical and clinical conditions. *Cardiovasc. Ther.* **2010**, *28*, 72–91. [CrossRef] [PubMed]

101. Salmi, A.A.; Zaki, N.M.; Zakaria, R.; Nor Aliza, A.G.; Rasool, A.H. Arterial stiffness, inflammatory and pro-atherogenic markers in gestational diabetes mellitus. *Vasa* **2012**, *41*, 96–104. [CrossRef] [PubMed]

102. Bugatto, F.; Quintero-Prado, R.; Visiedo, F.M.; Vilar-Sanchez, J.M.; Figueroa-Quinones, A.; Lopez-Tinoco, C.; Torrejon, R.; Bartha, J.L. The influence of lipid and proinflammatory status on maternal uterine blood flow in women with late onset gestational diabetes. *Reprod. Sci.* **2017**. [CrossRef] [PubMed]

103. McManus, R.; Summers, K.; de Vrijer, B.; Cohen, N.; Thompson, A.; Giroux, I. Maternal, umbilical arterial and umbilical venous 25-hydroxyvitamin D and adipocytokine concentrations in pregnancies with and without gestational diabetes. *Clin. Endocrinol.* **2014**, *80*, 635–641. [CrossRef] [PubMed]

104. Farhan, S.; Winzer, C.; Tura, A.; Quehenberger, P.; Bieglmaier, C.; Wagner, O.F.; Huber, K.; Waldhausl, W.; Pacini, G.; Kautzky-Willer, A. Fibrinolytic dysfunction in insulin-resistant women with previous gestational diabetes. *Eur. J. Clin. Investig.* **2006**, *36*, 345–352. [CrossRef] [PubMed]

105. Morimitsu, L.K.; Fusaro, A.S.; Sanchez, V.H.; Hagemann, C.C.; Bertini, A.M.; Dib, S.A. Fibrinolytic dysfunction after gestation is associated to components of insulin resistance and early type 2 diabetes in latino women with previous gestational diabetes. *Diabetes Res. Clin. Pract.* **2007**, *78*, 340–348. [CrossRef] [PubMed]

106. Lowe, L.P.; Metzger, B.E.; Lowe, W.L., Jr.; Dyer, A.R.; McDade, T.W.; McIntyre, H.D.; Group, H.S.C.R. Inflammatory mediators and glucose in pregnancy: Results from a subset of the Hyperglycemia and Adverse Pregnancy Outcome (HAPO) Study. *J. Clin. Endocrinol. Metab.* **2010**, *95*, 5427–5434. [CrossRef] [PubMed]

107. Cawyer, C.R.; Horvat, D.; Leonard, D.; Allen, S.R.; Jones, R.O.; Zawieja, D.C.; Kuehl, T.J.; Uddin, M.N. Hyperglycemia impairs cytotrophoblast function via stress signaling. *Am. J. Obstet. Gynecol.* **2014**, *211*, 541. [CrossRef] [PubMed]

108. Leipold, H.; Knoefler, M.; Gruber, C.; Klein, K.; Haslinger, P.; Worda, C. Plasminogen activator inhibitor 1 gene polymorphism and gestational diabetes mellitus. *Obstet. Gynecol.* **2006**, *107*, 651–656. [CrossRef] [PubMed]

109. Gingras, V.; Vigneault, J.; Weisnagel, S.J.; Tchernof, A.; Robitaille, J. Accelerometry-measured physical activity and inflammation after gestational diabetes. *Med. Sci. Sports Exerc.* **2013**, *45*, 1307–1312. [CrossRef] [PubMed]

110. Glueck, C.J.; Phillips, H.; Cameron, D.; Wang, P.; Fontaine, R.N.; Moore, S.K.; Sieve-Smith, L.; Tracy, T. The 4G/4G polymorphism of the hypofibrinolytic plasminogen activator inhibitor type 1 gene: An independent risk factor for serious pregnancy complications. *Metabolism* **2000**, *49*, 845–852. [CrossRef] [PubMed]

111. Greene, A.D.; Lang, S.A.; Kendziorski, J.A.; Sroga-Rios, J.M.; Herzog, T.J.; Burns, K.A. Endometriosis: Where are we and where are we going? *Reproduction* **2016**, *152*, R63–R78. [CrossRef] [PubMed]

112. Bruse, C.; Guan, Y.; Carlberg, M.; Carlstrom, K.; Bergqvist, A. Basal release of urokinase plasminogen activator, plasminogen activator inhibitor-1, and soluble plasminogen activator receptor from separated and cultured endometriotic and endometrial stromal and epithelial cells. *Fertil. Steril.* **2005**, *83*, 1155–1160. [CrossRef] [PubMed]

113. Gilabert-Estelles, J.; Estelles, A.; Gilabert, J.; Castello, R.; Espana, F.; Falco, C.; Romeu, A.; Chirivella, M.; Zorio, E.; Aznar, J. Expression of several components of the plasminogen activator and matrix metalloproteinase systems in endometriosis. *Hum. Reprod.* **2003**, *18*, 1516–1522. [CrossRef] [PubMed]

114. Ramon, L.A.; Gilabert-Estelles, J.; Cosin, R.; Gilabert, J.; Espana, F.; Castello, R.; Chirivella, M.; Romeu, A.; Estelles, A. Plasminogen activator inhibitor-1 (PAI-1) 4G/5G polymorphism and endometriosis. Influence of PAI-1 polymorphism on PAI-1 antigen and mRNA expression. *Thromb. Res.* **2008**, *122*, 854–860. [CrossRef] [PubMed]

115. Gentilini, D.; Vigano, P.; Castaldi, D.; Mari, D.; Busacca, M.; Vercellini, P.; Somigliana, E.; di Blasio, A.M. Plasminogen activator inhibitor-1 4G/5G polymorphism and susceptibility to endometriosis in the Italian population. *Eur. J. Obstet. Gynecol. Reprod. Biol.* **2009**, *146*, 219–221. [CrossRef] [PubMed]

116. Goncalves-Filho, R.P.; Brandes, A.; Christofolini, D.M.; Lerner, T.G.; Bianco, B.; Barbosa, C.P. Plasminogen activator inhibitor-1 4G/5G polymorphism in infertile women with and without endometriosis. *Acta Obstet. Gynecol. Scand.* **2011**, *90*, 473–477. [CrossRef] [PubMed]

117. Goodarzi, M.O.; Dumesic, D.A.; Chazenbalk, G.; Azziz, R. Polycystic ovary syndrome: Etiology, pathogenesis and diagnosis. *Nat. Rev. Endocrinol.* **2011**, *7*, 219–231. [CrossRef] [PubMed]

118. Koiou, E.; Tziomalos, K.; Katsikis, I.; Delkos, D.; Tsourdi, E.A.; Panidis, D. Disparate effects of pharmacotherapy on plasma plasminogen activator inhibitor-1 levels in women with the polycystic ovary syndrome. *Hormones* **2013**, *12*, 559–566. [CrossRef] [PubMed]

119. Koiou, E.; Tziomalos, K.; Dinas, K.; Katsikis, I.; Kandaraki, E.A.; Tsourdi, E.; Mavridis, S.; Panidis, D. Plasma plasminogen activator inhibitor-1 levels in the different phenotypes of the polycystic ovary syndrome. *Endocr. J.* **2012**, *59*, 21–29. [CrossRef] [PubMed]

120. Orio, F., Jr.; Palomba, S.; Cascella, T.; Tauchmanova, L.; Nardo, L.G.; di Biase, S.; Labella, D.; Russo, T.; Savastano, S.; Tolino, A.; et al. Is plasminogen activator inhibitor-1 a cardiovascular risk factor in young women with polycystic ovary syndrome? *Reprod. Biomed. Online* **2004**, *9*, 505–510. [CrossRef]

121. Tarkun, I.; Canturk, Z.; Arslan, B.C.; Turemen, E.; Tarkun, P. The plasminogen activator system in young and lean women with polycystic ovary syndrome. *Endocr. J.* **2004**, *51*, 467–472. [CrossRef] [PubMed]

122. Palomba, S.; Orio, F., Jr.; Falbo, A.; Russo, T.; Tolino, A.; Zullo, F. Plasminogen activator inhibitor 1 and miscarriage after metformin treatment and laparoscopic ovarian drilling in patients with polycystic ovary syndrome. *Fertil. Steril.* **2005**, *84*, 761–765. [CrossRef] [PubMed]

123. Fauser, B.C.; Tarlatzis, B.C.; Rebar, R.W.; Legro, R.S.; Balen, A.H.; Lobo, R.; Carmina, E.; Chang, J.; Yildiz, B.O.; Laven, J.S.; et al. Consensus on women's health aspects of polycystic ovary syndrome (PCOS): The Amsterdam ESHRE/ASRM-Sponsored 3rd PCOS Consensus Workshop Group. *Fertil. Steril.* **2012**, *97*, 28–38. [CrossRef] [PubMed]

124. Liu, Y.X. Plasminogen activator/plasminogen activator inhibitors in ovarian physiology. *Front. Biosci.* **2004**, *9*, 3356–3373. [CrossRef] [PubMed]

125. Atiomo, W.U.; Hilton, D.; Fox, R.; Lee, D.; Shaw, S.; Friend, J.; Wilkin, T.J.; Prentice, A.G. Immunohistochemical detection of plasminogen activator inhibitor-1 in polycystic ovaries. *Gynecol. Endocrinol.* **2000**, *14*, 162–168. [CrossRef] [PubMed]

126. Glueck, C.J.; Wang, P.; Fontaine, R.N.; Sieve-Smith, L.; Tracy, T.; Moore, S.K. Plasminogen activator inhibitor activity: An independent risk factor for the high miscarriage rate during pregnancy in women with polycystic ovary syndrome. *Metabolism* **1999**, *48*, 1589–1595. [CrossRef]

127. Meigs, J.B.; O'Donnell, C.J.; Tofler, G.H.; Benjamin, E.J.; Fox, C.S.; Lipinska, I.; Nathan, D.M.; Sullivan, L.M.; D'Agostino, R.B.; Wilson, P.W. Hemostatic markers of endothelial dysfunction and risk of incident type 2 diabetes: The Framingham Offspring Study. *Diabetes* **2006**, *55*, 530–537. [CrossRef] [PubMed]
128. Smith, A.; Patterson, C.; Yarnell, J.; Rumley, A.; Ben-Shlomo, Y.; Lowe, G. Which hemostatic markers add to the predictive value of conventional risk factors for coronary heart disease and ischemic stroke? The Caerphilly Study. *Circulation* **2005**, *112*, 3080–3087. [CrossRef] [PubMed]
129. Yarmolinsky, J.; Bordin Barbieri, N.; Weinmann, T.; Ziegelmann, P.K.; Duncan, B.B.; Ines Schmidt, M. Plasminogen activator inhibitor-1 and type 2 diabetes: A systematic review and meta-analysis of observational studies. *Sci. Rep.* **2016**, *6*, 17714. [CrossRef] [PubMed]
130. Sales, M.F.; Soter, M.O.; Candido, A.L.; Fernandes, A.P.; Oliveira, F.R.; Ferreira, A.C.; Sousa, M.O.; Ferreira, C.N.; Gomes, K.B. Correlation between plasminogen activator inhibitor-1 (PAI-1) promoter 4G/5G polymorphism and metabolic/proinflammatory factors in polycystic ovary syndrome. *Gynecol. Endocrinol.* **2013**, *29*, 936–939. [CrossRef] [PubMed]
131. Lee, Y.H.; Song, G.G. Plasminogen activator inhibitor-1 4G/5G and the MTHFR 677C/T polymorphisms and susceptibility to polycystic ovary syndrome: A meta-analysis. *Eur. J. Obstet. Gynecol. Reprod. Biol.* **2014**, *175*, 8–14. [CrossRef] [PubMed]

International Journal of
Molecular Sciences

Review

Role of Extracellular Matrix in Development and Cancer Progression

Cameron Walker [1],[†], Elijah Mojares [1],[†] and Armando del Río Hernández [1],*

Cellular and Molecular Biomechanics Laboratory, Department of Bioengineering, Imperial College London, London SW7 2AZ, UK; c.walker17@imperial.ac.uk (C.W.); e.mojares17@imperial.ac.uk (E.M.)
*   Correspondence: a.del-rio-hernandez@imperial.ac.uk; Tel.: +44 (0) 20-7594-5187
†   These authors contributed equally to this work.

Received: 14 August 2018; Accepted: 28 September 2018; Published: 4 October 2018

Abstract: The immense diversity of extracellular matrix (ECM) proteins confers distinct biochemical and biophysical properties that influence cell phenotype. The ECM is highly dynamic as it is constantly deposited, remodelled, and degraded during development until maturity to maintain tissue homeostasis. The ECM's composition and organization are spatiotemporally regulated to control cell behaviour and differentiation, but dysregulation of ECM dynamics leads to the development of diseases such as cancer. The chemical cues presented by the ECM have been appreciated as key drivers for both development and cancer progression. However, the mechanical forces present due to the ECM have been largely ignored but recently recognized to play critical roles in disease progression and malignant cell behaviour. Here, we review the ways in which biophysical forces of the microenvironment influence biochemical regulation and cell phenotype during key stages of human development and cancer progression.

Keywords: tumour microenvironment; cancer progression; extracellular matrix; matrix remodelling; fibrosis

---

1. Introduction

The extracellular matrix (ECM) is most commonly defined as the non-cellular component of tissue that provides both biochemical and essential structural support for its cellular constituents. Rather than serving simply as an intercellular filling, the ECM is a physiologically active component of living tissue, responsible for cell–cell communication, cell adhesion, and cell proliferation [1]. Fundamentally, the ECM is composed of and interlocking mesh of water, minerals, proteoglycans, and fibrous proteins secreted by resident cells. However, every organ has a unique composition of these elements to serve a particular tissue-specific purpose [1,2]. Indeed, this unique composition arises through dynamic biophysical and biochemical feedback between cellular components and their evolving microenvironment during tissue development [3,4]. For any specific tissue, components of the ECM are created and arranged by resident cells in accordance with the needs of the tissue. The production of essential fibrous proteins, such as collagen, elastin, and laminin are controlled by the ECM and adapt during various stages of embryonic development and disease progression. As a highly dynamic structure, the ECM is constantly undergoing a remodelling process, by which components are degraded and modified, facilitated primarily by ECM proteinases [5,6]. The balance between degradation and secretion of ECM, orchestrated by ECM-modifying cells, is responsible for tensional homeostasis and the properties of each organ, such as elasticity and compressive/tensile strength.

In vitro, most animal cells are known to only maintain viability when adhered to a substrate [7]. In this regard, cells rely heavily on their sense of touch to survive by protruding, adhering, and spatially interacting with the surrounding ECM. Various cellular growth factor receptors and adhesion molecules along the cell membrane, such as integrins, are responsible for the cell's ability to adhere and communicate with its environment [8,9]. Indeed, cells have been shown to transduce cues from the ECM, such as spatial context and mechanical rigidity, to coordinate crucial morphological organization and signalling events through regulation of gene transcription. This process in which a cell converts external mechanical stimuli into a downstream intracellular chemical signal is known as mechanotransduction [10]. The sensitivity by which cells respond to biophysical and biochemical cues of the ECM demonstrates the importance of tissue homeostasis in the maintenance of healthy resident cells. Accordingly, dysregulation of ECM remodelling has been shown to contribute significantly to cell fate through various fibrotic conditions, characterized by excess ECM deposition and increased rigidity [11]. Due to increased interstitial pressure, unresolved loss of tissue homeostasis has been linked to an elevated risk of various conditions, such as osteoarthritis, cardiovascular disease, and cancer [11]. In this review, we will discuss the role of the ECM in critical physiological processes, such as tissue development and cancer, and some potential targets for therapeutic intervention.

2. Primary Components of the Extracellular Matrix (ECM)

The ECM is composed of various proteins that give rise to different structures and properties that exist within it. The main components of the ECM include collagen, proteoglycans, laminin, and fibronectin. Even among these ECM components, there are subtypes that further specify their function in the overall structure and properties of the ECM. As structure dictates function, different subtypes and combinations of ECM molecules confer different functions that are essential for the whole body to function.

2.1. Collagen as the Basis of ECM Architecture

Collagen is the most significant component of the ECM and the most abundant protein in human tissue, with 28 unique subtypes discovered [12–15]. Each type is composed of homotrimers or heterotrimers of left handed helical α chains that are twisted to form a right handed triple helix structure [13,16]. The collagen superfamily is a large group of proteins that contain the Gly-X-Y motif, where X and Y are usually either proline or hydroxyproline [16,17]. Despite the large amounts of bulky proline, the right-hand helical structure is stabilized by the small glycine, interchained hydrogen bonds, and electrostatic interactions involving lysine and aspartate [17,18]. Fibrillar collagens form fibrous structures often found in tendons, cartilage, skin, and cornea [13,14]. Each collagen fibre is made up of several subtypes of collagen in response to its tissue location. The most abundant type of fibrillar collagen, type I collagen, and can be found in connective tissues ranging from skin and bone to tendon and cornea [19]. Collagen I is involved heavily in processes such as a wound repair and organ development.

All fibrillar collagens are first produced as precursors. The α chains are assembled together in the rough endoplasmic reticulum to form the triple helical structure. Proline and lysine are hydroxylated and the molecule is glycosylated to initiate the formation of the triple helical structure [20]. The procollagen is then brought to the Golgi apparatus where it is prepared for cellular export. Processing of the procollagen happens either during or after secretion in the ECM [21–24]. The C terminal propeptide is cleaved off by specific matrix metalloproteinases (MMPs) and if it is not removed, it leads to high solubility of collagen that prevents it from forming fibrils [25]. For collagen types I, II, and III, the N-propeptides are cleaved off, while for type V, XI, and other fibrillar collagens, the N-propeptides remain (Figure 1A). This modifies the shape and diameter of the fibril without affecting fibril formation [15,25–27]. The N-propeptides of type V and XI collagens protrude from the gaps between collagen molecules to prevent lateral growth via steric hindrance and charge interactions [25,26]. Type V and XI collagens are currently believed to be responsible for nucleating and modulating the fibril formation of collagen [25,26]. It has been shown that the deletion of collagen V in mice leads to failure of fibril assembly despite its low amounts in the total collagen content in most tissues [28].

Once the microfibrils are formed, these may bind with other microfibrils so that they will grow into larger fibres. This process is mediated by other ECM proteins (Figure 1C) [29]. Small leucine rich proteoglycans (SLRPs) such as decorin and biglycan have collagen binding motifs allowing them to modulate fibre growth, size, morphology, and content [15,29,30]. Another subfamily of collagen are fibril-associated collagens with interrupted helices (FACIT) that do not form fibrils themselves but are associated with the surface of collagen microfibrils [13]. Their primary function is to mediate the formation of a higher-order structure via binding with other extracellular matrix proteins such as SLRPs and proteoglycans [26,31]. The supramolecular assembly of collagen is further stabilized by lysyl oxidase (LOX), which leads to overall enhanced mechanical properties. The N terminal and C terminal ends of individual collagen molecules are covalently cross linked by LOX both within and between microfibres, contributing to the great tensile strength of collagen [31,32].

In addition to fibrillar and FACIT collagens, there also exist network forming collagens such as type IV, VIII, and X. These are found in the basal lamina of basement membranes (Figure 1B) [13]. Collagen IV forms a tetramer through their 7S N-terminal domain. Each of these collagen IV molecules is bound to another collagen IV molecule via their C-terminal NC1 domain of each α-chain, forming a hexamer [13]. These two domains of collagen IV allow it to form a stable collagen network that separates the basal lamina from the interstitial stroma [33]. Other ECM proteins such as laminin, nidogen, and perlecan can be found in the basal lamina that strengthens this barrier to effectively maintain the organization of the cells in the body (Figure 1B) [33,34].

Although different types of collagen are able to build various types of supramolecular structures that form the basis of the architecture of the ECM, the contribution of other ECM proteins such as proteoglycans, laminins, and fibronectin cannot be ignored. They largely influence the chemical and physical properties of the extracellular matrix such as through their growth factor binding motifs and innate chemical properties. Furthermore, they also serve as connectors between the cells and the ECM.

2.2. Proteoglycans as Functional Modifiers of the ECM

Proteoglycans are characterized as proteins that have glycosaminoglycans (GAGs) covalently bonded to them. These GAGs are long chains of negatively charged disaccharide repeats that can either be heparin sulphate, chondroitin/dermatan sulphate, hyaluronan, or keratin sulphate. Due to the negative charge of these GAGs, proteoglycans are able to sequester water and cations, which gives them their space-filling and lubrication functions [35]. For the purpose of this review, only transmembrane proteoglycans and those found in the pericellular and extracellular space will be discussed.

Of the 13 transmembrane proteoglycans, four of them are syndecans, proteins thought to act as co-receptors [35]. Syndecans have an intracellular domain, transmembrane domain, and ectodomain (Figure 1A). The GAGs, typically heparan sulphates, are found attached to the ectodomain, which can be shed through the action of MMPs [35,36]. The ectodomain of syndecans is intrinsically disordered, which allows it to interact with a wide variety of molecules to perform a broad range of biological functions (Box 1) [35]. Some of its functions involve binding to growth factors and morphogens, facilitating exosome uptake, and being co-receptors of receptor tyrosine kinases [36–39].

One of the proteoglycans found in the pericellular area of the basement membrane is perlecan. As a large heparan sulfate proteoglycan (HSPG), perlecan has multiple domains, each with different binding sites and functions (Figure 1A) [40]. These heparan sulfates can bind to a variety of molecules such as growth factors, growth factor receptors, collagen, and other ECM proteins. In the basement membrane, perlecan binds and links collagen IV, nidogen, and laminin in order to further strengthen the basement lamina (Figure 1B) [33,34,41].

Proteoglycans found in the extracellular space are classified into hyalectans and SLRPs. The structure of hyalectans are identical: the hyaluronic acid binding N terminal and lectin binding C terminal with GAGs are attached between the N and C terminal ends (Figure 1A). Hyalectans are encoded by 4 distinct genes: aggrecan, versican, neurocan, and brevican [35]. Aggrecan is found mostly in bone cartilage and the brain while neurocan and brevican are found in the central nervous system.

On the other hand, versican is found in the ECM of almost all tissues and organs [42]. They can serve as molecular bridges between the cell surface and the extracellular matrix [35]. Versican has been shown to bind to collagen type I and fibronectin, which are both substrates of integrins [43]. The binding of versican to fibronectin's RGD motif leads to loss of cell adhesion as it sequesters fibronectin from the cell's integrins [42,43].

SLRPs make up the largest family of proteoglycans due to its 18 distinct gene products each with multiple splice variants and processed forms [35]. These proteins have a relatively short protein core with a central region dominated by leucine-rich repeats (LRRs). They are expressed in the ECM during development of various tissue types, suggesting their critical involvement in directing organ size and shape during embryonic development and homeostasis [44,45]. Decorin and biglycan are SLRPs that have collagen-binding motifs and regulate collagen fibre assembly along with other proteoglycans (Figure 1C) [46].

In summary, proteoglycans vary in form and structure that confer different functions in the ECM. They are integral in the maintenance of a healthy ECM without which would lead to a non-functional ECM and a collapse of its structure.

Box 1. Sensing the extracellular matrix's (ECM) mechanical properties.

The ECM is sensed by the cell through transmembrane proteins such as integrins and syndecans and other glycoproteins. Integrins are one of the most versatile transmembrane proteins as various heterodimer combinations allows it to bind to fibronectin, laminin, and collagen [47]. Integrins themselves are mechanosensors. Stretching has been shown to increase integrin binding to the ECM via conversion of integrins to its high-affinity state in smooth muscle cells and fibroblasts [48]. Integrins also experience a conformational change in their cytoplasmic domains, allowing it to activate several signalling pathways such as mitogen-activated protein (MAP) kinases and Rho GTPases [49–51]. Syndecans can also bind to fibronectin, resulting in a synergy between integrin and syndecan to activate signalling cascades through focal adhesion kinase (FAK) and subsequent focal adhesion complex stabilization [52]. There are other receptors for other ECM components such as CD44 for hyaluronan, 67 kDa laminin receptor for laminin, and discoidin domain receptors (DDRs) for collagen [53–57].

Integrins and syndecans activate various pathways such as the MAPK and Rac1/RhoA pathways. The selective activation of these pathways leads to context-dependent regulation of cell survival, growth, proliferation, motility, spreading, or migration [52,58]. Integrins are connected to the actin cytoskeleton through vinculins, talins, and other scaffold proteins while syndecans are connected to the microfilaments through syntenin and through the actin cytoskeleton via a-actinin [52,58–61]. The adhesion complexes formed by integrins and syndecans have been found to be mechanosensitive [9,62]. The intracellular signalling and mechanotransduction through these receptors is still an active field of research. Much is still to be discovered about the pathways that facilitate ECM mediated cellular responses.

2.3. Connecting the Cell to the ECM through Laminin

Laminins are trimeric glycoproteins consisting of α, β, and γ chains that are often found in the basal lamina or some mesenchymal compartments [15]. The 12 mammalian α, β, and γ chains can theoretically create 60 unique laminins but only 16 combinations have been observed so far [34,63]. The α chains vary in size from 200 and 400 kDa while β and γ chains have sizes from 120 to 200 kDa. A trimer can then have a size varying from 400 to 800 kDa [63]. During rotary shadowing electron microscopy, laminins look like cross-shaped molecules [34,64,65]. The three chains form an α-helical coiled coil structure that forms the long arm of the cross while the three short arms are composed of one chain each (Figure 1A) [34]. At the end of the long arm are 5 laminin G-like (LG) domains from the α chain that serve as attachment sites for the cell. Integrins, dystroglycan, Lutheran glycoprotein, or sulfated glycolipids bind to these LG domains [63]. At the end of each short arm are laminin N-terminal (LN) domains that are important for laminin polymerization and basement membrane assembly (Figure 1B) [34].

Laminins have cell type-specific functions such as adhesion, differentiation, migration, phenotype maintenance, and apoptotic resistance [63]. Through binding of integrins, laminins are able to create a dynamic link between the cell and the ECM (Box 1) [63]. Unique heterotrimeric laminins will have unique integrin heterodimers binding partners to allow the induction of signalling pathways and organization of intracellular cytoskeleton [63,66]. Collagen IV deposition in the basement membrane is seen as the maturation of the basement membrane that is essential for structural stability later in development [34,67]. However, the exact mechanism by which laminins bind to collagen IV remains unclear. Initial studies indicated that nidogen binds to laminin through the LE domains of the γ1 chain and collagen IV, thus serving as an intermediary between the two networks found in the basement membrane. However, recent research has indicated that nidogen might not be the major bridge in connecting laminins and collagen IV [34]. It has been observed that the interaction between laminins and collagen IV is directly mediated by heparan sulfates [68]. Perlecan was thought to mediate this function, but genetic ablation of perlecan in mice did not result in collagen IV depletion [34,67]. It has since been postulated that agrin, another pericellular HSPG, serves as a compensating candidate. In this model, both perlecan and agrin would bind to the nidogen containing laminin network and to the collagen IV's 7S and NC1 domains (Figure 1B) [69,70]. Laminins serve crucial roles in both basement membrane assembly and ECM–cell interactions. Recent studies have indicated that basement membrane assembly is initialized through laminin polymerization [53–58]. Indeed, genetic ablation of either β1 or γ1 chains proved to be lethal due to the resultant failure of basement membrane assembly. While collagen, proteoglycans, and hyaluronic acid comprise the major structural component of the ECM, laminins are one of the molecules that bridge the interaction gap between the cells and the ECM [15].

2.4. Fibronectin as the Mechanosensitive Connection Between the Cell and ECM

Fibronectin is a multi-domain protein that interacts with the various previously described ECM components to connect the cell to the ECM [15]. It is encoded by a single gene, but it has 20 isoforms in humans as a result of alternative splicing of the mRNA [71,72]. Similar to collagen, fibronectin forms a fibrillar network in the ECM (Figure 1C) [71]. Fibronectin naturally exists as a dimer outside the cell, mediated by the two cysteine disulfide bonds, which is crucial for its ability to assemble in a fibrillar fashion (Figure 1A) [71,73]. Fibronectin matrix assembly is mediated by selective binding to α5β1 integrins through an RGD binding motif and a synergy site on the fibronectin molecule [59,62]. Through these integrins, the compact and soluble secreted fibronectin is unfolded revealing cryptic binding sites for other fibronectin molecules to form the fibronectin fibrillar network (Figure 1C) [1,71,74]. Anti-integrin and anti-fibronectin antibodies have been shown to prevent fibronectin fibril formation [71,75,76]. Fibronectin binding induces integrin clustering that provides local high concentrations of fibronectin at the cell surface. This phenomenon promotes fibronectin–fibronectin interactions through the N terminal assembly domains of each molecule [71].

Once fibronectin is tethered to the cell surface by integrins, the actin cytoskeleton can pull onto fibronectin molecules to change its conformation [71,72]. This will affect the C terminal regions of fibronectin, revealing cryptic binding sites for fibronectin, heparan sulfates, heparin, collagen, and other ECM proteins [77–81]. It is through strong non-covalent protein-protein interactions that the fibronectin network matures and becomes insoluble, although other ECM proteins may mediate mature lateral interactions between fibrils [71]. These interactions stabilize the relatively weak binding sites at individual sites. However, the turnover of the fibronectin matrix is still largely unexplored [71].

Due to fibronectin's multiple binding sites for other ECM proteins, it has been implicated in various functions, including a role in collagen type I assembly. It has been shown that in the absence of fibronectin, collagen fibrils do not accumulate, suggesting a role for fibronectin in collagen assembly [82,83]. However, this relationship may prove reciprocal as recent studies have also implicated that collagen has a role in enhancing fibronectin assembly [71,84,85].

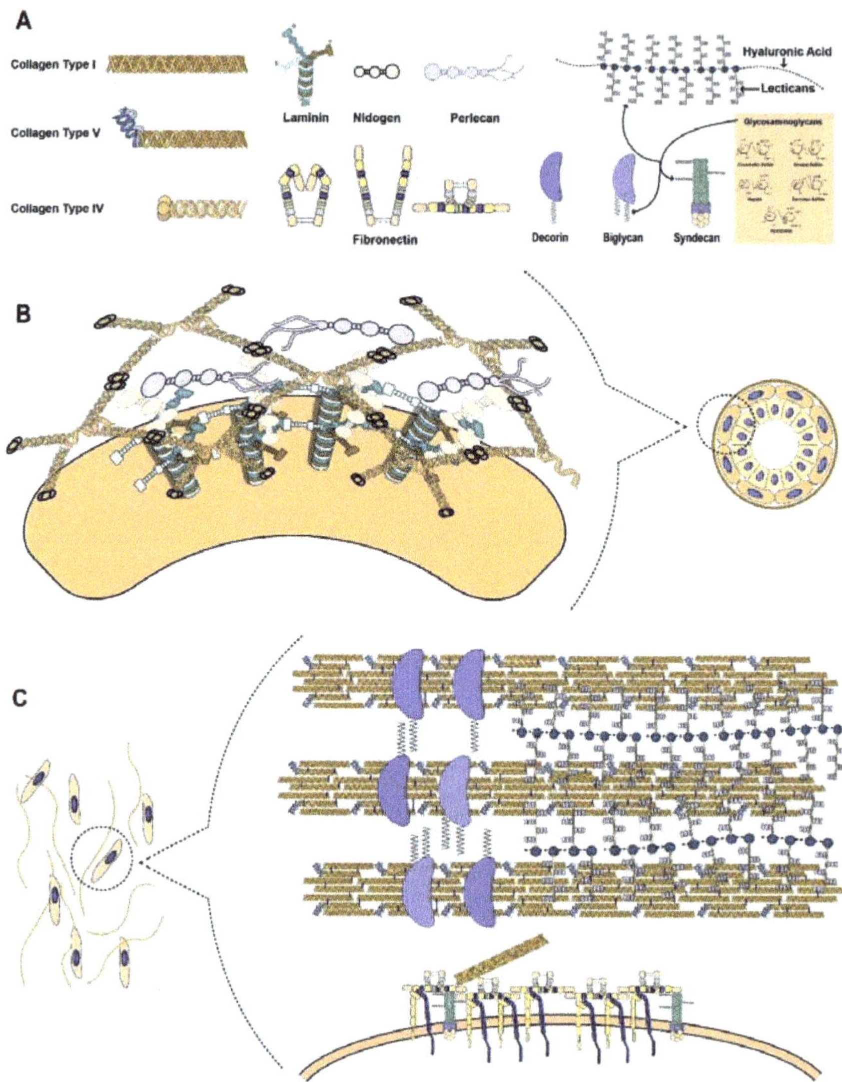

Figure 1. Unique ECM molecules and their organization in the basement membrane and interstitial stroma. Panel **A** (top) shows the unique components of the extracellular matrix. Panel **B** and **C** (middle) shows how these different collagens, proteoglycans, laminins, and fibronectin are organized within the basement membrane (**B**) and interstitial ECM (**C**). A breast acinus with epithelial cells is surrounded by myoepithelial cells and the basement membrane. In the basement membrane, the laminin is bound to the cell and forms a network through its long arms. It is then connected to the collagen IV network through nidogen and proteoglycans such as perlecan and agrin. Outside the basement membrane is the interstitial ECM where fibroblasts that produce and remodel the ECM can be found. In the interstitial stroma, collagen fibres are made up of fibrils composed of collagen I and collagen V. The different proteoglycans, such as decorin, biglycan, and hyalectans, holds the fibrils together to form a collagen fibre. Fibronectin is bound to the cell via integrins and syndecans. Once fibronectin is unfolded, it reveals cryptic binding sites for heparan sulfate proteoglycans (HSPGs) and collagen. Modified and combined figures from Mouw et al. 2014 [15] and Hohenester and Yurchenco 2013 [34].

3. Function of ECM

The plethora of unique ECM molecules serves several functions that influence biochemical and biophysical processes in the cell simultaneously (Figure 2). While the ECM has been considered for many years as an inert scaffold solely providing structure for the cells, its role in determining the functions and phenotypes of cells has clearly emerged in the last two decades. The ECM can serve as binding sites, controlling the adhesion and movement of cells [86]. This is emphasized in the complex structure and composition of the basement membrane that serves as a barrier between epithelial cells and the interstitial stroma [6,87]. In addition to structural integrity and anchorage, the ECM components have several binding sites for growth factors, controlling their release and presentation to target cells. This is especially important in morphogenesis as it establishes morphogen gradients [88]. Finally, the ECM transmits mechanical signals to the cells, which activates several intracellular signalling pathways and cytoskeletal machinery [89]. Indeed, the ECM serves several functions and here we review the function of the ECM in the context of development and maintenance of the stem cell niche.

The function of the ECM is best described in the context of development. The development of a mammalian embryo from a foetus to a fully developed organism is a well-orchestrated phenomenon that involves carefully controlled mechanisms. In such a relatively rapid process, the spatiotemporal composition, amount, and characteristics of the ECM must be tightly regulated. Several studies have shown that mutated ECM components lead to birth defects or embryonic lethality, which emphasizes its role in development [2,87]. The geometry, rigidity, and other physical properties of the ECM are sensed by the cells and ultimately direct their differentiation and the complex spatial and structural arrangements they form in tissues (Box 1).

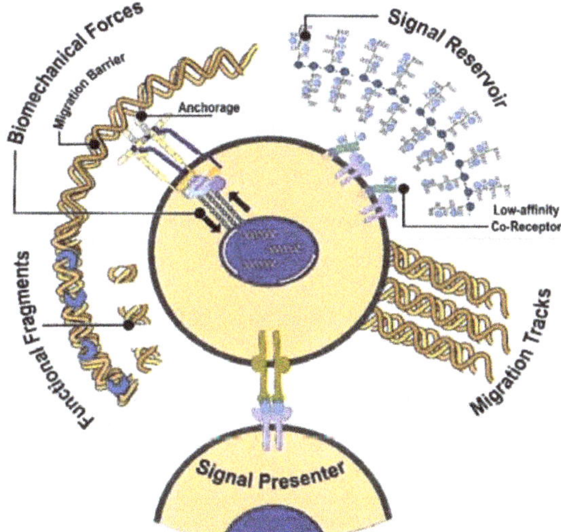

Figure 2. Functions of the ECM. The ECM serves as a point of anchorage for the cells that is essential for maintaining tissue polarity and asymmetric stem cell division. Depending on the context, it can impede or facilitate migration. It can sequester growth factors and prevent its free diffusion. Other ECM components can bind growth factors and can serve as co-receptors or signal presenters, which help determine the direction of cell-cell communication. Through the action of metalloproteinases (MMPs), fragments of the ECM can also influence cell behaviour. The physical properties of the ECM can be sensed by focal adhesion complexes, which lead to a variety of changes in cell phenotype such as reorganization of the 3D genome. Figure modified and adopted from Lu, Weaver, and Werb 2012 [90].

3.1. ECM as Tracks for Migration and Proliferation

Migration of cells is essential for tissue development and can be best illustrated by neural crest cells, which migrate from the periphery of the neural tube to different parts of the embryo to form parts of the heart, spinal nerve, skin, and cranium [91]. How these cells direct their migration and final destination is a complex question that has been extensively studied.

The ECM influences the migration track and speed of migrating cells through its topography, composition, and physical properties. The alignment of the underlying ECM has been shown to direct cell migration and proliferation. Sharma et al. [92] previously used aligned fibres to direct cell migration in the context of wound healing in vitro. Moreover, self-aligning materials have been recently used to create ECM constructs for brain tissue regeneration in vivo [93].

Cells migrate from regions with lower ECM concentration to higher ECM concentration due to an adhesion gradient in a type of cell migration known as haptotaxis [87]. However, this relationship is nuanced. If the concentration of the ECM is too high, the adhesion force experienced by the cell is too large from them to continue to migrate. Accordingly, the speed of migration is dependent on the ECM concentration as well. As migration is a coordinated interplay between adhesion and deadhesion of the cell onto the ECM, the speed of migration is characterized by a bell-shape function with respect to ECM concentration [94].

The speed of migration is also influenced by the composition of the ECM. Hartman et al. [95] have shown that fibroblasts cultured in rigidity gradients composed of fibronectin exhibited cell migration while those cultured on matrices covered with either laminin or a fibronectin/laminin mixture did not exhibit any migration. These results indicate that the ECM not only serves as a track for migration, but also dictates cell migration due to its mechanical properties. It was previously demonstrated by Wang et al. [96] that matrix stiffness and cell contractility also control RNA localization of genes responsible for cellular organization and signalling to cellular protrusions. Proteases that degrade the ECM also facilitate migration of cells through a process involving the interplay of MMPs, adamlysins, meprins, metalloproteinase inhibitors (MMPIs), and other enzymes [2]. It is important to note that constant ECM remodelling is occurring in development. The ECM components and their concentration are continuously modified to dictate the developmental program.

3.2. ECM as the Dynamic Blueprint for Development

The role of ECM in structural organization is best studied in branching morphogenesis, which involves epithelial buds and tubes invading the surrounding mesenchyme rich ECM [87]. This process can be seen in various parts of the body, including mammary glands, salivary glands, and kidneys [87]. Structural organization emphasizes the key functions played by the ECM and different ECM components such as glycosaminoglycans (GAGs), collagen, and proteoglycans. Furthermore, the ECM is constantly changing as this process occurs, highlighting the spatiotemporal control needed to facilitate the development of these organs.

Branching in the mammary gland occurs at the terminal end buds that have a thin hyaluronic acid-rich ECM and an accumulation of thick ECM composed of collagen IV, laminins 1 and 5, and HSPGs at the flanks [97,98]. Thick ECM provides a structure to maintain the tubular organization by serving as anchors for the cells while the reduced ECM at the end bud facilitates the migration of epithelial cells, specifically the cell-budding process [87,99,100]. However, fibrillar collagen does not only serve as a physical barrier. As it binds to discoidin domain receptors (DDRs) expressed by mammary gland cells, it prevents hyper-proliferation possibly to maintain the tubular structure [87,101].

The architecture of the ECM serves as a guide to control how branching occurs through local anisotropies in terms of tension as well [99]. Topographical variation in structure and elasticity of the ECM provides a blueprint for where cells can bend, twist, and break off to form complex morphologies that are essential for the different organs. Using micro-patterned organotypic cultures, Nelson et al. [102] showed that tissue geometry dictates the position of the branches. Furthermore, Gjorevski and Nelson [103] have shown that endogenous patterns of mechanical stress in the surrounding ECM specify the branching pattern. There are several techniques to study ECM architecture in 3D such as atomic force microscopy (AFM) combined with second-harmonic generation (SHG) [104]. Robinson et al. used image analyses algorithms to analyse AFM and SHG micrographs to monitor and analyse ECM remodelling of pancreatic stellate cells. This technique could similarly be used to study how the mechanical forces and ECM architecture continuously evolve during development.

As the end bud of the mammary gland grows, it continuously degrades the ECM, which in turn releases factors dictating the branching direction of the growing buds [2]. ECM degradation releases collagen fragments tumstatin and endostatin that regulate the migration, survival, and proliferation of these cells [105]. Furthermore, the ECM's binding sites for morphogens and growth factors, such as Wnt glycoproteins, epidermal growth factors (EGFs) and fibroblast growth factors (FGFs), allow them to sequester and control the release of these factors [35,106]. Through this, the ECM facilitates the formation of a morphogen gradient, which is required for diverse types of cells and structures to develop [88]. The growth of the bud is finally terminated through the deposition of inelastic ECM composed of sulphated GAGs (SGAGs) and collagen I [99].

Overall, the ECM modulates the growth of tissues to form complex structures that are required for these organs to function. The ECM provides structural organization not only through its action as a physical barrier to growing cells, but also by activating intracellular signalling in a time and context dependent manner. The ECM does this through growth factor distribution modulation, physical anisotropies, and anchorage.

3.3. ECM as the Driver for Cell Fate

The ECM influences cell fate through the previously discussed morphogenetic gradient in development. However, this is not the only mechanism by which ECM affects cell fate, as the physical properties of the ECM also play a critical role.

The role of ECM composition on cell fate is exemplified in mammary gland differentiation. Even with hormonal stimulation in vitro, mammary gland cells do not secrete milk proteins. However, upon exposure of these cells to laminin-1, they begin secreting milk proteins [107]. This activation is due to integrin binding by laminin-1 leading to phosphorylation of the prolactin receptor, an upstream regulator of STAT5. STAT5 then activates the transcription of milk proteins β-lactoglobulin and β-casein, indicating that an appropriate 3-D ECM microenvironment is critical for cells to function properly [102,107].

Using a simple yet elegant approach, Engler et al. [10] demonstrated that when mesenchymal stem cells (MSCs) are cultured on collagen matrices with various elasticities, the MSCs differentiated into osteocytes, myocytes, and neurons on the substrates that resembled their respective native tissue. The process by which these MSCs are able to sense the elasticity and stiffness mechanical environment before initiating an intracellular signalling cascade to dictate cell fate is known as mechanotransduction. This phenomenon allows the cells to sense the physical properties of the underlying matrix and activate appropriate intracellular pathways [89]. When myosin II, a key molecule in various mechanotransduction-signalling pathways, was blocked, MSCs became insensitive to the matrix elasticity mediated cell differentiation [10]. This study emphasized that the ECM's physical properties themselves have differentiating capability.

Engler's results indicate that environmental cues can be relayed to the nucleus biochemically or biophysically [89]. Indeed, often biochemical and biophysical relays work in conjunction with one another to transmit environmental information [89,108]. One prominent example of a mechanosensitive genetic regulator is the pair of the transcription co-activators, yes-associated protein (YAP) and transcriptional coactivator with PDZ-binding motif (TAZ). These functionally redundant transcriptional coactivators are known to play crucial roles in critical cellular processes, such as proliferation, wound healing, fibrosis, and other physiological processes that involve changes in biomechanical properties [109]. Importantly, these transcription co-activators are known to be sensitive to both biochemical and biophysical cues. The nuclear localization of YAP/TAZ is regulated by cell shape, stiffness, ECM topology, and shear stress [109–115]. The integrin complexes that sense this matrix stiffness are linked to the cytoskeleton, which in turn is linked to the nucleus through the linker of nucleoskeleton and cytoskeleton (LINC) complex, which is composed of nesprins, sun, and lamin proteins [116,117]. This allows direct transmission of mechanical cues from the extracellular matrix to the nucleus [117]. The extensive role of YAP/TAZ in cell processes elucidates the role of mechanotransduction in development and diseases, such as cancer [108].

In addition to matrix stiffness, the response of the cell to the ECM's physical properties has also been shown to be dependent on the ligand tether length. It has been shown that the focal adhesion sizes and cell-adhesion strengths were affected when the tether length of the ECM coating's ligand was varied [118]. Coating stiff ECM with RGD ligands with longer tether lengths lead to the cell sensing a softer ECM thereby controlling mechanotransduction mediated YAP/TAZ nuclear localization in cancer. This might be a novel effective treatment against cancer.

4. Tissue Homeostasis

The ECM is a highly dynamic structure. Even after development, ECM is constantly being deposited, degraded and modified to maintain tissue homeostasis. This is especially important in maintaining the phenotype of cells and in physiological processes such as wound healing, angiogenesis, and bone remodelling [6,119,120].

To maintain tissue homeostasis, the cells in contact with the ECM sense the properties of the ECM through receptors and focal adhesion complexes. In turn, the cell regulates the expression of ECM components and enzymes based on the signals of the ECM. This creates a feedback mechanism wherein the cell also influences the ECM, which results in a balance of deposition and degradation of ECM components [116].

The response of the cells to other stimuli, including shear stresses exerted by blood flow, are ultimately influenced by the ECM component. Chen and Tzima [121] showed that platelet endothelial cell adhesion molecule-1 (PECAM-1), a mechanosensitive molecule, is essential for vascular remodelling, which occurs in response to long-term changes in hemodynamic conditions. Furthermore, Collins et al. [122] recently demonstrated that the ability of platelet endothelial cell adhesion molecule-1 (PECAM-1) to respond to mechanical forces is influenced by the type of ECM they are adhered to. This exemplifies the complexity and importance of the feedback mechanism that exists between the ECM and the cell to maintain tissue homeostasis.

The importance of the ECM in maintaining tissue homeostasis is exemplified by the study performed by Weaver et al. [123] where they were able to revert the malignant breast cancer cell phenotype to the normal phenotype. They did this by culturing breast cancer cells onto basement membrane based 3-D substrates coated in integrin $\beta 1$ blocking antibodies [123]. This study confirmed that the ECM is able to override the mutations causing the cancer phenotype, emphasizing the ECM's role in maintaining the correct cell phenotype. The role of dysregulated ECM in cancer progression will be discussed further in the next section.

That an imbalance in the deposition and degradation of the ECM leads to diseases is a hallmark not just of cancer but also other prominent diseases including fibrosis [124–126]. Overall, the ECM's role in tissue homeostasis is to direct proper cell response and phenotype to maintain the tissue's mechanical integrity and function.

5. ECM in Cancer

5.1. Dysregulation of ECM Molecules in Cancer Progression

Traditional perspectives of cancer have shifted to reflect the important role of the ECM in regulating cell proliferation, migration, and apoptosis. On a microscopic level, the particular arrangement and orientation of ECM constituents form a tissue-specific microenvironment that plays a critical role in tumour progression [11,127,128]. It is now understood that the ECM not only undergoes continuous active remodelling, but also elicits biochemical and biophysical cues to influence cell adhesion and migration [129]. As such, small changes in microenvironment homeostasis can have significant effects on the proliferation of cancer cells. As the most significant ECM component, collagen dictates the primary functional properties of the matrix. Indeed, changes in the deposition or degradation of collagen can lead to the loss of ECM homeostasis [130,131].

As tumour cells proliferate, the surrounding ECM undergoes significant architectural changes in a dynamic interplay between the microenvironment and resident cells. These changes, including increased secretion of fibronectin and collagens I, III, and IV illustrate that tumour progression demand a continuous interaction between the ECM and tumour cells (Figure 3) [132]. Increased deposition of matrix proteins promotes tumour progression by interfering with cell–cell adhesion, cell polarity, and ultimately amplifying growth factor signalling [133]. However, the exact role of collagen deposition in tumour progression is nuanced. Recent studies have shown that increased collagen cross-linking and deposition leads to tumour progression via increased integrin signalling [134,135]. Interestingly, however, depletion of fibrillar collagens I and III can also promote malignant behaviour, indicating that biomechanical forces produced by collagen deposition can have both beneficial and deleterious effects on tumour progression [136,137].

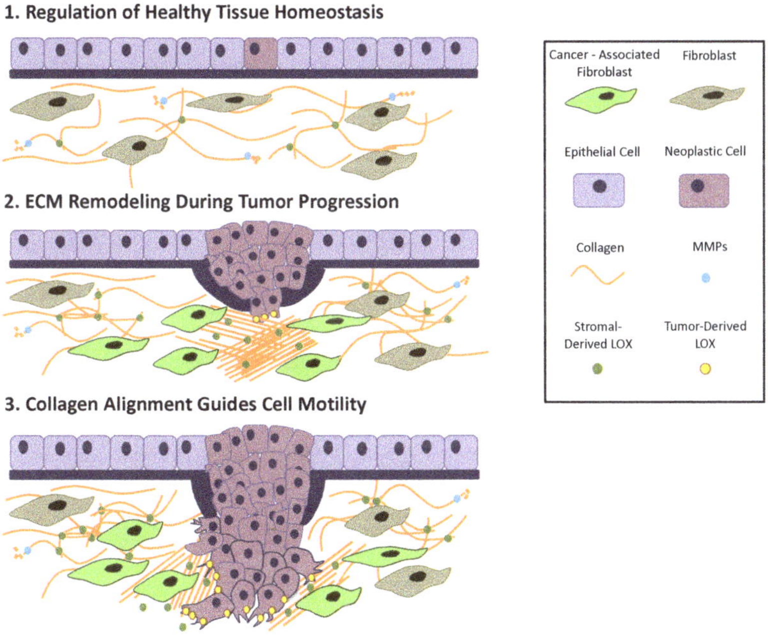

Figure 3. ECM remodelling during cancer progression and initiation. (**1**) Epithelial neoplastic cells proliferate rapidly, inducing strain on the basement membrane. (**2**) Basement membrane bulges due to mechanical strain. Adjacent cancer-associated fibroblasts increase deposition of collagen. Stromal-derived lysyl oxidase (LOX) aligns collagen. (**3**) Neoplastic cells breach membrane and migrate along aligned collagen. (Adapted from Lu et al. [6].)

Collagen cross-linking can occur in both an enzyme-mediated and non-enzyme-mediated fashion. Regulated collagen cross-linking is coordinated primarily by LOX and the LOX family of amine oxidase enzymes [1]. LOX, secreted by primary tumour cells, is responsible for catalysing the cross-linking of both collagen and elastin, which in turn increases matrix stiffness and total adjacent ECM volume. Increased ECM stiffness activates integrins and augments Rho-generated cytoskeletal tension to promote focal adhesion formation and cell motility [138]. Elevated LOX activity has been clinically associated with increased collagen cross-linking, fibrosis, and elevated risk of cancer metastasis [139]. Moreover, elevated LOX activity found on invasive edges of tumours has been noted to drive actin polymerization, cell contractility, and migration, providing a pathway for successive tumour cells to follow [130].

Visualization of surrounding epithelial tissue during tumour metastasis has revealed localized matrix organization and alignment along the leading edge of invasive tumours [131,140]. Indeed, local cell invasion of these tumours has been observed to be oriented along aligned collagen fibres, suggesting that the linearization of collagen fibres facilitates tumour invasion [141]. It is believed that these densely aligned collagen fibres act as tracks for proliferating neoplastic cells to migrate out of the tumour. Breast cancer serves as an important example of collagen alignment during tumour metastasis. Although collagen within epithelial structures is typically tangled and disorganized, collagenous tissue surrounding mammary tumours is frequently thickened, stiffened, and aligned perpendicularly to the tumour boundary [142]. Recent studies indicate that the topography of matrix fibres increases the efficiency of tumour migration by reducing the protrusions along the collagen fibre, and hence the distance travelled by the migrating cell [143].

Much like collagen and LOX, elevated levels of the glycosaminoglycan hyaluronic acid in the ECM correlates to increased likelihood of malignancy and poor prognosis [144]. As a naturally occurring omnipresent linear polysaccharide, hyaluronic acid is critical in determining the compressive properties of most biological tissues. The combination of tensile resistance due to collagen and compression compliance due to hyaluronic acid creates the ideal biophysical properties for tissue homeostasis [145]. In addition, it has been found that hyaluronic acid is both an induction signal for mesenchymal transition and a migration substrate [146]. Accordingly, hyaluronic acid is frequently used as a biomarker for prostate and breast cancer. While augmented levels of collagen and LOX directly promote ECM stiffness and mechanically drive cell motility and proliferation, the exact role of hyaluronic acid in cancer metastasis remains unclear. However, its dysregulation can serve as a key biomarker for metastasis and cancer invasion.

5.2. Protein Unfolding Mediates Mechanotransduction

ECM signalling is a crucial cellular process that drives cell proliferation, differentiation, and defers apoptosis [147]. In brief, if a cell cannot sense its mechanical environment, it cannot survive. Many studies have reported that cells are capable of sensing their microenvironment through chemical signalling, such as growth factors and metabolic precursors [148–150]. In order to detect ECM rigidity, it is believed that cells mechanically probe their microenvironment via lamellipodia and sense the mechanical feedback and resistance of their environment through integrin-based focal adhesions, triggering an intracellular signalling cascade [151]. The ability for cells to probe their microenvironment is attributed to the actin cytoskeleton, as inhibition of F-actin polymerization limits the ability of cells to generate force, which induces a biological effect similar to plating cells on a soft substrate. Specifically, the ability for cells to produce internal forces is derived from contractile actin bundles and their upstream regulators, such as Rho-associated protein kinase (ROCK), which are necessary to mechanically sense their environment [109]. While mechanical rigidity clearly has profound effects on cell behaviour, the mechanism that translates mechanical force into gene transcription is not fully understood.

Our recent work has illustrated the importance of protein unfolding in the transduction of mechanical force exerted by the ECM. Indeed, talin, a prominent molecule in focal adhesion complexes that couples focal adhesions to the actin cytoskeleton, has been shown to mechanically unfold during force transmission [152]. Deleted in liver cancer 1 (DLC1) is a negative regulator of RhoA and cell contractility that regulates cell behaviour when concentrated to focal-adhesion complexes bound to talin [153]. Mechanical clamping of the R8 domain of talin prevented mechanical unfolding of the molecule, interrupting downstream signalling of DLC1 and, consequently, cell behaviour [153].

Moreover, single molecule force microscopy revealed that every talin rod subdomain is susceptible to unfolding over a physiologically relevant range of forces between 10 and 40 pN [152]. Because the observed range of talin subdomain stabilities within the focal adhesion complex depend on small structural differences, it is possible the mechanical stability of talin rod bundles could be influenced by a few single point mutations. These mutations could lead to misinterpretation of ECM signals, altering cellular response. Incorrect interpretations of ECM information could influence the behaviour of cancer cells in the tumour microenvironment, potentially triggering DLC1 deactivation, increased cell contractility, and cell migration [152,153].

5.3. YAP & TAZ Mechanotransduction in Cancer Progression

As robust regulators of cell proliferation and survival, YAP and TAZ play critical roles in regulating organ development, cell differentiation, and progenitor cell self-renewal [109]. During these processes, the YAP/TAZ proteins actively shuttle between the nucleus and the cytoplasm. While in the cytoplasm, the YAP/TAZ proteins play a relatively passive role, regulating specific signalling cascades, such as the Wnt signalling pathway. Meanwhile, when in the nucleus, they readily interact with DNA-binding transcription factors, particularly TEA domain family members (TEAD), to regulate genetic expression associated with proliferation, a key hallmark of cancer [154,155]. Upon biochemical inhibition, YAP/TAZ accumulate in the cytoplasm, suggesting that the main functionality of YAP/TAZ is gene transcription regulation in the nucleus. Importantly, upon cell detachment from a substrate, YAP/TAZ activity is inhibited; suggesting that YAP/TAZ translocation to the nucleus can be regulated by the F-actin cytoskeleton and mechanical force [156]. Moreover, in mammalian systems, matrix elasticity and cell-spreading geometry are noted to heavily regulate YAP/TAZ nuclear transport and their corresponding physiological processes [157]. Taken together, these results suggest that focal adhesion and cytoskeleton-mediated cell signalling of mechanical rigidity is coupled to the YAP/TAZ pathway to induce metastasis and tumour invasion, indicating a direct chemical pathway linking mechanical force with malignant cellular behaviour.

Although cytoskeletal tension is sufficient for YAP/TAZ nuclear translocation, there exist multiple potential pathways and proteins that mediate YAP/TAZ nuclear translocation. For example, the heparan sulfate proteoglycan, agrin, is most commonly known for its role in the formation of neuromuscular junctions during embryogenesis. However, recent advances in the field have suggested that agrin may also serve as an ECM sensor that stabilizes focal adhesions and facilitates YAP/TAZ nuclear translocation through the lipoprotein-related receptor-4 (Lrp4) and muscle-specific kinase (MuSK) pathway [158,159]. Activation of Lrp4 and MuSK by agrin inhibits the Hippo tumour suppressor pathway, ultimately leading to elevated YAP/TAZ nuclear translocation [158,160]. Agrin depletion was shown to promote the inhibitory phosphorylation of YAP, which forced nuclear YAP to remain in the cytosol [161]. Contrarily, supplementary introduction of agrin into cells cultured on compliant matrices was sufficient for YAP activation [161]. Multiple junctional proteins, including the Angiomotin (AMOT) family of proteins regulate YAP/TAZ in combination with changes in actomyosin contractility [161,162]. AMOT proteins have been shown to directly bind to YAP, inhibiting its function. F-actin competitively binds with AMOT to disrupt YAP:AMOT complexes, releasing YAP from its inhibitory state to translocate into the nucleus [161]. Interestingly, agrin depletion elevated YAP:AMOT binding, which ultimately led to decreased YAP activity [159]. Moreover, recent work demonstrates that the Ras-related GTPase, Rap2, is also a key intracellular mediator that transduces ECM rigidity signals to influence YAP/TAZ nuclear translocation [163,164]. At low ECM stiffness, Rap2 is known to bind and activate MAP4K4, MAP4K6, MAP4K7, and ARHGAP29, which stimulate LATS1a and LATS2 while inhibiting YAP and TAZ nuclear translocation [163]. These findings demonstrate that ostensibly unrelated proteins, such as Rap2 and agrin, play significant roles in ECM sensing and regulation of YAP/TAZ activity.

5.4. ECM-Mediated Tumour Initiation and Migration

A crucial hallmark of carcinoma and other cancer cells is their ability to migrate through surrounding tissues, penetrating the adjacent basement membrane. This dense, highly cross-linked membrane of ECM serves not only as an anchor for epithelial cells to surrounding connective tissue, but also as a significant barrier to epithelial cell migration [165]. However, due to the need for cells to migrate within the body during healthy tissue homeostasis, cancer cells have adopted a few methods of traversing the collagenous barrier [166]. One such method is the use of mechanical force. Mechanical force has increasingly been seen as a compelling factor in triggering the breaching of the basement membrane. As epithelial cancer cells proliferate, they are spatially constrained by the bordering basement membrane. This burgeoning population of cancer cells significantly increases the mechanical stress along the membrane, ultimately causing rupture and allowing cells to escape their microenvironment (Figure 3) [167].

Another method of membrane navigation is anchor cell invasion, in which anchor cells breach the basement membrane using protrusive, F-actin rich subcellular structures called invadopodia [168,169]. Indeed, electron micrographs of invasive tumours have demonstrated that leading invasive cells extend a single protrusive arm into the basement membrane [170]. After initial breach by the invadopodia, the membrane fissure widens, allowing for subsequent cells to traverse the collagen boundary [171]. However, along these breaching sites, elevated levels of collagen IV degrading products have also been found, indicating a possible third factor in the migration of cancer cells [172]. Increased accumulation of MMPs along the basement membrane has led to the commonly held assumption that proteases were solely responsible for degradation of the basement membrane. However, staining of the membrane during invasion reveal that laminin and collagen IV are in fact pushed aside by the invadopodia rather than fully degraded [165,167]. These results indicate that MMPs may, rather, play a role in the initial breaching of the basement membrane or in softening the matrix while anchor cell invadopodia facilitate direct invasion [166].

5.5. Metalloproteinases (MMPs) in Tumour Progression

The role of MMPs in cancer cell invasion is multi-pronged: they not only assist in the degradation of surrounding ECM barriers, but also release active growth factors and promote tumour angiogenesis [148]. The ECM is known to promote cell proliferation primarily through contact with the integrin family of cell surface receptors. However, certain ECM binding sites responsible for cell proliferation and survival have been shown to be "cryptic" or partially hidden within the ECM. MMPs simply unmask these hidden binding sites by degrading and loosening surrounding collagen, allowing for integrins along the cell membrane to interact directly with the matrix [148,173].

In addition to removing physical barriers and revealing cryptic binding sites, MMP-mediated degradation of collagen also exposes signalling components embedded within the ECM [173]. Stored in an inactive state when embedded within collagen, various growth factors are activated upon ECM degradation and allowed to bind with their target receptor. For example, MMP-2 mediated ECM degradation is known to release the active form of transforming growth factor-β (TGF-β). Upon its release, TGF-β is able to modulate cell invasion, immune response, and cell proliferation [174–176]. In effect, MMPs not only physically manipulate the surrounding ECM to allow for cell migration, but also create a microenvironment conducive to tumour development through growth factor release and cryptic binding site exposure. Thus, targeting MMPs could serve as a promising therapeutic approach, despite a previous lack of success (Box 2).

Despite MMP-induced angiogenesis, vasculature networks in the tumour are frequently disorganized with inter-capillary regions often exceeding the diffusion distance of oxygen. As such, hypoxia, or the state in which cells are devoid of oxygen, is a hallmark of cancer. In fact, measurements of the partial pressure of oxygen in tumours reveal that poorly oxygenated tumours strongly correlate to increased malignancy [173]. It is believed that cancer cells are able to withstand oxygen-derived regions by altering the transcription of various genes associated with angiogenesis. Hypoxia-inducible factors (HIF) are known play a crucial part in the regulating this intracellular cancer cell response to hypoxia [177–179]. Recent studies have indicated that HIF-1α, a member of the HIF family of transcription factors, has been associated with increased MMP and collagen production [177–179]. Importantly, HIF-1α is known to increase LOX deposition, ultimately stiffening the surrounding matrix [180]. Finally, HIF-1α has also been shown to activate transcription factors associated with epithelial–mesenchymal transition (EMT), the process by which cells lose their polarity and adhesion with adjacent cells, augmenting the invasive behaviour of cancer cells [180].

Box 2. MMPs as a therapeutic target.

As the role of the ECM in tumour progression becomes more apparent, cancer therapeutic interventions have begun to target key elements of the ECM in an attempt to limit metastasis. One key ECM component that has been targeted is the MMP family of enzymes. Due to the significant role of MMPs during cancer progression, the pharmaceutical industry has worked to develop safe therapies to inhibit MMP activity [181,182]. Several groups of synthetic MMP inhibitors, such as Marimastat, Minocycline, and Matimastat, have been developed to target broad groups of MMPs and tested in stage III clinical trials in late stage cancer patients [183].

Unfortunately, most therapies specifically targeting MMP activity demonstrated poor outcomes during clinical trials [184]. A few possible explanations exist for the poor clinical outcomes. Firstly, patients selected to receive the MMP-inhibiting therapies were late-stage cancer patients. As previously discussed, MMPs are known to play a role in tumour initiation and progression. It is possible that MMP-inhibiting agents could be more effective in early stage patients. Moreover, it is known that specific MMPs play different roles during cancer progression. It is likely that synthetic inhibitors need to be developed to target unique MMP subgroups at specific timeframes during cancer progression.

5.6. Role of Mechanical Stress in Tumour Growth and Treatment

As cancer progresses, tumours rapidly grow in size and stiffen due to the increased appearance of structural components, such as ECM, cancer-associated fibroblasts (CAFs), and cancer cells. This rapid rise in the rigidity of tumours is indeed one of the only easily detectable mechanical features of tumours that aid physicians in predicting malignancy and prognosis [185–188]. As the tumour progresses and stiffens, internally generated forces allow the tumour to disarrange adjacent healthy tissue and migrate into surrounding spaces. Accordingly, tumour progression is directly facilitated by these intratumour-generated forces and forces arising from interactions with its microenvironment [189]. These mechanical forces induce two unique types of stress on tumour cells: fluid and solid stress [190].

Generally, solid stress is created by the non-fluid components of the tumour. Initial evidence for the existence of solid stress within tumours came from the realization that blood and lymphatic vessels are mechanically compressed during tumour formation [191]. Growth-induced solid stress accumulates within tumours as the cancer cells rapidly proliferate. During this rapid reproduction process, cells grow into one another and strain the tumour microenvironment, which ultimately strains the surrounding healthy tissue [192]. In addition to intratumour-generated solid stress, externally generated solid stress accrues due to the adjacent tissue, which attempts to resist tumour expansion. In brief, solid stresses directly influence tumour progression in two manners: they first apply direct mechanical stress on cancer cells to alter genetic expression and, therefore, increase malignancy and invasion [193]. Secondly, solid stress deforms blood and lymphatic vessels to induce hypoxia [194].

As the name suggests, fluid stresses stem from forces generated by the fluid elements of the tumour. This includes shear stresses created by blood and lymphatic flow within the vessels, microvasculature (capillaries), and interstitial fluid flow [195,196]. In fact, fluid stress and solid stress are highly intertwined, as compression of blood/lymphatic vessels by solid stress greatly influences the fluid stress exerted on the surrounding epithelial tissue [190,197,198]. Vessel constriction reduces the cross-sectional area of the vessel to increase resistance to lymphatic flow, which in turn increases shear stress, interstitial fluid volume, and decreases perfusion rates [199]. This decrease of perfusion rates and flow significantly limits the ability for lymphatic vessels to remove excess fluid from the tumour, which ultimately increases interstitial fluid pressure in adjacent tumour tissue. Moreover, compression of blood and lymphatic vessels has significant negative ramifications for the effectiveness of chemo and immunotherapies [200].

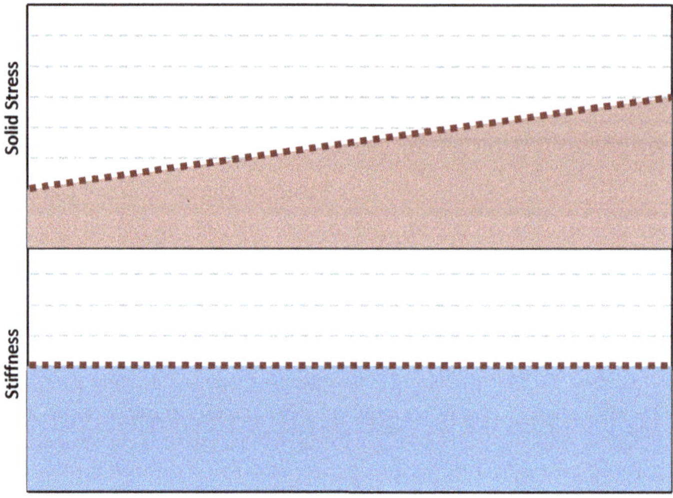

Figure 4. Solid stress and stiffness as a function of tumour diameter (adapted from Nia et al.) [201]. As rigidity of the ECM remains constant, an increase in tumour diameter is associated with increased solid stress within the tumour.

Elevated solid and fluid stress within tumours place cancer cells in an entirely unique physiological environment. Increases in tension and compression acting on the cells mechanically activates tumourigenic pathways, increases proliferation rates, and promotes collective migration [146,148]. In addition to elevated rigidity, cancer cells produce and therefore are exposed to an elevated level of force than adjacent tissues [202]. While the bulk rigidity of tumours is relatively simple to quantify, measuring solid stress within tumours has proven to be a much more elusive task. Researchers have recently begun to quantify solid stress in individual tumour cells. Nia et al. [202] has recently provided the experimental framework for creating in-situ two-dimensional mapping of solid stress.

Researchers accomplish this mapping by releasing the solid stress within tissues in a controlled method using predefined geometry that encapsulates the tumour in agarose gel and records deformation after a precise incision is made. Combination of mathematical modelling and experimental analysis has revealed a few important findings: solid stress increases linearly with tumour size while rigidity remains constant, and adjacent healthy tissue contributes significantly to the solid stress within the tumour. These findings suggest that rigidity of the tumour is decoupled from the solid stress implemented on tumour cells (Figure 4).

5.7. Quantification of Tumour Cell Mechanical Stress in vivo

Tissue development, growth, and regeneration are crucially dependent on spatiotemporal variations in microenvironment mechanics. However, most current techniques for stress quantification utilize two-dimensional analysis that can only be performed in vitro, limiting its application for determining microenvironment mechanics during tumour progression. To address this issue, researchers in the Campàs group recently developed a novel oil micro droplet technique to quantify local cell-generated mechanical stresses in tumours in a spatiotemporal manner [201,203–205].

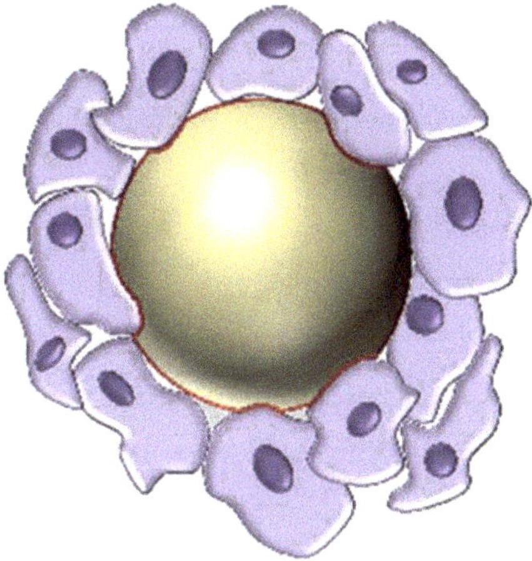

Figure 5. Schematic of oil micro droplet in vivo stress quantification described by Campàs et al. [203]. An oil droplet with calibrated surface tension is injected into living embryonic or cancerous tissue. As cells proliferate, they exert force onto the micro droplet and deform it. Deformations in curvature (red) of the oil droplet are used to calculate anisotropic stress within the tissue.

Fluorescent oil micro droplets with calibrated surface tensions are injected between tumour cells in living tissue while fluorescence microscopy is used to image localized oil droplet deformation. Provided droplet surface tension, measurements of curvature deformation along the oil droplet yield precise information regarding localized anisotropic mechanical stresses exerted by adjacent cells [201,203]. This technique of localized stress quantification shown above in Figure 5 revealed that the magnitude of cell-generated stress varies only weakly spatially during tumour progression, but increases dramatically over time [205]. Campàs et al. [203] further adapted the oil micro droplet technique to incorporate a biocompatible ferrofluid magnetic micro droplets to serve as mechanical actuators [206,207]. Using this technique, researchers are able to actively apply localized stress on tissues while observing tissue mechanical response. Indeed, this novel ferrofluid micro droplet allows for the simultaneous measurement of tissue mechanical properties and local cell-generated mechanical stress [207].

5.8. Role of ECM Mechanics in Behaviour of Myofibroblastic Cells

Many aggressive malignancies, such as pancreatic ductal adenocarcinoma (PDAC), are characterized by extensive desmoplasia and collagen deposition, which ultimately increases the rigidity of the tumour. Myofibroblast-like cells, such as pancreatic stellate cells (PSCs), are crucial mediators in the production of this fibrotic ECM [208]. When quiescent, PSCs are responsible for ECM turnover and remodelling through the production of MMPs. During wound repair, PSCs become activated by numerous soluble factors, including IL-1, IL-6, and TGF-β [141,142]. Alternatively, PSCs in the tumour desmoplasia of human pancreatic cancers behave erratically, become chronically activated, and create a microenvironment conducive to tumour growth [144]. It has been shown that pancreatic tumour cells are able to induce activation of PSCs through increased secretion of TGF-β1 and PDGF [141]. However, recent studies have indicated that PSCs may be able to sustain activation due to the mechanical properties of the microenvironment alone [209,210].

Our recent work has shown that matrix stiffness is sufficient for activation of PSCs. Upon activation, PSCs were found to mechanically sense the increased rigidity of the environment as they produce excess collagen [210,211]. This mechanosensing of tissue stiffness activates intracellular-signalling pathways within the PSC, encouraging the myofibroblast-like cell to produce excess deposits collagen.

This process of stiffness mechanosensing forms a positive-feedback loop, in which PSCs continue to secrete collagen as the matrix becomes stiffer and stiffer [210,212]. Moreover, we also found that matrix rigidity influences PSC migration, as PSCs migrate from adjacent soft tissue towards the stiff tumour microenvironment [212]. Thus, as the matrix is stiffened, distant PSCs are recruited towards the stiff tissue and become activated, further enhancing the positive feedback loop.

Inactivating the mechanosensing and remodelling capability of PSCs may be an effective therapeutic strategy. All-trans-retinoic acid (ATRA) has been shown to suppress PSC mechanosensing by downregulating MLC-2 actomyosin contractility. This leads to PSC inactivation and turns off the positive feedback loop of increased matrix rigidity and PSC activation [213]. By inactivating PSC's ability to sense the mechanical environment, ATRA reduces fibrosis and suppresses cancer invasion. Furthermore, inactivation of PSC's ECM remodelling capability prevents its ability to mechanically liberate TGFβ from Latent TGFβ Binding Protein (LTBP) [214].

These results indicate that the mechanical environment is a powerful regulator of PDAC progression via PSC activation and ECM remodelling, suggesting that reprogramming of PSCs and other resident cells may be a viable therapeutic target to alleviate tumour growth. A review of the role of mechanical rigidity and mechanical stress in tumour proliferation can be seen in Box 3.

Box 3. Rigidity and stress influence cell behaviour.

As discussed in the previous three sections, mechanical stiffness and stress of the tumour both play extremely important yet distinct roles in influencing cancer cell behaviour. Although there is significant overlap in effects of rigidity and mechanical stress, the points below highlight certain differences:

Tumour Rigidity

- ❖ Caused by elevated ECM deposition
- ❖ Mechanically activates pancreatic stellate cells (PSCs) to produce ECM
- ❖ Induces EMT in epithelial cells
- ❖ Amplifies growth-factor signalling

Tumour Stress (Solid and Fluid)

- ❖ Caused by increased cell proliferation and blood flow
- ❖ Induces hypoxia in tumours
- ❖ Augments cell proliferation
- ❖ Increases chemo resistance

5.9. Seed and Soil

For many years, the scope of study regarding cancer metastasis had primarily focused on cancer cells and their migration through the vasculature. However, as the complex interplay between the ECM and cancer progression became apparent, the perspective on cancer progression to distant organs evolved. In 1889, surgeon Stephen Paget posited that cancer metastasis is dependent on complex interactions between the migrating tumour cells (the "seed") and its microenvironment (the "soil") [208,209]. Despite doubt over the course of the twentieth century, Paget's hypothesis was strengthened in the 1970s when Isaiah Fidler's research demonstrated that successful tumour migration could only occur at certain organ sites [215].

Further research demonstrated that primary tumours have the ability to induce the formation of a microenvironment at distant organ sites that are conducive to tumour growth [209,213]. This newly formed distant microenvironment, known as the pre-metastatic niche (PMN), promotes tumour growth through a variety of methods, including increased inflammation, vascular permeability, immunosuppression, and tissue stiffness [215–217]. Studies have indicated a variety of molecules and mechanisms involved with the generation of the PMN, such as tumour-derived secreted factors (TDSFs) and tumour-derived extracellular vesicles (EVs) [218,219].

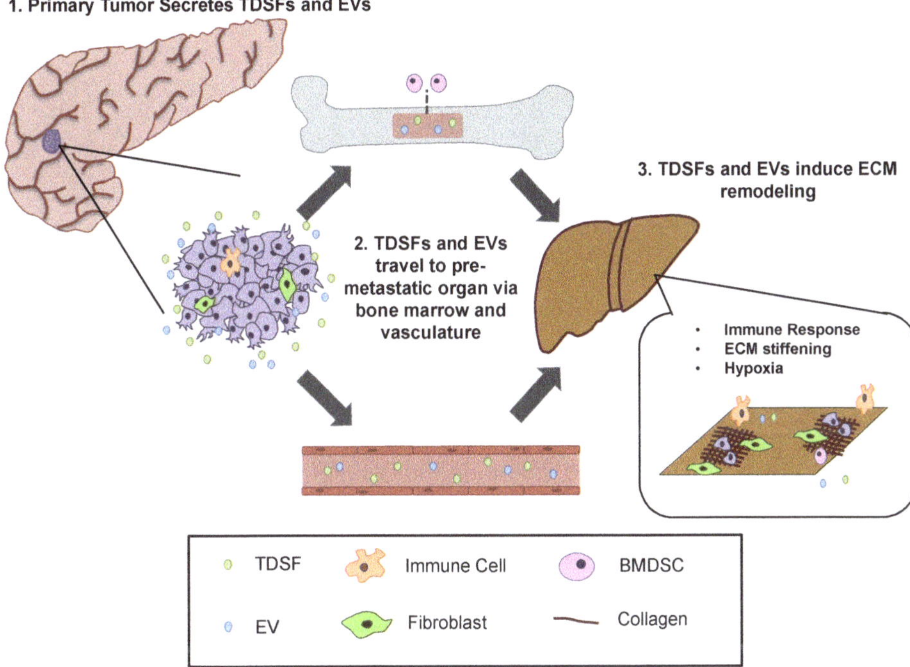

Figure 6. Illustration of pre-metastatic niche formation in the liver. (**1**) The primary tumour located in the pancreas emits tumour-derived secreted factors (TDSFs) and extracellular vesicles (EVs). (**2**) TDSFs and EVs migrate through the vasculature and bone marrow to the secondary organ. While in the bone marrow, TDSFs and EVs recruit bone marrow-derived stem cells (BDSCs), such as hematopoietic stem cells, to the secondary organ site. (**3**) TDSFs and EVs induce immune cell recruitment and ECM remodelling through LOX and cancer-associated fibroblasts at the pre-metastatic site. [220]

One primary form of tumour-derived EV is the exosome, which has a diameter ranging from 30–100 nm [218]. Exosomes containing proteins, mRNAs, and unique ECM-binding integrins are first secreted from the primary tumour and travel through the vasculature, promoting vessel leakiness in distant organ sites. EVs and TDSFs, such as TGF-β and MMP-9, then alter local resident cells and fibroblasts [219,221]. Fibroblasts altered by exosomes recruit more inflammatory cytokines, such as TGF-β, and dramatically increase collagen deposition to stiffen tissue in the PMN [222]. Just as in primary tumour tissue, elevated collagen deposition and inflammation increase interstitial stress within the tissue to induce hypoxia [220]. Finally, TDSF-mediated recruitment of non-resident cells, such as bone marrow-derived cells (BMDCs) and immune suppressor cells, ultimately attracts circulating tumour cells (CTCs) to the site (Figure 6) [223]. CTCs traveling through the vasculature are able to easily permeate into the PMN due elevated vasculature leakiness and collectively migrate through the organ toward the stiffer tissue, in a process known as durotaxis [224,225].

Interestingly, the notion of the PMN proposes that critical changes to the ECM, typically associated with primary tumour formation, can occur prior to the arrival of tumour cells [226,227]. The Seed and Soil process closely mirrors the process of primary tumour ECM remodelling, with one critical difference: it occurs in the absence of cancer cells. Furthermore, it specifies that TDSF and EV-mediated ECM remodelling is crucial in facilitating cancer metastasis. However, many aspects of the mechanism regarding PNM formation remain elusive.

Although exosomes have been known to play a crucial role in PMN development, why cancer cells only metastasize to specific organs, a process known as organotropism, has remained unclear. Recent research in the Lyden laboratory has demonstrated that exosomes secreted by primary tumour cells have specific integrin expression patterns that dictate organotropism. Indeed, exosome proteomics revealed that exosomal integrins $\alpha 6\beta 4$ and $\alpha 6\beta 1$ are closely linked with lung metastasis, while exosomal integrin $\alpha v\beta 5$ is associated with liver metastasis [220]. Targeting of these specific integrins decreased exosomal uptake by resident cells and decreased lung and liver metastasis, respectively. These results suggest that exosomal integrin expression could be used to predict organ-specific metastatic sites. Moreover, there is an implication that cancer therapies may be most beneficial if tailored to distinct metastatic sites (lung, liver, etc.) and each stages of cancer metastasis: pre-metastatic and post-metastatic [215].

6. Challenges and Future Perspectives

In this review we have discussed the complex and nuanced role of the ECM in tissue development and cancer progression. Over the past 20 years, studies have revealed the importance of the ECM in regulating crucial physiological processes such as stem cell lineage specification, cell migration, and proliferation [138–140]. Accordingly, perspectives have shifted to address cancer not only as a disease of uncontrolled cell proliferation, but also of dysregulation of the microenvironment. The ostensibly static ECM actively undergoes dynamic remodelling during all stages of cancer progression in a complex interplay between cancer cells, resident cells, and non-cellular components. Advances in understanding of the role of ECM in cancer progression have provided hope and revealed promising therapeutic targets for mitigating cancer's ability to metastasize. However, a lack of success in targeting broad ranges of proteins, such as MMPs and collagen, reveals the temporal sensitivity and specificity needed to effectively limit the spread of the tumour cells.

As neoplastic cells proliferate rapidly in tumours, they experience an increase in mechanical stress, which mechanically activates tumourigenic pathways, increases migration, and induces hypoxia. An essential region of future cancer research will be to ascertain the mechanism by which increased mechanical stress in tumours relates to malignant behaviour and angiogenesis. These signalling pathways relating to exogenous mechanical stress and malignant behaviour serve as an auspicious therapeutic target to resist cancer progression. Moreover, understanding the relationship between increased solid stress and angiogenesis mechanisms would elucidate possible advancements for drug delivery. Recent developments in in vivo cell stress quantification techniques may provide novel insights into the relationship between ECM-generated stress and hypoxia [213–216].

Funding: This work was supported by the ERC grant 282051 "ForceRegulation".

Conflicts of Interest: The authors declare no conflict of interest.

References

1. Frantz, C.; Stewart, K.M.; Weaver, V.M. The extracellular matrix at a glance. *J. Cell Sci.* **2010**, *123*, 4195–4200. [CrossRef] [PubMed]

2. Bonnans, C.; Chou, J.; Werb, Z. Remodelling the extracellular matrix in development and disease. *Nat. Rev. Mol. Cell Biol.* **2014**, *15*, 786–801. [CrossRef] [PubMed]

3. Kai, F.; Laklai, H.; Weaver, V.M. Force matters: Biomechanical regulation of cell invasion and migration in disease. *Trends Cell Biol.* **2016**, *26*, 486–497. [CrossRef] [PubMed]

4. Kim, S.H.; Turnbull, J.; Guimond, S. Extracellular matrix and cell signalling: The dynamic cooperation of integrin, proteoglycan and growth factor receptor. *J. Endocrinol.* **2011**, *209*, 139–151. [CrossRef] [PubMed]

5. Jablonska-Trypuc, A.; Matejczyk, M.; Rosochacki, S. Matrix metalloproteinases (mmps), the main extracellular matrix (ecm) enzymes in collagen degradation, as a target for anticancer drugs. *J. Enzyme Inhib. Med. Chem.* **2016**, *31*, 177–183. [CrossRef] [PubMed]

6.   Lu, P.; Takai, K.; Weaver, V.M.; Werb, Z. Extracellular matrix degradation and remodeling in development and disease. *Cold Spring Harb. Perspect. Biol.* **2011**, *3*, 1–24. [CrossRef] [PubMed]

7.   Gumbiner, B.M. Cell adhesion: The molecular basis of tissue architecture and morphogenesis. *Cell* **1996**, *84*, 345–357. [CrossRef]

8.   Chen, K.D.; Li, Y.S.; Kim, M.; Li, S.; Yuan, S.; Chien, S.; Shyy, J.Y. Mechanotransduction in response to shear stress. Roles of receptor tyrosine kinases, integrins, and shc. *J. Biol. Chem.* **1999**, *274*, 18393–18400. [CrossRef] [PubMed]

9.   Katsumi, A.; Orr, A.W.; Tzima, E.; Schwartz, M.A. Integrins in mechanotransduction. *J. Biol. Chem.* **2004**, *279*, 12001–12004. [CrossRef] [PubMed]

10.   Engler, A.J.; Sen, S.; Sweeney, H.L.; Discher, D.E. Matrix elasticity directs stem cell lineage specification. *Cell* **2006**, *126*, 677–689. [CrossRef] [PubMed]

11.   Cox, T.R.; Erler, J.T. Remodeling and homeostasis of the extracellular matrix: Implications for fibrotic diseases and cancer. *Dis. Model. Mech.* **2011**, *4*, 165–178. [CrossRef] [PubMed]

12.   Myllyharju, J.; Kivirikko, K.I. Collagens, modifying enzymes and their mutations in humans, flies and worms. *Trends Genet.* **2004**, *20*, 33–43. [CrossRef] [PubMed]

13.   Ricard-Blum, S. The collagen family. *Cold Spring Harb. Perspect. Biol.* **2011**, *3*, 1–19. [CrossRef] [PubMed]

14.   Ricard-Blum, S.; Ruggiero, F. The collagen superfamily: From the extracellular matrix to the cell membrane. *Pathol. Biol.* **2005**, *53*, 430–442. [CrossRef] [PubMed]

15.   Mouw, J.K.; Ou, G.; Weaver, V.M. Extracellular matrix assembly: A multiscale deconstruction. *Nat. Rev. Mol. Cell Biol.* **2014**, *15*, 771–785. [CrossRef] [PubMed]

16.   Shoulders, M.D.; Raines, R.T. Collagen structure and stability. *Annu. Rev. Biochem.* **2009**, *78*, 929–958. [CrossRef] [PubMed]

17.   Bella, J.; Eaton, M.; Brodsky, B.; Berman, H. Crystal and molecular structure of a collagen-like peptide at 1.9 a resolution. *Science* **1994**, *266*, 75–81. [CrossRef] [PubMed]

18.   Persikov, A.V.; Ramshaw, J.A.M.; Kirkpatrick, A.; Brodsky, B. Electrostatic interactions involving lysine make major contributions to collagen triple-helix stability. *Biochemistry* **2005**, *44*, 1414–1422. [CrossRef] [PubMed]

19.   Muiznieks, L.D.; Keeley, F.W. Molecular assembly and mechanical properties of the extracellular matrix: A fibrous protein perspective. *Biochim. Biophys. Acta* **2013**, *1832*, 866–875. [CrossRef] [PubMed]

20.   Myllyharju, J. Intracellular post-translational modifications of collagens. In *Collagen: Primer in Structure, Processing and Assembly*; Brinckmann, J., Notbohm, H., Müller, P.K., Eds.; Springer-Verlag Berlin Heidelberg: Heidelberg, Germany, 2005; pp. 115–147.

21.   Birk, D.E.; Zycband, E.I.; Winkelmann, D.A.; Trelstad, R.L. Collagen fibrillogenesis in situ: Fibril segments are intermediates in matrix assembly. *Proc. Natl. Acad. Sci. USA* **1989**, *86*, 4549–4553. [CrossRef] [PubMed]

22.   Canty, E.G.; Lu, Y.; Meadows, R.S.; Shaw, M.K.; Holmes, D.F.; Kadler, K.E. Coalignment of plasma membrane channels and protrusions (fibripositors) specifies the parallelism of tendon. *J. Cell Biol.* **2004**, *165*, 553–563. [CrossRef] [PubMed]

23.   Kalson, N.S.; Starborg, T.; Lu, Y.; Mironov, A.; Humphries, S.M.; Holmes, D.F.; Kadler, K.E. Nonmuscle myosin ii powered transport of newly formed collagen fibrils at the plasma membrane. *Proc. Natl. Acad. Sci. USA* **2013**, *110*, E4743–E4752. [CrossRef] [PubMed]

24.   Starborg, T.; Kalson, N.S.; Lu, Y.; Mironov, A.; Cootes, T.F.; Holmes, D.F.; Kadler, K.E. Using transmission electron microscopy and 3view to determine collagen fibril size and three-dimensional organization. *Nat. Protoc.* **2013**, *8*, 1433–1448. [CrossRef] [PubMed]

25.   Hulmes, D.J. Building collagen molecules, fibrils, and suprafibrillar structures. *J. Struct. Biol.* **2002**, *137*, 2–10. [CrossRef] [PubMed]

26.   Kadler, K.E.; Holmes, D.F.; Trotter, J.A.; Chapman, J.A. Collagen fibril formation. *Biochem. J.* **1996**, *316*, 1–11. [CrossRef] [PubMed]

27.   Hulmes, D.J.S. Collagen diversity, synthesis, and assembly. In *Collagen: Structure and Mechanics*, 1st ed.; Fratzl, P., Ed.; Springer: New York, NY, USA, 2008; pp. 15–47.

28.   Wenstrup, R.J.; Florer, J.B.; Brunskill, E.W.; Bell, S.M.; Chervoneva, I.; Birk, D.E. Type v collagen controls the initiation of collagen fibril assembly. *J. Biol. Chem.* **2004**, *279*, 53331–53337. [CrossRef] [PubMed]

29.   Bruckner, P. Suprastructures of extracellular matrices: Paradigms of functions controlled by aggregates rather than molecules. *Cell Tissue Res.* **2009**, *339*, 7–18. [CrossRef] [PubMed]

30. Ameye, L.; Young, M.F. Mice deficient in small leucine-rich proteoglycans: Novel in vivo models for osteoporosis, osteoarthritis, ehlers-danlos syndrome, muscular dystrophy, and corneal diseases. *Glycobiology* **2002**, *12*, 107R–116R. [CrossRef] [PubMed]

31. Molnar, J.; Fong, K.S.K.; He, Q.P.; Hayashi, K.; Kim, Y.; Fong, S.F.T.; Fogelgren, B.; Molnarne Szauter, K.; Mink, M.; Csiszar, K. Structural and functional diversity of lysyl oxidase and the lox-like proteins. *Biochim. Biophys. Acta* **2003**, *1647*, 220–224. [CrossRef]

32. Fratzl, P.; Misof, K.; Zizak, I.; Rapp, G.; Amenitsch, H.; Bernstorff, S. Fibrillar structure and mechanical properties of collagen. *J. Struct. Biol.* **1997**, *122*, 119–122. [CrossRef] [PubMed]

33. Hashmi, S.; Marinkovich, M.P. Molecular organization of the basement membrane zone. *Clin. Dermatol.* **2011**, *29*, 398–411. [CrossRef] [PubMed]

34. Hohenester, E.; Yurchenco, P.D. Laminins in basement membrane assembly. *Cell Adh. Migr.* **2013**, *7*, 56–63. [CrossRef] [PubMed]

35. Iozzo, R.V.; Schaefer, L. Proteoglycan form and function: A comprehensive nomenclature of proteoglycans. *Matrix Biol.* **2015**, *42*, 11–55. [CrossRef] [PubMed]

36. Leonova, E.I.; Galzitskaya, O.V. Structure and functions of syndecans in vertebrates. *Biochem. (Mosc.)* **2013**, *78*, 1071–1085. [CrossRef] [PubMed]

37. Miaczynska, M. Effects of membrane trafficking on signaling by receptor tyrosine kinases. *Cold Spring Harb. Perspect. Biol.* **2013**, *5*, 1–20. [CrossRef] [PubMed]

38. Christianson, H.C.; Svensson, K.J.; van Kuppevelt, T.H.; Li, J.P.; Belting, M. Cancer cell exosomes depend on cell-surface heparan sulfate proteoglycans for their internalization and functional activity. *Proc. Natl. Acad. Sci. USA* **2013**, *110*, 17380–17385. [CrossRef] [PubMed]

39. Fares, J.; Kashyap, R.; Zimmermann, P. Syntenin: Key player in cancer exosome biogenesis and uptake? *CellCell Adh. Migr.* **2017**, *11*, 124–126. [CrossRef] [PubMed]

40. Knox, S.M.; Whitelock, J.M. Perlecan: How does one molecule do so many things? *Cell Mol. Life Sci.* **2006**, *63*, 2435–2445. [CrossRef] [PubMed]

41. Farach-Carson, M.C.; Carson, D.D. Perlecan—A multifunctional extracellular proteoglycan scaffold. *Glycobiology* **2007**, *17*, 897–905. [CrossRef] [PubMed]

42. Wu, Y.J.; La Pierre, D.P.; Wu, J.; Yee, A.J.; Yang, B.B. The interaction of versican with its binding partners. *Cell Res.* **2005**, *15*, 483–494. [CrossRef] [PubMed]

43. Yamagata, M.; Yamada, K.M.; Yoneda, M.; Suzuki, S.; Kimata, K. Chondroitin sulfate proteoglycan (pg-m-like proteoglycan) is involved in the binding of hyaluronic acid to cellular fibronectin. *J. Biol. Chem.* **1986**, *261*, 13526–13535. [PubMed]

44. Iozzo, R.V. The biology of the small leucine-rich proteoglycans. *J. Biol. Chem.* **1999**, *274*, 18843–18846. [CrossRef] [PubMed]

45. Iozzo, R.V.; Karamanos, N. Proteoglycans in health and disease: Emerging concepts and future directions. *FEBS J.* **2010**, *277*, 3863. [CrossRef] [PubMed]

46. Kalamajski, S.; Oldberg, A. The role of small leucine-rich proteoglycans in collagen fibrillogenesis. *Matrix Biol.* **2010**, *29*, 248–253. [CrossRef] [PubMed]

47. Raab-Westphal, S.; Marshall, J.F.; Goodman, S.L. Integrins as therapeutic targets: Successes and cancers. *Cancers* **2017**, *9*, 1–28. [CrossRef] [PubMed]

48. Katsumi, A.; Naoe, T.; Matsushita, T.; Kaibuchi, K.; Schwartz, M.A. Integrin activation and matrix binding mediate cellular responses to mechanical stretch. *J. Biol. Chem.* **2005**, *280*, 16546–16549. [CrossRef] [PubMed]

49. O'Toole, T.; Katagiri, Y.; Faull, R.; Peter, K.; Tamura, R.; Quaranta, V.; Loftus, J.; Shattil, S.; Ginsberg, M. Integrin cytoplasmic domains mediate inside-out signal transduction. *J. Cell Biol.* **1994**, *124*, 1047–1059. [CrossRef] [PubMed]

50. Chiquet, M.; Renedo, A.S.; Huber, F.; Flück, M. How do fibroblasts translate mechanical signals into changes in extracellular matrix production? *Matrix Biol.* **2003**, *22*, 73–80. [CrossRef]

51. Orr, A.W.; Helmke, B.P.; Blackman, B.R.; Schwartz, M.A. Mechanisms of mechanotransduction. *Dev. Cell* **2006**, *10*, 11–20. [CrossRef] [PubMed]

52. Elfenbein, A.; Simons, M. Syndecan-4 signaling at a glance. *J. Cell Sci.* **2013**, *126*, 3799–3804. [CrossRef] [PubMed]

53. Banerji, S.; Wright, A.J.; Noble, M.; Mahoney, D.J.; Campbell, I.D.; Day, A.J.; Jackson, D.G. Structures of the cd44-hyaluronan complex provide insight into a fundamental carbohydrate-protein interaction. *Nat. Struct. Mol. Biol.* **2007**, *14*, 234–239. [CrossRef] [PubMed]

54. Misra, S.; Hascall, V.C.; Markwald, R.R.; Ghatak, S. Interactions between hyaluronan and its receptors (cd44, rhamm) regulate the activities of inflammation and cancer. *Front. Immunol.* **2015**, *6*, 1–31. [CrossRef] [PubMed]

55. Nelson, J.; McFerran, N.V.; Pivato, G.; Chambers, E.; Doherty, C.; Steele, D.; Timson, D.J. The 67 kda laminin receptor: Structure, function and role in disease. *Biosci. Rep.* **2008**, *28*, 33–48. [CrossRef] [PubMed]

56. Shrivastava, A.; Radziejewski, C.; Campbell, E.; Kovac, L.; McGlynn, M.; Ryan, T.E.; Davis, S.; Goldfarb, M.P.; Glass, D.J.; Lemke, G.; et al. An orphan receptor tyrosine kinase family whose members serve as nonintegrin collagen receptors. *Mol. Cell* **1997**, *1*, 25–34. [CrossRef]

57. Vogel, W.; Gish, G.D.; Alves, F.; Pawson, T. The discoidin domain receptor tyrosine kinases are activated by collagen. *Mol. Cell* **1997**, *1*, 13–23. [CrossRef]

58. Harburger, D.S.; Calderwood, D.A. Integrin signalling at a glance. *J. Cell Sci.* **2009**, *122*, 159–163. [CrossRef] [PubMed]

59. Zimmerman, P.; Tomatis, D.; Rosas, M.; Grootjans, J.; Leenaerts, I.; Degeest, G.; Reekmans, G.; Coomans, C.; David, G. Characterization of syntenin, a syndecan-binding pdz protein, as a component of cell adhesion sites and microfilaments. *Mol. Biol. Cell* **2001**, *13*, 339–350. [CrossRef] [PubMed]

60. Greene, D.K.; Tumova, S.; Couchman, J.R.; Woods, A. Syndecan-4 associates with alpha-actinin. *J. Biol. Chem.* **2003**, *278*, 7617–7623. [CrossRef] [PubMed]

61. Okina, E.; Grossi, A.; Gopal, S.; Multhaupt, H.A.; Couchman, J.R. Alpha-actinin interactions with syndecan-4 are integral to fibroblast-matrix adhesion and regulate cytoskeletal architecture. *Int. J. Biochem. Cell Biol.* **2012**, *44*, 2161–2174. [CrossRef] [PubMed]

62. 62. Bellin, R.M.; Kubicek, J.D.; Frigault, M.J.; Kamien, A.J.; Steward, R.L., Jr.; Barnes, H.M.; DiGlacomo, M.B.; Duncan, L.J.; Edgerly, C.K.; Morse, E.M.; et al. Defining the role of syndecan-4 in mechanotransduction using surface-modification approaches. *Proc. Natl. Acad. Sci. USA* **2009**, *106*, 22102–22107. [CrossRef] [PubMed]

63. Domogatskaya, A.; Rodin, S.; Tryggvason, K. Functional diversity of laminins. *Annu. Rev. Cell Dev. Biol.* **2012**, *28*, 523–553. [CrossRef] [PubMed]

64. Beck, K.; Hunter, I.; Engel, J. Structure and function of laminin: Anatomy of a multidomain glycoprotein. *FASEB J.* **1990**, *4*, 148–160. [CrossRef] [PubMed]

65. Engel, J.; Odermatt, E.; Engel, A.; Madri, J.A.; Furthmayr, H.; Rohde, H.; Timpl, R. Shapes, domain organizations and flexibility of laminin and fibronectin, two multifunctional proteins of the extracellular matrix. *J. Mol. Biol.* **1981**, *150*, 97–120. [CrossRef]

66. Berrier, A.L.; Yamada, K.M. Cell-matrix adhesion. *J. Cell Physiol.* **2007**, *213*, 565–573. [CrossRef] [PubMed]

67. Poschl, E.; Schlotzer-Schrehardt, U.; Brachvogel, B.; Saito, K.; Ninomiya, Y.; Mayer, U. Collagen iv is essential for basement membrane stability but dispensable for initiation of its assembly during early development. *Development* **2004**, *131*, 1619–1628. [CrossRef] [PubMed]

68. Behrens, D.T.; Villone, D.; Koch, M.; Brunner, G.; Sorokin, L.; Robenek, H.; Bruckner-Tuderman, L.; Bruckner, P.; Hansen, U. The epidermal basement membrane is a composite of separate laminin- or collagen iv-containing networks connected by aggregated perlecan, but not by nidogens. *J. Biol. Chem.* **2012**, *287*, 18700–18709. [CrossRef] [PubMed]

69. Oberbaumer, I.; Wiedemann, H.; Timpl, R.; Kuhn, K. Shape and assembly of type iv procollagen obtained from cell culture. *EMBO J.* **1982**, *1*, 805–810. [CrossRef] [PubMed]

70. Tsilibary, E.C.; Koliakos, G.G.; Charonis, A.S.; Vogel, A.M.; Raeger, L.A.; Furcht, L.T. Heparin type iv collagen interactions: Equilibrium binding and inhibition of type iv collagen self-assembly. *J. Biol. Chem.* **1988**, *263*, 19112–19116. [PubMed]

71. Singh, P.; Carraher, C.; Schwarzbauer, J.E. Assembly of fibronectin extracellular matrix. *Annu. Rev. Cell Dev. Biol.* **2010**, *26*, 397–419. [CrossRef] [PubMed]

72. Schwarzbauer, J.E.; DeSimone, D.W. Fibronectins, their fibrillogenesis, and in vivo functions. *Cold Spring Harb. Perspect. Biol.* **2011**, *3*, 1–20. [CrossRef] [PubMed]

73. Schwarzbauer, J.E. Identification of the fibronectin sequences required for assembly of a fibrillar matrix. *J. Cell Biol.* **1991**, *113*, 1463–1473. [CrossRef] [PubMed]

74. To, W.S.; Midwood, K.S. Plasma and cellular fibronectin: Distinct and independent functions during tissue repair. *Fibrogenesis Tissue Repair* **2011**, *4*, 1–17. [CrossRef] [PubMed]
75. Fogetry, F.J.; Akiyama, S.K.; Yamada, K.M.; Mosher, D.F. Inhibition of binding of fibronectin to matrix assembly sites by anti-integrin (α5β1) antibodies. *J. Cell Biol.* **1990**, *111*, 699–708.
76. McDonald, J.A.; Quade, B.J.; Broekelmann, T.J.; LaChance, R.; Forsman, K.; Hasegawa, E.; Akiyama, S. Fibronectin's cell-adhesive domain and an amino-terminal matrix assembly domain participate in its assembly into fibroblast pericellular matrix. *J. Biol. Chem.* **1987**, *262*, 2957–2967. [PubMed]
77. Chung, C.Y.; Erickson, H.P. Glycosaminoglycans modulate fibronectin matrix assembly and are essential for matrix incorporation of tenascin-c. *J. Cell Biol.* **1997**, *110*, 1413–1419.
78. Galante, L.L.; Schwarzbauer, J.E. Requirements for sulfate transport and the diastrophic dysplasia sulfate transporter in fibronectin matrix assembly. *J. Cell Biol.* **2007**, *179*, 999–1009. [CrossRef] [PubMed]
79. Morla, A.; Ruoslahti, E. A fibronectin self-assembly site involved in fibronectin matrix assembly: Reconstruction in a synthetic peptide. *J. Cell Biol.* **1992**, *118*, 421–429. [CrossRef] [PubMed]
80. Klass, C.M.; Couchman, J.R.; Woods, A. Control of extracellular matrix assembly by syndecan-2 proteoglycan. *J. Cell Sci.* **2000**, *113*, 493–506. [PubMed]
81. Woods, A. Syndecans: Transmembrane modulators of adhesion and matrix assembly. *J. Clin. Invest.* **2001**, *107*, 935–941. [CrossRef] [PubMed]
82. Dallas, S.L.; Sivakumar, P.; Jones, C.J.; Chen, Q.; Peters, D.M.; Mosher, D.F.; Humphries, M.J.; Kielty, C.M. Fibronectin regulates latent transforming growth factor-beta (tgf beta) by controlling matrix assembly of latent tgf beta-binding protein-1. *J. Biol. Chem.* **2005**, *280*, 18871–18880. [CrossRef] [PubMed]
83. Sottile, J.; Hocking, D.C. Fibronectin polymerization regulates the composition and stability of extracellular matrix fibrils and cell-matrix adhesions. *Mol. Biol. Cell* **2002**, *13*, 3546–3559. [CrossRef] [PubMed]
84. Dzamba, B.J.; Wu, H.; Jaenisch, R.; Peters, D.M. Fibronectin binding site in type i collagen regulates fibronectin fibril formation. *J. Cell Biol.* **1993**, *121*, 1165–1172. [CrossRef] [PubMed]
85. Colombi, M.; Zoppi, N.; De Petro, G.; Marchina, E.; Gardella, R.; Tavian, D.; Ferraboli, S.; Barlati, S. Matrix assembly induction and cell migration and invasion inhibition by a 13-amino acid fibronectin peptide. *J. Biol. Chem.* **2003**, *278*, 14346–14355. [CrossRef] [PubMed]
86. Brown, N.H. Extracellular matrix in development: Insights from mechanisms conserved between invertebrates and vertebrates. *Cold Spring Harb. Perspect Biol.* **2011**, *3*, 1–13. [CrossRef] [PubMed]
87. Rozario, T.; DeSimone, D.W. The extracellular matrix in development and morphogenesis: A dynamic view. *Dev. Biol.* **2010**, *341*, 126–140. [CrossRef] [PubMed]
88. Entchev, E.V.; Gonzalez-Gaitan, M.A. Morphogen gradient formation and vesicular trafficking. *Traffic* **2002**, *3*, 98–109. [CrossRef] [PubMed]
89. Uhler, C.; Shivashankar, G.V. Regulation of genome organization and gene expression by nuclear mechanotransduction. *Nat. Rev. Mol. Cell Biol.* **2017**, *18*, 717–727. [CrossRef] [PubMed]
90. Lu, P.; Weaver, V.M.; Werb, Z. The extracellular matrix: A dynamic niche in cancer progression. *J. Cell Biol.* **2012**, *196*, 395–406. [CrossRef] [PubMed]
91. Knecht, A.K.; Bronner-Fraser, M. Induction of the neural crest: A multigene process. *Nat. Rev. Genet.* **2002**, *3*, 453–461. [CrossRef] [PubMed]
92. Sharma, P.; Ng, C.; Jana, A.; Padhi, A.; Szymanski, P.; Lee, J.S.H.; Behkam, B.; Nain, A.S. Aligned fibers direct collective cell migration to engineer closing and nonclosing wound gaps. *Mol. Biol. Cell* **2017**, *28*, 2579–2588. [CrossRef] [PubMed]
93. Motalleb, R.; Berns, E.J.; Patel, P.; Gold, J.; Stupp, S.I.; Georg Kuhn, H. In vivo migration of endogenous brain progenitor cells guided by an injectable peptide amphiphile biomaterial. *J. Tissue Eng. Regen. Med.* **2018**, *12*, e2123–e2133. [CrossRef] [PubMed]
94. Palecek, S.P.; Loftus, J.C.; Ginsberg, M.H.; Lauffenburger, D.A.; Horwitz, A.F. Integrin-ligand binding properties govern cell migration speed through cell substratum adhesiveness. *Nature* **1997**, *385*, 537–540. [CrossRef] [PubMed]
95. Hartman, C.D.; Isenberg, B.C.; Chua, S.G.; Wong, J.Y. Extracellular matrix type modulates cell migration on mechanical gradients. *Exp. Cell Res.* **2017**, *359*, 361–366. [CrossRef] [PubMed]
96. Wang, T.; Hamilla, S.; Cam, M.; Aranda-Espinoza, H.; Mili, S. Extracellular matrix stiffness and cell contractility control rna localization to promote cell migration. *Nat. Commun.* **2017**, *8*, 1–16. [CrossRef] [PubMed]

97. Fata, J.E.; Werb, Z.; Bissell, M.J. Regulation of mammary gland branching morphogenesis by the extracellular matrix and its remodeling enzymes. *Breast Cancer Res.* **2004**, *6*, 1–11. [CrossRef] [PubMed]

98. Silberstein, G.; Strickland, P.; Coleman, S.; Daniel, C.W. Epithelium-dependent extracellular matrix synthesis in transforming-growth-factor β1-growth-inhibited mouse mammary gland. *J. Cell Biol.* **1990**, *110*, 2209–2219. [CrossRef] [PubMed]

99. Hinck, L.; Silberstein, G.B. Key stages in mammary gland development: The mammary end bud as a motile organ. *Breast Cancer Res.* **2005**, *7*, 245–251. [CrossRef] [PubMed]

100. Alford, D.; Baeckstrom, D.; Geyp, M.; Pitha, P.; Taylor-Papadimitriou, J. Integrin-matrix interactions affect the form of the structures developing from human mammary epithelial cells in collagen or fibrin gel. *J. Cell Sci.* **1998**, *111*, 521–532. [PubMed]

101. Vogel, W.F.; Aszodi, A.; Alves, F.; Pawson, T. Discoidin domain receptor 1 tyrosine kinase has an essential role in mammary gland development. *Mol. Cell Biol.* **2001**, *21*, 2906–2917. [CrossRef] [PubMed]

102. Nelson, C.M.; VanDujin, M.M.; Inman, J.L.; Fletcher, D.A.; Bissell, M.J. Tissue geometry determines sites of mammary branching morphogenesis in organotypic cultures. *Science* **2006**, *314*, 298–301. [CrossRef] [PubMed]

103. Gjorevski, N.; Nelson, C.M. Endogenous patterns of mechanical stress are required for branching morphogenesis. *Integr. Biol.* **2010**, *2*, 424–434. [CrossRef] [PubMed]

104. Robinson, B.K.; Cortes, E.; Rice, A.J.; Sarper, M.; Del Rio Hernandez, A. Quantitative analysis of 3d extracellular matrix remodelling by pancreatic stellate cells. *Biol Open* **2016**, *5*, 875–882. [CrossRef] [PubMed]

105. Ortega, N. New functional roles for non-collagenous domains of basement membrane collagens. *J. Cell Sci.* **2002**, *115*, 4201–4214. [CrossRef] [PubMed]

106. Sternlicht, M.D.; Kouros-Mehr, H.; Lu, P.; Werb, Z. Hormonal and local control of mammary branching morphogenesis. *Differentiation* **2006**, *74*, 365–381. [CrossRef] [PubMed]

107. Streuli, C.H.; Schmidhauser, C.; Bailey, N.; Yurchenco, P.D.; Skubitz, A.P.N.; Roskelley, C.; Bissell, M.J. Laminin mediates tissue-specific gene expression in mammary epithelia. *J. Cell Biol.* **1995**, *129*, 591–603. [CrossRef] [PubMed]

108. Panciera, T.; Azzolin, L.; Cordenonsi, M.; Piccolo, S. Mechanobiology of yap and taz in physiology and disease. *Nat. Rev. Mol. Cell Biol.* **2017**, *18*, 758–770. [CrossRef] [PubMed]

109. Dupont, S.; Morsut, L.; Aragona, M.; Enzo, E.; Giulitti, S.; Cordenonsi, M.; Zanconato, F.; Le Digabel, J.; Forcato, M.; Bicciato, S.; et al. Role of yap/taz in mechanotransduction. *Nature* **2011**, *474*, 179–183. [CrossRef]

110. Halder, G.; Dupont, S.; Piccolo, S. Transduction of mechanical and cytoskeletal cues by yap and taz. *Nat. Rev. Mol. Cell Biol.* **2012**, *13*, 591–600. [CrossRef]

111. Schroeder, M.C.; Halder, G. Regulation of the hippo pathway by cell architecture and mechanical signals. *Semin. Cell Dev. Biol.* **2012**, *23*, 803–811. [CrossRef] [PubMed]

112. Lee, H.J.; Diaz, M.F.; Price, K.M.; Ozuna, J.A.; Zhang, S.; Sevick-Muraca, E.M.; Hagan, J.P.; Wenzel, P.L. Fluid shear stress activates yap1 to promote cancer cell motility. *Nat. Commun.* **2017**, *8*, 1–14. [CrossRef] [PubMed]

113. Nakajima, H.; Yamamoto, K.; Agarwala, S.; Terai, K.; Fukui, H.; Fukuhara, S.; Ando, K.; Miyazaki, T.; Yokota, Y.; Schmelzer, E.; et al. Flow-dependent endothelial yap regulation contributes to vessel maintenance. *Dev. Cell* **2017**, *40*, 523–536. [CrossRef] [PubMed]

114. Wang, L.; Luo, J.Y.; Li, B.; Tian, X.Y.; Chen, L.J.; Huang, Y.; Liu, J.; Deng, D.; Lau, C.W.; Wan, S.; et al. Integrin-yap/taz-jnk cascade mediates atheroprotective effect of unidirectional shear flow. *Nature* **2016**, *540*, 579–582. [CrossRef] [PubMed]

115. Wang, K.C.; Yeh, Y.T.; Nguyen, P.; Limqueco, E.; Lopez, J.; Thorossian, S.; Guan, K.L.; Li, Y.J.; Chien, S. Flow-dependent yap/taz activities regulate endothelial phenotypes and atherosclerosis. *Proc. Natl. Acad. Sci. USA* **2016**, *113*, 11525–11530. [CrossRef] [PubMed]

116. Starr, D.A.; Fridolfsson, H.N. Interactions between nuclei and the cytoskeleton are mediated by sun-kash nuclear envelope bridges. *Annu. Rev. Cell Dev. Biol.* **2010**, *26*, 421–444. [CrossRef] [PubMed]

117. Lombardi, M.L.; Jaalouk, D.E.; Shanahan, C.M.; Burke, B.; Roux, K.J.; Lammerding, J. The interaction between nesprins and sun proteins at the nuclear envelope is critical for force transmission between the nucleus and cytoskeleton. *J. Biol. Chem.* **2011**, *286*, 26743–26753. [CrossRef] [PubMed]

118. Attwood, S.J.; Cortes, E.; Haining, A.W.; Robinson, B.; Li, D.; Gautrot, J.; Del Rio Hernandez, A. Adhesive ligand tether length affects the size and length of focal adhesions and influences cell spreading and attachment. *Sci. Rep.* **2016**, *6*, 34334. [CrossRef] [PubMed]

119. Gattazzo, F.; Urciuolo, A.; Bonaldo, P. Extracellular matrix: A dynamic microenvironment for stem cell niche. *Biochim. Biophys. Acta* **2014**, *1840*, 2506–2519. [CrossRef] [PubMed]

120. Rohani, M.G.; Parks, W.C. Matrix remodeling by mmps during wound repair. *Matrix Biol.* **2015**, *44–46*, 113–121. [CrossRef] [PubMed]

121. Chen, Z.; Tzima, E. Pecam-1 is necessary for flow-induced vascular remodeling. *Arterioscler. Thromb. Vasc. Biol.* **2009**, *29*, 1067–1073. [CrossRef] [PubMed]

122. Collins, C.; Osborne, L.D.; Guilluy, C.; Chen, Z.; O'Brien, E.T., 3rd; Reader, J.S.; Burridge, K.; Superfine, R.; Tzima, E. Haemodynamic and extracellular matrix cues regulate the mechanical phenotype and stiffness of aortic endothelial cells. *Nat. Commun.* **2014**, *5*, 1–12. [CrossRef] [PubMed]

123. Weaver, V.M.; Petersen, O.W.; Wang, F.; Larabell, C.A.; Briand, P.; Damsky, C.; Bissell, M.J. Reversion of the malignant phenotype of human breast cells in three-dimensional culture and in vivo by integrin blocking antibodies. *J. Cell Biol.* **1997**, *137*, 231–245. [CrossRef] [PubMed]

124. Burgess, J.K.; Mauad, T.; Tijin, G.; Karlsson, J.C.; Westergren-Thorsson, G. The extracellular matrix - the under-recognized element in lung disease? *J. Pathol.* **2016**, *240*, 397–409. [CrossRef] [PubMed]

125. Iredale, J.P.; Thompson, A.; Henderson, N.C. Extracellular matrix degradation in liver fibrosis: Biochemistry and regulation. *Biochim. Biophys. Acta* **2013**, *1832*, 876–883. [CrossRef] [PubMed]

126. Kolb, M.; Gauldie, J.; Bellaye, P.S. Editorial: Extracellular matrix: The common thread of disease progression in fibrosis? *Arthritis Rheumatol.* **2016**, *68*, 1053–1056. [PubMed]

127. Friedl, P.; Wolf, K. Tube travel: The role of proteases in individual and collective cancer cell invasion. *Cancer Res.* **2008**, *68*, 7247–7249. [CrossRef] [PubMed]

128. Gritsenko, P.G.; Ilina, O.; Friedl, P. Interstitial guidance of cancer invasion. *J. Pathol.* **2012**, *226*, 185–199. [CrossRef] [PubMed]

129. Geiger, B.; Yamada, K.M. Molecular architecture and function of matrix adhesions. *Cold Spring Harb Perspect Biol.* **2011**, *3*. [CrossRef] [PubMed]

130. Fang, M.; Yuan, J.; Peng, C.; Li, Y. Collagen as a double-edged sword in tumor progression. *Tumour Biol.* **2014**, *35*, 2871–2882. [CrossRef] [PubMed]

131. Provenzano, P.P.; Eliceiri, K.W.; Campbell, J.M.; Inman, D.R.; White, J.G.; Keely, P.J. Collagen reorganization at the tumor-stromal interface facilitates local invasion. *BMC Med.* **2006**, *4*, 38. [CrossRef] [PubMed]

132. Malik, R.; Lelkes, P.I.; Cukierman, E. Biomechanical and biochemical remodeling of stromal extracellular matrix in cancer. *Trends Biotechnol.* **2015**, *33*, 230–236. [CrossRef] [PubMed]

133. Paszek, M.J.; Zahir, N.; Johnson, K.R.; Lakins, J.N.; Rozenberg, G.I.; Gefen, A.; Reinhart-King, C.A.; Margulies, S.S.; Dembo, M.; Boettiger, D.; et al. Tensional homeostasis and the malignant phenotype. *Cancer Cell* **2005**, *8*, 241–254. [CrossRef] [PubMed]

134. Levental, K.R.; Yu, H.; Kass, L.; Lakins, J.N.; Egeblad, M.; Erler, J.T.; Fong, S.F.; Csiszar, K.; Giaccia, A.; Weninger, W.; et al. Matrix crosslinking forces tumor progression by enhancing integrin signaling. *Cell* **2009**, *139*, 891–906. [CrossRef] [PubMed]

135. Karagiannis, G.S.; Poutahidis, T.; Erdman, S.E.; Kirsch, R.; Riddell, R.H.; Diamandis, E.P. Cancer-associated fibroblasts drive the progression of metastasis through both paracrine and mechanical pressure on cancer tissue. *Mol. Cancer Res.* **2012**, *10*, 1403–1418. [CrossRef] [PubMed]

136. Ozdemir, B.C.; Pentcheva-Hoang, T.; Carstens, J.L.; Zheng, X.; Wu, C.C.; Simpson, T.R.; Laklai, H.; Sugimoto, H.; Kahlert, C.; Novitskiy, S.V.; et al. Depletion of carcinoma-associated fibroblasts and fibrosis induces immunosuppression and accelerates pancreas cancer with reduced survival. *Cancer Cell* **2014**, *25*, 719–734. [CrossRef] [PubMed]

137. Arnold, S.A.; Rivera, L.B.; Miller, A.F.; Carbon, J.G.; Dineen, S.P.; Xie, Y.; Castrillon, D.H.; Sage, E.H.; Puolakkainen, P.; Bradshaw, A.D.; et al. Lack of host sparc enhances vascular function and tumor spread in an orthotopic murine model of pancreatic carcinoma. *Dis. Model. Mech.* **2010**, *3*, 57–72. [CrossRef] [PubMed]

138. Xiao, Q.; Ge, G. Lysyl oxidase, extracellular matrix remodeling and cancer metastasis. *Cancer Microenviron* **2012**, *5*, 261–273. [CrossRef] [PubMed]

139. Erler, J.T.; Bennewith, K.L.; Nicolau, M.; Dornhofer, N.; Kong, C.; Le, Q.T.; Chi, J.T.; Jeffrey, S.S.; Giaccia, A.J. Lysyl oxidase is essential for hypoxia-induced metastasis. *Nature* **2006**, *440*, 1222–1226. [CrossRef] [PubMed]

140. Han, W.; Chen, S.; Yuan, W.; Fan, Q.; Tian, J.; Wang, X.; Chen, L.; Zhang, X.; Wei, W.; Liu, R.; et al. Oriented collagen fibers direct tumor cell intravasation. *Proc. Natl. Acad. Sci. USA* **2016**, *113*, 11208–11213. [CrossRef] [PubMed]

141. Conklin, M.W.; Eickhoff, J.C.; Riching, K.M.; Pehlke, C.A.; Eliceiri, K.W.; Provenzano, P.P.; Friedl, A.; Keely, P.J. Aligned collagen is a prognostic signature for survival in human breast carcinoma. *Am. J. Pathol.* **2011**, *178*, 1221–1232. [CrossRef] [PubMed]

142. Provenzano, P.P.; Inman, D.R.; Eliceiri, K.W.; Knittel, J.G.; Yan, L.; Rueden, C.T.; White, J.G.; Keely, P.J. Collagen density promotes mammary tumor initiation and progression. *BMC Med.* **2008**, *6*, 11. [CrossRef] [PubMed]

143. Riching, K.M.; Cox, B.L.; Salick, M.R.; Pehlke, C.; Riching, A.S.; Ponik, S.M.; Bass, B.R.; Crone, W.C.; Jiang, Y.; Weaver, A.M.; et al. 3d collagen alignment limits protrusions to enhance breast cancer cell persistence. *Biophys. J.* **2014**, *107*, 2546–2558. [CrossRef] [PubMed]

144. Josefsson, A.; Adamo, H.; Hammarsten, P.; Granfors, T.; Stattin, P.; Egevad, L.; Laurent, A.E.; Wikstrom, P.; Bergh, A. Prostate cancer increases hyaluronan in surrounding nonmalignant stroma, and this response is associated with tumor growth and an unfavorable outcome. *Am. J. Pathol.* **2011**, *179*, 1961–1968. [CrossRef] [PubMed]

145. Camenisch, T.D.; Spicer, A.P.; Brehm-Gibson, T.; Biesterfeldt, J.; Augustine, M.L.; Calabro, A., Jr.; Kubalak, S.; Klewer, S.E.; McDonald, J.A. Disruption of hyaluronan synthase-2 abrogates normal cardiac morphogenesis and hyaluronan-mediated transformation of epithelium to mesenchyme. *J. Clin. Invest.* **2000**, *106*, 349–360. [CrossRef] [PubMed]

146. McAtee, C.O.; Barycki, J.J.; Simpson, M.A. Emerging roles for hyaluronidase in cancer metastasis and therapy. *Adv. Cancer Res.* **2014**, *123*, 1–34. [PubMed]

147. Huveneers, S.; Danen, E.H. Adhesion signaling - crosstalk between integrins, src and rho. *J. Cell Sci.* **2009**, *122*, 1059–1069. [CrossRef] [PubMed]

148. Discher, D.E.; Mooney, D.J.; Zandstra, P.W. Growth factors, matrices, and forces combine and control stem cells. *Science* **2009**, *324*, 1673–1677. [CrossRef] [PubMed]

149. Calvo, F.; Ege, N.; Grande-Garcia, A.; Hooper, S.; Jenkins, R.P.; Chaudhry, S.I.; Harrington, K.; Williamson, P.; Moeendarbary, E.; Charras, G.; et al. Mechanotransduction and yap-dependent matrix remodelling is required for the generation and maintenance of cancer-associated fibroblasts. *Nat. Cell Biol.* **2013**, *15*, 637–646. [CrossRef] [PubMed]

150. Varelas, X. The hippo pathway effectors taz and yap in development, homeostasis and disease. *Development* **2014**, *141*, 1614–1626. [CrossRef] [PubMed]

151. Dupont, S. Role of yap/taz in cell-matrix adhesion-mediated signalling and mechanotransduction. *Exp. Cell Res.* **2016**, *343*, 42–53. [CrossRef] [PubMed]

152. Haining, A.W.; von Essen, M.; Attwood, S.J.; Hytonen, V.P.; Del Rio Hernandez, A. All subdomains of the talin rod are mechanically vulnerable and may contribute to cellular mechanosensing. *ACS Nano.* **2016**, *10*, 6648–6658. [CrossRef] [PubMed]

153. Haining, A.W.M.; Rahikainen, R.; Cortes, E.; Lachowski, D.; Rice, A.; von Essen, M.; Hytonen, V.P.; Del Rio Hernandez, A. Mechanotransduction in talin through the interaction of the r8 domain with dlc1. *PLoS Biol.* **2018**, *16*, e2005599. [CrossRef] [PubMed]

154. Wei, S.C.; Fattet, L.; Tsai, J.H.; Guo, Y.; Pai, V.H.; Majeski, H.E.; Chen, A.C.; Sah, R.L.; Taylor, S.S.; Engler, A.J.; et al. Matrix stiffness drives epithelial-mesenchymal transition and tumour metastasis through a twist1-g3bp2 mechanotransduction pathway. *Nat. Cell Biol.* **2015**, *17*, 678–688. [CrossRef] [PubMed]

155. Piccolo, S.; Dupont, S.; Cordenonsi, M. The biology of yap/taz: Hippo signaling and beyond. *Physiol. Rev.* **2014**, *94*, 1287–1312. [CrossRef] [PubMed]

156. Hong, W.; Guan, K.L. The yap and taz transcription co-activators: Key downstream effectors of the mammalian hippo pathway. *Semin. Cell Dev. Biol.* **2012**, *23*, 785–793. [CrossRef] [PubMed]

157. Low, B.C.; Pan, C.Q.; Shivashankar, G.V.; Bershadsky, A.; Sudol, M.; Sheetz, M. Yap/taz as mechanosensors and mechanotransducers in regulating organ size and tumor growth. *FEBS Lett.* **2014**, *588*, 2663–2670. [CrossRef] [PubMed]

158. Chakraborty, S.; Lakshmanan, M.; Swa, H.L.; Chen, J.; Zhang, X.; Ong, Y.S.; Loo, L.S.; Akincilar, S.C.; Gunaratne, J.; Tergaonkar, V.; et al. An oncogenic role of agrin in regulating focal adhesion integrity in hepatocellular carcinoma. *Nat. Commun.* **2015**, *6*, 6184. [CrossRef] [PubMed]

159. Chakraborty, S.; Njah, K.; Pobbati, A.V.; Lim, Y.B.; Raju, A.; Lakshmanan, M.; Tergaonkar, V.; Lim, C.T.; Hong, W. Agrin as a mechanotransduction signal regulating yap through the hippo pathway. *Cell Rep.* **2017**, *18*, 2464–2479. [CrossRef] [PubMed]

160. Tatrai, P.; Dudas, J.; Batmunkh, E.; Mathe, M.; Zalatnai, A.; Schaff, Z.; Ramadori, G.; Kovalszky, I. Agrin, a novel basement membrane component in human and rat liver, accumulates in cirrhosis and hepatocellular carcinoma. *Lab. Invest.* **2006**, *86*, 1149–1160. [CrossRef] [PubMed]

161. Chakraborty, S.; Hong, W. Linking extracellular matrix agrin to the hippo pathway in liver cancer and beyond. *Cancers (Basel)* **2018**, *10*, 45. [CrossRef] [PubMed]

162. Bassat, E.; Mutlak, Y.E.; Genzelinakh, A.; Shadrin, I.Y.; Baruch Umansky, K.; Yifa, O.; Kain, D.; Rajchman, D.; Leach, J.; Riabov Bassat, D.; et al. The extracellular matrix protein agrin promotes heart regeneration in mice. *Nature* **2017**, *547*, 179–184. [CrossRef] [PubMed]

163. Meng, Z.; Qiu, Y.; Lin, K.C.; Kumar, A.; Placone, J.K.; Fang, C.; Wang, K.C.; Lu, S.; Pan, M.; Hong, A.W.; et al. Rap2 mediates mechanoresponses of the hippo pathway. *Nature* **2018**, *560*, 655–660. [CrossRef] [PubMed]

164. Taira, K.; Umikawa, M.; Takei, K.; Myagmar, B.E.; Shinzato, M.; Machida, N.; Uezato, H.; Nonaka, S.; Kariya, K. The traf2- and nck-interacting kinase as a putative effector of rap2 to regulate actin cytoskeleton. *J. Biol. Chem.* **2004**, *279*, 49488–49496. [CrossRef] [PubMed]

165. Morrissey, M.A.; Hagedorn, E.J.; Sherwood, D.R. Cell invasion through basement membrane: The netrin receptor dcc guides the way. *Worm* **2013**, *2*, e26169. [CrossRef] [PubMed]

166. Kelley, L.C.; Lohmer, L.L.; Hagedorn, E.J.; Sherwood, D.R. Traversing the basement membrane in vivo: A diversity of strategies. *J. Cell Biol.* **2014**, *204*, 291–302. [CrossRef] [PubMed]

167. Hiramatsu, R.; Matsuoka, T.; Kimura-Yoshida, C.; Han, S.W.; Mochida, K.; Adachi, T.; Takayama, S.; Matsuo, I. External mechanical cues trigger the establishment of the anterior-posterior axis in early mouse embryos. *Dev. Cell* **2013**, *27*, 131–144. [CrossRef] [PubMed]

168. Hagedorn, E.J.; Ziel, J.W.; Morrissey, M.A.; Linden, L.M.; Wang, Z.; Chi, Q.; Johnson, S.A.; Sherwood, D.R. The netrin receptor dcc focuses invadopodia-driven basement membrane transmigration in vivo. *J. Cell Biol.* **2013**, *201*, 903–913. [CrossRef] [PubMed]

169. Linder, S.; Wiesner, C.; Himmel, M. Degrading devices: Invadosomes in proteolytic cell invasion. *Annu. Rev. Cell Dev. Biol.* **2011**, *27*, 185–211. [CrossRef] [PubMed]

170. Schoumacher, M.; Goldman, R.D.; Louvard, D.; Vignjevic, D.M. Actin, microtubules, and vimentin intermediate filaments cooperate for elongation of invadopodia. *J. Cell Biol.* **2010**, *189*, 541–556. [CrossRef] [PubMed]

171. Ihara, S.; Hagedorn, E.J.; Morrissey, M.A.; Chi, Q.; Motegi, F.; Kramer, J.M.; Sherwood, D.R. Basement membrane sliding and targeted adhesion remodels tissue boundaries during uterine-vulval attachment in caenorhabditis elegans. *Nat. Cell Biol.* **2011**, *13*, 641–651. [CrossRef] [PubMed]

172. Valastyan, S.; Weinberg, R.A. Tumor metastasis: Molecular insights and evolving paradigms. *Cell* **2011**, *147*, 275–292. [CrossRef] [PubMed]

173. Deryugina, E.I.; Quigley, J.P. Matrix metalloproteinases and tumor metastasis. *Cancer Metastasis Rev.* **2006**, *25*, 9–34. [CrossRef] [PubMed]

174. Imai, K.; Hiramatsu, A.; Fukushima, D.; Pierschbacher, M.D.; Okada, Y. Degradation of decorin by matrix metalloproteinases: Identification of the cleavage sites, kinetic analyses and transforming growth factor-β1 release. *Biochem. J.* **1997**, *322*, 809–814. [CrossRef] [PubMed]

175. Kessenbrock, K.; Wang, C.Y.; Werb, Z. Matrix metalloproteinases in stem cell regulation and cancer. *Matrix Biol.* **2015**, *44–46*, 184–190. [CrossRef] [PubMed]

176. Yang, L.; Pang, Y.; Moses, H.L. Tgf-beta and immune cells: An important regulatory axis in the tumor microenvironment and progression. *Trends Immunol.* **2010**, *31*, 220–227. [CrossRef] [PubMed]

177. Pozzi, A.; Moberg, P.E.; Miles, L.A.; Wagner, S.; Soloway, P.; Gardner, H.A. Elevated matrix metalloprotease and angiostatin levels in integrin alpha 1 knockout mice cause reduced tumor vascularization. *Proc. Natl. Acad. Sci. USA* **2000**, *97*, 2202–2207. [CrossRef] [PubMed]

178. Gilkes, D.M.; Semenza, G.L.; Wirtz, D. Hypoxia and the extracellular matrix: Drivers of tumour metastasis. *Nat. Rev. Cancer* **2014**, *14*, 430–439. [CrossRef] [PubMed]

179. Spill, F.; Reynolds, D.S.; Kamm, R.D.; Zaman, M.H. Impact of the physical microenvironment on tumor progression and metastasis. *Curr. Opin. Biotechnol.* **2016**, *40*, 41–48. [CrossRef] [PubMed]

180. Kumar, V.; Gabrilovich, D.I. Hypoxia-inducible factors in regulation of immune responses in tumour microenvironment. *Immunology* **2014**, *143*, 512–519. [CrossRef] [PubMed]

181. Konstantinopoulos, P.A.; Karamouzis, M.V.; Papatsoris, A.G.; Papavassiliou, A.G. Matrix metalloproteinase inhibitors as anticancer agents. *Int. J. Biochem. Cell Biol.* **2008**, *40*, 1156–1168. [CrossRef] [PubMed]

182. Cathcart, J.; Pulkoski-Gross, A.; Cao, J. Targeting matrix metalloproteinases in cancer: Bringing new life to old ideas. *Genes Dis.* **2015**, *2*, 26–34. [CrossRef] [PubMed]

183. Jiang, X.; Dutton, C.M.; Qi, W.-N.; Block, J.A.; Brodt, P.; Durko, M.; Scully, S.P. Inhibition of mmp-1 expression by antisense rna decreases invasiveness of human chondrosarcoma. *J. Orthop. Res.* **2003**, *21*, 1063–1070. [CrossRef]

184. Fingleton, B. Mmps as therapeutic targets—still a viable option? *Semin. Cell Dev. Biol.* **2008**, *19*, 61–68. [CrossRef] [PubMed]

185. Chaudhuri, O.; Koshy, S.T.; Branco da Cunha, C.; Shin, J.W.; Verbeke, C.S.; Allison, K.H.; Mooney, D.J. Extracellular matrix stiffness and composition jointly regulate the induction of malignant phenotypes in mammary epithelium. *Nat. Mater.* **2014**, *13*, 970–978. [CrossRef] [PubMed]

186. Jain, R.K.; Martin, J.D.; Stylianopoulos, T. The role of mechanical forces in tumor growth and therapy. *Annu. Rev. Biomed. Eng.* **2014**, *16*, 321–346. [CrossRef] [PubMed]

187. Tilghman, R.W.; Cowan, C.R.; Mih, J.D.; Koryakina, Y.; Gioeli, D.; Slack-Davis, J.K.; Blackman, B.R.; Tschumperlin, D.J.; Parsons, J.T. Matrix rigidity regulates cancer cell growth and cellular phenotype. *PLoS One* **2010**, *5*, e12905. [CrossRef] [PubMed]

188. Mpekris, F.; Angeli, S.; Pirentis, A.P.; Stylianopoulos, T. Stress-mediated progression of solid tumors: Effect of mechanical stress on tissue oxygenation, cancer cell proliferation, and drug delivery. *Biomech. Model. Mechanobiol.* **2015**, *14*, 1391–1402. [CrossRef] [PubMed]

189. Jain, R.K. An indirect way to tame cancer. *Sci. Am.* **2014**, *310*, 46–53. [CrossRef] [PubMed]

190. Stylianopoulos, T. The solid mechanics of cancer and strategies for improved therapy. *J. Biomech. Eng.* **2017**, *139*. [CrossRef] [PubMed]

191. Chauhan, V.P.; Boucher, Y.; Ferrone, C.R.; Roberge, S.; Martin, J.D.; Stylianopoulos, T.; Bardeesy, N.; DePinho, R.A.; Padera, T.P.; Munn, L.L.; et al. Compression of pancreatic tumor blood vessels by hyaluronan is caused by solid stress and not interstitial fluid pressure. *Cancer Cell* **2014**, *26*, 14–15. [CrossRef] [PubMed]

192. Cheng, G.; Tse, J.; Jain, R.K.; Munn, L.L. Micro-environmental mechanical stress controls tumor spheroid size and morphology by suppressing proliferation and inducing apoptosis in cancer cells. *PLoS One* **2009**, *4*, e4632. [CrossRef] [PubMed]

193. Northcott, J.M.; Dean, I.S.; Mouw, J.K.; Weaver, V.M. Feeling stress: The mechanics of cancer progression and aggression. *Front. Cell Dev. Biol.* **2018**, *6*, 17. [CrossRef] [PubMed]

194. Stylianopoulos, T.; Martin, J.D.; Chauhan, V.P.; Jain, S.R.; Diop-Frimpong, B.; Bardeesy, N.; Smith, B.L.; Ferrone, C.R.; Hornicek, F.J.; Boucher, Y.; et al. Causes, consequences, and remedies for growth-induced solid stress in murine and human tumors. *Proc. Natl. Acad. Sci. USA* **2012**, *109*, 15101–15108. [CrossRef] [PubMed]

195. Koumoutsakos, P.; Pivkin, I.; Milde, F. The fluid mechanics of cancer and its therapy. *Annu. Rev. Fluid Mech.* **2013**, *45*, 325–355. [CrossRef]

196. Mahadevan, N.R.; Zanetti, M. Tumor stress inside out: Cell-extrinsic effects of the unfolded protein response in tumor cells modulate the immunological landscape of the tumor microenvironment. *J. Immunol.* **2011**, *187*, 4403–4409. [CrossRef] [PubMed]

197. Martinez-Outschoorn, U.E.; Balliet, R.M.; Rivadeneira, D.B.; Chiavarina, B.; Pavlides, S.; Wang, C.; Whitaker-Menezes, D.; Daumer, K.M.; Lin, Z.; Witkiewicz, A.K.; et al. Oxidative stress in cancer associated fibroblasts drives tumor-stroma co-evolution: A new paradigm for understanding tumor metabolism, the field effect and genomic instability in cancer cells. *Cell Cycle* **2010**, *9*, 3256–3276. [CrossRef] [PubMed]

198. Sarntinoranont, M.; Rooney, F.; Ferrari, M. Interstitial stress and fluid pressure within a growing tumor. *Ann. Biomed. Eng.* **2003**, *31*, 327–335. [CrossRef] [PubMed]

199. Rofstad, E.K.; Gaustad, J.V.; Egeland, T.A.; Mathiesen, B.; Galappathi, K. Tumors exposed to acute cyclic hypoxic stress show enhanced angiogenesis, perfusion and metastatic dissemination. *Int. J. Cancer* **2010**, *127*, 1535–1546. [CrossRef] [PubMed]

200. Stylianopoulos, T.; Jain, R.K. Combining two strategies to improve perfusion and drug delivery in solid tumors. *Proc. Natl. Acad. Sci. USA* **2013**, *110*, 18632–18637. [CrossRef] [PubMed]

201. Nia, H.T.; Liu, H.; Seano, G.; Datta, M.; Jones, D.; Rahbari, N.; Incio, J.; Chauhan, V.P.; Jung, K.; Martin, J.D.; et al. Solid stress and elastic energy as measures of tumour mechanopathology. *Nat. Biomed. Eng.* **2016**, *1*. [CrossRef] [PubMed]

202. Wirtz, D.; Konstantopoulos, K.; Searson, P.C. The physics of cancer: The role of physical interactions and mechanical forces in metastasis. *Nat. Rev. Cancer* **2011**, *11*, 512–522. [CrossRef] [PubMed]

203. Campas, O.; Mammoto, T.; Hasso, S.; Sperling, R.A.; O'Connell, D.; Bischof, A.G.; Maas, R.; Weitz, D.A.; Mahadevan, L.; Ingber, D.E. Quantifying cell-generated mechanical forces within living embryonic tissues. *Nat. Methods* **2014**, *11*, 183–189. [CrossRef] [PubMed]

204. Lucio, A.A.; Mongera, A.; Shelton, E.; Chen, R.; Doyle, A.M.; Campas, O. Spatiotemporal variation of endogenous cell-generated stresses within 3d multicellular spheroids. *Sci. Rep.* **2017**, *7*, 12022. [CrossRef] [PubMed]

205. Lucio, A.A.; Ingber, D.E.; Campas, O. Generation of biocompatible droplets for in vivo and in vitro measurement of cell-generated mechanical stresses. *Methods Cell Biol.* **2015**, *125*, 373–390. [PubMed]

206. Rowghanian, P.; Meinhart, C.D.; Campàs, O. Dynamics of ferrofluid drop deformations under spatially uniform magnetic fields. *J. Fluid Mech.* **2016**, *802*, 245–262. [CrossRef]

207. Serwane, F.; Mongera, A.; Rowghanian, P.; Kealhofer, D.A.; Lucio, A.A.; Hockenbery, Z.M.; Campas, O. In vivo quantification of spatially varying mechanical properties in developing tissues. *Nat. Methods* **2017**, *14*, 181–186. [CrossRef] [PubMed]

208. Joyce, J.A.; Pollard, J.W. Microenvironmental regulation of metastasis. *Nat. Rev. Cancer* **2009**, *9*, 239–252. [CrossRef] [PubMed]

209. Kaplan, R.N.; Riba, R.D.; Zacharoulis, S.; Bramley, A.H.; Vincent, L.; Costa, C.; MacDonald, D.D.; Jin, D.K.; Shido, K.; Kerns, S.A.; et al. Vegfr1-positive haematopoietic bone marrow progenitors initiate the pre-metastatic niche. *Nature* **2005**, *438*, 820–827. [CrossRef] [PubMed]

210. Rice, A.J.; Cortes, E.; Lachowski, D.; Cheung, B.C.H.; Karim, S.A.; Morton, J.P.; Del Rio Hernandez, A. Matrix stiffness induces epithelial-mesenchymal transition and promotes chemoresistance in pancreatic cancer cells. *Oncogenesis* **2017**, *6*, e352. [CrossRef] [PubMed]

211. Lachowski, D.; Cortes, E.; Pink, D.; Chronopoulos, A.; Karim, S.A.; Morton, J.P.; Del Rio Hernandez, A.E. Substrate rigidity controls activation and durotaxis in pancreatic stellate cells. *Sci. Rep.* **2017**, *7*, 2506. [CrossRef] [PubMed]

212. Lachowski, D.; Cortes, E.; Robinson, B.; Rice, A.; Rombouts, K.; Del Rio Hernandez, A.E. Fak controls the mechanical activation of yap, a transcriptional regulator required for durotaxis. *FASEB J.* **2017**, *32*, 1099–1107. [CrossRef] [PubMed]

213. Chronopoulos, A.; Robinson, B.; Sarper, M.; Cortes, E.; Auernheimer, V.; Lachowski, D.; Attwood, S.; Garcia, R.; Ghassemi, S.; Fabry, B.; et al. Atra mechanically reprograms pancreatic stellate cells to suppress matrix remodelling and inhibit cancer cell invasion. *Nat. Commun.* **2016**, *7*, 1–12. [CrossRef] [PubMed]

214. Sarper, M.; Cortes, E.; Lieberthal, T.J.; Del Rio Hernandez, A. Atra modulates mechanical activation of tgf-beta by pancreatic stellate cells. *Sci.. Rep.* **2016**, *6*, 1–10. [CrossRef] [PubMed]

215. Psaila, B.; Lyden, D. The metastatic niche: Adapting the foreign soil. *Nat. Rev. Cancer* **2009**, *9*, 285–293. [CrossRef] [PubMed]

216. Quail, D.F.; Joyce, J.A. Microenvironmental regulation of tumor progression and metastasis. *Nat. Med.* **2013**, *19*, 1423–1437. [CrossRef] [PubMed]

217. Hiratsuka, S.; Watanabe, A.; Aburatani, H.; Maru, Y. Tumour-mediated upregulation of chemoattractants and recruitment of myeloid cells predetermines lung metastasis. *Nat. Cell Biol.* **2006**, *8*, 1369–1375. [CrossRef] [PubMed]

218. Sleeman, J.P. The lymph node pre-metastatic niche. *J. Mol. Med. (Berl)* **2015**, *93*, 1173–1184. [CrossRef] [PubMed]

219. Ordonez-Moran, P.; Huelsken, J. Complex metastatic niches: Already a target for therapy? *Curr. Opin. Cell Biol.* **2014**, *31*, 29–38. [CrossRef] [PubMed]

220. Hoshino, A.; Costa-Silva, B.; Shen, T.L.; Rodrigues, G.; Hashimoto, A.; Tesic Mark, M.; Molina, H.; Kohsaka, S.; Di Giannatale, A.; Ceder, S.; et al. Tumour exosome integrins determine organotropic metastasis. *Nature* **2015**, *527*, 329–335. [CrossRef] [PubMed]

221. Lin, E.Y.; Li, J.F.; Gnatovskiy, L.; Deng, Y.; Zhu, L.; Grzesik, D.A.; Qian, H.; Xue, X.N.; Pollard, J.W. Macrophages regulate the angiogenic switch in a mouse model of breast cancer. *Cancer Res.* **2006**, *66*, 11238–11246. [CrossRef] [PubMed]

222. Nozawa, H.; Chiu, C.; Hanahan, D. Infiltrating neutrophils mediate the initial angiogenic switch in a mouse model of multistage carcinogenesis. *Proc. Natl. Acad. Sci. USA* **2006**, *103*, 12493–12498. [CrossRef] [PubMed]

Int. J. Mol. Sci. **2018**, *19*, 3028

223. Cox, T.R.; Bird, D.; Baker, A.M.; Barker, H.E.; Ho, M.W.; Lang, G.; Erler, J.T. Lox-mediated collagen crosslinking is responsible for fibrosis-enhanced metastasis. *Cancer Res.* **2013**, *73*, 1721–1732. [CrossRef] [PubMed]

224. Cox, T.R.; Rumney, R.M.H.; Schoof, E.M.; Perryman, L.; Hoye, A.M.; Agrawal, A.; Bird, D.; Latif, N.A.; Forrest, H.; Evans, H.R.; et al. The hypoxic cancer secretome induces pre-metastatic bone lesions through lysyl oxidase. *Nature* **2015**, *522*, 106–110. [CrossRef] [PubMed]

225. Yan, H.H.; Pickup, M.; Pang, Y.; Gorska, A.E.; Li, Z.; Chytil, A.; Geng, Y.; Gray, J.W.; Moses, H.L.; Yang, L. Gr-1+cd11b+ myeloid cells tip the balance of immune protection to tumor promotion in the premetastatic lung. *Cancer Res.* **2010**, *70*, 6139–6149. [CrossRef] [PubMed]

226. Peinado, H.; Zhang, H.; Matei, I.R.; Costa-Silva, B.; Hoshino, A.; Rodrigues, G.; Psaila, B.; Kaplan, R.N.; Bromberg, J.F.; Kang, Y.; et al. Pre-metastatic niches: Organ-specific homes for metastases. *Nat. Rev. Cancer* **2017**, *17*, 302–317. [CrossRef] [PubMed]

227. Sceneay, J.; Smyth, M.J.; Moller, A. The pre-metastatic niche: Finding common ground. *Cancer Metastasis Rev.* **2013**, *32*, 449–464. [CrossRef] [PubMed]

International Journal of
Molecular Sciences

Review

Human Cancer and Platelet Interaction, a Potential Therapeutic Target

Shike Wang [1], Zhenyu Li [2] and Ren Xu [1,3,*

[1] Markey Cancer Center, University of Kentucky, Lexington, KY 40536, USA; Shike.Wang@uky.edu
[2] Division of Cardiovascular Medicine, Department of Internal Medicine, College of Medicine, University of Kentucky, 741 South Limestone Street, Lexington, KY 40536, USA; zhenyuli08@uky.edu
[3] Department of Molecular and Biomedical Pharmacology, University of Kentucky, Lexington, KY 40536, USA
*   Correspondence: ren.xu2010@uky.edu; Tel.: +1-859-323-7889; Fax: +1-859-257-6030

Received: 2 March 2018; Accepted: 16 April 2018; Published: 20 April 2018

Abstract: Cancer patients experience a four-fold increase in thrombosis risk, indicating that cancer development and progression are associated with platelet activation. Xenograft experiments and transgenic mouse models further demonstrate that platelet activation and platelet-cancer cell interaction are crucial for cancer metastasis. Direct or indirect interaction of platelets induces cancer cell plasticity and enhances survival and extravasation of circulating cancer cells during dissemination. In vivo and in vitro experiments also demonstrate that cancer cells induce platelet aggregation, suggesting that platelet-cancer interaction is bidirectional. Therefore, understanding how platelets crosstalk with cancer cells may identify potential strategies to inhibit cancer metastasis and to reduce cancer-related thrombosis. Here, we discuss the potential function of platelets in regulating cancer progression and summarize the factors and signaling pathways that mediate the cancer cell-platelet interaction.

Keywords: cancer metastasis; platelet; biomarker; cancer therapy

---

1. Introduction

During tumor progression, a small number of cancer cells invade into surrounding tissue from the primary lesion and get into the circulation system through the intravastation process [1]. These circulating tumor cells (CTCs) were first identified by Thomas Ashworth in 1869 [2]. Given the recent progress in CTC isolation, the association between CTC and cancer metastasis or prognosis has been identified in many types of cancer, including lung cancer [3,4], breast cancer [5], colon cancer [6] and castration-resistant prostate cancer [7]. In fact, multiple clinical trials have been done or are ongoing to test whether CTC counts can be used as a prognosis marker. The roles of CTCs in cancer metastasis and cancer relapse are well established in animal models [8,9]. Single cell RNA sequencing data show that CTCs exhibit the epithelial-to-mesenchymal transition (EMT) [10] and stem cell phenotypes [11,12], suggesting that CTCs are the driver of cancer metastasis.

CTCs directly interact with red blood cells [13], platelets, macrophages [14], and many other immune cells [15–17]. CTCs also encounter shear stress induced by blood flow [18]. These interactions play important roles in the colonization of CTC at distant organs. It has been shown that CTCs induce the differentiation of macrophages. Cytokines secreted by the differentiated macrophage, in turn, enhances CTC-inflammatory cell interaction, stroma breakdown, and CTC invasion [19,20]. Clinical data show that the number of CTC is negatively associated with CD3$^+$ T cells and cytotoxic (CD8$^+$) T cells [21], suggesting that T cell-mediated immunity is abnormal in patients with high CTC counts [16]. In addition, programmed death-ligand 1 (PD-L1) expression has been detected on the surface of CTCs, which may contribute to the immune escape from T cells and promote cancer metastasis [22].

Clinical evidence and mouse models demonstrate that platelet-cancer cell interaction is crucial for cancer metastasis [23]. Platelets, originally derived from megakaryocytes in the bone marrow [24], are the key regulator of thrombosis [25,26]. The major function of platelets is to prevent bleeding and reduce blood loss in case of vascular injury [27]. It has been reported that platelet count is associated with metastasis and poor prognosis in cancer patients [28,29]. Consistently, with the clinical evidence, the size and number of tumor nodules are reduced by halving the platelet count in the murine model of ovarian cancer [30]. In addition, long-term application of low-dose anti-platelet drugs, such as aspirin, inhibits cancer metastasis and significantly reduces cancer incidence [31,32]. Together these results suggest that platelet activation is a potential target and prognosis marker for cancer treatment [29,33,34].

In this review, we discuss the function and regulation of cancer cell-platelet interaction during cancer development and progression. We also summarize the factors and pathways mediating the interaction and potential targets to halt platelet-induced cancer progression.

2. Roles of Platelets in Cancer Development and Progression

2.1. Roles of Platelets in Tumor Development

Platelet activation by physiological agonists results in secretion of a variety of cytokines and growth factors in the platelet releasates (molecules released after platelets activating) [35,36]. Platelet releasates, induced by the agonists of the thrombin receptors, protease activated receptor-1 (PAR1) and PAR4 [37], promote the proliferation of MCF-7 and MDA-MB-231 breast cancer cells and angiogenesis via the phosphoinositide 3-kinase/protein kinase C (PI3K/PKC) pathway [38]. Platelet activation induced by other agonists, including the adenosine diphosphate (ADP) (through its receptor P2Y12 and P2Y1) also promotes tumor growth in ovarian cancer and pancreatic cancer [39,40]. Recently, the relationship between P2Y12 and cancer was reviewed by Ballerini et al. indicating the important role of P2Y12 in malignant cells [41].

Many of the platelet-derived factors involved in cancer progression are important components of tumor microenvironment, such as transforming growth factor beta (TGF-β), vascular endothelial growth factor (VEGF), and platelet-derived growth factor (PDGF) [42–44]. TGF-β1, a member of the TGF-β family, can be secreted during platelets activation [45]. A recent study showed that platelet-derived TGF-β1 promotes the growth of primary ovarian cancer in murine models [46]. Incubation of platelets with TGF-β1-blocking antibody or downregulation of TGF-βR1 receptor expression in cancer cells with siRNA inhibits proliferation in ovarian cancer cells [47]. It has been shown that platelet extracts induce hepatocellular carcinoma growth [48] by suppressing the expression of Krüppel-like factor 6 [49], a tumor suppressor in many cancers [50]. Protein levels of VEGF, PDGF and platelet factor 4 (PF4) in platelets are elevated in colorectal cancer patients compared to healthy control [51]. VEGF and PDGF are the well-characterized angiogenesis regulator [52,53]. It has been shown that platelets induce tumor angiogenesis by releasing platelet-derived growth factor D and VEGF, and subsequently promotes the tumor growth [54]. PF4 accelerates Kras-driven tumorigenesis in lung cancer [55]. Interestingly, PF4 has also been identified as a chemokine that exhibits anti-angiogenesis activity [56] and may inhibit tumor growth through anti-angiogenesis [57]. PF4 may bind to VEGF or basic fibroblast growth factor (bFGF), thereby inhibiting receptor binding and bFGF dimerization [58,59]. These results suggest that the function of PF4 in cancer development is context-dependent.

Platelet-derived microRNA has recently been identified as a regulator of tumor development. Lawrence E. Goldfinger showed that platelet-derived microparticles transfer miR-24 into cancer cells. Platelet miR-24 subsequently targets mt-Nd2 and Snora75, modulates mitochondrial function, and inhibits tumor growth [60]. Although most data support that platelets promote cancer progression, especially in metastatic dissemination [61,62], this study suggest that the platelets suppress tumor development at the initiation stage. Therefore, the function of platelets in cancer progression may be stage- and context-dependent.

2.2. Roles of Platelets in Cancer Metastasis

About 90% cancer related death is due to cancer metastasis [63]. Depletion of platelets or inhibition of platelet activation inhibits cancer metastases [64,65], indicating that platelets are required for cancer metastasis. During metastasis, cancer cells must detach from the primary tumor and intravasate into circulation, where tumor cells encounter immune cells and experience fluid shear stress. The shear force can sensitize both colon and prostate cancer cells to TNF-related apoptosis-inducing ligand (TRAIL)-induced apoptosis [66]. It has been speculated that binding of platelets to cancer cells protects cancer cells from shear-induced damage and facilitates cancer colonization [67].

EMT, characterized by disruption of cell-cell adhesion and expression of mesenchymal markers, provides cancer cells with the increased cell plasticity and stemness required for colonization and metastasis [68,69]. Platelet-cancer cell interaction promotes EMT in tumor cells and enhances the rate of tumor extravasation in vivo through the TGF-β/Smad and NF-κB pathways [70,71]. Platelet microparticles (PMPs) are the most abundant microparticles in the blood, which may transport miRNA and many other factors promoting EMT. For instance, miR-939 in PMPs promotes the EMT by downregulating E-cadherin and claudin by targeting the 3'UTR region of these genes [72]. In addition, the platelet receptor C-type lectin-like receptor 2(CLEC2) binds to Aggrus expressed in cancer cells and induces the EMT phenotype and cancer metastasis [73]. Tissue factor (TF) is a transmembrane receptor that initiates the extrinsic coagulation pathway. TF is highly expressed in many cancers, and the expression is associated with cancer metastasis [74]. Co-culturing patient-derived ovarian cancer cells with platelets increases the EMT/stemness biomarker and TF protein levels in cancer cells. TF further enhances platelet recruitment and tumorsphere formation [75]. Platelet-released PDGF can also enhance Cyclooxygenase (COX)-2 expression and induce the EMT markers [76]. These studies suggest that platelets promote the EMT process through multiple pathways.

Platelet activation and adhesion depend on integrin signaling [77]. Five integrins, including $\alpha2\beta1$, $\alpha5\beta1$, $\alpha6\beta1$, $\alpha IIb\beta3$ and $\alpha v\beta3$, have been identified in platelets, which bind preferentially to collagen, fibronectin, laminin, vitronectin, and fibrinogen, respectively [78]. It has been shown that platelet $\alpha6\beta1$ mediates the platelet-cancer cell interaction by binding to metalloproteinase (ADAM) 9 on tumor cells, and subsequently induces platelet activation and cancer cell extravasation. Deletion of integrin $\alpha6\beta1$ on platelets reduces lung metastasis [79]. Knockout mouse experiments show that platelet $\beta3$ integrin contributes to cancer bone metastasis [80]. Treatment with the $\alpha IIb\beta3$ antagonist significantly reduces the bone metastasis of breast cancer in mice though depletion of platelet derived lysophosphatidic acid (LPA) [81]. Interestingly, $\alpha IIb\beta3$ expression is also detected on tumor cells [82]; however, roles of the tumor cell $\alpha IIb\beta3$ in cancer metastasis remains unclear.

Anoikis is a programmed cell death induced by cell detachment [83]. Anoikis resistance is required for CTC survival and colonization in distant organs. Platelet interaction protects cancer cells from anoikis [84]. RhoA-(myosin phosphatase targeting subunit 1) MYPT1-protein phosphatase (PP1)-mediated Yes-associated protein 1 (YAP1) dephosphorylation and nuclear translocation are induced by platelets, resulting in apoptosis resistance [85]. Apoptosis signal-regulating kinase 1 (Ask1) is a stress-responsive Ser/Thr mitogen-activated protein kinase kinase kinase (MAP3K) in the Jun N-terminal kinases (JNK) and p38 pathways [86]. Once the Ask1 is deficient in platelet, activating phosphorylation of protein kinase B (Akt), JNK, and p38 is reduced, and tumor metastasis is attenuated [87].

Acid sphingomyelinase (Asm) is another secreted protein mediating the interaction between cancer cells and platelets. Asm released from activated platelets induces the production of ceramide. Ceramide activates the $\alpha5\beta1$ integrin on melanoma cells and promotes metastasis in vivo [88]. Treatment with exogenous Asm activates p38 kinase pathway in melanoma cells. Activation of p38 is required for tumor cell adhesion and metastasis in vivo [89]. This evidence suggests that platelets promote cancer metastasis through direct and indirect interactions.

2.3. Impact of Platelet on the Anti-Tumor Immunity

In order to survive in circulation, CTCs need to overcome not only the shear force-induced damage, but also attacks from immune cells. Antitumor immunity activity is well-characterized in NK cells [90]. Depletion of NK cells significantly promotes cancer metastasis in mouse [91]. It has been shown that binding of platelets protects CTCs from NK cells [92]. MHC class I is usually downregulated in tumor cells [93]. Platelet-derived MHC class I is transferred to tumor cells upon interaction, subsequently reducing the NK cells' antitumor reactivity [94]. In addition, TGF-β released by platelets inhibits the anti-tumor activity of NK cells by reducing the expression of natural-killer group 2, member D on NK cells [95].

Platelet-derived TGF-β has multiple functions in suppressing the antitumor immunity. TGF-β1 is required for converting conventional CD4[+] T (Tconv) cells into induced regulatory T (iTreg) cells [96]. In the hemophilia A mice, TGF-β1 along with other platelet contents induces Foxp3 expression in Tconv cells, and then converts them into functional iTreg cells [97]. Treg cells have the ability of killing activated T cells through a granzyme B (GzmB)-dependent mechanism [98,99]. In addition, platelets constitutively express the non-signaling TGF-β-docking receptor glycoprotein A repetitions predominant (GARP) [100]. Platelet-intrinsic GARP may facilitate TGF-β activation in tumor tissue and subsequently constrains the T cell function in the cancer microenvironment [101]. These data support the hypothesis that platelets promote cancer metastasis by repressing immune response.

3. Cancer Induces Platelet Activation

The interaction between platelets and cancer cells is bidirectional, and cancer cells have profound effects on platelet generation and activation. Cancer patients often have an abnormal platelet count and activation. More than five-fold increase of thrombosis and thromboembolism incidences have been detected in cancer patients compared with normal person [102]. Furthermore, the extracellular vesicles derived from breast cancer cell lines induce tissue factor-independent platelet activation and aggregation, providing a potential mechanism for cancer-associated thrombosis [103]. Fiorella Guadagni et al. showed that the cancer-associated oxidative stress also contributes to persistent platelet activation [104].

Cathepsins K (CAT K) is a protease up-regulated in many cancers [105,106]. It has been shown that platelet aggregation is induced by CAT K in a dose-dependent manner through proteolytically-activated receptors (PAR) 3 and 4. During this process, sonic hedgehog, osteoprotegerin, parathyroid hormone-related protein, and TGF-β are released, which, in turn, induce downstream signaling pathways in breast cancer cells [107]. This study further suggests that cancer cells have profound impacts on platelet activation.

Levels of lipid phosphate phosphatase 1 (LPP1), the key enzyme in phospholipid biosynthesis pathways, is reduced in platelet derived from ovarian cancer patients. The reduction of LLP1 may contribute to the increased risk of thrombosis in cancer patient [108]. However, plasma levels of β-thromboglobulin and PF-4, two markers of platelet α granule secretion and platelet aggregation, have little difference between ovarian cancer patient and patients with benign ovarian tumors. In addition, platelets derived from ovarian cancer patients do not exhibit an enhanced aggregation response to ADP or collagen [109]. These findings suggest that platelet hyperactivation in cancer patients is cancer-type dependent.

4. Platelets, a Potential Therapeutic Target and Biomarker for Cancer Treatment

4.1. Platelet Is a Potential Target to Suppresses Cancer Metastasis

Given the crucial roles of platelets in cancer progression, targeting cancer cell-platelet interaction is considered a promising strategy for cancer prevention and treatment. In fact, many compounds that target platelets exhibit anti-tumor activities. Aspirin is the traditional drug to reduce fever, pain, and inflammation [110]. It is also widely used in patients with a high risk of heart disease and thrombosis because of its unique ability to inhibit platelet COX-1 [111]. Treatment with aspirin suppresses the function of platelets in promoting cancer metastasis in mice [112]. Population and clinical studies also demonstrate that aspirin significantly reduces the risk of colon cancer development and inhibits cancer growth and invasiveness [113,114]. Tamoxifen is used widely as antiestrogen therapy for breast cancer [115]. Interestingly, one study shows that tamoxifen and its metabolite, 4-hydroxytamoxifen, directly inhibit platelet-mediated metastasis [116]. Specifically, delivery of the platelet aggregation inhibitor ticagrelor to tumor tissue also inhibits the EMT phenotypes and cancer metastasis in vivo [117].

The compound 2CP, a derivative of 4-*O*-benzoyl-3-methoxy-beta-nitrostyrene, binds to the CLEC-2 and inhibits the platelet aggregation and cancer metastasis in vivo [118]. Phosphodiesterases (PDEs) regulate cyclic nucleotide signaling by catalyzing cyclic adenosine monophosphate (cAMP) and cyclic guanosine monophosphate (cGMP) to the inactive form. PDE2, PDE3, and PDE5 expression is detected in platelets [119,120]. Selective PDE inhibitors, such as caffeine and theophylline, inhibit platelet aggregation and cancer cell invasion, and enhance anti-cancer drug efficiency in vivo [121–123]. Glycoprotein IIb/IIIa (GPIIb/IIIa) antagonists inhibit platelet aggregation, and a pre-clinical study shows that treatment with GPIIb/IIIa antagonists significantly decreases tumor nodules in lung metastasis [124].

4.2. Targeting Platelets Is a Potential Strategy to Overcome Drug Resistance

The chemotherapeutic response in human epidermal growth factor receptor 2-negative breast cancer is significantly associated with that platelets surround primary tumor [125]. Clinical data show that primary tumor cells surrounded with platelets are less responsive to neo-adjuvant chemotherapy. In addition, platelets promote EMT in cancer cells, which is associated with chemoresistance [126,127]. It has been reported that platelet-derived ADP and ATP increase the level of EMT inducer Slug and cytidine deaminase, and enhances gemcitabine resistance. The P2Y12 inhibitor abolishes the survival signal induced by platelet-derived ADP and ATP [39]. These data suggest that targeting platelets is a potential strategy to overcome chemoresistance.

4.3. Platelets in Anti-Cancer Drug Delivery

Platelets, and their secreted vesicles, are potential drug delivery vehicles [128]. Platelets have little effect on drug activity, and using them as drug delivery vehicles may reduce side effects [129]. The platelet-loaded drugs are protected from clearance and, thus, can circulate in blood for a relatively long time [130]. A recent study shows that platelet membrane-coated particles specifically deliver drugs to CTC and reduce cancer lung metastases in mice [131]. Current drug delivery systems depend on unique cancer markers and tumor-specific microenvironment cues, such as pH and hypoxia. However, the microenvironment of CTCs is different from the solid tumor. Platelets may provide an effective delivery system to target CTCs and inhibit cancer metastasis.

5. Conclusions and Future Direction

Platelet-cancer cell interaction promotes cancer cells metastasis by enhancing CTC survival and extravasation (Figure 1). Growth factors, metabolites, and microRNA released by activated platelets induce EMT and enhance cancer cell stemness, which is crucial for cancer cell colonization at the distant organs (Table 1). Importantly, cancer cells also induce platelet activation and aggregation, and subsequently elevate the risk of thrombosis. Therefore, targeting platelet-cancer cell interaction is a potential strategy to reduce both cancer metastasis and cancer-associated thrombosis. Nevertheless, targeting platelets has not been utilized for cancer therapy in the clinic because the cancer cell-platelet interaction is still not completely understood. For instance, the key factor that regulates cancer cell-platelet interaction has not been identified; roles of platelets in tumor initiation and primary tumor development remained to be determined. We believe that addressing these questions may help to achieve the goal of targeting platelet-cancer interaction for cancer therapy.

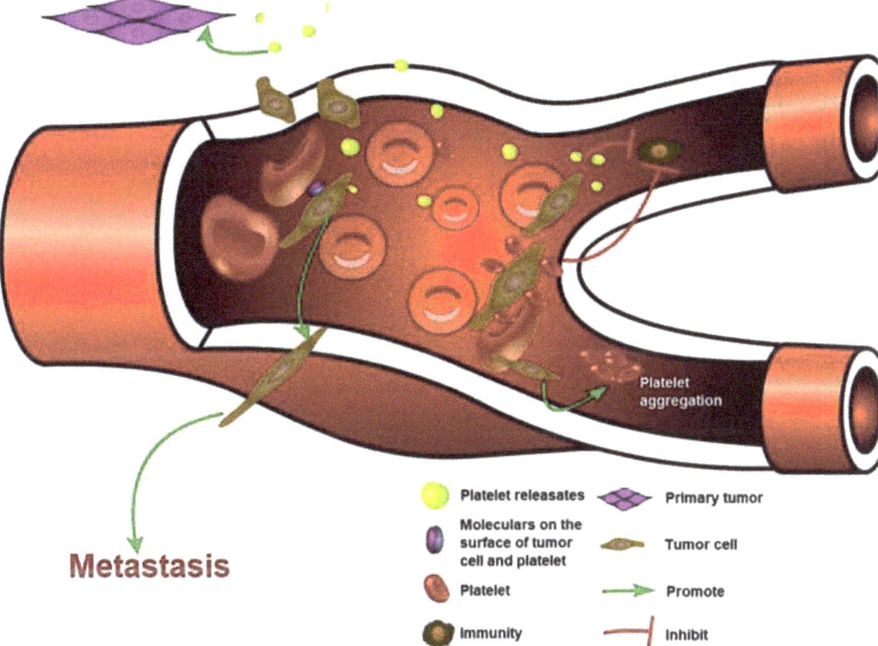

Figure 1. The interaction between cancer cell and platelet. Circulating tumor cells induce platelet activation and aggregation. Activated platelets release a variety of factors, which promote primary tumor growth and cancer metastasis. Binding of platelets also protects CTCs from flow shear force and immune cell attacks.

Table 1. Function of platelet–derived factors and proteins in cancer development and progression.

Platelet Related Factors	Function	Mechanism	Inhibitors	Ref.
TGF-β	Promote primary tumor growth,	TGF-β1 promotes cancer cell proliferation directly	SB431542, decorin	[46,47]
	Enhance EMT phenotype and promote tumor cell extravasation	TGF-β releasing induces the EMT phenotype depending on podoplanin		[70,71]
		Platelets and tumor cells contacts activate TGF-β/SMAD and NF-κb pathway		
	Downregulate reactivity of NK cell, inhibit antitumor immunity	TGF-β down-regulates the NKG2D expression, the activating immunoreceptor		[95,96]
		TGF-β downregulates inflammatory cytokine production		
VEGF	Promote the angiogenesis	Enhance endothelial cell growth		[52]
PDGF	Promote the tumorigenesis	Stimulate the cells in tumor stroma and promote angiogenesis	Olaratumab, imatinib, sunitinib, sorafenib, pazopa-nib, nilotinib, cediranib, trapidil	[53]
	Induce EMT markers	Upregulate the expression of COX-2		[76]
PF4	Inhibit tumor growth and metastasis	Inhibit endothelial proliferation in vitro and angiogenesis in vivo		[57]
	Promote Kras-driven tumorigenensis	Promote platelet production and modulate the tumor mocroenvironment to accelerate the tumor growth		[55]
P2Y12	Promote primary tumor growth	Recruits Gβγ subunits, causing phosphoinositide-3-kinase-dependent Akt phosphorylation and Rap1b activation	clopidogrel, ticagrelor, prasugrel	[40,41]
		Induce ERK1/2 and paxillin Ser83 phosphorylation		
MiRNA 24	Induce the tumor growth inhibition at early stage	Transfer to tumor cells, then induce the mitochondrial dysfunction and tumor cell apoptosis		[60]
MiRNA 939	promotes epithelial to mesenchymal transition	Transfer to tumor cells, downregulate E-cadherin and up-regulate vimentin		[72]
CLEC2	Promote EMT and tumor extravasation in mouse model	Bind with Aggrus, attenuate Aggrus-induced platelet aggregation	2A2B10, 2CP	[73]
Integrin (α6β1, αIIbβ3)	Promote metastasis	Bind with molecular on tumor cell surface, such as ADAM9	ML464, scFv Ab; A11, 7E3 F(ab′)2	[79,80]
LPA	Enhance bone metastasis	enhances the LPA-dependent production of IL-6 and IL-8 to stimulate osteoclast-mediated bone resorption		[81]
Asm	Promote tumor cell adhesion and metastasis	Activate α5β1 on melanoma cells		[88]
Ask1	Promote cancer metastasis	Protect the cancer cells from anoikis		[86,87]

Acknowledgments: This study was supported by start-up funding from Markey Cancer Center and funding support from NCI (1R01CA207772, 1R01CA215095, and 1R21CA209045 to Ren Xu) and the United States Department of Defense (W81XWH-15-1-0052 to Ren Xu).

Author Contributions: Shike Wang and Ren Xu wrote the manuscript, and Zhenyu Li read and revised the manuscript.

Conflicts of Interest: The authors declare no conflict of interest.

Abbreviations

VEGF	Vascular endothelial growth factor
PDGF	Platelet derived growth factor
P2Y12	Platelet factor 4 (PF4), ADP receptor
CLEC2	C-type lectin-like receptor 2
LPA	Lysophosphatidic acid
Asm	Acid sphingomyelinase
Ask1	Apoptosis signal-regulating kinase 1

References

1. Gupta, G.P.; Massague, J. Cancer metastasis: Building a framework. *Cell* **2006**, *127*, 679–695. [CrossRef] [PubMed]
2. Ashworth, T.R. A case of cancer in which cells similar to those in the tumors were seen in the blood after death. *Aust. Med. J.* **1869**, *14*, 146–149.
3. Aggarwal, C.; Wang, X.; Ranganathan, A.; Torigian, D.; Troxel, A.; Evans, T.; Cohen, R.B.; Vaidya, B.; Rao, C.; Connelly, M.; et al. Circulating tumor cells as a predictive biomarker in patients with small cell lung cancer undergoing chemotherapy. *Lung Cancer* **2017**, *112*, 118–125. [CrossRef] [PubMed]
4. Tartarone, A.; Lerose, R.; Rodriquenz, M.G.; Mambella, G.; Calderoni, G.; Bozza, G.; Aieta, M. Molecular characterization and prognostic significance of circulating tumor cells in patients with non-small cell lung cancer. *J. Thorac. Dis.* **2017**, *9* (Suppl. 13), S1359–S1363. [CrossRef] [PubMed]
5. Cristofanilli, M.; Budd, G.T.; Ellis, M.J.; Stopeck, A.; Matera, J.; Miller, M.C.; Reuben, J.M.; Doyle, G.V.; Allard, W.J.; Terstappen, L.W.; et al. Circulating tumor cells, disease progression, and survival in metastatic breast cancer. *N. Engl. J. Med.* **2004**, *351*, 781–791. [CrossRef] [PubMed]
6. Cohen, S.J.; Punt, C.J.; Iannotti, N.; Saidman, B.H.; Sabbath, K.D.; Gabrail, N.Y.; Picus, J.; Morse, M.A.; Mitchell, E.; Miller, M.C.; et al. Prognostic significance of circulating tumor cells in patients with metastatic colorectal cancer. *Ann. Oncol.* **2009**, *20*, 1223–1229. [CrossRef] [PubMed]
7. De Bono, J.S.; Scher, H.I.; Montgomery, R.B.; Parker, C.; Miller, M.C.; Tissing, H.; Doyle, G.V.; Terstappen, L.W.; Pienta, K.J.; Raghavan, D. Circulating tumor cells predict survival benefit from treatment in metastatic castration-resistant prostate cancer. *Clin. Cancer Res.* **2008**, *14*, 6302–6309. [CrossRef] [PubMed]
8. Zhang, S.; Wu, T.; Peng, X.; Liu, J.; Liu, F.; Wu, S.; Liu, S.; Dong, Y.; Xie, S.; Ma, S. Mesenchymal phenotype of circulating tumor cells is associated with distant metastasis in breast cancer patients. *Cancer Manag. Res.* **2017**, *9*, 691–700. [CrossRef] [PubMed]
9. Thiele, J.A.; Bethel, K.; Kralickova, M.; Kuhn, P. Circulating Tumor Cells: Fluid Surrogates of Solid Tumors. *Annu. Rev. Pathol.* **2017**, *12*, 419–447. [CrossRef] [PubMed]
10. Ting, D.T.; Wittner, B.S.; Ligorio, M.; Vincent Jordan, N.; Shah, A.M.; Miyamoto, D.T.; Aceto, N.; Bersani, F.; Brannigan, B.W.; Xega, K.; et al. Single-cell RNA sequencing identifies extracellular matrix gene expression by pancreatic circulating tumor cells. *Cell Rep.* **2014**, *8*, 1905–1918. [CrossRef] [PubMed]
11. Okumura, T.; Yamaguchi, T.; Watanabe, T.; Nagata, T.; Shimada, Y. Flow Cytometric Detection of Circulating Tumor Cells Using a Candidate Stem Cell Marker, p75 Neurotrophin Receptor (p75NTR). *Methods Mol. Biol.* **2017**, *1634*, 211–217. [PubMed]
12. Mirza, S.; Jain, N.; Rawal, R. Evidence for circulating cancer stem-like cells and epithelial-mesenchymal transition phenotype in the pleurospheres derived from lung adenocarcinoma using liquid biopsy. *Tumour Biol.* **2017**, *39*. [CrossRef] [PubMed]
13. Takeishi, N.; Imai, Y.; Yamaguchi, T.; Ishikawa, T. Flow of a circulating tumor cell and red blood cells in microvessels. *Phys. Rev. E* **2015**, *92*. [CrossRef] [PubMed]
14. Hamilton, G.; Rath, B. Circulating tumor cell interactions with macrophages: Implications for biology and treatment. *Transl. Lung Cancer Res.* **2017**, *6*, 418–430. [CrossRef] [PubMed]

15. Gruber, I.; Landenberger, N.; Staebler, A.; Hahn, M.; Wallwiener, D.; Fehm, T. Relationship between circulating tumor cells and peripheral T-cells in patients with primary breast cancer. *Anticancer Res.* **2013**, *33*, 2233–2238. [PubMed]

16. Mego, M.; Gao, H.; Cohen, E.N.; Anfossi, S.; Giordano, A.; Sanda, T.; Fouad, T.M.; De Giorgi, U.; Giuliano, M.; Woodward, W.A.; et al. Circulating Tumor Cells (CTC) Are Associated with Defects in Adaptive Immunity in Patients with Inflammatory Breast Cancer. *J. Cancer* **2016**, *7*, 1095–1104. [CrossRef] [PubMed]

17. Lin, M.; Liang, S.Z.; Shi, J.; Niu, L.Z.; Chen, J.B.; Zhang, M.J.; Xu, K.C. Circulating tumor cell as a biomarker for evaluating allogenic NK cell immunotherapy on stage IV non-small cell lung cancer. *Immunol. Lett.* **2017**, *191*, 10–15. [CrossRef] [PubMed]

18. McCarty, O.J.; Ku, D.; Sugimoto, M.; King, M.R.; Cosemans, J.M.; Neeves, K.B. Dimensional analysis and scaling relevant to flow models of thrombus formation: Communication from the SSC of the ISTH. *J. Thromb. Haemost.* **2016**, *14*, 619–622. [CrossRef] [PubMed]

19. Unsicker, K.; Spittau, B.; Krieglstein, K. The multiple facets of the TGF-beta family cytokine growth/differentiation factor-15/macrophage inhibitory cytokine-1. *Cytokine Growth Factor Rev.* **2013**, *24*, 373–384. [CrossRef] [PubMed]

20. Hamilton, G.; Rath, B.; Klameth, L.; Hochmair, M.J. Small cell lung cancer: Recruitment of macrophages by circulating tumor cells. *Oncoimmunology* **2016**, *5*, e1093277. [CrossRef] [PubMed]

21. Sun, W.W.; Xu, Z.H.; Lian, P.; Gao, B.L.; Hu, J.A. Characteristics of circulating tumor cells in organ metastases, prognosis, and T lymphocyte mediated immune response. *Onco Targets Ther.* **2017**, *10*, 2413–2424. [CrossRef] [PubMed]

22. Wang, X.; Sun, Q.; Liu, Q.; Wang, C.; Yao, R.; Wang, Y. CTC immune escape mediated by PD-L1. *Med. Hypotheses* **2016**, *93*, 138–139. [CrossRef] [PubMed]

23. Stone, J.P.; Wagner, D.D. P-selectin mediates adhesion of platelets to neuroblastoma and small cell lung cancer. *J. Clin. Investig.* **1993**, *92*, 804–813. [CrossRef] [PubMed]

24. Machlus, K.R.; Thon, J.N.; Italiano, J.E., Jr. Interpreting the developmental dance of the megakaryocyte: A review of the cellular and molecular processes mediating platelet formation. *Br. J. Haematol.* **2014**, *165*, 227–236. [CrossRef] [PubMed]

25. Didelot, M.; Docq, C.; Wahl, D.; Lacolley, P.; Regnault, V.; Lagrange, J. Platelet aggregation impacts thrombin generation assessed by calibrated automated thrombography. *Platelets* **2017**, 1–6. [CrossRef] [PubMed]

26. Davi, G.; Patrono, C. Platelet activation and atherothrombosis. *N. Engl. J. Med.* **2007**, *357*, 2482–2494. [CrossRef] [PubMed]

27. Versteeg, H.H.; Heemskerk, J.W.; Levi, M.; Reitsma, P.H. New fundamentals in hemostasis. *Physiol. Rev.* **2013**, *93*, 327–358. [CrossRef] [PubMed]

28. Zhang, M.; Huang, X.Z.; Song, Y.X.; Gao, P.; Sun, J.X.; Wang, Z.N. High Platelet-to-Lymphocyte Ratio Predicts Poor Prognosis and Clinicopathological Characteristics in Patients with Breast Cancer: A Meta-Analysis. *BioMed Res. Int.* **2017**, *2017*. [CrossRef] [PubMed]

29. Tjon-Kon-Fat, L.A.; Lundholm, M.; Schroder, M.; Wurdinger, T.; Thellenberg-Karlsson, C.; Widmark, A.; Wikstrom, P.; Nilsson, R.J.A. Platelets harbor prostate cancer biomarkers and the ability to predict therapeutic response to abiraterone in castration resistant patients. *Prostate* **2018**, *78*, 48–53. [CrossRef] [PubMed]

30. Stone, R.L.; Nick, A.M.; McNeish, I.A.; Balkwill, F.; Han, H.D.; Bottsford-Miller, J.; Rupairmoole, R.; Armaiz-Pena, G.N.; Pecot, C.V.; Coward, J.; et al. Paraneoplastic thrombocytosis in ovarian cancer. *N. Engl. J. Med.* **2012**, *366*, 610–618. [CrossRef] [PubMed]

31. Rothwell, P.M.; Wilson, M.; Price, J.F.; Belch, J.F.; Meade, T.W.; Mehta, Z. Effect of daily aspirin on risk of cancer metastasis: A study of incident cancers during randomised controlled trials. *Lancet* **2012**, *379*, 1591–1601. [CrossRef]

32. Patrignani, P.; Patrono, C. Aspirin and Cancer. *J. Am. Coll. Cardiol.* **2016**, *68*, 967–976. [CrossRef] [PubMed]

33. Fu, S.; Niu, Y.; Zhang, X.; Zhang, J.R.; Liu, Z.P.; Wang, R.T. Squamous cell carcinoma antigen, platelet distribution width, and prealbumin collectively as a marker of squamous cell cervical carcinoma. *Cancer Biomark.* **2017**, *21*, 317–321. [CrossRef] [PubMed]

34. Liu, W.; Ha, M.; Yin, N. Combination of platelet count and lymphocyte to monocyte ratio is a prognostic factor in patients undergoing surgery for non-small cell lung cancer. *Oncotarget* **2017**, *8*, 73198–73207. [CrossRef] [PubMed]

35. Sheu, J.R.; Fong, T.H.; Liu, C.M.; Shen, M.Y.; Chen, T.L.; Chang, Y.; Lu, M.S.; Hsiao, G. Expression of matrix metalloproteinase-9 in human platelets: Regulation of platelet activation in in vitro and in vivo studies. *Br. J. Pharmacol.* **2004**, *143*, 193–201. [CrossRef] [PubMed]

36. Pintucci, G.; Froum, S.; Pinnell, J.; Mignatti, P.; Rafii, S.; Green, D. Trophic effects of platelets on cultured endothelial cells are mediated by platelet-associated fibroblast growth factor-2 (FGF-2) and vascular endothelial growth factor (VEGF). *Thromb. Haemost.* **2002**, *88*, 834–842. [CrossRef] [PubMed]

37. Ma, L.; Perini, R.; McKnight, W.; Dicay, M.; Klein, A.; Hollenberg, M.D.; Wallace, J.L. Proteinase-activated receptors 1 and 4 counter-regulate endostatin and VEGF release from human platelets. *Proc. Natl. Acad. Sci. USA* **2005**, *102*, 216–220. [CrossRef] [PubMed]

38. Jiang, L.; Luan, Y.; Miao, X.; Sun, C.; Li, K.; Huang, Z.; Xu, D.; Zhang, M.; Kong, F.; Li, N. Platelet releasate promotes breast cancer growth and angiogenesis via VEGF-integrin cooperative signalling. *Br. J. Cancer* **2017**, *117*, 695–703. [CrossRef] [PubMed]

39. Elaskalani, O.; Falasca, M.; Moran, N.; Berndt, M.C.; Metharom, P. The Role of Platelet-Derived ADP and ATP in Promoting Pancreatic Cancer Cell Survival and Gemcitabine Resistance. *Cancers* **2017**, *9*, 142. [CrossRef] [PubMed]

40. Cho, M.S.; Noh, K.; Haemmerle, M.; Li, D.; Park, H.; Hu, Q.; Hisamatsu, T.; Mitamura, T.; Mak, S.L.C.; Kunapuli, S.; et al. Role of ADP receptors on platelets in the growth of ovarian cancer. *Blood* **2017**, *130*, 1235–1242. [CrossRef] [PubMed]

41. Ballerini, P.; Dovizio, M.; Bruno, A.; Tacconelli, S.; Patrignani, P. P2Y12 Receptors in Tumorigenesis and Metastasis. *Front. Pharmacol.* **2018**, *9*, 66. [CrossRef] [PubMed]

42. Waldmann, T.A. Cytokines in Cancer Immunotherapy. *Cold Spring Harb. Perspect. Biol.* **2017**. [CrossRef] [PubMed]

43. Musolino, C.; Allegra, A.; Innao, V.; Allegra, A.G.; Pioggia, G.; Gangemi, S. Inflammatory and Anti-Inflammatory Equilibrium, Proliferative and Antiproliferative Balance: The Role of Cytokines in Multiple Myeloma. *Mediat. Inflamm.* **2017**, *2017*. [CrossRef] [PubMed]

44. Lee, M.; Rhee, I. Cytokine Signaling in Tumor Progression. *Immune Netw.* **2017**, *17*, 214–227. [CrossRef] [PubMed]

45. Assoian, R.K.; Komoriya, A.; Meyers, C.A.; Miller, D.M.; Sporn, M.B. Transforming growth factor-beta in human platelets. Identification of a major storage site, purification, and characterization. *J. Biol. Chem.* **1983**, *258*, 7155–7160. [PubMed]

46. Hu, Q.; Hisamatsu, T.; Haemmerle, M.; Cho, M.S.; Pradeep, S.; Rupaimoole, R.; Rodriguez-Aguayo, C.; Lopez-Berestein, G.; Wong, S.T.C.; Sood, A.K.; et al. Role of Platelet-Derived Tgfbeta1 in the Progression of Ovarian Cancer. *Clin. Cancer Res.* **2017**, *23*, 5611–5621. [CrossRef] [PubMed]

47. Cho, M.S.; Bottsford-Miller, J.; Vasquez, H.G.; Stone, R.; Zand, B.; Kroll, M.H.; Sood, A.K.; Afshar-Kharghan, V. Platelets increase the proliferation of ovarian cancer cells. *Blood* **2012**, *120*, 4869–4872. [CrossRef] [PubMed]

48. Carr, B.I.; Cavallini, A.; D'Alessandro, R.; Refolo, M.G.; Lippolis, C.; Mazzocca, A.; Messa, C. Platelet extracts induce growth, migration and invasion in human hepatocellular carcinoma in vitro. *BMC Cancer* **2014**, *14*, 43. [CrossRef] [PubMed]

49. He, A.D.; Xie, W.; Song, W.; Ma, Y.Y.; Liu, G.; Liang, M.L.; Da, X.W.; Yao, G.Q.; Zhang, B.X.; Gao, C.J.; et al. Platelet releasates promote the proliferation of hepatocellular carcinoma cells by suppressing the expression of KLF6. *Sci. Rep.* **2017**, *7*, 3989. [CrossRef] [PubMed]

50. Narla, G.; Heath, K.E.; Reeves, H.L.; Li, D.; Giono, L.E.; Kimmelman, A.C.; Glucksman, M.J.; Narla, J.; Eng, F.J.; Chan, A.M.; et al. KLF6, a candidate tumor suppressor gene mutated in prostate cancer. *Science* **2001**, *294*, 2563–2566. [CrossRef] [PubMed]

51. Peterson, J.E.; Zurakowski, D.; Italiano, J.E., Jr.; Michel, L.V.; Connors, S.; Oenick, M.; D'Amato, R.J.; Klement, G.L.; Folkman, J. VEGF, PF4 and PDGF are elevated in platelets of colorectal cancer patients. *Angiogenesis* **2012**, *15*, 265–273. [CrossRef] [PubMed]

52. Zizzo, N.; Patruno, R.; Zito, F.A.; Di Summa, A.; Tinelli, A.; Troilo, S.; Misino, A.; Ruggieri, E.; Goffredo, V.; Gadaleta, C.D.; et al. Vascular endothelial growth factor concentrations from platelets correlate with tumor angiogenesis and grading in a spontaneous canine non-Hodgkin lymphoma model. *Leuk. Lymphoma* **2010**, *51*, 291–296. [CrossRef] [PubMed]

53. Heldin, C.H.; Lennartsson, J.; Westermark, B. Involvement of platelet-derived growth factor ligands and receptors in tumorigenesis. *J. Intern. Med.* **2018**, *283*, 16–44. [CrossRef] [PubMed]

54. Repsold, L.; Pool, R.; Karodia, M.; Tintinger, G.; Joubert, A.M. An overview of the role of platelets in angiogenesis, apoptosis and autophagy in chronic myeloid leukaemia. *Cancer Cell Int.* **2017**, *17*, 89. [CrossRef] [PubMed]

55. Pucci, F.; Rickelt, S.; Newton, A.P.; Garris, C.; Nunes, E.; Evavold, C.; Pfirschke, C.; Engblom, C.; Mino-Kenudson, M.; Hynes, R.O.; et al. PF4 Promotes Platelet Production and Lung Cancer Growth. *Cell Rep.* **2016**, *17*, 1764–1772. [CrossRef] [PubMed]

56. Maione, T.E.; Gray, G.S.; Petro, J.; Hunt, A.J.; Donner, A.L.; Bauer, S.I.; Carson, H.F.; Sharpe, R.J. Inhibition of angiogenesis by recombinant human platelet factor-4 and related peptides. *Science* **1990**, *247*, 77–79. [CrossRef] [PubMed]

57. Yamaguchi, K.; Ogawa, K.; Katsube, T.; Shimao, K.; Konno, S.; Shimakawa, T.; Yoshimatsu, K.; Naritaka, Y.; Yagawa, H.; Hirose, K. Platelet factor 4 gene transfection into tumor cells inhibits angiogenesis, tumor growth and metastasis. *Anticancer Res.* **2005**, *25*, 847–851. [PubMed]

58. Sato, Y.; Waki, M.; Ohno, M.; Kuwano, M.; Sakata, T. Carboxyl-terminal heparin-binding fragments of platelet factor 4 retain the blocking effect on the receptor binding of basic fibroblast growth factor. *Jpn. J. Cancer Res.* **1993**, *84*, 485–488. [CrossRef] [PubMed]

59. Gengrinovitch, S.; Greenberg, S.M.; Cohen, T.; Gitay-Goren, H.; Rockwell, P.; Maione, T.E.; Levi, B.Z.; Neufeld, G. Platelet factor-4 inhibits the mitogenic activity of VEGF121 and VEGF165 using several concurrent mechanisms. *J. Biol. Chem.* **1995**, *270*, 15059–15065. [CrossRef] [PubMed]

60. Michael, J.V.; Wurtzel, J.G.T.; Mao, G.F.; Rao, A.K.; Kolpakov, M.A.; Sabri, A.; Hoffman, N.E.; Rajan, S.; Tomar, D.; Madesh, M.; et al. Platelet microparticles infiltrating solid tumors transfer miRNAs that suppress tumor growth. *Blood* **2017**, *130*, 567–580. [CrossRef] [PubMed]

61. Labelle, M.; Begum, S.; Hynes, R.O. Platelets guide the formation of early metastatic niches. *Proc. Natl. Acad. Sci. USA* **2014**, *111*, E3053–E3061. [CrossRef] [PubMed]

62. Sharma, D.; Brummel-Ziedins, K.E.; Bouchard, B.A.; Holmes, C.E. Platelets in tumor progression: A host factor that offers multiple potential targets in the treatment of cancer. *J. Cell. Physiol.* **2014**, *229*, 1005–1015. [CrossRef] [PubMed]

63. Seyfried, T.N.; Huysentruyt, L.C. On the origin of cancer metastasis. *Crit. Rev. Oncog.* **2013**, *18*, 43–73. [CrossRef] [PubMed]

64. Gasic, G.J.; Gasic, T.B.; Stewart, C.C. Antimetastatic effects associated with platelet reduction. *Proc. Natl. Acad. Sci. USA* **1968**, *61*, 46–52. [CrossRef] [PubMed]

65. Gasic, G.J.; Gasic, T.B.; Galanti, N.; Johnson, T.; Murphy, S. Platelet-tumor-cell interactions in mice. The role of platelets in the spread of malignant disease. *Int. J. Cancer* **1973**, *11*, 704–718. [CrossRef] [PubMed]

66. Mitchell, M.J.; King, M.R. Fluid Shear Stress Sensitizes Cancer Cells to Receptor-Mediated Apoptosis via Trimeric Death Receptors. *New J. Phys.* **2013**, *15*. [CrossRef] [PubMed]

67. Egan, K.; Cooke, N.; Kenny, D. Living in shear: Platelets protect cancer cells from shear induced damage. *Clin. Exp. Metast.* **2014**, *31*, 697–704. [CrossRef] [PubMed]

68. Thiery, J.P.; Acloque, H.; Huang, R.Y.; Nieto, M.A. Epithelial-mesenchymal transitions in development and disease. *Cell* **2009**, *139*, 871–890. [CrossRef] [PubMed]

69. Yeung, K.T.; Yang, J. Epithelial-mesenchymal transition in tumor metastasis. *Mol. Oncol.* **2017**, *11*, 28–39. [CrossRef] [PubMed]

70. Takemoto, A.; Okitaka, M.; Takagi, S.; Takami, M.; Sato, S.; Nishio, M.; Okumura, S.; Fujita, N. A critical role of platelet TGF-beta release in podoplanin-mediated tumour invasion and metastasis. *Sci. Rep.* **2017**, *7*, 42186. [CrossRef] [PubMed]

71. Labelle, M.; Begum, S.; Hynes, R.O. Direct signaling between platelets and cancer cells induces an epithelial-mesenchymal-like transition and promotes metastasis. *Cancer Cell* **2011**, *20*, 576–590. [CrossRef] [PubMed]

72. Tang, M.; Jiang, L.; Lin, Y.; Wu, X.; Wang, K.; He, Q.; Wang, X.; Li, W. Platelet microparticle-mediated transfer of miR-939 to epithelial ovarian cancer cells promotes epithelial to mesenchymal transition. *Oncotarget* **2017**, *8*, 97464–97475. [CrossRef] [PubMed]

73. Fujita, N.; Takagi, S. The impact of Aggrus/podoplanin on platelet aggregation and tumour metastasis. *J. Biochem.* **2012**, *152*, 407–413. [CrossRef] [PubMed]

74. Ruf, W. Tissue factor and cancer. *Thromb. Res.* **2012**, *130* (Suppl. 1), S84–S87. [CrossRef] [PubMed]

75. Orellana, R.; Kato, S.; Erices, R.; Bravo, M.L.; Gonzalez, P.; Oliva, B.; Cubillos, S.; Valdivia, A.; Ibanez, C.; Branes, J.; et al. Platelets enhance tissue factor protein and metastasis initiating cell markers, and act as chemoattractants increasing the migration of ovarian cancer cells. *BMC Cancer* **2015**, *15*, 290. [CrossRef] [PubMed]

76. Dovizio, M.; Maier, T.J.; Alberti, S.; Di Francesco, L.; Marcantoni, E.; Munch, G.; John, C.M.; Suess, B.; Sgambato, A.; Steinhilber, D.; et al. Pharmacological inhibition of platelet-tumor cell cross-talk prevents platelet-induced overexpression of cyclooxygenase-2 in HT29 human colon carcinoma cells. *Mol. Pharmacol.* **2013**, *84*, 25–40. [CrossRef] [PubMed]

77. Xu, X.R.; Carrim, N.; Neves, M.A.; McKeown, T.; Stratton, T.W.; Coelho, R.M.; Lei, X.; Chen, P.; Xu, J.; Dai, X.; et al. Platelets and platelet adhesion molecules: Novel mechanisms of thrombosis and anti-thrombotic therapies. *Thromb. J.* **2016**, *14* (Suppl. 1), 29. [CrossRef] [PubMed]

78. Lavergne, M.; Janus-Bell, E.; Schaff, M.; Gachet, C.; Mangin, P.H. Platelet Integrins in Tumor Metastasis: Do They Represent a Therapeutic Target? *Cancers* **2017**, *9*, 133. [CrossRef] [PubMed]

79. Mammadova-Bach, E.; Zigrino, P.; Brucker, C.; Bourdon, C.; Freund, M.; De Arcangelis, A.; Abrams, S.I.; Orend, G.; Gachet, C.; Mangin, P.H. Platelet integrin alpha6beta1 controls lung metastasis through direct binding to cancer cell-derived ADAM9. *JCI Insight* **2016**, *1*, e88245. [CrossRef] [PubMed]

80. Bakewell, S.J.; Nestor, P.; Prasad, S.; Tomasson, M.H.; Dowland, N.; Mehrotra, M.; Scarborough, R.; Kanter, J.; Abe, K.; Phillips, D.; et al. Platelet and osteoclast beta3 integrins are critical for bone metastasis. *Proc. Natl. Acad. Sci. USA* **2003**, *100*, 14205–14210. [CrossRef] [PubMed]

81. Boucharaba, A.; Serre, C.M.; Gres, S.; Saulnier-Blache, J.S.; Bordet, J.C.; Guglielmi, J.; Clezardin, P.; Peyruchaud, O. Platelet-derived lysophosphatidic acid supports the progression of osteolytic bone metastases in breast cancer. *J. Clin. Investig.* **2004**, *114*, 1714–1725. [CrossRef] [PubMed]

82. Chopra, H.; Hatfield, J.S.; Chang, Y.S.; Grossi, I.M.; Fitzgerald, L.A.; O'Gara, C.Y.; Marnett, L.J.; Diglio, C.A.; Taylor, J.D.; Honn, K.V. Role of tumor cytoskeleton and membrane glycoprotein IRGpIIb/IIIa in platelet adhesion to tumor cell membrane and tumor cell-induced platelet aggregation. *Cancer Res.* **1988**, *48*, 3787–3800. [PubMed]

83. Frisch, S.M.; Screaton, R.A. Anoikis mechanisms. *Curr. Opin. Cell Biol.* **2001**, *13*, 555–562. [CrossRef]

84. Buchheit, C.L.; Weigel, K.J.; Schafer, Z.T. Cancer cell survival during detachment from the ECM: Multiple barriers to tumour progression. *Nat. Rev. Cancer* **2014**, *14*, 632–641. [CrossRef] [PubMed]

85. Haemmerle, M.; Taylor, M.L.; Gutschner, T.; Pradeep, S.; Cho, M.S.; Sheng, J.; Lyons, Y.M.; Nagaraja, A.S.; Dood, R.L.; Wen, Y.; et al. Platelets reduce anoikis and promote metastasis by activating YAP1 signaling. *Nat. Commun.* **2017**, *8*, 310. [CrossRef] [PubMed]

86. Ichijo, H.; Nishida, E.; Irie, K.; ten Dijke, P.; Saitoh, M.; Moriguchi, T.; Takagi, M.; Matsumoto, K.; Miyazono, K.; Gotoh, Y. Induction of apoptosis by ASK1, a mammalian MAPKKK that activates SAPK/JNK and p38 signaling pathways. *Science* **1997**, *275*, 90–94. [CrossRef] [PubMed]

87. Kamiyama, M.; Shirai, T.; Tamura, S.; Suzuki-Inoue, K.; Ehata, S.; Takahashi, K.; Miyazono, K.; Hayakawa, Y.; Sato, T.; Takeda, K.; et al. ASK1 facilitates tumor metastasis through phosphorylation of an ADP receptor P2Y12 in platelets. *Cell Death Differ.* **2017**, *24*, 2066–2076. [CrossRef] [PubMed]

88. Carpinteiro, A.; Becker, K.A.; Japtok, L.; Hessler, G.; Keitsch, S.; Pozgajova, M.; Schmid, K.W.; Adams, C.; Muller, S.; Kleuser, B.; et al. Regulation of hematogenous tumor metastasis by acid sphingomyelinase. *EMBO Mol. Med.* **2015**, *7*, 714–734. [CrossRef] [PubMed]

89. Carpinteiro, A.; Beckmann, N.; Seitz, A.; Hessler, G.; Wilker, B.; Soddemann, M.; Helfrich, I.; Edelmann, B.; Gulbins, E.; Becker, K.A. Role of Acid Sphingomyelinase-Induced Signaling in Melanoma Cells for Hematogenous Tumor Metastasis. *Cell. Physiol. Biochem.* **2016**, *38*, 1–14. [CrossRef] [PubMed]

90. Wargo, J.A.; Schumacher, L.Y.; Comin-Anduix, B.; Dissette, V.B.; Glaspy, J.A.; McBride, W.H.; Butterfield, L.H.; Economou, J.S.; Ribas, A. Natural killer cells play a critical role in the immune response following immunization with melanoma-antigen-engineered dendritic cells. *Cancer Gene Ther.* **2005**, *12*, 516–527. [CrossRef] [PubMed]

91. Shimaoka, H.; Takeno, S.; Maki, K.; Sasaki, T.; Hasegawa, S.; Yamashita, Y. A cytokine signal inhibitor for rheumatoid arthritis enhances cancer metastasis via depletion of NK cells in an experimental lung metastasis mouse model of colon cancer. *Oncol. Lett.* **2017**, *14*, 3019–3027. [CrossRef] [PubMed]

92. Nieswandt, B.; Hafner, M.; Echtenacher, B.; Mannel, D.N. Lysis of tumor cells by natural killer cells in mice is impeded by platelets. *Cancer Res.* **1999**, *59*, 1295–1300. [PubMed]

93. Zitvogel, L.; Tesniere, A.; Kroemer, G. Cancer despite immunosurveillance: Immunoselection and immunosubversion. *Nat. Rev. Immunol.* **2006**, *6*, 715–727. [CrossRef] [PubMed]

94. Placke, T.; Orgel, M.; Schaller, M.; Jung, G.; Rammensee, H.G.; Kopp, H.G.; Salih, H.R. Platelet-derived MHC class I confers a pseudonormal phenotype to cancer cells that subverts the antitumor reactivity of natural killer immune cells. *Cancer Res.* **2012**, *72*, 440–448. [CrossRef] [PubMed]

95. Kopp, H.G.; Placke, T.; Salih, H.R. Platelet-derived transforming growth factor-beta down-regulates NKG2D thereby inhibiting natural killer cell antitumor reactivity. *Cancer Res.* **2009**, *69*, 7775–7783. [CrossRef] [PubMed]

96. Fadok, V.A.; Bratton, D.L.; Konowal, A.; Freed, P.W.; Westcott, J.Y.; Henson, P.M. Macrophages that have ingested apoptotic cells in vitro inhibit proinflammatory cytokine production through autocrine/paracrine mechanisms involving TGF-beta, PGE2, and PAF. *J. Clin. Investig.* **1998**, *101*, 890–898. [CrossRef] [PubMed]

97. Haribhai, D.; Luo, X.; Chen, J.; Jia, S.; Shi, L.; Schroeder, J.A.; Weiler, H.; Aster, R.H.; Hessner, M.J.; Hu, J.; et al. TGF-beta1 along with other platelet contents augments Treg cells to suppress anti-FVIII immune responses in hemophilia A mice. *Blood Adv.* **2016**, *1*, 139–151. [CrossRef] [PubMed]

98. Grossman, W.J.; Verbsky, J.W.; Barchet, W.; Colonna, M.; Atkinson, J.P.; Ley, T.J. Human T regulatory cells can use the perforin pathway to cause autologous target cell death. *Immunity* **2004**, *21*, 589–601. [CrossRef] [PubMed]

99. Gondek, D.C.; Lu, L.F.; Quezada, S.A.; Sakaguchi, S.; Noelle, R.J. Cutting edge: Contact-mediated suppression by CD4+CD25+ regulatory cells involves a granzyme B-dependent, perforin-independent mechanism. *J. Immunol.* **2005**, *174*, 1783–1786. [CrossRef] [PubMed]

100. Tran, D.Q.; Andersson, J.; Wang, R.; Ramsey, H.; Unutmaz, D.; Shevach, E.M. GARP (LRRC32) is essential for the surface expression of latent TGF-beta on platelets and activated FOXP3+ regulatory T cells. *Proc. Natl. Acad. Sci. USA* **2009**, *106*, 13445–13450. [CrossRef] [PubMed]

101. Rachidi, S.; Metelli, A.; Riesenberg, B.; Wu, B.X.; Nelson, M.H.; Wallace, C.; Paulos, C.M.; Rubinstein, M.P.; Garrett-Mayer, E.; Hennig, M.; et al. Platelets subvert T cell immunity against cancer via GARP-TGFbeta axis. *Sci. Immunol.* **2017**, *2*, 11. [CrossRef] [PubMed]

102. Silverstein, M.D.; Heit, J.A.; Mohr, D.N.; Petterson, T.M.; O'Fallon, W.M.; Melton, L.J. Trends in the incidence of deep vein thrombosis and pulmonary embolism: A 25-year population-based study. *Arch. Intern. Med.* **1998**, *158*, 585–593. [CrossRef] [PubMed]

103. Gomes, F.G.; Sandim, V.; Almeida, V.H.; Rondon, A.M.R.; Succar, B.B.; Hottz, E.D.; Leal, A.C.; Vercoza, B.R.F.; Rodrigues, J.C.F.; Bozza, P.T.; et al. Breast-cancer extracellular vesicles induce platelet activation and aggregation by tissue factor-independent and -dependent mechanisms. *Thromb. Res.* **2017**, *159*, 24–32. [CrossRef] [PubMed]

104. Ferroni, P.; Santilli, F.; Cavaliere, F.; Simeone, P.; Costarelli, L.; Liani, R.; Tripaldi, R.; Riondino, S.; Roselli, M.; Davi, G.; et al. Oxidant stress as a major determinant of platelet activation in invasive breast cancer. *Int. J. Cancer* **2017**, *140*, 696–704. [CrossRef] [PubMed]

105. Lindeman, J.H.; Hanemaaijer, R.; Mulder, A.; Dijkstra, P.D.; Szuhai, K.; Bromme, D.; Verheijen, J.H.; Hogendoorn, P.C. Cathepsin K is the principal protease in giant cell tumor of bone. *Am. J. Pathol.* **2004**, *165*, 593–600. [CrossRef]

106. Chen, B.; Platt, M.O. Multiplex zymography captures stage-specific activity profiles of cathepsins K, L, and S in human breast, lung, and cervical cancer. *J. Transl. Med.* **2011**, *9*, 109. [CrossRef] [PubMed]

107. Andrade, S.S.; Gouvea, I.E.; Silva, M.C.; Castro, E.D.; de Paula, C.A.; Okamoto, D.; Oliveira, L.; Peres, G.B.; Ottaiano, T.; Facina, G.; et al. Cathepsin K induces platelet dysfunction and affects cell signaling in breast cancer -molecularly distinct behavior of cathepsin K in breast cancer. *BMC Cancer* **2016**, *16*, 173. [CrossRef] [PubMed]

108. Hu, Q.; Wang, M.; Cho, M.S.; Wang, C.; Nick, A.M.; Thiagarajan, P.; Aung, F.M.; Han, X.; Sood, A.K.; Afshar-Kharghan, V. Lipid profile of platelets and platelet-derived microparticles in ovarian cancer. *BBA Clin.* **2016**, *6*, 76–81. [CrossRef] [PubMed]

109. Feng, S.; Kroll, M.H.; Nick, A.M.; Sood, A.K.; Afshar-Kharghan, V. Platelets are not hyperreactive in patients with ovarian cancer. *Platelets* **2016**, *27*, 716–718. [CrossRef] [PubMed]

110. Cuzick, J.; Thorat, M.A.; Bosetti, C.; Brown, P.H.; Burn, J.; Cook, N.R.; Ford, L.G.; Jacobs, E.J.; Jankowski, J.A.; La Vecchia, C.; et al. Estimates of benefits and harms of prophylactic use of aspirin in the general population. *Ann. Oncol.* **2015**, *26*, 47–57. [CrossRef] [PubMed]

111. Vane, J. Towards a better aspirin. *Nature* **1994**, *367*, 215–216. [CrossRef] [PubMed]

112. Guillem-Llobat, P.; Dovizio, M.; Bruno, A.; Ricciotti, E.; Cufino, V.; Sacco, A.; Grande, R.; Alberti, S.; Arena, V.; Cirillo, M.; et al. Aspirin prevents colorectal cancer metastasis in mice by splitting the crosstalk between platelets and tumor cells. *Oncotarget* **2016**, *7*, 32462–32477. [CrossRef] [PubMed]

113. Lichtenberger, L.M.; Fang, D.; Bick, R.J.; Poindexter, B.J.; Phan, T.; Bergeron, A.L.; Pradhan, S.; Dial, E.J.; Vijayan, K.V. Unlocking Aspirin's Chemopreventive Activity: Role of Irreversibly Inhibiting Platelet Cyclooxygenase-1. *Cancer Prev. Res.* **2017**, *10*, 142–152. [CrossRef] [PubMed]

114. Sandler, R.S.; Halabi, S.; Baron, J.A.; Budinger, S.; Paskett, E.; Keresztes, R.; Petrelli, N.; Pipas, J.M.; Karp, D.D.; Loprinzi, C.L.; et al. A randomized trial of aspirin to prevent colorectal adenomas in patients with previous colorectal cancer. *N. Engl. J. Med.* **2003**, *348*, 883–890. [CrossRef] [PubMed]

115. Fisher, B.; Costantino, J.P.; Wickerham, D.L.; Cecchini, R.S.; Cronin, W.M.; Robidoux, A.; Bevers, T.B.; Kavanah, M.T.; Atkins, J.N.; Margolese, R.G.; et al. Tamoxifen for the prevention of breast cancer: Current status of the National Surgical Adjuvant Breast and Bowel Project P-1 study. *J. Natl. Cancer Inst.* **2005**, *97*, 1652–1662. [CrossRef] [PubMed]

116. Johnson, K.E.; Forward, J.A.; Tippy, M.D.; Ceglowski, J.R.; El-Husayni, S.; Kulenthirarajan, R.; Machlus, K.R.; Mayer, E.L.; Italiano, J.E., Jr.; Battinelli, E.M. Tamoxifen Directly Inhibits Platelet Angiogenic Potential and Platelet-Mediated Metastasis. *Arterioscler. Thromb. Vasc. Biol.* **2017**, *37*, 664–674. [CrossRef] [PubMed]

117. Zhang, Y.; Wei, J.; Liu, S.; Wang, J.; Han, X.; Qin, H.; Lang, J.; Cheng, K.; Li, Y.; Qi, Y.; et al. Inhibition of platelet function using liposomal nanoparticles blocks tumor metastasis. *Theranostics* **2017**, *7*, 1062–1071. [CrossRef] [PubMed]

118. Chang, Y.W.; Hsieh, P.W.; Chang, Y.T.; Lu, M.H.; Huang, T.F.; Chong, K.Y.; Liao, H.R.; Cheng, J.C.; Tseng, C.P. Identification of a novel platelet antagonist that binds to CLEC-2 and suppresses podoplanin-induced platelet aggregation and cancer metastasis. *Oncotarget* **2015**, *6*, 42733–42738. [CrossRef] [PubMed]

119. Gresele, P.; Momi, S.; Falcinelli, E. Anti-platelet therapy: Phosphodiesterase inhibitors. *Br. J. Clin. Pharmacol.* **2011**, *72*, 634–646. [CrossRef] [PubMed]

120. Rondina, M.T.; Weyrich, A.S. Targeting phosphodiesterases in anti-platelet therapy. *Handb. Exp. Pharmacol.* **2012**, *210*, 225–238.

121. Uzawa, K.; Kasamatsu, A.; Baba, T.; Usukura, K.; Saito, Y.; Sakuma, K.; Iyoda, M.; Sakamoto, Y.; Ogawara, K.; Shiiba, M.; et al. Targeting phosphodiesterase 3B enhances cisplatin sensitivity in human cancer cells. *Cancer Med.* **2013**, *2*, 40–49. [CrossRef] [PubMed]

122. Tzanakakis, G.N.; Agarwal, K.C.; Vezeridis, M.P. Prevention of human pancreatic cancer cell-induced hepatic metastasis in nude mice by dipyridamole and its analog RA-233. *Cancer* **1993**, *71*, 2466–24671. [CrossRef]

123. Desai, P.B.; Duan, J.; Sridhar, R.; Damle, B.D. Reversal of doxorubicin resistance in multidrug resistant melanoma cells in vitro and in vivo by dipyridamole. *Methods Find. Exp. Clin. Pharmacol.* **1997**, *19*, 231–239. [PubMed]

124. Amirkhosravi, A.; Mousa, S.A.; Amaya, M.; Blaydes, S.; Desai, H.; Meyer, T.; Francis, J.L. Inhibition of tumor cell-induced platelet aggregation and lung metastasis by the oral GpIIb/IIIa antagonist XV454. *Thromb. Haemost.* **2003**, *90*, 549–554. [CrossRef] [PubMed]

125. Ishikawa, S.; Miyashita, T.; Inokuchi, M.; Hayashi, H.; Oyama, K.; Tajima, H.; Takamura, H.; Ninomiya, I.; Ahmed, A.K.; Harman, J.W.; et al. Platelets surrounding primary tumor cells are related to chemoresistance. *Oncol. Rep.* **2016**, *36*, 787–794. [CrossRef] [PubMed]

126. Haslehurst, A.M.; Koti, M.; Dharsee, M.; Nuin, P.; Evans, K.; Geraci, J.; Childs, T.; Chen, J.; Li, J.; Weberpals, J.; et al. EMT transcription factors snail and slug directly contribute to cisplatin resistance in ovarian cancer. *BMC Cancer* **2012**, *12*, 91. [CrossRef] [PubMed]

127. Tsukasa, K.; Ding, Q.; Yoshimitsu, M.; Miyazaki, Y.; Matsubara, S.; Takao, S. Slug contributes to gemcitabine resistance through epithelial-mesenchymal transition in CD133(+) pancreatic cancer cells. *Hum. Cell* **2015**, *28*, 167–174. [CrossRef] [PubMed]

128. Sarkar, S.; Alam, M.A.; Shaw, J.; Dasgupta, A.K. Drug delivery using platelet cancer cell interaction. *Pharm. Res.* **2013**, *30*, 2785–2794. [CrossRef] [PubMed]

129. Xu, P.; Zuo, H.; Zhou, R.; Wang, F.; Liu, X.; Ouyang, J.; Chen, B. Doxorubicin-loaded platelets conjugated with anti-CD22 mAbs: A novel targeted delivery system for lymphoma treatment with cardiopulmonary avoidance. *Oncotarget* **2017**, *8*, 58322–58337. [CrossRef] [PubMed]

130. Xu, P.; Zuo, H.; Chen, B.; Wang, R.; Ahmed, A.; Hu, Y.; Ouyang, J. Doxorubicin-loaded platelets as a smart drug delivery system: An improved therapy for lymphoma. *Sci. Rep.* **2017**, *7*, 42632. [CrossRef] [PubMed]
131. Li, J.; Ai, Y.; Wang, L.; Bu, P.; Sharkey, C.C.; Wu, Q.; Wun, B.; Roy, S.; Shen, X.; King, M.R. Targeted drug delivery to circulating tumor cells via platelet membrane-functionalized particles. *Biomaterials* **2016**, *76*, 52–65. [CrossRef] [PubMed]

International Journal of
Molecular Sciences

Article

Overexpression of the Vitronectin V10 Subunit in Patients with Nonalcoholic Steatohepatitis: Implications for Noninvasive Diagnosis of NASH

Maria Del Ben [1,†] , Diletta Overi [2,†], Licia Polimeni [1], Guido Carpino [3], Giancarlo Labbadia [1],
Francesco Baratta [1,2], Daniele Pastori [1,2] , Valeria Noce [4] , Eugenio Gaudio [2],
Francesco Angelico [5] and Carmine Mancone [4,*]

[1]   Department of Internal Medicine and Medical Specialties, Sapienza University of Rome,
    Viale del Policlinico 155, 00161 Rome, Italy; maria.delben@uniroma1.it (M.D.B.);
    licia.polimeni@uniroma1.it (L.P.); giancarlo.labbadia@uniroma1.it (G.L.);
    francesco.baratta@uniroma1.it (F.B.); daniele.pastori@uniroma1.it (D.P.)
[2]   Department of Anatomical, Histological, Forensic Medicine and Orthopedics Sciences,
    Sapienza University of Rome, Via Borelli 50, 00161 Rome, Italy; diletta.overi@uniroma1.it (D.O.);
    eugenio.gaudio@uniroma1.it (E.G.)
[3]   Department of Movement, Human and Health Sciences, Division of Health Sciences,
    University of Rome "Foro Italico", Piazza Lauro De Bosis 6, 00135 Rome, Italy; guido.carpino@uniroma1.it
[4]   Department of Cellular Biotechnologies and Haematology, Sapienza University of Rome,
    Viale del Policlinico 155, 00161 Rome, Italy; noce@bce.uniroma1.it
[5]   Department of Public Health and Infectious Diseases, Sapienza University of Rome, Viale del Policlinico 155,
    00161 Rome, Italy; francesco.angelico@uniroma1.it
*   Correspondence: carmine.mancone@uniroma1.it; Tel.: +39-06-4997-9670
†   These authors contributed equally to this work.

Received: 30 January 2018; Accepted: 16 February 2018; Published: 18 February 2018

Abstract: Nonalcoholic steatohepatitis (NASH) is the critical stage of nonalcoholic fatty liver disease (NAFLD). The persistence of necroinflammatory lesions and fibrogenesis in NASH is the leading cause of liver cirrhosis and, ultimately, hepatocellular carcinoma. To date, the histological examination of liver biopsies, albeit invasive, remains the means to distinguish NASH from simple steatosis (NAFL). Therefore, a noninvasive diagnosis by serum biomarkers is eagerly needed. Here, by a proteomic approach, we analysed the soluble low-molecular-weight protein fragments flushed out from the liver tissue of NAFL and NASH patients. On the basis of the assumption that steatohepatitis leads to the remodelling of the liver extracellular matrix (ECM), NASH-specific fragments were in silico analysed for their involvement in the ECM molecular composition. The 10 kDa C-terminal fragment of the ECM protein vitronectin (VTN) was then selected as a promising circulating biomarker in discriminating NASH. The analysis of sera of patients provided these major findings: the circulating VTN fragment (i) is overexpressed in NASH patients and positively correlates with the NASH activity score (NAS); (ii) originates from the disulfide bond reduction between the V10 and the V65 subunits. In conclusion, V10 determination in the serum could represent a reliable tool for the noninvasive discrimination of NASH from simple steatosis.

Keywords: nonalcoholic fatty liver disease; nonalcoholic steatohepatitis; liver fibrosis; secretome

---

1. Introduction

Nonalcoholic fatty liver disease (NAFLD) is by now the most common liver disease in the developed and developing countries with a global estimated median prevalence of 20% [1]. Most NAFLD patients have simple steatosis or nonalcoholic fatty liver (NAFL) with low inflammation, tissue damage, or liver fibrosis. However, 13% to 31% of the cases develop definite nonalcoholic steatohepatitis (NASH), characterized by hepatic steatosis and inflammation with ballooning, with or without fibrosis [2]. The prognosis is poorer in patients with NASH, and NASH patients can progress to liver cirrhosis and ultimately hepatocellular carcinoma (HCC) [3]. Therefore, it is crucial to obtain a prompt diagnosis of NAFLD patients with NASH for an effective clinical management.

Currently, the differentiation of NAFL from definite steatohepatitis and liver fibrosis in the whole spectrum of NAFLD is still based on the histological examination of liver biopsies to assign a grade and stage through scoring systems [4]. The NAFLD Activity Score (NAS) is widely accepted and used in clinical practice [5]. To stage fibrosis, the NASH Clinical Research Network system is one of the most validated systems currently available and defines four fibrosis stages [5,6]. Nevertheless, although the histological evaluation of liver biopsies addresses the full spectrum of lesions of NAFLD, this procedure remains invasive and limited by sampling error, diagnostic accuracy, and hazard to the patients [7].

Serum biomarkers offer a noninvasive and cost-effective alternative to liver biopsy both for patients and clinicians. Previous studies on NAFLD subjects described some circulating proteins and metabolites with a potential for discriminating between NASH and NAFL [8–11]. Nevertheless, the feasibility, specificity, and sensitivity of these candidate biomarkers are relatively low or, in some cases, only reflect one aspect of NAFLD progression instead of the overall condition. There is, therefore, an increasingly pressing need to identify NAS-related serum biomarkers that can reflect the grade of the disease. In this regard, low-abundance peptides in the serum may represent a reservoir of NASH-specific products originating from leakage as a result of cell death, extracellular matrix remodelling, or damages [12]. Unfortunately, one of the challenges in proteomic-based serum marker discovery is that serum samples are dominated by high-abundance proteins whose presence obscures less abundant products. In particular, the high concentrations of albumin and immunoglobulins (Igs) in serum samples prevent the successful identification of low-abundance biomarkers in studies based on top-down peptidomic approaches. Moreover, methods developed for depleting the high-abundance proteins, while eliminating the most of albumin and Igs, lead to the impoverishment of minor products. Therefore, identifying potential circulating markers of NASH straight from serum samples still remains an ambitious challenge.

In this study, to overcome this issue, we firstly performed a proteomic analysis of soluble low-molecular-weight (LMW) polypeptides flushed out from liver biopsies of NAFLD patients. The 10 kDa subunit (V10) of vitronectin was identified as upregulated in the flush-out samples of the NASH liver sample group. Hence, our aim has been to provide a proof of concept regarding the possible correlation of serum vitronectin fragments with histopathology in patients affected by NAFLD and its putative role in the assessment of disease severity.

2. Results

2.1. Characteristics of the Patients

The group of NAFLD patients included 50 subjects (24 males and 26 females). Among them, 27 subjects did not present definite steatohepatitis (NAFL group) while 23 subjects displayed definite steatohepatitis (NASH group). The fibrosis stage in NASH patients was more advanced than in those affected by NAFL. The patient's age did not statistically differ between NAFL and NASH groups. The body mass index (BMI), aspartate aminotransferase (AST), and alanine aminotransferase (ALT) levels were significantly higher in subjects with NASH than in those with NAFL. Gamma-glutamyltransferase (GGT) levels and AST/ALT ratio were not significantly different between the groups. The clinical and serological characteristics of the patients are summarized in Table 1.

Table 1. Clinical characteristics of NAFL and NASH populations.

	NAFL	NASH	p Value
All	27	23	
Gender (male/female)	14/13	10/13	0.55
Nas	2.81 ± 1.04	5.43 ± 0.99	**<0.001**
Fibrosis 1/2/3/4	11/12/4/0	3/5/9/6	**<0.001**
AGE at enrolment (years)	52.3 ± 14.4	51.7 ± .9.8	0.44
Body mass index (kg/m^2)	28.7 ± 4.1	30.5 ± 3.9	0.067
AST (U/L)	37.6 ± 17.8	61.9 ± 40.9	**0.005**
ALT (U/L)	69.3 ± 36.1	97.1 ± 56.1	**0.024**
GGT (U/L)	62.4 ± 41.5	80.7 ± 86.7	0.18
AST/ALT ratio	0.56 ± 0.18	0.68 ± 0.31	0.07

Data are reported as Means ± Standard Deviation; p values <0.05 are reported in bold; AST (Aspartate transaminase); ALT (Alanine transaminase); GGT (Gamma-glutamyl transferase).

2.2. Identification of LMW Polypeptides in NAFL and NASH Samples

The LMW polypeptides released from the liver tissue are an essential part of the NASH microenvironment and represent a reservoir of early promising and specific biomarkers firstly detectable in the NASH-specific secretome and, subsequently, circulating in the blood as low-abundance products. Therefore, a comparative proteomic analysis of enriched LMW polypeptides flushed out from the liver biopsies of NAFLD patients could represent a powerful alternative strategy for the discovery of circulating subnanomolar biomarkers.

Secretomes from the liver biopsies were individually collected, and equal amounts of proteins from patients having a diagnosis of simple steatosis (NAFL group) and definite steatohepatitis (NASH group) were pooled. One hundred micrograms of proteins from each pool was separated by non-reducing SDS-PAGE with a buffer system specific for the resolution of low-molecular-weight polypeptides. Each gel lane was cut into seven sections ranging from 3 to 20 kDa, and the polypeptides in each gel section were digested and submitted to nLC–MS-based proteomic analysis. A total of 235 and 119 proteins in NASH and NAFL groups, respectively, were identified (Table S1). Aiming at selecting peptides originating from the fragmentation of whole proteins, we used the electrophoretically derived molecular weight (MWexp) of the protein as the identification constraint. On the basis of the assumption that the MWexp of a protein should correlate with its theoretical molecular weight (MWcal) obtained from the database search, 63 and 23 identifications resulted as protein fragments in NASH and NASL groups, respectively (Table S1). Interestingly, by comparing these two datasets, 7 protein fragments were identified in the NAFL group, 16 were commonly identified, and 47 were found exclusively in the NASH group, thus confirming that the disease progression leads to increased protein fragmentation (Figure 1).

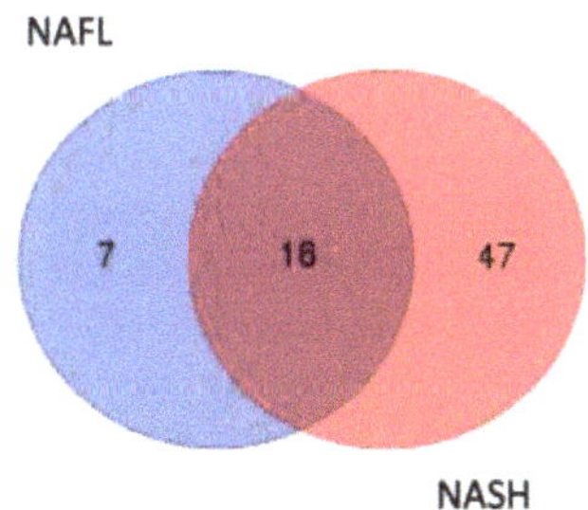

Figure 1. Venn diagram of the overlap of identified protein fragments in NASH and NAFL groups.

2.3. In Silico Analysis of NASH-Specific Fragments

In response to inflammatory stimuli, NASH leads to changes in the composition of the liver extracellular matrix (ECM), where high levels of ECM protein fragments are generated and released into the circulation at low concentrations [13]. Therefore, it is conceivable that liver-specific ECM circulating peptides may be the one of the most suitable candidates as circulating NAS-related biomarkers for assessing grade and stage in NASH patients.

The 47 proteins, whose fragments were found exclusively in the NASH group, were then analysed for their involvement in the ECM molecular composition by interrogating the MatrisomeDB 2.0 (http://matrisomeproject.mit.edu/proteins/), a searchable database that integrates experimental proteomic data on ECM and ECM-associated proteins from the ECM Atlas [14]. As shown in Table 2, seven proteins were found to be associated with liver ECM, five of which as minor components or regulators (i.e.: LGALS3, LGALS4, CTSB, SERPINB1, SERPINC1), and the vitronectin (VTN) and fibrinogen alpha chain (FGA) as main structural components. Since several reports demonstrated that the VTN is upregulated in the ECM of fibrotic liver [15–18], it is reasonable to hypothesize that this protein may be subjected to turnover and degradation. Accordingly, we decided to challenge the detection of the VTN fragment directly in the sera of patients.

Table 2. ECM and ECM-associated components analysis of NASH fragments by MatrisomeDB 2.0.

Gene Symbol	Name	Matrisome Division	Category
FGA	Fibrinogen alpha chain	Core matrisome	ECM Glycoproteins
VTN	Vitronectin	Core matrisome	ECM Glycoproteins
LGALS3	Galectin-3	Matrisome-associated	ECM-affiliated Proteins
LGALS4	Galectin-4	Matrisome-associated	ECM-affiliated Proteins
CTSB	Cathepsin B	Matrisome-associated	ECM Regulators
SERPINB1	Leukocyte elastase inhibitor	Matrisome-associated	ECM Regulators
SERPINC1	Antithrombin-III	Matrisome-associated	ECM Regulators

2.4. Correlation between Circulating VTN Fragments and Liver HistoMorphology

Mature vitronectin is a multifunctional plasma and extracellular matrix protein of 459 amino acids with a MWcal of 52 kDa; the observed MWexp of the glycosylated form is 75 kDa [19]. We identified the VTN fragment in the 12–7 kDa SDS-PAGE-displayed molecular weight range. Interestingly, by tandem mass spectrometry analysis, we found that the sequence of the tryptic peptides, originated from the digestion of the VTN fragment, matched with the C-terminal end of the protein (Table S2). To detect this fragment in the sera of our cohort of patients, a non-reducing western blotting analysis was performed by using an antibody that recognizes the vitronectin C-terminal end. As expected, the 75 kDa form of VTN (V75) was detected in the serum of NAFLD patients (Figure 2). Moreover, an additional signal of approximately 10 kDa (V10) was detected, thus confirming the existence of a circulating C-terminal fragment. Surprisingly, the V10 level of expression did not reflect the expression level of the circulating mature form. We then analysed the serum levels of V10 and V75 in the frame of NASH by densitometry. As shown in Figure 3, NASH patients displayed increased levels of V10 compared to the NAFL group ($p = 0.027$). On the contrary, lower amounts of V75 were measured in the serum of NASH patients ($p = 0.013$) and in patients with a fibrosis score >2 compared to patients with a fibrosis score =1 ($p < 0.01$). Then, we further extended the analysis by using the V10/V75 ratio (Figure 3). Interestingly, the V10/V75 ratio was significantly higher in NASH compared to NAFL patients ($p = 0.003$). By bivariate analysis (Figure 3), V10, V75, and V10/V75 ratio significantly correlated with the NAS score (respectively: $r = 0.311$, $p = 0.028$; $r = -0.318$, $p = 0.024$; $r = 0.399$, $p = 0.004$) but not with clinical–serological parameters (i.e., BMI, ALT, AST); only the V75 fragments correlated with the fibrosis score ($r = -0.424$; $p = 0.002$). To further assess whether these results could be actually due to the reported differences in BMI, ALT, and AST between the two groups (NAFL and NASH), a linear regression analysis was performed by two models using NAS or, alternatively, fibrosis as independent variables. The ratio V10/V75 resulted a predictor of NAS (beta = 0.354; p = 0.010) but not of fibrosis (beta= 0.166; p = 0.248), independently from the above-mentioned parameters (i.e., BMI, ALT, AST); moreover, the V75 fragment resulted a predictor of fibrosis (beta= −0.292; p = 0.034) but not of NAS (beta= −0.279; p = 0.051), independently from BMI, ALT, AST. Given the low beta value, we further tried to individuate a possible cut-off value for the V10/V75 ratio by analysing sera from healthy subjects (N = 6). The V10, V75, and V10/V75 ratio values resulted significantly higher in NAFLD patients compared to healthy subjects (p < 0.05). For the V10/V75 ratio, we assumed the highest value obtained in healthy subjects as a possible cut-off (0.27) to be used as a threshold for NAFLD patients. The binary logistic regression demonstrated that NASH diagnosis was associated with a V10/V75 ratio over (>0.27) the identified threshold (OR: 5.254; CI 95%: 1.142–24.163; p = 0.033) after adjustment for BMI, AST, ALT.

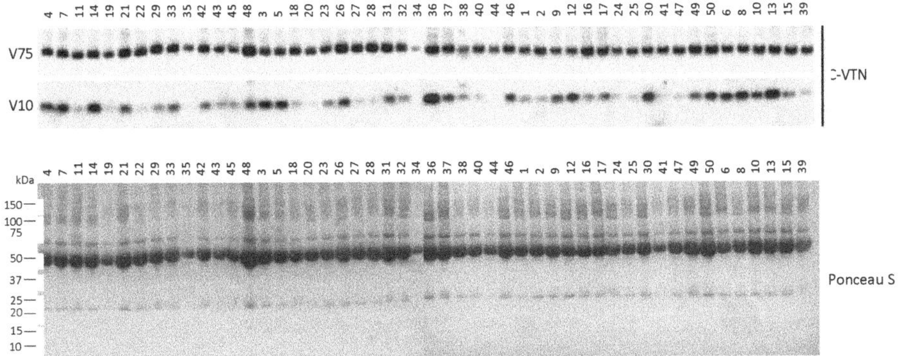

Figure 2. Western blotting analysis of 10 kDa VTN fragment levels in NAFLD patients. Dilutions (1:10) of the serum samples were separated on a non-reducing SDS polyacrylamide gel and probed with an antibody specific for the vitronectin C-terminal end (C-VTN). Total protein staining by Ponceau S on the nitrocellulose membrane is shown. Patient IDs are indicated. These images are representative of experiments carried out in triplicate

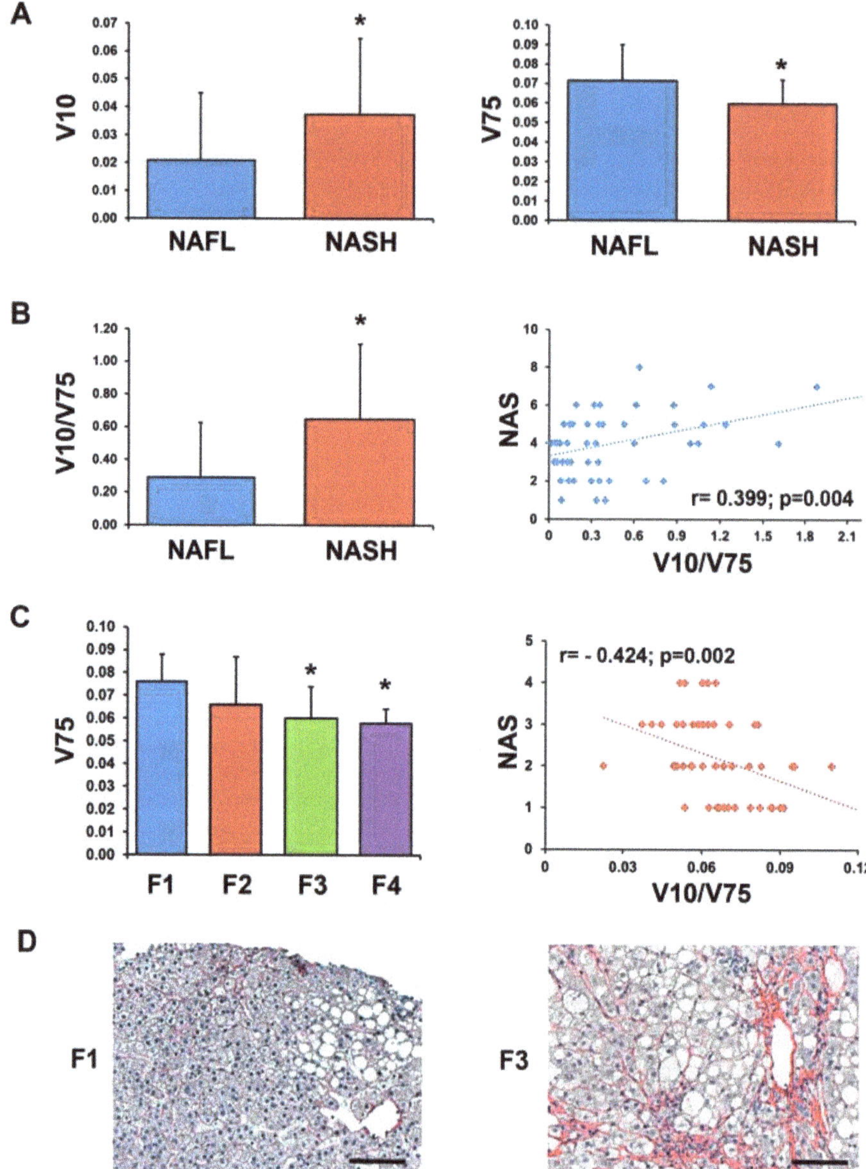

Figure 3. V10 and V75 levels are increased in NASH patients' serum. (**A**) Densitometry analysis of the V10 and V75 blots imaged in Figure 2. The total lane density detected from the Ponceau S staining of the transferred protein in the blots was used for normalization. Patients with definite steatohepatitis (NASH group) have higher V10 and lower V75 levels compared with patients without definite steatohepatitis (NAFL group); * = $p < 0.05$. (**B**) Histogram showing that the V10/V75 ratio was higher in NASH compared to NAFL group (left); * < 0.05. The scatter plot graph on the right shows the correlation between the V10/V75 ratio and the NAS score. (**C**) The histogram on the left shows V75 serum levels in patients divided according to the fibrosis (F) score; * < 0.01 versus F1 group. The scatter plot graph on the right shows the correlation between V75 levels and the fibrosis score. (**D**) Sirius red stains in liver biopsies are representative of F1 and F3 stages, respectively. Scale bar = 100 μm.

2.5. Circulating V10 Originates by the Reduction of VTN Clipped Form

VTN is produced and secreted by the hepatocytes in two molecular forms of 75 kDa: a single, non-cleaved chain and a furin-mediated clipped form composed of two chains of 65 and 10 kDa which are held together by a disulfide bond [19]. Consequently, we wondered whether the increased level of the C-terminal fragment of vitronectin in the sera of patients with NASH may be related to the degradation of vitronectin by proteases overproduced in NASH or to the reduction of the disulfide bond between the two vitronectin subunits.

The degradation of vitronectin occurs by means of matrix metalloproteinases (MMPs)-1, -2, -3, -7, and -9 [20]. Since MMP-2 and -9 are found extensively upregulated in NASH [21], we first sought to determine whether in vitro digestion of circulating VTN by these MMPs gave rise to increased levels of V10. As indicated in Figure 4A, the proteolytic activity generated fragments ranging from 37 to 45 kDa. Nevertheless, the 10 kDa signals were found at lower levels with respect to the control, thus suggesting that the MMP-2 and -9 are not responsible for the V10 production.

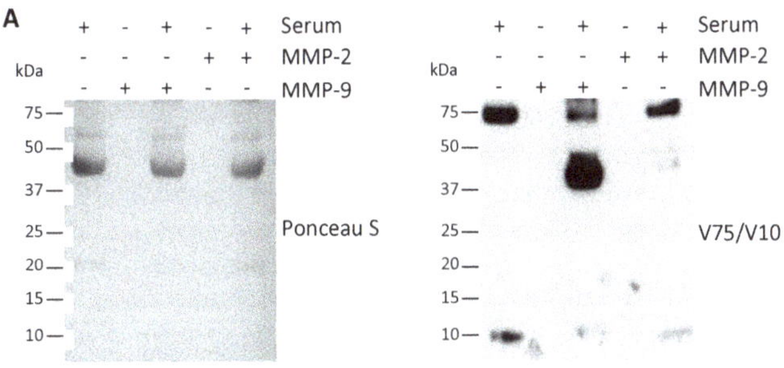

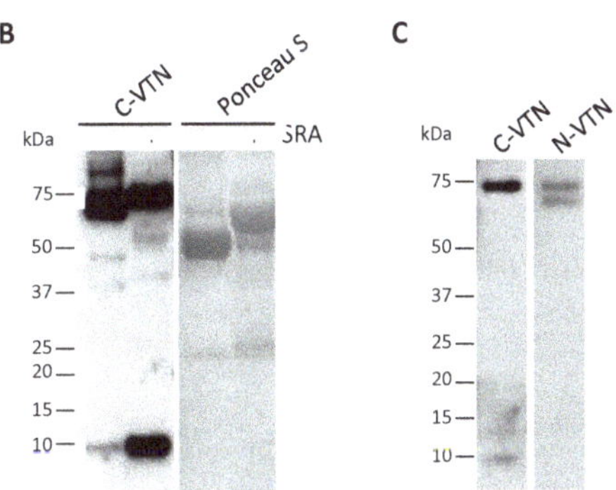

Figure 4. The 10 kDa VTN fragment originates from the reduction of the disulfide bond between the V65 and V10 subunits. (**A**) C-VTN non-reducing western blotting analysis of 1:10 dilutions of a NASH serum sample in the presence or absence of MMP-2 and -9. Total protein staining by Ponceau S on the nitrocellulose membrane is shown. (**B**) C-VTN western blotting analysis of 1:10 dilutions of a NASH serum sample in the presence or absence of a sample-reducing agent (SRA). Total protein staining by Ponceau S on the nitrocellulose membrane is shown. (**C**) Non-reducing western blotting analysis of 1:10 dilutions of a NASH serum sample probed with antibodies specific for the vitronectin C-terminal (C-VTN) and N-terminal (N-VTN) ends. These images are representative of experiments carried out in triplicate.

Then, we verified the hypothesis that V10 originated by the releasing of the 10 kDa subunit from the vitronectin clipped form by means of disulfide bond reduction. Firstly, we proved that the addition of a reducing agent generated a signal corresponding to the same molecular weight of V10 (Figure 4B). Then, we analysed the presence of the free circulating 65 kDa subunit (V65) by using an antibody that recognizes the vitronectin N-terminal end. As reported in Figure 4C, in addition to the 75 kDa signal, the immunoblot after non-reducing electrophoresis highlights the presence of a further band corresponding to the V65 subunit, thus suggesting that the circulating 10 kDa fragment could be the result of the disulfide bond reduction.

3. Discussion

Our analysis on sera of NAFLD patients showed that: (i) circulating V10 is overexpressed in NASH patients compared to NAFL patients and positively correlates with NAS; (ii) circulating V75 is underexpressed in NASH patients compared to NAFL patients and negatively correlates with NAS and liver fibrosis; (iii) the V10/V75 ratio is a predictor of NAS score, independently from BMI, ALT, and AST; (iv) circulating V10 originates from the disulfide bond reduction of the V75 clipped form.

The aim of the present study was to identify circulating serum markers which could be helpful in distinguishing patients with high NAFLD activity or with advanced fibrosis. For this purpose, we focused our attention on those peptides originated by the fragmentation of mature proteins that may reflect valuable disease-related information. However, the proteomic screening for these fragments was performed on liver biopsy secretomes to overcome complexity and detection problems associated with peptide analysis of serum or plasma. Serum and plasma proteins are present at concentrations that are likely to extend over 10 orders of magnitude [22]. In addition, the 22 major proteins, including albumin and immunoglobulins, representing 99% of total serum polypeptides, are preferentially sequenced by mass spectrometry [23]. Therefore, the identification of the remaining 1%, consisting of thousands of LMW proteins and peptides, remains challenging despite the proposed strategies for peptide extraction from serum [24,25]. Here, by enriching LMW polypeptides flushed out from the liver specimens of patients, we generated NAFL- and NASH-specific data sets of LMW protein fragments whose comparison highlighted an increased tissue protein turnover during the progression of the disease from the simple steatosis to the steatohepatitis. To gain insights on the specific valence of the protein fragments found in NASH, we focused our attention on the liver ECM specific products which may reflect the pathological matrix remodelling in steatohepatitis. Five ECM-derived fragments were found, and, among these, we restricted the experimental observations to the vitronectin 10 kDa C-terminal fragment.

Vitronectin has been extensively studied in the frame of liver fibrosis in chronic liver diseases such as viral hepatitis B and C infections and HCC. Particularly, the ECM-associated vitronectin is found at low concentrations in normal liver and markedly increases in the cirrhotic liver, while the circulating 75 kDa form follows an opposite trend [15–18].

Here, for the first time, we analysed the expression of circulating VTN in the frame of NAFLD. We demonstrated that in NASH patients, VTN undergoes molecular remodelling that releases the two subunits from the clipped form. Since the small subunit V10 was found at high concentrations in NASH, while the V75 form was found to be decreased, the ratio V10/V75 might be a suitable serological indicator helpful for discriminating the presence of steatohepatitis in NAFLD patients. A 10 kDa vitronectin C-terminal fragment has been previously identified as a serum marker of HCC [26]. Nevertheless, while in HCC the V10 is produced by the catalytic activity of MMP-2 on the V75 non-cleaved form, we demonstrated that in NASH it originates from the disulfide bond reduction of the clipped form. Oxidative stress has been extensively demonstrated to be a major factor in the development of NASH [27–29]. Particularly, it has been shown that the production of reactive oxygen and nitrogen species in a context of liver steatosis promotes lipid peroxidation which, in turn, supports the necroinflammatory milieu and leads to the stimulation of collagen synthesis in hepatic stellate cells [30,31]. To counteract the oxidative stress, mammalian cells activate thioredoxin (Trx), which maintains a reducing environment by catalysing an electron flux from nicotinamide adenine dinucleotide phosphate to Trx through Trx reductase, which reduces its target proteins using highly conserved thiol groups [32]. Therefore, it is conceivable that the oxidative stress-induced elevated serum [33] and tissue (Table S1) Trx levels in NASH patients may be responsible for the free V10 production. Moreover, since increased hepatic vitronectin expression favors fibrogenesis by recruiting lymphocytes within the inflamed liver tissue and promoting a wound-healing response [19,34], the oxidative stress-induced reduction of the functional V75 clipped form may be the consequence of a pathophysiological response for counteracting liver fibrosis and the disease progression. Further experiments may challenge the robustness of these insights.

The validation and clinical availability of serum biomarkers of NASH are desirable to aid clinicians in the discrimination of NAFLD patients with steatohepatitis from simple steatosis and for the noninvasive monitoring of disease progression and response to therapy. To date, the most promising proposed serum biomarkers (i.e.: cytokeratin-18 M30 fragment [8], fibroblast growth factor 21 [35], interleukin 1 receptor antagonist [36], pigment epithelium-derived factor [11], osteoprotegerin [37]), showed a potential in diagnosing NAFLD and NASH. However, taken individually, they reflect only one aspect of the NAFLD pathological scenery, i.e., hepatocyte apoptosis, inflammation, fibrosis, insulin sensitivity, and steatosis. Moreover, some of them, showing low sensitivity and accuracy, need to be validated in a larger cohort [38]. However, it has been demonstrated that tests performed by combining these biomarkers improved the accuracy in the diagnosis of NASH [38]. Thus, we believe that the measurement of V10 in blood could be a further useful tool for detecting NASH by means of a panel of markers. However, the present study represents a proof of concept regarding the possibility to detect vitronectin fragments in the serum of patients affected by NAFLD and a possible correlation with histomorphology. Thus, our results should be replicated in a larger cohort of NAFLD patients to validate their eventual clinical relevance. In addition, it has, in general, several limitations: (i) a sampling bias may have originated as liver biopsy was used as gold standard for assessing the use of V10; (ii) the samples were all selected within the Italian population; (iii) the lack of validation based on a gel-free quantitative immunoassay able to discriminate the C-terminal V10 from the V75; (iv) the specificity of the V10 fragment in relation to hepatocyte injury in NASH has not been challenged in patients with other liver diseases. Moreover, since the presence of threonine rather than methionine at position 381 was proposed to be responsible for the susceptibility of VTN to cleavage at Arg^{379}–Ala^{380} for subunits production [19], a genetic SNP analysis of the VTN gene may be considered for those NASH patients with low levels of circulating V10.

In conclusion, our findings suggest that the circulating VTN 10 kDa subunit may be a reliable tool in the discrimination of patients with NASH.

4. Materials and Methods

4.1. Patient Characteristics

The study was approved by the Institutional Ethic Committee of Sapienza University of Rome (prot. 873/11, Rif. 2277 approved on 13 October 2011) and conforms to the ethical guidelines of the 1975 Declaration of Helsinki. The population for the current study consisted of 50 well-characterized, biopsy-proven adult NAFLD patients. To be eligible for the study, NAFLD patients had to fulfill the following criteria: ultrasound evidence of fatty liver (defined according to Hamagouchi criteria), absence of current or past excessive alcohol drinking as defined by an average daily consumption of alcohol >20 g for women and >30 g for men; negative tests for the presence of hepatitis B surface antigen and antibody to hepatitis C virus. Percutaneous liver biopsy was performed under ultrasound guidance in fatty liver patients with clinical suspicion of NASH by their treating hepatologists. The decision to perform the biopsy was individualized and based on a persistent elevation of serum alanine aminotransferase levels (>1.5 above normal values) for more than 6 months and the presence of risk factors for NASH. A single operator performed ultrasound-guided liver biopsies. The pathologist who examined the biopsy specimens was blinded to patients' identity or clinical information. Diagnosis of definite steatohepatitis (i.e., NASH) was defined using standard criteria [4], and NAFLD activity score (NAS) was calculated on the basis of separate scores for steatosis, hepatocellular ballooning, and inflammation [5]. Fibrosis was scored on a scale of 0–4 [5]. Blood samples from six control (healthy) subjects was analysed. Control subjects (three male and three female) with age = 49 ± 13 years (mean $\pm$ standard deviation) were included; these subjects had a BMI < 25, did not have metabolic risk factors, (e.g., diabetes, hypertension), were negative for viral hepatitis markers, had normal liver serological tests, and had no sign of liver steatosis at ultrasound. A written informed consent was obtained from all subjects.

4.2. Liver Secretome, Protein Digestion, and Peptide Purification

Secreted proteins from fresh liver biopsies were obtained by overnight shaking (600 RPM) in Washing Buffer (0.5 M NaCl, 10 mM Tris Base pH 7.5, 1× protease inhibitor cocktail (Sigma Aldrich, St. Louis, MO, USA)) at 37 °C. After centrifugation at 13,000 rpm for 1 min, the supernatants were collected, and protein concentrations were measured by Bradford assay. Equal amounts of samples were individually collected to generate pools of 100 µg. The samples were then separated on 12% gels (Life technologies, Thermo Fisher Scientific, Waltham, MA, USA) by SDS-PAGE with MES running buffer (Life technologies). The gels were stained by Simply Blue Safe Stain (Life technologies), and seven sections for each gel lane were cut. Protein-containing gel pieces were washed with 100 µL of 0.1 M ammonium bicarbonate (5 min at RT). Then, 100 µL of 100% acetonitrile (ACN) was added to each tube, and the tubes were incubated for 5 min at RT. The liquid was discarded, the washing step repeated once more, and the gel plugs were shrunk by adding ACN. The dried gel pieces were reconstituted with 100 µL of 10 mM DTT/0.1 M ammonium bicarbonate and incubated for 40 min at 56 °C for cysteine reduction. The excess liquid was then discarded, and the cysteines were alkylated with 100 µL of 55 mM IAA/0.1 M ammonium bicarbonate (20 min at RT, in the dark). The liquid was discarded, the washing step was repeated once more, and the gel plugs were shrunk by adding ACN. The dried gel pieces were reconstituted with 12.5 ng/µL trypsin in 50 mM ammonium bicarbonate and digested overnight at 37 °C. The supernatant from the digestions was saved in a fresh tube, and 100 µL of 1% TFA/30% ACN was added to the gel pieces for an additional extraction of peptides. The extracted solution and digested mixture were then combined and vacuum-centrifuged for organic component evaporation. The peptides were resuspended with 40 µL of 2.5% ACN/0.1% TFA, desalted, filtered through a C18 microcolumn ZipTip, and eluted from the C18 bed using 10 µL of 80% ACN/0.1% TFA. The organic component was once again removed by evaporation in a vacuum centrifuge, and the peptides were resuspended in a suitable nanoLC injection volume (typically 3−10 µL) of 2.5% ACN/0.1% TFA.

4.3. NanoLC Analysis and Mass Spectrometry Analysis

An UltiMate 3000 RSLC nano-LC system (ThermoFisher Scientific, Waltham, MA, USA) equipped with an integrated nanoflow manager and microvacuum degasser was used for peptide separation. The peptides were loaded onto a 75 µm NanoSeries C18 column (ThermoFisher, P/N 164534) for multistep gradient elution (eluent A 0.05% TFA; eluent B 0.04% TFA in 80% ACN) from 5% to 20% eluent B within 10 min, from 20% to 50% eluent B within 45 min, and for further 5 min from 50% to 90% eluent B with a constant flow of 0.3 µL/min. After 5 min, the eluted sample fractions were continuously diluted with 1.2 µL/min a-cyano-4-hydroxycinnamic acid (CHCA) and spotted onto a MALDI target using a HTC-xt spotter (PAL SYSTEM) with an interval of 20 s resulting in 168 fractions for each gel slice. Mass Spectrometry Analysis MALDI-TOF MS spectra were acquired using a 5800 MALDI TOF/TOF Analyzer (Sciex, Concord, ON, Canada). The spectra were acquired in the positive reflector mode by 20 subspectral accumulations (each consisting of 50 laser shots) in an 800−4000 mass range, focus mass 2100 Da, using a 355 nm Nb:YAG laser with a 20 kV acceleration voltage. Peak labeling was automatically done by 4000 Series Explorer software Version 4.1.0 (Sciex) without any kind of smoothing of peaks or baseline, considering only peaks that exceeded a signal-to-noise ratio of 10 (local noise window 200 m/z) and a half maximal width of 2.9 bins. The calibration was performed using default calibration originated by five standard spots (Mass Standards kit for Calibration P/N 4333604). Only the MS/MS spectra of preselected peaks (out of peak pairs with a mass difference of 6.02, 10.01, 12.04, 16.03, and 20.02 Da) were integrated over 1000 laser shots in the 1 kV positive ion mode with the metastable suppressor turned on. Air at the medium gas pressure setting (1.25×10^{-6} Torr) was used as the collision gas in the CID-off mode. After smoothing and baseline subtractions, spectra were generated automatically by 4000 Series Explorer software. The MS and MS/MS spectra were processed by ProteinPilot Software 4.5 (Sciex) which acts as an interface between the Oracle database containing raw spectra and a local copy of the MASCOT search engine (Version 2.1, Matrix Science, Ltd.). The Paragon algorithm was used with identification as the Sample Type, iodacetamide as cysteine alkylation, with the search option "biological modifications" checked, and trypsin as the selected enzyme. MS/MS protein identification was performed against the Swiss-Prot database (number of protein sequences: 254757; released on 20121210) without taxon restriction using a confidence threshold of 95% (Proteinpilot Unused score $\geq$ 1.31). The monoisotopic precursor ion tolerance was set to 0.12 Da and the MS/MS ion tolerance to 0.3 Da. The minimum required peptide length was set to six amino acids.

4.4. Immunoblotting Analysis

Dilutions 1:10 of the serum samples were separated in 12% gels (Life technologies) by SDS-PAGE with MES running buffer (Life technologies) and electroblotted onto nitrocellulose (GE, Healthcare, Little Chalfont, UK) membranes. The blots were incubated with primary and secondary antibodies. The antibodies were revealed using ECL (Millipore). To control for equal protein loading and transfer, the membranes were stained with Ponceau S solution (Sigma). The following antibodies were used: anti-C-VTN (Abcam, Cambridge, UK, ab113700), anti-N-VTN (Santa Cruz Biotechnology, Dallas, Texas, USA, sc-15332); the secondary anti-rabbit peroxidase-conjugated antibody was from Jackson ImmunoResearch. The chemiluminescent blots were imaged with the ChemiDoc™ Touch Imaging System (Bio-Rad Laboratories, Hercules, CA, USA), and vitronectin band densities were quantified by ImageLab software version 5.1.2 (Bio-Rad). Total protein lane densities by Ponceau S staining were used for normalization.

4.5. In Vitro Degradation of Vitronectin by Metalloproteases

MMP-2 (ab81550) and MMP-9 (ab82955) were from Abcam. Before incubation with the serum sample, pro MMP-9 was activated by incubation with 4-aminophenylmercuric acetate 1 mM (Sigma, St. Louis, MO, USA) overnight at 37 °C. The digestion of vitronectin was carried out by incubation of the substrate at 37 °C for 24 h with MMPs in an enzyme-to-substrate ratio of 1:20 in 50 mMTris-HCl, pH 7.5, containing 0.15 M NaCl, $CaCl_2$ 10 mM, 0.05% Brij 35 and 0.02% NaN_3.

4.6. Statistics

Outcome measures for between-group comparisons of patient characteristics (Age, body mass, AST, ALT, GGT, V10, and V75) were reported as mean and standard deviation or median and interquartile range, as appropriate. The Student t test or Mann–Whitney U test were used to compare groups for normally or not normally distributed data, respectively. The Pearson correlation coefficient or the Rho Spearman nonparametric correlation test were used. All tests are two-tailed, and a p-value of <0.05 was considered as statistically significant.

Linear multiple regression analyses were performed to identify factors independently associated with NAS and liver fibrosis. A further multiple logistic regression analysis was carried out to test the independent factors associated with the V10/V75 ratio above the upper normal value (0.27) found in healthy controls. The analyses were performed using SPSS software v.23 (IBM, Milan, Italy).

Supplementary Materials: Supplementary materials can be found at www.mdpi.com/xxx/s1.

Acknowledgments: This work was supported by Sapienza University of Rome "Fondi di Ateneo". We are deeply grateful to Maria Teresa Catanese for the editing.

Author Contributions: Carmine Mancone conceived and designed the research; Carmine Mancone, Licia Polimeni, Diletta Overi, Guido Carpino, Giancarlo Labbadia, Francesco Baratta, Maria Del Ben, and Francesco Angelico, contributed to the acquisition, analysis, and interpretation of data; Carmine Mancone and Francesco Angelico contributed to draft the work; Eugenio Gaudio, Daniele Pastori, and Valeria Noce revised it critically for important intellectual content. All authors have read and approved the final manuscript.

Conflicts of Interest: The authors declare no conflict of interest.

References

1. Vernon, G.; Baranova, A.; Younossi, Z.M. Systematic review: The epidemiology and natural history of non-alcoholic fatty liver disease and non-alcoholic steatohepatitis in adults. *Aliment Pharmacol. Ther.* **2011**, *34*, 274–285. [CrossRef] [PubMed]

2. Bedossa, P. Pathology of non-alcoholic fatty liver disease. *Liver Int.* **2017**, *37*, 85–89. [CrossRef] [PubMed]

3. Dam-Larsen, S.; Franzmann, M.; Andersen, I.B.; Christoffersen, P.; Jensen, L.B.; Sørensen, T.I.; Becker, U.; Bendtsen, F. Long term prognosis of fatty liver: Risk of chronic liver disease and death. *Gut* **2004**, *53*, 750–755. [CrossRef] [PubMed]

4. Chalasani, N.; Younossi, Z.; Lavine, J.E.; Diehl, A.M.; Brunt, E.M.; Cusi, K.; Charlton, M.; Sanyal, A.J. The diagnosis and management of non-alcoholic fatty liver disease: Practice Guideline by the American Association for the Study of Liver Diseases, American College of Gastroenterology, and the American Gastroenterological Association. *Hepatology* **2012**, *55*, 2005–2023. [CrossRef] [PubMed]

5. Kleiner, D.E.; Brunt, E.M.; van Natta, M.; Behling, C.; Contos, M.J.; Cummings, O.W.; Ferrell, L.D.; Liu, Y.C.; Torbenson, M.S.; Unalp-Arida, A.; et al. Design and validation of a histological scoring system for nonalcoholic fatty liver disease. *Hepatology* **2005**, *41*, 1313–1321. [CrossRef] [PubMed]

6. Sanyal, A.J.; Brunt, E.M.; Kleiner, D.E.; Kowdley, K.V.; Chalasani, N.; Lavine, J.E.; Ratziu, V.; McCullough, A. Endpoints and clinical trial design for nonalcoholic steatohepatitis. *Hepatology* **2011**, *54*, 344–353. [CrossRef] [PubMed]

7. Ratziu, V.; Charlotte, F.; Heurtier, A.; Gombert, S.; Giral, P.; Bruckert, E.; Grimaldi, A.; Capron, F.; Poynard, T.; LIDO Study Group. Sampling variability of liver biopsy in nonalcoholic fatty liver disease. *Gastroenterology* **2005**, *128*, 1898–1906. [CrossRef] [PubMed]

8. Feldstein, A.E.; Wieckowska, A.; Lopez, A.R.; Liu, Y.C.; Zein, N.N.; McCullough, A.J. Cytokeratin-18 fragment levels as noninvasive biomarkers for nonalcoholic steatohepatitis: A multicenter validation study. *Hepatology* **2009**, *50*, 1072–1078. [CrossRef] [PubMed]

9. Goh, G.B.; Issa, D.; Lopez, R.; Dasarathy, S.; Sargent, R.; Hawkins, C.; Pai, R.K.; Yerian, L.; Khiyami, A.; Pagadala, M.R.; et al. The development of a non-invasive model to predict the presence of non-alcoholic steatohepatitis in patients with non-alcoholic fatty liver disease. *J Gastroenterol. Hepatol.* **2016**, *31*, 995–1000. [CrossRef] [PubMed]

10. Dong, S.; Zhan, Z.Y.; Cao, H.Y.; Wu, C.; Bian, Y.Q.; Li, J.Y.; Cheng, G.H.; Liu, P.; Sun, M.Y. Urinary metabolomics analysis identifies key biomarkers of different stages of nonalcoholic fatty liver disease. *World J. Gastroenterol.* **2017**, *23*, 2771–2784. [CrossRef]

11. Yilmaz, Y.; Eren, F.; Ayyildia, T.; Colak, Y.; Kurt, R.; Senates, E.; Tuncer, I.; Dolar, E.; Imeryuz, N. Serum pigment epithelium-derived factor levels are increased in patients with biopsy-proven nonalcoholic fatty liver disease and independently associated with liver steatosis. *Clin. Chim. Acta* **2011**, *412*, 2296–2299. [CrossRef]

12. Drake, R.R.; Cazares, L.; Semmes, O.J. Mining the low molecular weight proteome of blood. *Proteomics Clin. Appl.* **2007**, *1*, 758–768. [CrossRef] [PubMed]

13. Karsdal, M.A.; Manon-Jensen, T.; Genovese, F.; Kristensen, J.H.; Nielsen, M.J.; Sand, J.M.; Hansen, N.U.; Bay-Jensen, AC.; Bager, C.L.; Krag, A.; et al. Novel insights into the function and dynamics of extracellular matrix in liver fibrosis. *Am. J. Physiol. Gastrointest. Liver Physiol.* **2015**, *308*, G807–G830. [CrossRef] [PubMed]

14. Naba, A.; Clauser, K.R.; Ding, H.; Whittaker, C.A.; Carr, S.A.; Hynes, R.O. The Extracellular Matrix: Tools and Insights for the "Omics" Era. *Matrix Biol.* **2016**, *49*, 10–24. [CrossRef] [PubMed]

15. Koukoulis, G.K.; Shen, J.; Virtanen, I.; Gould, V.E. Vitronectin in the cirrhotic liver: An immunomarker of mature fibrosis. *Hum. Pathol.* **2001**, *32*, 1356–1362. [CrossRef] [PubMed]

16. Kobayashi, J.; Yamada, S.; Kawasaki, H. Distribution of vitronectin in plasma and liver tissue: Relationship to chronic liver disease. *Hepatology* **1994**, *20*, 1412–1417. [CrossRef] [PubMed]

17. Inuzuka, S.; Ueno, T.; Torimura, T.; Tamaki, S.; Sakata, R.; Sata, M.; Yoshida, H.; Tanikawa, K. Vitronectin in liver disorders: Biochemical and immunohistochemical studies. *Hepatology* **1992**, *15*, 629–636. [CrossRef] [PubMed]

18. Baiocchini, A.; Montaldo, C.; Conigliaro, A.; Grimaldi, A.; Correani, V.; Mura, F.; Ciccosanti, F.; Rotiroti, N.; Brenna, A.; Montalbano, M.; et al. Extracellular Matrix Molecular Remodeling in Human Liver Fibrosis Evolution. *PLoS ONE* **2016**, *11*, e0151736. [CrossRef] [PubMed]

19. Schvartz, I.; Seger, D.; Shaltiel, S. Vitronectin. *Int. J. Biochem. Cell. Biol.* **1999**, *31*, 539–544. [CrossRef]

20. Imai, K.; Shikata, H.; Okada, Y. Degradation of vitronectin by matrix metalloproteinases-1, -2, -3, -7 and -9. *FEBS Lett.* **1995**, *369*, 249–251. [CrossRef]

21. Ljumovic, D.; Diamantis, I.; Alegakis, A.K.; Kouroumalis, E.A. Differential expression of matrix metalloproteinases in viral and non-viral chronic liver diseases. *Clin. Chim. Acta* **2004**, *349*, 203–211. [CrossRef] [PubMed]

22. Finoulst, I.; Pinkse, M.; van Dongen, W.; Verhaert, P. Sample preparation techniques for the untargeted LC-MS-based discovery of peptides in complex biological matrices. *J. Biomed. Biotechnol.* **2011**, *2011*, 245291. [CrossRef] [PubMed]

23. Anderson, N.L.; Anderson, N.G. The human plasma proteome: History, character, and diagnostic prospects. *Mol. Cell. Proteom.* **2002**, *1*, 845–867. [CrossRef] [PubMed]

24. Kawashima, Y.; Fukutomi, T.; Tomonaga, T.; Takahashi, H.; Nomura, F.; Maeda, T.; Kodera, Y. High-yield peptide-extraction method for the discovery of subnanomolar biomarkers from small serum samples. *J. Proteome Res.* **2010**, *9*, 1694–1705. [CrossRef] [PubMed]

25. Chen, L.; Zhai, L.; Li, Y.; Li, N.; Zhang, C.; Ping, L.; Chang, L.; Wu, J.; Li, X.; Shi, D.; Xu, P. Development of gel-filter method for high enrichment of low molecular weight proteins from serum. *PLoS ONE* **2015**, *10*, e0115862. [CrossRef] [PubMed]

26. Paradis, V.; Degos, F.; Dargère, D.; Pham, N.; Belghiti, J.; Degott, C.; Janeau, J.L.; Bezeaud, A.; Delforge, D.; Cubizolles, M.; et al. Identification of a new marker of hepatocellular carcinoma by serum protein profiling of patients with chronic liver diseases. *Hepatology* **2005**, *41*, 40–47. [CrossRef] [PubMed]

27. Tariq, Z.; Green, C.J.; Hodson, L. Are oxidative stress mechanisms the common denominator in the progression from hepatic steatosis towards non-alcoholic steatohepatitis (NASH)? *Liver Int.* **2014**, *34*, e180–e190; [CrossRef] [PubMed]

28. Polimeni, L.; del Ben, M.; Baratta, F.; Perri, L.; Albanese, F.; Pastori, D.; Violi, F.; Angelico, F. Oxidative stress: New insights on the association of non-alcoholic fatty liver disease and atherosclerosis. *World J. Hepatol.* **2015**, *7*, 1325–1336. [CrossRef] [PubMed]

29. Del Ben, M.; Polimeni, L.; Baratta, F.; Bartimoccia, S.; Carnevale, R.; Loffredo, L.; Pignatelli, P.; Violi, F.; Angelico, F. Serum Cytokeratin-18 Is Associated with NOX2-Generated Oxidative Stress in Patients with Nonalcoholic Fatty Liver. *Int. J. Hepatol.* **2014**, *2014*, 784985. [CrossRef] [PubMed]

30. Rolo, A.P.; Teodoro, J.S.; Palmeira, C.M. Role of oxidative stress in the pathogenesis of nonalcoholic steatohepatitis. *Free Radic. Biol. Med.* **2012**, *52*, 59–69. [CrossRef] [PubMed]

31. Lee, K.S.; Buck, M.; Houglum, K.; Chojkier, M. Activation of hepatic stellate cells by TGFα and collagen type I is mediated by oxidative stress through c-myb expression. *J. Clin. Invest.* **1995**, *96*, 2461–2468. [CrossRef] [PubMed]

32. Lee, S.; Kim, S.M.; Lee, R.T. Thioredoxin and thioredoxin target proteins: From molecular mechanisms to functional significance. *Antioxid. Redox Signal.* **2013**, *18*, 1165–1207. [CrossRef] [PubMed]

33. Sumida, Y.; Nakashima, T.; Yoh, T.; Furutani, M.; Hirohama, A.; Kakisaka, Y.; Nakajima, Y.; Ishikawa, H.; Mitsuyoshi, H.; Okanoue, T.; et al. Serum thioredoxin levels as a predictor of steatohepatitis in patients with nonalcoholic fatty liver disease. *J. Hepatol.* **2003**, *38*, 32–38. [CrossRef]

34. Edwards, S.; Lalor, P.F.; Tuncer, C.; Adams, D.H. Vitronectin in human hepatic tumours contributes to the recruitment of lymphocytes in an alpha v beta3-independent manner. *Br. J. Cancer* **2006**, *95*, 1545–1554. [CrossRef] [PubMed]

35. Xu, J.; Lloyd, D.J.; Hale, C.; Stanislaus, S.; Chen, M.; Sivits, G.; Vonderfecht, S.; Hecht, R.; Li, Y.S.; Lindberg, R.A.; et al. Fibroblast growth factor 21 reverses hepatic steatosis, increases energy expenditure, and improves insulin sensitivity in diet-induced obese mice. *Diabetes* **2009**, *58*, 250–259. [CrossRef] [PubMed]
36. Petrasek, J.; Bala, S.; Csak, T.; Lippai, D.; Kodys, K.; Menashy, V.; Barrieau, M.; Min, S.Y.; Kurt-Jones, E.A.; Szabo, G. IL-1 receptor antagonist ameliorates inflammasome-dependent alcoholic steatohepatitis in mice. *J. Clin. Invest.* **2012**, *122*, 3476–3489. [CrossRef] [PubMed]
37. An, J.J.; Han, D.H.; Kim, D.M.; Kim, S.H.; Rhee, Y.; Lee, E.J.; Lim, S.K. Expression and regulation of osteoprotegerin in adipose tissue. *Yonsei Med. J.* **2007**, *48*, 765–772. [CrossRef] [PubMed]
38. Yang, M.; Xu, D.; Liu, Y.; Guo, X.; Li, W.; Guo, C.; Zhang, H.; Gao, Y.; Mao, Y.; Zhao, J. Combined Serum Biomarkers in Non-Invasive Diagnosis of Non-Alcoholic Steatohepatitis. *PLoS ONE* **2015**, *10*, e0131664. [CrossRef] [PubMed]

International Journal of
Molecular Sciences

Article

Porcine Breast Extracellular Matrix Hydrogel for Spatial Tissue Culture

Girdhari Rijal [1], Jing Wang [2,†], Ilhan Yu [3,†], David R. Gang [2], Roland K. Chen [3] and Weimin Li [1,*]

[1] Department of Biomedical Sciences, Elson S. Floyd College of Medicine, Washington State University, Spokane, WA 99202, USA; girdhari.rijal@wsu.edu
[2] Tissue Imaging and Proteomics Laboratory, Washington State University, Pullman, WA 99164, USA; jing.wang6@wsu.edu (J.W.); gangd@wsu.edu (D.R.G.)
[3] School of Mechanical and Materials Engineering, Washington State University, Pullman, WA 99164, USA; ilhan.yu@wsu.edu (I.Y.); roland.chen@wsu.edu (R.K.C.)
* Correspondence: weimin.li@wsu.edu; Tel.: +1-509-368-6625
† These authors contributed equally to this work.

Received: 16 September 2018; Accepted: 22 September 2018; Published: 25 September 2018

Abstract: Porcine mammary fatty tissues represent an abundant source of natural biomaterial for generation of breast-specific extracellular matrix (ECM). Here we report the extraction of total ECM proteins from pig breast fatty tissues, the fabrication of hydrogel and porous scaffolds from the extracted ECM proteins, the structural properties of the scaffolds (tissue matrix scaffold, TMS), and the applications of the hydrogel in human mammary epithelial cell spatial cultures for cell surface receptor expression, metabolomics characterization, acini formation, proliferation, migration between different scaffolding compartments, and in vivo tumor formation. This model system provides an additional option for studying human breast diseases such as breast cancer.

Keywords: porcine; breast cancer; extracellular matrix; hydrogel; tissue matrix scaffold; scaffold; 3D culture

1. Introduction

Hydrogel is a gel formed by networks of polymer chains, which have the ability to retain the water and gelatinize easily through various cross-linking processes. It has a wide range of applications ranging from biomedical research, tissue engineering, and clinical utilization to drug delivery [1–4]. Hydrogels are commonly classified into two major categories, synthetic and natural, based on the origin and biochemical properties of the source materials used to produce the gel. Although synthetic hydrogels have been used in research fields for decades owing to many advantages, including long service life, high capacity of water absorption, high gel strength, easy availability, low cost, comparably simple fabrication process, adjustable signaling inputs by integration of different extracellular matrix (ECM) proteins or polypeptides, and experimental reproducibility [3], a growing trend of using hydrogels derived from natural, especially tissue-specific native biomaterials in biomedical and bioengineering applications has become increasingly robust recently [5].

Among native tissue-derived hydrogels, collagen and the laminin-rich ECM (lrECM or Matrigel) [6,7] have been extensively used in various bioassays and experiments for phenotypic and mechanistic studies of human diseases as well as for tissue engineering [8–11]. These studies have contributed substantially to our understanding of cell biology on substrata that resemble those existing in native environments, which are beyond the supporting capacity of plastic and synthetic polymeric surfaces. The compositional differences of the substrata that cells live on are important for cell adhesion, migration, and growth since different cell surface receptors are expressed in response to their interacting proteins or particles within the matrices for extracellular and intracellular signaling and other biological functions. To this end, collagen and the compositionally under-defined mouse sarcoma lrECM may induce cellular phenotypes different than those seen in cells grown on tissue-specific matrices since the levels and types of collagen and other components within ECM of different organs or tissues are quite distinct [12–14].

In an effort to promote tissue-specific scaffolding tools available to study breast cancers, we recently reported a tissue matrix scaffold (TMS) system generated using mouse mammary fat pad (MFP) ECM [15]. Both hydrogel and porous TMS were derived from the same ECM source and used in spatial breast tumor modeling and drug testing, representing an ideal platform for consistent studies from in vitro to in vivo without switching to different culturing matrices. In this current work, we present the biochemical composition of the ECM of female porcine breast fatty tissues and introduce the fabrication of hydrogel and porous scaffold using the purified ECM. Finally, select applications of the scaffolds in breast cancer research are exhibited.

2. Results and Discussion

2.1. Extraction and Identification of Porcine Breast Tissue ECM Proteins

Proper handling of fresh tissues is critical for minimal degradation and maximal identification of native proteins within the tissues. Homogenization of the fresh porcine breast fatty tissues directly obtained from local slaughter house was conducted in a precooled homogenizer, with the container holding the tissues embedded in ice and sliced tissues mixed with ice-cold water as described in the materials and methods. The homogenized tissues were then decellularized with the non-ionic detergent triton X-100 or the zwitterionic detergent CHAPSO to remove the cellular contents while preserving the native and active states of the ECM proteins. Since porcine breast tissues contain rich fat that is difficult to remove compared to that within the mammary tissues of mice and human, we applied lipase in the detergent solution to maximize lipid removal and the decellularization efficiency (Figure 1a). The detergents and lipase were removed with multiple rounds of washing with ddH$_2$O. The resulting ECM dry weight was about 2% of the total fresh tissues used for the extraction. Whole ECM protein extraction from the ECM was carried out using a gradient of concentrations of urea and thiourea solutions to solubilize proteins that could be dissolved under different urea solution conditions due to their sizes and native conformational states. After dialysis and concentration of the protein extract, the total protein amount within the extract was measured at about 90% of the ECM used for the extraction.

The proteomic composition of the porcine ECM was identified using liquid chromatography coupled with tandem mass spectrometry (LC-MS/MS). The majority of ECM proteins detected were different types of collagen, representing about 70% of the major proteins listed (Figure 1b), with collagen I and III being the most abundant (~77% based on single chain ratios) among the total collagen content. While the sum of glycoproteins and proteoglycans accounted for about 5% of the ECM major proteins, the content of myosin and tropomyosin was about 19%, with other ECM proteins comprising the remaining 6%. Interestingly, when we compared these data with the proteomic profiles of mouse normal MFP ECM that we reported previously [15], the ECM of rat mammary tissues [12] and the ECM of human breast tissues adjacent (considered to be normal) to tumors [16], it appeared that the ECM total collagen content of the human breast tissue ECM (about 80–85%) was in between of that of the porcine breast ECM and the rodent mammary ECM (about 85–90%). Another intriguing result was that higher myosin and tropomyosin content was detected in the porcine breast ECM compared to that of the rodent mammary ECM (less than 0.5%) and of the human breast ECM (about 1%) [16]. The amounts of the overall glycoproteins and proteoglycans within the ECM of porcine, rodent, and human were similar at about 5–7%. These ECM compositional differences of the porcine, mouse, and human breast tissues could be due to the anatomical and physiological natures of the native tissues that are related to the functions of the tissues in each specific species. It is noteworthy that the reproductive cycle seems to play a role in mammary tissue ECM compositional changes [12,17].

Additionally, the differences in the buffers and methods used to extract native ECM proteins clearly have an impact on the types and amounts of proteins identified [12,15,16,18]. Future proteomic analyses of mammary ECM compositions of the aforementioned and additional species using the same extraction and analytical methods will potentially identify novel discrepancies upon cross comparing the data sets obtained from the same experiments.

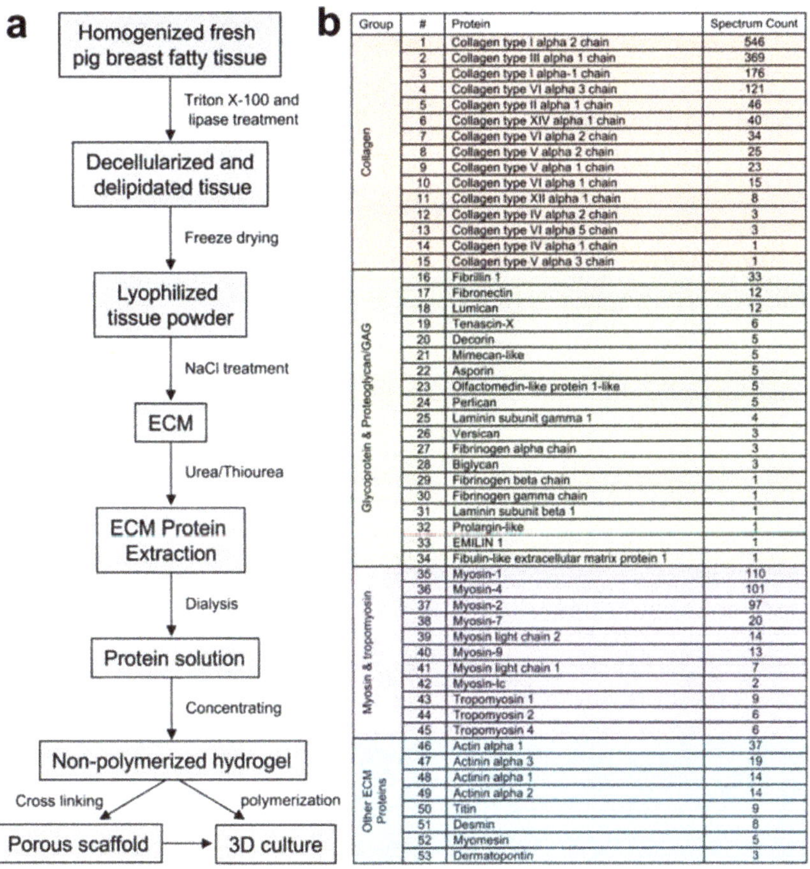

Group	#	Protein	Spectrum Count
Collagen	1	Collagen type I alpha 2 chain	546
	2	Collagen type III alpha 1 chain	369
	3	Collagen type I alpha-1 chain	176
	4	Collagen type VI alpha 3 chain	121
	5	Collagen type II alpha 1 chain	46
	6	Collagen type XIV alpha 1 chain	40
	7	Collagen type VI alpha 2 chain	34
	8	Collagen type V alpha 2 chain	25
	9	Collagen type V alpha 1 chain	23
	10	Collagen type VI alpha 1 chain	15
	11	Collagen type XII alpha 1 chain	8
	12	Collagen type IV alpha 2 chain	3
	13	Collagen type VI alpha 5 chain	3
	14	Collagen type IV alpha 1 chain	1
	15	Collagen type V alpha 3 chain	1
Glycoprotein & Proteoglycan/GAG	16	Fibrillin 1	33
	17	Fibronectin	12
	18	Lumican	12
	19	Tenascin-X	6
	20	Decorin	5
	21	Mimecan-like	5
	22	Asporin	5
	23	Olfactomedin-like protein 1-like	5
	24	Perlican	5
	25	Laminin subunit gamma 1	4
	26	Versican	3
	27	Fibrinogen alpha chain	3
	28	Biglycan	3
	29	Fibrinogen beta chain	1
	30	Fibrinogen gamma chain	1
	31	Laminin subunit beta 1	1
	32	Prolargin-like	1
	33	EMILIN 1	1
	34	Fibulin-like extracellular matrix protein 1	1
Myosin & tropomyosin	35	Myosin-1	110
	36	Myosin-4	101
	37	Myosin-2	97
	38	Myosin-7	20
	39	Myosin light chain 2	14
	40	Myosin-9	13
	41	Myosin light chain 1	7
	42	Myosin-Ic	2
	43	Tropomyosin 1	9
	44	Tropomyosin 2	6
	45	Tropomyosin 4	6
Other ECM Proteins	46	Actin alpha 1	37
	47	Actinin alpha 3	19
	48	Actinin alpha 1	14
	49	Actinin alpha 2	14
	50	Titin	9
	51	Desmin	8
	52	Myomesin	5
	53	Dermatopontin	3

Figure 1. Extracellular matrix (ECM) protein extraction and identification from porcine breast fatty tissues. (**a**) Outline of the major steps of the breast ECM protein extraction and downstream applications. (**b**) The major ECM proteins of porcine breast fatty tissues.

2.2. Generation and Characterization of the Porcine Breast ECM Hydrogel Scaffold

To assess the impact of storage temperature of the ECM hydrogel on polymerization, the gels (8 mg/mL) directly stored at 4, −20, −80 °C, and flash froze in liquid nitrogen followed by storing at −80 °C were thawed on ice and dropped onto the bottom of a 100 mm tissue culture dish, which was then placed in a 37 °C incubator for polymerization. After 30 min of incubation, the gels from the different stocking conditions all gelatinized (Figure 2a). When the polymerized gels were submersed in 1× PBS or DMEM, they retained their initial shapes and remained undissolved during the 10-day testing period (Figure 2b). To maximize the preservation of the ECM proteins in their native forms and avoid protein degradation, we used liquid nitrogen flash freezing followed by storing at −80 °C as a standard approach for long-term stocking of the hydrogel solutions at different concentrations.

Short-term storage on ice or at 4 °C for up to four weeks did not seem to affect the polymerization of the hydrogel and its performance in experiments as described below.

Generation of porous TMS using the ECM hydrogel was achieved using a combination of a gas foaming method [19] and a freeze-gelation approach [20] with modifications as described in the methods. This technique not only induced gelation of the hydrogel but also allowed production of interconnecting porous structures within the solidified gel. The sustainability of the fabricated porous scaffolds under regular tissue culture conditions (37 °C, 5% CO_2) was evaluated in 1× PBS or DMEM for 7 days. Our results showed that the porous scaffolds were stable and retained their shape during the period of testing (Figure 2c,d). Hematoxylin and eosin (H&E) staining of the cross sections of the scaffolds showed inter-connective pores at the sizes of about 100–200 μm (Figure 2e, left panel), highly resembling those of the decellularized porcine breast fatty tissue ECM (Figure 2e, right panel).

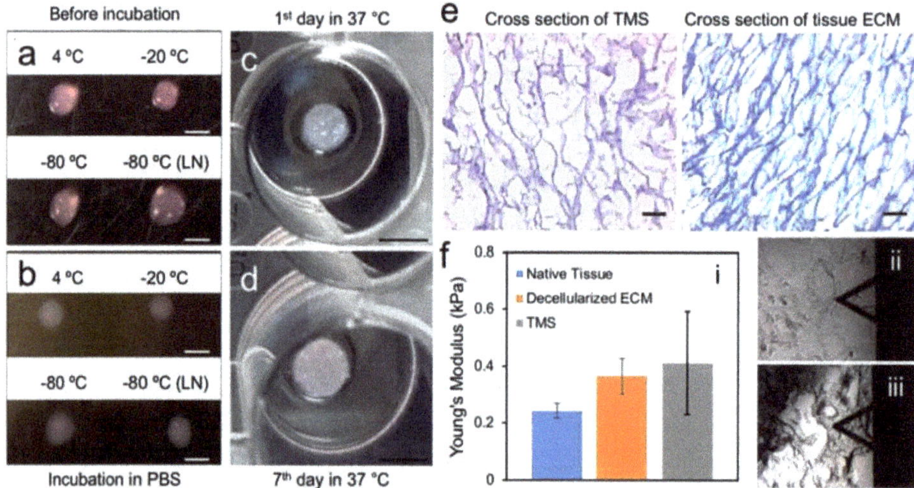

Figure 2. Physical and structurally properties of the porcine breast ECM hydrogel and porous TMS. (**a**) The ice-cold ECM hydrogel stored under different conditions (hydrogels kept in −20 °C and −80 °C were thawed in ice at 4 °C) dispensed on the clean and dry plate before polymerization tested and subjected for polymerization at 37 °C for 30 min. (**b**) The polymerized hydrogel was immersed in 1× PBS or DMEM and placed in 37 °C incubator (5% CO_2) for stability test. Scale bars, 4 mm for (**a**,**b**). (**c**,**d**) The sustainability of the porous scaffolds generated from the ECM hydrogel was assessed by submersing the scaffolds in PBS or DMEM and incubation at 37 °C (5% CO_2) for 7 days. Scale Bars, 6 mm. (**e**) H&E staining of the cross sections of the porous scaffolds and decellularized porcine breast fatty tissue ECM. Scale bars, 100 μm. (**f**) AFM measurement of the mechanical properties of porcine native breast tissues, decellularized breast ECM, and porous TMS: (**i**) Young's moduli of the samples; (**ii**) and (**iii**) Illustrations of the positions of the probe tips on the decellularized ECM and TMS samples, respectively.

In order to evaluate the mechanical resemblance of the reconstituted scaffolds to that of the matrix of native tissues, we used AFM to measure native porcine breast tissues, decellularized porcine breast ECM, and porous scaffolds generated using porcine breast ECM hydrogel. The AFM testing method on the stiffness of biological tissue structures, cells, and specific regions of ECM is not standardized. Thus, previous reports have shown different results depending on their individual testing conditions [21,22]. On a large scale, measurements of decellularized organ matrices with AFM have been reported [23,24]. On a fine scale, mechanical properties of breast cancer cells and their structures have been investigated using AFM [25]. The probe tip size, indentation rate, and maximum force applied to samples should be optimized based on the mechanical strength of the samples. Thus, we optimized our testing parameters specifically for the breast tissues as described in the methods. The positioning of the probe tip for indention on the decellularized ECM and TMS has been illustrated in Figure 2f-ii,f-iii.

Our AFM results showed that the average Young's modulus of the native porcine breast tissues was 0.243 ± 0.027 kPa (Figure 2f–i), which corresponded to a similar range of the compliance of human normal breast tissues [25,26] and is about 45–50% higher than that of mouse normal mammary tissues [27,28]. The Young's modulus of the decellularized porcine breast ECM was 0.366 ± 0.061 kPa (Figure 2f–i), which is about 50% higher than that of the porcine breast native tissues and stiffer than decellularized mouse normal mammary fatty tissues [28]. A similar trend of higher Young's modulus in decellularized tissues than in native tissues was observed in AFM measurement of human liver samples [29]. Porous TMS generated using ECM hydrogel at the concentration of 40 mg/mL (data for the concentrations of 20 mg/mL and 60 mg/mL were not shown) demonstrated an average Young's modulus of 0.411 ± 0.180 kPa (Figure 2f–i), which is close to that of the decellularized porcine breast ECM. These data collectively indicate that the structural and mechanical properties of the reconstituted porous TMS highly resemble those of the decellularized native ECM and are suitable for spatial tissue cultures that closely mimic native breast tissue microenvironment.

2.3. ECM Support of Cell Surface Receptor Expression and Metabolomes in Spatial Culture

Our proteomics data showed that collagen I and III were the major protein components, whereas the main basement membrane protein laminin only accounted for a minimal amount of the porcine breast tissue ECM. Therefore, we carried out immunofluorescence (IF) staining of integrin β1 and β4 plasma membrane receptors for collagen I/III and laminin, respectively, in MM231 cells grown on the breast ECM hydrogel-coated glass coverslips for 24 h. The expression of focal adhesion kinase (FAK) on the surface of the cells was also immunostained to serve as an indicator of the adhesion sites of the cells. Our confocal fluorescence microscopy results showed that high levels of integrin β1 and low levels of integrin β4 were observed in the cells cultured on the hydrogel matrix (Figure 3a,b), indicating that the cells attached to the matrix through integrin β1 and collagen I/III interactions.

To assess the capability of the porcine breast ECM hydrogel in capturing the metabolites secreted from cells grown on it, 4% porcine breast ECM hydrogel was added into the wells of 96-well plates at a thickness of 4 mm, followed by inserting porous TMS illustrated in Figure 2c (2-mm thick, 6-mm diameter, 100-μm pore size) into the surface section of the gel before its polymerization in a 37 °C incubator. A total of 1.0×10^4 MM231 cells per well were seeded on the top of the porous scaffold and cultured in RPMI 1640 medium (Corning; $1\times$; 2 g/L of D-glucose; 10% FBS) under optimal conditions (37 °C, 5% CO_2) for 3 days. The hydrogel samples underneath the porous scaffolds were collected, cross sectioned (Figure 3c, left panel) and processed for mass spectrometry (MS) analysis of the metabolites collected within the gels as described in the methods.

Based on our MS spectral data, metabolite distribution patterns and relative abundance levels were determined using the SCiLS Lab MS imaging analysis software, which also grouped compounds (detected as ions with specific m/z values) into distinct ion distribution patterns across the scaffolds for comparison. A total of 41 distinct chemical features (i.e., the mass over charge ratios detected in MS) were selected based on their abundance distribution profiles and the positive scores (close to 1) resulted from the receiver operating characteristic (ROC) curve analyses in the MM231 cell culture-laden hydrogel samples compared to the medium-containing blank hydrogel samples. As illustrated by one of the forty-one results, the compound with m/z 663.024 was absent from the blank scaffold (Figure 3c, right panel, top) but showed high accumulation with a distinct pattern of being more abundant near the edges of the ECM hydrogel scaffold than in the center. This compound had a ROC value of close to 1 (highly positive; Figure 3d, top panel). The box plot for this compound (Figure 3d, bottom panel) further demonstrated its preferential accumulation within the hydrogel samples of the MM231 cell cultures. Similar results were found in all 41 ions (data not shown), i.e., higher specific chemical abundances were detected in the hydrogel samples of the MM231 cell cultures with high ROC scores. Four box plots for the ions with top ROC scores from the 41 individual groups were shown in Figure S1, which had m/z ratios at 663.024, 531.060, 839.043, and 768.371 with the ROC scores of 0.995, 0.996, 0.994 and 0.988, respectively. Together, these data demonstrate that the porcine breast ECM hydrogel scaffolds support matrix-associated cell membrane receptor expression and secretion of metabolites from the cells grown in the tissue-mimicking 3D space.

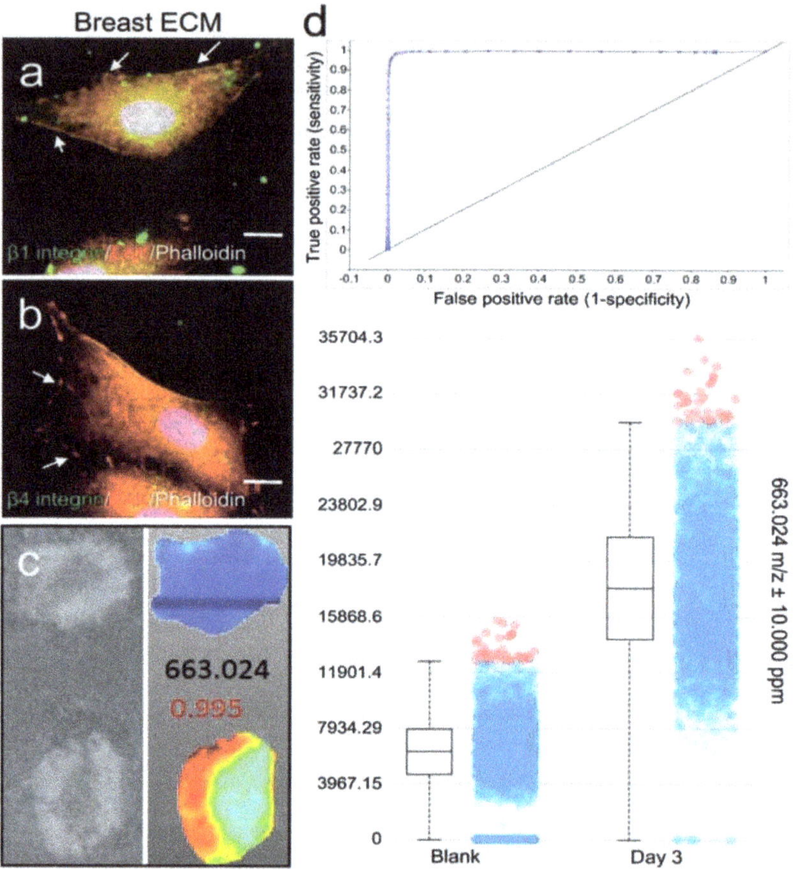

Figure 3. Membrane surface receptor expression and metabolomic analysis of cells grown on porcine breast fatty tissue ECM hydrogel. Integrin β1 (**a**) or β4 (**b**) receptor (green) and FAK (red) expression in MM231 cells cultured on the ECM hydrogel-coated coverslips were assessed by IF. Phalloidin staining (white) of F-actin was used to contour the cells. Scale bars, 10 μm. (**c**) MS imaging of cross sections of the medium-conditioned blank (upper panels) and MM231 cell culture-laden (bottom panels) hydrogel scaffolds. (**d**) Examples of ROC (upper) and box-and-whiskers (bottom) plots from the MS imaging analysis, showing data for a compound with m/z 663.024, where differences between the hydrogel samples cultured with or without MM231 cells are clearly evident.

2.4. Applying Porcine Breast ECM Scaffold in Support of Spatial Cell Proliferation, Coculture of Cancer Cells and Stromal Cells, and Tumor Formation

To evaluate the proliferation of mammary epithelial cells on porcine breast ECM hydrogel, normal MCF10A or MM231 cells were seeded in 96-well plates coated with or without the porcine breast ECM hydrogel, collagen, and Matrigel, and cultured under optimal conditions (37 °C; 5% CO_2) for 7 days. Cell proliferation was measured using WST-1 reagent (Sigma-Aldrich, St. Louis, MO, USA) at different time points (Day 1, 3, 5, 7). Our results showed that both type of the cells proliferated faster on Matrigel compared to those on collagen and ECM hydrogel (Figure 4a), possibly due to the compositional nature of Matrigel, which contains growth factors and underdefined cellular components derived from Engelbreth–Holm–Swarm (*EHS*) *sarcoma* sources [15,30]. In contrast, the growth of the cells cultured on the porcine breast ECM hydrogel is even slower than those on collagen but at comparable levels, suggesting certain degree of similarities of the two types of gels on supporting cell population expansion.

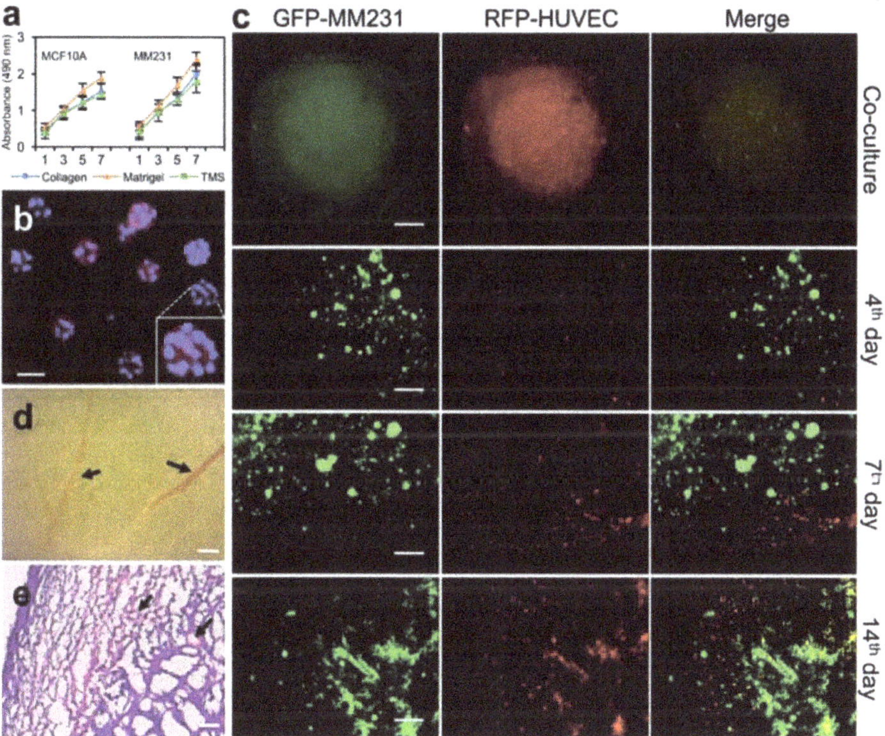

Figure 4. Applications of porcine breast ECM hydrogel and porous scaffolds in proliferation, migration, and tumor formation experiments. (**a**) Proliferation of MCF10A or MM231 cells on the porcine breast ECM hydrogel. Error bars, mean ± SD. (**b**) MCF10A cell acini formation within the ECM hydrogel. (**c**) Compartmental culture of MM231 cells and HUVECs to assess cross-compartmental migration of different types of cells. (**d**) The vasculature of a whole mount cross section of breast tumors from mice MFP implanted with MM231 cell-laden porous TMS derived from the porcine ECM hydrogel. (**e**) H&E staining of the cross sections of mice breast tumors developed from the cancer cell-laden TMS. Black arrows in d and e indicate blood vessels. Scale bars, 50 μm for (**b**); 1 mm—upper first panel (**c**) & 500 μm—lower three panels (**c**); 200 μm for (**d,e**).

The spatial growth and proliferation of mammary epithelial cells within the breast ECM hydrogel were further assessed using an acini formation assay. Briefly, 1×10^4 MCF10A cells were mixed into 200 μL of 2% porcine breast ECM hydrogel and cultured on 8-well chamber slide under optimal conditions as described above for 7 days. Acini formation was assessed using light microscopy, and the structures of the acini were characterized with phalloidin and DAPI IF staining followed by confocal microscopy. Our results showed that MCF10A cells formed acini at different sizes, with larger ones having hollow centers (Figure 4b). Compared to the traditional acini formation assay using Matrigel, the current method applies fewer matrix materials, simplified procedures, and shorter culture times. Importantly, the purified porcine total ECM hydrogel contains neither growth factors nor tumor cell products and has well-defined ECM proteins, lending promise for low background tissue cultures and adjustable options for adding desired culturing components within culture medium.

In order to observe the spatial expansion of cancer cell population and recruitment of stromal cells, we have devised a coculture system using the porcine breast ECM hydrogel and porous TMS (Figure 2c,d) derived from the hydrogel. 1×10^4 GFP-MM231 cells were seeded on the porous scaffold (round, 2 mm thick, 4 mm diameter) placed in a well of 96-well plates. A layer of the ECM hydrogel (8 mg/mL) containing 1×10^4 RFP-HUVECs (human umbilical vein endothelial cells) was covered on top of the porous scaffold (Figure 4c, top panels). After polymerization of the gel, 100 µL of 1× DMEM containing 10% FBS was added into the well (replicate samples were prepared). The two types of cells within the different yet mutually accessible compartments were cultured for 14 days. The distribution and migration of the cells within the co-culture system were imaged over time using confocal microscopy. Our data showed that both MM231 cells and HUVECs progressively increased their numbers and migrated out of their initial living compartment into adjacent areas (Figure 4, middle and bottom panels). The interaction of the two types of cells was also increased, as exhibited by the yellow overlapping regions. Clearly, this compartmental culture approach using hydrogel and porous scaffold derived from the same native tissue ECM allows for the observation of certain cellular phenotypes, such as spatial cell migration and interaction, in an advanced tissue-mimicking environment that could be difficult to be captured in live tissues or other nontissue-specific culture models.

We next tested the efficiency of the porcine breast ECM scaffold in supporting tumor formation in vivo. 1×10^5 MM231 cells were seeded on a spherical porous TMS (4 mm diameter) and cultured for 24 h under optimal conditions as described before. Triplicate samples were prepared for the experiment. Then, the scaffolds were implanted separately into the mammary fat pad (MFP, 4th nipple region from the rostral side) of eight-week-old nulliparous NOD/SCID female mice. Tumor development was observed over a period of 4 weeks. The sizes of the tumors were dynamically measured using a caliper and reached an average of about 2-cm diameter at the end of week 4 post-implantation, at which point the tumors were collected, formalin fixed, cross-sectioned (10 µm thick), and H&E stained for histological examination as we reported previously [15]. Our histological data showed that visible blood vessels had been grown into the tumors in most of the regions of the outer half of the tumor bodies (Figure 4d,e, arrows), which were filled with mixtures of cancer cells and stromal cells clustered within or overlapped with ECM structures (Figure 4e). This data indicates that the porcine breast ECM-based scaffold represents another tissue-specific platform, in addition to mice ECM-based scaffolds [15], to support consistent formation of breast tumors in animals.

3. Materials and Methods

3.1. Tissue Collection and Decellularization

The fresh mammary tissues from female pigs were collected aseptically from a local slaughter house, where the research purpose of using the designated tissues was informed. The animal use protocol (#04965) has been approved (01/30/2017) by WSU Institutional Animal Care and Use Committee (IACUC). The tissues on ice were immediately transferred to the lab, sliced into small pieces, and homogenized in sterilized ice-cold deionized distilled water (ddH$_2$O). After centrifugation (10,000 RPM/17,000 RCF; 30 min) at 42 °C (Pig fat is relatively sticky and melts only above 37 °C) of the homogenized tissues, the supernatant together with the fat were discarded. The centrifugation process was repeated for a couple of rounds to maximally remove the fat until there was visible oil droplets on the surface of the supernatant. The homogenized tissues were then mixed with a sufficient amount (at least 10 times more than the homogenized tissue volume) of 0.1% Triton X-100 or 4% CHAPSO with constant agitation at room temperature for 12 h and centrifuged as described above. This decellularization process was repeated one more time. The pelleted ECM was transferred into 50 mL conical tubes containing 0.1% Triton X-100, cOmplete mini protease inhibitor cocktail (Sigma-Aldrich, St. Louis, MO, USA), and lipase (1 mg/g of ECM), and incubated at 37 °C with constant rotation for 12 h, followed by several rounds of washing with ddH$_2$O and centrifugation to ensure complete removal of Triton X-100, protease inhibitors, and lipase from the samples. The ECM was lyophilized and assessed for decellularization efficiency using the removal of DNA content (<50 ng per milligram of dried ECM) and the maximal retention of collagen and glycosaminoglycan (GAG) within the ECM as parameter references as described previously [15].

3.2. Extraction of Porcine Breast Tissue ECM Proteins

The lyophilized ECM was pulverized by grinding in liquid nitrogen, followed by treating the ECM powder twice with 3.4 M NaCl buffer (99.25 g NaCl, 6.25 mL of 2 M Tris Base, 0.75 g EDTA, and distilled water to 500 mL, final pH 7.4) at 4 °C for 15 min. The ECM was pelleted by centrifugation and homogenized in 2 M urea buffer (60 g Urea, 3.025 g Tris Base, 4.5 g NaCl, distilled water to 500 mL, pH 7.4) at 4 °C overnight. The sample was then centrifuged at 13,000 RPM/28,720 RCF for 30 min. The supernatant was collected and kept on ice. Further homogenization of the remaining ECM was followed using 4 M and 6 M urea buffer, respectively, and the supernatant from each extraction was collected and stored as described above. The remaining insoluble ECM was treated with increasing concentration of urea/thiourea (6 M/0.5 M; 6 M/2 M; 7 M/0.5 M; and 7 M/2 M) for 12 h at 4 °C, and supernatant collected as before. The insoluble ECM sediment was further homogenized with 8 M urea, and then with 2% of n-octyl β-D-glucopiranoside (OG) overnight at 4 °C. Again, the supernatant was collected after centrifugation at 13,000 RPM/28,720 RCF for 30 min. The urea concentrations of the different batches of the supernatants were brought to 2 M. Then the tissue ECM protein extracts were pooled together and dialyzed in cold TBS (6.05 g Tris Base, 9.0 g NaCl, total volume of 1 L, pH 7.4, with 5 mL of chloroform) for at least 2 h. Dialysis was repeated twice in cold TBS without chloroform for 12 h. Further dialysis with serum free 1× DMEM medium is optional. The sterile ECM protein solution was then concentrated using polyethylene glycol (PEG) and stored at −80 °C for future use.

3.3. Identification of the ECM Proteins and Data Analysis

Identification of the extracted ECM proteins was conducted as reported previously [15] . Briefly, the ECM proteins were solubilized in 8 M urea solutions containing 1% ProteaseMAX (Promega), subjected to Dithiothreitol (DTT) reduction, iodoacetamide (IAA) alkylation, trypsin digestion, followed by LC-MS/MS. The raw data was converted to mgf files, which were used to search against *Sus Scrofa* amino acid sequence database with decoy reverse entries and a list of common contaminants (81,280 total entries with 40,602 pig proteins from UniProt database downloaded 11_28_2017) using in-house Mascot search engine 2.2.07 (Matrix Science, Boston, MA, USA) with variable methionine and proline oxidation, and with asparagine and glutamine deamidation. Peptide mass tolerance was set at 15 ppm and fragment mass at 0.6 Da. Protein annotations, significance of identification and spectral based quantification was performed with the help of Scaffold software (version 4.4.1, Proteome Software Inc., Portland, OR, USA). Protein identifications were accepted if they could be established at greater than 80.0% probability within 1% false discovery rate and contained at least two identified peptides. Protein probabilities were assigned by the ProteinProphet algorithm [31].

3.4. Generation of Porous Scaffold Using the ECM Hydrogel

The porcine breast ECM hydrogel at the concentration of 40 mg/mL was mixed with sodium bicarbonate (final concentration 1 mg/mL), filled in desired culture vessels (for example 96-well plates), flash-frozen in liquid nitrogen, and stored in −80 °C overnight. The frozen hydrogel blocks were then immersed into precooled (−80 °C) ethanol bath containing 0.1 M acetic acid until there was no bubbles arising from the solidified gel (the processing time depends on the size of the frozen gel).

The ethanol was replenished, and the scaffolds remained immersed for at least 24 h. Before cell culture, the scaffolds were washed three times with $1\times$ PBS or tissue culture medium.

3.5. Atomic Force Microscopy (AFM)

AFM (Dimension Icon ScanAsyst, Santa Barbara, CA, USA) was used to determine the Young's moduli of the samples. Three different types of samples were prepared for this study: porcine breast native fatty tissues, decellularized ECM of porcine breast native fatty tissues, and porous TMSs generated using hydrogel extracted from the decellularized ECM of porcine breast native fatty tissues as described above. Samples were embedded in optimal cutting temperature (OCT) compound and frozen in ethanol-dry ice bath, cross sectioned into 20 µm of thickness, mounted on silane-coated slides, and rinsed with $1\times$ PBS and ddH$_2$O to remove OCT. The samples were kept at 4 °C until AFM measurement. Before AFM, the samples were allowed to warm up to room temperature in a sealed plastic bag to prevent drying. AFM measurement of the mechanical properties of the samples was conducted through borosilicate spherical-shaped tips (Novascan, Boone, IA, USA) with a size of 5 µm and 0.06 N/m spring constant [25]. The spring constant of cantilever was matched with the stiffness of the porcine breast tissues or decellularized ECM samples to be tested. During AFM measurement, the maximum force was set to 2 nN. The total indentation depth was 1 µm with an indentation rate of 20 µm/s. For the native tissue samples, data points were collected by every 10 µm along the x-axis. For decellularized ECM and TMS, due to the porous structures of the samples, the data points were collected by manually selecting areas of interest to avoid probing inside pores. At least 7 different points were measured for each sample. The data was processed using Hertz's model with Poisson's ratio of 0.5 to determine the Young's moduli of the samples.

3.6. Cells and Tissue Cultures

Human breast cancer epithelial cells MDA-MB-231 (MM231) and normal breast epithelial cells MCF10A were purchased from ATCC, and GFP-MM231 cells and RFP-HUVECs were from Cell Biolabs and ANGIO-PROTEOMIE, respectively. The cells were cultured in optimal medium under 37 °C and 5% CO$_2$ conditions as described previously [15,32].

3.7. Immunofluorescence Staining (IF)

IF staining of the cells cultured on coverslips coated with porcine breast tissue ECM hydrogel and confocal microscopy were carried out as described previously [33]. Integrin β1 (#MAB17782), integrin β4 (MAB4060), and FAK (#NBP1-84750) antibodies were purchased from Novus Biologicals (Littleton, CO, USA). Alexa Fluor 663 Phalloidin probe was purchased from ThermoFisher Scientific (#A22284) (Waltham, MA, USA).

3.8. Metabolome Analysis

The hydrogel samples collected from the tissue cultures were flash frozen, cross sectioned (10-µm thick) and mounted onto the cold indium tin oxide (ITO)-coated microscope slides (Figure 3c, left panel). Three experimental replicates were prepared, and three consecutive sections from an individual replicate were collected. A matrix solution of 2,5-dihydroxybenzoic acid (DHBA, 20 mg/mL) dissolved in water and methanol mixture (50:50 in volume) was applied to the tissue sections on the different slides using a TM-Sprayer. The DHBA-coated slides were analyzed using a Bruker SolariX 9.4T MALDI FTICR mass spectrometer (Bruker Daltonics, Billerica, MA, USA) to scan ions within a mass range of 150–3000 Da. Imaging acquisition was carried out using FlexImaging (Bruker Daltonics). The spectral analyses were performed using DataAnalysis (Bruker Daltonics) and SciLS Lab (SciLS, Bremen, Germany).

3.9. Cell Proliferation Assay

The growth and proliferation of the cells on TMS scaffolds were measured using WST-1 reagent as described previously [19,32]. Briefly, 2% of porcine breast ECM hydrogel, 100 μL was dispensed into each well of a 96-well culture plate and incubated at 37 °C for cross linking as described above, forming TMS scaffolds. MCF10A or MDA-MB-231 suspended in the respective culture media were seeded on the scaffolds (1×10^4 per scaffold) and allowed to attach for 45 min as we described in our previous paper [19]. 100 μL of respective culture medium was then added and cultured under the optimal conditions (37 °C, 5% CO_2), replacing media in every alternate day. The proliferation of the cells grown on the scaffolds was measured using WST-1 (Sigma-Aldrich) at the points indicated (1st, 3rd, 5th and 7th day). WST-1 solution (10 μL) was added at a 1:10 dilution into the cultures and incubated for 2 h. The supernatants of the cultures were collected and the colorimetric reactions that reflect the proliferation status were measured using a Synergy 2 microplate reader (BioTek, Winooski, VT, USA) for the absorbance at 490 nm. Error bars represent standard deviations (SD) of the means of three independent experiments.

3.10. Data Availability

The MS proteomics and metabolomics data have been deposited to the ProteomeXchange Consortium via the PRIDE partner repository with the data set identifier PXD011011 (proteomics) and PXD010960 (metabolomics), respectively.

4. Conclusions

Porcine mammary ECM contains abundant collagen as well as other structural or compositional proteins similar to those in mouse [15] and human breast ECM [16]. Additionally, the physical structures of the decellularized mammary ECM of these different native tissue materials are also close to each other. These physicochemical resemblances of the native ECM across the different mammalian species potentially benefit our research involving the applications of the ECM in phenotypic, mechanistic, or pharmacological response studies of human diseases relevant to the ECM source organs or tissues. The tissue-specific native ECM-based studies will provide valuable and clinically relevant insights into the development of human cancers and other diseases. We expect to see a steady growth of using tissue-specific biomatrices to address disease-specific questions in the future.

Supplementary Materials: Supplementary materials are available online at http://www.mdpi.com/1422-0067/19/10/2912/s1.

Author Contributions: W.L. and G.R. conceptualized the project. W.L. wrote the manuscript. G.R., J.W., and I.Y. contributed to the writing. D.R.G., R.K.C. and W.L. contributed to editing the manuscript. G.R., J.W., and I.Y. performed the experiments.

Funding: This project was supported by a WSU Startup Fund to W.L.

Acknowledgments: The authors thank the colleagues in WSU E.S.F. College of Medicine, School of Mechanical and Materials Engineering, and Tissue Imaging and Proteomics Laboratory for their technical support and discussions. We also thank Grzegorz Sabat in the Biotechnology Center (Proteomics-Mass Spectrometry) of the University of Wisconsin-Madison for technical support on the ECM proteomics data analysis. The authors apologize to the scientists whose relevant publications were not cited because of space limitations.

Conflicts of Interest: G.R. and W.L. are authors on an international patent related to this work (PCT/US2017/039135). The authors declare no other conflict of interest.

References

1. Kopecek, J. Hydrogel biomaterials: A smart future? *Biomaterials* **2007**, *28*, 5185–5192. [CrossRef] [PubMed]
2. Rijal, G.; Li, W. 3D scaffolds in breast cancer research. *Biomaterials* **2016**, *81*, 135–156. [CrossRef] [PubMed]
3. Ahmed, E.M. Hydrogel: Preparation, characterization, and applications: A review. *J. Adv. Res.* **2015**, *6*, 105–121. [CrossRef] [PubMed]

4. Rijal, G.; Kim, B.S.; Pati, F.; Ha, D.H.; Kim, S.W.; Cho, D.W. Robust tissue growth and angiogenesis in large-sized scaffold by reducing H_2O_2-mediated oxidative stress. *Biofabrication* **2017**, *9*, 015013. [CrossRef] [PubMed]

5. Rijal, G.; Li, W. Native-mimicking in vitro microenvironment: An elusive and seductive future for tumor modeling and tissue engineering. *J. Biol. Eng.* **2018**, *12*, 20. [CrossRef] [PubMed]

6. Kleinman, H.K.; McGarvey, M.L.; Hassell, J.R.; Star, V.L.; Cannon, F.B.; Laurie, G.W.; Martin, G.R. Basement membrane complexes with biological activity. *Biochemistry* **1986**, *25*, 312–318. [CrossRef] [PubMed]

7. Kleinman, H.K.; Martin, G.R. Matrigel: Basement membrane matrix with biological activity. *Semin. Cancer Biol.* **2005**, *15*, 378–386. [CrossRef] [PubMed]

8. Egeblad, M.; Rasch, M.G.; Weaver, V.M. Dynamic interplay between the collagen scaffold and tumor evolution. *Cur. Opin. Cell Biol.* **2010**, *22*, 697–706. [CrossRef] [PubMed]

9. Hakkinen, K.M.; Harunaga, J.S.; Doyle, A.D.; Yamada, K.M. Direct comparisons of the morphology, migration, cell adhesions, and actin cytoskeleton of fibroblasts in four different three-dimensional extracellular matrices. *Tissue Eng. Part A* **2011**, *17*, 713–724. [CrossRef] [PubMed]

10. Debnath, J.; Muthuswamy, S.K.; Brugge, J.S. Morphogenesis and oncogenesis of MCF-10A mammary epithelial acini grown in three-dimensional basement membrane cultures. *Methods* **2003**, *30*, 256–268. [CrossRef]

11. Glowacki, J.; Mizuno, S. Collagen scaffolds for tissue engineering. *Biopolymers* **2008**, *89*, 338–344. [CrossRef] [PubMed]

12. Goddard, E.T.; Hill, R.C.; Barrett, A.; Betts, C.; Guo, Q.; Maller, O.; Borges, V.F.; Hansen, K.C.; Schedin, P. Quantitative extracellular matrix proteomics to study mammary and liver tissue microenvironments. *Int. J. Biochem. Cell. Biol.* **2016**, *81 Pt A*, 223–232. [CrossRef]

13. Gilbert, T.W.; Sellaro, T.L.; Badylak, S.F. Decellularization of tissues and organs. *Biomaterials* **2006**, *27*, 3675–3683. [CrossRef] [PubMed]

14. DeQuach, J.A.; Mezzano, V.; Miglani, A.; Lange, S.; Keller, G.M.; Sheikh, F.; Christman, K.L. Simple and high yielding method for preparing tissue specific extracellular matrix coatings for cell culture. *PLoS ONE* **2010**, *5*, e13039. [CrossRef] [PubMed]

15. Rijal, G.; Li, W. A versatile 3D tissue matrix scaffold system for tumor modeling and drug screening. *Sci. Adv.* **2017**, *3*, e1700764. [CrossRef] [PubMed]

16. Naba, A.; Pearce, O.M.T.; Del Rosario, A.; Ma, D.; Ding, H.; Rajeeve, V.; Cutillas, P.R.; Balkwill, F.R.; Hynes, R.O. Characterization of the Extracellular Matrix of Normal and Diseased Tissues Using Proteomics. *J. Proteome Res.* **2017**, *16*, 3083–3091. [CrossRef] [PubMed]

17. Schedin, P.; Mitrenga, T.; McDaniel, S.; Kaeck, M. Mammary ECM composition and function are altered by reproductive state. *Mol. Carcinog.* **2004**, *41*, 207–220. [CrossRef] [PubMed]

18. O'Brien, J.H.; Vanderlinden, L.A.; Schedin, P.J.; Hansen, K.C. Rat mammary extracellular matrix composition and response to ibuprofen treatment during postpartum involution by differential GeLC-MS/MS analysis. *J. Proteome Res.* **2012**, *11*, 4894–4905. [CrossRef] [PubMed]

19. Rijal, G.; Bathula, C.; Li, W. Application of Synthetic Polymeric Scaffolds in Breast Cancer 3D Tissue Cultures and Animal Tumor Models. *Int. J. Biomater.* **2017**, *2017*, 8074890. [CrossRef] [PubMed]

20. Ho, M.H.; Kuo, P.Y.; Hsieh, H.J.; Hsien, T.Y.; Hou, L.T.; Lai, J.Y.; Wang, D.M. Preparation of porous scaffolds by using freeze-extraction and freeze-gelation methods. *Biomaterials* **2004**, *25*, 129–138. [CrossRef]

21. Sicard, D.; Fredenburgh, L.E.; Tschumperlin, D.J. Measured pulmonary arterial tissue stiffness is highly sensitive to AFM indenter dimensions. *J. Mech. Behav. Biomed. Mater.* **2017**, *74*, 118–127. [CrossRef] [PubMed]

22. Luo, Q.; Kuang, D.; Zhang, B.; Song, G. Cell stiffness determined by atomic force microscopy and its correlation with cell motility. *Biochim. Biophys. Acta* **2016**, *1860*, 1953–1960. [CrossRef] [PubMed]

23. Melo, E.; Cardenes, N.; Garreta, E.; Luque, T.; Rojas, M.; Navajas, D.; Farre, R. Inhomogeneity of local stiffness in the extracellular matrix scaffold of fibrotic mouse lungs. *J. Mech. Behav. Biomed. Mater.* **2014**, *37*, 186–195. [CrossRef] [PubMed]

24. Klaas, M.; Kangur, T.; Viil, J.; Maemets-Allas, K.; Minajeva, A.; Vadi, K.; Antsov, M.; Lapidus, N.; Jarvekulg, M.; Jaks, V. The alterations in the extracellular matrix composition guide the repair of damaged liver tissue. *Sci. Rep.* **2016**, *6*, 27398. [CrossRef] [PubMed]

25. Acerbi, I.; Cassereau, L.; Dean, I.; Shi, Q.; Au, A.; Park, C.; Chen, Y.Y.; Liphardt, J.; Hwang, E.S.; Weaver, V.M. Human breast cancer invasion and aggression correlates with ECM stiffening and immune cell infiltration. *Integr. Biol.* **2015**, *7*, 1120–1134. [CrossRef] [PubMed]

26. Tilghman, R.W.; Cowan, C.R.; Mih, J.D.; Koryakina, Y.; Gioeli, D.; Slack-Davis, J.K.; Blackman, B.R.; Tschumperlin, D.J.; Parsons, J.T. Matrix rigidity regulates cancer cell growth and cellular phenotype. *PLoS ONE* **2010**, *5*, e12905. [CrossRef] [PubMed]

27. Paszek, M.J.; Zahir, N.; Johnson, K.R.; Lakins, J.N.; Rozenberg, G.I.; Gefen, A.; Reinhart-King, C.A.; Margulies, S.S.; Dembo, M.; Boettiger, D.; et al. Tensional homeostasis and the malignant phenotype. *Cancer Cell* **2005**, *8*, 241–254. [CrossRef] [PubMed]

28. Seo, B.R.; Bhardwaj, P.; Choi, S.; Gonzalez, J.; Andresen Eguiluz, R.C.; Wang, K.; Mohanan, S.; Morris, P.G.; Du, B.; Zhou, X.K.; et al. Obesity-dependent changes in interstitial ECM mechanics promote breast tumorigenesis. *Sci. Transl. Med.* **2015**, *7*, 301ra130. [CrossRef] [PubMed]

29. Mazza, G.; Al-Akkad, W.; Telese, A.; Longato, L.; Urbani, L.; Robinson, B.; Hall, A.; Kong, K.; Frenguelli, L.; Marrone, G.; et al. Rapid production of human liver scaffolds for functional tissue engineering by high shear stress oscillation-decellularization. *Sci. Rep.* **2017**, *7*, 5534. [CrossRef] [PubMed]

30. Kibbey, M.C. Maintenance of the EHS sarcoma and Matrigel preparation. *J. Tissue Cult. Methods* **1994**, *16*, 227–230. [CrossRef]

31. Nesvizhskii, A.I.; Keller, A.; Kolker, E.; Aebersold, R. A statistical model for identifying proteins by tandem mass spectrometry. *Anal. Chem.* **2003**, *75*, 4646–4658. [CrossRef] [PubMed]

32. Li, W.; Petrimpol, M.; Molle, K.D.; Hall, M.N.; Battegay, E.J.; Humar, R. Hypoxia-induced endothelial proliferation requires both mTORC1 and mTORC2. *Circ. Res.* **2007**, *100*, 79–87. [CrossRef] [PubMed]

33. Li, W.; Laishram, R.S.; Ji, Z.; Barlow, C.A.; Tian, B.; Anderson, R.A. Star-PAP control of BIK expression and apoptosis is regulated by nuclear PIPKIalpha and PKCdelta signaling. *Mol. Cell* **2012**, *45*, 25–37. [CrossRef] [PubMed]

International Journal of
Molecular Sciences

Microfabrication-Based Three-Dimensional (3-D) Extracellular Matrix Microenvironments for Cancer and Other Diseases

Kena Song, Zirui Wang, Ruchuan Liu, Guo Chen * and Liyu Liu *

College of Physics, Chongqing University, Chongqing 401331, China; kenasong@cqu.edu.cn (K.S.);
cquphywzr@cqu.edu.cn (Z.W.); phyliurc@cqu.edu.cn (R.L.)
* Correspondence: wezer@cqu.edu.cn (G.C.); lyliu@cqu.edu.cn (L.L.); Tel.: +86-023-6567-8392 (G.C.)

Received: 23 December 2017; Accepted: 19 January 2018; Published: 21 March 2018

Abstract: Exploring the complicated development of tumors and metastases needs a deep understanding of the physical and biological interactions between cancer cells and their surrounding microenvironments. One of the major challenges is the ability to mimic the complex 3-D tissue microenvironment that particularly influences cell proliferation, migration, invasion, and apoptosis in relation to the extracellular matrix (ECM). Traditional cell culture is unable to create 3-D cell scaffolds resembling tissue complexity and functions, and, in the past, many efforts were made to realize the goal of obtaining cell clusters in hydrogels. However, the available methods still lack a precise control of cell external microenvironments. Recently, the rapid development of microfabrication techniques, such as 3-D printing, microfluidics, and photochemistry, has offered great advantages in reconstructing 3-D controllable cancer cell microenvironments in vitro. Consequently, various biofunctionalized hydrogels have become the ideal candidates to help the researchers acquire some new insights into various diseases. Our review will discuss some important studies and the latest progress regarding the above approaches for the production of 3-D ECM structures for cancer and other diseases. Especially, we will focus on new discoveries regarding the impact of the ECM on different aspects of cancer metastasis, e.g., collective invasion, enhanced intravasation by stress and aligned collagen fibers, angiogenesis regulation, as well as on drug screening.

Keywords: microfabrication; extracellular matrix; cancer; metastasis

---

1. Introduction

The lethality of cancer lies in its ability to form metastases that accounts for about 90% of cancer deaths according to the available statistics [1,2]. The phenomenon of cancer metastasis has been investigated extensively in the last decade [3–5], and the neighboring microenvironment of cancer cells, i.e., the extracellular matrix (ECM), has been found to significantly impact tumor and metastasis development [6–10].

Cancer cells are not isolated, and their complicated cell–cell communications, development, metastases, and functions are always closely connected with the ECM microenvironment [11–13], e.g., tumor cells must break through the ECM, a critical step for cancer metastasis, to be able to reach the lymphatic or vascular system [14]. Therefore, an in-depth understanding of the interactions between cancer cells and ECM, from both physical and biological perspectives, is necessary to uncover the mechanism of cancer metastasis. This may also help to find potential therapeutic strategies to control malignant cancer. To achieve this goal, constructing a realistic in vitro cell culture system, particularly involving cell proliferation, migration, invasion, and apoptosis in relation to the ECM, becomes imperative.

In fact, the structure of the ECM in vivo is a complicated system, especially around a neoplastic tissue. The 3-D structure of the ECM in healthy, perilesional, and neoplastic tissues is different. The ECM in a healthy area shows a homogeneous distribution of structure, proteins, and glycoproteins, with collagen fibers intersecting to form a random network. Conversely, the ECM in perilesional and neoplastic areas shows a heterogeneous distribution of the structure, with a dense matrix, irregular shape, and asymmetric profile. The heterogeneous degree of glycoproteins distribution and the parallel degree of collagen fibrils become more obvious closer to the neoplastic tissue [15,16]. In addition, the degree of stiffness of the ECM is an important parameter related to the occurring lesions. The increased stiffness of perilesional areas may represent a new predictive marker of invasion [17].

Traditionally, cells have been cultured in Petri dishes that can only provide a two-dimensional (2-D) extracellular environment: cells can only attach to the surface of the medium and cannot form any 3-D scaffolds to mimic the real tissue complexity and functions [18–20]. Although some important discoveries have been made by using 2-D cancer cell culture systems, they are still insufficient for understanding the complex interactions between cancer cells and the ECM. Numerous studies have indicated that cell morphology, signaling patterns, and cellular functions are different in 3-D tissue microenvironments in vitro compared to 2-D Petri-dish systems [21,22], e.g., 2-D cell cultures do not fully support the recovery of the cellular phenotypes found in tissues in vivo [23]; also, when addressing drug toxicity effects, pharmacokinetic studies performed in 2-D polarized intestinal cells showed distinct features compared to those from toxicology screening tests conducted in a 3-D system composed of interconnected channels and chambers representative of distinct tissue types [24,25].

Realizing a 3-D tissue microenvironment similar to the one found in vivo, is one of the major challenges, but also the key factor to bridge the gap. In the past, many efforts were made to mimic the complex 3-D tissue microenvironment and particularly its influence on cell proliferation, migration, invasion, and apoptosis in relation to the ECM, for example by embedding cell clusters in a low-density Matrigel to form a lumina to mimic the in vivo epithelial layer [26], or by aggregating cells into 3-D spheroidal structures on a low-adhesion surface of a 2-D culture system [27]. However, a precise control of cell morphology (including size, density, and shape) is still lacking, and quantitatively adjustments of the external microenvironment, such as the medium and nutrition gradient, cannot be achieved [28].

In this review, we will discuss some commonly used, modern microfabrication techniques for the effective reconstruction of 3-D cell culture microenvironments in vitro that mimic real tissue structures or systems in vivo. On the basis of the available literature and the latest progress, we will elaborate on how these specifically designed 3-D cells–ECM microenvironments in vitro are applicable to the study of cancer, including metastasis, tumor angiogenesis, and drug screening, as well as of other diseases.

2. Microfabrication Techniques in the Effective Reconstruction of 3-D ECM Microenvironments In Vitro

Recently, the rapid development of microfabrication techniques, such as 3-D printing [29–32], microfluidics [33,34], photopolymerization, photochemistry, photoreaction [35–40], mold-based Diskoid In Geometrically Micropatterned ECM (DIGME) [41,42], as well as the corresponding cell culture approaches, have made it possible to precisely design and construct sophisticated controllable tissue models in vitro that mimic the 3-D architectures and physiological conditions present in vivo [28,43]. Consequently, various biofunctionalized hydrogels, such as collagen, fibronectin, and Matrigel, have become ideal candidates [12,44–50]. Furthermore, combined with advanced modern biomicroscopic tools, the direct and clear observation of cell morphology, motility, and other behaviors has become much easier, thereby helping the researchers acquire new insights into various diseases. Some commonly used methods and materials to engineer 3-D cancer or other disease models and the strengths and limitations of each technique are summarized in Tables 1 and 2, respectively.

Table 1. Methods and materials used to engineer 3-D cancer or other disease models.

Technique	Method	Material
3-D bioprinting	BIP (bioink printing) [29] Extrusion printing [51] Laser-assisted bioprinting [37,38] Microvalve printing [52–55]	Hydrogel Biomolecular ink [29] Matrigel alginate [51] Chitosan [52] Gelatin [52] Alginate [54] Fibrinogen [55]
microfluidics	Soft lithography [33,35,36,56–59]	Matrigel [58] Collagen [18,33,35,57] Gelatin [57] Calcium alginate [58,59]
photochemistry	FLDW (femtosecond laser direct writing) [38] DMD-PP process(digital micromirror device-based projection printing) [37] UV-light exporsure [39]	BSA [38] Fibronectin [38] Polyethylene glycol (PEG) [37] Poly(ethylene glycol) diacrylate (PEGDA) [37] Methacrylamide-modified gelatin or hydrogel [39]

Table 2. Strengths and limitations of each technique.

Technique	Strength	Limitation
3-D bioprinting	Flexible	Unable to achieve accuracy at less than 50 μm (the highest accuracy is inkjet printing in publishing). The temperature is not easy to control precisely. High cost.
microfluidics	High precision Stable	Structures are limited, e.g., multilayers, lumen.
photochemistry	Curing fast High precision	Limited by optical characteristics. Must be combined with other technologies, e.g., 3-D printing, to achieve high accuracy and flexible structure.

2.1. Microfabrication Techniques Contribute to the Construction of Special Structures, Such as Interface, Regions of Oriented Collagen Fibers, and Vascular Structures in the ECM

Metastasis usually involves the infiltration of cancer cells through aligned collagen fibers in the ECM and their subsequent penetration through the basement membrane to reach the blood or lymph vessels [14]. Moreover, characteristic changes in the orientation of the collagen fibers in the ECM are usually accompanied by the initiation and progression of tumors and are labeled as tumor-associated collagen signature. This collagen signature has served as an early diagnostic marker in various pathological processes related to cancer [60]. To mimic such processes, such as the ECM infiltration and collagen signature in in vitro cell culture systems, a successful reconstruction of the special structures of the ECM becomes significantly important. Microfabrication techniques seem to be ideal tools to realize this goal.

Zhu et al. [34] created a 3-D heterogeneous Matrigel structure, i.e., an ECM system with an interface inside it, to simulate the nonhomogeneous ECM microenvironment in vivo by curing two Matrigel sections of identical concentration at different times. Brin M. Gillette et al. [61] also fabricated a stable collagen fiber-mediated interface by micropatterning and gelling 3-D ECMs of flexible composition to mimic the nonhomogeneous and anisotropic properties of the native tissue.

Besides the realization of interfaces, microfabrication provides a new way to establish aligned collagen fiber structures. Liu et al. [14] created a 3-D sandwiched ECM microenvironment in a microfluidic chip with pillar-like inner structures by injecting collagen I in a fluid state into the Matrigel. Collagen I showed a specific orientation, perpendicular to the heterogeneous interface, induced by the internally developing strain during the solidification of the composite ECM, as shown in Figure 1A. Such aligned collagen fibers might resemble the highly oriented collagen fiber bundles found in cancer patients. Alexandra et al. [62] also varied the alignment of individual fibers and their network by using a microfabrication-based 3-D culture approach to investigate the dynamics of ECM alignment around tissues of defined geometry.

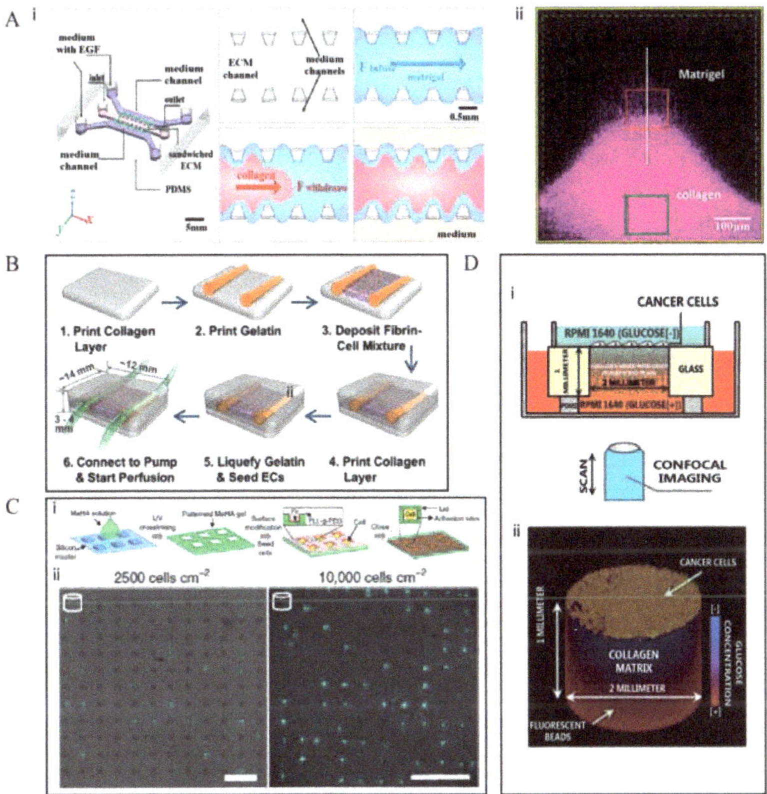

Figure 1. Diverse microfabrication techniques to recreate 3-D ECM microenvironments in vitro. (**A**) Microfluidics is used to build 3-D sandwiched ECM formed by collagen and Matrigel (i). An aligned collagen fiber zone is established at the surface (ii) [14]; (**B**) A multiscale vascular system with collagen as ECM is produced by using the 3-D bioprinting technology [63]; (**C**) Photochemistry is used to build 3-D patterned hydrogels as scaffolds with a range of different geometries (i). The size and shape of the cells is controlled inside microniche chambers (ii). Scale bar: 100 μm [23]; (**D**) A glucose gradient is established in the microfluidic chip, with collagen as the ECM. The system is used to study the trend of collective invasion of cancer cells. The cartoon of device is shown in (i), and the corresponding 3-D view is shown in (ii) [64]. All figures shown here are reproduced with permission from copyright owner.

Other complex structures, such as in vascular or multicomponent systems, can also be reconstructed with modern microfabrication techniques. Lee et al. [63] developed a 3-D bioprinting platform, as illustrated in Figure 1B, to construct large scale fluidic vascular channels with an adjacent capillary network within a 3-D hydrogel through a natural maturation process, as seen for angiogenic endothelial cells sprouting from a large channel edge and branching out into the collagen matrix. Their model can help to engineer desired 3-D vascular niches or thick vascularized tissues. Yong Da Sie et al. [38] adopted multiphoton excitation photochemistry to fabricate a complex 3-D multicomponent microstructure by inserting human fibronectin at specific locations on the 3-D scaffold formed by bovine serum albumin (BSA). Their model can be used to test the role of single factors in cell growth and metastasis.

2.2. Advanced Microfabrication Techniques Can Realize a Systematic and Quantitative Regulation of Cell Geometry, Such as Density, Volume, Size, Shape, and Location, in a Complex 3-D ECM Setting for Long Periods of Culture

Geometric cues are also considered to affect cellular functions, e.g., cell survival, self-renewal, and differentiation, in 3-D ECM microenvironments [23,65]. A precise control of cellular geometry in a complex 3-D scaffold, for long periods of culture, can be achieved effectively through the modern microfabrication techniques.

Lee et al. [66] fabricated a 3-D spheroid culture-based system with transparent and cell-repulsive polyacrylamide hydrogel, for testing nanoparticle toxicity. In their system, the sizes and shapes of the cells were accurately controlled by constraining the cells inside the pores. In 2011, Cheri Y. Li et al. [67] fabricated a 3-D cell-laden hydrogel microtissue with a photocurable hydrogel as the ECM. This system could control the cell size as well as the cell type. They made the microtissues functional by surface-encoding DNA. Through this system, they achieved multiple types of microtissues with multiplexed patterns. Several years later, Bao et al. [23] provided a more robust method to encapsulate single cells within 3-D matrix structures using the photopolymerization technique, for example in hydrogel microniches, which allows both cell adhesion and nutritional permeability, as shown in Figure 1C. These microniches were prism-shaped, with a controlled geometry of the bottom plane, and had precisely defined volumes, which could constrain the cell size and geometry in a systematic and quantitative manner. This device could also be easily combined with a confocal microscope to acquire images rapidly.

2.3. Modern Microfabrication Techniques Can Help to Control Precisely the Concentration Gradient of the Medium, with Respect to Growth Factosr, Nutrients, and Drugs, in the ECM

Metastasis is an energy-consuming process, and cancer cell invasion cannot be achieved without a driving force, such as growth factors, nutrition, or drugs [66,68]. Microfabrication techniques can help us to setup a concentration gradient inside the ECM to further investigate the complex metastasis process in vitro.

Liu et al. [64] designed a 3-D mesoscopic ecology by drilling a hole into a glass slide and filling it with 4.7 mg/ml collagen solution. A stable glucose gradient was established through the gel with low-glucose medium at the top and high-glucose medium at the bottom of the device, as demonstrated in Figure 1D. This device can help to test chemotaxis and trophotaxis of metastatic cells towards glucose or growth factors in a 3-D homogeneous ECM microenvironment. Ashley A. Jaeger et al. [56] also fabricated a 3-D bioreactor using Matrigel as the ECM. Their system could quantitatively control the oxygen gradient mimicking the tumor microenvironment, and could potentially be applied to study gas exchanges in cancer cell cultures.

Multiple concentration gradients can also be realized. Fan et al. [18] constructed a 3-D microfluidic system with an array of hollow microchambers in collagen I for cell culture, by employing the micromoulding and microfluidic techniques. This delicately designed microfluidic system could realize up to four controlled biochemical and drug gradients, simultaneously, through four independent microchannels encompassing the central collagen platform, and offered a robust platform for high-throughput drug and biochemical screening studies.

In conclusion, on the basis of the advanced microfabrication techniques, e.g., 3-D printing, microfluidics, and photochemistry, reconstructing a 3-D cells–ECM microenvironment in vitro, which perfectly mimics tissue morphology and functions in vivo, becomes an attainable goal. Especially, these techniques prove to be very useful in forming a special morphology, such as interface, oriented collagen fiber region, and vascular structures, or in quantitatively regulating cellular geometry and precisely controlling concentration gradients in the ECM.

3. New Discoveries about Cancer and Other Diseases, Based on 3-D ECM Systems In Vitro

Because of the rapid development of advanced microfabrication techniques, numerous novel discoveries, related to cancer and other diseases, have arisen in the last decade. In this section of the review, we will discuss some important studies and latest progress in the field of cancer and other diseases, based on 3-D in vitro cell–ECM microenvironments. We will especially focus on cancer metastasis—new discoveries of the impact of the ECM, e.g., collective invasion, intravasation enhancement by stress and aligned collagen fibers, angiogenesis regulation, as well as drug screening.

3.1. The Impact of Tumor Biotic Factors and Secretion on Normal Tissues and Cells via 3-D ECM Microenvironments

Cancer cells are reported to be able to affect normal tissues and cells by secreting biotic factors, which are transferred through the neighboring ECM. Therefore, a comprehensive understanding of the complicated interactions between cancer cells and normal tissues via a 3-D ECM would be of great benefit for a better control of malignant cancers.

Eliza Li Shan Fong et al. [69] established a 3-D Ewing sarcoma model using an electrospinning apparatus ex vivo. They found that Ewing sarcoma cells cultured in porous 3-D electrospun poly(ε-caprolactone) exhibited remarkable differences in the expression pattern of the insulin-like growth factor-1 receptor/mammalian target of rapamycin pathway, and they were more resistant to traditional cytotoxic drugs compared to the cells grown in a 2-D monolayer culture system. Jia et al. [29] fabricated a 3-D ECM-like substrate with arbitrary micropatterns by combining inkjet printing and electrospinning. They found a close spatiotemporal interaction between breast cancer cells and stromal cells through transforming growth factor (TGF)-β1, which is secreted by breast cancer cells and can induce spatial differentiation of fibroblasts, as shown in Figure 2A1,A2.

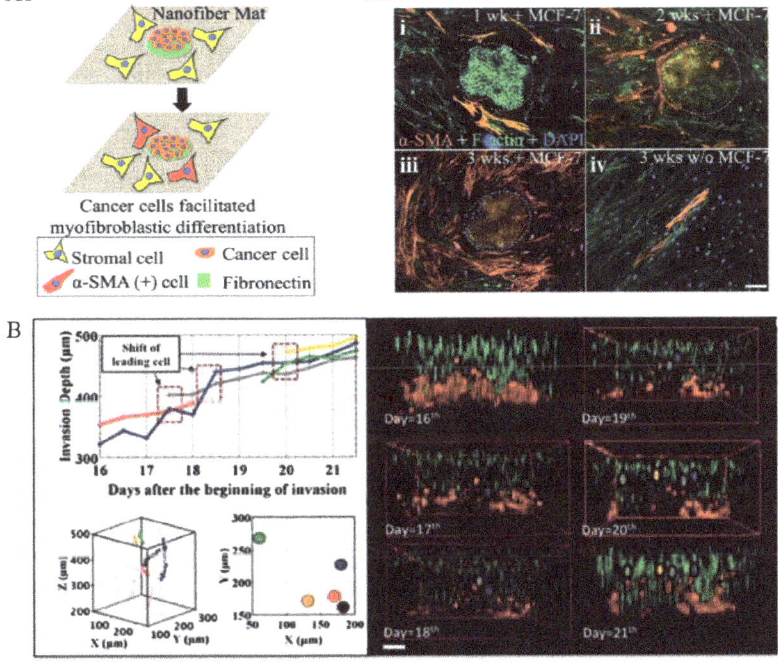

Figure 2. *Cont.*

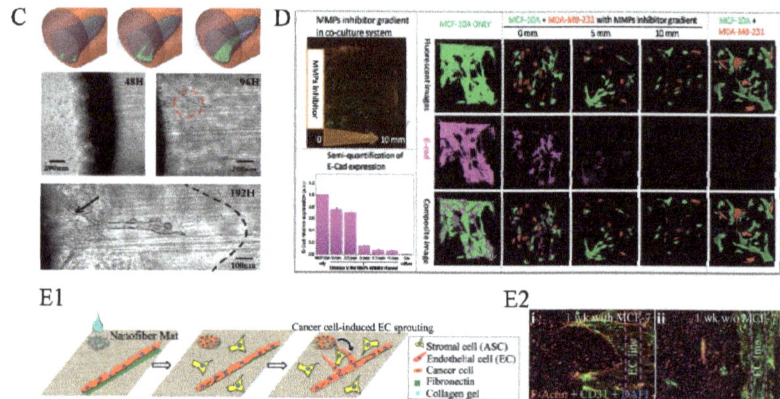

Figure 2. New discoveries about cancer and other diseases based on 3-D ECM systems in vitro. (**A1**) Spatiotemporal interactions between breast cancer cells and stromal cells are studied in a coculture system established by 3-D bioprinting technology. Cancer cells could recruit the stromal cells by differentiating into α-SMA positive cells, dashed line circle indicates the colony formed by cancer cells. (**A2**) Cancer cells (MCF-7) and stromal cells are co-cultured with high seeding density on the polycaprolactone nanofiber matrices. (i-iii) show the fluorescence images after co-cultured for 1 week to 3 weeks. The control seeding, without cancer cells, shows that only a small part of stromal cells became α-SMA-positive cells (iv) Scale bar: 100 μm [29]; (**B**) Using a 3D ECM of collagen, a glucose gradient is established for studying the collective invasion of cancer cells. Scale bar: 150 μm [64]; (**C**) A 3-D funnel-like Matrigel interface is created inside the ECM microenvironment. Through the complex and heterogeneous 3-D ECM microenvironment, ECM heterogeneity is proved to be an essential element in controlling collective cell invasive behaviors, red dashed circle indicates the network and connection of invading cells with other invading branches, and black arrow shows the cell plane. [34]; (**D**) A complex concentration gradient of specific biological molecules is built in a 3-D microfluidic system including microchamber arrays. Matrix Metalloproteinases (MMPs) produced by cancer cells are proven to play a dominant role in determining cellular behavior through the control of E-cad expression [18]; (**E**) Using a system made by bioprinting, cancer cells are shown to regulate angiogenesis through secretion. Cancer cells and endothelial cells are seeded far from each other (**E1**). Endothelial cells are found to sprout toward the cancer clusters. (**E2**) The fluorescence images of coculture of cancer cells and endothelial cells (i), and the control group of endothelial cells only (ii). Scale bar: 100 μm [29]. All figures shown here are reproduced with permission from copyright owner.

3.2. The Collective Invasion of Cancer Cells into a Homogeneous ECM System Driven by Medium, Nutrients, and Growth Factor Gradients

During metastasis in vivo, cancer cells are commonly found to migrate to distant locations collectively. Extensive investigations have indicated that the intercellular and cell–ECM interactions have significant impacts on this collective invasion behavior.

Liu et al. [64] constructed a 3-D mesoscopic ecology to investigate the process of invasion of metastatic cells into an elastic medium in the presence of a glucose gradient. They showed that human breast cancer cells seeded on the top surface of the gel with low-glucose medium could invade into the 3-D collagen matrix cooperatively from the top to the bottom, where the glucose medium concentration was high. This collective invasion of cancer cells was found to be similar to that seen in the case of exchanging leaders in the invading front, as shown in Figure 2B. John Casey et al. [40] fabricated 3-D microwell arrays of defined geometry and dimension using photoreactive hydrogels to study cell migration out from cell aggregates over time. They concluded that spheroids within PEG 20 kDa microwells had the narrowest diameters as the cells tended to condense and cluster together.

Brendon M. Baker et al. [57] used the sacrificial channel template approach to fabricate another ECM structure, composed of collagen gel, which contained microfluidically ported microchannels. A stable diffusion gradient of soluble angiogenic growth factors dependent on the structure of the microfluidic network, was established. They showed that human umbilical vein endothelial cell (HUVEC) invasion and angiogenic sprouting occurred at the locations with the strongest gradients.

3.3. Intravasation of Cancer Cells into a Heterogeneous ECM Structure with the Help of Gradients, Stress, or Aligned Collagen Fibers

The intravasation of cancer cells into the lymphatic or vascular systems involves the key process of the cancer cells breaking through the basement membrane to get into the vessels. With modern technology, the important and puzzling question concerning tumor metastasis, i.e., how the heterogeneous ECM structures, such as aligned collagen or interface, influence cell infiltration and the subsequent break into the basement membrane during metastasis, is slowly getting answered. In 2012, Ioannis K. Zervantonakis et al. [70] designed and constructed a 3-D tumor–endothelial intravasation microfluidic-based assay by interconnecting two independently addressable microchannels via a 3-D ECM hydrogel. Tumor and endothelial cells were seeded, respectively, in these two channels to mimic tumor cell invasion through 3-D ECM, in response to externally applied growth factor, such as epidermal growth factor (EGF), or biochemical gradients. Their experiments, aimed at addressing how tumor cells interact with normal epithelial cells, showed that breast carcinoma cells could invade through a HUVEC monolayer in the presence of macrophages, and the endothelial barrier impairment was associated with a higher number and faster dynamics of tumor–endothelial interactions. Several years later, Zhu et al. [34] used their system of 3-D heterogeneous ECM microenvironment in vitro to mimic the environment in vivo. Breast cancer cells were found to exhibit strikingly different invasive behaviors in homogenous and heterogeneous Matrigel with interior interfaces. As shown in Figure 2C, the interface in a heterogeneous ECM microenvironment guided cell invasion at an early stage, and the cells showed strong collective finger-like invasive behaviors. Liu et al. [14] used their 3-D ECM structure, as introduced in Section 2.1, to study cancer cell infiltration. They showed that the aligned collagen fibers were able to significantly enhance tumor metastatic potential, since metastatic breast cancer cells could accelerate their infiltration into rigid Matrigel by following the alignment direction. Epithelial–mesenchymal transition (EMT) is also a prominent signal of deterioration of tumors. Once a cancer cell undergoes a phenotypic change through EMT, it will obtain greater migratory and invasive ability, directing cells to penetrate into the vessels [71]. Sharmistha Saha et al. [72] combined the technology of electrospun and microfabrication to create fibrous scaffolds with random or aligned fiber orientation mimicking the complex ECM microenvironment in tumors. They found that the aligned fiber scaffold can guide EMT in cancer cells accompanied by the upregulation of TGF β-1.

3.4. Biophysical and Mechanical Properties of the ECM, such as Geometry, Density, Stiffness, Contractility, and Crosslinking of Collagen Fibers, Influence Tumor Angiogenesis Both Directly and Indirectly

Recently, ECM stiffness has progressively become a significant mechanical cue that precedes disease and drives its pathological progression by altering cellular behavior or influencing tumor angiogenesis [73]. Abnormal vessel growth and function are usually the hallmarks of cancer or inflammatory diseases and contribute to disease progression [74]. Tumor cells and tissues need to establish a blood supply, primarily through angiogenesis, from a pre-existing vascular network, to satisfy their increasing demand for nutrients and oxygen to perform metabolic functions, similar to the normal tissues [74,75]. Angiogenic programming of tumor tissues is a multidimensional process, regulated by cancer cells, stromal cells, their bioactive products, as well as ECM microenvironments [76].A few years ago, Mason et al. [77] took advantage of the tunable mechanical properties of collagen-based scaffolds to investigate the effects of matrix stiffness on 3-D microenvironments. They observed a significant increase in cellular spreading and extension with increasing matrix stiffness, and these effects lasted throughout the entire course of the cell culture period. Recently, Jia et al. [29] showed cancer-regulated angiogenesis by factors secreted from the cancer cells, which guided the sprouting of endothelial cells toward cancer clusters, by using their 3-D ECM structure with arbitrary micropatterns, as shown in Figure 2E1,E2.

3.5. 3-D Microenvironments for Drug Screening

Conventionally, pharmacokinetic studies are performed in rodents or in 2-D polarized intestinal cells before clinical administration, nevertheless, several compounds have still proven toxic to humans, despite such testing [9]. Advanced microfabrication techniques have proven their suitability in creating 3-D cell–ECM microenvironments in vitro that can be perfectly used for screening drugs.

Lee et al. [66] showed that the toxicity of CdTe and Au was significantly reduced in their system with a 3-D microenvironment, compared to that in 2-D cultures. In 2017, Fan et al. [18] used their complex gradient microfluidic system, which allowed incorporating multiple chemical, temporal, and spatial gradients within a 3-D micropatterned collagen scaffold in vitro, to test the role of matrix metalloproteinases (MMPs) in inhibiting cancer cells aggregation. As shown in Figure 2D, MMPs inhibitors were found to depress the abilities of the cancer cells to disrupt the structure of the surrounding lumen-like cell clusters; the down regulation of E-cad expression, due to the MMPs produced by invasive breast cancer cells, played a dominant role in determining the cellular behavior.

3.6. Other Diseases (Wound Healing, Clonal Acinar Development, Differentiation of Embryoid Bodies, Neurodevelopmental Processes, Cartilage Defects) Related to 3-D Cells–ECM Microenvironments

Besides cancer metastasis, 3-D cell–ECM microenvironments fabricated by modern microfabrication techniques can also be applied in the study of many other diseases or biomedical processes, including, but not limited to, wound healing, clonal acinar development, differentiation of embryoid bodies, neurodevelopmental processes, and so on.

The idea of the 3-D bioprinting of skin cells was considered adequate for transitioning the technology to clinical settings. For example, Skardal, A. et al. [78] used the bioprinting technology to deposit two layers of a fibrin–collagen gel containing endothelial cells to treat full-thickness skin wounds in mice.

In 2015, Dolega et al. [79] constructed a 3-D cell culture system based on a flow-focusing microfluidic chip that encapsulates cells in Matrigel microbeads. Each individual microbead acts as a single-cell culture compartment, generating one acinus per bead, on an average. They used this 3-D cell culture system to investigate acinar development, by tracing the microbead under temporally and spatially controlled conditions. They found that acinar development proceeds through clonal growth from single cells, which are self-sufficient to differentiate in a Matrigel environment.

In studying the differentiation of embryoid bodies, Fung et al. [80] presented a 3-D microfluidic platform, using biocompatible materials, based on a Y-channel device with two inlets for two different culturing media, to study the differentiation of embryoid bodies by in vitro aggregation of embryonic stem cells. They found that the differentiation of the same embryonic stem cells into different specialized cell lineages can be achieved by applying the microfluidic technology.

In the treatment of cartilage defects, the differentiation of stem cells into chondrocytes is an important cue [81]. Many physical and mechanical characteristics of the ECM, such as hypoxia, stiffness, topography, and pore size, are closely related to the interdifferentiation process of chondrocytes and stem cells [82]. Kuan-Han Wu et al. [81] used gelatin as ECM in a microfabrication process to create a scaffold with different sizes of microbubbles and pores. They found that the size of microbubbles and pores can impact chondrogenesis, and a larger pore size is required for inducing mature chondrogenesis. Xiaobin Huang et al. [83] chose pH-degradable Poly(vinyl alcohol) (PVA) hydrogel as an ECM scaffold to study the effects of oxygen tension on chondrogenesis. They found that the degree of hypoxia affects the expression of many cytokines (e.g., MMP13, ALPL, COL10A1, ELISA), which has a further influence on chondrogenesis. For example, hypoxia in 2.5% O_2 would promote cartilage matrix production.

Recently, Lancaster et al. [84] developed a 3-D cerebral organoid culture system for deriving brain tissues to study human neurodevelopmental processes in vitro. This method recapitulated not only the fundamental mechanisms of mammalian neurodevelopment, but also displayed a wide range of characteristics of human brain development.

4. Conclusions

The significant progress of microfabrication in the last decade has advanced our understanding of cancer and other diseases. In this review, we have discussed some important findings and latest developments of the above approaches with respect to cancer and other diseases in 3-D ECM scaffolds. We stress that these 3-D ECM microenvironments are closely related to the process of cancer metastasis that presents multiple aspects, such as collective invasion and intravasation, tumor angiogenesis regulation, as well as drug screening.

Acknowledgments: Grants from the National Natural Science Foundation of China (Grant No. 11604030, No. 11474345, No. 11674043), Natural Science Foundation Project of CQ CSTC (Grant No. 2015jcyjA00042) and the Fundamental Research Funds for the Central Universities (Project No. 106112017CDJXY300001).

Author Contributions: Kena Song and Zirui Wang contributed equally to this work. Guo Chen, Ruchuan Liu, and Liyu Liu designed the study; Kena Song, Zirui Wang, and Guo Chen investigated the literature; Kena Song, Zirui Wang, Ruchuan Liu, Guo Chen, and Liyu Liu provided advice; Kena Song and Guo Chen wrote the paper.

Conflicts of Interest: The authors declare no conflict of interest.

Abbreviations

3-D	Three dimensional
2-D	Two dimensional
ECM	Extracellular matrix
BSA	Bovine serum albumin
TGF	Transforming growth factor
HUVEC	Human umbilical vein endothelial cells
MMPs	Matrix metalloproteinases
EGF	Epidermal growth factor
DIGME	Diskoid In Geometrically Micropatterned ECM
EMT	Epithelial-mesenchymal transition
PEG	Polyethylene glycol
PEGDA	Poly(ethylene glycol) diacrylate
DMD-PP	Digital micromirror device-based projection printing
FLDW	Femtosecond laser direct writing
MMPs	Matrix Metalloproteinases
PVA	Poly(vinyl alcohol)

References

1. Sporn, M.B. The war on cancer. *Lancet* **1996**, *347*, 1377. [CrossRef]
2. Sleeman, J.P.; Steeg, P.S. Cancer metastasis as a therapeutic target. *Eur. J. Cancer* **2010**, *46*, 1177–1180. [CrossRef] [PubMed]
3. Mowers, E.E.; Sharifi, M.N.; Macleod, K.F. Autophagy in cancer metastasis. *Oncogene* **2017**, *36*, 1619–1630. [CrossRef] [PubMed]
4. Seyfried, T.N.; Huysentruyt, L.C. On the Origin of Cancer Metastasis. *Crit. Rev. Oncog.* **2013**, *18*, 43–73. [CrossRef] [PubMed]
5. Wells, A.; Grahovac, J.; Wheeler, S.; Ma, B.; Lauffenburger, D.A. Targeting tumor cell motility as a strategy against invasion and metastasis. *Trends Pharmacol. Sci.* **2013**, *34*, 283–289. [CrossRef] [PubMed]
6. Butcher, D.T.; Alliston, T.; Weaver, V.M. A tense situation: Forcing tumour progression. *Nat. Rev. Cancer* **2009**, *9*, 108–122. [CrossRef] [PubMed]

7.	Sinkus, R.; Lorenzen, J.; Schrader, D.; Lorenzen, M.; Dargatz, M.; Holz, D.J. High-resolution tensor MR elastography for breast tumour detection. *Phys. Med. Biol.* **2000**, *45*, 1649–1664. [CrossRef] [PubMed]

8.	Paszek, M.J.; Zahir, N.; Johnson, K.R.; Lakins, J.N.; Rozenberg, G.I.; Gefen, A.; Reinhartking, C.A.; Margulies, S.S.; Dembo, M.; Boettiger, D. Tensional homeostasis and the malignant phenotype. *Cancer Cell.* **2005**, *3*, 241–254. [CrossRef] [PubMed]

9.	Broekelmann, T.J.; Limper, A.H.; Colby, T.V.; Mcdonald, J.A. Transforming growth factor beta 1 is present at sites of extracellular matrix gene expression in human pulmonary fibrosis. *Proc. Natl. Acad. Sci. USA* **1991**, *88*, 6642–6646. [CrossRef] [PubMed]

10.	Qi, Z.; Atsuchi, N.; Ooshima, A.; Takeshita, A.; Ueno, H. Blockade of type β transforming growth factor signaling prevents liver fibrosis and dysfunction in the rat. *Proc. Natl. Acad. Sci. USA* **1999**, *96*, 2345–2349. [CrossRef] [PubMed]

11.	Choi, J.; Lee, E.K.; Choo, J.; Yuh, J.; Hong, J.W. Micro 3D cell culture systems for cellular behavior studies: Culture matrices, devices, substrates, and in-situ sensing methods. *Biotech. J.* **2015**, *10*, 1682–1688. [CrossRef] [PubMed]

12.	Kolacna, L.; Bakesova, J.; Varga, F.; Kost'akova, E.; Planka, L.; Necas, A.; Lukas, D.; Amler, E.; Pelouch, V. Biochemical and Biophysical Aspects of Collagen Nanostructure in the Extracellular Matrix. *Physiol. Res.* **2007**, *56*, S51–S60. [PubMed]

13.	Gupta, G.P.; Massague, J. Cancer Metastasis: Building a Framework. *Cell* **2006**, *127*, 679–695. [CrossRef] [PubMed]

14.	Han, W.; Chen, S.; Yuan, W.; Fan, Q.; Tian, J.; Wang, X.; Chen, L.; Zhang, X.; Wei, W.; Liu, R.; et al. Oriented collagen fibers direct tumor cell intravasation. *Proc. Natl. Acad. Sci. USA* **2016**, *113*, 11208–11213. [CrossRef] [PubMed]

15.	Genovese, L.; Zawada, L.; Tosoni, A.; Ferri, A.; Zerbi, P.; Allevi, R.; Nebuloni, L.; Alfano, M. Cellular Localization, Invasion, and Turnover Are Differently Influenced by Healthy and Tumor-Derived Extracellular Matrix. *Tissue Eng. Part A* **2014**, *20*, 2005. [CrossRef] [PubMed]

16.	Alfano, M.; Nebuloni, M.; Allevi, R.; Zerbi, P.; Longhi, E.; Luciano, R.; Locatelli, I.; Pecoraro, A.; Indrieri, M.; Speziali, C.; et al. Linearized texture of three-dimensional extracellular matrix is mandatory for bladder cancer cell invasion. *Sci. Rep.* **2016**, *6*, 36128. [CrossRef] [PubMed]

17.	Nebuloni, M.; Albarello, L.; Andolfo, A.; Magagnotti, C.; Genovese, L.; Locatelli, I.; Tonon, G.; Longhi, E.; Zerbi, P.; Allevi, R.; et al. Insight on Colorectal Carcinoma Infiltration by Studying Perilesional Extracellular Matrix. *Sci Rep.* **2016**, *6*, 22522. [CrossRef] [PubMed]

18.	Fan, Q.; Liu, R.; Jiao, Y.; Tian, C.; Farrell, J.D.; Diao, W.; Wang, X.; Zhang, F.; Yuan, W.; Han, H. A novel 3-D bio-microfluidic system mimicking in vivo heterogeneous tumour microstructures reveals complex tumour-stroma interactions. *Lab Chip* **2017**. [CrossRef] [PubMed]

19.	Sung, K.E.; Beebe, D.J. Microfluidic 3D models of cancer. *Adv. Drug Deliv. Rev.* **2014**, 68–78. [CrossRef] [PubMed]

20.	Muthuswamy, S.K. 3D culture reveals a signaling network. *Breast Cancer Res.* **2011**, *13*, 103. [CrossRef] [PubMed]

21.	Patil, P.U.; Dambrosio, J.; Inge, L.J.; Mason, R.W.; Rajasekaran, A.K. Carcinoma cells induce lumen filling and EMT in epithelial cells through soluble E-cadherin-mediated activation of EGFR. *J. Cell Sci.* **2015**, *128*, 4366–4379. [CrossRef] [PubMed]

22.	Tanner, K.; Gottesman, M.M. Beyond 3D culture models of cancer. *Sci. Transl. Med.* **2015**, *7*. [CrossRef] [PubMed]

23.	Bao, M.; Xie, J.; Piruska, A.; Wts, H. 3D microniches reveal the importance of cell size and shape. *Nature Commun.* **2017**. [CrossRef] [PubMed]

24.	Ebrahimkhani, M.R.; Neiman, J.A.; Raredon, M.S.; Hughes, D.J.; Griffith, L.G. Bioreactor Technologies to Support Liver Function in vitro. *Adv. Drug Deliv. Rev.* **2014**, 132–157. [CrossRef] [PubMed]

25.	Esch, M.B.; King, T.L.; Shuler, M.L. The Role of Body-on-a-Chip Devices in Drug and Toxicity Studies. *Annu. Rev. Biomed. Eng.* **2011**, *13*, 55–72. [CrossRef] [PubMed]

26.	Wong, A.P.; Perezcastillejos, R.; Love, J.C.; Whitesides, G.M. Partitioning microfluidic channels with hydrogel to construct tunable 3-D cellular microenvironments. *Biomaterials* **2008**, *29*, 1853–1861. [CrossRef] [PubMed]

27.	Hu, W.; Ishii, K.S.; Fan, Q.; Ohta, A.T. Hydrogel microrobots actuated by optically generated vapour bubbles. *Lab Chip* **2012**, *12*, 3821–3826. [CrossRef] [PubMed]

28. Hahan, N.P.; Dolega, M.E.; Liguori, L.; Marquette, C.; Gac, S.L.; Gidrol, X.; Martin, D.K. A 3D Toolbox to Enhance Physiological Relevance of Human Tissue Models. *Trends Biotechnol.* **2016**, *34*, 757–769.

29. Jia, C.; Luo, B.; Wang, H.; Bian, Y.; Li, X.; Li, S.; Wang, H. Precise and Arbitrary Deposition of Biomolecules onto Biomimetic Fibrous Matrices for Spatially Controlled Cell Distribution and Functions. *Adv. Mater.* **2017**, 1701154. [CrossRef] [PubMed]

30. Loo, Y.; Hauser, C.A. Bioprinting synthetic self-assembling peptide hydrogels for biomedical applications. *Biomed. Mater.* **2015**, *11*. [CrossRef] [PubMed]

31. Murphy, S.V.; Atala, A. 3D bioprinting of tissues and organs. *Nat. Biotechnol.* **2014**, *32*, 773–785. [CrossRef] [PubMed]

32. Ozbolat, I.T. Bioprinting scale-up tissue and organ constructs for transplantation. *Trends Biotechnol.* **2015**, *33*, 395–400. [CrossRef] [PubMed]

33. Boghaert, E.; Gleghorn, J.P.; Lee, K.; Gjorevski, N.; Radisky, D.C.; Nelson, C.M. Host epithelial geometry regulates breast cancer cell invasiveness. *Proc. Natl. Acad. Sci. USA* **2012**, *109*, 19632–19637. [CrossRef] [PubMed]

34. Zhu, J.; Liang, L.; Jiao, Y.; Liu, L. Enhanced Invasion of Metastatic Cancer Cells via Extracellular Matrix Interface. *PLoS ONE* **2015**, *10*. [CrossRef] [PubMed]

35. Lee, J.S.; Romero, R.; Han, Y.M.; Kim, H.C.; Kim, C.J.; Hong, J.S.; Huh, D. Placenta-on-a-chip: A novel platform to study the biology of the human placenta. *J. Matern. Fetal Neonatal Med.* **2016**, *29*, 1046–1054. [CrossRef] [PubMed]

36. Kraning-Rush, C.M.; Carey, S.P.; Lampi, M.C.; Reinhart-King, C.A. Microfabricated collagen tracks facilitate single cell metastatic invasion in 3D. *Integr. Biol.* **2013**, *5*, 606–616. [CrossRef] [PubMed]

37. Soman, P.; Kelber, J.A.; Lee, J.W.; Wright, T.N.; Vecchio, K.S.; Klemke, R.L.; Chen, S. Cancer cell migration within 3D layer-by-layer microfabricated photocrosslinked PEG scaffolds with tunable stiffness. *Biomaterials* **2012**, *3*, 7064–7070. [CrossRef] [PubMed]

38. Sie, Y.D.; Li, Y.C.; Chang, N.S.; Campagnola, P.J.; Chen, S.J. Fabrication of three-dimensional multi-protein microstructures for cell migration and adhesion enhancement. *Biomed. Opt. Express* **2015**, *6*, 480–490. [CrossRef] [PubMed]

39. Markovic, M.; Van Hoorick, J.; Holzl, K.; Tromayer, M.; Gruber, P.M.; Nurnberger, S.; Dubruel, P.; Vlierberghe, S.V.; Liska, R.; Ovsianikov, A. Hybrid Tissue Engineering Scaffolds by Combination of Three-Dimensional Printing and Cell Photoencapsulation. *J. Nanotechnol. Eng. Med.* **2015**, *6*, 210011–210017. [CrossRef] [PubMed]

40. Casey, J.; Yue, X.; Nguyen, T.D.; Acun, A.; Zellmer, V.R.; Zhang, S.; Zorlutuna, P. 3D hydrogel-based microwell arrays as a tumor microenvironment model to study breast cancer growth. *Biomed. Mater.* **2017**, *12*, 025009. [CrossRef] [PubMed]

41. Alobaidi, A.A.; Sun, B. Probing three-dimensional collective cancer invasion with DIGME. *Cancer Converg.* **2017**, *1*. [CrossRef]

42. Alobaidi, A.A.; Xu, Y.; Chen, S.; Jiao, Y.; Sun, B. Probing cooperative force generation in collective cancer invasion. *Phys. Biol.* **2017**, *14*, 045005. [CrossRef] [PubMed]

43. Bissell, M.J.; Radisky, D.C. Putting tumours in context. *Nat. Rev. Cancer* **2001**, *1*, 46–54. [CrossRef] [PubMed]

44. Nelson, C.M.; VanDuijn, M.M.; Inman, J.L.; Fletcher, D.A.; Bissell, M.J. Tissue Geometry Determines Sites of Mammary Branching Morphogenesis in Organotypic Cultures. *Science* **2006**, *314*, 298–300. [CrossRef] [PubMed]

45. Khademhosseini, A.; Langer, R. Microengineered hydrogels for tissue engineering. *Biomaterials* **2007**, *28*, 5087–5092. [CrossRef] [PubMed]

46. Lee, K.Y.; Mooney, D.J. Hydrogels for Tissue Engineering. *Chem. Rev.* **2001**, *101*, 1869–1879. [CrossRef] [PubMed]

47. Willerth, S.M.; Arendas, K.J.; Gottlieb, D.I.; Sakiyama-Elbert, S.E. Optimization of Fibrin Scaffolds for Differentiation of Murine Embryonic Stem Cells into Neural Lineage Cells. *Biomaterials* **2006**, *27*, 5990–6003. [CrossRef] [PubMed]

48. Raeber, G.P.; Lutolf, M.P.; Hubbell, J.A. Molecularly engineered PEG hydrogels: A novel model system for proteolytically mediated cell migration. *Biophys. J.* **2005**, *89*, 1374–1388. [CrossRef] [PubMed]

49. Ma, P.X.; Choi, J. Biodegradable Polymer Scaffolds with Well-Defined Interconnected Spherical Pore Network. *Tissue Eng.* **2001**, *7*, 23–33. [CrossRef] [PubMed]

50. Kuntz, R.M.; Saltzman, W.M. Neutrophil motility in extracellular matrix gels: Mesh size and adhesion affect speed of migration. *Biophys. J.* **1997**, *72*, 1472–1480. [CrossRef]

51. Fedorovich, N.E.; Wijnberg, H.M.; Dhert, W.J.; Alblas, J. Distinct Tissue Formation by Heterogeneous Printing of Osteo- and Endothelial Progenitor Cells. *Tissue Eng. Part A* **2011**, *17*, 2113–2121. [CrossRef] [PubMed]

52. Yan, K.C.; Nair, K.; Sun, W. Three dimensional multi-scale modelling and analysis of cell damage in cell-encapsulated alginate constructs. *J. Biomech.* **2010**, *43*, 1031–1038. [PubMed]

53. Cheng, J.; Lin, F.; Liu, H.; Yan, Y.; Wang, X.; Zhang, R.; Xiong, Z. Rheological Properties of Cell-Hydrogel Composites Extruding Through Small-Diameter Tips. *J. Manuf. Sci. Eng.* **2008**, *130*, 021014. [CrossRef]

54. Yan, Y.; Wang, X.; Pan, Y.; Liu, H.; Cheng, J.; Xiong, Z.; Lin, F.; Wu, R.; Zhang, R.; Lu, Q. Fabrication of viable tissue-engineered constructs with 3D cell-assembly technique. *Biomaterials* **2005**, *26*, 5864–5871. [CrossRef] [PubMed]

55. Xu, W.; Wang, X.; Yan, Y.; Zheng, W.; Xiong, Z.; Lin, F.; Wu, R.; Zhang, R. Rapid Prototyping Three-Dimensional Cell/Gelatin/Fibrinogen Constructs for Medical Regeneration. *J. Bioact. Compat. Polym.* **2007**, *22*, 363–377. [CrossRef]

56. Jaeger, A.A.; Das, C.K.; Morgan, N.Y.; Pursley, R.; Mcqueen, P.G.; Hall, M.D.; Pohida, T.J.; Gottesman, M.M. Microfabricated polymeric vessel mimetics for 3-D cancer cell culture. *Biomaterials* **2013**, *34*, 8301–8313. [CrossRef] [PubMed]

57. Baker, B.M.; Trappmann, B.; Stapleton, S.C.; Toro, E.; Chen, C.S. Microfluidics embedded within extracellular matrix to define vascular architectures and pattern diffusive gradients. *Lab Chip* **2013**, *13*, 3246–3252. [CrossRef] [PubMed]

58. Kim, S.H.; Chi, M.; Yi, B.; Kim, S.H.; Oh, S.; Kim, Y.; Park, S.; Sung, J.H. Three-dimensional intestinal villi epithelium enhances protection of human intestinal cells from bacterial infection by inducing mucin expression. *Integr. Biol.* **2014**, *6*, 1122–1131. [CrossRef] [PubMed]

59. Sung, J.H.; Yu, J.; Luo, D.; Shuler, M.L.; March, J.C. Microscale 3-D hydrogel scaffold for biomimetic gastrointestinal (GI) tract model. *Lab Chip* **2011**, *11*, 389–392. [CrossRef] [PubMed]

60. Pavithra, V.; Sowmya, S.V.; Rao, R.S.; Patil, S.; Augustine, D.; Haragannavar, V.C.; Nambiar, S.K. Tumor-associated Collagen Signatures: An Insight. *World J. Dent.* **2017**, *8*, 224–230. [CrossRef]

61. Gillette, B.M.; Jensen, J.A.; Tang, B.; Yang, G.J.; Bazarganlari, A.; Zhong, M.; Sia, S.K. In situ collagen assembly for integrating microfabricated three-dimensional cell-seeded matrices. *Nat. Mater.* **2008**, *7*, 636–640. [CrossRef] [PubMed]

62. Alexandra, S.P.; Nerger, B.A.; Wolf, A.E.; Sundaresan, S.; Nelson, C.M. Dynamics of Tissue-Induced Alignment of Fibrous Extracellular Matrix. *Biophys. J.* **2017**, *113*, 702–713.

63. Lee, V.K.; Lanzi, A.M.; Ngo, H.; Yoo, S.; Vincent, P.A.; Dai, G. Generation of Multi-Scale Vascular Network System within 3D Hydrogel using 3D Bio-Printing Technology. *Cell. Mol. Bioeng.* **2014**, *7*, 460–472. [CrossRef] [PubMed]

64. Liu, L.; Duclos, G.; Sun, B.; Lee, J.; Wu, A.; Kam, Y.; Sontag, E.D.; Stone, H.A.; Sturm, J.C.; Gatenby, R.A.; Austin, R.H. Minimization of thermodynamic costs in cancer cell invasion. *Proc. Natl. Acad. Sci. USA* **2013**, *110*, 1686–1691. [CrossRef] [PubMed]

65. Scadden, D.T. The stem-cell niche as an entity of action. *Nature* **2006**, *441*, 1075–1079. [CrossRef] [PubMed]

66. Lee, J.; Lilly, G.D.; Doty, R.C.; Podsiadlo, P.; Kotov, N.A. In vitro toxicity testing of nanoparticles in 3D cell culture. *Small* **2009**, *5*, 1213–1221. [CrossRef] [PubMed]

67. Li, C.Y.; Wood, D.K.; Hsu, C.M.; Bhatia, S.N. DNA-templated assembly of droplet-derived PEG microtissues. *Lab Chip* **2011**, *11*, 2967–2975. [CrossRef] [PubMed]

68. Vazquez, A.; Liu, J.; Zhou, Y.; Oltvai, Z.N. Catabolic efficiency of aerobic glycolysis: The Warburg effect revisited. *BMC Syst. Biol.* **2010**, *4*, 58. [CrossRef] [PubMed]

69. Fong, E.L.; Lamhamedicherradi, S.E.; Burdett, E.; Ramamoorthy, V.; Lazar, A.J.; Kasper, F.K.; Farachcarson, M.C.; Vishwamitra, D.; Demicco, E.G.; Menegaz, B.A.; et al. Modeling Ewing sarcoma tumors in vitro with 3D scaffolds. *Proc. Natl. Acad. Sci. USA* **2013**, *110*, 6500–6505. [CrossRef] [PubMed]

70. Zervantonakis, I.K.; Hughesalford, S.K.; Charest, J.L.; Condeelis, J.; Gertler, F.B.; Kamm, R.D. Three-dimensional microfluidic model for tumor cell intravasation and endothelial barrier function. *Proc. Natl. Acad. Sci. USA* **2012**, *109*, 13515–13520. [CrossRef] [PubMed]

71. Bryony, S.W.; Werb, Z. Stromal Effects on Mammary Gland Development and Breast Cancer. *Science* **2002**, *296*, 1046–1049.

72. Saha, S.; Duan, X.; Wu, L.; Lo, P.; Chen, H.; Wang, Q. Electrospun fibrous scaffolds promote breast cancer cell alignment and epithelial-mesenchymal transition. *Langmuir* **2012**, *28*, 2028–2034. [CrossRef] [PubMed]

73. Marsha, C.L.; Cynthia, A. Reinhart-King, Targeting extracellular matrix stiffness to attenuate disease: From molecular mechanisms to clinical trials. *Sci Transl. Med.* **2018**, *10*, eaao0475. [CrossRef]

74. Potente, M.; Gerhardt, H.; Carmeliet, P. Basic and Therapeutic Aspects of Angiogenesis. *Cell* **2011**, *146*, 873–887. [CrossRef] [PubMed]

75. Lagory, E.L.; Giaccia, A.J. The ever-expanding role of HIF in tumour and stromal biology. *Nat. Cell Biol.* **2016**, *18*, 356–365. [CrossRef] [PubMed]

76. Michele, D.P.; Biziato, D.; Petrova, T.V. Microenvironmental regulation of tumour angiogenesis. *Nat. Rev. Cancer* **2017**, *17*. [CrossRef]

77. Mason, B.N.; Starchenko, A.; Williams, R.M.; Bonassar, L.J.; Reinhartking, C.A. Tuning three-dimensional collagen matrix stiffness independently of collagen concentration modulates endothelial cell behavior. *Acta Biomater.* **2013**, *9*, 4635–4644. [CrossRef] [PubMed]

78. Skardal, A.; Mack, D.L.; Kapetanovic, E.; Atala, A.; Jackson, J.D.; Yoo, J.J.; Soker, S. Bioprinted Amniotic Fluid-Derived Stem Cells Accelerate Healing of Large Skin Wounds. *Stem Cell. Transl. Med.* **2012**, *1*, 792–802. [CrossRef] [PubMed]

79. Dolega, M.E.; Abeille, F.; Picolletdhahan, N.; Gidrol, X. Controlled 3D culture in Matrigel microbeads to analyze clonal acinar development. *Biomaterials* **2015**, *52*, 347–357. [CrossRef] [PubMed]

80. Fung, W.; Beyzavi, A.; Abgrall, P.; Nguyen, N.; Li, H. Microfluidic platform for controlling the differentiation of embryoid bodies. *Lab Chip* **2009**, *9*, 2591–2595. [CrossRef] [PubMed]

81. Wu, K.; Mei, C.; Lin, C.; Yang, K.; Yu, J. The influence of bubble size on chondrogenic differentiation of adipose-derived stem cells in gelatin microbubble scaffolds. *J. Mater. Chem. B* **2018**, *6*, 125–132. [CrossRef]

82. Mahadik, B.P.; Wheeler, T.D.; Skertich, L.J.; Kenis, P.J.; Harley, B.A. Microfluidic generation of gradient hydrogels to modulate hematopoietic stem cell culture environment. *Adv. Healthc. Mater.* **2014**, *3*, 449–458. [CrossRef] [PubMed]

83. Huang, X.; Hou, Y.; Zhong, L.; Huang, D.; Qian, H.; Karperien, M., Chen, W. Promoted Chondrogenesis of Cocultured Chondrocytes and Mesenchymal Stem Cells under Hypoxia Using In-situ Forming Degradable Hydrogel Scaffolds. *Biomacromolecules* **2018**, *19*, 94–102. [CrossRef] [PubMed]

84. Lancaster, M.A.; Renner, M.; Martin, C.; Wenzel, D.; Bicknell, L.S.; Hurles, M.E.; Homfray, T.; Penninger, J.M.; Jackson, A.P.; Knoblich, J.A. Cerebral organoids model human brain development and microcephaly. *Nature* **2013**, *501*, 373–379. [CrossRef] [PubMed]

International Journal of
Molecular Sciences

Review

Mast Cells: Key Contributors to Cardiac Fibrosis

Scott P. Levick * and Alexander Widiapradja

Kolling Institute for Medical Research, University of Sydney, St Leonards 2065, Australia;
alexander.widiapradja@sydney.edu.au
* Correspondence: scott.levick@sydney.edu.au; Tel.: +61-2-9926-4911

Received: 27 November 2017; Accepted: 22 December 2017; Published: 12 January 2018

Abstract: Historically, increased numbers of mast cells have been associated with fibrosis in numerous cardiac pathologies, implicating mast cells in the development of cardiac fibrosis. Subsequently, several approaches have been utilised to demonstrate a causal role for mast cells in animal models of cardiac fibrosis including mast cell stabilising compounds, rodents deficient in mast cells, and inhibition of the actions of mast cell-specific proteases such as chymase and tryptase. Whilst most evidence supports a pro-fibrotic role for mast cells, there is evidence that in some settings these cells can oppose fibrosis. A major gap in our current understanding of cardiac mast cell function is identification of the stimuli that activate these cells causing them to promote a pro-fibrotic environment. This review will present the evidence linking mast cells to cardiac fibrosis, as well as discuss the major questions that remain in understanding how mast cells contribute to cardiac fibrosis.

Keywords: heart; protease; tryptase; chymase; TNF-α; collagen; extracellular matrix; histamine

1. Introduction

Mast cells (MCs) are non-circulating immune cells that develop only once bone marrow-derived precursors have reached their target tissue. These tissue MCs then go through several stages of maturation driven primarily by the c-kit ligand, stem cell factor, with the final MC phenotype being highly dependent on the microenvironment in which they reside. There is now a substantial amount of evidence supporting a role for MCs in cardiac remodeling and heart failure [1]. In fact, MC numbers increase in the heart both in humans and in experimental animals in a wide variety of cardiac pathologies including myocardial infarction (MI) [2–4], hypertension [5–7], transplantation [8], myocarditis [9], volume overload [10,11], and in the failing heart [12]. Many of the effects of cardiac MCs involve regulation of the extracellular matrix (ECM), whether it be inducing ECM degradation or promoting increased ECM synthesis. The latter characterises cardiac fibrosis, which is a characteristic of almost all cardiac pathologies and is the focus of this review article. Cardiac MCs can influence fibrosis by direct effects on fibroblasts, as well as indirect effects, both brought about by the many proteases, cytokines, growth factors, and other products manufactured by these multi-faceted cells. This review article will discuss the evidence supporting a role for MCs in cardiac fibrosis by presenting studies that have utilised MC stabilizing compounds, rodents deficient in MCs, and specific inhibitors of MC proteases. This review will also discuss important unanswered questions in the field, including the elusive mediators that activate cardiac MCs causing them to promote fibrosis.

2. Studies Associating Mast Cells with Cardiac Fibrosis

MCs were first linked to cardiac fibrosis more than 50 years ago with the observation that these cells were increased in human hearts with endocardial fibrosis [13]. Since then, there have been numerous other observations that have associated MCs with cardiac fibrosis arising from multiple etiologies. In 1988, increased MCs were found to be associated with areas of fibrosis in biopsies obtained from 92 human diseased hearts by Turlington et al. [14]. In 1989, Olivetti et al. [6] observed an increased number of MCs in the right ventricle (RV) of rats following constriction of the pulmonary artery, a technique that results in RV fibrosis. Subsequently, Li et al. [8] reported that MCs were increased in human hearts following transplantation and that MC number correlated with fibrosis ($r = 0.63$). Strengthening this relationship was the observation that patients with high numbers of MCs at two weeks post-transplantation showed a 17% increase in fibrosis by week 3, whilst those patients with lesser numbers of MCs had only a 3.5% increase in fibrosis. Perhaps not surprisingly, patients in the high MC group also scored higher on the rejection scale.

Levels of the MC-specific amine, histamine, were reported to be elevated in experimental Chagas' disease induced by infection of mice with Trypanosoma cruzi virus [15], with MCs in these mice appearing in areas of fibrosis [16]. Further, MC degranulation occurs soon after infection of mice with experimental myocarditis induced by coxsackievirus [17]. MC density also increases [18] in myocarditis and very strongly correlates with collagen volume fraction ($r = 0.946$) [19].

MCs were also linked to fibrosis in the hypertensive left ventricle (LV) when Panizo et al. [5] observed an increase in MC density in the LV of spontaneously hypertensive rats (SHR) that strongly correlated with collagen volume fraction ($r = 0.87$). Shiota et al. [7] also reported increased MC densities across the lifespan of the SHR. Even in stenotic aortic valves, MCs contained increased cathepsin G, which correlated with expression levels of collagen I and III [20]. More recently, Luitel et al. [21] confirmed in mice the earlier findings of Olivetti [6] in rats that MC density and degranulation increase in the RV following constriction of the pulmonary artery. Whilst these studies clearly show a strong association between MCs and fibrosis in the heart from varying etiologies, these associations do not establish causality. These studies are summarized in Table 1.

Table 1. Summary of in vivo studies associating mast cells (MCs) with cardiac fibrosis.

Species	Model/Pathology	Outcome	Heart Chamber	References
Human	Fibrosis	↑ MC	LV	[13,14]
Rat	Pulmonary hypertension	↑ MC	RV	[6]
Human	Transplantation	MC number correlated with fibrosis	LV	[8]
Mouse	Myocarditis	↑ histamine, ↑ MC correlated with fibrosis	LV	[15,19]
Rat	Hypertension	↑ MC correlated with fibrosis	LV	[5]
Mouse	Pulmonary hypertension	↑ MC	RV	[21]

LV = Left ventricle, RV = Right ventricle.

3. Evidence for the Causal Involvement of Mast Cells in Cardiac Fibrosis

3.1. Studies with Mast Cell Stabilizers

MC stabilizers prevent the release of MC mediators (e.g., histamine). This stabilization may ultimately involve blocking calcium channels, without which MC granules cannot fuse to the cell membrane and be exuded. Several studies have used this approach to examine the role of MCs in cardiac fibrosis. Palaniyandi et al. [19] demonstrated that the MC stabilizer disodium cromoglycate (also known as cromolyn) could dramatically reduce fibrosis in a model of myocarditis, whereby rats were injected with porcine cardiac myosin emulsified with complete freund's adjuvant with *Mycobacterium tuberculosis* H37RA. A subsequent study confirmed the anti-fibrotic effect of cromolyn in myocarditis in rats [22]. We provided the first causal evidence that MCs play a role in cardiac fibrosis in the hypertensive heart [23]. SHR were treated with the MC stabilizing compound nedocromil (30 mg/kg/day) from 8 weeks of age (prior to the development of fibrosis) through to 24 weeks of age. This resulted in complete prevention of fibrosis in the LV, as determined by collagen volume fraction (Figure 1A). This included the observation that MC stabilization prevented macrophage recruitment and normalized cytokine profiles (IFN-γ, IL-4, IL-6 and IL-10). Interestingly, we found that IL-10 was dramatically decreased in untreated SHR, and was returned to normal after MC inhibition. In a previous study, Palaniyandi et al. [24] had demonstrated that IL-10 inhibited acute myocarditis-induced pathological changes in the heart, and that this likely involved the inhibition of MCs since histamine levels and MC density were reduced by IL-10. Thus, IL-10 may represent an endogenous MC inhibitor, with a loss of IL-10 leaving MCs susceptible to activation stimuli. Confirming the pro-fibrotic role of MCs in the pressure overloaded heart, Kanellakis et al. [25] showed that cromolyn prevented LV fibrosis in mice with transaortic constriction. Similarly in the atria, the MC stabilizer, cromolyn, prevented fibrosis following transaortic constriction-induced pressure overload on the heart [26]. Even in STZ-induced diabetic hearts, nedocromil was able to reduce cardiac fibrosis [27]. More recently, Li et al. [28] found that nedocromil (30 mg/kg/day) prevented fibrosis from developing in rats following five weeks of transaortic constriction. Thus, the MC stabilzer studies strongly argue for a role for MCs in cardiac fibrosis. However, one must be aware of possible off target effects of these compounds, such as inhibition of sensory nerves. These studies are summarized in Table 2.

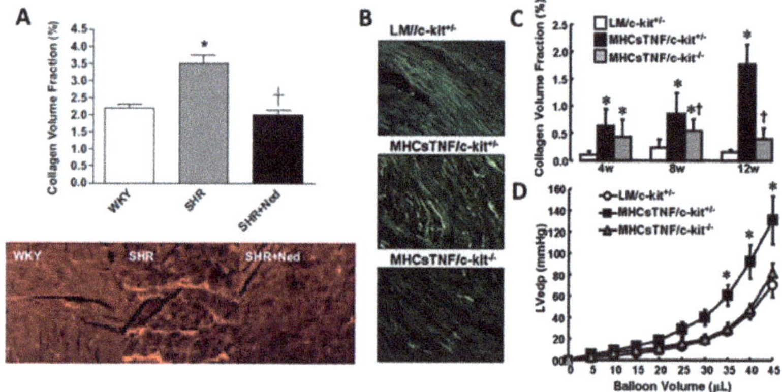

Figure 1. MC stabilization with nedocromil, or MC deficiency prevents cardiac fibrosis. (**A**) Quantification and representative picrosirius red-stained images (20× magnification) for left ventricle (LV) collagen volume fraction for Wistar Kyoto rats (WKY), spontaneously hypertensive rats (SHR), and SHR treated with the MC stabilizer, nedocromil (Ned, 30 mg/kg/day), * = $p < 0.05$ vs WKY, † = $p < 0.05$ vs SHR; (**B**) representative images of picrosirius red stained LV collagen in control mice (LM/c-kit$^{+/-}$), TNF-α overexpressing mice (MHCsTNF/c-kit$^{+/-}$), and TNF-α overexpressing mice crossed with MC-deficient mice (MHCsTNF/c-kit$^{-/-}$); (**C**) quantification of collagen volume fraction in control LM//c-kit$^{+/-}$ mice, MHCsTNF/c-kit$^{+/-}$ mice, and MHCsTNF/c-kit$^{-/-}$ mice; and (**D**) LV pressure-volume relationship for control LM//c-kit$^{+/-}$ mice, MHCsTNF/c-kit$^{+/-}$ mice, and MHCsTNF/c-kit$^{-/-}$ mice. * = $p < 0.05$ vs LM/c-kit$^{+/-}$, † = $p < 0.05$ vs MHCsTNF/c-kit$^{+/-}$. (Copied with permission from Levick et al., Hypertension, 2009;53:1041–1047 (**A**); and Zhang et al., Circulation, 2011;124:2106–2116 (**B–D**)).

Table 2. In vivo studies establishing cause and effect between MCs and cardiac fibrosis.

Species	Intervention	Pathology	Outcome	Heart Chamber	Reference
MC Stabilizers					
Rat	Cromolyn	Myocarditis	↓ fibrosis	LV	[19,22]
Rat	Nedocromil	Hypertension	↓ fibrosis	LV	[23]
Mouse	Cromolyn	Transaortic constriction	↓ fibrosis	LV	[25]
Mouse	Cromolyn	Transaortic constriction	↓ fibrosis	Atria	[26]
Mouse	Nedocromil	STZ-induced diabetes	↓ fibrosis	LV	[27]
Rat	Nedocromil	Transaortic constriction	↓ fibrosis	LV	[28]
MC-deficient Rodents					
Mouse	$Kit^{W/Wv}$	Abdominal aortic banding	↓ perivascular fibrosis	LV	[29]
Mouse	$Kit^{W/Wv}$	Transaortic constriction	↓ fibrosis	Atria	[26]
Mouse	$Kit^{W/W-sh}$	TNF-α overexpression	↓ fibrosis, ↓ diastolic dysfunction	LV	[30]
Rat	Ws/Ws	Hyperhomocysteinemia	↑ fibrosis	LV	[31]
Rat	Ws/Ws	Radiation	↑ fibrosis	LV	[32]

Table 2. *Cont.*

Species	Intervention	Pathology	Outcome	Heart Chamber	Reference
Targeting Proteases					
Canine	Chymase inhibitor (SUNC8257)	Pacing-induced heart failure	↓ collagen I and III mRNA, ↓ fibrosis	LV	[33]
Rat	Chymase inhibitor (NK3201)	Myocardial infarction	↓ collagen I and III mRNA, ↓ fibrosis, ↓ E/A	LV	[34]
Mouse	Chymase inhibitor (NK3201)	Intermittent hypoxia	↓ perivascular fibrosis	LV	[35]
Rat	PAR-2 antagonist (FSLLRY, tryptase)	Hypertension	↓ fibrosis	LV	[36]
Mouse	$H2R^{-/-}$	Transaortic constriction	↓ fibrosis, ↑ systolic function	LV	[37]

LV = Left ventricle, RV = Right ventricle, PAR-2 = Protease activated receptor-24. MC products.

3.2. Studies with Mast Cell-Deficient Rodents

3.2.1. Types of Mast Cell-Deficient Rodents

There are several mutant mice that have been utilized as models of MC-deficiency to study MC function in vivo including $Kit^{W/W-v}$ and $Kit^{W/W-sh}$ mice [38–41]. These mouse strains carry spontaneous loss of function mutations at both alleles of the dominant *white spotting* locus (*W*, i.e., *c-kit*). This mutation exhibits a significant reduction in c-kit tyrosine kinase-dependent signalling, hence disrupting normal MC development and survival [39,42]. The different mutant alleles of c-kit reflect the variable non-MC-related effects that these mice display. The mutated *W* allele gives rise to truncated c-kit without the transmembrane domain, resulting in no protein expression on the cell surface [43]. Alternatively, the W^v mouse has a point mutation at the tyrosine kinase-encoding domain of c-kit. Unlike the *W* mouse, $Kit^{W/W-v}$ mice still express the c-kit protein on the cell surface, although with reduced tyrosine kinase activity [43,44]. $Kit^{W/W-v}$ mice have no detectable MCs in multiple organs by the time they reach 6 to 8 weeks of age [39]. However, due to malfunction of the c-kit protein, these mice display phenotypic abnormalities such as macrocytic anaemia, infertility, impaired melanogenesis, lack of intestinal cells of Cajal, spontaneous dermatitis, gastric ulcers and duodenum dilatation [45–53]. This strain has traditionally been the most popular strain used to study MC-deficiency.

The *W-sash* ($Kit^{W/W-sh}$) mutation is an inversion mutation in the transcriptional regulatory elements, upstream of the c-kit transcription start site of mouse chromosome 5, resulting in functionally impaired c-kit protein [54]. This mutation was first described 23 years ago from crossing two inbred strain mice (C3H/HEH × 101/H), although it is only fairly recently that this strain has gained popularity as a model of MC-deficiency in vivo [40,55]. Adult $Kit^{W/W-sh}$ mice are MC-deficient at multiple anatomical sites [56], however, unlike the $Kit^{W/W-v}$ mouse, they are fertile and not anaemic [40,57]. They also exhibit normal levels of myeloid cells, B cells, T cells, dendritic cells, and basophils [58]. Importantly, like the $Kit^{W/W-v}$ mouse, $Kit^{W/W-sh}$ mice can be successfully reconstituted with bone marrow derived mast cells with normal c-kit expression by adoptive transfer via intraperitoneal, intradermal or intravenous injection [55,56,58]. A comprehensive study by Grimbaldeston et al. [58] details the advantageous of $Kit^{W/W-sh}$ mice over $Kit^{W/W-v}$ mice.

Another mutant mouse that can be used to study MC biology is the *steel-Dickie* (Sl^d) mouse. This mutation occurs due to a 4.0 kilobase intragenic deletion and truncates the *Sl* coding sequence. Mast cell growth factor (MGF) is encoded by the *Sl* gene, hence this mutation results in soluble truncated growth factor that lacks both transmembrane and cytoplasmic domains [59]. Sl^d mice carry a homozygous mutation in the *Sl* gene as a complete deletion of *Sl* alleles, resulting in the complete absence of MGF and is lethal [60–62]. Phenotypically, Sl^d mice exhibit melanocytes defects, severe anaemia and sterility [63].

There is also a MC-deficient rat. The *Ws/Ws* rat has a 12-base deletion in the tyrosine kinase domain of c-kit, the receptor for stem cell factor. The *Ws/Ws* phenotype is otherwise normal except for white spotting of the skin and macrocytic anemia that is spontaneously ameliorated by 10 weeks of age [41]. Therefore, these rats do not exhibit the severe anemia seen in some MC-deficient mouse strains.

3.2.2. Pro-Fibrotic Role for Mast Cells

Hara et al. [29] were the first to use MC-deficient mice to evaluate cardiac fibrosis in any form. They used male $Kit^{W/Wv}$ mice exposed to abdominal aortic banding for 15 weeks and found that whilst perivascular fibrosis occurred in WT mice with banding, collagen levels were normal in $Kit^{W/Wv}$ mice. To definitively confirm the role of MCs in MC-deficient mice, MCs must be replaced by adoptive transfer. Unfortunately, Hara et al. were unsuccessful in their attempt to replenish MCs in $Kit^{W/Wv}$ mice, thus, could not confirm that a lack of MCs was solely responsible for the resistance to adverse remodelling. The inability to reconstitute MCs to the hearts of these mice may have been due to beginning reconstitution just 2 days before initiating banding. Typically, it takes 6 to 8 weeks for injected MCs to reconstitute to the heart. Liao et al. [26] used the $Kit^{W/Wv}$ mouse to investigate the contribution of MCs to atrial fibrosis following transaortic constriction. These authors reported that mice deficient in MCs were protected against atrial fibrosis, however, they also did not perform MC reconstitution experiments. In an interesting study, Zhang et al. [30] investigated the role of MCs in cardiac fibrosis in a model of TNF-α overexpression. Mice with cardiac-restricted TNF-α overexpression were crossed with $Kit^{W/W-sh}$ mice. While fibrosis developed in the hearts of TNF-α overexpressing mice, this did not occur in MC-deficient TNF-α overexpressing hearts, indicating that MCs mediate the pro-fibrotic actions of TNF-α in this setting (Figure 1B,C). Further, the lack of MCs restored normal diastolic function as indicated by normalisation of the LV pressure volume relationship (Figure 1D). This study shows that in the setting of elevated TNF-α, MCs mediate the pro-fibrotic actions of TNF-α and that MC-mediated cardiac fibrosis contributes to diastolic dysfunction. This study offers probably the most conclusive evidence to date that MCs contribute to cardiac fibrosis, however, the caveat here is that this is an artificial up-regulation of TNF-α, which may or may not be relevant to conditions such as hypertension. These studies are summarized in Table 2.

3.2.3. Anti-Fibrotic Role of Mast Cells

Despite the evidence that MCs are pro-fibrotic in the heart, there does appear to be some exceptions. Martin Hauer-Jensen's laboratory has used the *Ws/Ws* MC-deficient rat to identify some of these exceptions. In one study, Joseph et al. [31] fed *Ws/Ws* rats a diet high in homocysteine for 10 weeks to induce hyperhomocysteinemia that causes perivascular and interstitial cardiac fibrosis, without concomitant cardiomyocyte hypertrophy. Both forms of fibrosis were further increased in *Ws/Ws* rats fed homocysteine, indicating that MCs were protective in this setting. In another study on cardiac injury caused by radiation (single dose at 18 Gy), Boerma et al. [32] found that MC-competent rats had increased collagen III compared to rats deficient in MCs. However, this was not the case for collagen I. These studies are summarized in Table 2.

The reasons why MCs appear to be pro-fibrotic except in the instances of hyperhomocysteinemia and radiation are not clear. One tempting explanation is the use of MC-deficient rats in the studies indicating an anti-fibrotic role versus MC-deficient mice in the studies supporting a pro-fibrotic role. However, MC stabilisers in rats appear to be anti-fibrotic, arguing against this hypothesis. It may be that MCs are directed to take on a pro-fibrotic phenotype when the stimulus involves altered cardiovascular hemodynamics (e.g., hypertension) or infection (e.g., myocarditis), whereas remodelling not related to these types of stimuli invoke an anti-fibrotic MC phenotype. Deeper analysis of the types of mediators released by MCs in each of these settings will be required to confirm or refute this hypothesis.

4. MC products

Figure 2 indicates specific mediators released by MCs and their potential contributions to cardiac fibrosis. These mediators are discussed below.

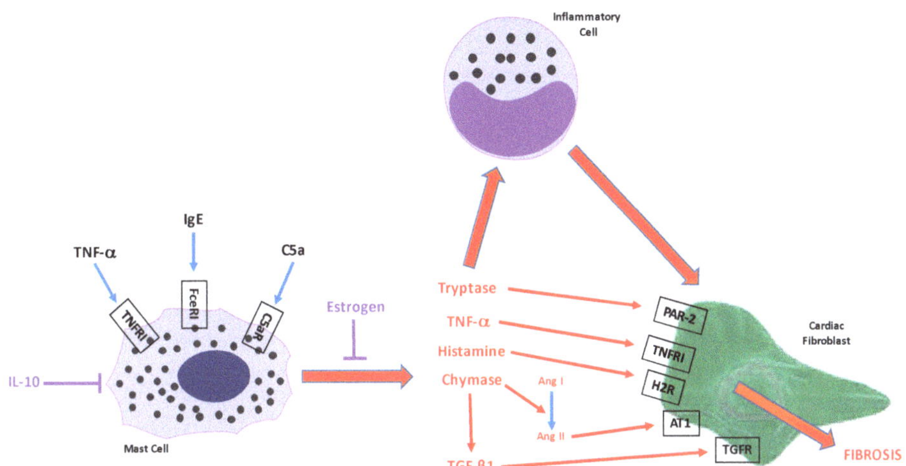

Figure 2. Schematic depicting potential MC activation stimuli and interactions with other cell types that lead to cardiac fibrosis. Candidates for cardiac MC activation include IgE, TNF-α, and C5a. These then cause the release of MC mediators including the proteases tryptase and chymase, TNF-α, histamine, and TGF-β1. These mediators can then have direct effects on cardiac fibroblasts, but may also contribute to an inflammatory response that then activates cardiac fibroblasts via numerous cytokines. IL-10 and estrogen likely oppose cardiac MC activation/degranulation.

4.1. Proteases

Cardiac MC phenotype has not been well studied. However, cardiac MCs, like MCs in general, store large amounts of specific proteases. Cardiac MCs fall under the connective tissue type MC phenotype since they contain both chymase and tryptase (Figure 3). On the other hand, mucosal MCs are tryptase$^+$, but chymase$^-$. Most of the work pertaining to cardiac MC products that contribute to fibrosis has focused on these proteases.

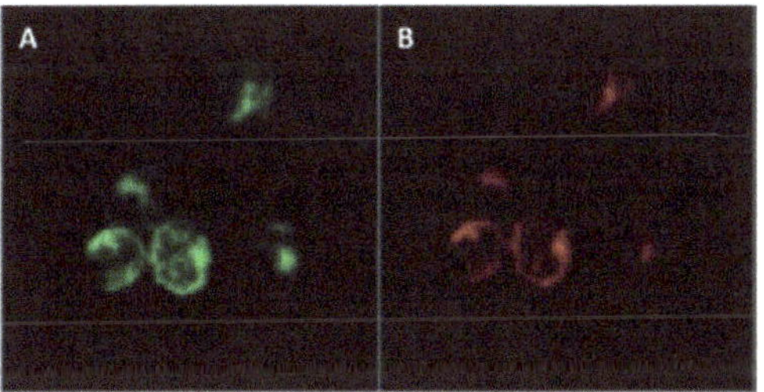

Figure 3. Immunolabelling of chymase (**A**, green) and tryptase (**B**, red) in cardiac mast cells isolated from rats (400× magnification; Copied with permission from Morgan et al., Inflamm Res, 2008;57:1–6).

4.1.1. Chymase

Chymase is a chymotrypsin-like serine protease stored in the granules of MCs. In a mouse model of transaortic constriction, Hara et al. [29] observed an increase in *Mcp5* in the heart, one of the mouse genes for chymase. Chymase activity in the heart was reported to increase 5.2-fold in hamsters with hypertension induced by the two-kidney, one-clip approach [64]. Similarly, there is evidence of chymase up-regulation in humans, with chymase mRNA and protein increased in patients with aortic stenosis undergoing valve replacement surgery [65].

The first study to demonstrate a causal role for chymase in cardiac fibrosis was by Matsumoto et al. [33]. In this study, heart failure was induced by rapid pacing in beagles (270 bpm, 22 days). Dogs with heart failure were treated with the chymase inhibitor SUNC8257 (10 mg/kg, orally twice a day), with chymase inhibition reducing collagen I and III mRNA levels and fibrosis as determined by picrosirius red staining (Figure 4). In the setting of MI-induced remodelling in rats, the chymase inhibitor, NK3201 reduced collagen I and III levels as well as fibrosis following 4 weeks post-MI, although it was unclear whether this analysis only included the infarct region or the entire LV [34]. Intriguingly, whilst there was a small improvement in diastolic function with chymase inhibition, as determined by E/A ratio, LV dilatation and systolic function were not improved. In an interesting study, Matsumoto et al. [35] investigated the role of chymase in cardiac remodelling caused by intermittent hypoxia mimicking sleep apnoea. Mice were placed in chambers that delivered intermittent hypoxia (30 s of 4.5% to 5.5% O_2 followed by 30 s of 21% O_2 for 8 h/day during the daytime) or normoxic conditions for 10 days. In addition to other remodelling parameters, perivascular fibrosis was increased by hypoxia and reduced by the chymase inhibitor, NK3201.

Figure 4. Picrosirius red-stained images of perivascular fibrosis in normal, paced (vehicle), and paced plus the chymase inhibitor SUNC8257 (Chy-I) dog hearts (100× magnification; Copied with permission from Matsumoto et al., Circulation, 2003;107:2555–2558).

In the Matsumoto study in dogs, TGF-β1 mRNA was reduced by chymase inhibition whereas ACE mRNA was not. Further, Shimizu et al. [66] observed that aortic banding in male Syrian hamsters induced an increase in chymase activity and a decrease in angiotensin converting enzyme (ACE) activity, while causing cardiac fibrosis. The authors concluded that this indicated that chymase was primarily responsible for angiotensin II formation in this setting. These studies reflect the two known mechanisms likely to underlie chymases role in promoting fibrosis (i.e., angiotensin II and TGF-β1). Evidence has shown that chymase plays a role in the formation of angiotensin II in a non-canonical pathway of the renin angiotensin system (RAS). Chymase acted as an angiotensin-(1–12) converting enzyme in generating angiotensin II in SHR neonatal myocytes [67]. Chymase was also found to be responsible for this bio-conversion in the left atria of patients undergoing surgery for treatment of resistant atrial fibrillation or LV of normal patients dying from motor vehicle accidents [68,69]. Further supporting this, cardiac angiotensin II formation was reduced by the chymase inhibitor NK3201 in a mouse model of intermittent hypoxia [35]. Thus, one mechanism by which chymase exerts pro-fibrotic actions is via angiotensin II. Although, one needs to be aware of species differences in the contribution of chymase versus ACE to the cardiac angiotensin II pool. Balcells et al. [70] have noted that the human heart has dramatically more angiotensin II than other species, with dog being second, followed by mouse, rabbit, and rat. Human cardiac angiotensin II was almost entirely accounted for by chymase activity, as was also the case in the dog [70]. Approximately 25% of cardiac angiotensin II formation in the rat was attributed to chymase, whereas the contribution was less again in the rabbit and mouse. Akasu et al. [71] found similar results. They reported that angiotensin II was almost entirely accounted for by chymase in human, hamster, dog, and marmoset hearts. Pig and rabbit hearts showed ACE as the primary mechanism for angiotensin II synthesis. Interestingly, and in contrast to Balcells et al. [70], Akasu et al. [71] found that chymase was the primary determinant of angiotensin II formation in the rat heart. The mouse heart was not investigated.

The other pathway by which chymase exerts pro-fibrotic actions involves TGF-β1. At a cellular level, treatment of isolated neonatal cardiac fibroblasts with chymase (15–30 ng/mL) for 24 h resulted in cell proliferation [72]. This was accompanied by increased mRNA and protein levels of TGF-β1. Chymase up-regulation of TGF-β1 was independent of angiotensin II since blockade of AT_1 and AT_2 receptors did not alter TGF-β1 production. The proliferative and collagen producing effects of chymase could be reduced by a neutralising antibody to TGF-β1. The downstream mediator of TGF-β1 was Smad2/3 and not p38 or ERK pathways. In vivo, cromolyn reduced TGF-β1 levels in rats with myocarditis [19]. Shiota et al. [7] demonstrated that activated MCs were a major source of TGF-β1 and were localised to areas of fibrosis in 12 month and 20 month old SHR that were in advanced stages of LV hypertrophy and heart failure, respectively. In a mouse model of TNF-α overexpression, crossing these mice with MC-deficient mice reduced TGF-β1 levels as well TGF-β receptors in the heart [30].

4.1.2. Tryptase

Tryptase is a serine protease also stored in the granules of MCs. Fewer efforts have focused on the role of tryptase in cardiac fibrosis, even though tryptase levels increase in fibrotic hearts [23,28]. Mice with encephalomyocarditis virus-induced myocarditis show up-regulated mRNA levels of *Mcp6*, the gene for tryptase, 14 days after infection, which tracked with an increase in collagen I gene expression [18]. Using an in vitro approach, we demonstrated that tryptase could increase ECM synthesis by cardiac fibroblasts after 72 h [23]. Interestingly, proliferation was induced much more rapidly (24 h), suggesting that a fibroblast proliferative response might be the primary action of tryptase. In a follow-up study, we demonstrated that the effects of tryptase on cardiac fibroblasts were mediated by protease activated receptor-2 (PAR-2), which induced selective MAP kinase pathways with ERK1/2 mediating the pro-fibrotic actions of tryptase on cardiac fibroblasts, with no involvement from p38 or JNK [36]. This pathway also mediated cardiac fibroblast conversion to the myofibroblast phenotype. Critically, blockade of PAR-2 with FSLLRY (10 µg/kg/day) in SHR prevented fibrosis from occurring, independent of blood pressure. Thus, the role of tryptase appears to be more direct than is the case for chymase. In another follow-up study, we identified an autocrine/paracrine response by cardiac MCs mediating their own protease release. Inhibition of tryptase with nafamostat mesilate (5 mg/kg/day) reduced plasma chymase levels in rats with transaortic constriction [28]. To investigate this further, sections of rat LV were cultured in a novel tissue culture system and treated with tryptase. Tryptase caused the release of chymase into the media and a concomitant increase in collagen production that could be reduced by the chymase inhibitor chymostatin. These results suggest that tryptase also acts in an autocrine/paracrine manner to induce chymase release from MCs, and subsequent fibrosis.

4.2. Other Mast Cell Products

MCs release many products other than proteases that are capable of influencing the ECM. Several of these will be discussed below. However, it is important to recognize that many of these products can be produced by other cell types, and the relative contribution of the MCs to the overall pool of some of these products is unclear.

4.2.1. Histamine

The human heart contains considerable amounts of histamine (1035 ± 65 ng/g of atrial tissue) [73]. Histamine is the classic MC product mediating hypersensitivity reactions, and this is also true in the heart, which participates in anaphylaxis. However, histamine can also contribute to cardiac remodelling [74]. In the most direct assessment to date, Zeng et al. [37] performed transaortic constriction on H2 histamine receptor deficient mice (H2R$^{-/-}$, Table 2). After four weeks,

H2R$^{-/-}$ mice showed reduced cardiac fibrosis and slightly improved systolic function, indicating a role for histamine in cardiac fibrosis. The investigators performed additional studies in isolated cardiac fibroblasts and determined that both histamine and the H2R-selective agonist amthamine dihydrobromide increased protein levels of calcineurin; this was prevented by the H2R antagonist famotidine. Importantly, H2R activation also up-regulated myofibroblast conversion, fibronectin production, and procollagen I and III up-regulation at the gene level. Calcineurin was subsequently shown to mediate fibroblast proliferation, fibronectin production, and collagen gene regulation in response to H2R activation. This study clearly shows the capability of histamine to have direct effects on cardiac fibroblasts. Interestingly though, the actions of histamine may extend beyond this. Of the four known histamine receptors, three (H1, H2 and H3) are found in the heart. Due to their localisation, the actions of these receptors in the heart are extremely complex. H1R and H2R are present in the sinoatrial and atrioventricular nodes of the heart, suggesting regulation of heart rate [75]. H1R modulates cardiac autonomic nerve function [76,77]. Interestingly, histamine enables noradrenaline release in the rat heart via H2R [78]. Given the pro-fibrotic properties of noradrenaline, this raises the possibility that MC-mediated release of noradrenaline via H2R could promote fibrosis. In fact, a recent clinical study linked H2R antagonist use to a 62% reduced risk of heart failure [79]. Cardiac sensory nerves possess H3R. Our recent findings that the sensory nerve neuropeptide substance P plays a critical role in cardiac fibrosis in the hypertensive heart raises the possibility that MC histamine could be responsible for its release [80]. If this is the case it sets up an interesting feed forward mechanism since we have data indicating that substance P and its cognate receptor the neurokinin-1 receptor are responsible for the increase in MC density observed in the hypertensive heart, but neurokinin-1 receptors do not contribute to MC activation in this setting. Thus, other stimuli activate MCs, potentially resulting in histamine release and amplification of the substance P response.

4.2.2. Components of the Renin Angiotensin System

Renin is the first enzyme in the RAS, its role being to cleave angiotensinogen to angiotensin I, which in turn can be cleaved by ACE or chymase to active angiotensin II. In 2004, Roberto Levi's group demonstrated that rat cardiac MCs contain renin [81]. Extrapolating the relevance of this finding to humans, Levi's group also showed that the human MC line HMC-1 produced active renin that could convert angiotensinogen to angiotensin I. Subsequently, Hara et al. [82] confirmed the presence of renin mRNA in HMC-1 cells and further identified angiotensinogen mRNA in these cells. Interestingly, these authors identified pre-formed angiotensin II in HMC-1 cells that was released in response to calcitonin gene-related peptide. Although MC-derived renin contributes to ischaemia-induced arrhythmias in the heart [83], no specific evidence shows directly the role of MC-derived components of the RAS in cardiac fibrosis. However, given the known pro-fibrotic effect of angiotensin II, it is reasonable to assume that MC RAS contributes at least to some degree to cardiac fibrosis.

4.2.3. TNF-α

Immunolabelling indicates that MCs are likely the main source of TNF-α in the heart [84,85]. MC stabilizers such as ketotifen and cromoglycate prevented TNF-α release in hearts undergoing ischemia reperfusion, further supporting this supposition [86]. MC-derived TNF-α has been shown to stimulate collagen production by dermal fibroblasts [87], however, the extent to which MC-derived TNF-α contributes to cardiac fibrosis has not yet been investigated.

4.2.4. TGF-β

The role of MCs in generating TGF-β via chymase has already been discussed in Section 4.1.1, however, direct production of TGF-β by MCs could also contribute to organ fibrosis. Inhibition of TGF-β1 has been shown to mediate the pro-fibrotic effects of MCs on dermal fibroblasts in culture [87]. Although this suggests effects of MC-derived TGF-β, the alternate possibility cannot be ruled out that chymase induced activation of TGF-β produced by fibroblasts. There are no studies directly investigating the contribution of MC-derived TGF-β to cardiac fibrosis.

4.2.5. Matrix Metalloproteinases

Investigation of MCs in airways of dogs found that these cells produce matrix metalloproteinase (MMP)-2 and -9 (gelatinase A and B) [88]. Stem cell factor selectively up-regulated MMP-9 [88]. Interestingly, TGF-β opposed the actions of stem cell factor on MMP-9. The significance of MMPs is their ability to regulate the ECM, whether it be by initiating degradation, or alternatively, inducing pro-fibrotic responses [89]. While MC regulation of MMPs has been established in cardiac volume overload [10,90], this results in ECM degradation in that setting. The contribution of MCs to MMPs has not been investigated in relation to cardiac fibrosis. However, conditioned media from MCs was able to increase MMP-2 activation in neonatal cardiac fibroblasts [91]. Additionally, chymase inhibition prevented MMP-9 activation in pigs undergoing ischemia reperfusion [92].

5. Important Questions

While there is accumulating evidence that MCs contribute to cardiac fibrosis, there are several important questions that remain unanswered.

5.1. What Activates Cardiac Mast Cells?

Probably the most pressing question is what are the stimuli that activate these cells causing them to promote fibrosis? There are several good candidates still to be investigated. These are depicted in Figure 2.

5.1.1. Immunoglobulin E

Like all MCs, cardiac MCs possess the IgE receptor, FcεRI. This is evidenced by their degranulation in response to antibody against FcεRI in vitro [93]. Interestingly, FcεRI activation causes release of histamine and tryptase, as well as leukotriene C_4 and prostaglandin D_2 by isolated human cardiac MCs [94]. As discussed earlier, MC histamine and tryptase are linked to cardiac fibrosis. Surprisingly, the role of IgE and FcεRI in causing cardiac fibrosis has not yet been investigated.

5.1.2. TNF-α

TNF-α receptor I (TNFRI) and TNF-α receptor II (TNFRII) mediate the actions of TNF-α. $Tnfrl^{-/-}$ mice have improved cardiac remodelling responses (including fibrosis) following MI, while $TnfrII^{-/-}$ mice have worse fibrosis. Thus, TNFRI exacerbates remodelling leading to heart failure, whereas TNFRII has cardioprotective actions. The evidence implicating TNF-α and TNFRI comes from studies that crossed TNF-α overexpressing mice with MC-deficient mice. TNF-α overexpressing mice develop fibrosis; this fibrosis is reduced in mice lacking MCs. This indicates that MCs mediate the pro-fibrotic effects of TNF-α, implying that TNF-α plays a role in activating MCs. What is not clear from that study is whether this is direct activation of MCs by TNF-α via TNFRI or whether the activation is indirect with TNF-α up-regulating other mediators that then activate cardiac MCs. This question requires MC-specific deletion of TNFRI. Also, whether MC-derived TNF-α represents a viable treatment target is questionable given the failure of targeting TNF-α in heart failure patients.

5.1.3. Complement 5a

Patella et al., demonstrated more than 20 years ago that isolated human cardiac MCs degranulate in response to the complement factor, C5a [93]. The response to C5a was more rapid than to IgE and reached the same maximal response as IgE. However, Füreder et al. [95] subsequently reported that only 5% to 15% of human cardiac MCs that they examined possessed the C5a receptor and that human cardiac MCs did not release histamine in response to C5a. Nevertheless, coronary infusion of C5a

(500 ng) in pigs undergoing MI led to an increase in coronary histamine levels indicating MC activation by C5a [96]. It will require MC-specific deletion of the C5a receptor to help resolve this question.

5.2. What Are the Specific Mechanisms by Which Mast Cells Cause Cardiac Fibrosis?

Despite clear evidence from multiple animal models that MCs are involved in cardiac fibrosis and that MC proteases mediate these actions along with possible contributions from other MC mediators, we know very little about the specifics. One example is the temporal involvement of MCs in cardiac fibrosis. Our own belief is that these cells are important in initiating fibrosis, but may not be involved continually throughout the process. This somewhat stems from our experience in MC regulation of the ECM in volume overload models where MC density increases in the first few days following initiation of volume overload, before returning to normal [90]. This suggests that MCs are important early to initiate processes that then continue throughout the remodelling process. Supporting this, MCs density is increased in the young SHR [7] and continued to be increased at 8 and 12 weeks of age, when fibrosis develops, before returning to normal levels by 16 weeks of age once fibrosis is established. Interestingly, MC density increases again around the age that heart failure develops in the SHR [7].

Also undetermined are the specifics of MC interactions with cardiac fibroblasts. In vitro it has been shown that MCs can act directly on cardiac fibroblasts to induce myofibroblast conversion, proliferation, and excess collagen synthesis [36,72], however, do these direct interactions actually occur in vivo, and to what extent? While cardiac MC number does vary between species, overall there are very few MCs in the heart. Clearly, although few in number they have a significant impact, however, it does raise the question of how many cardiac fibroblasts and MCs can actually interact in a paracrine manner in vivo. This we do not know. We do not even know whether released MC products circulate in the heart, giving them the opportunity to reach fibroblasts from remote parts of the heart, or whether they just act locally. Figure 5 depicts local versus distant interactions between MCs and cardiac fibroblasts.

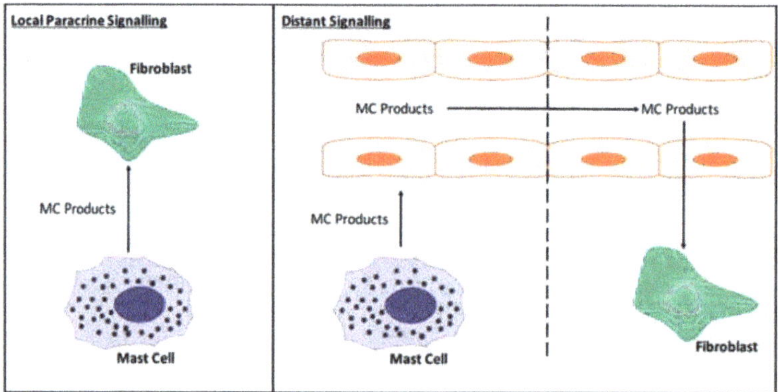

Figure 5. Schematic depicting the possible interactions between MCs and fibroblasts in the heart. (**Left**) MCs may act in a paracrine manner to signal only to fibroblasts in their local area. This would limit direct MC-fibroblast interactions as a mechanism by which MCs cause fibrosis; (**Right**) Alternatively, MC products may be taken up in the general coronary circulation allowing their products to be distributed to fibroblasts throughout the heart. This would allow for greater MC-fibroblast interactions.

Also likely important in MC modulation of cardiac fibrosis are interactions with other inflammatory cells (Figure 2), especially if direct MC-fibroblast interactions are minimal. Conditioned media from bone marrow-derived MCs activated with the calcium ionophore A23187 (500 ng/mL) dramatically up-regulated VCAM-1, ICAM-1, P-selectin, and E-selectin in mouse heart endothelial cells, suggesting a role for MCs in inflammatory cell recruitment [97]. In support of this, we had observed reduced numbers of macrophages in the SHR LV following treatment with nedocromil compared to untreated SHR [23]. This qualitative assessment suggests a role for MCs in the recruitment of macrophages to the fibrotic heart, however, quantitative assessment of this effect is still lacking. Should MCs regulate macrophage recruitment, then it should be determined as to whether this is a direct effect on macrophages, or whether other cells are involved (i.e., MC activation of endothelial cells). We also reported that MC stabilization reduced IFN-γ and IL-4 production, demonstrating the contribution of MCs to the overall cytokine pool, whether directly or indirectly.

Very little work has investigated MC-cardiomyocyte interactions. In our own studies, we did not observe any significant effect on cardiac hypertrophy, either by MC stabilization with nedocromil or inhibition of tryptase with FSLLRY [23,36]. Chymase appears to be able to induce cardiomyocyte death by entering these cells and inducing translocation of nuclear receptor subfamily 4A1 (NR4A1) from the nucleus to the cytoplasm. This was identified in ischemia reperfusion injury; the significance of this to cardiac fibrosis has not been explored. Media from cultured MCs can invoke death in cultured neonatal cardiomyocytes, with a neutralizing antibody against chymase opposing this effect [98]. Cardiomyocyte death is a stimulus for cardiac fibrosis. Thus, MC interactions with cardiomyocytes could be a stimulus for fibrosis, however, this is yet to be established.

5.3. What Is the Cardiac Mast Cell Phenotype and Are There Gender Differences?

A large amount of the work contributing to our understanding of the role of MCs in adverse cardiac remodeling has come from the laboratory of Joseph Janicki. Recently, the Janicki lab produced evidence indicating that differences in MC phenotype may underlie cardioprotection in pre-menopausal females. Intact and ovariectomised (ovx) female rats were exposed to pressure overload induced by transaortic constriction. LV MC density did not increase in intact females, but did increase in oxv rats [99]. Further, whilst there was a small increase in LV chymase levels in intact rats, the increase was greater in ovx animals. This was also the case for TGF-β1, presumably related to the increase in chymase. When oxv female rats were treated with estrogen, MC density, LV chymase, and TGF-β1 were reduced, as was fibrosis. This suggests that estrogen in pre-menopausal females provides a level of protection from cardiac fibrosis by reducing the ability of MCs to either: (1) respond to activation stimuli and therefore reduce levels of MC proteases that contribute to fibrosis in males; or (2) reduce production of mediators responsible for promoting fibrosis. Essentially, phenotypic differences exist between male and female MCs and this may underlie pre-menopausal cardioprotection in females (Figure 2). This is an extremely interesting concept that needs to be further explored.

These potential phenotypic differences between male and female MCs leads to another gap in our cardiac MC understanding; very little is known about the cardiac MC phenotype in general. This is an area that has been badly neglected. We know that cardiac MCs are tryptase^{+}/chymase^{+} (Figure 3), as well as containing histamine and TNF-α [84,94,100]. Patella et al. [94] have demonstrated that isolated human cardiac MCs produce leukotrienes in response to activation by C5a. Beyond these few mediators, essentially nothing is known about cardiac MC phenotype. Whether cardiac MC sub-populations exist, as is the case for macrophages and T cells, is currently unknown, but should become a focus of investigation.

6. Conclusions

There is strong experimental evidence that MCs contribute to cardiac fibrosis, at least in part through the release of MC-specific proteases. However, the specifics of how MCs promote fibrosis is not clear, including the role of non-protease MC products, the stimuli that activate cardiac MCs, how cardiac MCs and fibroblasts interact in vivo, and specifics of cardiac MC phenotype (e.g., sub-populations and gender differences). These questions must be answered if targeting MCs is to eventually become a therapeutic approach in humans.

Int. J. Mol. Sci. **2018**, *19*, 231

Acknowledgments: This work was supported by the National Heart Lung and Blood Institute at the National Institutes of Health [HL093215, HL132908 to Scott P. Levick]; Greater Milwaukee Foundation—Elsa Schoeneich Medical Research Fund to Scott P. Levick; the George and Mary Thomson Fellowship to Scott P. Levick; and an MCW Presidential Postdoctoral Fellow Award to Alexander Widiapradja.

Conflicts of Interest: The authors declare no conflict of interest. The funding sponsors had no role in the writing of this manuscript.

Abbreviations

MC	Mast cell
MI	Myocardial Infarction
RV	Right Ventricle
LV	Left Ventricle
SHR	Spontaneously Hypertensive Rat
ECM	Extracellular Matrix
Sl^d	Steel Dickie
ACE	Angiotensin Converting Enzyme
PAR-2	Protease Activated Receptor 2
H1R	Histamine Receptor 1
H2R	Histamine Receptor 2
H3R	Histamine Receptor 3
H4R	Histamine Receptor 4
RAS	Renin Angiotensin System
HMC-1	Human Mast Cell Line 1
MMP	Matrix Metalloproteinase
TNFRI	Tumor Necrosis Factor Receptor I
TNFRII	Tumor Necrosis Factor Receptor II
C5a	Complement Factor 5a
Ovx	Ovariectomised

References

1. Levick, S.P.; Melendez, G.C.; Plante, E.; McLarty, J.L.; Brower, G.L.; Janicki, J.S. Cardiac mast cells: The centrepiece in adverse myocardial remodelling. *Cardiovasc. Res.* **2011**, *89*, 12–19. [CrossRef] [PubMed]
2. Engels, W.; Reiters, P.H.; Daemen, M.J.; Smits, J.F.; van der Vusse, G.J. Transmural changes in mast cell density in rat heart after infarct induction in vivo. *J. Pathol.* **1995**, *177*, 423–429. [CrossRef] [PubMed]
3. Frangogiannis, N.G.; Perrard, J.L.; Mendoza, L.H.; Burns, A.R.; Lindsey, M.L.; Ballantyne, C.M.; Michael, L.H.; Smith, C.W.; Entman, M.L. Stem cell factor induction is associated with mast cell accumulation after canine myocardial ischemia and reperfusion. *Circulation* **1998**, *98*, 687–698. [CrossRef] [PubMed]
4. Somasundaram, P.; Ren, G.; Nagar, H.; Kraemer, D.; Mendoza, L.; Michael, L.H.; Caughey, G.H.; Entman, M.L.; Frangogiannis, N.G. Mast cell tryptase may modulate endothelial cell phenotype in healing myocardial infarcts. *J. Pathol.* **2005**, *205*, 102–111. [CrossRef] [PubMed]
5. Panizo, A.; Mindán, F.J.; Galindo, M.F.; Cenarruzabeitia, E.; Hernández, M.; Díez, J. Are mast cells involved in hypertensive heart disease? *J. Hypertens.* **1995**, *13*, 1201–1208. [CrossRef] [PubMed]
6. Olivetti, G.; Lagrasta, C.; Ricci, R.; Sonnenblick, E.H.; Capasso, J.M.; Anversa, P. Long-term pressure-induced cardiac hypertrophy: Capillary and mast cell proliferation. *AJP-Heart Circ. Phys.* **1989**, *257*, H1766–H1772. [CrossRef] [PubMed]
7. Shiota, N.; Rysä, J.; Kovanen, P.T.; Ruskoaho, H.; Kokkonen, J.O.; Lindstedt, K.A. A role for cardiac mast cells in the pathogenesis of hypertensive heart disease. *J. Hypertens.* **2003**, *21*, 1823–1825. [CrossRef]
8. Li, Q.Y.; Raza-Ahmad, A.; MacAulay, M.A.; Lalonde, L.D.; Rowden, G.; Trethewey, E.; Dean, S. The relationship of mast cells and their secreted products to the volume of fibrosis in posttransplant hearts. *Transplantation* **1992**, *53*, 1047–1051. [CrossRef] [PubMed]
9. Estensen, R.D. Eosinophilic myocarditis: A role for mast cells? *Arch. Pathol. Lab. Med.* **1984**, *108*, 358–359. [PubMed]

10. Brower, G.L.; Chancey, A.L.; Thanigaraj, S.; Matsubara, B.B.; Janicki, J.S. Cause and effect relationship between myocardial mast cell number and matrix metalloproteinase activity. *Am. J. Physiol. Heart Circ. Physiol.* **2002**, *283*, H518–H525. [CrossRef] [PubMed]
11. Stewart, J.A.; Wei, C.C.; Brower, G.L.; Rynders, P.E.; Hankes, G.H.; Dillon, A.R.; Lucchesi, P.A.; Janicki, J.S.; Dell'Italia, L.J. Cardiac mast cell- and chymase-mediated matrix metalloproteinase activity and left ventricular remodeling in mitral regurgitation in the dog. *J. Mol. Cell. Cardiol.* **2003**, *35*, 311–319. [CrossRef]
12. Batlle, M.; Roig, E.; Perez-Villa, F.; Lario, S.; Cejudo-Martin, P.; Garcia-Pras, E.; Ortiz, J.; Roque, M.; Orus, J.; Rigol, M.; et al. Increased expression of the renin-angiotensin system and mast cell density but not of angiotensin-converting enzyme II in late stages of human heart failure. *J. Heart Lung Transplant.* **2006**, *25*, 1117–1125. [CrossRef] [PubMed]
13. Fernex, M.; Sternby, N.H. Mast cells and coronary heart disease. Relationship between number of mast cells in the myocardium, severity of coronary atherosclerosis and myocardial infarction in an autopsy series of 672 cases. *Acta Pathol. Microbiol. Scand.* **1964**, *62*, 525–538. [CrossRef] [PubMed]
14. Turlington, B.S.; Edwards, W.D. Quantitation of mast cells in 100 normal and 92 diseased human hearts. Implications for interpretation of endomyocardial biopsy specimens. *Am. J. Cardiovasc. Pathol.* **1988**, *2*, 151–157. [PubMed]
15. Pires, J.G.; Milanez, M.C.; Pereira, F.E. Histamine levels in tissues of Trypanosoma cruzi-infected mice. *Agents Actions Suppl.* **1992**, *36*, 96–98. [PubMed]
16. Postan, M.; Correa, R.; Ferrans, V.J.; Tarleton, R.L. In vitro culture of cardiac mast cells from mice experimentally infected with Trypanosoma cruzi. *Int. Arch. Allergy Immunol.* **1994**, *105*, 251–257. [CrossRef] [PubMed]
17. Fairweather, D.; Frisancho-Kiss, S.; Gatewood, S.; Njoku, D.; Steele, R.; Barrett, M.; Rose, N.R. Mast cells and innate cytokines are associated with susceptibility to autoimmune heart disease following coxsackievirus B3 infection. *Autoimmunity* **2004**, *37*, 131–145. [CrossRef] [PubMed]
18. Kitaura-Inenaga, K.; Hara, M.; Higuchi, K.; Yamamoto, K.; Yamaki, A.; Ono, K.; Nakano, A.; Kinoshita, M.; Sasayama, S.; Matsumori, A. Gene expression of cardiac mast cell chymase and tryptase in a murine model of heart failure caused by viral myocarditis. *Circ. J.* **2003**, *67*, 881–884. [CrossRef] [PubMed]
19. Palaniyandi, S.S.; Watanabe, K.; Ma, M.; Tachikawa, H.; Kodama, M.; Aizawa, Y. Involvement of mast cells in the development of fibrosis in rats with postmyocarditis dilated cardiomyopathy. *Biol. Pharm. Bull.* **2005**, *28*, 2128–2132. [CrossRef]
20. Helske, S.; Syväranta, S.; Kupari, M.; Lappalainen, J.; Laine, M.; Lommi, J.; Turto, H.; Mayranpaa, M.; Werkkala, K.; Kovanen, P.T.; et al. Possible role for mast cell-derived cathepsin G in the adverse remodelling of stenotic aortic valves. *Eur. Heart J.* **2006**, *27*, 1495–1504. [CrossRef] [PubMed]
21. Luitel, H.; Sydykov, A.; Schymura, Y.; Mamazhakypov, A.; Janssen, W.; Pradhan, K.; Wietelmann, A.; Kosanovic, D.; Dahal, B.K.; Weissmann, N.; et al. Pressure overload leads to an increased accumulation and activity of mast cells in the right ventricle. *Physiol. Rep.* **2017**, *5*, e13146. [CrossRef] [PubMed]
22. Mina, Y.; Rinkevich-Shop, S.; Konen, E.; Goitein, O.; Kushnir, T.; Epstein, F.H.; Feinberg, M.S.; Leor, J.; Landa-Rouben, N. Mast cell inhibition attenuates myocardial damage, adverse remodeling, and dysfunction during fulminant myocarditis in the rat. *J. Cardiovasc. Pharmacol. Ther.* **2013**, *18*, 152–161. [CrossRef] [PubMed]
23. Levick, S.P.; McLarty, J.L.; Murray, D.B.; Freeman, R.M.; Carver, W.E.; Brower, G.L. Cardiac mast cells mediate left ventricular fibrosis in the hypertensive rat heart. *Hypertension* **2009**, *53*, 1041–1047. [CrossRef] [PubMed]
24. Palaniyandi, S.S.; Watanabe, K.; Ma, M.; Tachikawa, H.; Kodama, M.; Aizawa, Y. Inhibition of mast cells by interleukin-10 gene transfer contributes to protection against acute myocarditis in rats. *Eur. J. Immunol.* **2004**, *34*, 3508–3515. [CrossRef] [PubMed]
25. Kanellakis, P.; Ditiatkovski, M.; Kostolias, G.; Bobik, A. A pro-fibrotic role for interleukin-4 in cardiac pressure overload. *Cardiovasc. Res.* **2012**, *95*, 77–85. [CrossRef] [PubMed]
26. Liao, C.H.; Akazawa, H.; Tamagawa, M.; Ito, K.; Yasuda, N.; Kudo, Y.; Yamamoto, R.; Ozasa, Y.; Fujimoto, M.; Wang, P.; et al. Cardiac mast cells cause atrial fibrillation through PDGF-A-mediated fibrosis in pressure-overloaded mouse hearts. *J. Clin. Investig.* **2010**, *120*, 242–253. [CrossRef] [PubMed]
27. Huang, Z.G.; Jin, Q.; Fan, M.; Cong, X.L.; Han, S.F.; Gao, H.; Shan, Y. Myocardial remodeling in diabetic cardiomyopathy associated with cardiac mast cell activation. *PLoS ONE* **2013**, *8*, e60827. [CrossRef] [PubMed]

28. Li, J.; Jubair, S.; Levick, S.P.; Janicki, J.S. The autocrine role of tryptase in pressure overload-induced mast cell activation, chymase release and cardiac fibrosis. *IJC Metab. Endocr.* **2016**, *10*, 16–23. [CrossRef] [PubMed]

29. Hara, M.; Ono, K.; Hwang, M.W.; Iwasaki, A.; Okada, M.; Nakatani, K.; Sasayama, S.; Matsumori, A. Evidence for a role of mast cells in the evolution to congestive heart failure. *J. Exp. Med.* **2002**, *195*, 375–381. [CrossRef] [PubMed]

30. Zhang, W.; Chancey, A.L.; Tzeng, H.P.; Zhou, Z.; Lavine, K.J.; Gao, F.; Sivasubramanian, N.; Barger, P.M.; Mann, D.L. The development of myocardial fibrosis in transgenic mice with targeted overexpression of tumor necrosis factor requires mast cell-fibroblast interactions. *Circulation* **2011**, *124*, 2106–2116. [CrossRef] [PubMed]

31. Joseph, J.; Kennedy, R.H.; Devi, S.; Wang, J.; Joseph, L.; Hauer-Jensen, M. Protective role of mast cells in homocysteine-induced cardiac remodeling. *AJP-Heart Circ. Phys.* **2005**, *288*, H2541–H2545. [CrossRef] [PubMed]

32. Boerma, M.; Wang, J.; Wondergem, J.; Joseph, J.; Qiu, X.; Kennedy, R.H.; Hauer-Jensen, M. Influence of mast cells on structural and functional manifestations of radiation-induced heart disease. *Cancer Res.* **2005**, *65*, 3100–3107. [CrossRef] [PubMed]

33. Matsumoto, T.; Wada, A.; Tsutamoto, T.; Ohnishi, M.; Isono, T.; Kinoshita, M. Chymase inhibition prevents cardiac fibrosis and improves diastolic dysfunction in the progression of heart failure. *Circulation* **2003**, *107*, 2555–2558. [CrossRef] [PubMed]

34. Kanemitsu, H.; Takai, S.; Tsuneyoshi, H.; Nishina, T.; Yoshikawa, K.; Miyazaki, M.; Ikeda, T.; Komeda, M. Chymase inhibition prevents cardiac fibrosis and dysfunction after myocardial infarction in rats. *Hypertens. Res.* **2006**, *29*, 57–64. [CrossRef] [PubMed]

35. Matsumoto, C.; Hayashi, T.; Kitada, K.; Yamashita, C.; Miyamura, M.; Mori, T.; Ukimura, A.; Ohkita, M.; Jin, D.; Takai, S.; et al. Chymase plays an important role in left ventricular remodeling induced by intermittent hypoxia in mice. *Hypertension* **2009**, *54*, 164–171. [CrossRef] [PubMed]

36. McLarty, J.L.; Melendez, G.C.; Brower, G.L.; Janicki, J.S.; Levick, S.P. Tryptase/Protease-activated receptor 2 interactions induce selective mitogen-activated protein kinase signaling and collagen synthesis by cardiac fibroblasts. *Hypertension* **2011**, *58*, 264–270. [CrossRef] [PubMed]

37. Zeng, Z.; Shen, L.; Li, X.; Luo, T.; Wei, X.; Zhang, J.; Cao, S.; Huang, X.; Fukushima, Y.; Bin, J.; et al. Disruption of histamine H2 receptor slows heart failure progression through reducing myocardial apoptosis and fibrosis. *Clin. Sci.* **2014**, *127*, 435–448. [CrossRef] [PubMed]

38. Geissler, E.N.; McFarland, E.C.; Russell, E.S. Analysis of pleiotropism at the dominant white-spotting (W) locus of the house mouse: A description of ten new W alleles. *Genetics* **1981**, *97*, 337–361. [PubMed]

39. Kitamura, Y.; Go, S.; Hatanaka, K. Decrease of mast cells in W/Wv mice and their increase by bone marrow transplantation. *Blood* **1978**, *52*, 447–452. [PubMed]

40. Lyon, M.F.; Glenister, P.H. A new allele sash (Wsh) at the W-locus and a spontaneous recessive lethal in mice. *Genet. Res.* **1982**, *39*, 315–322. [CrossRef] [PubMed]

41. Niwa, Y.; Kasugai, T.; Ohno, K.; Morimoto, M.; Yamazaki, M.; Dohmae, K.; Nishimune, Y.; Kondo, K.; Kitamura, Y. Anemia and mast cell depletion in mutant rats that are homozygous at "white spotting (Ws)" locus. *Blood* **1991**, *78*, 1936–1941. [PubMed]

42. Kitamura, Y. Heterogeneity of mast cells and phenotypic change between subpopulations. *Annu. Rev. Immunol.* **1989**, *7*, 59–76. [CrossRef] [PubMed]

43. Nocka, K.; Tan, J.C.; Chiu, E.; Chu, T.Y.; Ray, P.; Traktman, P.; Besmer, P. Molecular bases of dominant negative and loss of function mutations at the murine c-kit/white spotting locus: W37, Wv, W41 and W. *EMBO J.* **1990**, *9*, 1805–1813. [PubMed]

44. Reith, A.D.; Rottapel, R.; Giddens, E.; Brady, C.; Forrester, L.; Bernstein, A. W mutant mice with mild or severe developmental defects contain distinct point mutations in the kinase domain of the c-kit receptor. *Genes Dev.* **1990**, *4*, 390–400. [CrossRef] [PubMed]

45. Huizinga, J.D.; Thuneberg, L.; Kluppel, M.; Malysz, J.; Mikkelsen, H.B.; Bernstein, A. W/kit gene required for interstitial cells of Cajal and for intestinal pacemaker activity. *Nature* **1995**, *373*, 347–349. [CrossRef] [PubMed]

46. Nakano, T.; Waki, N.; Asai, H.; Kitamura, Y. Different repopulation profile between erythroid and nonerythroid progenitor cells in genetically anemic W/Wv mice after bone marrow transplantation. *Blood* **1989**, *74*, 1552–1556. [PubMed]

47. Puddington, L.; Olson, S.; Lefrancois, L. Interactions between stem cell factor and c-Kit are required for intestinal immune system homeostasis. *Immunity* **1994**, *1*, 733–739. [CrossRef]

48. Tsai, M.; Tam, S.Y.; Wedemeyer, J.; Galli, S.J. Mast cells derived from embryonic stem cells: A model system for studying the effects of genetic manipulations on mast cell development, phenotype, and function in vitro and in vivo. *Int. J. Hematol.* **2002**, *75*, 345–349. [CrossRef] [PubMed]

49. Galli, S.J.; Arizono, N.; Murakami, T.; Dvorak, A.M.; Fox, J.G. Development of large numbers of mast cells at sites of idiopathic chronic dermatitis in genetically mast cell-deficient WBB6F1-W/Wv mice. *Blood* **1987**, *69*, 1661–1666. [PubMed]

50. Galli, S.J.; Zsebo, K.M.; Geissler, E.N. The kit ligand, stem cell factor. *Adv. Immunol.* **1994**, *55*, 1–96. [PubMed]

51. Shimada, M.; Kitamura, Y.; Yokoyama, M.; Miyano, Y.; Maeyama, K.; Yamatodani, A.; Takahashi, Y.; Tatsuta, M. Spontaneous stomach ulcer in genetically mast-cell depleted W/Wv mice. *Nature* **1980**, *283*, 662–664. [CrossRef] [PubMed]

52. Yokoyama, M.; Tatsuta, M.; Baba, M.; Kitamura, Y. Bile reflux: A possible cause of stomach ulcer in nontreated mutant mice of W/WV genotype. *Gastroenterology* **1982**, *82 Pt 1*, 857–863. [PubMed]

53. Kitamura, Y.; Yokoyama, M.; Matsuda, H.; Shimada, M. Coincidental development of forestomach papilloma and prepyloric ulcer in nontreated mutant mice of W/Wv and SI/SId genotypes. *Cancer Res.* **1980**, *40*, 3392–3397. [PubMed]

54. Nagle, D.L.; Kozak, C.A.; Mano, H.; Chapman, V.M.; Bucan, M. Physical mapping of the Tec and Gabrb1 loci reveals that the Wsh mutation on mouse chromosome 5 is associated with an inversion. *Hum. Mol. Genet.* **1995**, *4*, 2073–2079. [CrossRef] [PubMed]

55. Mallen-St Clair, J.; Pham, C.T.; Villalta, S.A.; Caughey, G.H.; Wolters, P.J. Mast cell dipeptidyl peptidase I mediates survival from sepsis. *J. Clin. Investig.* **2004**, *113*, 628–634. [CrossRef] [PubMed]

56. Wolters, P.J.; Mallen-St Clair, J.; Lewis, C.C.; Villalta, S.A.; Baluk, P.; Erle, D.J.; Caughey, G.H. Tissue-selective mast cell reconstitution and differential lung gene expression in mast cell-deficient Kit(W-sh)/Kit(W-sh) sash mice. *Clin. Exp. Allergy* **2005**, *35*, 82–88. [CrossRef] [PubMed]

57. Tono, T.; Tsujimura, T.; Koshimizu, U.; Kasugai, T.; Adachi, S.; Isozaki, K.; Nishikawa, S.; Morimoto, M.; Nishimune, Y.; Nomura, S.; et al. c-kit Gene was not transcribed in cultured mast cells of mast cell-deficient Wsh/Wsh mice that have a normal number of erythrocytes and a normal c-kit coding region. *Blood* **1992**, *80*, 1448–1453. [PubMed]

58. Grimbaldeston, M.A.; Chen, C.C.; Piliponsky, A.M.; Tsai, M.; Tam, S.Y.; Galli, S.J. Mast cell-deficient W-sash c-kit mutant kitW-sh/W-sh mice as a model for investigating mast cell biology in vivo. *Am. J. Pathol.* **2005**, *167*, 835–848. [CrossRef]

59. Brannan, C.I.; Lyman, S.D.; Williams, D.E.; Eisenman, J.; Anderson, D.M.; Cosman, D.; Bedell, M.A.; Jenkins, N.A.; Copeland, N.G. Steel-Dickie mutation encodes a c-kit ligand lacking transmembrane and cytoplasmic domains. *Proc. Natl. Acad. Sci. USA* **1991**, *88*, 4671–4674. [CrossRef] [PubMed]

60. Copeland, N.G.; Gilbert, D.J.; Cho, B.C.; Donovan, P.J.; Jenkins, N.A.; Cosman, D.; Anderson, D.; Lyman, S.D.; Williams, D.E. Mast cell growth factor maps near the steel locus on mouse chromosome 10 and is deleted in a number of steel alleles. *Cell* **1990**, *63*, 175–183. [CrossRef]

61. Huang, E.; Nocka, K.; Beier, D.R.; Chu, T.Y.; Buck, J.; Lahm, H.W.; Wellner, D.; Leder, P.; Besmer, P. The hematopoietic growth factor KL is encoded by the Sl locus and is the ligand of the c kit receptor, the gene product of the W locus. *Cell* **1990**, *63*, 225–233. [CrossRef]

62. Zsebo, K.M.; Williams, D.A.; Geissler, E.N.; Broudy, V.C.; Martin, F.H.; Atkins, H.L.; Hsu, R.Y.; Birkett, N.C.; Okino, K.H.; Murdock, D.C.; et al. Stem cell factor is encoded at the Sl locus of the mouse and is the ligand for the c-kit tyrosine kinase receptor. *Cell* **1990**, *63*, 213–224. [CrossRef]

63. Bernstein, S. Steel Dickie. *Mouse News Lett.* **1960**, *23*, 33–34.

64. Shiota, N.; Jin, D.; Takai, S.; Kawamura, T.; Koyama, M.; Nakamura, N.; Miyazaki, M. Chymase is activated in the hamster heart following ventricular fibrosis during the chronic stage of hypertension. *FEBS Lett.* **1997**, *406*, 301–304. [CrossRef]

65. Helske, S.; Lindstedt, K.A.; Laine, M.; Mäyränpää, M.; Werkkala, K.; Lommi, J.; Turto, H.; Kupari, M.; Kovanen, P.T. Induction of local angiotensin II-producing systems in stenotic aortic valves. *J. Am. Coll. Cardiol.* **2004**, *44*, 1859–1866. [CrossRef] [PubMed]

66. Shimizu, M.; Tanaka, R.; Fukuyama, T.; Aoki, R.; Orito, K.; Yamane, Y. Cardiac remodeling and angiotensin II-forming enzyme activity of the left ventricle in hamsters with chronic pressure overload induced by ascending aortic stenosis. *J. Vet. Med. Sci.* **2006**, *68*, 271–276. [CrossRef] [PubMed]

67. Ahmad, S.; Varagic, J.; Westwood, B.M.; Chappell, M.C.; Ferrario, C.M. Uptake and metabolism of the novel peptide angiotensin-(1-12) by neonatal cardiac myocytes. *PLoS ONE* **2011**, *6*, e15759. [CrossRef] [PubMed]

68. Ahmad, S.; Simmons, T.; Varagic, J.; Moniwa, N.; Chappell, M.C.; Ferrario, C.M. Chymase-dependent generation of angiotensin II from angiotensin-(1-12) in human atrial tissue. *PLoS ONE* **2011**, *6*, e28501. [CrossRef] [PubMed]

69. Ahmad, S.; Wei, C.C.; Tallaj, J.; Dell'Italia, L.J.; Moniwa, N.; Varagic, J.; Ferrario, C.M. Chymase mediates angiotensin-(1-12) metabolism in normal human hearts. *J. Am. Soc. Hypertens.* **2013**, *7*, 128–136. [CrossRef] [PubMed]

70. Balcells, E.; Meng, Q.C.; Johnson, W.H., Jr.; Oparil, S.; Dell'Italia, L.J. Angiotensin II formation from ACE and chymase in human and animal hearts: Methods and species considerations. *Am. J. Physiol.* **1997**, *273 Pt 2*, H1769–H1774. [CrossRef]

71. Akasu, M.; Urata, H.; Kinoshita, A.; Sasaguri, M.; Ideishi, M.; Arakawa, K. Differences in tissue angiotensin II-forming pathways by species and organs in vitro. *Hypertension* **1998**, *32*, 514–520. [CrossRef] [PubMed]

72. Zhao, X.Y.; Zhao, L.Y.; Zheng, Q.S.; Su, J.L.; Guan, H.; Shang, F.J.; Niu, X.L.; He, Y.P.; Lu, X.L. Chymase induces profibrotic response via transforming growth factor-β1/Smad activation in rat cardiac fibroblasts. *Mol. Cell. Biochem.* **2008**, *310*, 159–166. [CrossRef] [PubMed]

73. Gristwood, R.W.; Lincoln, J.C.; Owen, D.A.; Smith, I.R. Histamine release from human right atrium. *Br. J. Pharmacol.* **1981**, *74*, 7–9. [CrossRef] [PubMed]

74. Asanuma, H.; Minamino, T.; Ogai, A.; Kim, J.; Asakura, M.; Komamura, K.; Sanada, S.; Fujita, M.; Hirata, A.; Wakeno, M.; et al. Blockade of histamine H2 receptors protects the heart against ischemia and reperfusion injury in dogs. *J. Mol. Cell. Cardiol.* **2006**, *40*, 666–674. [CrossRef] [PubMed]

75. Matsuda, N.; Jesmin, S.; Takahashi, Y.; Hatta, E.; Kobayashi, M.; Matsuyama, K.; Kawakami, N.; Sakuma, I.; Gando, S.; Fukui, H.; et al. Histamine H1 and H2 receptor gene and protein levels are differentially expressed in the hearts of rodents and humans. *J. Pharmacol. Exp. Ther.* **2004**, *309*, 786–795. [CrossRef] [PubMed]

76. Powers, M.J.; Peterson, B.A.; Hardwick, J.C. Regulation of parasympathetic neurons by mast cells and histamine in the guinea pig heart. *Auton. Neurosci.* **2001**, *87*, 37–45. [CrossRef]

77. Hardwick, J.C.; Kotarski, A.F.; Powers, M.J. Ionic mechanisms of histamine-induced responses in guinea pig intracardiac neurons. *Am. J. Physiol. Regul. Integr. Comp. Physiol.* **2006**, *290*, R241–R250. [CrossRef] [PubMed]

78. Fuder, H.; Ries, P.; Schwarz, P. Histamine and serotonin released from the rat perfused heart by compound 48/80 or by allergen challenge influence noradrenaline or acetylcholine exocytotic release. *Fundam. Clin. Pharmacol.* **1994**, *8*, 477–490. [CrossRef] [PubMed]

79. Leary, P.J.; Tedford, R.J.; Bluemke, D.A.; Bristow, M.R.; Heckbert, S.R.; Kawut, S.M.; Krieger, E.V.; Lima, J.A.; Masri, C.S.; Ralph, D.D.; et al. Histamine H2 receptor antagonists, left ventricular morphology, and heart failure risk: The MESA study. *J. Am. Coll. Cardiol.* **2016**, *67*, 1544–1552. [CrossRef] [PubMed]

80. Dehlin, H.M.; Manteufel, E.J.; Monroe, A.L.; Reimer, M.H., Jr.; Levick, S.P. Substance P acting via the neurokinin-1 receptor regulates adverse myocardial remodeling in a rat model of hypertension. *Int. J. Cardiol.* **2013**, *168*, 4643–4651. [CrossRef] [PubMed]

81. Silver, R.B.; Reid, A.C.; Mackins, C.J.; Askwith, T.; Schaefer, U.; Herzlinger, D.; Levi, R. Mast cells: A unique source of renin. *Proc. Natl. Acad. Sci. USA* **2004**, *101*, 13607–13612. [CrossRef] [PubMed]

82. Hara, M.; Ono, K.; Wada, H.; Sasayama, S.; Matsumori, A. Preformed angiotensin II is present in human mast cells. *Cardiovasc. Drugs Ther.* **2004**, *18*, 415–420. [CrossRef] [PubMed]

83. Mackins, C.J.; Kano, S.; Seyedi, N.; Schafer, U.; Reid, A.C.; Machida, T.; Silver, R.B.; Levi, R. Cardiac mast cell-derived renin promotes local angiotensin formation, norepinephrine release, and arrhythmias in ischemia/reperfusion. *J. Clin. Investig.* **2006**, *116*, 1063–1070. [CrossRef] [PubMed]

84. Levick, S.P.; Gardner, J.D.; Holland, M.; Hauer-Jensen, M.; Janicki, J.S.; Brower, G.L. Protection from adverse myocardial remodeling secondary to chronic volume overload in mast cell deficient rats. *J. Mol. Cell. Cardiol.* **2008**, *45*, 56–61. [CrossRef] [PubMed]

85. Frangogiannis, N.G.; Lindsey, M.L.; Michael, L.H.; Youker, K.A.; Bressler, R.B.; Mendoza, L.H.; Spengler, R.N.; Smith, C.W.; Entman, M.L. Resident cardiac mast cells degranulate and release preformed TNF-{α}, initiating the cytokine cascade in experimental canine myocardial ischemia/reperfusion. *Circulation* **1998**, *98*, 699–710. [CrossRef] [PubMed]

86. Gilles, S.; Zahler, S.; Welsch, U.; Sommerhoff, C.P.; Becker, B.F. Release of TNF-α during myocardial reperfusion depends on oxidative stress and is prevented by mast cell stabilizers. *Cardiovasc. Res.* **2003**, *60*, 608–616. [CrossRef] [PubMed]

87. Gordon, J.R.; Galli, S.J. Promotion of mouse fibroblast collagen gene expression by mast cells stimulated via the Fc epsilon RI. Role for mast cell-derived transforming growth factor β and tumor necrosis factor alpha. *J. Exp. Med.* **1994**, *180*, 2027–2037. [CrossRef] [PubMed]

88. Fang, K.C.; Wolters, P.J.; Steinhoff, M.; Bidgol, A.; Blount, J.L.; Caughey, G.H. Mast cell expression of gelatinases A and B is regulated by kit ligand and TGF-β. *J. Immunol.* **1999**, *162*, 5528–5535. [PubMed]

89. Iyer, R.P.; Patterson, N.L.; Fields, G.B.; Lindsey, M.L. The history of matrix metalloproteinases: Milestones, myths, and misperceptions. *Am. J. Physiol. Heart Circ. Phys.* **2012**, *303*, H919–H930. [CrossRef] [PubMed]

90. Chancey, A.L.; Brower, G.L.; Janicki, J.S. Cardiac mast cell-mediated activation of gelatinase and alteration of ventricular diastolic function. *Am. J. Physiol. Heart Circ. Physiol.* **2002**, *282*, H2152–H2158. [CrossRef] [PubMed]

91. De Almeida, A.; Mustin, D.; Forman, M.F.; Brower, G.L.; Janicki, J.S.; Carver, W. Effects of mast cells on the behavior of isolated heart fibroblasts: Modulation of collagen remodeling and gene expression. *J. Cell. Physiol.* **2002**, *191*, 51–59. [CrossRef] [PubMed]

92. Oyamada, S.; Bianchi, C.; Takai, S.; Chu, L.M.; Sellke, F.W. Chymase inhibition reduces infarction and matrix metalloproteinase-9 activation and attenuates inflammation and fibrosis after acute myocardial ischemia/reperfusion. *J. Pharmacol. Exp. Ther.* **2011**, *339*, 143–151. [CrossRef] [PubMed]

93. Patella, V.; Marino, I.; Lamparter, B.; Arbustini, E.; Adt, M.; Marone, G. Human heart mast cells. Isolation, purification, ultrastructure, and immunologic characterization. *J. Immunol.* **1995**, *154*, 2855–2865. [PubMed]

94. Patella, V.; de Crescenzo, G.; Ciccarelli, A.; Marinò, I.; Adt, M.; Marone, G. Human heart mast cells: A definitive case of mast cell heterogeneity. *Int. Arch. Allergy Immunol.* **1995**, *106*, 386–393. [CrossRef] [PubMed]

95. Fureder, W.; Agis, H.; Willheim, M.; Bankl, H.C.; Maier, U.; Kishi, K.; Muller, M.R.; Czerwenka, K.; Radaszkiewicz, T.; Butterfield, J.H.; et al. Differential expression of complement receptors on human basophils and mast cells. Evidence for mast cell heterogeneity and CD88/C5aR expression on skin mast cells. *J. Immunol.* **1995**, *155*, 3152–3160. [PubMed]

96. Ito, B.R.; Engler, R.L.; del Balzo, U. Role of cardiac mast cells in complement C5a-induced myocardial ischemia. *Am. J. Physiol.* **1993**, *264 Pt 2*, H1346–H1354. [CrossRef]

97. Zhang, J.; Alcaide, P.; Liu, L.; Sun, J.; He, A.; Luscinskas, F.W.; Shi, G.P. Regulation of endothelial cell adhesion molecule expression by mast cells, macrophages, and neutrophils. *PLoS ONE* **2011**, *6*, e14525. [CrossRef] [PubMed]

98. Hara, M.; Matsumori, A.; Ono, K.; Kido, H.; Hwang, M.W.; Miyamoto, T.; Iwasaki, A.; Okada, M.; Nakatani, K.; Sasayama, S. Mast cells cause apoptosis of cardiomyocytes and proliferation of other intramyocardial cells in vitro. *Circulation* **1999**, *100*, 1443–1449. [CrossRef] [PubMed]

99. Li, J.; Jubair, S.; Janicki, J.S. Estrogen inhibits mast cell chymase release to prevent pressure overload induced adverse cardiac remodeling. *Hypertension* **2015**, *65*, 328–334. [CrossRef] [PubMed]

100. Marone, G.; Triggiani, M.; Cirillo, R.; Giacummo, A.; Hammarstrom, S.; Condorelli, M. IgE-mediated activation of human heart in vitro. *Agents Actions* **1986**, *18*, 194–196. [CrossRef] [PubMed]

International Journal of
Molecular Sciences

Review

Thrombospondins: A Role in Cardiovascular Disease

Dimitry A. Chistiakov [1,*] [ID], **Alexandra A. Melnichenko [2]**, **Veronika A. Myasoedova [2]** [ID], **Andrey V. Grechko [3]** and **Alexander N. Orekhov [2,4]** [ID]

[1] Department of Fundamental and Applied Neurobiology, Serbsky Federal Medical Research Center of Psychiatry and Narcology, 119991 Moscow, Russia

[2] Laboratory of Angiopathology, Institute of General Pathology and Pathophysiology, Russian Academy of Medical Sciences, 125315 Moscow, Russia; zavod@ifarm.ru (A.A.M.); myika@yandex.ru (V.A.M.)

[3] Federal Research and Clinical Center of Intensive Care Medicine and Rehabilitology, 109240 Moscow, Russia; avg-2007@yandex.ru

[4] Institute for Atherosclerosis Research, Skolkovo Innovative Center, Moscow 121609, Russia; a.h.opexob@gmail.com

* Correspondence: dimitry.chistiakov@lycos.com; Tel.: +7-495-637-5055; Fax: +7-495-637-4000

Received: 26 June 2017; Accepted: 13 July 2017; Published: 17 July 2017

Abstract: Thrombospondins (TSPs) represent extracellular matrix (ECM) proteins belonging to the TSP family that comprises five members. All TSPs have a complex multidomain structure that permits the interaction with various partners including other ECM proteins, cytokines, receptors, growth factors, etc. Among TSPs, TSP1, TSP2, and TSP4 are the most studied and functionally tested. TSP1 possesses anti-angiogenic activity and is able to activate transforming growth factor (TGF)-β, a potent profibrotic and anti-inflammatory factor. Both TSP2 and TSP4 are implicated in the control of ECM composition in hypertrophic hearts. TSP1, TSP2, and TSP4 also influence cardiac remodeling by affecting collagen production, activity of matrix metalloproteinases and TGF-β signaling, myofibroblast differentiation, cardiomyocyte apoptosis, and stretch-mediated enhancement of myocardial contraction. The development and evaluation of TSP-deficient animal models provided an option to assess the contribution of TSPs to cardiovascular pathology such as (myocardial infarction) MI, cardiac hypertrophy, heart failure, atherosclerosis, and aortic valve stenosis. Targeting of TSPs has a significant therapeutic value for treatment of cardiovascular disease. The activation of cardiac TSP signaling in stress and pressure overload may be therefore beneficial.

Keywords: thrombospondins; cardiac remodeling; cardiac hypertrophy; heart failure; atherosclerosis; myocardial infarction; cardiac fibrosis

---

1. Introduction

The function of the extracellular matrix (ECM) is not only limited by providing structural support and immobilization of cells. Functional significance of the ECM also relies in mediating cell-cell and cell-matrix contacts, signaling conduction, and triggering cell adhesion, motility, and differentiation. The composition and functional properties of the ECM vary depending on the cell type and tissue/organ specificity. For example, in the cardiovascular system, the ECM is involved in maintaining structural continuity of the heart and blood vessels, providing an essential scaffold for cell attachment and functioning, control of cell growth, viability and death, regulation of diastolic stiffness, and performing tissue repair/remodeling in a case of cardiovascular damage and inflammation [1]. In the ECM, structural changes induced by local microenvironment can lead to functional matrix alterations. Changes in the matrix may be then conducted to adjacent cells and affect their activity and behavior. For example, the cardiovascular ECM mediates blood flow-induced mechanotransduction [2,3] and adaptive responses of vascular cells and cardiomyocytes to various stress stimuli [4,5].

In cardiovascular pathology such as atherosclerosis, arterial restenosis or heart failure, ECM-associated responses are frequently maladaptive and may lead to adverse tissue remodeling and fibrosis [6,7]. Inflammatory, profibrotic, prooxidant, hypoxic, and other pathological stimuli may induce substantial modifications and impair matrix turnover, which in turn may cause qualitative and quantitative changes in matrix architecture and composition by increasing content of certain ECM proteins and decreasing amount of other matrix components [8,9]. For instance, in failing hearts of patients with dilated cardiomyopathy, significant changes in the content of some non-fibrillar matrix and matricellular proteins were found. Implementation of the mechanical unloading of the left ventricle by left ventricular assist device resulted in the restoration of the fibrillar ECM and basement membrane and improved clinical outcome [9].

Matricellular proteins comprise non-structural ECM proteins that modulate cell function and behavior. Matricellular proteins include thrombospondins (TSPs), tenascins, periostin, osteopontin, CCN proteins, and osteonectin/secreted protein acidic and rich in cysteine (SPARC) [10]. Tenascins contain three members (tenascin C, tenascin R, and tenascin X) that are especially abundant in developing embryonic tissues like cartilage, tendon, bone, and nervous system where these proteins promote migration, proliferation, and differentiation of stem and lineage-specific progenitor cells [11,12]. Periostin plays multiple physiological and pathogenic roles including regulation of mesenchymal differentiation in the developing heart, tissue repair, and involvement in cancer and valvular heart disease [13–15]. Osteopontin triggers biomineralization, bone remodeling, and immunity as well as pathological ectopic calcification [16–18]. Osteonectin is a Ca^{2+}-binding protein whose primary function is to contribute to osteogenesis by conducting biomineralization of the bone and cartilage [19]. CCN proteins include at least six members (CCN1–6) with a complex multidomain structure that provides an option to bind numerous ligands and participate in a variety of biological processes such as angiogenesis, inflammation, tissue repair, fibrosis, and carcinogenesis [20]. Like CNN proteins, TSPs have several functional domains and indeed are able to interact with multiple partners. TSPs are abundantly distributed in various tissues and organs including the cardiovascular system. At steady state, TSP expression is low but can be up-regulated in response to wounding. TSPs are likely to contribute to post-injury tissue repair and remodeling [21]. In this review, we consider structural and functional aspects of the TSP protein family in relation to cardiovascular disease.

2. Thrombospondins Structure

The TSP family contains five members (TSP1–5) that represent multimeric glycoproteins, which bind Ca^{2+}, interact with other ECM proteins, and contribute to the associations between cells and between cells and ECM. TSMs divided to two subgroups: trimeric subgroup A (TSP1 and TSP2) and pentameric subgroup B (TSP3, TSP4, and TSP5). TSPs have a complex multidomain structure (Figure 1). The C-terminal domain, type III repeats and epidermal growth factor (EGF)-like repeats are present in all TSPs and underline the TSP family. The oligomerization domain can be also found in all family members but it is more variable compared with other shared structures [22]. The subgroup A has three EGF-like repeats and type I repeats (also known as thrombospondin repeats; TSRs), von Willebrand factor type C (vWC) domains, and the N-terminal domain. The subgroup B contains four EGF-like repeats but vWC domains and TSRs are missing. While TSP3 and TSP4 have the N-terminal domain, TSP5 has not [23].

The evolutionary analysis showed that the subgroup B is evolutionary younger than the subgroup A [24]. In fish, a TSP4-like sequence, an ortholog of the tetrapod *TSP5* gene is found [25]. The TSP5-coding sequence evolved more quickly from the TSP4-like sequence as an innovation in the tetrapod lineage. Thus, all *TSP* genes show conservation of synteny between fish and tetrapods. In humans, the *TSP1*, *TSP3*, *TSP4* and *TSP5* genes reside within paralogous regions and are the result of gene duplications [24].

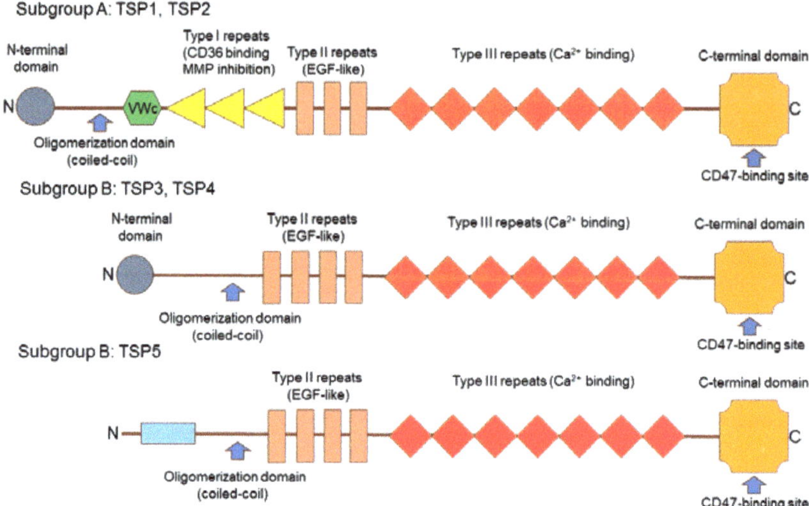

Figure 1. Structure of thrombospondins (TPSs). TSP family includes five members: TSP1–5. Subgroup A comprises TSP1 and TSP2 that form pentamers while subgroup B contains trimeric TSP3–5. TSPs have a complex multidomain architecture that provides an option to bind various ligands. For example, C-terminal domain contains CD47-binding site. Type III repeats are involved in Ca^{2+} binding while Type I repeats are responsible for the interaction with CD36, a receptor for TSP1 and TSP2, and inhibition of matrix proteinases (MMPs). Type II EGF-like domains are involved in the regulation of various signaling pathways such Notch and others. C-terminal domain, Type III repeats and Type II epidermal growth factor (EGF)-like repeats share high homology in all TSPs and represent a signature of the TSP family. Von Willebrand factor type C (VWc) domain is cysteine-rich and is implicated in binding members of the transforming growth factor-β (TGF-β) superfamily. The oligomerization (coiled-coil) domain drives formation of TSH homooligomers. The N-terminal domain, which is present in TSP1–4 and absent in TSP5, is less conservative. This domain regulates structure and stability of the coiled-coil region and binds heparin.

Due to the availability of various structural domains, TSPs can interact with different surface receptors and ECM proteins. In the TSP molecule, EGF-like domains are employed for binding of integrins and Ca^{2+}, TSPs are needed to bind transforming growth factor (TGF)-β and CD36, while the N-terminal domain is required for binding heparin and integrins [26]. The C-terminal domain contains a binding site for CD47, an essential TSP receptor [27]. TSP-mediated effects indeed depend on the availability of the binding partner and local microenvironment that explains cell- and tissue-specific actions of TSPs. In addition, TSPs display different cellular distributions, different temporal expression profiles and have distinct functional responsibilities and modes of transcriptional regulation.

TSP1 that was identified first represents the most studied thrombospondin. TSP1 and TSP2 are expressed by several cell types in response to damage or during remodeling [28,29]. TSP1 is known due to the functional significance in the control of angiogenesis and thrombosis and capacity to increase bioavailability TGF-β by liberating this cytokine from its latent form [23]. In the heart, TSP2 contributes to maintaining of cardiac matrix integrity via its actions on matrix metalloproteinases (MMPs) [30].

The highest expression levels of TSP3 and TSP5 were detected in the vascular wall and tendon [31]. However, both these TSPs are the least studied among TSP family members. TSP4 has a role in vascular inflammation [31], regulation of myocyte contractility, angiogenesis, and ECM remodeling [32,33].

In our understanding of TSP functions, the most profound recent progress was made in studying the cardiovascular system. Genetic studies showed association between single nucleotide polymorphisms (SNPs) in *TSP1*, *TSP2*, and *TSP4* genes with cardiovascular pathology [34–39]. All TSP disease-associated SNPs are functional. For example, the A387P polymorphism of *TSP4* and N700S polymorphism of *TSP1* alter Ca^{2+}-binding sites [40]. Ca^{2+} binding is essential for TSP structure and function. The TSP1 S700 variant had significantly less capacity to bind Ca^{2+} compared to the N700 allele. In fact, the P387 variant of *TSP4* represents a gain-of-function allele because it acquired an additional Ca^{2+}-binding site absent in the A387 allele [40]. A recent meta-analysis showed that the N700S polymorphism of *TSP1* is associated with coronary artery disease (CAD) especially in Asian populations (heterozygote model: odds ratio (OR) = 1.57 [95% confidence interval (CI): 1.01–2.44]; dominant model: OR = 1.56 [95% CI: 1.00–2.43]). The *TSP4* A387P polymorphism is associated with increased CAD risk in American population (homozygote model: OR = 1.29 [95% CI: 1.04–1.61]; recessive model: OR = 1.27 [95% CI: 1.02–1.58]). No association was shown for the THS2 3′ (untranslated region) UTR polymorphism and higher CAD risk [41].

3. Cardiac Integrity

Expression of all TSP family members was found in the heart. In cardiac remodeling, TSP1, TSP2, and TSP4 are up-regulated [42–45]. In pressure overload, myocardial expression of TSP3 and TSP5 was shown to be also increased [46].

TSP1, TSP2, and TSP4 are involved in cardiac fibrosis but possess opposite effects. While TSP1 and TSP2 promote fibrosis [47,48], TSP4 inhibits profibrotic mechanisms as was shown in animal models [49,50] and human heart allografts [43,48]. In human allografts, increased levels of TSP1 and TSP2 suggested for induction of both the fibrotic response and allograft rejection [43,48]. In TSP1 or TSP2-deficient murine cardiac remodeling models (i.e., those affected with doxorubicin-induced cardiomyopathy [51], diabetic cardiomyopathy [52,53], dilated cardiomyopathy [54], or MI [44,55,56] the profibrotic role of both TSPs was confirmed. Mechanistically, profibrotic effects of TSP1 and TSP2 in the heart lead to the stimulation of TGF-β, a key inducer of cardiac fibrosis [48,57,58], suppression of MMPs [48,59], and inhibition of angiogenesis [43,47,55]. Profibrotic activity of TSP1 can be also mediated the calreticulin/low density lipoprotein receptor-related protein 1 (LRP1) complex whose stimulation results in the activation of prosurvival protein kinases such as PI3K and migration of fibroblasts [60,61]. The N-terminal domain of TSP1 has the calreticulin-binding site to stimulate association of calreticulin with LRP1 to the signal focal adhesion disassembly and providing signal into the cytoplasm [62].

TSP2 expression was observed only in biopsy specimens from the hypertrophic hearts of rats that rapidly developed heart failure suggesting for a possible value of TSP2 to serve as a marker of early onset of heart failure [59]. TSP2-deficient mice were extremely vulnerable to rapid progression from angiotensin II-induced cardiac hypertrophy to cardiac failure and fatal rupture since 70% of animals died from cardiac rupture whereas the rest of them progressed to heart failure [59]. These data therefore indicate a role of TSP2 as an essential regulator of cardiac integrity.

In the heart, lack of TSP4 leads to advanced fibrosis [49,50] suggesting for the anti-fibrotic role. TSP4 also triggers heart stress adaptation by increasing intracellular Ca^{2+} content in cardiac muscle cells and enhancing contractility [63]. TSP4-induced adaptive response against ER stress also protects cardiomyocytes from pressure overload [1]. In TSP4-deficient mice, tendon collagen fibrils were found to be significantly larger than in wild-type mice suggesting for the negative TSP4-dependent regulation of collagen synthesis in ligaments [64]. However, it is unknown whether this TSP inhibit cardiac collagen production.

The reason of profound differences in the actions of TSP4 and TSP1/TSP2 on myocardial fibrosis may rely on the structural differences between these TSPs. TSP4 lacks domains responsible for the control of MMP function, angiogenesis and TGF-β stimulation. These domains are present in TSP1 and TSP2 [23]. Thus, in order to recognize effects of every TSP in heart remodeling, it is necessary to monitor expression of each TSP in various steps of remodeling and adaptive reaction to heart damage.

Expression of TSP1–4 was found in human aortic valves [65]. In fibrosclerotic and stenotic valves, TSP2 production was increased. TSP2 was up-regulated in myofibroblasts and some endothelial cells (ECs) and was associated with myofibroblast proliferation and neovascularization. TSP2 activation was followed by suppression of Akt and NF-κB [65]. Cardiac-related expression of TSP4 was highest in the valves suggesting for a potential activity in these regions [32,50].

4. TSPs in Angiogenesis

Angiogenesis is an essential physiological process involved in the developmental vasculogenesis and tissue repair after injury. In pathology, formation of neovessels occurs in vascular proliferative diseases such as atherosclerosis and in tumors. The anti-angiogenic properties of TSP1 and TSP2 are established and confirmed in different models of angiogenesis [66,67]. The anti-angiogenic activity of TSP1 and TSP2 may be especially attractive in the context of anti-tumor therapy because both TSPs are able to inhibit tumor-associated angiogenesis and suppress tumor growth [68,69]. Studying of TSP1 effects on tumor neovessel formation provided new insights into a specific role and a power of this TSP in preventing angiogenesis. It became obvious that tissue expression of TSP1 is able to define fate of angiogenesis even without influencing by pro-angiogenic signals [70].

A phenomenon of dormant tumors that have a microscopic size and do not expand is associated with high expression and inhibitory effects of TSP1 and tissue inhibitor of matrix proteinases TIMP-1 [71]. Down-regulation of TSP1 and reduced tumor sensitivity to angiostatin leads to proangiogenic switch and induction of rapid growth and tumor expansion [72]. MicroRNA (miR)-467 was reported to function as a negative regulator of TSP1 expression [73]. This miRNA is induced by high glucose and leads to the sequestration of TSP1 mRNA in the non-polysomal fraction of tumor cells and induction of angiogenesis. Inhibition of miR-467 suppresses tumor growth and angiogenesis [74].

The anti-angiogenic activity of TSP1 is attributed to a structural domain known as the TSP type I repeat [75]. This domain serves as a single Ca^{2+}-binding site for endothelial receptor CD36 that is essential for mediating anti-angiogenic effects of TSP1 and TSP2 [76,77]. Histidine-rich glycoprotein (HRGP), a circulating protein, contains a CD36 homology domain and blocks TSP-dependent anti-angiogenic effects by binding to either TSP1 or TSP2 [75,78]. Another TSP-1-dependent anti-angiogenic mechanism is consisted of the engagement of CD47 that disrupts CD47 interaction with vascular endothelial growth factor (VEGF) receptor 2 (VEGFR2) and blunts VEGFR2-mediated proangiogenic signaling associated with activation of endothelial NO synthase (eNOS) and soluble guanylate cyclase (sGC) [79]. In silico analysis showed that TSP1 binding to CD47 can also enhance VEGFR2 degradation [80]. By contrast, in TSP1-deficient mice, phosphorylation of endothelial VEGFR2 is up-regulated thereby providing a stimulatory signal to Akt or Src that in turn activate eNOS through phosphorylation [79]. Another suppressive mechanism, by which TSP1 can inhibit myristic acid-stimulated eNOS-dependent signaling that leads to the induction of increased adhesion properties of ECs and vascular smooth muscle cells (VSMCs), is dampening of the CD36-mediated uptake of free fatty acids or engagement of CD47 [81,82].

TSP1 can also diminish the sGC/3′, 5′-cyclic GMP (cGMP) signaling by limiting cGMP-dependent activation of the downstream cGMP-activated kinase (PKG) [83]. The inhibitory effect of TSP1 on eNOS/cGMP signaling in ECs are more potent than that of TSP2 suggesting for a role of TSP1 as a dominant regulator of NO/cGMP signaling pathway through CD47 [84].

NO is essential for activation sGC that contains a heme responsible for NO binding [85]. NO-dependent stimulation of sGC leads to intensive production of cGMP, a signaling messenger that is involved in the vascular tone regulation by relaxation of VSMC contractility [86], inhibiting platelet aggregation [87] and blood cell adhesion to the endothelium [88]. Except for limiting NO bioavailability, inhibitory effects of TSP1 on sGC activity and cGMP production can also involve suppression of the hydrogen sulfide (H_2S)-mediated signaling through blocking activity of H_2S-biosynthesiting enzymes, cystathionine β-synthase (CBS) and cystathionine γ-lyase (CSE) [89]. Another inhibitory mechanism may involve TSP1-dependent stimulation of reactive oxygen (ROS) and nitrogen species production [90], which in turn inactivate sGC via covalent enzyme modification [91] or heme nitrosylation [92].

Compared to TSP1 and TSP2, TSP4 exerts proangiogenic properties [93]. TSP4 was detected in the lumen of neovessels. TSP4-deficient mice have diminished angiogenesis in comparison with wild-type mice. Mice transgenic for the CAD-associated human *TSP4* P387 variant displayed more intensive angiogenesis compared with mice bearing the A387 allele of *TSP4*. Pulmonary ECs derived from TSP-lacking mice exhibited reduced adhesion and migratory properties in contrast to wild-type ECs. In addition, recombinant TSP4 was shown to stimulate vessel development and EC motility/proliferation through binding to integrin-α2 and gabapentin receptor α2δ-1 [93].

5. TSPs in Atherosclerotic Blood Vessels

Vascular expression was shown for all TSPs [31]. In blood vessels, TSP1, TSP2, and TSP4 are involved in the functional and structural regulation of the vascular wall and interactions with blood-borne cells. In diabetic rats, TSP1 levels were increased after vascular wounding [94]. In rats with balloon-induced injury, blockade of TSP1 in carotid artery with antibody resulted in enhanced reendothelization and decreased neointima formation [95]. In apolipoprotein E (ApoE)-deficient mice, an atherosclerotic animal model, inhibition of TSP1 and TSP4 caused delayed atherosclerosis progression and less inflammatory conditions. However, at late atherogenic stages, the plaques developed proinflammatory content and had the same size as lesions in control mice [31,96]. TSP-1 or TSP4 did not alter intraplaque lipid content but induced dramatic changes in macrophages counts within the plaque by influencing either function (TSP1) or infiltration (TSP4) of macrophages to the lesion. In advanced lesions, TSP2 plays an anti-atherogenic role by stimulating phagocytic function of macrophages needed to perform clearance of apoptotic and necrotic cells and cell debris [96].

In ApoE-deficient mice, TSP4 depletion does not affect lesional matrix deposition [31] while TSP1 deficiency leads to the formation of more fibrotic lesions of less size but also increases inflammation associated with enhanced activity of macrophages [96]. These macrophages are involved in intensive degradation of the fibrous cap associated with increased accumulation of MMP-9 in the cap but display altered phagocytosis that finally results in increase of the necrotic core and plaque destabilization. It appears that TSP1 promotes atherogenesis in early stages through induction of endothelial dysfunction, stimulation of VSMC proliferation and inhibiting collagen deposition. However, in late stages, TSP1 switches this role to the anti-atherogenic function by repressing lesional maturation via stimulation of the phagocytic activity of macrophages and reducing necrosis [97].

In ApoE-deficient mice fed on high-fat diet, strong plaque TSP1 expression was detected in fibrous cap-associated VSMCs and inflammatory cells in the shoulder of the plaque and foam cells. Weak TSP1 expression was found in the adventitia and media of the atherosclerotic wall [96]. Expression of TSP2 was found in human arterial VSMCs [98] but was not detectable in the endothelial plaque lining and intraplaque neovessels [99]. TSP3, TSP4, and TSP5 are all also linked to atherosclerotic plaques. Expression of TSP3 was detected in the tunica media and tunica adventitia, on the luminal endothelial surface, and in the plaques in ApoE-deficient mice [31]. TSP4 exerts proatherosclerotic and proinflammatory effects in vessels since TSP4-knockout mice developed less inflammatory lesions with lowered macrophage content, diminished activation of ECs, and reduced production of proinflammatory cytokines [31]. TSP4 stimulates adhesion and movement of macrophages and neutrophils in an integrin $\alpha_v\beta_3$-dependent manner [13,40]. The CAD risk-associated *TSP4* P387 variant was shown to enhance leukocyte attachment to ECs and motility and promote proinflammatory signaling in vascular and blood-borne cells likely due the ability to bind more Ca^{2+}. This ability leads to conformational changes in the mutant TSP4 molecule and provides better interaction with cell surface receptors [13,40].

TSP5 protein was histochemically detected in normal and affected (i.e., atherosclerotic and stenotic) human arteries where it is produced by VSMCs [100]. In ApoE-deficient mice, TSP5 expression is associated with tunica media and a few plaque cells [31]. TSP5 is involved in maintaining VSMC quiescence and contractile phenotype via interaction with integrin $\alpha_7\beta_1$ [101]. A disintegrin and metalloproteinase with thrombospondin motifs 7 (ADAMTS7) that is also expressed in VSMCs can degrade TSP5 and promote VSMC migration and recruitment for neointima formation [102].

CD47 (also known as the integrin-associated receptor; IAP) is the receptor for TSP1 [103]. After stimulation with TSP1, CD47-mediated pathway is involved in the control of leukocyte functions, vascular resistance, and intracellular signaling in ECs and VSMCs [104]. TSP1 binding to CD47 has global functional consequences by inhibiting endothelial nitric oxide (NO) production, controlling vascular tone, and maintaining systemic hemodynamics and cardiac dynamics in stressful conditions [105–108]. The interaction between TSP1 and CD47 also regulates thrombosis/hemostasis, immune responses, and mitochondrial function [109]. CD47-binding capacity can be shared between all TSP family members since CD47-binding site is located in the C-terminal region, which is homologous in all TSHs.

TSP1-mediated overactivation of NADPH oxidase may exert important pathogenic effects in cardiovascular pathology. TSP1 is able to stimulate Nox1 and Nox4 through CD47-dependent signaling. In VSMCs, TSP1 binding to CD47 leads to phospholipase C (PLC)-catalyzed biosynthesis of diacyl glycerol, a stimulator of protein kinase C (PKC) that in turn phosphorylates the NADPH oxidase core subunit p47^{phox} followed by activation of Nox1. Nox1 overactivity enhances ROS formation that further inhibits VSMC-dependent vasorelaxation and induces vascular dysfunction by promoting oxidative stress [90]. In ischemic VSMCs, TSP1 can also increase ROS generation via stimulation of the cell surface receptor signal-regulatory protein-α (SIRP-α) and subsequent recruitment of the p47^{phox} subunit [110].

TSP1-induced Nox4 up-regulation stimulates ROS-dependent proliferation of VSMCs and neointimal formation, a hallmark of the proatherogenic arterial remodeling [111,112]. In addition, TSP1/CD47-dependent stimulation of Nox1 enhances micropinocytosis of non-modified low density lipoprotein (LDL) by macrophages and promotes their transformation to foam cells, another key characteristics of atherogenesis. In macrophages, Nox1 overactivity induces dephosphorylation of actin-binding protein cofilin, PI3K-dependent activation of myotubularin-related protein 6 (MTMR6) followed by cytoskeletal rearrangements and increased LDL uptake [113]. Thus, TSP1-mediated stimulation of ROS-dependent signaling and oxidative stress have numerous pathogenic consequences, which promote atherogenesis.

6. TSPs in Myocardial Infarction

Myocardial injury initiates the post-MI tissue repair response aimed to restore cardiac conduction and contractility, blood supply, and replace necrotic cardiomyocytes in the infarct with a scar [114]. After MI, cardiomyocyte necrosis occurs early in post-MI remodeling of the infarct area while apoptosis occurs in the infarct and distant cardiac regions both in the early and late stages of remodeling [115]. In the early stage of cardiac repair, infiltrated inflammatory cells release MMPs, particularly MMP-2/9, to degrade ECM in the injured myocardial area and adjacent regions [116]. In parallel with ECM destruction, transformation of cardiac fibroblasts to myofibroblasts associated with their proliferation and migration to the site of injury begins. In response to profibrotic signals, myofibroblasts produce collagen and other ECM components, which are deposited in the cell-free infarcted zone that was cleared by macrophages from necrotic cells [117]. In the proliferative stage of myocardial repair, the collagen amount quickly rises in the injured region while collagen fibers undergo cross-linking in the maturation stage [118]. Collagen can be also accumulated in the non-infarcted area that can initiate reactive myocardial hypertrophy [116].

Compared to pressure overload, post-MI reperfusion in TSP1-deficient mice leads to more intensive and long-term heart inflammation in the infarct border zone and excessive remodeling [55]. These observations allow to suggest that TSP1 exerts a barrier function in the infarct border zone to limit propagation of inflammation and fibrosis into the non-injured cardiac regions. The mechanisms of this effect are not well studied. TGF-β-dependent up-regulation of TSP1 production may be involved in this process [55,56]. TSP1 also enhances apoptosis of activated T cells through CD47-dependent activation of proapoptotic Bcl-2 family member BNIP3 that primarily links it to inflammation [119,120].

TSP1 inhibits both eNOS-dependent NO production and NO-mediated signaling that stimulates vascular relaxation and angiogenesis [107]. These data may suggest for a putative contribution of TSP1 to heart ischemia and/or MI [121]. However, TSP1 also exhibit cardioprotective effects by activating TGF-β, an anti-inflammatory cytokine [56]. Indeed, inhibition of CD47 represents a more attractive therapeutic target than inhibition of TSP1. In ischemic mouse models, exposure to CD47-blocking agents results in significant improvement of tissue survival and decreased vasculopathy, an evidence of the therapeutic value of CD47 suppression to treat cardiovascular disease [122]. In MI or ischemia, down-regulation of CD47 may have unpleasant sequela such as diminished apoptosis of inflammatory macrophages, which can result in elevated levels of proinflammatory cytokines [123]. Enhanced TSP1 production may be involved in NO resistance observed in aging and ischemic heart disease [122]. Indeed, therapeutic targeting of vessel NO signaling through TSP1/CD47 can be valuable [121].

Compared with TSP1, our knowledge of a role of TSP2 in MI is constrained. TSP2 and TSP4 do not involved to the control of NO signaling [106]. However, TSP2 shares with TSP1 many anti-angiogenic properties, promotes CD36-mediated apoptosis of ECs, and initiates cell cycle arrest [68]. TSP2-deficient mice exhibited increased angiogenesis that was associated with MMP-9 up-regulation [123]. Similarly, TSP2 deficiency was shown to induce enhanced angiogenesis and delayed skin wound contraction accompanied with increased MMP-2/9 and soluble VEGF production [124]. These findings indicate that anti-angiogenic effects TSP2 can be released through multiple mechanisms. TSP2 deletion in mouse was shown to induce increased vascularity and defects in connective tissue formation due to impaired collagen fibrillogenesis [125,126] indicating that cardiac TSP2 may be involved in post-injury heart remodeling and repair through the control of fibrillogenesis [127]. In favor of the involvement of TSP2 into cardiac repair, the ability of TSP2 to positively modulate function of human cardiomyocyte progenitor cells (hCMPCs) in hypoxic conditions was demonstrated [128]. In mice, short-term exposure to hypoxia stimulates migratory and invasive properties of hCMPCs while prolonged exposure activates proliferation, angiogenesis, and blocks migration likely due to TSP2 actions [129]. Limitation of migratory activity of hCMPCs is necessary to induce proliferation and differentiation of progenitor cells into cardiomyocytes in the infarcted region.

Post-MI and in cardiac hypertrophy, TSP4 expression was shown to be chronically up-regulated in the heart, especially in the left ventricle [45]. TSP4 mRNA levels directly correlated with the rate of left ventricular remodeling indicating the role of TSP4 in post-MI cardiac remodeling [130]. TSP4 plays a cardioprotective role by diminishing cardiac ER stress. Furthermore, cardiac TSP4 overproduction is protective against MI [63]. Further studies showed that the type III repeat domain and the C-terminal domain of TSP2 are involved in Atf6α binding and regulation of ER stress response [131]. ATF6α is an ER stress-regulated transcription factor that drives expression of ER chaperons needed to initiate the unfolded protein response and prevent ER stress [132]. TSP4 cardiac-specific transgenic mice was resistant to myocardial infarction (MI) while TSP4-deficient mice exert cardiac maladaptation.

In the pilot genetic study, an association of the *TMP1* N700S and *TMP4* A387P variants with higher risk familial premature MI was demonstrated. The *TMP2* T/G 3′UTR polymorphism was associated with lower risk of familial MI [34]. A global meta-analysis confirmed association with CAD only for the *TMP1* N700S and *TMP4* A387P polymorphisms but not for the *TMP2* T/G 3′UTR [41]. The *TSP4* A387P polymorphism was associated with increased coronary risk in post-MI subjects who had elevated levels of high density lipoprotein (HDL) cholesterol and C-reactive protein (CRP), an inflammatory marker [39]. Accordingly, the TT genotype of the *TMP2* T/G 3′UTR variant showed association with plaque erosion independently of age, gender, and cigarette smoking in cases of sudden death [133]. Lesional erosion that is induced by intimal injury and does not lead to plaque rupture is a frequent cause of sudden death [134]. However, the results of case-control studies were contradictory since no significant association of the *TMP1* N700S, *TMP2* T/G 3′UTR, and *TMP4* A387P genetic variants, and with both CAD and MI were found in other studies [135–138]. Possible reasons of such an inconsistency may be referred to the different patients' selection criteria, insufficient size of the population samples tested, racial differences, etc. For example, the frequency of the *TSP1* N700S variant was reported to be extremely low in the Chinese Han population [138]. This therefore underlines the need to recruit case-control cohorts of a larger size to provide a sufficient statistical power that is critical to the success of genetic association studies to detect causal genes of human complex diseases such as CAD.

7. TMPs in Cardiac Hypertrophy

Cardiac hypertrophy can be induced by chronic pressure overload (for instance, by essential hypertension) and is characterized by extensive growth of cardiac muscle cells, proliferation of cardiac fibroblasts, increased ECM deposition (i.e., fibrosis), and intensive cell death. Heart fibrosis occurs due to the massive production and deposition of collagens type I and type III, which exceeds their degradation. Fibrosis is resulted from the up-regulation of collagen synthesis, down-regulation of collagen destruction, or both [139]. In heart hypertrophy, cardiac matrix composition is altered due to collagen redistribution and increased cross-linking that can lead to changes in ECM functional properties [140]. MMPs is the most frequent type of enzymes involved in matrix remodeling. Except for matrix degradation, MMPs also promote ECM synthesis by liberating growth factors and other profibrotic messengers from the matrix [141]. In the heart, chronic hypertension stimulates apoptosis of cardiomyocytes [142] and inflammation [143].

In hypertensive cardiac disease, levels of TSP1, TSP2, and TSP4 are elevated [45,59,144]. Pressure overload stimulates heart expression of TSP1 and TSP4 [45,144]. TSP2 up-regulation was observed in hypertrophic hearts of rats that overexpressed renin and further progressed to heart failure [59].

In mice, TSP1 deficiency led to early onset of heart hypertrophy and enhanced late dilatation in response to pressure overload. Degenerative morphological changes in cardiomyocytes were observed due to the sarcomeric loss and rupture of sarcolemma. Cardiac remodeling was abnormal and accompanied with abundant infiltration of defective fibroblasts. The fibroblast-to-myofibroblast transdifferentiation was impaired. Collagen synthesis was reduced due to the perturbed TGF-β signaling. Furthermore, myocardial production of MMP-3 and MMP-9 was up-regulated [144]. Indeed, TSP1 loss in the heart leads to adverse consequences by impairing response to pressure overload and inducing aberrant tissue remodeling. TSP1 activation in the pressure-overloaded myocardium is critical in the control of the fibroblast phenotype and heart remodeling through up-regulation of TGF-β-dependent pathway and cardiac matrix preservation through inhibition of MMPs. However, no significant changes in inflammatory responses was detected [144] that is rather paradoxical since TGF-β exerts anti-inflammatory properties. In an ischemia-reperfusion model, TSP1 deletion was associated with prolonged post-MI inflammatory response and increased release of proinflammatory factors such as chemokine (C-C motif) ligand 2 (CCL2), chemokine (C-X-C motif) ligand 10 (CXCL10), interleukin (IL)-1β, IL-6, macrophage inflammatory protein-1α (MIP-1α) [55]. In experimental diabetic cardiomyopathy complicated with abdominal aortic coartaction, implementation of LKSL, a peptide that antagonizes TSP1-dependent TGF-β activation, had beneficial actions on the myocardium by inhibiting TGF-β-driven fibrosis [52]. Hence, from the pharmacological point of view, antagonizing TSP1-dependent activation of TGF-β looks more attractive and efficient than blockade of TSP1 signaling in heart hypertrophy.

Reduced vascularity because of decreased angiogenesis is supposed to promote progression of heart hypertrophy to heart failure [145]. The inhibitory actions of TSP1 on cancer angiogenesis were broadly investigated. However, there are contradictory results of studies that examine microvascular effects of TSP1 in the heart. Global deletion of TSP1 was reported to induce the development of dense cardiac capillary network and higher cardiac mass [146]. These findings were not confirmed [144]. Up-regulation of MMPs in TSP1-deficient mice may substantially contribute to enhanced angiogenesis [144].

In response to vasoactive stress, TSP1-deficient mice responded by increased heart rate and changes in blood pressure by elevation of central diastolic and mean arterial blood pressure and reduction of peripheral blood pressure and pulse pressure. In response to epinephrine, the hypertensive response was diminished in both TSP1-deficient or CD47-deficient mice [105]. These data indicate an important role of TSP1 and its receptor CD47 in the acute regulation of blood pressure, which possess vasoconstrictor effects to hold global hemodynamics under vasoactive stress. By inhibiting NO-dependent vasorelaxation, TSP1 maintains blood pressure under stressful conditions [107]. Since CD47 is crucially involved in TSP1-mediated regulation of blood pressure, pharmaceutical targeting of the TSP1/CD47 mechanism may be useful for treatment of hypertension.

In cardiac hypertrophy, TSP2 and TSP4 activities are rather oriented to the control of matrix composition. In pressure overload, TSP2-lacking mice developed cardiac rupture or heart failure accompanied with higher activities of MMP-2/9 indicating that TSP2 contributes to the control of cardiac integrity in hypertrophic hearts [59]. The up-regulation of TSP2 was recognized as a useless effort to rescue the integrity in pressure overload. In response to experimentally induced transverse aortic constriction (TAC), TSP4-deficient mice exhibited advanced heart hypertrophy and fibrosis along with left ventricular dilation, decreased systolic and impaired diastolic function. These cardiac changes resembled age-dependent hypertrophy and fibrosis [50]. TAC-induced interstitial fibrosis was accompanied with up-regulated collagen production, increased MMP expression, lowered microvessel density and was not associated with apoptosis of cardiac muscle cells and inflammation [147]. However, it is unclear whether decrease in myocardial capillary density is the effect of fibrosis or angiogenesis.

TSP4-lacking hearts failed the ability to respond properly to acute pressure overload by enhanced heart contractility or by activation of stretch-response pathways (Akt- and ERK1/2-dependent). In addition to missing capacity to reply normally to acute pressure overload, TSP4-deficient mice also failed the ability to perform an appropriate adaptive response to chronic press overload that induces cardiac dilation, greater myocardial mass, and decline in heart function. However, no changes in interstitial fibrosis was detected. Pressure overload affected cardiac contractility of a whole cardiac muscle, not in separate cardiomyocytes [49]. Therefore, TSP4 can serve as a cardiomyocyte-interstitial mechano-sensing molecule, which regulates adaptive myocardial contractile reactions in response to acute stress. Stable and stressed ECM is likely to negatively regulate function of cardiac muscle cells, which is confronted by TSP4 in normal conditions [49].

8. TSPs in Heart Failure

The aberrant remodeling leads to myocardial overwork that if untreated can progress to heart failure. Structural changes associated with right or left heart failure are differentiated [148]. Right failure is characterized by intensive collagen degradation and cross-linking disruption and off-center hypertrophy. In left failure, cardiomyocyte hypertrophy is concentric while interstitial collagen content and cross-linking is increased [149]. Heart remodeling is accompanied by substantial cardiomyocyte loss, which is supposed to represent one of the major mechanisms of heart failure progression. Apoptosis, necrosis, and autophagy mediate the cardiomyocyte death [150,151]. Cardiac inflammation also contributes to heart failure through enhanced production of TNF-α and other inflammatory cytokines that exhibit adverse effects on the myocardium [152].

TSP1 levels were decreased in human failing hearts and positively correlated with TGF-β levels indicating that cardiac fibrosis is an attribute of early stages of heart failure [153]. In rats with artificially induced heart failure, TSP1 production was up-regulated [154,155]. The TSP1/CD47 signaling seems to play a central role in promoting left ventricular hypertrophy and heart failure [155]. Up-regulation of histone deacetylase 3 (HDAC3) and Ca^{2+}/calmodulin protein kinase II (CaMKII) stimulates hypertrophy of the left ventricle. Accordingly, blockade of either CD47, HDAC3 or CaMKII has beneficial cardiac effects by reducing hypertrophy and softening heart failure [155] thereby underlining a value of the TSP1/CD47 pathway in the pathogenesis of heart failure.

In a murine model of age-related heart failure, a modulatory role was shown for miR-18/19, both are members of the aging-associated miRNA cluster 17–92 that targets TSP1 and connective tissue growth factor (CTGF) [156]. In age-related heart failure, expression of cluster 17–92 miRNA members was down-regulated. Accordingly, TSP1 and CTGF were increased. Importantly, these expression changes occurred only in cardiomyocytes, not in fibroblasts. Indeed, miR-18/19 protect heart against age-related heart failure. With aging, expression of the cluster 17–92 declines while expression of TSP1 and GTCF grows, a phenomenon that predisposes to heart failure [156].

In aged mice, TSP2 production is also increased and is attributed to the ECM surrounding cardiac muscle cells [54]. In older animals, TSP2 deficiency was associated with severe dilated cardiomyopathy, cardiac fibrosis, altered systolic function, and inflammation. Cardiomyocytes were subjected to increased cellular stress and death. Cardiac capillary density was not affected. MMP-2 expression was up-regulated and tissue transglutaminase-2 was down-regulated that led to aberrant cross-linking. Cardiac expression of the TSP2 transgene prevented dilated cardiomyopathy in aged rats [54].

The cardioprotective action of TSP2 was confirmed in a model of doxorubicin-induced cardiomyopathy. Deletion of TSP2 in mice caused higher death rate in response to doxorubicin while survived animals has diminished heart function associated with intensive apoptosis of cardiac muscle cells and ECM destruction because of MMP-1/9 activation [51]. As shown in a model of viral myocarditis-induced heart failure, TSP2 deficiency also stimulates heart inflammation suggesting for the involvement of TSP2 in the regulation of inflammatory response [157]. Lack of TSP2 is related to reduced activation of regulatory T cells, advanced necrosis and fibrosis, heart dilation, and diminished systolic function. TSP2 overexpression prevents heart failure and reduces mortality through attenuating heart inflammation, inflammation, and virus-induced cardiac death. TSP2 was also up-regulated in the myocardial biopsy samples from subjects affected with viral myocarditis [157]. Overall, these data demonstrate protective effects of TSP2 against heart failure.

The up-regulation of cardiac TSP4 expression was observed in the heart of rats with pressure overload-induced heart failure [45,158] and in rats with heart failure induced by the volume overload after aortocaval fistula [159]. Interestingly, there were no significant changes in systolic function and expression of genes responsible for Ca^{2+} homeostasis, neurohumoral regulation, contractility, and cytoskeleton organization during transition from left ventricular hypertrophy to heart failure [158]. Only, expression of ECM proteins such as TSP4 and matrix Gla protein was elevated indicating that progression from hypertrophy to heart failure is regulated by ECM remodeling. In the heart of TSP4-deficient mice subjected to TAC to increase left ventricle load, massive ECM depositions, higher cardiac mass, decreased microvessel density, abnormal heart function, and inflammation, but no signs of apoptosis were observed [50]. These observations show that increase of cardiac TSP4 expression is an adaptive response to pressure overload. TSP4 display cardioprotective effects by regulating myocardial remodeling in pressure overload to prevent progression to heart failure.

9. TSPs in Calcific Aortic Valve Disease

The ECM and factors that control ECM composition and remodeling are implicated in the pathogenesis of calcific aortic valve disease (CAVD). Inflammation and angiogenesis, both are regulated by TSPs, also closely linked to the pathogenesis of aortic stenosis [160]. CAVD is characterized by increased proliferation of myofibroblasts, neovasculogenesis, and valvular calcification. Expression of TSPs 1–4 was detected in stenotic valves, with up-regulation of TSP2 [65]. TSP2 expression was permanently increased in neovessels during progression from early valve remodeling to adverse stenosis indicating a role of TSP2 in the control of CAVD-associated neovascularization [65]. Further studies are required to discover a precise mechanism underlining a role of TSP2 in calcified aortic valves.

10. TSPs in other Pathologies

The involvement of TSP1 to hypoxia-induced pulmonary hypertension was shown. This disorder is characterized by increased pressure in the pulmonary artery, pulmonary vein, and lung vasculature. Pulmonary hypertension is accompanied by narrowing of lung-associated vessels due to thickening of the tunica intima and tunica media as a result of pathogenic vascular remodeling. Increased workload of the heart causes hypertrophy of the right ventricle that can finally progress to the right heart failure [161].

Chronic lung ischemia induces overexpression of TSP1 in the pulmonary artery as was shown in a pig model of experimental pulmonary hypertension. TSP1 overactivity correlated with increased death of ECs and endothelial dysfunction likely due to the proapoptotic effects of TSP1 [162]. Similarly, up-regulated levels of various matricellular proteins including TSP1, TSP2, and TSP4 were detected in the right ventricle of monocrotaline-induced pulmonary hypertensive rats [163]. In humans affected with pulmonary hypertension, elevated levels of circulating TSH1 were also reported [164]. Deletion of TSP1 in mice was related to increased arterial VSMC hyperplasia, proliferation, and growth, less advanced vascular remodeling, lowered right ventricular hypertrophy and right ventricle systolic pressure compared with wild-type counterparts exposed to chronic hypoxia. In fact, TSP1-deficient animals showed increased resistance to hypoxia-induced pulmonary hypertension [165].

Mechanistically, TSP1-induced activation of TGF-β promotes VSMC hyperplasia and proliferation in the pulmonary artery and lung arteries of less caliber [166]. In hypoxia-induced human pulmonary artery VSMCs, TSP1 activation also up-regulates production of the NADPH oxidase subunit, Nox4, that can be inhibited by the peroxisome proliferator-activated receptor γ (PPARγ) or its agonist, rosiglitazone [167]. In hypoxic VSMCs, TGF-β acts in an autocrine manner stimulating insulin-like growth factor binding protein-3 (IGFBP-3) activation via the Akt/PI3K mechanism. IGFBP-3 then increases Nox4 expression, which induces VSMC proliferation and transformation of cardiac fibroblasts to myofibroblasts through ROS-dependent signaling and therefore aggravates lung arterial thickening and right ventricular hypertrophy [168–170]. Cardiovascular protective effects of PPARγ are mediated through inhibition of hypoxia-induced binding of the transcription factor NF-κB to the *Nox4* promoter that prevents transcription [171]. On the other hand, hypoxia was shown to down-regulate PPARγ in pulmonary arterial VSMCs via the ERK1/2- NF-κB-Nox4 mechanism [172]. In summary, these observations suggest for a pathogenic role of both TSP1 and TGF-β in pulmonary hypertension that cooperate in induction of abnormal tissue remodeling associated with increased arterial VSMC hyperplasia/proliferation and hypertrophy of the cardiac right ventricle.

In addition to the involvement of coronary atherosclerosis and post-MI cardiac remodeling, a role of TSPs in cerebrovascular ischemic disease and ischemic stroke was reported. After stroke, increased production of TSP1 and TSP2 was observed in experimental models of ischemic stroke [173,174]. The post-stroke activation of TSPs is necessary to promote an adaptive response to brain injury in order to the recover synaptic plasticity and motor function [175] and regulate angiogenic and platelet-mediated prothrombotic mechanisms [176]. In the resolution phase of the repair of brain infarct, TSP-1/CD36 interaction is important to activate clearance of dead and apoptotic cells by macrophages in response to stimulation by IL-4 or monocyte colony-stimulating factor (M-CSF) [174,177]. Thus, these data indicate protective effects of TSP1 and TSP2 on brain function during healing of ischemic cerebral injury.

TSP1 plays a protective role in non-ischemic neurological pathology such as Alzheimer's disease (AD), fragile X syndrome, and Down syndrome, both are associated with serious mental impairments and reduced synaptic plasticity. Decreased expression of TSP1 was shown in a subset of cortical pyramidal neurons and astrocytes that are prone to AD [178,179]. TSP1 was shown to protect neurons against β-amyloid-induced synaptic degeneration [179]. On the other hand, β-amyloid inhibits release of TSP1 by astrocytes that in turn attenuates expression of synaptic proteins such as synaptophysin and PSD95 followed by aberrations in the morphology of dendritic spines and reduction of synaptic plasticity [180]. Similar abnormalities such as spine malformations and reduced synaptic density were observed in Down syndrome astrocytes, astrocytes from animal models of fragile X syndrome, and astrocytes from TSP-deficient mice indicating a pathological role of TSP1 deficits [181,182]. In the AD brain, prostaglandin E2, an inflammatory messenger, reduces astrocytic expression of TSP1 by induction of miR-135 that targets the TSP1 mRNA. Binding of prostaglandin E2 to its receptor EP4 leads to protein kinase A (PKA)-dependent stimulation of the CCAAT/enhancer-binding protein δ (CEBPD), a transcriptional coactivator that up-regulates expression of miR-135 in astrocytes [183].

In summary, these findings suggest for a key role of TSP1 in astrogenesis and maturation, spine development, and synaptogenesis. Thus, decreased TSP1 activity is linked to neurodegenerative, neurodevelopmental, and mental pathology associated with dysfunction of dendritic spines and aberrations in synaptic plasticity.

Thrombospondins may contribute to the pathogenesis of congenital heart defects (i.e., congenital heart disease) associated with structural heart anomalies presented at birth. In children with congenital ventricular septal defect characterized by the perforation of the ventricular septum, serum levels of TSP1 were dramatically increased and showed positive correlation with the risk of ventricular septal defect [184] thereby suggesting for a potential value for early diagnosis of this cardiac defect. A pathologic role of TSP1 up-regulation in the ventricular septal defect is unclear but may be related to the constitutive impairment of the TGF-β signaling associated with alterations of migration of neural crest cells [185] that contribute to the formation of the septum part, which separates the pulmonary circulation from the aorta [186].

Depletion of the Hect domain E3 ubiquitin ligase Nedd4 in mice resulted in detrimental abnormalities in heart development associated with the formation of double-outlet right ventricle and atrioventricular cushion defects that was fatal for developing embryos [187]. TSP1 expression was markedly up-regulated in Nedd4-deficient mice. Nedd4 is involved in the ubiquitination of a variety of ion channels, membrane transporters, growth factors and their receptors [188] followed by proteosomal degradation. Interestingly, VEGFR2 is a Nedd4 substrate [189] whose degradation is promoted by TSP1. Indeed, Nedd4-deficient mice experience adverse problems in cardiogenesis and vasculogenesis associated with letal developmental deviations likely due to the defects in protein trafficking machinery, ER stress, and heart malformations due to enhanced and deregulated growth factor signaling. In addition, Nedd4 deficiency may induce the premature control loss of the activity of sodium channels such as cardiac voltage-gated channel $Na_v1.5$ [190] and peripheral neuronal channels $Na_v1.2$ and $Na_v1.7$ [191], which is rather lethal because of the inability to support heart conduction/contractility and cardiac/neuronal connectivity in a proper manner.

Some cardiac congenital aberrations such as for example the Holt-Oram syndrome are accompanied by alterations in electrical conduction [192]. The role of TSPs in the regulation of myocardial electrophysiology and contraction is widely unclear. Direct evidence for the involvement to the cardiac muscle contractility was obtained only for TSP4 that modulates heart contraction in response to stress induced by enhanced blood flow [49]. In TSP1-deficient mice, no difference in heart contractility was observed compared to the wild-type counterparts [193]. Although TSPs are able to bind many Ca^{2+} cations, this property may be primarily essential for Ca^{2+}-dependent signaling. However, there are some evidence in favor of a potential involvement of TPSs in the control of muscle contraction. As mentioned above, the TSP1/CD47 mechanism is implicated in CaKMII-mediated cardiac hypertrophy [155]. In differentiated SMCs, CaKMII controls cell contractility [194]. In peripheral sensory nerves, painful nerve injury interrupts Ca^{2+} signaling by stimulating plasma membrane Ca^{2+}-ATPase (PMCA) activity and inhibiting sarco-endoplasmic reticulum Ca^{2+}-ATPase (SERCA) activity that results in depletion of ER-associated Ca^{2+} depots and increase of cytoplasmic Ca^{2+} levels. Injury-induced TSP4 up-regulation resembles effects of painful injury on Ca^{2+} homeostasis by reducing Ca^{2+} current (I_{Ca}) via high-voltage-activated Ca^{2+} channels and stimulating I_{Ca} through low-voltage-activated Ca^{2+} channels in dorsal root ganglion neurons. This leads to the PMCA activation, SERCA depression, increase of store-operated Ca^{2+} influx, and Ca^{2+} signaling disruption through TSP4-dependent stimulation of the voltage-gated Ca^{2+} channel $\alpha_2\delta_1$ subunit ($Ca_v\alpha_2\delta_1$) and PKC-mediated signaling [195]. It would be interesting to examine whether TPSs contribute to the regulation of cardiomyocyte-specific Ca^{2+} handling and activity of SERCA and CaKMII that are primarily involved in the myocardial contractility and electric conductivity.

11. Therapeutic Potential of Thrombospondins

The multidomain structural architecture of TSP molecules defines their diverse functions and pleiotropic actions. TSPs have a capability to bind a variety of proteins such as cytokines, growth factors, receptors, and proteases that emphasizes their function in a tissue- and cell type-specific manner [23]. So far, the antiangiogenic activity of TSP1 and TSP2 in cancers inflamed an interest to use these molecules for anti-cancer therapy [196]. Peptides mimicking the anti-angiogenic domains of TSPs and recombinant proteins were developed [197,198].

As known, in TSP1 and TSP2, the anti-angiogenic function is related to the properdin (type-1) repeats located at the N-terminal stalk region [22]. Small peptides derived from this region exhibited only weak inhibitory effects on angiogenesis. However, a single D-amino acid substitution (D-isoleucine) of a particular properdin-region heptapeptide was found to strengthen the anti-angiogenic activity by 1000-fold [199]. Finally, a potent anti-angiogenic TSP1 mimetic nonapeptide analog of this substituted heptapeptide named ABT-510 was constructed [200]. In preclinical studies, ABT-510 showed an ability to efficiently inhibit VEGF-induced migration of microvascular ECs and exhibited anti-angiogenic and anti-tumor activity in several mouse and human xenograft models [201]. In humans, ABT-510 alone or in combination with cytotoxic agents was tested in several Phase I clinical trials to treat a variety of advanced solid and soft cancers [202–207]. Overall, ABT-510 administration was safe and well-tolerated, with modest adverse effects with injection-site reactions and fatigue as the most frequent. The anti-tumor effect of ABT-510 varied depending on the cancer type. In principal, treatment with ABT-510 alone had a modest efficiency. However, combinational therapy with cytotoxic agents improved the efficiency of anti-cancer therapy. In Phase II trials [208–210], ABT-510 monotherapy led to stabilization of tumor growth and inhibition of tumor expression of proangiogenic factors. In overall, treatment with ABT-510 alone showed a moderate efficiency in Phase II clinical studies by providing only a modest prolongation of overall survival of patients. Thus, based on the results of Phase II trials, clinical application of ABT-510 in combination with other anti-cancer agents was recommended.

In the context of cardiovascular therapy, anti-angiogenic approaches tested in tumors may be helpful in graft atherosclerosis. Mapping of salutary and deleterious effects to different TSP domains will provide an option to construct other TSP-targeting agents for widespread cardiovascular pathology such as MI and heart failure. In heart hypertrophy, atherosclerosis, heart failure, and MI, the down-regulation of the TSP/CD47 axis to enhance angiogenesis and restore NO-dependent signaling would be beneficial [121]. CD47 blockade with a monoclonal antibody was preclinically tested in animal models of ischemia resulted in improvement of angiogenesis and great increases in tissue survival [109,211].

So far, assessment of therapeutic effects CD47 blockade with a monoclonal antibody undergoes transition from the preclinical phase to clinical evaluation. The main purpose of these Phase I clinical trials is to check biosafety/tolerability of a CD47 antibody CC-90002 and find an optimal dose for treatment of advanced hematological neoplasms in combination with Rituximab, an anti-CD20 monoclonal antibody (trial NCT02367196) or for monotherapy of acute myeloid leukemia and high-risk myelodysplastic syndrome (trial NCT02641002). Another humanized anti-CD47 monoclonal antibody, Hu5F9-G4, will be clinically tested alone for treatment of recurrent/refractory acute myeloid leukemia (trial NCT02678338) and advanced solid malignancy or lymphoma (trial NCT02216409) [212]. At present, patients are enrolled for these clinical studies. Expected beneficial effects of this immunotherapy involve the inhibition of tumor angiogenesis and invasion, decrease of tumor-induced macrophage apoptosis and functional impairment, and depletion of CD47-expressing cancer stem cells that are key contributors to tumor relapse and chemoresistance [213]. In a case of the evident tolerability to antibodies, these trials will proceed to the Phase II.

In order to target a profibrotic activity of TSP1 through the activation of TGF-β, the activation sequence (LKSL) in the TSP1 molecule essential for the interaction with the latency-associated peptide (LAP) was mapped [214] and an LKSL peptide was developed [215]. The peptide antagonizes TSH1 binding to LAP and inhibits TGF-β liberation from the latent complex with LAT. The anti-fibrotic activity of the LKSL peptide was demonstrated in various animal models including experimental models of liver fibrosis [216], unilateral ureteral obstruction [217], diabetic nephropathy [218], and post-hemorrhagic hydrocephalus [219]. Regarding cardiovascular pathology, cardioprotective effects of LKSL peptide-mediated inhibition of TGF-β activation were observed in TAC-induced cardiomyopathy in type 1 diabetic rats. Diabetic rats treated with the LKSL peptide did not develop cardiac fibrosis and had improved heart function [52]. However, in a recent study, detrimental effects of implication of this peptide to treat angiotensin II-induced abdominal aortic aneurysm in ApoE-deficient mice were observed [220]. LKSL-dependent suppression of TGF-β activation further aggravated abdominal aortic aneurysm associated with increase of aortic diameter, adverse atherosclerosis within the aortic arch, and aortic elastin fragmentation due to down-regulation of the TGF-β-target gene lysyl oxidase-like 1 (LOXL1), an enzyme involved in cross-linking of elastin [221]. These data suggest for protective role of TGF-β against aortic aneurism. Inhibition of TGF-β signaling may therefore have deleterious consequences by impairing vascular ECM repair and promoting aortic infiltration of inflammatory cells. However, in overall, LKSL-mediated suppression of TSH1-dependent TGF-β activation showed beneficial results in inhibition of tissue fibrosis and should be further explored to prevent or diminish the advanced profibrotic response in hypertrophic hearts or in post-MI cardiac repair.

Since the regenerative potential of human heart is limited, MI-induced loss of cardiomyocytes can result in heart failure and death. Stem cell therapy has emerged as a promising strategy for healing cardiac injury, directly or indirectly, and seems to offer functional benefits to patients. Cardiac stem cell therapy involves using of hematopoietic, mesenchymal, and cardiac stem cells for regenerative purposes. However, a common challenge is to increase the retention and survival of engrafted cells at the injured site in order to strengthen their chances for proliferation and differentiation to functional cardiomyocytes [222]. To enhance the regenerative and prosurvival capacity, stem cells are subjected to ischemic/pharmacological preconditioning before transplantation. For example, regenerative properties of CD34+ hematopoietic progenitor cells from diabetic patients with atherosclerosis are frequently reduced and impaired. Treatment of CD34+ progenitors with TSP1-derived peptide RFYVVMWK promotes expression of TSP-1, integrins, and P-selectin that in turn increases adhesion capability and retention of the autologous progenitor cell engraft although do not affect apoptosis and viability [223]. Hypoxic exposure of adipose tissue-derived mesenchymal cells from aged mice improve their functionality by decreasing expression of anti-angiogenic, prothrombotic, and profibrotic molecules such as TSP-1, plasminogen activator inhibitor-1 (PAI-1), and TGF-β [224]. Preconditioning of mesenchymal stem cells with oxytocin significantly stimulates their therapeutic potential, angiogenic properties, and resistance to hypoxia and apoptosis through induction of various prosurvival and anti-apoptotic factors including TSP-1 [225].

12. Conclusions

Effects of TSPs on the mechanisms of cardiac remodeling are represented in Figure 2. Table 1 recapitulates TSP-dependent actions in the cardiovascular system. In summary, TSP1, TSP2, and TSP4 possess protective properties against heart hypertrophy since their deletion in animal models of pressure overload results in aberrant remodeling [49,50,59,144,147]. Lack of TSP1 and TSP2 activates MMP production and causes dilation of the left ventricle whereas altered TSP1-dependent TGF-β activation disturbs conversion of fibroblasts to myofibroblasts and down-regulates cardiac matrix synthesis [144]. In a murine pressure overload model, loss of TSP4 resulted in increased cardiac mass and fibrosis [50]. TSP4 also protects from abnormal heart remodeling through induction of stretch-mediated enhancement of myocardial contraction in pressure overload and prevention of the ER stress in cardiomyocytes [49,63]. On the basis of the role in heart hypertrophy, TSP1 and TSP2 would be suggested to influence post-MI remodeling but their effects are needed to be evaluated.

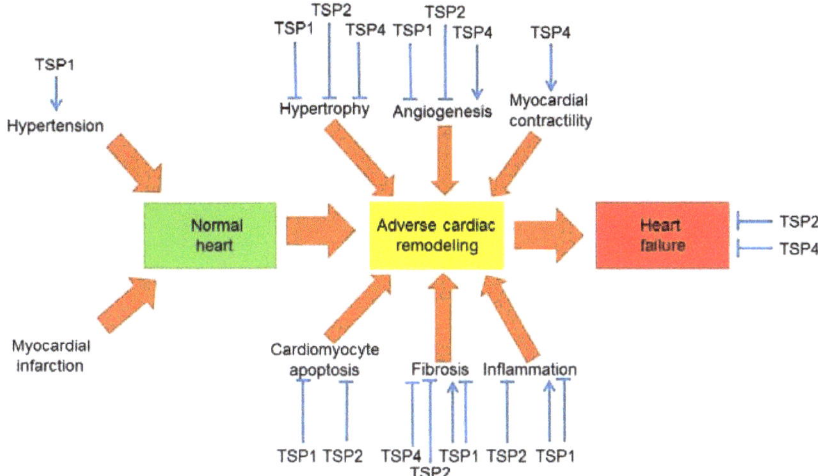

Figure 2. The role of thrombospondins (TSPs) in cardiovascular pathology. TSP1, TSP2, and TSP4 are the best preclinically studied TSPs in experimental models of cardiovascular pathology involving knockout or overexpression of these TSPs. Sharp arrows define stimulatory effects. Other type of arrows defines inhibitory effects.

Table 1. The role of thrombospondins in cardiovascular physiology and pathology

Characteristics	TSP1	TSP2	TSP3	TSP4	TSP5
Expression in the vascular wall	Yes	Yes	Yes	Yes	Yes
Expression in the atherosclerotic plaque	Yes	Yes	Yes	Yes	Yes
Cardiac expression	Yes	Yes	No?	Yes	Unknown
Angiogenesis in the myocardium	Inhibition	Inhibition	Unknown	Activation	Unknown
Up-regulated expression in cardiac remodeling	Yes	Yes	Yes	Yes	Yes
Inhibition of MMP-2/3/9	Yes	Yes	No	No	No
Cardiac fibrosis	Activation/ Inhibition	Inhibition	Unknown	Inhibition	Unknown
VSMC proliferation/ hyperplasia	Activation	Activation	Unknown	No effect	Inhibition
Blood pressure	Vasoconstriction	Vasoconstriction	Unknown	Unknown	Unknown
Inflammation	Activation/ Inhibition	Inhibition	Unknown	Activation (moderate)	Unknown
Effects on macrophages	Stimulation of phagocytosis Foam cell formation	Unknown	Unknown	Recruitment to the plaque	Unknown
Plaque progression	Activation	Unknown	Unknown	Activation	Unknown
Oxidative stress	Activation	Unknown	Unknown	Unknown	Unknown
Cardiomyocyte apoptosis	Inhibition	Inhibition	Unknown	Unknown	Unknown
Cardiac contractility	No effect	Unknown	Unknown	Activation	Unknown
Cardiac hypertrophy	Inhibition	Inhibition	Unknown	Inhibition	Unknown
Heart failure	Inhibition?	Inhibition	Unknown	Inhibition	Unknown

Abbreviations: MMP, matrix metalloproteinase; TSP, thrombospondin; VSMC, vascular smooth muscle cell.

In failing hearts, TSP2 and TSP4 protect cardiac ECM from adverse remodeling. In TSP-deficient ageing heart, systolic function is altered while fibrinogenesis and cardiac dilatation are activated [54]. Also, an advanced cardiac muscle cell death, inflammation, MMP-2 activation and aberrant collagen cross-linking was observed [54]. TSP2 possess cardioprotective properties against heart failure in viral myocarditis by repressing inflammation, fibrotic response, and cardiomyocyte death [157]. In doxorubicin-induced cardiomyopathy, the anti-hypertrophic activity of TSP2 is related to the inhibition of apoptosis of cardiac muscle cells and preserving ECM from the damage [51]. As in a case of TSP2, TSP4 deletion results in fibrosis, altered diastolic function, and depressed systolic function denoting the involvement of TSP4 in the regulation of ECM composition [50]. Finally, TSP4 appears to possess the proatherosclerotic activity since ApoE-deficient mice lacking TSP4 had reduced macrophage accumulation in the plaques due to reduced proinflammatory activation of ECs and decreased recruitment of inflammatory leukocytes to the endothelium [31].

So far, the most comprehensive functional assessment of the physiological and pathological roles was done only for TSP1, a founder member of the TSP family. Less complete results were obtained for TSP2 and TSP4. From the scientific literature, there is a profound data deficit about the function of TSP3 and TSP5. Although TSP3 is expressed by VSMCs at significant levels, its expression in the homeostatic myocardium seems to be absent [42]. However, stimulation of cardiac fibroblasts with a peptide matricryptin generated by a limited proteolysis of collagen Iα1 by MMP-2/9 results in the induction of TSP3 expression. Along with TSP3, matricryptin induces production of many ECM structural and regulatory proteins that contribute to post-MI cardiac repair and promote scar formation and angiogenesis [226]. This effect of matricryptin has a therapeutic promise and therefore should be further evaluated. Since TSP1 and TSP2 inhibit proteolytic activation of MMP-2 and MMP-9 [227], these TSPs can potentially suppress cardiac expression of TSP3. Therefore, assessing reciprocal TSP regulation would be intriguing. Thus, future prospects in the thrombospondin-related research may ultimately concern investigation of TSP3 and TSP5 functions. Further, TSPs have a variety of binding partners and the number of TSP ligands is growing. However, regulatory mechanisms of binding these ligands are widely unknown. Actually, ligand binding is spatially and temporally regulated, and it would be of great interest to reveal these regulatory patterns and recognize their functional significance.

Overall, studies involving knockout mice indicate that deficiency of TSP1, TSP2, and TSP4 appears to be deleterious in cardiovascular pathology. Therefore, enhancing of cardiac TSP-dependent signaling in stressful settings such as pressure overload may be profitable. Precise analysis of the relationship between the TSP structure and function, identification of new receptors, and functional mapping of various domains will be useful for the development of novel drugs to target TSPs and promote the gain in TSP-dependent signaling pathways.

Acknowledgments: This work was supported by Russian Science Foundation (Grant # 14–15–00112).

Author Contributions: Dimitry A. Chistiakov, Alexandra A. Melnichenko, and Veronika A. Myasoedova drafted the paper. Andrey V. Grechko and Alexander N. Orekhov corrected the paper. Alexander N. Orekhov finally approved the paper. All authors read the final version and approved submission.

Conflicts of Interest: All authors state that they have no conflict of interests.

References

1. Chistiakov, D.A.; Sobenin, I.A.; Orekhov, A.N. Vascular extracellular matrix in atherosclerosis. *Cardiol. Rev.* **2013**, *21*, 270–288. [CrossRef] [PubMed]

2. Lewinsohn, A.D.; Anssari-Benham, A.; Lee, D.A.; Taylor, P.M.; Chester, A.H.; Yacoub, M.H.; Screen, H.R. Anisotropic strain transfer through the aortic valve and its relevance to the cellular mechanical environment. *Proc. Inst. Mech. Eng. H* **2011**, *225*, 821–830. [CrossRef] [PubMed]

3. Okech, W.; Abberton, K.M.; Kuebel, J.M.; Hocking, D.C.; Sarelius, I.H. Extracellular matrix fibronectin mediates an endothelial cell response to shear stress via the heparin-binding, matricryptic RWRPK sequence of FNIII1H. *Am. J. Physiol. Heart Circ. Physiol.* **2016**, *311*, H1063–H1071. [CrossRef] [PubMed]

4. Casals, G.; Fernández-Varo, G.; Melgar-Lesmes, P.; Marfà, S.; Reichenbach, V.; Morales-Ruiz, M.; Jiménez, W. Factors involved in extracellular matrix turnover in human derived cardiomyocytes. *Cell Physiol. Biochem.* **2013**, *32*, 1125–1136. [CrossRef] [PubMed]

5. Luu, N.T.; Glen, K.E.; Egginton, S.; Rainger, G.E.; Nash, G.B. Integrin-substrate interactions underlying shear-induced inhibition of the inflammatory response of endothelial cells. *Thromb. Haemost.* **2013**, *109*, 298–308. [CrossRef] [PubMed]

6. Von Bary, C.; Makowski, M.; Preissel, A.; Keithahn, A.; Warley, A.; Spuentrup, E.; Buecker, A.; Lazewatsky, J.; Cesati, R.; Onthank, D.; et al. MRI of coronary wall remodeling in a swine model of coronary injury using an elastin-binding contrast agent. *Circ. Cardiovasc. Imaging* **2011**, *4*, 147–155. [CrossRef] [PubMed]

7. Li, J.; Philip, J.L.; Xu, X.; Theccanat, T.; Abdur Razzaque, M.; Akhter, S.A. β-Arrestins regulate human cardiac fibroblast transformation and collagen synthesis in adverse ventricular remodeling. *J. Mol. Cell Cardiol.* **2014**, *76*, 73–83. [CrossRef] [PubMed]

8. Mujumdar, V.S.; Tyagi, S.C. Temporal regulation of extracellular matrix components in transition from compensatory hypertrophy to decompensatory heart failure. *J. Hypertens.* **1999**, *17*, 261–270. [CrossRef] [PubMed]

9. Sakamuri, S.S.; Takawale, A.; Basu, R.; Fedak, P.W.; Freed, D.; Sergi, C.; Oudit, G.Y.; Kassiri, Z. Differential impact of mechanical unloading on structural and nonstructural components of the extracellular matrix in advanced human heart failure. *Transl. Res.* **2016**, *172*, 30–44. [CrossRef] [PubMed]

10. Bornstein, P. Thrombospondins as matricellular modulators of cell function. *J. Clin. Investig.* **2001**, *107*, 929–934. [CrossRef] [PubMed]

11. Kurihara, K.; Sato, I. Distribution of tenascin-C and -X, and soft X-ray analysis of the mandibular symphysis during mandible formation in the human fetus. *Okajimas Folia Anat. Jpn.* **2004**, *81*, 49–55. [CrossRef] [PubMed]

12. Huang, W.; Zhang, L.; Niu, R.; Liao, H. Tenascin-R distinct domains modulate migration of neural stem/progenitor cells in vitro. *In Vitro Cell Dev. Biol. Anim.* **2009**, *45*, 10–14. [CrossRef] [PubMed]

13. Gillan, L.; Matei, D.; Fishman, D.A.; Gerbin, C.S.; Karlan, B.Y.; Chang, D.D. Periostin secreted by epithelial ovarian carcinoma is a ligand for $\alpha_V\beta_3$ and alphavbeta5 integrins and promotes cell motility. *Cancer Res.* **2002**, *62*, 5358–5364. [PubMed]

14. Hoersch, S.; Andrade-Navarro, M.A. Periostin shows increased evolutionary plasticity in its alternatively spliced region. *BMC Evol. Biol.* **2010**, *10*, 30. [CrossRef] [PubMed]

15. Hakuno, D.; Kimura, N.; Yoshioka, M.; Mukai, M.; Kimura, T.; Okada, Y.; Yozu, R.; Shukunami, C.; Hiraki, Y.; Kudo, A.; et al. Periostin advances atherosclerotic and rheumatic cardiac valve degeneration by inducing angiogenesis and MMP production in humans and rodents. *J. Clin. Investig.* **2010**, *120*, 2292–2306. [CrossRef] [PubMed]

16. Steitz, S.A.; Speer, M.Y.; McKee, M.D.; Liaw, L.; Almeida, M.; Yang, H.; Giachelli, C.M. Osteopontin inhibits mineral deposition and promotes regression of ectopic calcification. *Am. J. Pathol.* **2002**, *161*, 2035–2046. [CrossRef]

17. McKee, M.D.; Addison, W.N.; Kaartinen, M.T. Hierarchies of extracellular matrix and mineral organization in bone of the craniofacial complex and skeleton. *Cell. Tissues Organs* **2005**, *181*, 176–188. [CrossRef] [PubMed]

18. Wang, K.X.; Denhardt, D.T. Osteopontin: Role in immune regulation and stress responses. *Cytokine Growth Factor Rev.* **2008**, *19*, 333–345. [CrossRef] [PubMed]

19. Midura, R.J.; Midura, S.B.; Su, X.; Gorski, J.P. Separation of newly formed bone from older compact bone reveals clear compositional differences in bone matrix. *Bone* **2011**, *49*, 1365–1374. [CrossRef] [PubMed]

20. Chen, C.C.; Lau, L.F. Functions and mechanisms of action of CCN matricellular proteins. *Int. J. Biochem. Cell Biol.* **2009**, *41*, 771–783. [CrossRef] [PubMed]

21. Dobaczewski, M.; Gonzalez-Quesada, C.; Frangogiannis, N.G. The extracellular matrix as a modulator of the inflammatory and reparative response following myocardial infarction. *J. Mol. Cell. Cardiol.* **2010**, *48*, 504–511. [CrossRef] [PubMed]

22. Carlson, C.B.; Lawler, J.; Mosher, D.F. Structures of thrombospondins. *Cell Mol. Life Sci.* **2008**, *65*, 672–686. [CrossRef] [PubMed]

23. Adams, J.C.; Lawler, J. The thrombospondins. *Cold Spring Harb. Perspect. Biol.* **2011**, *3*, a009712. [CrossRef] [PubMed]

24. McKenzie, P.; Chadalavada, C.; Bohrer, J.; Adams, J.C. Phylogenomic analysis of vertebrate thrombospondins, reveals fish-specific paralogues, ancestral gene relationships and a tetrapod innovation. *BMC Evol. Biol.* **2006**, *6*, 33. [CrossRef] [PubMed]

25. Bentley, A.A.; Adams, J.C. The evolution of thrombospondins and their ligand-binding activities. *Mol. Biol. Evol.* **2010**, *27*, 2187–2197. [CrossRef] [PubMed]

26. Murphy-Ullrich, J.E.; Iozzo, R.V. Thrombospondins in physiology and disease: New tricks for old dogs. *Matrix Biol.* **2012**, *31*, 152–154. [CrossRef] [PubMed]

27. Rath, G.M.; Schneider, C.; Dedieu, S.; Rothhut, B.; Soula-Rothhut, M.; Ghoneim, C.; Sid, B.; Morjani, H.; El Btaouri, H.; Martiny, L. The C-terminal CD47/IAP-binding domain of thrombospondin-1 prevents camptothecin- and doxorubicin-induced apoptosis in human thyroid carcinoma cells. *Biochim. Biophys. Acta* **2006**, *1763*, 1125–1134. [CrossRef] [PubMed]

28. Bornstein, P.; Armstrong, L.C.; Hankenson, K.D.; Kyriakides, T.R.; Yang, Z. Thrombospondin 2, a matricellular protein with diverse functions. *Matrix Biol.* **2000**, *19*, 557–568. [CrossRef]

29. Kyriakides, T.R.; Bornstein, P. Matricellular proteins as modulators of wound healing and the foreign body response. *Thromb. Haemost.* **2003**, *90*, 986–992. [CrossRef] [PubMed]

30. Schellings, M.W.; Van Almen, G.C.; Sage, E.H.; Heymans, S. Thrombospondins in the heart: Potential functions in cardiac remodeling. *J. Cell Commun. Signal* **2009**, *3*, 201–213. [CrossRef] [PubMed]

31. Frolova, E.G.; Pluskota, E.; Krukovets, I.; Burke, T.; Drumm, C.; Smith, J.D.; Blech, L.; Febbraio, M.; Bornstein, P.; Plow, E.F.; et al. Thrombospondin-4 regulates vascular inflammation and atherogenesis. *Circ. Res.* **2010**, *107*, 1313–1325. [CrossRef] [PubMed]

32. Mustonen, E.; Ruskoaho, H.; Rysä, J. Thrombospondins, potential drug targets for cardiovascular diseases. *Basic Clin. Pharmacol. Toxicol.* **2013**, *112*, 4–12. [CrossRef] [PubMed]

33. Muppala, S.; Xiao, R.; Krukovets, I.; Verbovetsky, D.; Yendamuri, R.; Habib, N.; Raman, P.; Plow, E.; Stenina-Adognravi, O. Thrombospondin-4 mediates TGF-β-induced angiogenesis. *Oncogene* **2017** (in press). [CrossRef] [PubMed]

34. Topol, E.J.; McCarthy, J.; Gabriel, S.; Moliterno, D.J.; Rogers, W.J.; Newby, L.K.; Freedman, M.; Metivier, J.; Cannata, R.; O'Donnell, C.J.; et al. Single nucleotide polymorphisms in multiple novel thrombospondin genes may be associated with familial premature myocardial infarction. *Circulation* **2001**, *104*, 2641–2644. [CrossRef] [PubMed]

35. Cui, J.; Randell, E.; Renouf, J.; Sun, G.; Han, FY.; Younghusband, B.; Xie, Y.G. Gender dependent association of thrombospondin-4 A387P polymorphism with myocardial infarction. *Arterioscler. Thromb. Vasc. Biol.* **2004**, *24*, e183–e1844. [CrossRef] [PubMed]

36. Cui, J.; Randell, E.; Renouf, J.; Sun, G.; Green, R.; Han, F.Y.; Xie, Y.G. Thrombospondin-4 1186G>C (A387P) is a sex-dependent risk factor for myocardial infarction: A large replication study with increased sample size from the same population. *Am. Heart J.* **2006**, *152*, 543.e1–543.e5. [CrossRef] [PubMed]

37. Wessel, J.; Topol, E.J.; Ji, M.; Meyer, J.; McCarthy, J.J. Replication of the association between the thrombospondin-4 A387P polymorphism and myocardial infarction. *Am. Heart J.* **2004**, *147*, 905–909. [CrossRef] [PubMed]

38. Zwicker, J.I.; Peyvandi, F.; Palla, R.; Lombardi, R.; Canciani, M.T.; Cairo, A.; Ardissino, D.; Bernardinelli, L.; Bauer, K.A.; Lawler, J.; et al. The thrombospondin-1 N700S polymorphism is associated with early myocardial infarction without altering von Willebrand factor multimer size. *Blood* **2006**, *108*, 1280–1283. [CrossRef] [PubMed]

39. Corsetti, J.P.; Ryan, D.; Moss, A.J.; McCarthy, J.; Goldenberg, I.; Zareba, W.; Sparks, C.E. Thrombospondin-4 polymorphism (A387P) predicts cardiovascular risk in postinfarction patients with high HDL cholesterol and C-reactive protein levels. *Thromb. Haemost.* **2011**, *106*, 1170–1178. [CrossRef] [PubMed]

40. Stenina, O.I.; Ustinov, V.; Krukovets, I.; Marinic, T.; Topol, E.J.; Plow, E.F. Polymorphisms A387P in thrombospondin-4 and N700S in thrombospondin-1 perturb calcium binding sites. *FASEB J.* **2005**, *19*, 1893–1895. [CrossRef] [PubMed]

41. Zhang, X.J.; Wei, C.Y.; Li, W.B.; Zhang, L.L.; Zhou, Y.; Wang, Z.H.; Tang, M.X.; Zhang, W.; Zhang, Y.; Zhong, M. Association between single nucleotide polymorphisms in thrombospondins genes and coronary artery disease: A meta-analysis. *Thromb. Res.* **2015**, *136*, 45–51. [CrossRef] [PubMed]

42. Adolph, K.W. Relative abundance of thrombospondin 2 and thrombospondin 3 mRNAs in human tissues. *Biochem. Biophys. Res. Commun.* **1999**, *258*, 792–796. [CrossRef] [PubMed]

43. Zhao, X.M.; Hu, Y.; Miller, G.G.; Mitchell, R.N.; Libby, P. Association of thrombospondin-1 and cardiac allograft vasculopathy in human cardiac allografts. *Circulation* **2001**, *103*, 525–531. [CrossRef] [PubMed]
44. Sezaki, S.; Hirohata, S.; Iwabu, A.; Nakamura, K.; Toeda, K.; Miyoshi, T.; Yamawaki, H.; Demircan, K.; Kusachi, S.; Shiratori, Y.; et al. Thrombospondin-1 is induced in rat myocardial infarction and its induction is accelerated by ischemia/reperfusion. *Exp. Biol. Med.* **2005**, *230*, 621–630.
45. Mustonen, E.; Aro, J.; Puhakka, J.; Ilves, M.; Soini, Y.; Leskinen, H.; Ruskoaho, H.; Rysä, J. Thrombospondin-4 expression is rapidly upregulated by cardiac overload. *Biochem. Biophys. Res. Commun.* **2008**, *373*, 186–191. [CrossRef] [PubMed]
46. Stenina-Adognravi, O. Thrombospondins: Old players, new games. *Curr. Opin. Lipidol.* **2013**, *24*, 401–419. [CrossRef] [PubMed]
47. Reinecke, H.; Robey, T.E.; Mignone, J.L.; Muskheli, V.; Bornstein, P.; Murry, C.E. Lack of thrombospondin-2 reduces fibrosis and increases vascularity around cardiac cell grafts. *Cardiovasc. Pathol.* **2013**, *22*, 91–95. [CrossRef] [PubMed]
48. Vanhoutte, D.; van Almen, G.C.; Van Aelst, L.N.; Van Cleemput, J.; Droogné, W.; Jin, Y.; Van de Werf, F.; Carmeliet, P.; Vanhaecke, J.; Papageorgiou, A.P.; et al. Matricellular proteins and matrix metalloproteinases mark the inflammatory and fibrotic response in human cardiac allograft rejection. *Eur. Heart J.* **2013**, *34*, 1930–1941. [CrossRef] [PubMed]
49. Cingolani, O.H.; Kirk, J.A.; Seo, K.; Koitabashi, N.; Lee, D.I.; Ramirez-Correa, G.; Bedja, D.; Barth, A.S.; Moens, A.L.; Kass, D.A. Thrombospondin-4 is required for stretch-mediated contractility augmentation in cardiac muscle. *Circ. Res.* **2011**, *109*, 1410–1414. [CrossRef] [PubMed]
50. Frolova, E.G.; Sopko, N.; Blech, L.; Popovic, Z.B.; Li, J.; Vasanji, A.; Drumm, C.; Krukovets, I.; Jain, M.K.; Penn, M.S.; Plow, E.F.; et al. Thrombospondin-4 regulates fibrosis and remodeling of the myocardium in response to pressure overload. *FASEB J.* **2012**, *26*, 2363–2373. [CrossRef] [PubMed]
51. Van Almen, G.C.; Swinnen, M.; Carai, P.; Verhesen, W.; Cleutjens, J.P.; D'hooge, J.; Verheyen, F.K.; Pinto, Y.M.; Schroen, B.; Carmeliet, P.; et al. Absence of thrombospondin-2 increases cardiomyocyte damage and matrix disruption in doxorubicin-induced cardiomyopathy. *J. Mol. Cell Cardiol.* **2011**, *51*, 318–328. [CrossRef] [PubMed]
52. Belmadani, S.; Bernal, J.; Wei, C.C.; Pallero, M.A.; Dell'italia, L.; Murphy-Ullrich, J.E.; Berecek, K.H. A thrombospondin-1 antagonist of transforming growth factor-β activation blocks cardiomyopathy in rats with diabetes and elevated angiotensin II. *Am. J. Pathol.* **2007**, *171*, 777–789. [CrossRef] [PubMed]
53. Gonzalez-Quesada, C.; Cavalera, M.; Biernacka, A.; Kong, P.; Lee, D.W.; Saxena, A.; Frunza, O.; Dobaczewski, M.; Shinde, A.; Frangogiannis, N.G. Thrombospondin-1 induction in the diabetic myocardium stabilizes the cardiac matrix in addition to promoting vascular rarefaction through angiopoietin-2 upregulation. *Circ. Res.* **2013**, *113*, 1331–1344. [CrossRef] [PubMed]
54. Swinnen, M.; Vanhoutte, D.; Van Almen, G.C.; Hamdani, N.; Schellings, M.W.; D'hooge, J.; Van der Velden, J.; Weaver, M.S.; Sage, E.H.; Bornstein, P.; et al. Absence of thrombospondin-2 causes age-related dilated cardiomyopathy. *Circulation* **2009**, *120*, 1585–1597. [CrossRef] [PubMed]
55. Frangogiannis, N.G.; Ren, G.; Dewald, O.; Zymek, P.; Haudek, S.; Koerting, A.; Winkelmann, K.; Michael, L.H.; Lawler, J.; Entman, M.L. Critical role of endogenous thrombospondin-1 in preventing expansion of healing myocardial infarcts. *Circulation* **2005**, *111*, 2935–2942. [CrossRef] [PubMed]
56. Chatila, K.; Ren, G.; Xia, Y.; Huebener, P.; Bujak, M.; Frangogiannis, N.G. The role of the thrombospondins in healing myocardial infarcts. *Cardiovasc. Hematol. Agents Med. Chem.* **2007**, *5*, 21–27. [CrossRef] [PubMed]
57. Cai, H.; Yuan, Z.; Fei, Q.; Zhao, J. Investigation of thrombospondin-1 and transforming growth factor-β expression in the heart of aging mice. *Exp. Ther. Med.* **2012**, *3*, 433–436. [PubMed]
58. Sun, H.; Zhao, Y.; Bi, X.; Li, S.; Su, G.; Miao, Y.; Ma, X.; Zhang, Y.; Zhang, W.; Zhong, M. Valsartan blocks thrombospondin/transforming growth factor/Smads to inhibit aortic remodeling in diabetic rats. *Diagn. Pathol.* **2015**, *10*, 18. [CrossRef] [PubMed]
59. Schroen, B.; Heymans, S.; Sharma, U.; Blankesteijn, W.M.; Pokharel, S.; Cleutjens, J.P.; Porter, J.G.; Evelo, C.T.; Duisters, R.; van Leeuwen, R.E.; Janssen, B.J.; et al. Thrombospondin-2 is essential for myocardial matrix integrity: Increased expression identifies failure-prone cardiac hypertrophy. *Circ. Res.* **2004**, *95*, 515–522. [CrossRef] [PubMed]

60. Orr, A.W.; Elzie, C.A.; Kucik, D.F.; Murphy-Ullrich, J.E. Thrombospondin signaling through the calreticulin/LDL receptor-related protein co-complex stimulates random and directed cell migration. *J. Cell Sci.* **2003**, *116*, 2917–2927. [CrossRef] [PubMed]

61. Pallero, M.A.; Elzie, C.A.; Chen, J.; Mosher, D.F.; Murphy-Ullrich, J.E. Thrombospondin 1 binding to calreticulin-LRP1 signals resistance to anoikis. *FASEB J.* **2008**, *22*, 3968–3979. [CrossRef] [PubMed]

62. Yan, Q.; Murphy-Ullrich, J.E.; Song, Y. Structural insight into the role of thrombospondin-1 binding to calreticulin in calreticulin-induced focal adhesion disassembly. *Biochemistry* **2010**, *49*, 3685–3694. [CrossRef] [PubMed]

63. Lynch, J.M.; Maillet, M.; Vanhoutte, D.; Schloemer, A.; Sargent, M.A.; Blair, N.S.; Lynch, K.A.; Okada, T.; Aronow, B.J.; Osinska, H.; et al. A thrombospondin-dependent pathway for a protective ER stress response. *Cell* **2012**, *149*, 1257–1268. [CrossRef] [PubMed]

64. Frolova, E.G.; Drazba, J.; Krukovets, I.; Kostenko, V.; Blech, L.; Harry, C.; Vasanji, A.; Drumm, C.; Sul, P.; Jenniskens, G.J.; et al. Control of organization and function of muscle and tendon by thrombospondin-4. *Matrix Biol.* **2014**, *37*, 35–48. [CrossRef] [PubMed]

65. Pohjolainen, V.; Mustonen, E.; Taskinen, P.; Näpänkangas, J.; Leskinen, H.; Ohukainen, P.; Peltonen, T.; Aro, J.; Juvonen, T.; Satta, J.; et al. Increased thrombospondin-2 in human fibrosclerotic and stenotic aortic valves. *Atherosclerosis* **2012**, *220*, 66–71. [CrossRef] [PubMed]

66. Streit, M.; Riccardi, L.; Velasco, P.; Brown, L.F.; Hawighorst, T.; Bornstein, P.; Detmar, M. Thrombospondin-2: A potent endogenous inhibitor of tumor growth and angiogenesis. *Proc. Natl. Acad. Sci. USA* **1999**, *96*, 14888–14893. [CrossRef] [PubMed]

67. Lawler, J. Thrombospondin-1 as an endogenous inhibitor of angiogenesis and tumor growth. *J. Cell. Mol. Med.* **2002**, *6*, 1–12. [CrossRef] [PubMed]

68. Lawler, J.; Detmar, M. Tumor progression: The effects of thrombospondin-1 and -2. *Int. J. Biochem. Cell Biol.* **2004**, *36*, 1038–1045. [CrossRef] [PubMed]

69. Zhang, X.; Lawler, J. Thrombospondin-based antiangiogenic therapy. *Microvasc. Res.* **2007**, *74*, 90–99. [CrossRef] [PubMed]

70. Almog, N.; Henke, V.; Flores, L.; Hlatky, L.; Kung, A.L.; Wright, R.D.; Berger, R.; Hutchinson, L.; Naumov, G.N.; Bender, E.; et al. Prolonged dormancy of human liposarcoma is associated with impaired tumor angiogenesis. *FASEB J.* **2006**, *20*, 947–949. [CrossRef] [PubMed]

71. Almog, N. Molecular mechanisms underlying tumor dormancy. *Cancer Lett.* **2010**, *294*, 139–146. [CrossRef] [PubMed]

72. Almog, N.; Ma, L.; Raychowdhury, R.; Schwager, C.; Erber, R.; Short, S.; Hlatky, L.; Vajkoczy, P.; Huber, P.E.; Folkman, J.; et al. Transcriptional switch of dormant tumors to fast-growing angiogenic phenotype. *Cancer Res.* **2009**, *69*, 836–844. [CrossRef] [PubMed]

73. Bhattacharyya, S.; Sul, K.; Krukovets, I.; Nestor, C.; Li, J.; Adognravi, O.S. Novel tissue-specific mechanism of regulation of angiogenesis and cancer growth in response to hyperglycemia. *J. Am. Heart Assoc.* **2012**, *1*, e005967. [CrossRef] [PubMed]

74. Krukovets, I.; Legerski, M.; Sul, P.; Stenina-Adognravi, O. Inhibition of hyperglycemia-induced angiogenesis and breast cancer tumor growth by systemic injection of microRNA-467 antagonist. *FASEB J.* **2015**, *29*, 3726–3736. [CrossRef] [PubMed]

75. Silverstein, R.L.; Febbraio, M. CD36-TSP-HRGP interactions in the regulation of angiogenesis. *Curr. Pharm. Des.* **2007**, *13*, 3559–3567. [CrossRef] [PubMed]

76. Frieda, S.; Pearce, A.; Wu, J.; Silverstein, R.L. Rec mbinant GST/CD36 fusion proteins define a thrombospondin binding domain. Evidence for a single calcium-dependent binding site on CD36. *J. Biol. Chem.* **1995**, *270*, 2981–2986. [PubMed]

77. Simantov, R.; Silverstein, R.L. CD36: A critical anti-angiogenic receptor. *Front. Biosci.* **2003**, *8*, s874–s882. [CrossRef] [PubMed]

78. Simantov, R.; Febbraio, M.; Silverstein, R.L. The antiangiogenic effect of thrombospondin-2 is mediated by CD36 and modulated by histidine-rich glycoprotein. *Matrix Biol.* **2005**, *24*, 27–34. [CrossRef] [PubMed]

79. Kaur, S.; Martin-Manso, G.; Pendrak, M.L.; Garfield, S.H.; Isenberg, J.S.; Roberts, D.D. Thrombospondin-1 inhibits VEGF receptor-2 signaling by disrupting its association with CD47. *J. Biol. Chem.* **2010**, *285*, 38923–38932. [CrossRef] [PubMed]

80. Bazzazi, H.; Isenberg, J.S.; Popel, A.S. Inhibition of VEGFR2 activation and its downstream signaling to ERK1/2 and calcium by thrombospondin-1 (TSP1): In silico investigation. *Front. Physiol* **2017**, in press.

81. Isenberg, J.S.; Ridnour, L.A.; Dimitry, J.; Frazier, W.A.; Wink, D.A.; Roberts, D.D. CD47 is necessary for inhibition of nitric oxide-stimulated vascular cell responses by thrombospondin-1. *J. Biol. Chem.* **2006**, *281*, 26069–26080. [CrossRef] [PubMed]

82. Isenberg, J.S.; Jia, Y.; Fukuyama, J.; Switzer, C.H.; Wink, D.A.; Roberts, D.D. Thrombospondin-1 inhibits nitric oxide signaling via CD36 by inhibiting myristic acid uptake. *J. Biol. Chem.* **2007**, *282*, 15404–15415. [CrossRef] [PubMed]

83. Isenberg, J.S.; Hyodo, F.; Matsumoto, K.; Romeo, M.J.; Abu-Asab, M.; Tsokos, M.; Kuppusamy, P.; Wink, D.A.; Krishna, M.C.; Roberts, D.D. Thrombospondin-1 limits ischemic tissue survival by inhibiting nitric oxide-mediated vascular smooth muscle relaxation. *Blood* **2007**, *109*, 1945–1952. [CrossRef] [PubMed]

84. Isenberg, J.S.; Annis, D.S.; Pendrak, M.L.; Ptaszynska, M.; Frazier, W.A.; Mosher, D.F.; Roberts, D.D. Differential interactions of thrombospondin-1, -2, and -4 with CD47 and effects on cGMP signaling and ischemic injury responses. *J. Biol. Chem.* **2009**, *284*, 1116–1125. [CrossRef] [PubMed]

85. Zhao, Y.; Brandish, P.E.; Ballou, D.P.; Marletta, M.A. A molecular basis for nitric oxide sensing by soluble guanylate cyclase. *Proc. Natl. Acad. Sci. USA* **1999**, *96*, 14753–14758. [CrossRef] [PubMed]

86. Murad, F.; Rapoport, R.M.; Fiscus, R. Role of cyclic-GMP in relaxations of vascular smooth muscle. *J. Cardiovasc. Pharmacol.* **1985**, *7*, S111–S118. [CrossRef] [PubMed]

87. Isenberg, J.S.; Romeo, M.J.; Yu, C.; Yu, C.K.; Nghiem, K.; Monsale, J.; Rick, M.E.; Wink, D.A.; Frazier, W.A.; Roberts, D.D. Thrombospondin-1 stimulates platelet aggregation by blocking the antithrombotic activity of nitric oxide/cGMP signaling. *Blood* **2008**, *111*, 613–623. [CrossRef] [PubMed]

88. Koom, Y.K.; Kim, J.M.; Kim, S.Y.; Koo, J.Y.; Oh, D.; Park, S.; Yun-Choi, H.S. Elevated plasma concentration of NO and cGMP may be responsible for the decreased platelet aggregation and platelet leukocyte conjugation in platelets hypo-responsive to catecholamines. *Platelets* **2009**, *20*, 555–565.

89. Miller, T.W.; Kaur, S.; Ivins-O'Keefe, K.; Roberts, D.D. Thrombospondin-1 is a CD47-dependent endogenous inhibitor of hydrogen sulfide signaling in T cell activation. *Matrix Biol.* **2013**, *32*, 316–324. [CrossRef] [PubMed]

90. Csányi, G.; Yao, M.; Rodríguez, A.I.; Al Ghouleh, I.; Sharifi-Sanjani, M.; Frazziano, G.; Huang, X.; Kelley, E.E.; Isenberg, J.S.; Pagano, P.J. Thrombospondin-1 regulates blood flow via CD47 receptor-mediated activation of NADPH oxidase 1. *Arterioscler. Thromb. Vasc. Biol.* **2012**, *32*, 2966–2973. [CrossRef] [PubMed]

91. Mayer, B.; Kleschyov, A.L.; Stessel, H.; Russwurm, M.; Münzel, T.; Koesling, D.; Schmidt, K. Inactivation of soluble guanylate cyclase by stoichiometric S-nitrosation. *Mol. Pharmacol.* **2009**, *75*, 886–891. [CrossRef] [PubMed]

92. Tsai, A.L.; Berka, V.; Sharina, I.; Martin, E. Dynamic ligand exchange in soluble guanylyl cyclase (sGC): Implications for sGC regulation and desensitization. *J. Biol. Chem.* **2011**, *286*, 43182–43192. [CrossRef] [PubMed]

93. Muppala, S.; Frolova, E.; Xiao, R.; Krukovets, I.; Yoon, S.; Hoppe, G.; Vasanji, A.; Plow, E.; Stenina-Adognravi, O. Proangiogenic properties of thrombospondin-4. *Arterioscler. Thromb. Vasc. Biol.* **2015**, *35*, 1975–1986. [CrossRef] [PubMed]

94. Stenina, O.I.; Krukovets, I.; Wang, K.; Zhou, Z.; Forudi, F.; Penn, M.S.; Topol, E.J.; Plow, E.F. Increased expression of thrombospondin-1 in vessel wall of diabetic Zucker rat. *Circulation* **2003**, *107*, 3209–3215. [CrossRef] [PubMed]

95. Chen, D.; Asahara, T.; Krasinski, K.; Witzenbichler, B.; Yang, J.; Magner, M.; Kearney, M.; Frazier, W.A.; Isner, J.M.; Andrés, V. Antibody blockade of thrombospondin accelerates reendothelialization and reduces neointima formation in balloon-injured rat carotid artery. *Circulation* **1999**, *100*, 849–854. [CrossRef] [PubMed]

96. Moura, R.; Tjwa, M.; Vandervoort, P.; Van Kerckhoven, S.; Holvoet, P.; Hoylaerts, M.F. Thrombospondin-1 deficiency accelerates atherosclerotic plaque maturation in ApoE$^{-/-}$ mice. *Circ. Res.* **2008**, *103*, 1181–1189. [CrossRef] [PubMed]

97. Stenina, O.I.; Plow, E.F. Counterbalancing forces: What is thrombospondin-1 doing in atherosclerotic lesions? *Circ. Res.* **2008**, *103*, 1053–1055. [CrossRef] [PubMed]

98. Siegel-Axel, D.I.; Runge, H.; Seipel, L.; Riessen, R. Effects of cerivastatin on human arterial smooth muscle cell growth and extracellular matrix expression at varying glucose and low-density lipoprotein levels. *J. Cardiovasc. Pharmacol.* **2003**, *41*, 422–433. [CrossRef] [PubMed]

99. Reed, M.J.; Iruela-Arispe, L.; O'Brien, E.R.; Truong, T.; LaBell, T.; Bornstein, P.; Sage, E.H. Expression of thrombospondins by endothelial cells. Injury is correlated with TSP-1. *Am. J. Pathol.* **1995**, *147*, 1068–1080. [PubMed]

100. Riessen, R.; Fenchel, M.; Chen, H.; Axel, D.I.; Karsch, K.R.; Lawler, J. Cartilage oligomeric matrix protein (thrombospondin-5) is expressed by human vascular smooth muscle cells. *Arterioscler. Thromb. Vasc. Biol.* **2001**, *21*, 47–54. [CrossRef] [PubMed]

101. Wang, L.; Wang, X.; Kong, W. ADAMTS-7, a novel proteolytic culprit in vascular remodeling. *Sheng Li Xue Bao* **2010**, *62*, 285–294. [PubMed]

102. Wang, L.; Zheng, J.; Bai, X.; Liu, B.; Liu, C.J.; Xu, Q.; Zhu, Y.; Wang, N.; Kong, W.; Wang, X. ADAMTS-7 mediates vascular smooth muscle cell migration and neointima formation in balloon-injured rat arteries. *Circ. Res.* **2009**, *104*, 688–698. [CrossRef] [PubMed]

103. Gao, A.G.; Lindberg, F.P.; Finn, M.B.; Blystone, S.D.; Brown, E.J.; Frazier, W.A. Integrin-associated protein is a receptor for the C-terminal domain of thrombospondin. *J. Biol. Chem.* **1996**, *271*, 21–24. [CrossRef] [PubMed]

104. Yao, M.; Roberts, D.D.; Isenberg, J.S. Thrombospondin-1 inhibition of vascular smooth muscle cell responses occurs via modulation of both cAMP and cGMP. *Pharmacol. Res.* **2011**, *63*, 13–22. [CrossRef] [PubMed]

105. Isenberg, J.S.; Frazier, W.A.; Krishna, M.C.; Wink, D.A.; Roberts, D.D. Enhancing cardiovascular dynamics by inhibition of thrombospondin-1/CD47 signaling. *Curr. Drug Targets* **2008**, *9*, 833–841. [CrossRef] [PubMed]

106. Isenberg, J.S.; Qin, Y.; Maxhimer, J.B.; Sipes, J.M.; Despres, D.; Schnermann, J.; Frazier, W.A.; Roberts, D.D. Thrombospondin-1 and CD47 regulate blood pressure and cardiac responses to vasoactive stress. *Matrix Biol.* **2009**, *28*, 110–119. [CrossRef] [PubMed]

107. Bauer, E.M.; Qin, Y.; Miller, T.W.; Bandle, R.W.; Csanyi, G.; Pagano, P.J.; Bauer, P.M.; Schnermann, J.; Roberts, D.D.; Isenberg, J.S. Thrombospondin-1 supports blood pressure by limiting eNOS activation and endothelial-dependent vasorelaxation. *Cardiovasc. Res.* **2010**, *88*, 471–481. [CrossRef] [PubMed]

108. Soto-Pantoja, D.R.; Kaur, S.; Roberts, D.D. CD47 signaling pathways controlling cellular differentiation and responses to stress. *Crit. Rev. Biochem. Mol. Biol.* **2015**, *50*, 212–230. [CrossRef] [PubMed]

109. Roberts, D.D.; Miller, T.W.; Rogers, N.M.; Yao, M.; Isenberg, J.S. The matricellular protein thrombospondin-1 globally regulates cardiovascular function and responses to stress via CD47. *Matrix Biol.* **2012**, *31*, 162–169. [CrossRef] [PubMed]

110. Yao, M.; Rogers, N.M.; Csányi, G.; Rodriguez, A.I.; Ross, M.A.; St Croix, C.; Knupp, H.; Novelli, E.M.; Thomson, A.W.; Pagano, P.J.; et al. Thrombospondin-1 activation of signal-regulatory protein-α stimulates reactive oxygen species production and promotes renal ischemia reperfusion injury. *J. Am. Soc. Nephrol.* **2014**, *25*, 1171–1186. [CrossRef] [PubMed]

111. Tong, X.; Khandelwal, A.R.; Qin, Z.; Wu, X.; Chen, L.; Ago, T.; Sadoshima, J.; Cohen, R.A. Role of smooth muscle Nox4-based NADPH oxidase in neointimal hyperplasia. *J. Mol. Cell Cardiol.* **2015**, *89*, 185–194. [CrossRef] [PubMed]

112. Tong, X.; Khandelwal, A.R.; Wu, X.; Xu, Z.; Yu, W.; Chen, C.; Zhao, W.; Yang, J.; Qin, Z.; Weisbrod, R.M.; et al. Pro-atherogenic role of smooth muscle Nox4-based NADPH oxidase. *J. Mol. Cell Cardiol.* **2016**, *92*, 30–40. [CrossRef] [PubMed]

113. Csányi, G.; Feck, D.M.; Ghoshal, P.; Singla, B.; Lin, H.; Nagarajan, S.; Meijles, D.N.; Al Ghouleh, I.; Cantu-Medellin, N.; Kelley, E.E.; et al. CD47 and Nox1 mediate dynamic fluid-phase macropinocytosis of native LDL. *Antioxid. Redox. Signal* **2017**, *26*, 886–901. [CrossRef] [PubMed]

114. Tiyyagura, S.R.; Pinney, S.P. Left ventricular remodeling after myocardial infarction: Past, present, and future. *Mt. Sinai J. Med.* **2006**, *73*, 840–851. [PubMed]

115. Buja, L.M.; Vela, D. Cardiomyocyte death and renewal in the normal and diseased heart. *Cardiovasc Pathol.* **2008**, *17*, 349–374. [CrossRef] [PubMed]

116. Jugdutt, B.I. Ventricular remodeling after infarction and the extracellular collagen matrix: When is enough enough? *Circulation* **2003**, *108*, 1395–1403. [CrossRef] [PubMed]

117. Chistiakov, D.A.; Orekhov, A.N.; Bobryshev, Y.V. The role of cardiac fibroblasts in post-myocardial heart tissue repair. *Exp. Mol. Pathol.* **2016**, *101*, 231–240. [CrossRef] [PubMed]

118. Holmes, J.W.; Borg, T.K.; Covell, J.W. Structure and mechanics of healing myocardial infarcts. *Annu. Rev. Biomed. Eng.* **2005**, *7*, 223–253. [CrossRef] [PubMed]

119. Lamy, L.; Ticchioni, M.; Rouquette-Jazdanian, A.K.; Samson, M.; Deckert, M.; Greenberg, A.H.; Bernard, A. CD47 and the 19 kDa interacting protein-3 (BNIP3) in T cell apoptosis. *J. Biol. Chem.* **2003**, *278*, 23915–23921. [CrossRef] [PubMed]

120. Lamy, L.; Foussat, A.; Brown, E.J.; Bornstein, P.; Ticchioni, M.; Bernard, A. Interactions between CD47 and thrombospondin reduce inflammation. *J. Immunol.* **2007**, *178*, 5930–5939. [CrossRef] [PubMed]

121. Isenberg, J.S.; Roberts, D.D.; Frazier, W.A. CD47: A new target in cardiovascular therapy. *Arterioscler. Thromb. Vasc. Biol.* **2008**, *28*, 615–621. [CrossRef] [PubMed]

122. Isenberg, J.S.; Hyodo, F.; Pappan, L.K.; Abu-Asab, M.; Tsokos, M.; Krishna, M.C.; Frazier, W.A.; Roberts, D.D. Blocking thrombospondin-1/CD47 signaling alleviates deleterious effects of aging on tissue responses to ischemia. *Arterioscler. Thromb. Vasc. Biol.* **2007**, *27*, 2582–2588. [CrossRef] [PubMed]

123. Mirochnik, Y.; Kwiatek, A.; Volpert, O.V. Thrombospondin and apoptosis: Molecular mechanisms and use for design of complementation treatments. *Curr. Drug Targets* **2008**, *9*, 851–862. [CrossRef] [PubMed]

124. Krady, M.M.; Zeng, J.; Yu, J.; MacLauchlan, S.; Skokos, E.A.; Tian, W.; Bornstein, P.; Sessa, W.C.; Kyriakides, T.R. Thrombospondin-2 modulates extracellular matrix remodeling during physiological angiogenesis. *Am. J. Pathol.* **2008**, *173*, 879–891. [CrossRef] [PubMed]

125. Maclauchlan, S.; Skokos, E.A.; Agah, A.; Zeng, J.; Tian, W.; Davidson, J.M.; Bornstein, P.; Kyriakides, T.R. Enhanced angiogenesis and reduced contraction in thrombospondin-2-null wounds is associated with increased levels of matrix metalloproteinases-2 and -9, and soluble VEGF. *J. Histochem. Cytochem.* **2009**, *57*, 301–313. [CrossRef] [PubMed]

126. Kyriakides, T.R.; Zhu, Y.H.; Smith, L.T.; Bain, S.D.; Yang, Z.; Lin, M.T.; Danielson, K.G.; Iozzo, R.V.; LaMarca, M.; McKinney, C.E.; et al. Mice that lack thrombospondin 2 display connective tissue abnormalities that are associated with disordered collagen fibrillogenesis, an increased vascular density, and a bleeding diathesis. *J. Cell Biol.* **1998**, *140*, 419–430. [CrossRef] [PubMed]

127. Kyriakides, T.R.; Leach, K.J.; Hoffman, A.S.; Ratner, B.D.; Bornstein, P. Mice that lack the angiogenesis inhibitor, thrombospondin 2, mount an altered foreign body reaction characterized by increased vascularity. *Proc. Natl. Acad. Sci. USA* **1999**, *96*, 4449–4454. [CrossRef] [PubMed]

128. Bornstein, P.; Kyriakides, T.R.; Yang, Z.; Armstrong, L.C.; Birk, D.E. Thrombospondin 2 modulates collagen fibrillogenesis and angiogenesis. *J. Investig. Dermatol. Symp. Proc.* **2000**, *5*, 61–66. [CrossRef] [PubMed]

129. Van Oorschot, A.A.; Smits, A.M.; Pardali, E.; Doevendans, P.A.; Goumans, M.J. Low oxygen tension positively influences cardiomyocyte progenitor cell function. *J. Cell Mol. Med.* **2011**, *15*, 2723–2734. [CrossRef] [PubMed]

130. Mustonen, E.; Leskinen, H.; Aro, J.; Luodonpää, M.; Vuolteenaho, O.; Ruskoaho, H.; Rysä, J. Metoprolol treatment lowers thrombospondin-4 expression in rats with myocardial infarction and left ventricular hypertrophy. *Basic Clin. Pharmacol. Toxicol.* **2010**, *107*, 709–717. [CrossRef] [PubMed]

131. Brody, M.J.; Schips, T.G.; Vanhoutte, D.; Kanisicak, O.; Karch, J.; Maliken, B.D.; Blair, N.S.; Sargent, M.A.; Prasad, V.; Molkentin, J.D. Dissection of thrombospondin-4 domains involved in intracellular adaptive endoplasmic reticulum stress-responsive signaling. *Mol. Cell Biol.* **2015**, *36*, 2–12. [PubMed]

132. Jin, J.K.; Blackwood, E.A.; Azizi, K.; Thuerauf, D.J.; Fahem, A.G.; Hofmann, C.; Kaufman, R.J.; Doroudgar, S.; Glembotski, C.C. ATF6 decreases myocardial ischemia/reperfusion damage and links ER stress and oxidative stress signaling pathways in the heart. *Circ. Res.* **2017**, *120*, 862–875. [CrossRef] [PubMed]

133. Burke, A.; Creighton, W.; Tavora, F.; Li, L.; Fowler, D. Decreased frequency of the 3′UTR T>G single nucleotide polymorphism of thrombospondin-2 gene in sudden death due to plaque erosion. *Cardiovasc. Pathol.* **2010**, *19*, e45–e49. [CrossRef] [PubMed]

134. Farb, A.; Burke, A.P.; Tang, A.L.; Liang, T.Y.; Mannan, P.; Smialek, J.; Virmani, R. Coronary plaque erosion without rupture into a lipid core. A frequent cause of coronary thrombosis in sudden coronary death. *Circulation* **1996**, *93*, 1354–1363. [CrossRef] [PubMed]

135. Koch, W.; Hoppmann, P.; de Waha, A.; Schömig, A.; Kastrati, A. Polymorphisms in thrombospondin genes and myocardial infarction. A case-control study and a meta-analysis of available evidence. *Hum. Mol. Genet.* **2008**, *17*, 1120–1126. [CrossRef] [PubMed]

136. Ashokkumar, M.; Anbarasan, C.; Saibabu, R.; Kuram, S.; Raman, S.C.; Cherian, K.M. An association study of thrombospondin 1 and 2 SNPs with coronary artery disease and myocardial infarction among South Indians. *Thromb. Res.* **2011**, *128*, e49–e53. [CrossRef] [PubMed]

137. Zhou, X.; Huang, J.; Chen, J.; Zhao, J.; Yang, W.; Wang, X.; Gu, D. Thrombospondin-4 A387P polymorphism is not associated with coronary artery disease and myocardial infarction in the Chinese Han population. *Clin. Sci.* **2004**, *106*, 495–500. [CrossRef] [PubMed]

138. Zhou, X.; Huang, J.; Chen, J.; Zhao, J.; Ge, D.; Yang, W.; Gu, D. Genetic association analysis of myocardial infarction with thrombospondin-1 N700S variant in a Chinese population. *Thromb. Res.* **2004**, *113*, 181–186. [CrossRef] [PubMed]

139. Burlew, B.S.; Weber, K.T. Connective tissue and the heart. Functional significance and regulatory mechanisms. *Cardiol. Clin.* **2000**, *18*, 435–442. [CrossRef]

140. Fomovsky, G.M.; Thomopoulos, S.; Holmes, J.W. Contribution of extracellular matrix to the mechanical properties of the heart. *J. Mol. Cell Cardiol.* **2010**, *48*, 490–496. [CrossRef] [PubMed]

141. Spinale, F.G. Matrix metalloproteinases: Regulation and dysregulation in the failing heart. *Circ. Res.* **2002**, *90*, 520–530. [CrossRef] [PubMed]

142. Dorn, G.W., 2nd. Apoptotic and non-apoptotic programmed cardiomyocyte death in ventricular remodelling. *Cardiovasc. Res.* **2009**, *81*, 465–473. [CrossRef] [PubMed]

143. Xia, Y.; Lee, K.; Li, N.; Corbett, D.; Mendoza, L.; Frangogiannis, N.G. Characterization of the inflammatory and fibrotic response in a mouse model of cardiac pressure overload. *Histochem. Cell Biol.* **2009**, *131*, 471–481. [CrossRef] [PubMed]

144. Xia, Y.; Dobaczewski, M.; Gonzalez-Quesada, C.; Chen, W.; Biernacka, A.; Li, N.; Lee, D.W.; Frangogiannis, N.G. Endogenous thrombospondin 1 protects the pressure-overloaded myocardium by modulating fibroblast phenotype and matrix metabolism. *Hypertension* **2011**, *58*, 902–911. [CrossRef] [PubMed]

145. Rohini, A.; Agrawal, N.; Koyani, C.N.; Singh, R. Molecular targets and regulators of cardiac hypertrophy. *Pharmacol. Res.* **2010**, *61*, 269–280. [CrossRef] [PubMed]

146. Malek, M.H.; Olfert, I.M. Global deletion of thrombospondin-1 increases cardiac and skeletal muscle capillarity and exercise capacity in mice. *Exp. Physiol.* **2009**, *94*, 749–760. [CrossRef] [PubMed]

147. Moens, A.L.; Cingolani, O.; Arkenbout, E.; Kass, D.A. Exacerbated cardiac remodeling to pressure-overload in mice lacking thrombospondin-4. *Eur. J. Heart Fail.* **2008**, *7*, 149–150. [CrossRef]

148. Bronzwaer, J.G.; Paulus, W.J. Diastolic and systolic heart failure: Different stages or distinct phenotypes of the heart failure syndrome? *Curr. Heart Fail. Rep.* **2009**, *6*, 281–286. [CrossRef] [PubMed]

149. Chatterjee, K.; Massie, B. Systolic and diastolic heart failure: Differences and similarities. *J. Card. Fail.* **2007**, *13*, 569–576. [CrossRef] [PubMed]

150. Diwan, A.; Dorn, G.W., 2nd. Decompensation of cardiac hypertrophy: Cellular mechanisms and novel therapeutic targets. *Physiology* **2007**, *22*, 56–64. [CrossRef] [PubMed]

151. Nishida, K.; Otsu, K. Cell death in heart failure. *Circ. J.* **2008**, *72*, A17–A21. [CrossRef]

152. Shan, K.; Kurrelmeyer, K.; Seta, Y.; Wang, F.; Dibbs, Z.; Deswal, A.; Lee-Jackson, D.; Mann, D.L. The role of cytokines in disease progression in heart failure. *Curr. Opin. Cardiol.* **1997**, *12*, 218–223. [CrossRef]

153. Batlle, M.; Pérez-Villa, F.; Lázaro, A.; García-Pras, E.; Vallejos, I.; Sionis, A.; Castel, M.A.; Roig, E. Decreased expression of thrombospondin-1 in failing hearts may favor ventricular remodeling. *Transplant. Proc.* **2009**, *41*, 2231–2233. [CrossRef]

154. Muñoz-Pacheco, P.; Ortega-Hernández, A.; Caro-Vadillo, A.; Casanueva-Eliceiry, S.; Aragoncillo, P.; Egido, J.; Fernández-Cruz, A.; Gómez-Garre, D. Eplerenone enhances cardioprotective effects of standard heart failure therapy through matricellular proteins in hypertensive heart failure. *J. Hypertens.* **2013**, *31*, 2309–2319. [CrossRef]

155. Sharifi-Sanjani, M.; Shoushtari, A.H.; Quiroz, M.; Baust, J.; Sestito, S.F.; Mosher, M.; Ross, M.; McTiernan, C.F.; St Croix, C.M.; Bilonick, R.A.; et al. Cardiac CD47 drives left ventricular heart failure through Ca^{2+}-CaMKII-regulated induction of HDAC3. *J. Am. Heart Assoc.* **2014**, *3*, e000670. [CrossRef]

156. Van Almen, G.C.; Verhesen, W.; Van Leeuwen, R.E.; Van de Vrie, M.; Eurlings, C.; Schellings, M.W.; Swinnen, M.; Cleutjens, J.P.; Van Zandvoort, M.A.; Heymans, S.; et al. MicroRNA-18 and microRNA-19 regulate CTGF and TSP-1 expression in age-related heart failure. *Aging Cell* **2011**, *10*, 769–779. [CrossRef]

157. Papageorgiou, A.P.; Swinnen, M.; Vanhoutte, D.; VandenDriessche, T.; Chuah, M.; Lindner, D.; Verhesen, W.; De Vries, B.; D'hooge, J.; Lutgens, E.; et al. Thrombospondin-2 prevents cardiac injury and dysfunction in viral myocarditis through the activation of regulatory T-cells. *Cardiovasc. Res.* **2012**, *94*, 115–124. [CrossRef]

158. Rysä, J.; Leskinen, H.; Ilves, M.; Ruskoaho, H. Distinct upregulation of extracellular matrix genes in transition from hypertrophy to hypertensive heart failure. *Hypertension* **2005**, *45*, 927–933. [CrossRef]

159. Melenovsky, V.; Benes, J.; Skaroupkova, P.; Sedmera, D.; Strnad, H.; Kolar, M.; Vlcek, C.; Petrak, J.; Benes, J., Jr.; Papousek, F.; et al. Metabolic characterization of volume overload heart failure due to aorto-caval fistula in rats. *Mol. Cell Biochem.* **2011**, *354*, 83–96. [CrossRef]

160. Yetkin, E.; Waltenberger, J. Molecular and cellular mechanisms of aortic stenosis. *Int. J. Cardiol.* **2009**, *135*, 4–13. [CrossRef]

161. Rawat, D.K.; Alzoubi, A.; Gupte, R.; Chettimada, S.; Watanabe, M.; Kahn, A.G.; Okada, T.; McMurtry, I.F.; Gupte, S.A. Increased reactive oxygen species, metabolic maladaptation, and autophagy contribute to pulmonary arterial hypertension-induced ventricular hypertrophy and diastolic heart failure. *Hypertension* **2014**, *64*, 1266–1274. [CrossRef] [PubMed]

162. Sage, E.; Mercier, O.; Van den Eyden, F.; De Perrot, M.; Barlier-Mur, A.M.; Dartevelle, P.; Eddahibi, S.; Herve, P.; Fadel, E. Endothelial cell apoptosis in chronically obstructed and reperfused pulmonary artery. *Respir. Res.* **2008**, *9*, 19. [CrossRef] [PubMed]

163. Imoto, K.; Okada, M.; Yamawaki, H. Expression profile of matricellular proteins in hypertrophied right ventricle of monocrotaline-induced pulmonary hypertensive rats. *J. Vet. Med. Sci.* **2017**, *79*, 1096–1102. [CrossRef] [PubMed]

164. Kaiser, R.; Frantz, C.; Bals, R.; Wilkens, H. The role of circulating thrombospondin-1 in patients with precapillary pulmonary hypertension. *Respir. Res.* **2016**, *17*, 96. [CrossRef] [PubMed]

165. Ochoa, C.D.; Yu, L.; Al-Ansari, E.; Hales, C.A.; Quinn, D.A. Thrombospondin-1 null mice are resistant to hypoxia-induced pulmonary hypertension. *J. Cardiothorac. Surg.* **2010**, *5*, 32. [CrossRef] [PubMed]

166. Kumar, R.; Mickael, C.; Kassa, B.; Gebreab, L.; Robinson, J.C.; Koyanagi, D.E.; Sanders, L.; Barthel, L.; Meadows, C.; Fox, D.; et al. TGF-β activation by bone marrow-derived thrombospondin-1 causes Schistosoma- and hypoxia-induced pulmonary hypertension. *Nat. Commun.* **2017**, *8*, 15494. [CrossRef] [PubMed]

167. Green, D.E.; Kang, B.Y.; Murphy, T.C.; Hart, C.M. Peroxisome proliferator-activated receptor gamma (PPARγ) regulates thrombospondin-1 and Nox4 expression in hypoxia-induced human pulmonary artery smooth muscle cell proliferation. *Pulm. Circ.* **2012**, *2*, 483–491. [CrossRef] [PubMed]

168. Cucoranu, I.; Clempus, R.; Dikalova, A.; Phelan, P.J.; Ariyan, S.; Dikalov, S.; Sorescu, D. NAD(P)H oxidase 4 mediates transforming growth factor-beta1-induced differentiation of cardiac fibroblasts into myofibroblasts. *Circ. Res.* **2005**, *97*, 900–907. [CrossRef] [PubMed]

169. Sturrock, A.; Cahill, B.; Norman, K.; Huecksteadt, T.P.; Hill, K.; Sanders, K.; Karwande, S.V.; Stringham, J.C.; Bull, D.A.; Gleich, M.; Kennedy, T.P.; Hoidal, J.R. Transforming growth factor-beta1 induces Nox4 NAD(P)H oxidase and reactive oxygen species-dependent proliferation in human pulmonary artery smooth muscle cells. *Am. J. Physiol. Lung Cell Mol. Physiol.* **2006**, *290*, L661–L673. [CrossRef] [PubMed]

170. Ismail, S.; Sturrock, A.; Wu, P.; Cahill, B.; Norman, K.; Huecksteadt, T.; Sanders, K.; Kennedy, T.; Hoidal, J. NOX4 mediates hypoxia-induced proliferation of human pulmonary artery smooth muscle cells: The role of autocrine production of transforming growth factor-β-1 and insulin-like growth factor binding protein-3. *Am. J. Physiol. Lung Cell. Mol. Physiol.* **2009**, *296*, L489–L499. [CrossRef] [PubMed]

171. Lu, X.; Murphy, T.C.; Nanes, M.S.; Hart, C.M. PPARγ regulates hypoxia-induced Nox4 expression in human pulmonary artery smooth muscle cells through NF-κB. *Am. J. Physiol. Lung Cell Mol. Physiol.* **2010**, *299*, L559–L566. [CrossRef] [PubMed]

172. Lu, X.; Bijli, K.M.; Ramirez, A.; Murphy, T.C.; Kleinhenz, J.; Hart, C.M. Hypoxia downregulates PPARγ via an ERK1/2-NF-κB-Nox4-dependent mechanism in human pulmonary artery smooth muscle cells. *Free Radic. Biol. Med.* **2013**, *63*, 151–160. [CrossRef] [PubMed]

173. Lin, T.N.; Kim, G.M.; Chen, J.J.; Cheung, W.M.; He, Y.Y.; Hsu, C.Y. Differential regulation of thrombospondin-1 and thrombospondin-2 after focal cerebral ischemia/reperfusion. *Stroke* **2003**, *34*, 177–186. [CrossRef] [PubMed]

174. Woo, M.S.; Yang, J.; Beltran, C.; Cho, S. Cell surface CD36 protein in monocyte/macrophage contributes to phagocytosis during the resolution phase of ischemic stroke in mice. *J. Biol. Chem.* **2016**, *291*, 23654–23661. [CrossRef] [PubMed]

175. Liauw, J.; Hoang, S.; Choi, M.; Eroglu, C.; Choi, M.; Sun, G.H.; Percy, M.; Wildman-Tobriner, B.; Bliss, T.; Guzman, R.G.; Barres, B.A.; Steinberg, G.K. Thrombospondins 1 and 2 are necessary for synaptic plasticity and functional recovery after stroke. *J. Cereb. Blood Flow Metab.* **2008**, *28*, 1722–1732. [CrossRef] [PubMed]

176. Jurk, K.; Jahn, U.R.; Van Aken, H.; Schriek, C.; Droste, D.W.; Ritter, M.A.; Bernd Ringelstein, E.; Kehrel, B.E. Platelets in patients with acute ischemic stroke are exhausted and refractory to thrombin, due to cleavage of the seven-transmembrane thrombin receptor (PAR-1). *Thromb. Haemost.* **2004**, *91*, 334–344. [CrossRef] [PubMed]

177. Yesner, L.M.; Huh, H.Y.; Pearce, S.F.; Silverstein, R.L. Regulation of monocyte CD36 and thrombospondin-1 expression by soluble mediators. *Arterioscler. Thromb. Vasc. Biol.* **1996**, *16*, 1019–1025. [CrossRef] [PubMed]

178. Buée, L.; Hof, P.R.; Roberts, D.D.; Delacourte, A.; Morrison, J.H.; Fillit, H.M. Immunohistochemical identification of thrombospondin in normal human brain and in Alzheimer's disease. *Am. J. Pathol.* **1992**, *141*, 783–788. [PubMed]

179. Son, S.M.; Nam, D.W.; Cha, M.Y.; Kim, K.H.; Byun, J.; Ryu, H.; Mook-Jung, I. Thrombospondin-1 prevents amyloid beta-mediated synaptic pathology in Alzheimer's disease. *Neurobiol. Aging* **2015**, *36*, 3214–3227. [CrossRef] [PubMed]

180. Rama Rao, K.V.; Curtis, K.M.; Johnstone, J.T.; Norenberg, M.D. Amyloid-β inhibits thrombospondin 1 release from cultured astrocytes: Effects on synaptic protein expression. *J. Neuropathol. Exp. Neurol.* **2013**, *72*, 735–744. [PubMed]

181. Garcia, O.; Torres, M.; Helguera, P.; Coskun, P.; Busciglio, J. A role for thrombospondin-1 deficits in astrocyte-mediated spine and synaptic pathology in Down's syndrome. *PLoS ONE* **2010**, *5*, e14200. [CrossRef] [PubMed]

182. Cheng, C.; Lau, S.K.; Doering, L.C. Astrocyte-secreted thrombospondin-1 modulates synapse and spine defects in the fragile X mouse model. *Mol. Brain* **2016**, *9*, 74. [CrossRef] [PubMed]

183. Ko, C.Y.; Chu, Y.Y.; Narumiya, S.; Chi, J.Y.; Furuyashiki, T.; Aoki, T.; Wang, S.M.; Chang, W.C.; Wang, J.M. CCAAT/enhancer-binding protein δ/miR135a/thrombospondin 1 axis mediates PGE2-induced angiogenesis in Alzheimer's disease. *Neurobiol. Aging* **2015**, *36*, 1356–1368. [CrossRef] [PubMed]

184. Long, J.; Liu, S.; Zeng, X.; Yang, X.; Huang, H.; Zhang, Y.; Chen, J.; Xu, Y.; Huang, D.; Qiu, X. Population study confirms serum proteins' change and reveals diagnostic values in congenital ventricular septal defect. *Pediatr. Cardiol* **2017**, in press. [CrossRef] [PubMed]

185. Wang, J.; Nagy, A.; Larsson, J.; Dudas, M.; Sucov, H.M.; Kaartinen, V. Defective ALK5 signaling in the neural crest leads to increased postmigratory neural crest cell apoptosis and severe outflow tract defects. *BMC Dev. Biol.* **2006**, *6*, 51. [CrossRef] [PubMed]

186. Gao, Z.; Kim, G.H.; Mackinnon, A.C.; Flagg, A.E.; Bassett, B.; Earley, J.U.; Svensson, E.C. Ets1 is required for proper migration and differentiation of the cardiac neural crest. *Development* **2010**, *137*, 1543–1551. [CrossRef] [PubMed]

187. Fouladkou, F.; Lu, C.; Jiang, C.; Zhou, L.; She, Y.; Walls, J.R.; Kawabe, H.; Brose, N.; Henkelman, R.M.; Huang, A.; Bruneau, B.G.; Rotin, D. The ubiquitin ligase Nedd4–1 is required for heart development and is a suppressor of thrombospondin-1. *J. Biol. Chem.* **2010**, *285*, 6770–6780. [CrossRef] [PubMed]

188. Yang, B.; Kumar, S. Nedd4 and Nedd4–2: Closely related ubiquitin-protein ligases with distinct physiological functions. *Cell Death Differ.* **2010**, *17*, 68–77. [CrossRef] [PubMed]

189. Murdaca, J.; Treins, C.; Monthouël-Kartmann, M.N.; Pontier-Bres, R.; Kumar, S.; Van Obberghen, E.; Giorgetti-Peraldi, S. Grb10 prevents Nedd4-mediated vascular endothelial growth factor receptor-2 degradation. *J. Biol. Chem.* **2004**, *279*, 26754–26761. [CrossRef] [PubMed]

190. Van Bemmelen, M.X.; Rougier, J.S.; Gavillet, B.; Apothéloz, F.; Daidié, D.; Tateyama, M.; Rivolta, I.; Thomas, M.A.; Kass, R.S.; Staub, O.; et al. Cardiac voltage-gated sodium channel Nav1.5 is regulated by Nedd4–2 mediated ubiquitination. *Circ. Res.* **2004**, *95*, 284–291. [CrossRef] [PubMed]

191. Fotia, A.B.; Ekberg, J.; Adams, D.J.; Cook, D.I.; Poronnik, P.; Kumar, S. Regulation of neuronal voltage-gated sodium channels by the ubiquitin-protein ligases Nedd4 and Nedd4–2. *J. Biol. Chem.* **2004**, *279*, 28930–28935. [CrossRef] [PubMed]

192. Moskowitz, I.P.; Pizard, A.; Patel, V.V.; Bruneau, B.G.; Kim, J.B.; Kupershmidt, S.; Roden, D.; Berul, C.I.; Seidman, C.E.; Seidman, J.G. The T-Box transcription factor Tbx5 is required for the patterning and maturation of the murine cardiac conduction system. *Development* **2004**, *131*, 4107–4116. [CrossRef] [PubMed]

193. Murhy-Ullrich, J.E.; Sage, E.H. Revisiting the matricellular concept. *Matrix Biol.* **2014**, *37*, 1–14. [CrossRef] [PubMed]

194. Marganski, W.A.; Gangopadhyay, S.S.; Je, H.D.; Gallant, C.; Morgan, K.G. Targeting of a novel Ca+2/calmodulin-dependent protein kinase II is essential for extracellular signal-regulated kinase-mediated signaling in differentiated smooth muscle cells. *Circ. Res.* **2005**, *97*, 541–549. [CrossRef] [PubMed]

195. Guo, Y.; Zhang, Z.; Wu, H.E.; Luo, Z.D.; Hogan, Q.H.; Pan, B. Increased thrombospondin-4 after nerve injury mediates disruption of intracellular calcium signaling in primary sensory neurons. *Neuropharmacology* **2017**, *117*, 292–304. [CrossRef] [PubMed]

196. Patra, D.; Sandell, L.J. Antiangiogenic and anticancer molecules in cartilage. *Expert Rev. Mol. Med.* **2012**, *14*, e10. [CrossRef] [PubMed]

197. Rusnati, M.; Urbinati, C.; Bonifacio, S.; Presta, M.; Taraboletti, G. Thrombospondin-1 as a paradigm for the development of antiangiogenic agents endowed with multiple mechanisms of action. *Pharmaceuticals* **2010**, *3*, 1241–1278. [CrossRef] [PubMed]

198. Henkin, J.; Volpert, O.V. Therapies using anti-angiogenic peptide mimetics of thrombospondin-1. *Expert Opin. Ther. Targets* **2011**, *15*, 1369–1386. [CrossRef] [PubMed]

199. Dawson, D.W.; Volpert, O.V.; Pearce, S.F.; Schneider, A.J.; Silverstein, R.L.; Henkin, J.; Bouck, N.P. Three distinct D-amino acid substitutions confer potent antiangiogenic activity on an inactive peptide derived from a thrombospondin-1 type 1 repeat. *Mol. Pharmacol.* **1999**, *55*, 332–338. [PubMed]

200. Westphal, J.R. Technology evaluation: ABT-510, Abbott. *Curr. Opin. Mol. Ther.* **2004**, *6*, 451–457. [PubMed]

201. Amato, R.J. Renal cell carcinoma: Review of novel single-agent therapeutics and combination regimens. *Ann. Oncol.* **2005**, *16*, 7–15. [CrossRef] [PubMed]

202. Hoekstra, R.; de Vos, F.Y.; Eskens, F.A.; Gietema, J.A.; Van der Gaast, A.; Groen, H.J.; Knight, R.A.; Carr, R.A.; Humerickhouse, R.A.; Verweij, J.; et al. Phase I safety, pharmacokinetic, and pharmacodynamic study of the thrombospondin-1-mimetic angiogenesis inhibitor ABT-510 in patients with advanced cancer. *J. Clin. Oncol.* **2005**, *23*, 5188–5197. [CrossRef] [PubMed]

203. Hoekstra, R.; de Vos, F.Y.; Eskens, F.A.; de Vries, E.G.; Uges, D.R.; Knight, R.; Carr, R.A.; Humerickhouse, R.; Verweij, J.; Gietema, J.A. Phase I study of the thrombospondin-1-mimetic angiogenesis inhibitor ABT-510 with 5-fluorouracil and leucovorin: A safe combination. *Eur. J. Cancer* **2006**, *42*, 467–472. [CrossRef] [PubMed]

204. Gietema, J.A.; Hoekstra, R.; De Vos, F.Y.; Uges, D.R.; Van der Gaast, A.; Groen, H.J.; Loos, W.J.; Knight, R.A.; Carr, R.A.; Humerickhouse, R.A.; et al. A phase I study assessing the safety and pharmacokinetics of the thrombospondin-1-mimetic angiogenesis inhibitor ABT-510 with gemcitabine and cisplatin in patients with solid tumors. *Ann. Oncol.* **2006**, *17*, 1320–1327. [CrossRef] [PubMed]

205. Gordon, M.S.; Mendelson, D.; Carr, R.; Knight, R.A.; Humerickhouse, R.A.; Iannone, M.; Stopeck, A.T. A phase 1 trial of 2 dose schedules of ABT-510, an antiangiogenic, thrombospondin-1-mimetic peptide, in patients with advanced cancer. *Cancer* **2008**, *113*, 3420–3429. [CrossRef] [PubMed]

206. Nabors, L.B.; Fiveash, J.B.; Markert, J.M.; Kekan, M.S.; Gillespie, G.Y.; Huang, Z.; Johnson, M.J.; Meleth, S.; Kuo, H.; Gladson, C.L.; et al. A phase 1 trial of ABT-510 concurrent with standard chemoradiation for patients with newly diagnosed glioblastoma. *Arch. Neurol.* **2010**, *67*, 313–319. [CrossRef] [PubMed]

207. Uronis, H.E.; Cushman, S.M.; Bendell, J.C.; Blobe, G.C.; Morse, M.A.; Nixon, A.B.; Dellinger, A.; Starr, M.D.; Li, H.; Meadows, K.; et al. A phase I study of ABT-510 plus bevacizumab in advanced solid tumors. *Cancer Med.* **2013**, *2*, 316–324. [CrossRef] [PubMed]

208. Markovic, S.N.; Suman, V.J.; Rao, R.A.; Ingle, J.N.; Kaur, J.S.; Erickson, L.A.; Pitot, H.C.; Croghan, G.A.; McWilliams, R.R.; Merchan, J.; et al. A phase II study of ABT-510 (thrombospondin-1 analog) for the treatment of metastatic melanoma. *Am. J. Clin. Oncol.* **2007**, *30*, 303–309. [CrossRef] [PubMed]

209. Ebbinghaus, S.; Hussain, M.; Tannir, N.; Gordon, M.; Desai, A.A.; Knight, R.A.; Humerickhouse, R.A.; Qian, J.; Gordon, G.B.; Figlin, R. Phase 2 study of ABT-510 in patients with previously untreated advanced renal cell carcinoma. *Clin. Cancer Res.* **2007**, *13*, 6689–6695. [CrossRef] [PubMed]

210. Baker, L.H.; Rowinsky, E.K.; Mendelson, D.; Humerickhouse, R.A.; Knight, R.A.; Qian, J.; Carr, R.A.; Gordon, G.B.; Demetri, G.D. Randomized, phase II study of the thrombospondin-1-mimetic angiogenesis inhibitor ABT-510 in patients with advanced soft tissue sarcoma. *J. Clin. Oncol.* **2008**, *26*, 5583–5588. [CrossRef] [PubMed]

211. Isenberg, J.S.; Romeo, M.; Abu-Asab, M.; Tsokos, M.; Oldenborg, A.; Pappan, L.; Wink, D.A.; Frazier, W.A.; Roberts, D.D. Increasing survival of ischemic tissue by targeting CD47. *Circ. Res.* **2007**, *100*, 712–720. [CrossRef] [PubMed]

212. Liu, J.; Wang, L.; Zhao, F.; Tseng, S.; Narayanan, C.; Shura, L.; Willingham, S.; Howard, M.; Prohaska, S.; Volkmer, J.; et al. Pre-clinical development of a humanized anti-CD47 antibody with anti-cancer therapeutic potential. *PLoS ONE* **2015**, *10*, e0137345. [CrossRef] [PubMed]

213. Naujokat, C. Monoclonal antibodies against human cancer stem cells. *Immunotherapy* **2014**, *6*, 290–308. [CrossRef] [PubMed]

214. Ribeiro, S.M.; Poczatek, M.; Schultz-Cherry, S.; Villain, M.; Murphy-Ullrich, J.E. The activation sequence of thrombospondin-1 interacts with the latency-associated peptide to regulate activation of latent transforming growth factor-beta. *J. Biol. Chem.* **1999**, *274*, 13586–13593. [CrossRef] [PubMed]

215. Murphy-Ullrich, J.E.; Poczatek, M. Activation of latent TGF-β by thrombospondin-1: Mechanisms and physiology. *Cytokine Growth Factor Rev.* **2000**, *11*, 59–69. [CrossRef]

216. Kondou, H.; Mushiake, S.; Etani, Y.; Miyoshi, Y.; Michigami, T.; Ozono, K. A blocking peptide for transforming growth factor-β1 activation prevents hepatic fibrosis in vivo. *J. Hepatol.* **2003**, *39*, 742–748. [CrossRef]

217. Xie, X.S.; Li, F.Y.; Liu, H.C.; Deng, Y.; Li, Z.; Fan, J.M. LSKL, a peptide antagonist of thrombospondin-1, attenuates renal interstitial fibrosis in rats with unilateral ureteral obstruction. *Arch. Pharm. Res.* **2010**, *33*, 275–284. [CrossRef] [PubMed]

218. Lu, A.; Miao, M.; Schoeb, T.R.; Agarwal, A.; Murphy-Ullrich, J.E. Blockade of TSP1-dependent TGF-β activity reduces renal injury and proteinuria in a murine model of diabetic nephropathy. *Am. J. Pathol.* **2011**, *178*, 2573–2586. [CrossRef] [PubMed]

219. Liao, F.; Li, G.; Yuan, W.; Chen, Y.; Zuo, Y.; Rashid, K.; Zhang, J.H.; Feng, H.; Liu, F. LSKL peptide alleviates subarachnoid fibrosis and hydrocephalus by inhibiting TSP1-mediated TGF-β1 signaling activity following subarachnoid hemorrhage in rats. *Exp. Ther. Med.* **2016**, *12*, 2537–2543. [CrossRef] [PubMed]

220. Krishna, S.M.; Seto, S.W.; Jose, R.J.; Biros, E.; Moran, C.S.; Wang, Y.; Clancy, P.; Golledge, J. A peptide antagonist of thrombospondin-1 promotes abdominal aortic aneurysm progression in the angiotensin II-infused apolipoprotein-E-deficient mouse. *Arterioscler. Thromb. Vasc. Biol.* **2015**, *35*, 389–398. [CrossRef] [PubMed]

221. Liu, X.; Zhao, Y.; Gao, J.; Pawlyk, B.; Starcher, B.; Spencer, J.A.; Yanagisawa, H.; Zuo, J.; Li, T. Elastic fiber homeostasis requires lysyl oxidase-like 1 protein. *Nat. Genet.* **2004**, *36*, 178–182. [CrossRef] [PubMed]

222. Hao, M.; Wang, R.; Wang, W. Cell therapies in cardiomyopathy: Current status of clinical trials. *Anal. Cell Pathol.* **2017**, *2017*, 9404057. [CrossRef] [PubMed]

223. Cointe, S.; Rhéaume, É.; Martel, C.; Blanc-Brude, O.; Dubé, E.; Sabatier, F.; Dignat-George, F.; Tardif, J.C.; Bonnefoy, A. Thrombospondin-1-derived peptide RFYVVMWK improves the adhesive phenotype of CD34[+] cells from atherosclerotic patients with type 2 diabetes. *Cell Transplant.* **2017**, *26*, 327–337. [CrossRef] [PubMed]

224. Efimenko, A.; Starostina, E.; Kalinina, N.; Stolzing, A. Angiogenic properties of aged adipose derived mesenchymal stem cells after hypoxic conditioning. *J. Transl. Med.* **2011**, *9*, 10. [CrossRef] [PubMed]

225. Noiseux, N.; Borie, M.; Desnoyers, A.; Menaouar, A.; Stevens, L.M.; Mansour, S.; Danalache, B.A.; Roy, D.C.; Jankowski, M.; Gutkowska, J. Preconditioning of stem cells by oxytocin to improve their therapeutic potential. *Endocrinology* **2012**, *153*, 5361–5372. [CrossRef] [PubMed]

226. Lindsey, M.L.; Iyer, R.P.; Zamilpa, R.; Yabluchanskiy, A.; DeLeon-Pennell, K.Y.; Hall, M.E.; Kaplan, A.; Zouein, F.A.; Bratton, D.; Flynn, E.R.; et al. A novel collagen matricryptin reduces left ventricular dilation post-myocardial infarction by promoting scar formation and angiogenesis. *J. Am. Coll. Cardiol.* **2015**, *66*, 1364–1374. [CrossRef] [PubMed]

227. Bein, K.; Simons, M. Thrombospondin type 1 repeats interact with matrix metalloproteinase 2. Regulation of metalloproteinase activity. *J. Biol. Chem.* **2000**, *275*, 32167–32173. [CrossRef] [PubMed]

Article

Increased Serum Levels of Fetal Tenascin-C Variants in Patients with Pulmonary Hypertension: Novel Biomarkers Reflecting Vascular Remodeling and Right Ventricular Dysfunction?

Ilonka Rohm [1], Katja Grün [1], Linda Marleen Müller [1], Daniel Kretzschmar [1], Michael Fritzenwanger [1], Atilla Yilmaz [2], Alexander Lauten [3], Christian Jung [4], P. Christian Schulze [1], Alexander Berndt [5] and Marcus Franz [1,*]

[1] Department of Internal Medicine I, Division of Cardiology, Angiology, Pneumology and Intensive Medical Care, Jena University Hospital, Friedrich-Schiller-University, 07747 Jena, Germany; Ilonka.Rohm@med.uni-jena.de (I.R.); Katja.Gruen@med.uni-jena.de (K.G.); Linda.Mueller@med.uni-jena.de (L.M.M.); DANIEL.KRETZSCHMAR@med.uni-jena.de (D.K.); Michael.Fritzenwanger@med.uni-jena.de (M.F.); Christian.Schulze@med.uni-jena.de (P.C.S.)

[2] Department of Internal Medicine II, Division of Cardiology, Elisabeth Klinikum Schmalkalden, 98574 Schmalkalden, Germany; Atilla.Yilmaz@elisabeth-klinikum.de

[3] Department of Cardiology, Charité–Universitätsmedizin Berlin, 12203 Berlin, Germany; Alexander.Lauten@charite.de

[4] Department of Internal Medicine, Division of Cardiology, University Hospital Düsseldorf, Heinrich Heine University, 40225 Düsseldorf, Germany; christian.jung@med.uni-duesseldorf.de

[5] Institute of Pathology, Jena University Hospital, Friedrich-Schiller-University, 07743 Jena, Germany; Alexander.Berndt@med.uni-jena.de

* Correspondence: Marcus.Franz@med.uni-jena.de; Tel.: +49-(0)-3641-9324127; Fax: +49-(0)-3641-9324102

Received: 8 October 2017; Accepted: 4 November 2017; Published: 8 November 2017

Abstract: Pulmonary vascular remodeling is a pathophysiological feature that common to all classes of pulmonary hypertension (PH) and right ventricular dysfunction, which is the major prognosis-limiting factor. Vascular, as well as cardiac tissue remodeling are associated with a re-expression of fetal variants of cellular adhesion proteins, including tenascin-C (Tn-C). We analyzed circulating levels of the fetal Tn-C splicing variants B^+ and C^+ Tn-C in serum of PH patients to evaluate their potential as novel biomarkers reflecting vascular remodeling and right ventricular dysfunction. Serum concentrations of B^+ and C^+ Tn-C were determined in 80 PH patients and were compared to 40 healthy controls by enzyme-linked immunosorbent assay. Clinical, laboratory, echocardiographic, and functional data were correlated with Tn-C levels. Serum concentrations of both Tn-C variants were significantly elevated in patients with PH ($p < 0.05$). Significant correlations could be observed between Tn-C and echocardiographic parameters, including systolic pulmonary artery pressure (B^+ Tn-C: $r = 0.31$, $p < 0.001$, C^+ Tn-C: $r = 0.26$, $p = 0.006$) and right atrial area (B^+ Tn-C: $r = 0.46$, $p < 0.001$, C^+ Tn-C: $r = 0.49$, $p < 0.001$), and laboratory values like BNP (B^+ Tn-C: $r = 0.45$, $p < 0.001$, C^+ Tn-C: $r = 0.42$, $p < 0.001$). An inverse correlation was observed between Tn-C variants and 6-minute walk distance as a functional parameter (B^+ Tn-C: $r = -0.54$, $p < 0.001$, C^+ Tn-C: $r = -0.43$, $p < 0.001$). In a multivariate analysis, B^+ Tn-C, but not C^+ Tn-C, was found to be an independent predictor of pulmonary hypertension. Both fetal Tn-C variants may represent novel biomarkers that are capable of estimating both pulmonary vascular remodeling and right ventricular load. The potential beneficial impact of Tn-C variants for risk stratification in patients with PH needs further investigation.

Keywords: fetal tenascin-C; pulmonary hypertension; vascular remodeling; right ventricular dysfunction

---

1. Introduction

Pulmonary hypertension (PH) is a clinical entity consisting of different conditions with elevated mean pulmonary arterial pressure (PAP) that is above 25 mmHg and is quantified by invasive measurements [1]. Elevation of PAP is associated with increased morbidity and mortality rates [2]. According to the classification used by current guidelines [1], five entities of PH are distinguished according to etiology. Group 1, pulmonary arterial hypertension (PAH), mainly includes the idiopathic and hereditary forms, as well as PH that is associated with connective tissue disease and drug intake. Group II is defined as PH due to left heart disease. Group III covers PH due to lung diseases and/or hypoxia. Group IV includes those conditions where chronic thromboembolic pulmonary events result in PH (CTEPH). Lastly, Group V covers miscellaneous disorders that lead to an increase in the pulmonary arterial pressure. However, significant overlap in clinical features and etiology exists [1].

Despite the diversity in PH etiology, certain pathophysiological processes occur in all forms of pulmonary hypertension. These include vasoconstriction, microthrombi formation, and pulmonary vascular remodeling, including structural and functional rebuilding of the extracellular matrix (ECM) [3]. Pathological tissue remodeling is known to be accompanied by the re-occurrence of fetal variants of matrix proteins, like tenascin-C (Tn-C), which are absent in non-diseased mature tissues. This effect has been repeatedly demonstrated for remodeling of cardiac tissue [4–9], but also occurs during vascular remodeling [10]. The cell adhesion molecule Tn-C is known to play a pivotal role in the regulation of cell adhesion, activation, differentiation, and migration in inflammation and tissue remodeling [11]. Different so-called fetal variants of this protein are generated by alternative splicing of the pre-mRNA, in particular B and C-domain, containing Tn-C (B^+ and C^+ Tn-C) [12,13]. These variants are expressed during embryonic development and under pathological conditions, while being virtually absent in healthy adult organs [14].

Because of the close association of fetal Tn-C variants and tissue, as well as vascular remodeling, the aim of the present study was to investigate the serum concentrations of B^+ and C^+ Tn-C in patients with pulmonary hypertension as compared to healthy control persons. Additionally, the assessment of echocardiographic, standard laboratory and functional parameters were performed to enable correlation analyses with circulating Tn-C to elucidate the possible role of these protein variants as novel biomarkers for diagnosis and risk stratification in patients with PH.

2. Results

2.1. Baseline Characteristics

For these investigations, 80 patients with PH and 40 apparently healthy subjects were recruited. Baseline characteristics of the included PH patients, as well as controls are reported in Table 1. PH patients had more co-morbidities, such as hyperlipidemia, coronary artery disease, diabetes mellitus, and chronic kidney disease. Medical therapy of the PH patients included more frequent intake of statins and glucocorticoids. Laboratory analyses revealed higher BNP, CRP, and creatinine, and lower LDL cholesterol and hemoglobin. The lower LDL cholesterol is at least in part explained by the higher intake of statins in the PH patients but also the chronic inflammatory state.

Table 1. Clinical data: Cardiovascular risk factors and medication of the study groups. In (**A**) presented for patients with pulmonary hypertension irrespective the etiology and control patients, in (**B**) presented for patients with pulmonary hypertension that are subdivided according to the etiology and control patients.

(A)

Clinical Parameter	Control Persons (n = 40)	PH Patients (n = 80)	p-Value
Age (years)	66 ± 7	70 ± 13	n.s.
BMI (kg/m^2)	27.9 ± 4.7	28.6 ± 6.0	n.s.
Gender, male (%)	33	39	n.s.
Systolic BP (mmHg)	146 ± 33	146 ± 33	n.s.
Diastolic BP (mmHg)	81 ± 17	78 ± 13	n.s.
Functional class	1.5 ± 0.6	2.6 ± 0.8	<0.001
Laboratory:			
BNP (pg/mL)	56 ± 70	445 ± 584	<0.001
CRP (mg/L)	2.7 ± 2.6	11.7 ± 17.5	0.002
Creatinine (μmol/l)	76 ± 16	111 ± 50	<0.001
LDL (mmol/l)	3.6 ± 1.0	2.7 ± 1.0	<0.001
Haemoglobin (mmol/l)	8.7 ± 0.8	7.9 ± 1.3	<0.001
Leukocytes (Gpt/l)	7.1 ± 1.4	7.5 ± 2.2	n.s.
Co-Morbidities:			
Hypertension (%)	95	83	0.03
Hyperlipidemia (%)	88	56	<0.001
Obesity (%) (BMI>30 kg/m^2)	42	38	n.s.
CAD (%)	0	26	<0.001
CKD (%) (GFR < 50 mL/min)	8	50	<0.001
Diabetes (%)	20	50	0.002
Smoking (%)	29	52	0.027
Medication:			
ASA (%)	25	18	n.s.
Beta blocker (%)	63	60	n.s.
ACE Inhibitor/Sartans (%)	85	78	n.s.
Statins (%)	38	59	0.026
Prednisolon (%)	0	11	0.027
ICS (%)	0	21	0.002

(B)

Characteristics	Control (n = 40)	PH Class I (n = 13)	PH Class II (n = 30)	PH Class III (n = 11)	PH Class IV (n = 12)	PH Class II & III (n = 14)	p-Value between Different PH-Classes
Age (years)	66.0 ± 6.8	64.5 ± 11.5	75.1 ± 8.1	58.5 ± 22.3	69.8 ± 11.5	74.9 ± 7.9	<0.001
BMI (kg/m^2)	28.0 ± 4.7	29.0 ± 8.4	27.9 ± 3.9	25.9 ± 7.5	30.6 ± 5.2	29.9 ± 6.5	n.s.
Functional class	1.5 ± 0.6	2.6 ± 0.8	2.6 ± 0.7	2.4 ± 1.1	2.3 ± 0.7	2.8 ± 0.7	n.s.
Laboratory:							
BNP (pg/mL)	56 ± 70	113 ± 82	635 ± 678	469 ± 469	111 ± 87	627 ± 731	0.021
CRP (mg/L)	2.7 ± 2.6	7.8 ± 10.7	13.8 ± 21.0	10.5 ± 12.9	5.2 ± 5.8	16.1 ± 22.0	n.s.
Creatinine (μmol/L)	76 ± 16	88 ± 44	127 ± 59	93 ± 24	99 ± 40	121 ± 47	n.s.
LDL (mmol/L)	3.6 ± 1.0	2.5 ± 1.0	2.5 ± 1.0	3.0 ± 1.0	3.0 ± 0.9	2.7 ± 1.2	n.s.
Haemoglobin (mmol/L)	8.7 ± 0.8	7.7 ± 1.5	7.5 ± 1.2	8.5 ± 1.0	8.9 ± 1.4	7.6 ± 1.6	0.012
Leukocytes (Gpt/L)	7.1 ± 1.4	6.7 ± 2.4	7.6 ± 1.7	8.3 ± 1.5	7.4 ± 1.8	7.7 ± 3.4	n.s.

Data are presented as mean ± standard deviation or percentage. ACE = angiotensin-converting enzyme, ASA = acetyl-salicylic acid, BMI = Body mass index, BNP = brain natriuretic peptide, CHD = coronary heart disease, CKD = chronic kidney disease, CRP = C-reactive protein, LDL = low-density lipoprotein, n.s. = not significant.

2.2. Serum Levels of B^+ and C^+ Tn-C

Both fetal splicing variants of Tn-C were higher in patients with PH (B^+ Tn-C: 752 (CI 520–1110) ng/mL versus 368 (CI 238–616) ng/mL, p < 0.001, Figure 1A; C^+ Tn-C: 79 (CI 48–125) ng/mL versus 66 (CI 44–82) ng/mL, p = 0.034, Figure 1B).

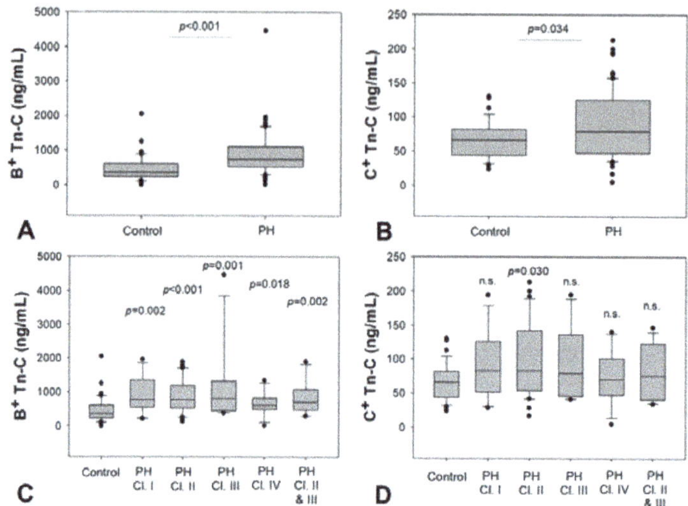

Figure 1. Increased serum levels of B⁺ and C⁺ Tn-C in patients with pulmonary hypertension (PH) compared to healthy controls. (**A**) shows the results for B⁺ Tn-C; (**B**) for C⁺ Tn-C; (**C,D**) show the subgroup analyses with respect to the PH etiology, PH classes are named according to the classification of the current guidelines. The box plots indicate the median (line inside the box), 25 and 75 percentile (lower and upper boundary of the box), 10 and 90 percentile (whiskers outside the box) as well as outlier values (dots). Cl. = class, Tn-C = Tenascin C.

For B⁺ Tn-C, analysis with respect to the PH etiology revealed significant elevations compared to healthy controls for all classes investigated: class I (780 (CI 559–1350) ng/mL, $p = 0.002$), class II (778 (CI 530–1191) ng/mL, $p < 0.001$), class III PH (811 (CI 447–1324) ng/mL, $p = 0.001$), and class IV (624 (CI 489–829) ng/mL, $p = 0.018$). Similar results were found for PH of mixed genesis, including PH resulting from lung and cardiac diseases (class II and class III) (706 (CI 478–1089) ng/mL, $p = 0.002$) (Figure 1C). For C⁺ Tn-C, significantly elevated concentrations were observed in patients with PH when compared to controls, but subgroup analysis only revealed a significant difference in patients with PH class II as compared to healthy controls (PH class II: 83 (CI 54–142), $p = 0.030$) (Figure 1D).

2.3. Correlation of Tn-C Serum Levels with Echocardiographic Parameters

For both patients' groups, echocardiographic parameters were obtained (Table 2) and were correlated with the serum concentrations of Tn-C. For both variants of Tn-C, a significant correlation between the serum concentration and the systolic PAP was observed (B⁺ Tn-C: $r = 0.31$, $p < 0.001$, C⁺ Tn-C: $r = 0.26$, $p = 0.006$). Additionally, a significant correlation between the serum levels of B⁺ and C⁺ Tn-C and the right atrial area was found (B⁺ Tn-C: $r = 0.46$, $p < 0.001$, C⁺ Tn-C: $r = 0.49$, $p < 0.001$). These correlations are demonstrated in Figure 2. No significant correlation was observed between Tn-C variants and TAPSE or TDI of the RV as markers of right heart dysfunction.

Table 2. Echocardiographic parameters of the study groups.

Parameter	Control (*n* = 40)	PH Class I (*n* = 13)	PH Class II (*n* = 30)	PH Class III (*n* = 11)	PH Class IV (*n* = 12)	PH Class II & III (*n* = 14)
LV-EF (%)	68 ± 7	63 ± 7	56 ± 13	63 ± 9	60 ± 6	52 ± 12
IVSDd (mm)	12 ± 2	12 ± 3	13 ± 3	13 ± 2	11 ± 2	12 ± 2
RVd basal (mm)	35 ± 2	41 ± 9	46 ± 7	47 ± 10	44 ± 7	49 ± 9
TAPSE (mm)	25 ± 2	22 ± 5	15 ± 3	19 ± 7	18 ± 3	16 ± 4
TDI-S′ (RV) (cm/s)	14 ± 2	12 ± 1	8 ± 1	9 ± 0	11 ± 1	9 ± 2
RA area (cm^2)	15.4 ± 2.3	20.8 ± 8.3	27.9 ± 9.6	19.7 ± 8.8	22.2 ± 5.2	31.0 ± 13.7
PAP sys (mmHg)	21 ± 4	59 ± 22	52 ± 16	57 ± 22	50 ± 20	56 ± 16

Data are presented as mean ± standard deviation. IVSDd = intraventricular septum diameter in diastole, LV-EF = left ventricular ejection fraction, PAP sys = systolic pulmonary arterial pressure, RA = right atrium, RVd = right ventricular diameter in diastole, TAPSE = tricuspid annular plane systolic excursion, TDI-S′ (RV) = tissue doppler imaging, right ventricle. PH classes given in this table correspond to the PH groups defined in the current guidelines.

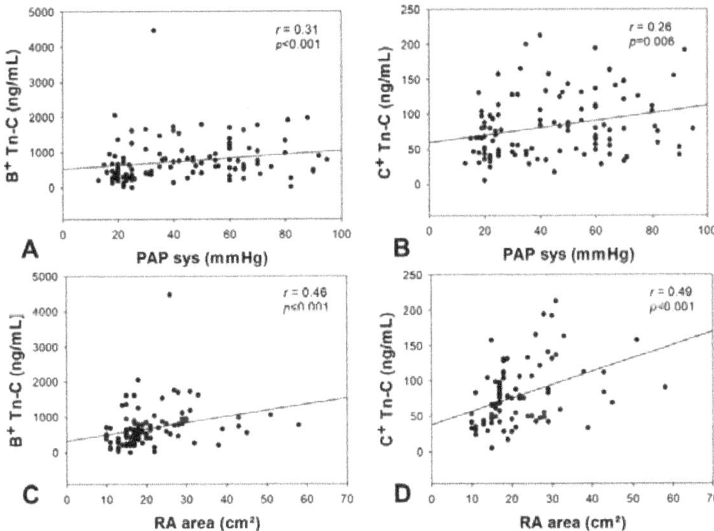

Figure 2. Correlation analyses between the serum concentration of B$^+$ and C$^+$ Tn-C and echocardiographic parameters. The correlation analysis graphs demonstrate significant correlations between B$^+$ and C$^+$ Tn-C and the systolic pulmonary artery pressure (**A**,**B**), as well as the area of the right atrium (**C**,**D**). PAP sys = systolic pulmonary arterial pressure, RA = right atrium, Tn-C = tenascin C, *p*-value = level of significance, r-value = correlation coefficient.

2.4. Correlation of Tn-C Serum Levels with BNP

Serum concentration of Tn-C correlated with BNP as a standard laboratory parameter of heart failure. For both Tn-C variants, a significant correlation could be demonstrated (B$^+$ Tn-C: *r* = 0.45, *p* < 0.001; C$^+$ Tn-C: *r* = 0.42, *p* < 0.001, Figure 3).

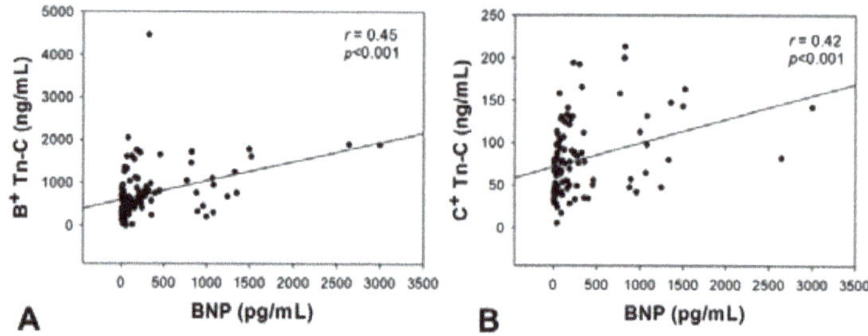

Figure 3. Correlation analyses between the serum concentration of B+ (**A**) and C+ (**B**) Tn-C and the brain natriuretic peptide (BNP). BNP = brain natriuretic peptide, Tn-C = tenascin C, *p*-value = level of significance, *r*-value = correlation coefficient.

2.5. Correlation of Tn-C Serum Concentrations with the 6-Minute Walk Distance

An inverse correlation was observed between serum levels of the two Tn-C variants and the 6-minute walk distance (B+ Tn-C: $r = -0.54$, $p < 0.001$; C+ Tn-C: $r = -0.43$, $p < 0.001$) as a parameter reflecting functional capacity (Figure 4).

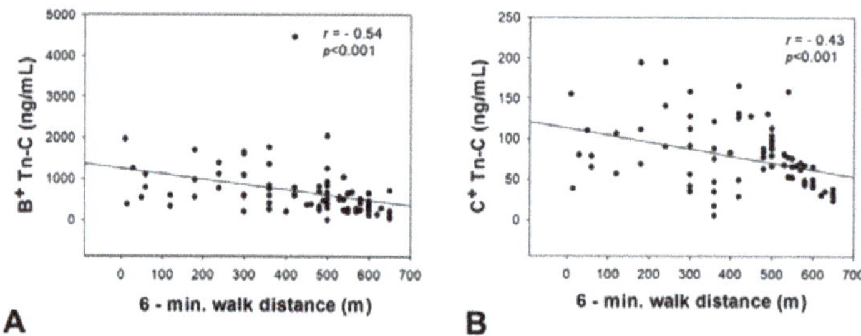

Figure 4. Correlation analyses between the serum concentration of B+ (**A**) and C+ (**B**) Tn-C and the 6-min walk distance. Tn-C = tenascin C, *p*-value = level of significance, *r*-value = correlation coefficient.

2.6. Impact of Clinical Parameters on Tn-C Serum Levels

Because of significant clinical differences in the patient groups that were enrolled in our study, we performed further analyses investigating possible confounding variables. To verify the predictive value of Tn-C serum concentration for the probability of pulmonary hypertension, a multivariate analysis was performed. Age, diabetes mellitus, CAD, CKD, statin medication, C-reactive protein levels, and the serum concentration of the two Tn-C variants were entered into the analysis as independent variables. After backward elimination, only CKD (Wald: 9.529, OR: 12.316, 95% CI: 2.501–60.647, $p = 0.002$), C-reactive protein (Wald: 4.414, OR: 1.199, 95% CI: 1.012–1.420, $p = 0.036$), and B+ Tn-C (Wald: 7.854, OR: 1.002, 95% CI: 1.001–1.004, $p = 0.005$), but not age, diabetes mellitus, CAD, statin medication, and C+ Tn-C were found to be independent predictors of pulmonary hypertension.

3. Discussion

Pathological tissue and vascular remodeling is associated with a variety of cardiovascular disorders results in the re-occurrence of fetal variants of extracellular matrix and cell adhesion modulating proteins like Tn-C, which are virtually absent in non-diseased adult organs [14]. The re-expression of fetal Tn-C variants like B+ and T+ Tn-C was repeatedly demonstrated to reflect the extent of cardiovascular remodeling and disease severity for several heart diseases [6,7,9]. Therefore, these molecules have been suggested as novel biomarkers not only for diagnosis and prognosis estimation, but also for therapeutic surveillance since serum levels also reflect reverse remodeling [4–6,9,15,16]. Fetal Tn-C re-occurrence has also been demonstrated with special regard to vascular remodeling processes [17,18].

In an animal model using monocrotaline-injected rats, Rabinovitch and colleagues demonstrated that the development of pulmonary arterial medial hypertrophy was accompanied by Tn-C expression in co-localization with proliferating smooth muscle cells [17]. Additionally, it could be shown that iloprost inhalation results in a reversion of the remodeling process reflected by a decrease in Tn-C occurrence [18]. Furthermore, Correira-Pinto et al. also demonstrated an overexpression of Tn-C in cardiac tissue of monocrotaline-induced pulmonary hypertension in an animal model reflecting PH-induced cardiac, and in particular, right ventricular pressure overload [19]. However, these observations have only been made for Tn-C found in tissue, not serum, and it is well known that lung tissue from PH patients is not available in daily routine because taking biopsies is not indicated in these patients and will not be an option, even in the future, due to the high risk associated with this invasive procedure.

In the present study, we investigated fetal Tn-C variants as serum biomarkers of tissue and vascular remodeling in human PH. We demonstrate that fetal Tn-C levels are increased in PH when compared to healthy controls. Elevated Tn-C concentrations in patients suffering from PH have previously been described by Schumann et al. [20]. The authors, however, did not analyze different fetal Tn-C variants [20]. Therefore, our data are the first to demonstrate both B$^+$ and C$^+$ Tn-C as functionally most important fetal splicing variants of the protein with prognostic significance in PH.

Subgroup analyses of the present study revealed that an elevated serum concentration of B$^+$ Tn-C occurs in PH due to different etiologies as compared to healthy controls. For C$^+$ Tn-C, subgroup analysis revealed significant differences only between PH class II and the control group. It has to be mentioned, that in this PH class, the highest number of patients has been included. Besides this, the intake of glucocorticoids might diminish statistical significance in PH due to lung diseases. An influence of glucocorticoids on Tn-C expression has been previously described in the literature [21–23].

Interestingly, Tn-C serum concentrations could be shown to correlate with echocardiographic parameters, such as systolic PAP and the area of the right atrium. This demonstrates significant correlations between the fetal variants of the matrix glycoprotein and parameters reflecting right ventricular load. This finding is supported by former studies that found PH-induced cardiac remodeling with an increased Tn-C occurrence [19]. This idea is supported by a publication investigating a patient cohort with acute pulmonary thromboembolism describing a correlation between Tn-C and systolic PAP [24]. Additionally, similar effects have been described in an animal model. Monocrotaline-induced right ventricular failure was shown to be associated with an up-regulation of Tn-C gene expression and results in significantly elevated plasma levels. However, in this model, a significant correlation could be observed for right ventricular ejection fraction and Tn-C [25]. In our present study, as markers of right ventricular function, TAPSE and TDI were echocardiographically measured but no significant correlation between these parameters and Tn-C were observed.

Another interesting finding of the present study was a significant positive correlation between serum BNP and both, B$^+$ and C$^+$ Tn-C levels. This reflects prior findings in human studies of patients with acute PH due to pulmonary thromboembolism [24], as well as animal models [25]. Besides PH, this correlation has also repeatedly been described in other cardiovascular diseases [16]. However, the present study was not only able to demonstrate a correlation between the fetal Tn-C variants and laboratory parameters, but also with functional data reflecting the physical capability of PH patients assessed by the 6-min walk test. This association was found for both Tn-C variants and is a novel finding of the current study.

There are certain limitations of our study. First, due to the small numbers of patients, statistical analyses underestimate the differences. Second, heterogeneity with respect to the clinical data of PH patients and the controls has to be mentioned. Moreover, significant differences in the subgroup analyses can be explained by the PH etiology. The highest mean BNP value is found in the PH class II, reflecting PH due to left heart disease. This is possibly due to the increased cardiac load not only of the right heart, but also the left heart occurring in the course of the underlying disease.

4. Material and Methods

4.1. Patients

We enrolled 80 patients with PH admitted to the Department of Internal Medicine I, Jena University Hospital, Friedrich Schiller University of Jena, Germany. 40 apparently healthy control subjects were recruited within the same department. Controls were enrolled after invasive exclusion of coronary artery disease [26]. Further exclusion criteria were malignant or autoimmune disease, hyperthyroidism, infection, history of pulmonary embolism or stroke, peripheral artery disease, and medical treatment, including corticosteroids or immunosuppressive agents. All of the patients underwent transthoracic echocardiography and a 6-minute walk test. Blood samples were collected for routine laboratory analyses, including serum brain natriuretic peptide (BNP) levels. Serum was stored using special low binding tubes (Protein LoBind Tubes, Eppendorf AG, Hamburg, Germany) and stored at $-80\ ^\circ$C after snap freezing in liquid nitrogen to reduce artificial protein degradation. Moreover, repeated freeze-thaw cycles were strictly avoided.

The investigation conforms with the principles outlined in the *Declaration of Helsinki* (*Br Med J 1964*; ii: 177) and was approved by the local ethics committee (registration number: 4732-03/16, 11 April 2016). All patients gave written informed consent before inclusion into the study.

4.2. Quantification of Serum Tn-C Levels and Standard Laboratory Values

Serum levels of B^+ and C^+ Tn-C were determined using enzyme-linked immunosorbent assay (ELISA). Well-established and validated ELISA assays are commercially available (Tenascin-C Large (FNIII-C) ELISA and Tenascin C Large (FNIII-B) ELISA, both IBL International GmbH, Hamburg, Germany). Routine standard laboratory parameters were measured according to standard hospital procedures.

4.3. Investigation of Echocardiographic Parameters and 6-Minute Walk Test

For each patient, a transthoracic echocardiography was performed to determine standard parameters, including left ventricular ejection fraction (EF), diastolic diameter of the interventricular septum (IVSDd), or relevant valve abnormalities. Moreover, special effort was made to carefully assess the parameters representing right heart morphology and dysfunction. Here, especially tricuspid annular plane systolic excursion (TAPSE), tissue doppler imaging of the right ventricle (TDI RV), the area of the right atrium (RA), and systolic pulmonary arterial pressure were measured. Moreover, 6-minute walk distance was documented for each patient.

4.4. Statistical Analysis

Statistical analysis was performed using SPSS (version 20.0, IBM Inc., Armonk, NY, USA) and SigmaPlot (version 12.0, Systat Software Inc., San Jose, CA, USA). The Kolmogorov Smirnov test was used to test the normal distribution of all the variables. When normally distributed, values are reported as mean ± standard deviation. When not normally distributed, the values are reported as median (25–75% Confidence Interval). The non-parametric Mann-Whitney Rank Sum Test was used to compare the number of different cells between two different study groups. To compare more than two groups, the Kruskal Wallis test was used. Bivariate correlations between parametric variables were assessed by the Spearman rank correlation test. To test the predictive value of Tn-C concentrations on the probability of the occurrence of pulmonary hypertension, a multivariate regression analysis was performed using a binary logistic model (backward elimination method: Wald). The presence of pulmonary hypertension was defined as the dependent variable. Age, diabetes mellitus, CAD, CKD,

C-reactive protein, statin treatment, and the two Tn-C variants were included into the first step. Then, multistep backward elimination (removal threshold $p > 0.10$) of independent variables was carried out. $p < 0.05$ was considered statistically significant.

5. Conclusions

In conclusion, the present study demonstrates that there is an increase of serum of B[+] and C[+] Tn-C in patients with PH. For B[+] Tn-C, this is visible for all PH classes. This is an important novel finding, because most of the studies in the field of PH have been conducted in PAH, which is a rare disease [27]. The prevalence of especially class II and class III PH is more frequent. For this reason, research in this field is necessary to improve diagnostic and therapeutic approaches.

Based on our findings, both the investigated Tn-C splicing variants, but especially B[+] Tn-C, can be suggested as promising novel biomarkers in human PH, which merits further investigation and validation in larger patient cohorts. Fetal Tn-C variants are functionally involved in vascular and tissue remodeling associated with PH probably irrespective of the particular etiology. This should be further investigated both in vivo and in vitro and raises the question, whether functional blocking strategies might be a therapeutic approach for PH treatment in the future. In this context, the availability of human recombinant antibodies specific to fetal Tn-C variants might be of certain interest since these antibodies can serve as vehicles for targeted delivery of both, bioactive molecules, such as immunocytokines and antibody-drug-conjugates. Further, this might have diagnostic potential for the development of novel radionuclides or molecular imaging strategies [28–30] to visualize pathologic lung tissue and vascular remodeling.

Acknowledgments: We thank Annett Schmidt for her technical assistance. This work was supported by a research grant from Actelion Pharmaceuticals, Germany.

Author Contributions: Ilonka Rohm, Katja Grün, Linda Marleen Müller, Alexander Berndt, Christian Jung and Marcus Franz designed the study, performed the experiments, analysed the data; Ilonka Rohm and Marcus Franz wrote the manuscript; Daniel Kretzschmar, Michael Fritzenwanger, Atilla Yilmaz, Alexander Lauten, P. Christian Schulze and Marcus Franz analysed the data and revised the manuscript.

Conflicts of Interest: The authors declare no conflict of interest

Abbreviations

ACE	Angiotensin-converting enzyme
ASA	Acetyl-salicylic acid
BMI	Body mass index
BNP	Brain natriuretic peptide
CAD	Coronary artery disease
CKD	Chronic kidney disease
CRP	C-reactive protein
ECM	Extracellular matrix
IVSDd	Intraventricular septum diameter in diastole
LDL	Low-density lipoprotein
LV-EF	Left ventricular ejection fraction
PH	Pulmonary hypertension
PAH	Pulmonary arterial hypertension
PAP	Pulmonary arterial pressure
RA	Right atrium
RV	Right ventricle
TAPSE	Tricuspid annular plane systolic excursion
TDI	Tissue doppler imaging
Tn-C	Tenascin C

References

1. Galie, N.; Humbert, M.; Vachiery, J.L.; Gibbs, S.; Lang, I.; Torbicki, A.; Simonneau, G.; Peacock, A.; Vonk Noordegraaf, A.; Beghetti, M.; et al. 2015 ESC/ERS Guidelines for the diagnosis and treatment of pulmonary hypertension: The Joint Task Force for the Diagnosis and Treatment of Pulmonary Hypertension of the European Society of Cardiology (ESC) and the European Respiratory Society (ERS): Endorsed by: Association for European Paediatric and Congenital Cardiology (AEPC), International Society for Heart and Lung Transplantation (ISHLT). *Eur. Respir. J.* **2015**, *46*, 903–975. [PubMed]
2. Benza, R.L.; Miller, D.P.; Barst, R.J.; Badesch, D.B.; Frost, A.E.; McGoon, M.D. An evaluation of long-term survival from time of diagnosis in pulmonary arterial hypertension from the REVEAL Registry. *Chest* **2012**, *142*, 448–456. [CrossRef] [PubMed]
3. Huber, L.C.; Bye, H.; Brock, M.; Swiss Society of Pulmonary Hypertension. The pathogenesis of pulmonary hypertension—An update. *Swiss Med. Wkly.* **2015**, *145*, w14202. [CrossRef] [PubMed]
4. Imanaka-Yoshida, K.; Hiroe, M.; Nishikawa, T.; Ishiyama, S.; Shimojo, T.; Ohta, Y.; Sakakura, T.; Yoshida, T. Tenascin-C modulates adhesion of cardiomyocytes to extracellular matrix during tissue remodeling after myocardial infarction. *Lab. Investig.* **2001**, *81*, 1015–1024. [CrossRef] [PubMed]
5. Franz, M.; Berndt, A.; Altendorf-Hofmann, A.; Fiedler, N.; Richter, P.; Schumm, J.; Fritzenwanger, M.; Figulla, H.R.; Brehm, B.R. Serum levels of large tenascin-C variants, matrix metalloproteinase-9, and tissue inhibitors of matrix metalloproteinases in concentric versus eccentric left ventricular hypertrophy. *Eur. J. Heart Fail.* **2009**, *11*, 1057–1062. [CrossRef] [PubMed]
6. Sarli, B.; Topsakal, R.; Kaya, E.G.; Akpek, M.; Lam, Y.Y.; Kaya, M.G. Tenascin-C as predictor of left ventricular remodeling and mortality in patients with dilated cardiomyopathy. *J. Investig. Med.* **2013**, *61*, 728–732. [CrossRef] [PubMed]
7. Franz, M.; Brehm, B.R.; Richter, P.; Gruen, K.; Neri, D.; Kosmehl, H.; Hekmat, K.; Renner, A.; Gummert, J.; Figulla, H.R.; et al. Changes in extra cellular matrix remodelling and re-expression of fibronectin and tenascin-C splicing variants in human myocardial tissue of the right atrial auricle: Implications for a targeted therapy of cardiovascular diseases using human SIP format antibodies. *J. Mol. Histol.* **2010**, *41*, 39–50. [PubMed]
8. Baldinger, A.; Brehm, B.R.; Richter, P.; Bossert, T.; Gruen, K.; Hekmat, K.; Kosmehl, H.; Neri, D.; Figulla, H.R.; Berndt, A.; et al. Comparative analysis of oncofetal fibronectin and tenascin-C expression in right atrial auricular and left ventricular human cardiac tissue from patients with coronary artery disease and aortic valve stenosis. *Histochem. Cell Biol.* **2011**, *135*, 427–441. [CrossRef] [PubMed]
9. Franz, M.; Berndt, A.; Neri, D.; Galler, K.; Grun, K.; Porrmann, C.; Reinbothe, F.; Mall, G.; Schlattmann, P.; Renner, A.; et al. Matrix metalloproteinase-9, tissue inhibitor of metalloproteinase-1, B$^+$ tenascin-C and ED-A$^+$ fibronectin in dilated cardiomyopathy: Potential impact on disease progression and patients' prognosis. *Int. J. Cardiol.* **2013**, *168*, 5344–5351. [CrossRef] [PubMed]
10. Wallner, K.; Sharifi, B.G.; Shah, P.K.; Noguchi, S.; DeLeon, H.; Wilcox, J.N. Adventitial remodeling after angioplasty is associated with expression of tenascin mRNA by adventitial myofibroblasts. *J. Am. Coll. Cardiol.* **2001**, *37*, 655–661. [CrossRef]
11. Midwood, K.S.; Hussenet, T.; Langlois, B.; Orend, G. Advances in tenascin-C biology. *Cell. Mol. Life Sci.* **2011**, *68*, 3175–3199. [CrossRef] [PubMed]
12. Borsi, L.; Balza, E.; Gaggero, B.; Allemanni, G.; Zardi, L. The alternative splicing pattern of the tenascin-C pre-mRNA is controlled by the extracellular pH. *J. Biol. Chem.* **1995**, *270*, 6243–6245. [CrossRef] [PubMed]
13. Jones, P.L.; Jones, F.S. Tenascin-C in development and disease: Gene regulation and cell function. *Matrix Biol.* **2000**, *19*, 581–596. [CrossRef]
14. Franz, M.; Jung, C.; Lauten, A.; Figulla, H.R.; Berndt, A. Tenascin-C in cardiovascular remodeling: Potential impact for diagnosis, prognosis estimation and targeted therapy. *Cell Adhes. Migr.* **2015**, *9*, 90–95. [CrossRef] [PubMed]
15. Hessel, M.H.; Bleeker, G.B.; Bax, J.J.; Henneman, M.M.; den Adel, B.; Klok, M.; Schalij, M.J.; Atsma, D.E.; van der Laarse, A. Reverse ventricular remodelling after cardiac resynchronization therapy is associated with a reduction in serum tenascin-C and plasma matrix metalloproteinase-9 levels. *Eur. J. Heart Fail.* **2007**, *9*, 1058–1063. [CrossRef] [PubMed]

16. Franz, M.; Berndt, A.; Grun, K.; Kuethe, F.; Fritzenwanger, M.; Figulla, H.R.; Jung, C. Serum levels of tenascin-C variants in congestive heart failure patients: Comparative analysis of ischemic, dilated, and hypertensive cardiomyopathy. *Clin. Lab.* **2014**, *60*, 1007–1013. [CrossRef] [PubMed]

17. Rabinovitch, M. Elastase and the pathobiology of unexplained pulmonary hypertension. *Chest* **1998**, *114* (Suppl. 3), 213S–224S. [CrossRef] [PubMed]

18. Schermuly, R.T.; Yilmaz, H.; Ghofrani, H.A.; Woyda, K.; Pullamsetti, S.; Schulz, A.; Gessler, T.; Dumitrascu, R.; Weissmann, N.; Grimminger, F.; et al. Inhaled iloprost reverses vascular remodeling in chronic experimental pulmonary hypertension. *Am. J. Respir. Crit. Care Med.* **2005**, *172*, 358–363. [CrossRef] [PubMed]

19. Correia-Pinto, J.; Henriques-Coelho, T.; Roncon-Albuquerque, R., Jr.; Lourenco, A.P.; Melo-Rocha, G.; Vasques-Novoa, F.; Gillebert, T.C.; Leite-Moreira, A.F. Time course and mechanisms of left ventricular systolic and diastolic dysfunction in monocrotaline-induced pulmonary hypertension. *Basic Res. Cardiol.* **2009**, *104*, 535–545. [CrossRef] [PubMed]

20. Schumann, C.; Lepper, P.M.; Frank, H.; Schneiderbauer, R.; Wibmer, T.; Kropf, C.; Stoiber, K.M.; Rudiger, S.; Kruska, L.; Krahn, T.; et al. Circulating biomarkers of tissue remodelling in pulmonary hypertension. *Biomarkers* **2010**, *15*, 523–532. [CrossRef] [PubMed]

21. Fassler, R.; Sasaki, T.; Timpl, R.; Chu, M.L.; Werner, S. Differential regulation of fibulin, tenascin-C, and nidogen expression during wound healing of normal and glucocorticoid-treated mice. *Exp. Cell Res.* **1996**, *222*, 111–116. [CrossRef] [PubMed]

22. Gratchev, A.; Kzhyshkowska, J.; Utikal, J.; Goerdt, S. Interleukin-4 and dexamethasone counterregulate extracellular matrix remodelling and phagocytosis in type-2 macrophages. *Scand. J. Immunol.* **2005**, *61*, 10–17. [CrossRef] [PubMed]

23. Ekblom, M.; Fassler, R.; Tomasini-Johansson, B.; Nilsson, K.; Ekblom, P. Downregulation of tenascin expression by glucocorticoids in bone marrow stromal cells and in fibroblasts. *J. Cell Biol.* **1993**, *123*, 1037–1045. [CrossRef] [PubMed]

24. Celik, A.; Kocyigit, I.; Calapkorur, B.; Korkmaz, H.; Doganay, E.; Elcik, D.; Ozdogru, I. Tenascin-C may be a predictor of acute pulmonary thromboembolism. *J. Atheroscler. Thromb.* **2011**, *18*, 487–493. [CrossRef] [PubMed]

25. Hessel, M.; Steendijk, P.; den Adel, B.; Schutte, C.; van der Laarse, A. Pressure overload-induced right ventricular failure is associated with re-expression of myocardial tenascin-C and elevated plasma tenascin-C levels. *Cell. Physiol. Biochem.* **2009**, *24*, 201–210. [CrossRef] [PubMed]

26. Task Force, M.; Montalescot, G.; Sechtem, U.; Achenbach, S.; Andreotti, F.; Arden, C.; Budaj, A.; Bugiardini, R.; Crea, F.; Cuisset, T.; et al. 2013 ESC guidelines on the management of stable coronary artery disease: The Task Force on the management of stable coronary artery disease of the European Society of Cardiology. *Eur. Heart J.* **2013**, *34*, 2949–3003.

27. Ling, Y.; Johnson, M.K.; Kiely, D.G.; Condliffe, R.; Elliot, C.A.; Gibbs, J.S.; Howard, L.S.; Pepke-Zaba, J.; Sheares, K.K.; Corris, P.A.; et al. Changing demographics, epidemiology, and survival of incident pulmonary arterial hypertension: Results from the pulmonary hypertension registry of the United Kingdom and Ireland. *Am. J. Respir. Crit. Care Med.* **2012**, *186*, 790–796. [CrossRef] [PubMed]

28. Bootz, F.; Neri, D. Immunocytokines: A novel class of products for the treatment of chronic inflammation and autoimmune conditions. *Drug Discov. Today* **2016**, *21*, 180–189. [CrossRef] [PubMed]

29. Casi, G.; Neri, D. Antibody-drug conjugates: Basic concepts, examples and future perspectives. *J. Control. Release* **2012**, *161*, 422–428. [CrossRef] [PubMed]

30. Rybak, J.N.; Trachsel, E.; Scheuermann, J.; Neri, D. Ligand-based vascular targeting of disease. *ChemMedChem* **2007**, *2*, 22–40. [CrossRef] [PubMed]

International Journal of
Molecular Sciences

Article

The ADAMTS5 Metzincin Regulates Zebrafish Somite Differentiation

Carolyn M. Dancevic [1,2], Yann Gibert [1,2], Joachim Berger [3], Adam D. Smith [1,2], Clifford Liongue [1,2], Nicole Stupka [1,2], Alister C. Ward [1,2,*] and Daniel R. McCulloch [1,2]

[1]   School of Medicine, Deakin University, Waurn Ponds, Victoria 3216, Australia;
     carolyn.dancevic@deakin.edu.au (C.M.D.); y.gibert@deakin.edu.au (Y.G.);
     adam.smith@bio-strategy.com (A.D.S.); c.liongue@deakin.edu.au (C.L.);
     nicole.stupka@deakin.edu.au (N.S.); daniel.mcculloch@uq.net.au (D.R.M.)
[2]   Centre for Molecular and Medical Research, Deakin University, Waurn Ponds, Victoria 3216, Australia
[3]   Australian Regenerative Medicine Institute, Monash University, Clayton, Victoria 3800, Australia;
     joachim.berger@monash.edu
*   Correspondence: award@deakin.edu.au

Received: 25 January 2018; Accepted: 1 March 2018; Published: 7 March 2018

Abstract: The ADAMTS5 metzincin, a secreted zinc-dependent metalloproteinase, modulates the extracellular matrix (ECM) during limb morphogenesis and other developmental processes. Here, the role of ADAMTS5 was investigated by knockdown of zebrafish *adamts5* during embryogenesis. This revealed impaired Sonic Hedgehog (Shh) signaling during somite patterning and early myogenesis. Notably, synergistic regulation of *myod* expression by ADAMTS5 and Shh during somite differentiation was observed. These roles were not dependent upon the catalytic activity of ADAMTS5. These data identify a non-enzymatic function for ADAMTS5 in regulating an important cell signaling pathway that impacts on muscle development, with implications for musculoskeletal diseases in which ADAMTS5 and Shh have been associated.

Keywords: metalloproteinase; extracellular matrix; ADAMTS; somite; muscle; zebrafish

---

1. Introduction

The A Disintegrin-like and Metalloproteinase domain with Thrombospondin-1 motifs (ADAMTS) metalloproteinases have important functions during developmental morphogenesis and are also implicated in chronic disease. The proteoglycanase subfamily of ADAMTS1, 4, 5, 8, 9, 15 and 20 have broad functions, many attributed to their ability to remodel extracellular matrix (ECM) components, such as the chondroitin sulphate proteoglycans versican and aggrecan. For example, *Adamts20* deficient *bt/bt* mice have defects in melanoblast survival [1] and *Adamts9* haplo-insufficient mice on an *Adamts20* deficient (*bt/bt*) background present with a secondary cleft palate [2], in each case associated with reduced versican proteolysis. Furthermore, ADAMTS1 has been implicated in promoting atherosclerosis [3] and ADAMTS15 acts as a tumor suppressor in breast carcinoma [4], potentially through proteoglycan proteolysis. However, non-enzymatic roles for several ADAMTS family members have been described [5–7].

ADAMTS5 has been implicated in classic morphogenesis during development as well as in chronic diseases such as arthritis and atherosclerosis. For example, combinatorial knockout of *Adamts5*, *Adamts9* and *Adamts20* in mice prevented generation of bioactive fragments of versican that are necessary for interdigital tissue apoptosis during development [8,9]. *Adamts5* knockout mice also developed myxomatous heart valves [10]. Furthermore, ADAMTS5 is considered one of the most important aggrecan-degrading enzymes in arthritis [11,12] and may also promote lipoprotein binding in atherosclerosis [13].

ECM remodeling is crucial to many developmental and disease processes, in part due to its role in controlling cell signaling. Heparan sulphate proteoglycans bind fibroblast growth factors (FGFs), thereby regulating their bioavailability to their receptors (FGFRs) [14] during developmental processes such as myogenesis [15], as well as acting as co-receptors for Sonic Hedgehog (Shh) signaling [16]. A recent study identified *Adamts9* as necessary for umbilical cord vascular development due, at least in part, on its facilitation of Shh signaling [17]. Furthermore, levels of Hedgehog (Hh) signaling correlate with the severity of osteoarthritis, which is potentially mediated by a pathway involving ADAMTS5 [18]. Combined, these studies are suggestive of a complex interplay between the ECM and crucial cell signaling pathways that involves ADAMTS proteoglycanases.

This study identifies a role for ADAMTS5 during zebrafish embryogenesis. Abrogation of *adamts5* expression disrupted Shh signaling during somite differentiation and reduced the expression of the myogenic regulator *myod*. Importantly, somite differentiation was synergistically dependent upon Shh and ADAMTS5. Moreover, these functions of ADAMTS5 were independent of catalytic function. These data indicate that ADAMTS5 plays an important non-enzymatic role in regulating the Shh pathway during embryogenesis that impacts on muscle development. This may be relevant in conditions where ADAMTS proteins interact with the Shh signaling pathway, such as osteoarthritis and umbilical cord vascular complications, as well as disorders where the myogenic program is disrupted, such as muscular dystrophies.

2. Results

2.1. The Secreted Metalloproteinase ADAMTS5 Is Expressed in Zebrafish Embryos

We have previously elucidated a role for ADAMTS5 during myoblast fusion in post-natal skeletal muscle from $Adamts5^{-/-}$ mice [19]. To investigate this further, zebrafish was employed as a highly manipulable model of vertebrate development, which possesses a strongly conserved *adamts5* gene that is maternally inherited and then dynamically expressed in early-stage embryos [20]. To obtain a detailed understanding of ADAMTS5 protein expression in zebrafish, whole-mount immunohistochemistry (IHC) was performed with a previously described anti-ADAMTS5 antibody directed to its pro-domain [21], which is highly conserved in ADAMTS5 across vertebrates [20,22]. At 8 h post fertilization (hpf) (~80% epiboly), ADAMTS5 was strongly expressed in the dorsal mesoendoderm at the animal pole with variable expression ventrally at the vegetal pole (Figure 1A). At 18 and 24 hpf, after the commencement of somitogenesis, ADAMTS5 was expressed in the rostral neural tube (floor plate) and bilaterally in the prosencephalon (Figure 1A).

2.2. Silencing of ADAMTS5 Expression

To explore *adamts5* function, the gene was targeted using two independent morpholino antisense oligonucleotides (MOs) that were directed to either the AUG translation start site (AUG-MO) or the splice site at the exon 2/3 boundary (2/3-MO) (Figure 1B), since exon 3 encodes for the catalytic domain of ADAMTS5 in human, mouse and zebrafish [22]. ADAMTS5 protein expression was found to be reduced upon *adamts5* AUG-MO injection as shown by IHC and immunoblotting (Figure 1C). To confirm altered splicing of *adamts5* transcripts after administration of the 2/3-MO, RT-PCR was performed followed by sequencing analysis (Figure 1D). This indicated a 71% reduction of correctly spliced *adamts5* transcript and identified an alternate *adamts5* transcript retaining the 569-bp intron between exons 2 and 3 that results in inclusion of several premature stop codons (Figure 1D). The AUG-MO was subsequently used throughout the study to ensure translation of the entire gene was disrupted, as well as to guarantee the maternal transcripts for this gene [20] were also affected; however, similar data was obtained with the *adamts5* 2/3-MO [23].

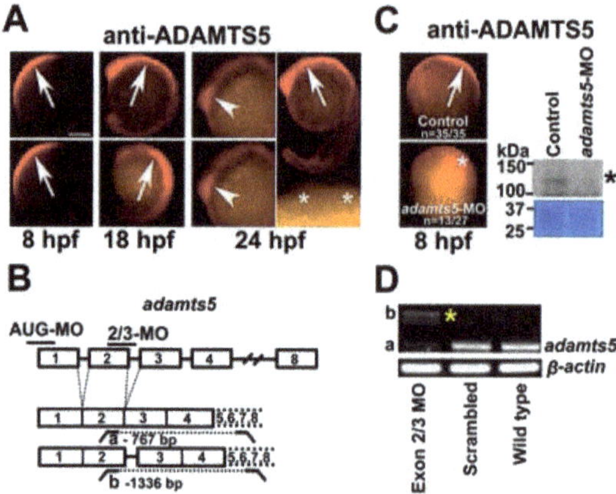

Figure 1. Expression and silencing of *adamts5* in zebrafish embryos. (**A**) ADAMTS5 expression in 8, 18 and 24 hpf wild-type embryos. Note strong early expression in the dorsal mesoendoderm (8 hpf, arrows) and variable expression ventrally (8 hpf, arrowhead), with later expression in the floor plate of the neural tube (18 and 24 hpf, arrows) and bilaterally in the prosencephalon (24 hpf, arrowheads). Asterisks = prosencephalon in no primary antibody control. Scale bar = 250 µm; (**B**) Schematic representation of the *adamts5* gene structure targeted with antisense morpholino oligonucleotides (MO), and its subsequent splicing, indicating the primers used for RT-PCR and the size of the resultant products; (**C**) Reduced ADAMTS5 expression is seen in *adamts5* AUG-MO injected embryos (asterisk) versus control (arrow) by whole-mount antibody labelling (left-hand panel) and Western blot (right-hand panel) showing the 120 kDa ADAMTS5 species (asterisk) with a region of the Coomassie blue stained gel shown below, demonstrating even loading; (**D**) RT-PCR of *adamts5* mRNA obtained from 24 hpf embryos following injection of the *adamts5* 2/3-MO at the 1-cell stage, showing amplicons a and b (asterisk). *β-actin* was used as a house-keeping gene.

2.3. Notochord Morphology Is Perturbed in adamts5 Morphant Embryos

Shh signaling from the notochord has been previously demonstrated to be important for adaxial and paraxial mesoderm formation and *myod* expression during myogenesis [24], while *no tail* (*ntl*) is an independent marker for axial mesoderm (notochord) [25]. Expression of *shh* and *ntl* remained unchanged in 12 hpf *adamts5* morphants compared to controls [23]. However, at 18 hpf the pattern of *shh* (Figure 2A,D) and *ntl* (Figure 2B,E) staining was altered revealing disrupted notochord morphology.

2.4. Skeletal Muscle Formation Is Disrupted in adamts5 Morphant Embryos

Notochord perturbation is linked with defective somitic muscle formation and morphogenesis [24]. Therefore, the disrupted notochord morphology in the *adamts5* morphants suggested that skeletal muscle development might be affected. This is also consistent with previous observations indicating a skeletal muscle developmental defect in *Adamts5* knockout mice [19]. Reduced or absent paraxial mesodermal *myod* expression was also observed at 18 hpf (Figure 2C,F). To analyze potential myofiber defects, *adamts5* AUG-MO was administered to double-transgenic embryos, in which myofiber thin filaments were labeled with Lifeact-GFP whereas the sarcolemma and t-tubules of the myofiber were marked with mCherryCaaX via the CaaX-tag [26]. In control injected 3 dpf double-transgenic larvae, Lifeact-GFP revealed the typical striation of the highly organized myofibril and mCherryCaaX indicated regularly spaced t-tubules and ordered fiber membranes within chevron-shaped somites (Figure 2Ga–a'''). In contrast, the somites of *adamts5* morphants were U-shaped, which resembled a phenotype previously reported in *shh* mutant embryos [27] (Figure 2G(b)), confirming *shh* availability as a potential cause. In addition, myofibril striation within myofibers of *adamts5* morphants was partially lost and the sarcolemma appeared irregular, indicating disrupted muscle organization (Figure 2Gb–b′′′).

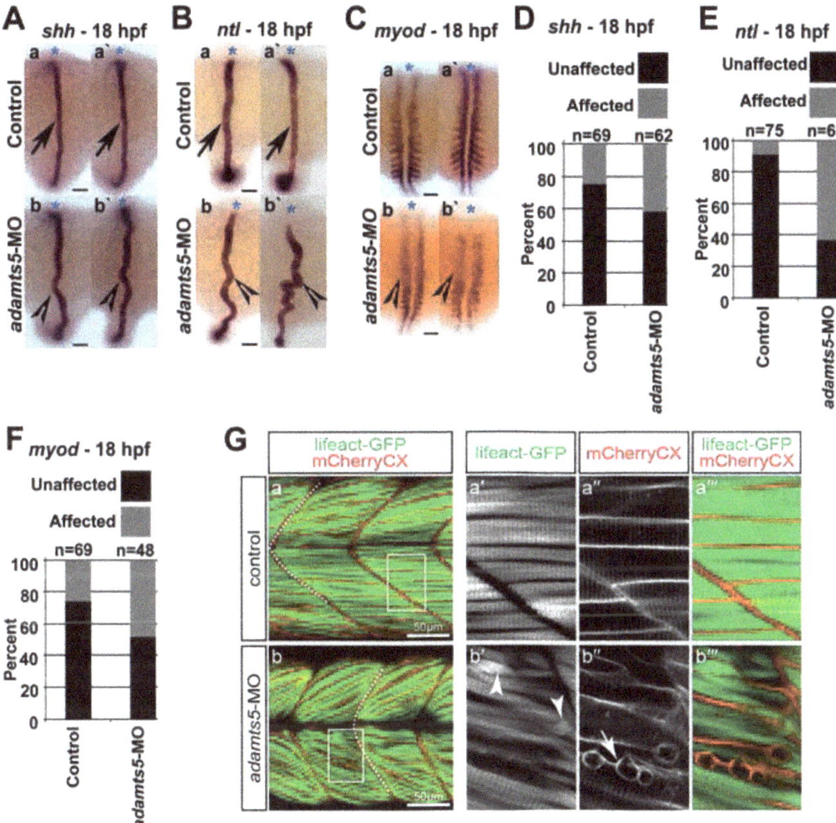

Figure 2. Notochord morphogenesis and muscle fiber formation is perturbed in *adamts5* morphant embryos. (**A**) Expression of *shh* in the notochord of 18 hpf control (**a**,**a′**, arrows) and *adamts5* morphant (**b**,**b′**, open arrowheads) embryos, with medio-lateral deviation in the *adamts5* morphants (**b**,**b′**), with anterior indicated (*); (**B**) Expression of *ntl* in the notochord of 18 hpf control (**a**,**a′**, arrows) and *adamts5* morphant (**b**,**b′**, open arrowheads) embryos, with medio-lateral deviation in the *adamts5* morphants with anterior indicated (*); (**C**) Expression of *myod* in adaxial and paraxial mesoderm of 18 hpf control embryos (**a**,**a′**) and its perturbation in *adamts5* morphants (**b**,**b′**, open arrowheads) with anterior indicated (*); (**D**) Quantitation of affected notochords in control and *adamts5* morphant embryos demarcated by *shh* in Figure 2A; (**E**) Quantitation of affected notochords in control and *adamts5* morphant embryos demarcated by *ntl* in Figure 2B; (**F**) Quantitation of embryos with perturbed *myod* expression in control and *adamts5* morphant embryos demarcated in Figure 2C; (**G**) Double-transgenic *Tg(acta1:lifeact-GFP)/Tg(acta1:mCherryCaaX)* embryos, in which thin filaments are marked green and sarcolemma red, reveal loss of muscle integrity in 3 dpf *adamts5* morphants. Muscle fibers of control injected larvae feature the typical striation of the myofibril and regular myofibers within chevron-shape somites, indicated by a dashed line (**a**). The boxed area in **a** is magnified in **a′–a′′′**. Myofibril striation is partially lost within *adamts5* morphants (arrowhead in **b′**) and the sarcolemma of the myofibers disrupted (arrow in **b′′**). The boxed area in **b** is magnified in **b′–b′′′**. Scale bar = 50 μm.

To further analyze myofiber differentiation, *myod* expression was examined. Reduced or absent paraxial mesodermal *myod* expression was observed at 12 hpf (Figure 3A(a,b)), whereas expression of adaxial mesodermal *myod* was largely unaffected (Figure 3A(a,b)). Similar observations were made with the *adamts5* 2/3-MO (Supplementary Figure S1) or upon co-injection of a *p53* morpholino with the *adamts5* AUG-MO (Supplementary Figure S2B). To ensure that the specificity of the phenotype was due to reduced *adamts5* expression, mRNA encoding either wild-type or catalytically-inactive (E^{411}A) ADAMTS5 were co-injected, with both able to partially rescue the reduced paraxial mesodermal *myod* expression (Figure 3A(c,d), respectively, and Figure 3B). This indicated that the enzymatic function of ADAMTS5 was not necessary to induce the reduced *myod* expression.

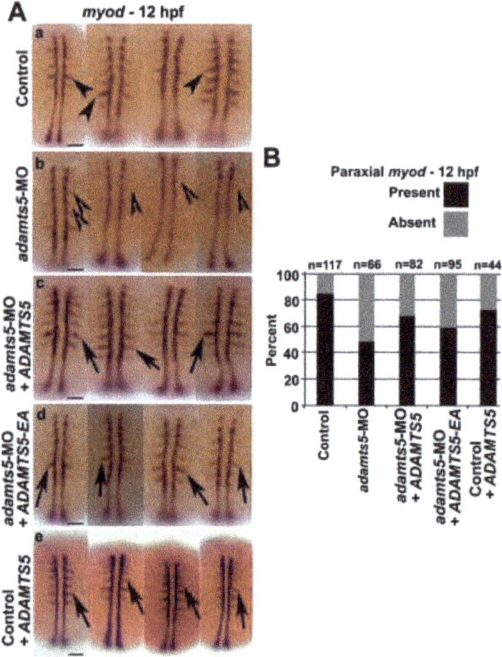

Figure 3. Loss of paraxial mesodermal *myod* expression in *adamts5* morphant embryos. (**A**) Expression of adaxial and paraxial *myod* in 12 hpf embryos injected with control MO (**a**, arrowheads), with *adamts5* morphant embryos showing substantial loss of paraxial expression (**b**, open arrowheads), as well as mild loss of paraxial *myod* expression (**b**, open arrowheads). Rescue of paraxial *myod* expression in *adamts5* morphants co-injected with mRNA encoding wild-type (**c**, arrows) or catalytically-inactive E^{411}A (**d**, arrows) ADAMTS5. Control embryos injected with *ADAMTS5* mRNA encoding wild-type ADAMTS5 show unaffected *myod* expression in paraxial mesoderm (**e**, arrows). Scale bar = 100 μm; (**B**) Quantitation of embryos showing present or absent *myod* patterning represented in (**A**).

2.5. Receptor-Mediated Sonic Hedgehog Signaling Is Affected in adamts5 Morphants

We hypothesized that reduced ADAMTS5 could lead to an altered extracellular environment that might disrupt Shh signaling, and that since adaxial mesoderm is in closer proximity to the notochord it might be less disrupted compared to the paraxial mesoderm. Therefore, cyclopamine, an antagonist of Smoothened (Smo), a receptor in the Shh signaling pathway [28] was used to understand whether Shh signaling through Smo was impaired in *adamts5* morphants. The presence of 5 μM cyclopamine did not affect adaxial *myod* expression at 12 hpf in wild-type embryos (Figure 4A(g–I),B). However, treatment of *adamts5* morphants with 5 μM cyclopamine severely affected adaxial expression of *myod* (Figure 4A(j–l),B) compared to untreated *adamts5* morphant embryos (Figure 4A(d–f),B). In a reciprocal experiment, the Smo agonist, SAG, was used to confirm the dependency of Shh signaling on *adamts5* expression. Administration of SAG on wild-type embryos disrupted paraxial *myod* expression in a similar manner to *adamts5* morphants (Figure 5A(d–I),B). However, the same concentration of SAG partially rescued the loss of paraxial *myod* patterning in the *adamts5* morphants (Figure 5A(g–l),B). These experiments collectively suggest an interaction between ADAMTS5 and Shh, such that they act synergistically to stimulate *myod* expression in adaxial mesoderm (Figure 6).

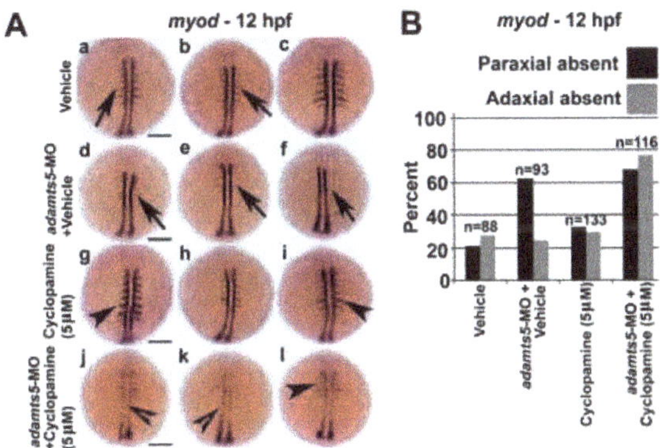

Figure 4. Combinatorial inhibition of Shh signaling and *adamts5* disrupt paraxial and adaxial *myod* expression. (**A**) Adaxial and paraxial *myod* expression in 12 hpf embryos treated with vehicle control (**a–c**, arrows denote paraxial *myod* expression), vehicle + *adamts5*-MO (**d–f**, arrows denote absent paraxial *myod* expression), 5 μM cyclopamine (**g–i**, arrowheads represent similar paraxial *myod* staining compared to control group) and 5 μM cyclopamine + *adamts5*-MO (**j–l**, open arrowheads represent absent adaxial *myod* expression and arrowhead represents absent paraxial *myod* staining compared to *adamts5* MO group). Scale bar = 200 μm; (**B**) Quantitation of embryos showing present or absent *myod* patterning represented in (**A**).

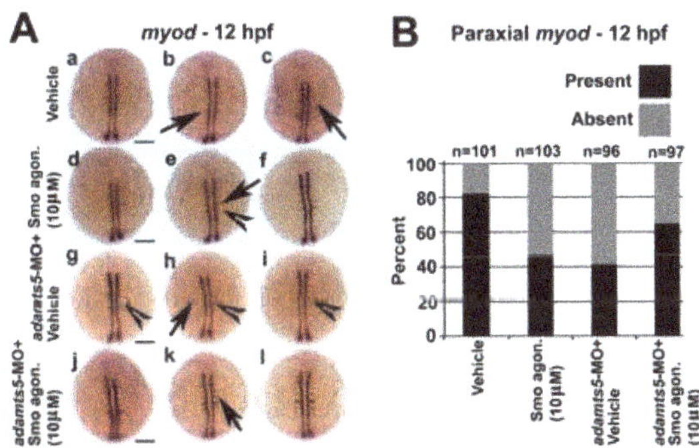

Figure 5. Combinatorial activation of Shh signaling and inhibition of *adamts5* rescues paraxial *myod* expression. (**A**) Adaxial and paraxial *myod* expression in 12 hpf embryos treated with vehicle control (**a–c**, arrows), 10 μM Smoothened agonist (Smo agon.) (**d–f**, arrow/open arrowhead represent present/absent *myod* expression in paraxial mesoderm), vehicle + *adamts5*-MO (**g–i**, arrow/open arrowhead represents present/absent *myod* expression in paraxial mesoderm) and 10 μM Smo agon. + *adamts5*-MO (**j–l**, arrows indicate *myod* expression present in paraxial mesoderm). Scale bar = 100 μm; (**B**) Quantitation of embryos showing present or absent *myod* patterning represented in (**A**).

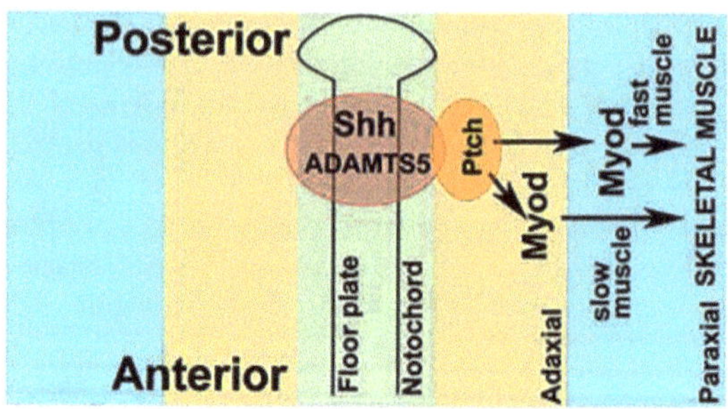

Figure 6. Model of interaction between Shh and ADAMTS5. Hypothetical model showing ADAMTS5 and Shh expression synergistically regulate downstream activation of MyoD in adaxial and paraxial mesoderm.

3. Discussion

Zebrafish myogenesis is controlled by multiple pathways [29–31]. Signaling from the notochord specifies slow-twitch muscle precursors in the adaxial mesoderm [24], which migrate laterally after somite formation to the most lateral muscle layer [32]. Myotomes develop following elongation and fusion of somitic cells and their attachment to the somite boundary, with the boundaries between myotome forming the critical myotendenous junctions that are the primary sites of force generation [33]. ECM–cell adhesion has been shown to be essential for multiple steps in this process [34]. This study has identified a novel non-catalytic function of the ECM protein ADAMTS5 in regulating Sonic Hedgehog signaling that impacted on somite differentiation, with reduced expression of *myod* in the paraxial mesoderm and disrupted myotome boundaries.

The phenotypes induced by the *adamts5* morphants were rescued with mRNA encoding both wild-type and catalytically-inactive ADAMTS5. Although unexpected, there is some precedence for ADAMTS family members demonstrating non-catalytic functions as reviewed recently [35]. For texample, ADAMTS1 has been shown to bind to VEGF through its C-terminal thrombospondin repeats and spacer domain to block VEGFR2 activation [5]. Moreover, both wild-type and catalytically-inactive ($E^{363}A$) ADAMTS15 were able to reduce breast cancer cell migration on matrices of fibronectin or laminin [6]. Furthermore, enzymatic activity was not required for enhancement of neurite outgrowth by ADAMTS4, which was instead dependent upon MAP kinase cascade activation [7]. ADAMTSL family members, which are structurally similar to ADAMTS family members but lack the N-terminal propeptide and catalytic domain, may also offer some important insights into non-catalytic functions of ADAMTS family members. Most notably, mutations of human ADAMTSL2 have been causally linked to the musculoskeletal disorder Geleophysic Dysplasia [36], where patients present with severe short stature, joint immobility and cardiac valvular abnormalities. Collectively, this suggests a role of ADAMTS5 in zebrafish muscle development is likely not related to its enzymatic function. However, mouse studies have highlighted considerable redundancy amongst ADAMTS members [35] suggesting that combinatorial targeting might be required to identify additional functions that may be dependent on enzymatic activity.

Shh is an important regulator of musculoskeletal development, given its role in somite and neural tube patterning. Duplication, and presumed overexpression, of Shh is associated with congenital muscular hypertrophy in humans [37]. Shh also enables the formation of the cranial musculature [38] and polarizes the limb during early morphogenesis [39,40]. Shh has also been demonstrated to mediate the patterning of somites [41,42]. Shh has the ability to activate myogenesis in vitro and in vivo [43] with expression and secretion of Shh from the notochord able to induce slow muscle fiber formation in vivo via *myod* [24]. The *adamts5* morphants displayed altered *myod* expression in the paraxial—but not adaxial—mesoderm despite levels of *shh* expression in the notochord being unaffected. This might be explained by reduced bioavailability of Shh in the absence of ADAMTS5. Since the adaxial *myod*-positive cells represent the slow muscle precursors that subsequently move through the fast muscle region where they impact on fast muscle differentiation [44], it would be of interest to examine the relative distribution of slow and fast twitch muscle fibers in the *adamts5* morphants.

The results obtained using agonists and antagonists of the downstream Smo pathway suggest that ADAMTS may work both upstream, as well as in parallel with Shh signals. Wnt/β-catenin signaling has been shown to act in co-operation with Shh (and BMPs) in embryonic myogenesis [31]. This could be mediated, at least partially, via ADAMTS5 since Wnt/β-catenin has been shown to act upstream of ADAMTS5 in other developmental situations, such as chondrocyte maturation and function [45]. Similarly, defects in Delta/Notch can affect somite boundary formation [46], with this pathway also shown to induce ADAMTS5 in joint cartilage, providing another potential upstream regulator of ADAMTS5 during somite differentiation.

Defective notochords have been identified in mutants of ECM components, such as fibrillin [47], collagen [48], the basement membrane proteins laminin alpha [49], beta and gamma [50], as well cell-associated molecules such as integrins [46]. A number of these defects are due to disrupted morphogenesis that results from perturbed ECM-cell interactions [49]. This suggests that altered morphogenesis as well as disrupted patterning may contribute to the perturbed notochord in *adamts5* morphants. In addition, U-shaped myotome boundaries have also been observed in mutants of ECM components, such as fibronectin [51] and laminin [52], or the alternative ECM processing enzyme MMP-11 [53], providing precedence for ADAMTS5 impacting on the myotome boundary.

This study has identified a new function for the metzincin ADAMTS5. By exploring the role of ADAMTS5 in zebrafish, understanding has been gained of a potential non-catalytic function in the regulation of muscle development and maintenance via interaction with the Sonic Hedgehog signaling pathway. Since both ADAMTS5 and Shh have independent—as well as potential combinatorial—roles during musculoskeletal development, the complex interplay between ADAMTS and Shh could be relevant to the development of musculoskeletal diseases, such as muscular dystrophies and arthritis. Further biochemical and functional characterization of potential interactions between ADAMTS5 and Shh in such diseases may reveal new insights into the development and progression of these diseases. Given that treatment options for these diseases are limited, this knowledge could then be applied to the development of novel therapeutics that specifically modulate this interaction to slow the progression of these debilitating conditions.

4. Materials and Methods

4.1. Zebrafish Lines and Maintenance

Wild-type and *Tg(acta1:lifeact-GFP)/Tg(acta1:mCherryCaax)* [26] zebrafish were maintained, raised and staged according to standard protocols [54]. Embryos were obtained by mating trios or using a mass embryo production system (MEPS) (Aquatic Habitats) and raised at 28.5 °C. Experiments were approved by the Deakin University Animal Ethics Committees (G14/2013, 15/05/2013).

4.2. Embryo Microinjection and Other Treatments

Morpholino antisense oligonucleotides (MOs; Gene Tools) targeting the ATG start codon (5′-atgctgtcgaaattacaggagtttggcgcgtat) and exon 2/3 splice site (5′-ctatcattgaggacgacggcctgcacgctg ccttcactgtggctcatgagatc) of zebrafish *adamts5* (GenBank: JF778846.1) were used to ablate the *adamts5* gene. MOs were solubilized in 1× Danieau buffer and 1 nL injected at a concentration of 1 mg/mL into one-cell stage embryos, as previously described [55]. Alternatively-spliced *adamts5* species were confirmed by Sanger sequencing (Australian Genome Research Facility, Melbourne, Australia) of RT-PCR amplicons generated with flanking primers (5′-ggcggatgtaggaactgtgt and 5′-ttacgcacctcacactgctc). Capped RNA encoding full-length wild-type or catalytically-inactive (E^{411}A)

ADAMTS5 [21] were synthesized using the T7 mMessage mMachine kit (Thermofisher Scientific, Scoresby, Victoria, Australia) and 40 pg was microinjected into one-cell stage embryos.

For other studies, injected embryos were treated with 5 µM cyclopamine (Sigma-Aldrich, St. Louis, MO, USA) in DMSO or 10 µM Smoothened agonist SAG (CAS 364590-63-6) (Merck Millipore, Darmstadt, Germany) in water, along with the corresponding vehicle control at 5.5 hpf and fixed at 12 hpf in 4% PFA/PBS.

4.3. Whole-Mount In Situ Hybridization and Immunofluorescence

Whole-mount in situ hybridization was performed as described [56]. The following antisense digoxigenin-labelled mRNA probes were synthesized by in vitro transcription: *shha* and *myod* [57], and *no-tail* (*ntl*) [58]. Immunofluorescence performed on whole embryos with polyclonal rabbit anti-propeptide ADAMTS5 (Cat# ab39203-100, Abcam, Pak Shak Kok, New Territories, Hong Kong, China) at 1:200 followed by anti-rabbit Alexa fluor 594 secondary antibody at 1:500 (Life Technologies, Carlsbad, CA, USA). Histochemical methods were performed as previously described [59].

4.4. Western Blotting

Cell lysates were extracted from 24 hpf embryos and subjected to Western blotting as described previously [60], using the polyclonal rabbit antibody against propeptide ADAMTS5 described above at 1:5000 followed by an anti-rabbit HRP antibody (Cell Signaling Technologies, Danvers, MA, USA) at 1:10,000. Protein concentrations were measured using the Bradford assay and equal loading of protein confirmed by Coomassie blue staining on duplicate SDS-PAGE gels.

4.5. Statistics

Two-tailed paired *t*-tests were performed between all treatment groups compared to the respective control groups with scoring performed blind. Significance was achieved at a *p*-value ≤ 0.05, with Gaussian distribution assumed in all cases.

Supplementary Materials: Supplementary materials can be found at http://www.mdpi.com/1422-0067/19/3/766/s1.

Acknowledgments: The authors thank Thomas Hall and Peter Currie for supplying cDNA constructs and Christopher Kintakas for gross morphological imaging of *adamts5* morphants. Additional thanks go to Suneel Apte and Sumeda Nandadasa for their helpful discussions regarding this work.

Author Contributions: Carolyn M. Dancevic and Daniel R. McCulloch conceived and designed the experiments; Carolyn M. Dancevic, Yann Gibert, Joachim Berger, Adam D. Smith, Clifford Liongue and Nicole Stupka performed the experiments; Carolyn M. Dancevic, Yann Gibert, Joachim Berger, Alister C. Ward and Daniel R. McCulloch analyzed the data; Carolyn M. Dancevic, Alister C. Ward and Daniel R. McCulloch wrote the paper.

Conflicts of Interest: The authors declare no conflict of interest.

References

1. Silver, D.L.; Hou, L.; Somerville, R.; Young, M.E.; Apte, S.S.; Pavan, W.J. The secreted metalloprotease ADAMTS20 is required for melanoblast survival. *PLoS Genet.* **2008**, *4*, e1000003. [CrossRef] [PubMed]
2. Enomoto, H.; Nelson, C.M.; Somerville, R.P.; Mielke, K.; Dixon, L.J.; Powell, K.; Apte, S.S. Cooperation of two ADAMTS metalloproteases in closure of the mouse palate identifies a requirement for versican proteolysis in regulating palatal mesenchyme proliferation. *Development* **2010**, *137*, 4029–4038. [CrossRef] [PubMed]
3. Jonsson-Rylander, A.C.; Nilsson, T.; Fritsche-Danielson, R.; Hammarstrom, A.; Behrendt, M.; Andersson, J.O.; Lindgren, K.; Andersson, A.K.; Wallbrandt, P.; Rosengren, B.; et al. Role of ADAMTS-1 in atherosclerosis: Remodeling of carotid artery, immunohistochemistry, and proteolysis of versican. *Arterioscler. Thromb. Vasc. Biol.* **2005**, *25*, 180–185. [PubMed]
4. Porter, S.; Span, P.N.; Sweep, F.C.; Tjan-Heijnen, V.C.; Pennington, C.J.; Pedersen, T.X.; Johnsen, M.; Lund, L.R.; Romer, J.; Edwards, D.R. ADAMTS8 and ADAMTS15 expression predicts survival in human breast carcinoma. *Int. J. Cancer* **2006**, *118*, 1241–1247. [CrossRef] [PubMed]

5. Iruela-Arispe, M.L.; Carpizo, D.; Luque, A. ADAMTS1: A matrix metalloprotease with angioinhibitory properties. *Ann. N. Y. Acad. Sci.* **2003**, *995*, 183–190. [CrossRef] [PubMed]

6. Kelwick, R.; Wagstaff, L.; Decock, J.; Roghi, C.; Cooley, L.S.; Robinson, S.D.; Arnold, H.; Gavrilovic, J.; Jaworski, D.M.; Yamamoto, K.; et al. Metalloproteinase-dependent and -independent processes contribute to inhibition of breast cancer cell migration, angiogenesis and liver metastasis by a disintegrin and metalloproteinase with thrombospondin motifs-15. *Int. J. Cancer* **2014**, *136*, E14–E26. [CrossRef] [PubMed]

7. Hamel, M.G.; Ajmo, J.M.; Leonardo, C.C.; Zuo, F.; Sandy, J.D.; Gottschall, P.E. Multimodal signaling by the ADAMTSs (a disintegrin and metalloproteinase with thrombospondin motifs) promotes neurite extension. *Exp. Neurol.* **2008**, *210*, 428–440. [CrossRef] [PubMed]

8. McCulloch, D.R.; Nelson, C.M.; Dixon, L.J.; Silver, D.L.; Wylie, J.D.; Lindner, V.; Sasaki, T.; Cooley, M.A.; Argraves, W.S.; Apte, S.S. ADAMTS metalloproteases generate active versican fragments that regulate interdigital web regression. *Dev. Cell* **2009**, *17*, 687–698. [CrossRef] [PubMed]

9. Dubail, J.; Aramaki-Hattori, N.; Bader, H.L.; Nelson, C.M.; Katebi, N.; Matuska, B.; Olsen, B.R.; Apte, S.S. A new Adamts9 conditional mouse allele identifies its non-redundant role in interdigital web regression. *Genesis* **2014**, *52*, 702–712. [CrossRef] [PubMed]

10. Dupuis, L.E.; McCulloch, D.R.; McGarity, J.D.; Bahan, A.; Wessels, A.; Weber, D.; Diminich, A.M.; Nelson, C.M.; Apte, S.S.; Kern, C.B. Altered versican cleavage in ADAMTS5 deficient mice; a novel etiology of myxomatous valve disease. *Dev. Biol.* **2011**, *357*, 152–164. [CrossRef] [PubMed]

11. Stanton, H.; Rogerson, F.M.; East, C.J.; Golub, S.B.; Lawlor, K.E.; Meeker, C.T.; Little, C.B.; Last, K.; Farmer, P.J.; Campbell, I.K.; et al. ADAMTS5 is the major aggrecanase in mouse cartilage in vivo and in vitro. *Nature* **2005**, *434*, 648–652. [CrossRef] [PubMed]

12. Glasson, S.S.; Askew, R.; Sheppard, B.; Carito, B.; Blanchet, T.; Ma, H.L.; Flannery, C.R.; Peluso, D.; Kanki, K.; Yang, Z.; et al. Deletion of active ADAMTS5 prevents cartilage degradation in a murine model of osteoarthritis. *Nature* **2005**, *434*, 644–648. [CrossRef] [PubMed]

13. Didangelos, A.; Mayr, U.; Monaco, C.; Mayr, M. Novel role of ADAMTS-5 protein in proteoglycan turnover and lipoprotein retention in atherosclerosis. *J. Biol. Chem.* **2012**, *287*, 19341–19345. [CrossRef] [PubMed]

14. Yayon, A.; Klagsbrun, M.; Esko, J.D.; Leder, P.; Ornitz, D.M. Cell surface, heparin-like molecules are required for binding of basic fibroblast growth factor to its high affinity receptor. *Cell* **1991**, *64*, 841–848. [CrossRef]

15. Rapraeger, A.C.; Krufka, A.; Olwin, B.B. Requirement of heparan sulfate for bFGF-mediated fibroblast growth and myoblast differentiation. *Science* **1991**, *252*, 1705–1708. [CrossRef] [PubMed]

16. Witt, R.M.; Hecht, M.L.; Pazyra-Murphy, M.F.; Cohen, S.M.; Noti, C.; van Kuppevelt, T.H.; Fuller, M.; Chan, J.A.; Hopwood, J.J.; Seeberger, P.H.; et al. Heparan sulfate proteoglycans containing a glypican 5 core and 2-*O*-sulfo-iduronic acid function as sonic hedgehog co-receptors to promote proliferation. *J. Biol. Chem.* **2013**, *288*, 26275–26288. [CrossRef] [PubMed]

17. Nandadasa, S.; Nelson, C.M.; Apte, S.S. ADAMTS9-mediated extracellular matrix dynamics regulates umbilical cord vascular smooth muscle differentiation and rotation. *Cell Rep.* **2015**, *11*, 1519–1528. [CrossRef] [PubMed]

18. Lin, A.C.; Seeto, B.L.; Bartoszko, J.M.; Khoury, M.A.; Whetstone, H.; Ho, L.; Hsu, C.; Ali, S.A.; Alman, B.A. Modulating hedgehog signaling can attenuate the severity of osteoarthritis. *Nat. Med.* **2009**, *15*, 1421–1425. [CrossRef] [PubMed]

19. Stupka, N.; Kintakas, C.; White, J.D.; Fraser, F.W.; Hanciu, M.; Aramaki Hattori, N.; Martin, S.; Coles, C.; Collier, F.; Ward, A.C.; et al. Versican processing by a disintegrin-like and metalloproteinase domain with thrombospondin-1 repeats proteinases-5 and -15 facilitates myoblast fusion. *J. Biol. Chem.* **2013**, *288*, 1907–1917. [CrossRef] [PubMed]

20. Brunet, F.G.; Fraser, F.W.; Binder, M.J.; Smith, A.D.; Kintakas, C.; Dancevic, C.M.; Ward, A.C.; McCulloch, D.R. The evolutionary conservation of the A Disintegrin-like and Metalloproteinase domain with Thrombospondin-1 motif metzincins across vertebrate species and their expression in teleost zebrafish. *BMC Evol. Biol.* **2015**, *15*, 22. [CrossRef] [PubMed]

21. Longpre, J.M.; McCulloch, D.R.; Koo, B.H.; Alexander, J.P.; Apte, S.S.; Leduc, R. Characterization of proADAMTS5 processing by proprotein convertases. *Int. J. Biochem. Cell Biol.* **2009**, *41*, 1116–1126. [CrossRef] [PubMed]

22. Brunet, F.; Kintakas, C.; Smith, A.D.; McCulloch, D.R. The function of the hyalectan class of proteoglycans and their binding partners during vertebrate development. In *Advances in Medicine and Biology*; Nova Science Publishers Inc.: Hauppauge, NY, USA, 2012; Volume 52, pp. 49–96.

23. Dancevic, C.M. Deakin University, Waurn Ponds, Victoria, Australia. *Unpublished data*, 2015.

24. Blagden, C.S.; Currie, P.D.; Ingham, P.W.; Hughes, S.M. Notochord induction of zebrafish slow muscle mediated by Sonic hedgehog. *Genes Dev.* **1997**, *11*, 2163–2175. [CrossRef] [PubMed]

25. Halpern, M.E.; Ho, R.K.; Walker, C.; Kimmel, C.B. Induction of muscle pioneers and floor plate is distinguished by the zebrafish no tail mutation. *Cell* **1993**, *75*, 99–111. [CrossRef]

26. Berger, J.; Tarakci, H.; Berger, S.; Li, M.; Hall, T.E.; Arner, A.; Currie, P.D. Loss of Tropomodulin4 in the zebrafish mutant trage causes cytoplasmic rod formation and muscle weakness reminiscent of nemaline myopathy. *Dis. Model. Mech.* **2014**, *7*, 1407–1415. [CrossRef] [PubMed]

27. Schauerte, H.E.; van Eeden, F.J.; Fricke, C.; Odenthal, J.; Strahle, U.; Haffter, P. Sonic hedgehog is not required for the induction of medial floor plate cells in the zebrafish. *Development* **1998**, *125*, 2983–2993. [PubMed]

28. Lewis, C.; Krieg, P.A. Reagents for developmental regulation of Hedgehog signaling. *Methods* **2014**, *66*, 390–397. [CrossRef] [PubMed]

29. Stickney, H.L.; Barresi, M.J.; Devoto, S.H. Somite development in zebrafish. *Dev. Dyn.* **2000**, *219*, 287–303. [CrossRef]

30. Rida, P.C.; Le Minh, N.; Jiang, Y.J. A Notch feeling of somite segmentation and beyond. *Dev. Biol.* **2004**, *265*, 2–22. [CrossRef] [PubMed]

31. Von Maltzahn, J.; Chang, N.C.; Bentzinger, C.F.; Rudnicki, M.A. Wnt signaling in myogenesis. *Trends Cell Biol.* **2012**, *22*, 602–609. [CrossRef] [PubMed]

32. Devoto, S.H.; Melancon, E.; Eisen, J.S.; Westerfield, M. Identification of separate slow and fast muscle precursor cells in vivo, prior to somite formation. *Development* **1996**, *122*, 3371–3380. [PubMed]

33. Long, J.H.; Adcock, B.; Root, R.G. Force transmission via axial tendons in undulating fish: A dynamic analysis. *Comp. Biochem. Physiol. A Mol. Integr. Physiol.* **2002**, *133*, 911–929. [CrossRef]

34. Goody, M.F.; Sher, R.B.; Henry, C.A. Hanging on for the ride: Adhesion to the extracellular matrix mediates cellular responses in skeletal muscle morphogenesis and disease. *Dev. Biol.* **2015**, *401*, 75–91. [CrossRef] [PubMed]

35. Dancevic, C.M.; McCulloch, D.R.; Ward, A.C. The ADAMTS hyalectanase family: Biological insights from diverse species. *Biochem. J.* **2016**, *473*, 2011–2022. [CrossRef] [PubMed]

36. Le Goff, C.; Morice-Picard, F.; Dagoneau, N.; Wang, L.W.; Perrot, C.; Crow, Y.J.; Bauer, F.; Flori, E.; Prost-Squarcioni, C.; Krakow, D.; et al. ADAMTSL2 mutations in geleophysic dysplasia demonstrate a role for ADAMTS-like proteins in TGF-β bioavailability regulation. *Nat. Genet.* **2008**, *40*, 1119–1123. [CrossRef] [PubMed]

37. Kroeldrup, L.; Kjaergaard, S.; Kirchhoff, M.; Kock, K.; Brasch-Andersen, C.; Kibaek, M.; Ousager, L.B. Duplication of 7q36.3 encompassing the Sonic Hedgehog (SHH) gene is associated with congenital muscular hypertrophy. *Eur. J. Med. Genet.* **2012**, *55*, 557–560. [CrossRef] [PubMed]

38. Balczerski, B.; Zakaria, S.; Tucker, A.S.; Borycki, A.G.; Koyama, E.; Pacifici, M.; Francis-West, P. Distinct spatiotemporal roles of hedgehog signalling during chick and mouse cranial base and axial skeleton development. *Dev. Biol.* **2012**, *371*, 203–214. [CrossRef] [PubMed]

39. Riddle, R.D.; Johnson, R.L.; Laufer, E.; Tabin, C. Sonic hedgehog mediates the polarizing activity of the ZPA. *Cell* **1993**, *75*, 1401–1416. [CrossRef]

40. Chiang, C.; Litingtung, Y.; Harris, M.P.; Simandl, B.K.; Li, Y.; Beachy, P.A.; Fallon, J.F. Manifestation of the limb prepattern: Limb development in the absence of sonic hedgehog function. *Dev. Biol.* **2001**, *236*, 421–435. [CrossRef] [PubMed]

41. Fan, C.M.; Tessier-Lavigne, M. Patterning of mammalian somites by surface ectoderm and notochord: Evidence for sclerotome induction by a hedgehog homolog. *Cell* **1994**, *79*, 1175–1186. [CrossRef]

42. Johnson, R.L.; Laufer, E.; Riddle, R.D.; Tabin, C. Ectopic expression of Sonic hedgehog alters dorsal-ventral patterning of somites. *Cell* **1994**, *79*, 1165–1173. [CrossRef]

43. Duprez, D.; Fournier-Thibault, C.; Le Douarin, N. Sonic Hedgehog induces proliferation of committed skeletal muscle cells in the chick limb. *Development* **1998**, *125*, 495–505. [PubMed]

44. Henry, C.A.; Amacher, S.L. Zebrafish slow muscle cell migration induces a wave of fast muscle morphogenesis. *Dev. Cell* **2004**, *7*, 917–923. [CrossRef] [PubMed]

45. Tamamura, Y.; Otani, T.; Kanatani, N.; Koyama, E.; Kitagaki, J.; Komori, T.; Yamada, Y.; Costantini, F.; Wakisaka, S.; Pacifici, M.; et al. Developmental regulation of Wnt/β-catenin signals is required for growth plate assembly, cartilage integrity, and endochondral ossification. *J. Biol. Chem.* **2005**, *280*, 19185–19195. [CrossRef] [PubMed]

46. Julich, D.; Geisler, R.; Holley, S.A. Tübingen 2000 Screen Consortium. Integrinalpha5 and delta/notch signaling have complementary spatiotemporal requirements during zebrafish somitogenesis. *Dev. Cell* **2005**, *8*, 575–586. [PubMed]

47. Gansner, J.M.; Madsen, E.C.; Mecham, R.P.; Gitlin, J.D. Essential role for fibrillin-2 in zebrafish notochord and vascular morphogenesis. *Dev. Dyn.* **2008**, *237*, 2844–2861. [CrossRef] [PubMed]

48. Gray, R.S.; Wilm, T.P.; Smith, J.; Bagnat, M.; Dale, R.M.; Topczewski, J.; Johnson, S.L.; Solnica-Krezel, L. Loss of col8a1a function during zebrafish embryogenesis results in congenital vertebral malformations. *Dev. Biol.* **2014**, *386*, 72–85. [CrossRef] [PubMed]

49. Parsons, M.J.; Pollard, S.M.; Saude, L.; Feldman, B.; Coutinho, P.; Hirst, E.M.; Stemple, D.L. Zebrafish mutants identify an essential role for laminins in notochord formation. *Development* **2002**, *129*, 3137–3146. [PubMed]

50. Pollard, S.M.; Parsons, M.J.; Kamei, M.; Kettleborough, R.N.; Thomas, K.A.; Pham, V.N.; Bae, M.K.; Scott, A.; Weinstein, B.M.; Stemple, D.L. Essential and overlapping roles for laminin alpha chains in notochord and blood vessel formation. *Dev. Biol.* **2006**, *289*, 64–76. [CrossRef] [PubMed]

51. Snow, C.J.; Peterson, M.T.; Khalil, A.; Henry, C.A. Muscle development is disrupted in zebrafish embryos deficient for fibronectin. *Dev. Dyn.* **2008**, *237*, 2542–2553. [CrossRef] [PubMed]

52. Peterson, M.T.; Henry, C.A. Hedgehog signaling and laminin play unique and synergistic roles in muscle development. *Dev. Dyn.* **2010**, *239*, 905–913. [CrossRef] [PubMed]

53. Jenkins, M.H.; Alrowaished, S.S.; Goody, M.F.; Crawford, B.D.; Henry, C.A. Laminin and Matrix metalloproteinase 11 regulate Fibronectin levels in the zebrafish myotendinous junction. *Skelet. Muscle* **2016**, *6*, 18. [CrossRef] [PubMed]

54. Kimmel, C.B.; Ballard, W.W.; Kimmel, S.R.; Ullmann, B.; Schilling, T.F. Stages of embryonic development of the zebrafish. *Dev. Dyn.* **1995**, *203*, 253–310. [CrossRef] [PubMed]

55. Nasevicius, A.; Ekker, S.C. Effective targeted gene 'knockdown' in zebrafish. *Nat Genet* **2000**, *26*, 216–220. [CrossRef] [PubMed]

56. Thisse, C.; Thisse, B. High-resolution in situ hybridization to whole-mount zebrafish embryos. *Nat. Protoc.* **2008**, *3*, 59–69. [CrossRef] [PubMed]

57. Gibert, Y.; Gajewski, A.; Meyer, A.; Begemann, G. Induction and prepatterning of the zebrafish pectoral fin bud requires axial retinoic acid signaling. *Development* **2006**, *133*, 2649–2659. [CrossRef] [PubMed]

58. Begemann, G.; Schilling, T.F.; Rauch, G.J.; Geisler, R.; Ingham, P.W. The zebrafish neckless mutation reveals a requirement for raldh2 in mesodermal signals that pattern the hindbrain. *Development* **2001**, *128*, 3081–3094. [PubMed]

59. Lange, M.; Norton, W.; Coolen, M.; Chaminade, M.; Merker, S.; Proft, F.; Schmitt, A.; Vernier, P.; Lesch, K.P.; Bally-Cuif, L. The ADHD-susceptibility gene lphn3.1 modulates dopaminergic neuron formation and locomotor activity during zebrafish development. *Mol. Psychiatry* **2012**, *17*, 946–954. [CrossRef] [PubMed]

60. Dancevic, C.M.; Fraser, F.W.; Smith, A.D.; Stupka, N.; Ward, A.C.; McCulloch, D.R. Biosynthesis and expression of a Disintegrin-like and Metalloproteinase domain with Thrombospondin-1 repeats 15: A novel versican-cleaving proteoglycanase. *J. Biol. Chem.* **2013**, *288*, 37267–37276. [CrossRef] [PubMed]

International Journal of
Molecular Sciences

Article

Glucocorticoids Improve Myogenic Differentiation In Vitro by Suppressing the Synthesis of Versican, a Transitional Matrix Protein Overexpressed in Dystrophic Skeletal Muscles

Natasha McRae [1] , Leonard Forgan [1], Bryony McNeill [1], Alex Addinsall [1], Daniel McCulloch [2], Chris Van der Poel [3] and Nicole Stupka [1,*]

1   School of Medicine, Deakin University, Waurn Ponds, VIC 3216, Australia; nmcrae@deakin.edu.au (N.M.); leonard.forgan@deakin.edu.au (L.F.); bryony.mcneill@deakin.edu.au (B.M.); aaddinsa@deakin.edu.au (A.A.)
2   Faculty of Law, The University of Queensland, Brisbane, QLD 4072, Australia; daniel.mcculloch@uq.net.au
3   Department of Physiology, Anatomy and Microbiology, La Trobe University, Bundoora, VIC 3086, Australia; c.vanderpoel@latrobe.edu.au
*   Correspondence: nicole.stupka@deakin.edu.au; Tel.: +61-3-5227-3160

Received: 8 November 2017; Accepted: 27 November 2017; Published: 6 December 2017

Abstract: In Duchenne muscular dystrophy (DMD), a dysregulated extracellular matrix (ECM) directly exacerbates pathology. Glucocorticoids are beneficial therapeutics in DMD, and have pleiotropic effects on the composition and processing of ECM proteins in other biological contexts. The synthesis and remodelling of a transitional versican-rich matrix is necessary for myogenesis; whether glucocorticoids modulate this transitional matrix is not known. Here, versican expression and processing were examined in hindlimb and diaphragm muscles from *mdx* dystrophin-deficient mice and C57BL/10 wild type mice. V0/V1 versican (*Vcan*) mRNA transcripts and protein levels were upregulated in dystrophic compared to wild type muscles, especially in the more severely affected *mdx* diaphragm. Processed versican (versikine) was detected in wild type and dystrophic muscles, and immunoreactivity was highly associated with newly regenerated myofibres. Glucocorticoids enhanced C2C12 myoblast fusion by modulating the expression of genes regulating transitional matrix synthesis and processing. Specifically, *Tgfβ1*, *Vcan* and hyaluronan synthase-2 (*Has2*) mRNA transcripts were decreased by 50% and *Adamts1* mRNA transcripts were increased three-fold by glucocorticoid treatment. The addition of exogenous versican impaired myoblast fusion, whilst glucocorticoids alleviated this inhibition in fusion. In dystrophic *mdx* muscles, versican upregulation correlated with pathology. We propose that versican is a novel and relevant target gene in DMD, given its suppression by glucocorticoids and that in excess it impairs myoblast fusion, a process key for muscle regeneration.

Keywords: Duchenne muscular dystrophy; fibrosis; glucocorticoids; myogenesis; *mdx* mouse; versican

---

1. Introduction

Duchenne muscular dystrophy (DMD) is a fatal hereditary disease affecting ~1:3500 boys, with glucocorticoid therapy being the only treatment with clinical efficacy [1]. DMD is caused by mutations in the dystrophin (*DMD*) gene, which renders dystrophic skeletal muscles vulnerable to ongoing contraction-induced injury, resulting in excessive inflammation, impaired regeneration and fibrosis [2]. Whilst fibrosis is usually thought of as a disease endpoint, it is important to note that endomysial extracellular matrix (ECM) accumulation precedes overt muscle degeneration in DMD [3], and is thought to actively contribute to the degeneration of dystrophic muscles [4–6].

The composition and processing of the ECM influences global cell behaviour, including cellular processes necessary for effective muscle repair [7–9]. Aberrant ECM synthesis and processing is observed in dystrophic muscles from patients with DMD [10] and in *mdx* mice [11], compromising regenerative myogenesis and exacerbating inflammatory processes [12]. TGF-β is considered to be a key cytokine driving fibrosis in DMD [13], and its levels are elevated in dystrophic muscles and in circulation [14].

The mature ECM of normal skeletal muscle is composed of glycoproteins, collagens and proteoglycans containing heparan sulphate and chondroitin sulphate/dermatan sulphate glycosaminoglycan (GAG) side chains [15]. Endomysial fibrosis in DMD is associated with the increased expression of not only mature ECM proteins [16,17], but also transitional ECM proteins such as hyaluronan and versican [18]. These transitional matrix proteins, through their synthesis and remodelling, regulate cell behaviour during normal development and regeneration, as well as functioning as a scaffold for mature ECM deposition [8,19]. ECM proteases are also upregulated in dystrophic muscles [20]. Strategies to limit aberrant ECM synthesis and remodelling in DMD are of therapeutic interest. Given the importance of a transitional matrix in tissue repair, versican is an especially relevant ECM protein [18].

Versican is a chondroitin sulphate proteoglycan (CSPG) [19], localised to pericellular regions of the basement membranes and interstitial matrices [9,21,22]. In skeletal muscle, the V0 and V1 splice variants of versican are the most abundant [9]. V0/V1 versican is comprised of the G1 and G3 globular domains at the N- and C-terminus respectively, with each of their core proteins sharing a common GAG-β domain and V0 versican containing an additional GAG-α domain. Chondroitin sulphate moieties on V0/V1 versican bind growth factors, cytokines and adhesion molecules, such as CD44 (cluster of differentiation 44), to regulate downstream signalling pathways [23]. The C- and N-termini of versican bind various ECM molecules [24], including hyaluronan [25], a large non-proteinaceous GAG of variable size which has been linked to myogenesis and muscle growth [8,26]. Hyaluronan is synthesised by hyaluronan synthases (HAS) and degraded by hyaluronidases (HYAL), with HAS2, HYAL1 and HYAL2 being the predominant skeletal muscle isoforms [26].

Recent studies have highlighted the role of versican in myoblast proliferation and myotube formation, processes critical for regenerative myogenesis [9,27]. In developing chick skeletal muscle, versican is synthesised early in myogenesis [27] and is localised to the pericellular matrix of developing myotubes [28]. V1 versican is cleaved at the Glu^{441}-Ala^{442} peptide bond within the GAG-β domain by specific A Disintegrin-like And Metalloproteinase Domain with ThromboSpondin-1 repeats (ADAMTS) proteoglycanases [29], which include ADAMTS1, -4, -5, -9, -15 and -20 [30], and presumably ADAMTS8, however this has not yet been proven [31]. This produces the bioactive G1-DPEAAE fragment known as versikine [19], which in other biological contexts can be pro-apoptotic [32] or pro-inflammatory [33]. Using C2C12 myoblasts as an in vitro model of regenerative myogenesis, we have shown that the processing of a versican and hyaluronan rich transitional, pericellular matrix by ADAMTS5 or -15 facilitated myotube formation [9].

Glucocorticoids delay disease progression in DMD by improving muscle strength, respiratory function, and increasing ambulation by up to four years [34,35]. These beneficial effects may be due to membrane stabilisation, decreased muscle necrosis and fibrosis, modulation of inflammation, and/or improved regeneration [12,36]. In dystrophic *mdx* mice, high dose treatment with the glucocorticoid deflazacort increased the proliferation and/or fusion of muscle precursor cells during myotube formation following crush injury, as well as enhancing the growth of intact myotubes [37]. More recently, when *mdx* mice were treated with prednisone or VBP15 (vamorolone; a dissociative glucocorticoid [38]) TGF-β related networks were suppressed, this included reduced gene expression of various collagen isoforms (1A1, 3A1 and 6A1), leading to improved muscle repair [12]. In vitro, glucocorticoids also enhanced myotube formation in primary wild type and dystrophic *mdx* myoblasts, as well as in C2C12 cells [39,40].

Given the importance of transitional matrix synthesis and remodelling for myofibre formation and that glucocorticoids can enhance myogenic differentiation [1,8,41], the effects of glucocorticoids on the ECM during myogensis and myoblast differentiation should be better characterised. In other disease models associated with a heightened pro-inflammatory state, glucocorticoids do modulate ECM composition and degradation with the specific effects being tissue and context dependent. In cultured rat mesangial cells, glucocorticoids decreased secreted and cell associated chondroitin sulphate and dermatan sulphate proteoglycan content by 50% [42]. In cultured airway fibroblasts stimulated with serum, glucocorticoids decreased versican gene and protein expression by 50% [43]. Hyaluronan content in skin and dermal fibroblasts was also reduced following glucocorticoid treatment [44], due to decreased *Has2* mRNA transcription and stability, and thus reduced hyaluronan synthesis [45]. Given their clinical usage in DMD and the ECM expansion observed in dystrophic muscles, understanding the transitional ECM gene targets regulated by glucocorticoids in skeletal muscle cells is an important step towards improving therapeutic outcomes for strategies targeting fibrosis in dystrophy.

Here, we show that V0/V1 versican mRNA transcript abundance and protein levels are elevated in diaphragm and hindlimb tibialis anterior (TA) muscles from dystrophic *mdx* mice when compared to C57BL/10 wild type mice. We identify a novel mechanism mediating the glucocorticoid stimulated increase in myoblast fusion and myotube formation in C2C12 cells. Specifically, in differentiating myoblasts, low doses of glucocorticoids (25 or 100 nM dexamethasone) modulated the expression of genes associated with the synthesis and processing of a versican–hyaluronan rich transitional matrix. This effect is dependent on versican, as glucocorticoid treatment improved myotube formation in the presence of excess versican. As a whole, these findings offer novel insight into the relevance of versican in dystrophic skeletal muscle pathology.

2. Results

2.1. Versican Expression Is Increased in Dystrophic Muscles and Correlate with the Severity of Pathology

The pathology of TA muscles from adult *mdx* mice is moderate compared to the human disease. Specifically, in hindlimb muscles from *mdx* mice there is minimal fibrosis, lower levels of inflammation and more effective regeneration as indicated by the presence of centrally nucleated fibres [46] (Figure 1B); although, muscle strength is compromised [47]. The pathology of *mdx* diaphragm muscles is more representative of DMD, with greatly impaired contractile function and high levels of endomysial fibrosis [48,49] (Figure 1D). *V0/V1 Vcan* mRNA transcript abundance was greater in dystrophic compared to wild type muscles (Figure 1I). To support this gene expression data, the immunoreactivity of full length V0/V1 versican and its cleaved bioactive fragment, versikine, was assessed in TA and diaphragm muscle cross-sections from *mdx* and wild type mice (Figure 2).

V0/V1 versican protein levels were upregulated in TA and diaphragm muscles from *mdx* mice (Figure 2B,D) compared to wild type mice (Figure 2A,C, and as quantified in Figure 2I). In concordance with the more severe pathology, versican immunoreactivity was greatest in the *mdx* diaphragm. In dystrophic muscles, versican staining was localised to regions of mononuclear infiltrate, which includes myoblasts, inflammatory cells and fibroblasts (Figure 2B,D; as indicated with a white asterisk). Versican staining was also associated with endomysial fibrosis in TA (Figure 3A–D) and diaphragm muscles (Figure 2D).

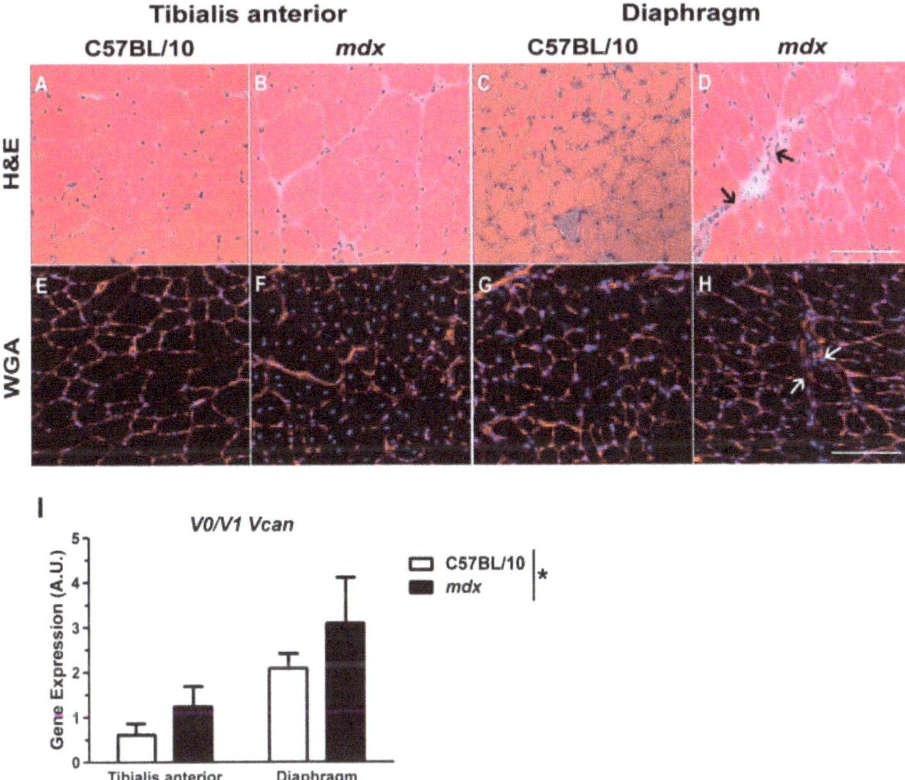

Figure 1. Muscle architecture and fibrosis in wild type and *mdx* mice. (**A–D**) Muscle architecture can be determined from the H and E-stained muscle cross-sections, and (**E–H**) areas of fibrosis can be observed by examining wheat germ agglutinin (WGA) stained cross-sections. (**I**) Versican (*V0/V1 Vcan*) mRNA transcript abundance was increased in *mdx* compared to wild type mice (* $p = 0.026$; main effect genotype; 2-way general linear model (GLM) ANOVA). Arrows denote areas of fibrosis and mononuclear infiltrate. Gene expression analysis was determined from $n = 3$ wild type mice and $n = 3$ *mdx* mice. Scale bar = 100 μm. Error bars = S.E.

Remodelling of V1 versican by ADAMTS proteoglycanases (e.g., ADAMTS1, -5 and -15) yields the bioactive versikine fragment [19]. Versikine immunoreactivity was predominantly localised to the pericellular region of myofibres, and did not differ between dystrophic and wild type TA or diaphragm muscles (Figure 2E–H, and as quantified in Figure 2J). In *mdx* TA and diaphragm muscle cross-sections, versikine staining was also associated with regions of mononuclear infiltrate (Figure 2F,H; as indicated with a white asterisk). When this association of versikine with mononuclear infiltrate was further interrogated in *mdx* TA muscle cross-sections by specifically assessing areas of regeneration, high levels of versikine immunoreactivity were co-localised with desmin positive, newly regenerated myofibres (Figure 3E–H). Furthermore, nuclear localisation of versikine (but not versican) was observed in muscle fibres and mononuclear infiltrate (Figure 3I,J).

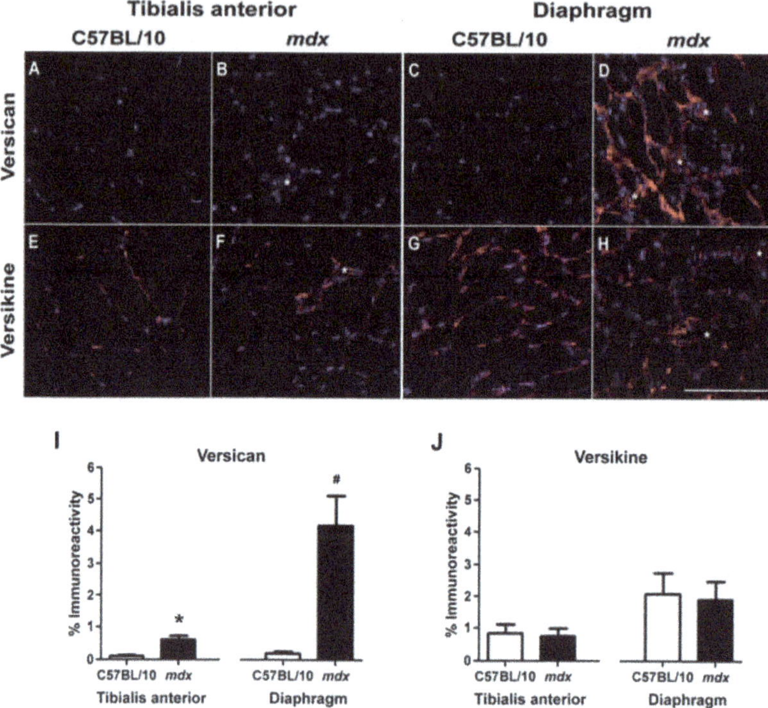

Figure 2. Versican and versikine expression in TA and diaphragm muscles from wild type and *mdx* mice. Representative images of (**A–D**) versican and (**E–H**) versikine immunoreactivity from TA and diaphragm muscles from wild type and *mdx* mice. (**I**) Quantification of versican immunoreactivity revealed an upregulation in TA (* $p = 0.001$) and diaphragm muscles (# $p = 0.0001$) from *mdx* mice when compared to wild type mice. (**J**) Versikine immunoreactivity was similar in *mdx* and wild type TA or diaphragm muscles. White asterisks denote areas of mononuclear infiltration. Immunoreactivity analysis was determined from $n = 5$ wild type mice and $n = 5$ *mdx* mice. Scale bar = 100 μm. Error bars = S.E.

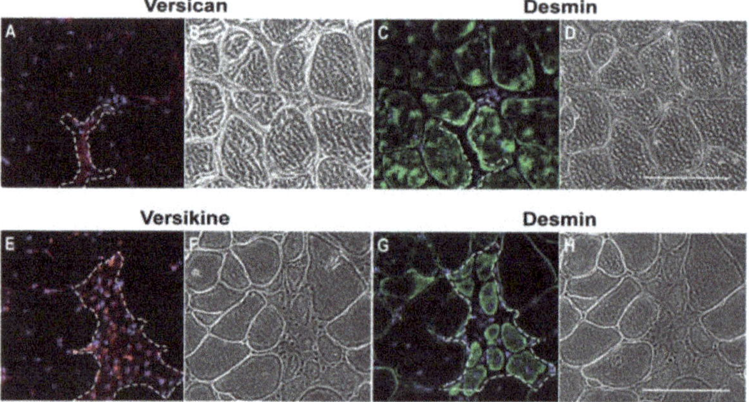

Figure 3. *Cont.*

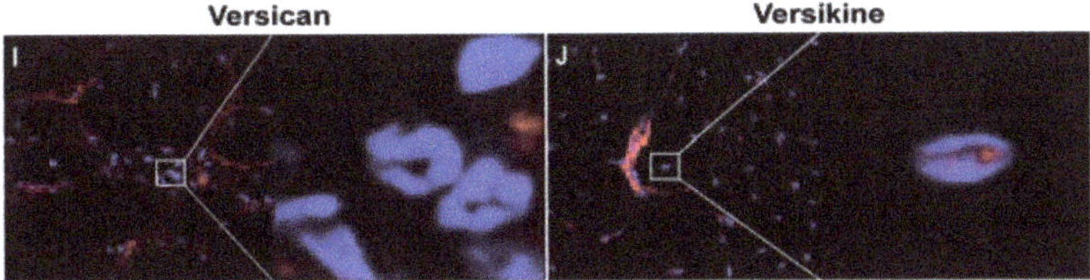

Figure 3. Versikine is localised to regenerating myofibres and mononuclear infiltrate, as well as within nuclei in *mdx* TA muscles. Serial cross-sections were stained with versican or versikine and desmin, phase images were captured to confirm localisation and tissue orientation. (**A–D**) Versican was localised to interstitium between myofibres. (**E–H**) Versikine was highly expressed in regenerating muscle, as indicated by its association with small desmin positive, centrally nucleated myofibres. (**I,J**) In dystrophic muscle cross-sections, nuclear localisation of versikine, but not versican, was also observed. Scale bars = 100 μm.

2.2. Glucocorticoids Enhance Myoblast Fusion and Myotube Formation

C2C12 myoblasts were used to investigate glucocorticoid mediated effects on transitional matrix synthesis and remodelling during myogenic differentiation. Similar to skeletal muscle development in vivo, the myogenic differentiation of C2C12 myoblasts is associated with an upregulation of transitional matrix genes, such as *V1/V0 Vcan*, *Adamts1*, *Adamts5*, *Adamts15*, *Pcsk6* [9], *Has2* and *Hyal2* [8,41]. The two-fold increase in the gene expression of the myogenic differentiation marker creatine kinase muscle (*Ckm*) (Figure 4A) and the associated dose dependent increase in total creatine kinase (CK) enzyme activity (Figure 4B), indicate that low dose glucocorticoid treatment enhanced myogenic differentiation. Essential to myogenic differentiation is the fusion of myoblasts into multinucleated myotubes, a multistep process involving migration, alignment, adhesion and membrane coalescence to form nascent myotubes [50]. Subsequent growth of these nascent myotubes occurs through the incorporation of additional myoblasts via secondary fusion [51,52]. In accordance with previously published findings [39,40], the fusion index, which is the proportion of nuclei fused into multinucleated myotubes compared to total nuclei, was increased in a dose dependent manner following treatment with 25 nM and 100 nM dexamethasone (Figure 4D). Myotube formation and maturation is considered to be a two-step process, with distinct signaling pathways contributing to the formation of nascent myotubes and the growth of mature myotubes [53–55]. Here, nascent myotubes containing 3–4 myonuclei and growing, mature myotubes undergoing secondary fusion and containing ≥5 myonuclei, were quantified as previously described [9]. In concordance with the fusion index data, there was an increase in the number of nascent myotubes (with 3–4 myonuclei) following treatment with 100 nM dexamethasone (Figure 4E), as well as a dose dependent increase in the number of mature myotubes with ≥5 myonuclei indicating enhanced secondary myoblast fusion (Figure 4F).

2.3. Glucocorticoids Regulate the Expression of Genes Associated with Transitional Matrix Synthesis and Processing during Myogenic Differentiation

Glucocorticoids may enhance myogenic differentiation by regulating the expression of genes associated with transitional matrix synthesis and processing. Specifically, 72 h of treatment with 25 nM and 100 nM of dexamethasone reduced *Tgfb1* mRNA transcript abundance by 50% and 54%, respectively (Figure 5A). The decrease in *Tgfb1* mRNA transcript abundance was associated with a reduction in versican (*V0/V1 Vcan*) gene expression by up to 56% (Figure 5B) and *Has2* gene expression by up to 58% (Figure 5C). *Has2* is the primary *Has* gene involved in hyaluronan synthesis in skeletal muscle [26]. *Adamts1* mRNA transcripts were increased up to three-fold in dexamethasone treated C2C12 cells (Figure 5D), whereas *Adamts5* or *Adamts15* mRNA levels were not affected by glucocorticoid treatment (Figure 5E,F). ADAMTS proteoglycanases are synthesised as inactive zymogens and become activated upon proteolytic processing by Furin and/or Pace4 [56]. Treatment with 25 and 100 nM dexamethasone decreased *Pcsk6* (Pace4) gene expression by 65% and 68% respectively (Figure 5G), whilst *Pcsk3* (Furin) mRNA levels were not significantly altered (Figure 5H). Lastly, dexamethasone had no effect on *Hyal2* mRNA transcript abundance (Figure 5I), the enzyme necessary for hyaluronan degradation [26].

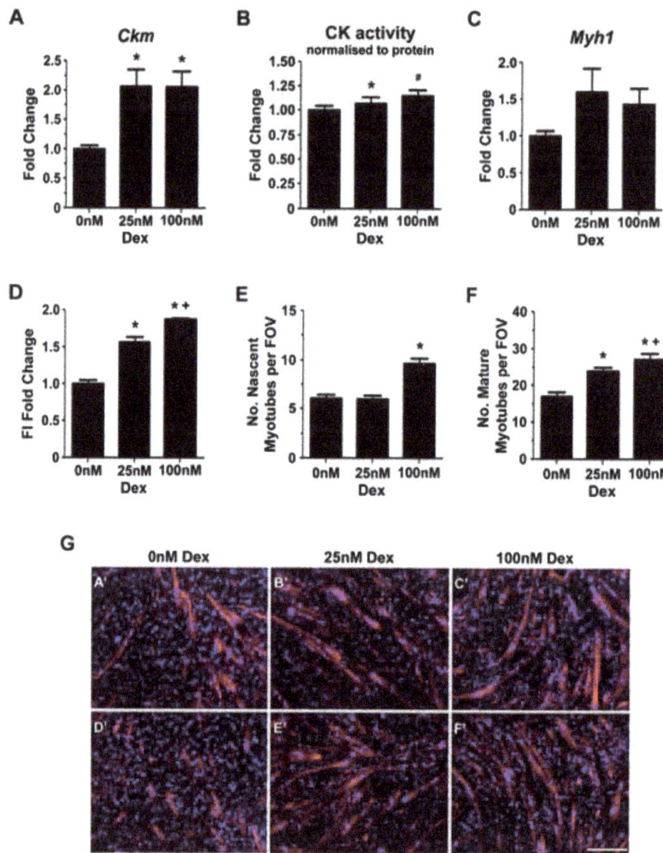

Figure 4. Low dose dexamethasone treatment for 72 h increased myogenic differentiation efficacy in C2C12 myoblasts. (**A**) The gene expression of myogenic differentiation marker creatine kinase (*Ckm*) was increased two-fold in cells treated with 25 nM and 100 nM dexamethasone (Dex) (* $p < 0.01$); (**B**) CK enzyme activity in cell lysates was also increased following treatment with 25 nM (* $p = 0.02$) and 100 nM dexamethasone (# $p = 0.01$); (**C**) Dexamethasone (Dex) did not significantly increase myosin heavy chain 1 (*Myh1*) mRNA transcripts; (**D**) Fusion index (FI; * $p < 0.001$) was greater in C2C12 cells treated with 25 nM and 100 nM dexamethasone compared to untreated control cells, and this increase was dose dependent (+ $p < 0.05$); (**E**) Treatment with 100 nM dexamethasone increased the formation of nascent myotubes containing 3–4 myonuclei (* $p < 0.001$); (**F**) The number of mature myotubes with ≥5 nuclei was greater in cultures treated with 25 nM or 100 nM versus 0 nM dexamethasone (* $p = 0.017$), and this increase was dose dependent (+ $p = 0.03$); (**G**) Representative images of C2C12 myotubes. Differentiation C2C12 myoblasts were treated with 0 nM dex (**A′,D′**), 25nM dex (**B′,E′**) or 100 nM dex (**C′,F′**) for 72 h. Myotubes were then stained with phalloidin for F-actin (red) and DAPI for nuclei (blue). CK enzyme activity was calculated from $n = 3$ biological replicates performed in quadruplicate. The fusion index and myotube number were calculated from $n = 3$ biological replicates performed in duplicate. Gene expression was determined from $n = 3$ biological replicates in triplicate. Scale bar = 200 μm. Error bars = S.E.

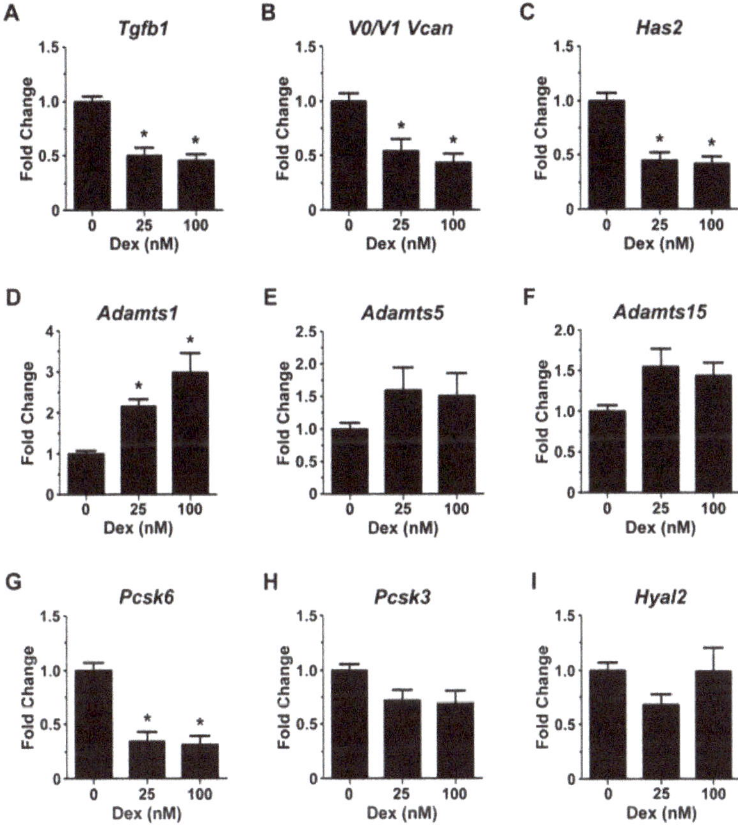

Figure 5. The expression of genes associated with a versican–hyaluronan rich transitional matrix is modulated by dexamethasone during myoblast differentiation. (**A–C**) Compared to untreated control cells, 25 nM or 100 nM dexamethasone treatment decreased *Tgfb1* (* $p < 0.0001$ and * $p = 0.0001$ respectively), *Vcan* (* $p = 0.005$ and * $p = 0.0006$, respectively) and *Has2* mRNA transcripts by approximately two-fold (* $p < 0.0001$ and * $p = 0.0001$, respectively); (**D**) *Adamts1* gene expression was increased up to three-fold in response to 25 nM and 100 nM dexamethasone treatment (* $p = 0.03$ and * $p < 0.001$, respectively); (**E,F**) *Adamts5* and *Adamts15* mRNA transcripts were not significantly increased; (**G,H**) *Pcsk6*, but not *Pcsk3*, mRNA transcripts were decreased approximately two-fold following treatment with 25 and 100 nM dexamethasone (* $p < 0.0001$). (**I**) *Hyal2* mRNA levels were not altered by dexamethasone treatment. Gene expression was determined from $n = 3$ biological replicates in triplicate. Error bars = S.E.

The effect of glucocorticoids on versican gene expression was confirmed by western blotting. Dexamethasone reduced protein levels of full length V0/V1 versican in a dose dependent manner by up to 50% (Figure 6B). ADAMTS dependent remodelling of versican during myogenic differentiation appeared not to be altered by glucocorticoids, as indicated by similar protein levels of versikine following treatment with 25 nM and 100 nM dexamethasone (Figure 6C).

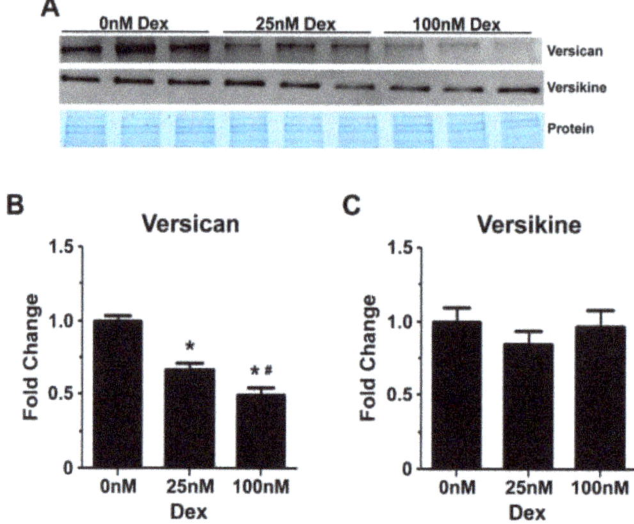

Figure 6. Versican and versikine protein expression in differentiating C2C12 myoblasts following dexamethasone treatment. (**A**) Representative western blots to assess versican and versikine expression in C2C12 cell lysates, with the respective stain free protein gel image to demonstrate even protein loading; (**B**) Decreased versican protein expression following treatment with 25 nM (* p = 0.00002) and 100 nM dexamethasone (* p = 0.0000001); and this decrease was dose dependent (# p = 0.03); (**C**) Versikine protein levels were not altered by dexamethasone treatment. Versican and versikine protein expression analysis was calculated from n = 3 biological replicates performed in quadruplicate. Error bars = S.E.

2.4. Glucocorticoids Rescue Myotube Formation in Differentiating Myoblasts Treated with Exogenous Versican and Versikine

In vitro, versican processing facilitates myoblast fusion and myotube formation, whilst an excess of versican appears to be detrimental [9]. Therefore, it is possible that reduced versican synthesis may contribute to the positive effects of glucocorticoids on regenerative myogenesis in dystrophic muscles. To test this hypothesis, differentiating C2C12 myoblasts were treated with V1 versican, versikine or empty vector conditioned media supplemented with 0 nM or 100 nM dexamethasone. The addition of conditioned media made the experimental conditions more challenging, with greater variability in fusion between biological replicates and a blunted response to dexamethasone. Nonetheless, excess full length or processed versican decreased myoblast fusion (main effect conditioned media; 2-way GLM ANOVA; * p < 0.001 for versican treated cells and ** p < 0.02 for versikine treated cells; Figure 7C). This decrease in fusion was ameliorated with glucocorticoid treatment (main effect dexamethasone; 2-way GLM ANOVA; # p < 0.001 for empty vector or versican treated cells and ## p < 0.001 for empty vector or versikine treated cells; Figure 7C).

As further evidence that versican impairs myoblast fusion, in cells treated with 0 nM dexamethasone, the versican conditioned media decreased the number of nascent myotubes compared to empty vector conditioned media (interaction; 2-way GLM ANOVA; * p < 0.001). Following treatment with 100 nM dexamethasone, the number of nascent myotubes was similar in cells treated with the versican or empty vector conditioned media (Figure 7D). Versikine had no effect on the number of nascent myotubes formed (Figure 7D). Unexpectedly, dexamethasone decreased the number of nascent myotube in cells treated with versikine or empty vector conditioned media (main effect dexamethasone; 2-way GLM ANOVA; # p < 0.01) (Figure 7D).

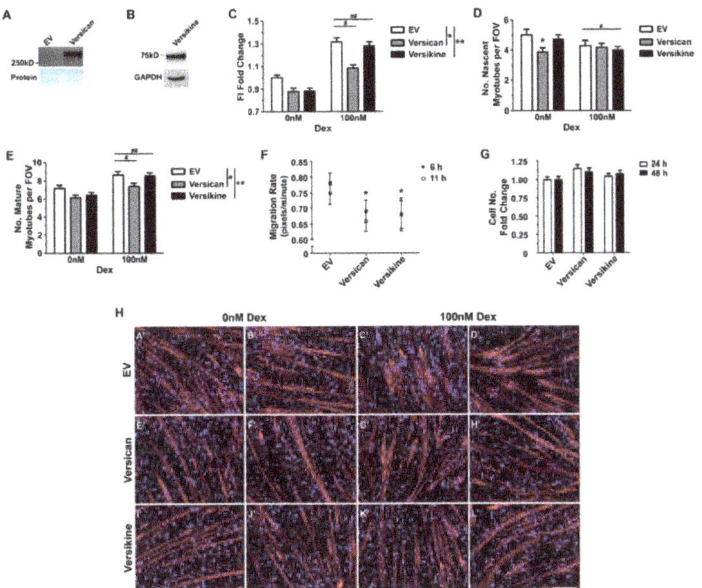

Figure 7. Dexamethasone ameliorated the impairment in myogenic differentiation associated with excess versican. (**A**,**B**) Representative western blots of versican, versikine or empty vector (EV) conditioned media, with the respective images of stain free protein gels or GAPDH as a loading controls; (**C**) The addition of versican and versikine conditioned media compromised myoblast differentiation, as assessed by fusion index (main effect conditioned media; 2-way GLM ANOVA; * $p < 0.001$ for versican conditioned media and ** $p < 0.02$ for versikine conditioned media relative to cells treated with the empty vector conditioned media). Whilst, glucocorticoids enhanced myoblast fusion (main effect dexamethasone; 2-way GLM ANOVA; # $p < 0.001$ for empty vector or versican treated cells and ## $p < 0.001$ empty vector or versikine treated cells); (**D**) In the absence of dexamethasone, excess versican reduced the formation of nascent myotubes (interaction; 2-way GLM ANOVA; * $p < 0.001$). In response to 100 nM dexamethasone, the number of nascent myotubes was similar in cells treated with versican or empty vector conditioned media. Versikine had no effect on nascent myotube formation, whilst dexamethasone decreased nascent myotube number in cells treated with versikine or empty vector conditioned media (main effect dexamethasone; 2-way GLM ANOVA; # $p < 0.01$); (**E**) Versican or versikine reduced the number of mature myotube formed per field of view (FOV) (main effect conditioned media; 2-way GLM ANOVA; * $p < 0.0001$ for versican treated cells and ** $p < 0.0001$ versikine treated cells), and 100 nM dexamethasone ameliorated this decrease in myotube number (main effect dexamethasone; 2 way GLM ANOVA; # $p = 0.0092$ for empty vector or versican treated cells and ## $p = 0.0061$ empty vector or versikine treated cells); (**F**) Versican or versikine conditioned media reduced the migration rate of C2C12 myoblasts compared to empty vector conditioned media (* $p = 0.01$ and * $p = 0.04$ respectively); (**G**) Myoblast cell number was not different following 24 h and 48 h of treatment with versican or versikine conditioned media compared to empty vector conditioned media; (**H**) Representative images of C2C12 myotubes. Differentiating C2C12 myoblasts were treated with 0 nM dex and EV conditioned media (**A′**,**B′**), 100 nM dex and EV conditioned media (**C′**,**D′**), 0 nM dex and versican conditioned media (**E′**,**F′**), 100 nM dex and versican conditioned media (**G′**,**H′**), 0 nM dex and versikine conditioned media (**I′**,**J′**), or 100 nM dex and versikine conditioned media (**K′**,**L′**) for 72 h. Myotubes were then stained with phalloidin for F-actin (red) and DAPI for nuclei (blue). Fusion index (FI) and myotube number were calculated from $n = 5$ biological replicates performed in duplicate. Migration rate was measured from $n = 5$ biological replicates performed in duplicate or triplicate. Myoblast proliferation was assessed from $n = 3$ biological replicates performed in 8 wells. Scale bar = 200 μm. Error bars = S.E.

With regards to the effects of versican, versikine and glucocorticoids on secondary fusion, the number of mature myotubes was reduced in cells treated with versican or versikine conditioned media (main effect conditioned media; 2-way GLM ANOVA; * $p < 0.0001$ for versican treated cells and ** $p < 0.0001$ versikine treated cells; Figure 7E). When differentiating myoblasts were treated with 100 nM dexamethasone, the number of mature myotubes increased (main effect dexamethasone; 2-way GLM ANOVA; # $p = 0.0092$ for versican treated cells and ## $p = 0.0061$ versikine treated cells; Figure 7E).

Alignment of myoblasts is essential for fusion, and this depends on carefully regulated migration [50,57,58]. Versican is known to modulate cell migration and depending on the biological context the effects can be stimulatory [59,60] or inhibitory [61]. The effects of versikine on cell migration have not been well characterised. In C2C12 cells treated with versican or versikine conditioned media for up to 11 h, myoblast migration rate was reduced by 12% and 13%, respectively (Figure 7F). Thus, excess versican, both the full-length protein and the cleaved bioactive fragment, may also impair regenerative myogenesis through a reduction in myoblast migration.

Myoblast viability and number can be a confounding factor in determining the efficacy of myogenic differentiation. Versican has been shown to increase proliferation in various biological contexts [62,63], including primary turkey myoblasts [27]. In contrast, versikine has been associated with apoptosis during interdigital web regression [32]. In actively proliferating C2C12 myoblast cultures, exogenous versican or versikine had no effect on cell number (Figure 6G).

3. Discussion

In dystrophic skeletal muscles, excess synthesis and inappropriate processing of ECM proteins lead to degeneration, fibrosis and compromised contractile function [17,64]. Similarities in the mechanisms of ECM expansion in patients with DMD and *mdx* mice have been observed, and contribute to the dystrophic pathology of these muscles [12]. The significance of versican in the generation and remodelling of a transitional matrix during skeletal muscle development and regeneration is continuing to gain recognition [8,9,27]. We propose that the carefully regulated synthesis and processing of a versican rich transitional matrix is also an important factor in differentiating between successful regenerative myogenesis or degeneration and fibrosis. A better understanding of versican function in muscular dystrophy is needed if progress is to be made in targeting the dysregulated ECM, which is a hallmark of DMD pathology.

Here, we report that the expression of full-length versican is increased in dystrophic *mdx* diaphragm and hindlimb muscles compared to wild type muscles, with the highest level of versican expression observed in the more severely affected *mdx* diaphragm muscles. These observations are in concordance with human data showing increased versican expression in muscle biopsies from patients with DMD compared to healthy controls, as assessed by immunohistochemistry [65] and microarray gene expression analysis [66]. Furthermore, deposition of chondroitin sulphate GAG side chains is upregulated in DMD [15], and V0/V1 versican is a significant source of chondroitin sulphate GAG chains in skeletal muscle. V0 versican is the most highly glycosylated isoform, followed by the V1 variant [67]. Versican is secreted and synthesised by activated satellite cells and myoblasts [27,68], newly formed myotubes [28], inflammatory cells [69] and fibroblasts [70].

Versican is transiently upregulated in myoblasts and newly formed myotubes during development and regeneration [28,71], whilst in healthy, mature skeletal muscle full length versican expression is quite low. Versican remodelling has been implicated in various developmental processes, and remodelling by specific ADAMTS proteoglycanases generates versikine [19,32]. Interestingly, in *mdx* TA muscles, versikine immunoreactivity was associated with small, recently regenerated myofibres. Furthermore, in dystrophic muscles, the nuclear localisation of versikine was observed in both muscle fibres and in mononuclear infiltrate. This is in line with observations by Carthy et al., who using the same anti-DPEAAE neo-epitope antibody (ThermoFisher Scientific, PA1-1748A, Waltham, MA, USA) detected nuclear versikine staining in vascular smooth muscle cells and proposed a potential role in mitotic spindle organization during cell division [63]. Versikine is further degraded by various ECM proteases. This hypothesis is supported by the observation that versikine immunoreactivity in developing mouse hindlimb muscles at E13.5 days is much higher than in mature muscles at 3 weeks of age [9]. This further degradation of versikine may account for the lack of difference in protein levels between diaphragm and TA muscles from *mdx* and wild type mice, despite increased V0/V1 versican expression.

In vivo, centrally nucleated fibres are indicative of recent damage and repair, with myoblast fusion being essential for effective regeneration. In diaphragm muscles from *mdx* mice, the proportion of centrally nucleated fibres is much lower, up to 2–3 folds, when compared to dystrophic TA muscles [48,72]. We hypothesise that excess versican accumulation contributes to the impaired regenerative capacity of *mdx* diaphragm muscles. As such, we propose that versican reduction could be a potential strategy to ameliorate the pathology of dystrophic muscles. In vitro evidence that processing of versican by ADAMTS5 or -15 facilitates myoblast fusion supports this hypothesis [9]. Furthermore, others have shown that the formation of multinucleated myotubes is associated with a reduction in chondroitin sulphate GAG sidechains, which also suggests a potential role for versican processing [73]. It is worth noting that versikine does not contain chondroitin sulphate GAG side chains [74].

Glucocorticoids improve muscle function in patients with DMD through various cellular mechanisms [36,75,76]. Of particular interest, are the effects of glucocorticoids on ECM synthesis and remodelling [77], on TGF-β [78] and TGF-β centred signalling networks, as these are highly relevant to regenerative myogenesis and fibrosis [12]. Our observation of a concentration dependent increase in myoblast fusion and myotube formation following low dose, 25 nM and 100 nM, dexamethasone treatment is in concordance with a number of studies reporting positive effects of glucocorticoids on myogenesis in vitro [39,40] and muscle regeneration in vivo [12,37]. In contrast to our findings, Ma et al. [79] reported inhibition of myogenic differentiation of C2C12 and primary mouse myoblasts following glucocorticoid treatment through the activation of glycogen synthase kinase 3β (GSK-3β). However, the concentration of dexamethasone used (10 μM) was 100–400 folds higher than our low dose treatment [79]. Furthermore, with lower concentrations of 10 nM and 100 nM dexamethasone, no significant decrease in myogenin or myosin heavy chain protein expression was reported, whilst the fusion index was not assessed [79].

The increase in myotube formation following glucocorticoid treatment was associated with decreased V0/V1 versican protein and *V0/V1 Vcan*, *Has2* and *Tgfb1* mRNA transcript abundance. This reduction in *Tgfb1* gene expression in response to glucocorticoids is in agreement with in vivo studies in *mdx* mice [12,78], and in vitro studies using hepatic stellate cells [80] and fetal lung fibroblasts [81]. In differentiating myoblasts, glucocorticoids appear to have specific effects on the various *Adamts* proteoglycanase isoforms and genes involved in ADAMTS activation, as dexamethasone treatment was also associated with an increase in *Adamts1*, but not *Adamts5* and -15, gene expression, as well as a decrease in *Pcsk6* (but not *Pcsk3*) mRNA transcript abundance. Altogether, these data suggest that glucocorticoids may attenuate the synthesis of a transitional, pericellular matrix, thus facilitating membrane coalescence during fusion in differentiating C2C12 myoblasts [9]. The effects of glucocorticoids on a versican and hyaluronan rich transitional matrix have been described in other biological contexts. Specifically, glucocorticoid induced skin atrophy is associated with reduced proteoglycan [82] and hyaluronan synthesis [44,45]. Glucocorticoids have also been reported to decrease versican expression in cultured rat mesangial cells and airway fibroblasts [43,79].

Myoblast fusion was reduced when differentiating C2C12 myoblasts were treated with versican conditioned media, supporting our hypothesis that excess versican impairs regenerative myogenesis. This decrease in myoblast fusion was ameliorated, but not fully reversed, with dexamethasone. Versikine conditioned media also impaired myoblast fusion, and this impairment was also ameliorated by glucocorticoid treatment. Hyaluronan can bind to versican, and perhaps also versikine, via the G1 N-terminus link module [25], thus contributing to pericellular matrix expansion. We have previously shown that expansion and inadequate processing of a versican–hyaluronan rich pericellular matrix impairs myoblast fusion and myotube formation [9]. This decrease in fusion was interrogated further by quantifying the number of nascent and mature myotubes following versican, versikine and/or dexamethasone treatment. An excess of versican decreased the number of nascent myotubes, and this decrease was ameliorated by dexamethasone treatment. Whereas, an excess of versikine did not compromise nascent myotube formation. Myotube hypertrophy and nuclear accretion occurs via secondary fusion, which involves distinct signalling pathways compared to the formation of nascent myotubes [51]. Excess versican and versikine reduced the formation of mature myotubes. Dexamethasone rescued this impairment and increased the number of mature myotubes formed.

Excess versican or versikine may impair myoblast migration, thus potentially contributing to the observed decrease in myoblast fusion and myotube number. The effects of versican on cell migration are context dependent, as during development, versican has been suggested to deter muscle cell migration and thus contribute to the patterning of the limb skeleton and joints [83]. The underlying mechanism by which versican and versikine regulate cell migration during myogenesis remains to be determined, but may involve CD44 signalling [57,84,85]. Full length versican, through its chondroitin sulphate side chains, can bind to the CD44 receptor directly [84]. Through interactions with the link module on the G1 domain, versican and versikine bind hyaluronan [19], which is also a known ligand for the CD44 receptor [85]. Altogether, these observations suggest a novel mechanism by which excess versican could compromise regenerative myogenesis in muscular dystrophy.

We propose that excess accumulation of versican in dystrophin-deficient muscles compromises regeneration and exacerbates fibrosis. Dadgar et al. [12] have recently confirmed that TGF-β centred signalling networks are key drivers for fibrosis and failed regeneration in muscular dystrophy. TGFβ stimulates V0/V1 versican synthesis in various biological contexts [59,86–88]. Furthermore, the interaction between TGF-β and versican is bidirectional, with versican potentiating TGFβ signalling [89] and regulating bioavailability [90]. When dystrophic *mdx* mice were treated with glucocorticoids, these TGF-β centered networks were suppressed and the dystrophic pathology was ameliorated [12]. Therefore, our in vitro observations that glucocorticoids reduce V0/V1 versican (*Vcan*) and *Tgfb1* expression in differentiating myoblasts highlight the relevance of these genes to regenerative myogenesis, especially in the context of dystrophy.

To fully characterise the role of versican in regenerative myogenesis in dystrophic skeletal muscles in vivo studies using genetic and/or pharmacological approaches are needed. The advantage of such in vivo studies is that they will allow the effects of versican on muscle repair and function to be investigated in the presence of an expanded ECM and increased inflammation. This is important, given the emerging role of versican [91,92] and versikine [33] in regulating inflammation in various biological contexts.

4. Materials and Methods

4.1. Mouse Models

All animal studies were approved by the La Trobe University and Deakin University Animal Ethics Committees, in accordance with the National Health and Medical Research Council (NH&MRC) guidelines, under the ethics numbers of AEC16-08 for La Trobe University (approved: 1 January 2016), and G35-2013 and A79/2011 (approved: 1 January 2015 and 1 January 2012, respectively) for Deakin University. Between 3 and 6 months of age, C57BL/10 (wild type) and *mdx* mice were deeply anaesthetised with sodium pentobarbitone (60 mg/kg) and killed by cardiac excision. TA and diaphragm muscles were collected for immunohistochemical analysis by embedding the blotted tissues in optimal cutting temperature compound (OCT) and freezing in thawing 2-methylbutane cooled in liquid nitrogen. TA and diaphragm muscles were also snap frozen in liquid nitrogen for gene expression analysis.

4.2. Skeletal Muscle Immunohistochemistry and Histology

Transverse frozen sections were cut from the mid-belly of the TA or diaphragm muscle strips at a thickness of 8 μm, mounted on slides and stored at $-80\ ^{\circ}$C until analysis. Immunohistochemistry for ADAMTS1 (Origene, TA317919, Rockville, MD, USA), ADAMTS5, ADAMTS15 (Abcam, ab45047, Cambridge, MA, USA), V0/V1 versican (anti-GAGβ; Merck Millipore, AB1033, Bayswater, VIC, Australia) or versikine (anti-DPEAAE neo-epitope; Thermo Fisher Scientific, PA1-1748A, Scoresby, VIC, Australia) were performed as previously described [9,32]. An anti-desmin rabbit polyclonal antibody (Abcam, ab15200, Cambridge, MA, USA) together with an anti-rabbit Alexa Fluor 488 secondary antibody (Thermo Fisher Scientific, R37116,) were used to detect myoblasts and newly regenerated myofibres [93,94]. Representative wild type and *mdx* TA and diaphragm muscle cross-sections were H and E stained for muscle architecture, and wheat germ agglutinin to assess fibrosis [95]. For analysis of V0/V1 versican and versikine immunoreactivity, four non-overlapping representative digital images were captured with a confocal microscope of each muscle section at 600× magnification (Olympus Fluoview FV10i) and analysed for area of immunoreactivity using Image-Pro Plus software (Version 7, Media Cybernetics, Silver Spring, MD, USA).

4.3. Cell Culture and Expression Constructs

HEK293T cells were grown in Dulbecco's Modified Eagle Medium (DMEM; 25 mM glucose) containing 10% fetal bovine serum (FBS) in atmospheric O_2 and 5% CO_2 at 37 $^{\circ}$C. Cells were transfected using Lipofectamine 2000 (Thermo Fisher Scientific) with constructs encoding the V1 versican construct (kindly provided by Dieter Zimmermann), the bioactive G1-DPEAAE versikine fragment was produced by the insertion of a stop codon in the V1 versican construct after the Glu^{441}-Ala^{442} peptide bond cleavage site [32], and empty vector control (pcDNA3.1MycHisA+ (Thermo Fisher Scientific)). Serum-free conditioned medium was collected for use in the myoblast differentiation experiments, as previously described [9], and underwent western blotting for confirmation of V1 versican and versikine protein expression. C2C12 myoblasts, a well characterized in vitro model of myoblast fusion and regenerative myogenesis [96,97], were maintained in growth medium (25 mM glucose DMEM plus 10% FBS) in atmospheric O_2 and 5% CO_2 at 37 $^{\circ}$C.

4.4. Glucocorticoid Treatment of Differentiating C2C12 Cells

To determine the effects of glucocorticoids on myoblast differentiation, cells were seeded at 25,000 cells/cm^2; in duplicate wells for fusion index determination, and in triplicate wells for gene and protein expression and creatine kinase activity analyses. Following 48 h of proliferation (and at >90% confluence), myoblasts were treated with differentiation medium (25 mM glucose DMEM plus 2% horse serum (HS)) supplemented with 0 nM, 25 nM or 100 nM dexamethasone (Prednisolone F, Sigma-Aldrich, D1756, Castle Hill, NSW, Australia) for 72 h (refreshed every 24 h). The latter is a low glucocorticoid concentration which has been shown to increase myoblast fusion efficiency in vitro [39]. Following this, the cells were harvested for biochemical analyses (see below) or fixed with 4% paraformaldehyde, and stained with Alexa-Fluor 568 phalloidin (1:50; Life Technologies) and DAPI (1:20; Life Technologies) for 20 min.

Fusion index was used as a proxy readout to assess myoblast differentiation efficacy. Nascent myotubes with 3–4 myonuclei and mature myotubes with ≥5 myonuclei were quantified, as previously described [9,50]. For the fusion index, 3 biological replicates (at different passages) in duplicate were performed. For the control cells treated with 0 nM dexamethasone, the % fusion index for each of the three biological replicates ranged from 17–20. For each experimental condition and biological replicate, 8897 ± 222 nuclei were counted to assess fusion index and 270 ± 23 myotubes were classified and counted to assess myotube maturation.

The fusion index data are supported by gene markers of myoblast differentiation and a commercially available creatine kinase enzyme activity assay kit, as per manufacturer's instructions (ab155901, Abcam) [98]. The cell lysates from the creatine kinase enzyme activity assay were also used for analysis of versican and versikine protein expression.

4.5. RNA Extraction, Reverse Transcription, and Quantitative RT-PCR

Cells were harvested in triplicate wells, collected in TRIzol (Sigma-Aldrich) and stored at -80 °C. TA and diaphragm muscles were mechanically homogenised in TRIzol. Upon thawing, RNA was extracted and 1 µg of total RNA was reverse-transcribed using an iScript cDNA synthesis kit (Bio-Rad, Gladesville, NSW, Australia). Quantitative RT-PCR was performed using iQ SYBR Green Supermix (Bio-Rad) and oligonucleotide primers for the murine genes of interest (Table 1). Relative changes in mRNA levels to untreated myotubes were calculated using the ΔC_t method. Real time data was normalised to cDNA content, as determined using a Quant-iT Oligreen ssDNA reagent kit (Life Technologies).

Table 1. List of primer sequences used for quantitative RT-PCR.

Accession Number	Name	Forward Sequence (5′-3′)	Reverse Sequence (5′-3′)
NM_009621.5	*Adamts1*	CCTGTGAAGCCAAAGGCATTG	TGCACACAGACAGAGGTAGAGT
NM_011782.2	*Adamts5*	GCTACTGCACAGGGAAGAGG	GCCAGGACACCTGCATATTT
NM_001329420.1	*Adamts15*	GCTCATCTGCCGAGCCAAT	CAGCCAGCCTTGATGCACTT
NM_007710.2	*Ckm* [1]	CCGTGTCACCTCTGCTGCTG	TCCTTCATATTGCCTCCCTTCTCC
XM_011250822.2	*Pcsk3* [2]	CAGCGAGACCTGAATGTGAA	CAGGGTCATAATTGCCTGCT
NM_008216.3	*Has2*	GGGACCTGGTGAGACAGAAG	ATGAGGCAGGGTCAAGCATA
XM_006511645.3	*Hyal2*	AGCCGCAACTTTGTCAGTTT	GAGTCCTCGGGTGTATGTGG
XM_017314318.1	*Myh1* [3]	GCTCAAAGCCCTGTGTTACC	CATAGACGGCTTTGGCTAGG
NM_001291184.1	*Pcsk6* [4]	ATTTCCCCAACCTCGTCTCT	AGCTGAGTCCTTGCCACCTA
NM_011577.2	*Tgfb1*	GCCTGAGTGGCTGTCTTTTGA	CACAAGAGCAGTGAGCGCTGAA
NM_001081249.1 (V0) NM_019389.2 (V1)	*V0/V1 Vcan* [5]	ACCAAGGAGAAGTTCGAGCA	CTTCCCAGGTAGCCAAATCA

[1] Creatine kinase muscle, [2] Furin, [3] Myosin heavy chain 1 (fast IIx/d isoform), [4] Pace4, [5] Versican (primers detect the V0 and V1 isoforms).

4.6. Versican or Versikine Treatment and C2C12 Myoblast Differentiation

To assess the effects of glucocorticoids on myotube formation in the presence of excess versican or versikine, C2C12 cells were seeded at 20,000 cells/cm^2 in duplicate wells, 24 h or 48 h later the differentiation medium was added for 4 or 3 days, respectively. Depending on the experimental conditions, the differentiation medium was supplemented with 0 nM or 100 nM dexamethasone and serum-free versican, versikine or empty vector conditioned media (diluted 1:4; refreshed daily). Fusion index and myotube number were determined and analysed as described above. Five biological replicates (at different passages) in duplicate were performed, with a total of 3156 $\pm$ 96 nuclei and 110 $\pm$ 3 myotubes counted per experimental condition for each biological replicate. For the control cells treated with empty vector conditioned media and 0 nM dexamethasone, the % fusion index for each of three biological replicates ranged from 19–27%. This greater variability in % fusion compared to the dexamethasone dose response experiments reflects the more challenging experimental conditions.

4.7. Versican or Versikine Treatment and C2C12 Myoblast Migration and Proliferation

To assess the specific effects of versican and versikine on myoblast migration, a process essential for effective differentiation [50,57], cells were seeded at 8500 cells/cm^2 in triplicate wells and 72 h later (when 100% confluent) a scratch wound assay was performed [58]. Growth media was supplemented with serum-free versican, versikine or empty vector conditioned media (diluted 1:4) for up to 11 h. To determine migration rate, digital images (three per well) were digitally captured at 0 h, 6 h and 11 h post-wounding. The distance between the edges of the scratches was measured using Adobe Photoshop CS6 (Adobe Systems), with the migration rate calculated in pixels/min.

To assess the effects of versican and versikine on myoblast proliferation, cells were seeded at 10,000 cells/cm^2 in growth media, supplemented with serum-free versican, versikine or empty vector conditioned media (diluted 1:4; refreshed daily) for 48 h. Myoblast number was assessed using the WST-1 cell proliferation reagent (Roche Life Science, Sydney, NSW, Australia), as per the manufacturer's directions.

4.8. Western Blot

To improve detection by the anti-V0/V1 versican antibody, the GAG side chains were removed. Specifically, 1 µL of chondroitinase ABC (Seikagaku, Tokyo, Japan) was added to the versican conditioned media and to the cell lysates from the creatine kinase enzyme assay for 2 h at 37 °C. The serum free conditioned media containing versikine was subjected to the western blotting as described by Dancevic et al. [30]. The versican conditioned media and C2C12 cell lysates underwent SDS-PAGE using pre-cast Mini-PROTEAN TGX Stain-Free Protein Gels (Bio-Rad), which were transferred onto PVDF membranes. Proteins were visualised by chemiluminescence on a Chemidoc XRS+ (Bio-Rad) and analysed using ImageLab software (Bio-Rad). Versican and versikine protein levels were normalised to total optical density of all protein bands on the TGX Stain-Free Protein Gel. Primary antibodies used were anti-GAGβ (V0/V1 versican) (1:200, Merck Millipore), anti-V0/V1 DPEAAE neo-epitope (versikine) (1:1000, Thermo Fisher Scientific, PA1-1748A), and anti-GAPDH (1:10,000, Merck Millipore, AB1033). Secondary antibodies used were peroxidase AffiniPure goat anti-rabbit IgG (1:5000, Jackson ImmunoResearch Laboratories) and anti-mouse IR680 (1:10,000, Sigma Aldrich).

4.9. Statistical Analyses

For the versican gene expression data, 2-way general linear model (GLM) ANOVA was performed with the factors being muscle type (diaphragm versus TA) and strain (C57BL/10 versus *mdx*). For the quantitation of versican and versikine immunoreactivity in TA and diaphragm muscles from *mdx* and wild type mice, independent *t*-tests were used to assess differences in immunoreactivity between C57BL/10 and *mdx* mice for a given muscle type. For the cell culture experiments, independent *t*-tests, 1-way or 2-way GLM ANOVA were performed as appropriate and followed by Tukey's post-hoc analysis where required. All data are presented as mean ± S.E. and were considered statistically significant when $p < 0.05$.

Acknowledgments: The authors wish to thank Alister Ward for critical reading of this manuscript. This work was supported, in whole or in part, by the Molecular and Medical Research SRC (Strategic Research Centre) and The Financial Markets Foundation for Children Grant 162-2010 (to Daniel McCulloch and Nicole Stupka).

Author Contributions: Nicole Stupka, Natasha McRae and Daniel McCulloch conceived and coordinated the study. Natasha McRae and Nicole Stupka wrote the paper. Chris Van der Poel and Nicole Stupka performed all mouse experiments and prepared the muscle cryosections. Natasha McRae performed experiments shown in Figures 1–7, analysed most data presented throughout, and prepared all figures and tables throughout the manuscript. Leonard Forgan and Bryony McNeill contributed to analysis of data presented in Figure 2I,J. Alex Addinsall contributed to experiments shown in Figure 6C,D. All authors reviewed the results, and approved the final version of the manuscript.

Conflicts of Interest: The authors declare no conflict of interest.

Abbreviations

ADAMTS	A disintegrin-like and metalloproteinase domain with thrombospondin-1 repeats
CSPG	Chondroitin sulphate proteoglycan
CK	Creatine kinase
Ckm	Creatine kinase muscle
Dex	Dexamethasone
DMEM	Dulbecco's modified eagle medium
DMD	Duchenne muscular dystrophy
ECM	Extracellular matrix

Int. J. Mol. Sci. **2017**, *18*, 2629

EV	Empty vector
FI	Fusion index
FOV	Field of view
GAG	Glycosaminoglycan
GLM	General linear model
GSK-3β	Glycogen synthase kinase-3β
Has	Hyaluronan synthase
HS	Horse serum
Hyal	Hyaluronidase
Myh1	Myosin heavy chain 1
OCT	Optimum cutting temperature
TA	Tibialis anterior
TGF-β	Transforming growth factor-β
Vcan	Versican

References

1. Beytia, M.; Vry, J.; Kirschner, J. Drug treatment of duchenne muscular dystrophy: Available evidence and perspectives. *Acta Myol.* **2012**, *31*, 4–8.
2. Gumerson, J.D.; Michele, D.E. The dystrophin-glycoprotein complex in the prevention of muscle damage. *J. Biomed. Biotechnol.* **2011**, *2011*, 210797. [CrossRef] [PubMed]
3. Duance, V.C.; Stephens, H.R.; Dunn, M.; Bailey, A.J.; Dubowitz, V. A role for collagen in the pathogenesis of muscular dystrophy? *Nature* **1980**, *284*, 470–472. [CrossRef] [PubMed]
4. Zhou, L.; Lu, H. Targeting fibrosis in duchenne muscular dystrophy. *J. Neuropathol. Exp. Neurol.* **2010**, *69*, 771–776. [CrossRef] [PubMed]
5. Desguerre, I.; Mayer, M.; Leturcq, F.; Barbet, J.P.; Gherardi, R.K.; Christov, C. Endomysial fibrosis in duchenne muscular dystrophy: A marker of poor outcome associated with macrophage alternative activation. *J. Neuropathol. Exp. Neurol.* **2009**, *68*, 762–773. [CrossRef] [PubMed]
6. Carvajal Monroy, P.L.; Grefte, S.; Kuijpers-Jagtman, A.M.; Helmich, M.P.; Wagener, F.A.; von den Hoff, J.W. Fibrosis impairs the formation of new myofibers in the soft palate after injury. *Wound Repair Regen.* **2015**, *23*, 866–873. [CrossRef] [PubMed]
7. Calve, S.; Simon, H.G. Biochemical and mechanical environment cooperatively regulate skeletal muscle regeneration. *FASEB J.* **2012**, *26*, 2538–2545. [CrossRef] [PubMed]
8. Calve, S.; Odelberg, S.J.; Simon, H.G. A transitional extracellular matrix instructs cell behavior during muscle regeneration. *Dev. Biol.* **2010**, *344*, 259–271. [CrossRef] [PubMed]
9. Stupka, N.; Kintakas, C.; White, J.D.; Fraser, F.W.; Hanciu, M.; Aramaki-Hattori, N.; Martin, S.; Coles, C.; Collier, F.; Ward, A.C.; et al. Versican processing by a disintegrin-like and metalloproteinase domain with thrombospondin-1 repeats proteinases-5 and -15 facilitates myoblast fusion. *J. Biol. Chem.* **2013**, *288*, 1907–1917. [CrossRef] [PubMed]
10. Pescatori, M.; Broccolini, A.; Minetti, C.; Bertini, E.; Bruno, C.; D'Amico, A.; Bernardini, C.; Mirabella, M.; Silvestri, G.; Giglio, V.; et al. Gene expression profiling in the early phases of DMD: A constant molecular signature characterizes DMD muscle from early postnatal life throughout disease progression. *FASEB J.* **2007**, *21*, 1210–1226. [CrossRef] [PubMed]
11. Marotta, M.; Ruiz-Roig, C.; Sarria, Y.; Peiro, J.L.; Nunez, F.; Ceron, J.; Munell, F.; Roig-Quilis, M. Muscle genome-wide expression profiling during disease evolution in *mdx* mice. *Physiol. Genom.* **2009**, *37*, 119–132. [CrossRef] [PubMed]
12. Dadgar, S.; Wang, Z.; Johnston, H.; Kesari, A.; Nagaraju, K.; Chen, Y.W.; Hill, D.A.; Partridge, T.A.; Giri, M.; Freishtat, R.J.; et al. Asynchronous remodeling is a driver of failed regeneration in duchenne muscular dystrophy. *J. Cell Biol.* **2014**, *207*, 139–158. [CrossRef] [PubMed]
13. Bernasconi, P.; Torchiana, E.; Confalonieri, P.; Brugnoni, R.; Barresi, R.; Mora, M.; Cornelio, F.; Morandi, L.; Mantegazza, R. Expression of transforming growth factor-β 1 in dystrophic patient muscles correlates with fibrosis. Pathogenetic role of a fibrogenic cytokine. *J. Clin. Investig.* **1995**, *96*, 1137–1144. [CrossRef] [PubMed]

14. Ishitobi, M.; Haginoya, K.; Zhao, Y.; Ohnuma, A.; Minato, J.; Yanagisawa, T.; Tanabu, M.; Kikuchi, M.; Iinuma, K. Elevated plasma levels of transforming growth factor β1 in patients with muscular dystrophy. *Neuroreport* **2000**, *11*, 4033–4035. [CrossRef] [PubMed]

15. Negroni, E.; Henault, E.; Chevalier, F.; Gilbert-Sirieix, M.; Van Kuppevelt, T.H.; Papy-Garcia, D.; Uzan, G.; Albanese, P. Glycosaminoglycan modifications in duchenne muscular dystrophy: Specific remodeling of chondroitin sulfate/dermatan sulfate. *J. Neuropathol. Exp. Neurol.* **2014**, *73*, 789–797. [CrossRef] [PubMed]

16. Stephens, H.R.; Duance, V.C.; Dunn, M.J.; Bailey, A.J.; Dubowitz, V. Collagen types in neuromuscular diseases. *J. Neurol. Sci.* **1982**, *53*, 45–62. [CrossRef]

17. Klingler, W.; Jurkat-Rott, K.; Lehmann-Horn, F.; Schleip, R. The role of fibrosis in duchenne muscular dystrophy. *Acta Myol.* **2012**, *31*, 184–195. [PubMed]

18. Wight, T.N. Provisional matrix: A role for versican and hyaluronan. *Matrix Biol.* **2016**. [CrossRef] [PubMed]

19. Nandadasa, S.; Foulcer, S.; Apte, S.S. The multiple, complex roles of versican and its proteolytic turnover by ADAMTS proteases during embryogenesis. *Matrix Biol.* **2014**, *35*, 34–41. [CrossRef] [PubMed]

20. Li, H.; Mittal, A.; Makonchuk, D.Y.; Bhatnagar, S.; Kumar, A. Matrix metalloproteinase-9 inhibition ameliorates pathogenesis and improves skeletal muscle regeneration in muscular dystrophy. *Hum. Mol. Genet.* **2009**, *18*, 2584–2598. [CrossRef] [PubMed]

21. Macri, L.; Silverstein, D.; Clark, R.A. Growth factor binding to the pericellular matrix and its importance in tissue engineering. *Adv. Drug Deliv. Rev.* **2007**, *59*, 1366–1381. [CrossRef] [PubMed]

22. Wight, T.N.; Kinsella, M.G.; Evanko, S.P.; Potter-Perigo, S.; Merrilees, M.J. Versican and the regulation of cell phenotype in disease. *Biochim. Biophys. Acta* **2014**, *1840*, 2441–2451. [CrossRef] [PubMed]

23. Sugahara, K.; Mikami, T.; Uyama, T.; Mizuguchi, S.; Nomura, K.; Kitagawa, H. Recent advances in the structural biology of chondroitin sulfate and dermatan sulfate. *Curr. Opin. Struct. Biol.* **2003**, *13*, 612–620. [CrossRef] [PubMed]

24. Keller, K.E.; Sun, Y.Y.; Vranka, J.A.; Hayashi, L.; Acott, T.S. Inhibition of hyaluronan synthesis reduces versican and fibronectin levels in trabecular meshwork cells. *PLoS ONE* **2012**, *7*, e48523. [CrossRef] [PubMed]

25. Naso, M.F.; Morgan, J.L.; Buchberg, A.M.; Siracusa, L.D.; Iozzo, R.V. Expression pattern and mapping of the murine versican gene (*Cspg2*) to chromosome 13. *Genomics* **1995**, *29*, 297–300. [CrossRef] [PubMed]

26. Calve, S.; Isaac, J.; Gumucio, J.P.; Mendias, C.L. Hyaluronic acid, *HAS1*, and *HAS2* are significantly upregulated during muscle hypertrophy. *Am. J. Physiol. Cell Physiol.* **2012**, *303*, C577–C588. [CrossRef] [PubMed]

27. Velleman, S.G.; Sporer, K.R.; Ernst, C.W.; Reed, K.M.; Strasburg, G.M. Versican, matrix gla protein, and death-associated protein expression affect muscle satellite cell proliferation and differentiation. *Poult. Sci.* **2012**, *91*, 1964–1973. [CrossRef] [PubMed]

28. Carrino, D.A.; Sorrell, J.M.; Caplan, A.I. Dynamic expression of proteoglycans during chicken skeletal muscle development and maturation. *Poult. Sci.* **1999**, *78*, 769–777. [CrossRef] [PubMed]

29. Sandy, J.D.; Westling, J.; Kenagy, R.D.; Iruela-Arispe, M.L.; Verscharen, C.; Rodriguez-Mazaneque, J.C.; Zimmermann, D.R.; Lemire, J.M.; Fischer, J.W.; Wight, T.N.; et al. Versican V1 proteolysis in human aorta in vivo occurs at the Glu441-Ala442 bond, a site that is cleaved by recombinant ADAMTS-1 and ADAMTS-4. *J. Biol. Chem.* **2001**, *276*, 13372–13378. [CrossRef] [PubMed]

30. Dancevic, C.M.; Fraser, F.W.; Smith, A.D.; Stupka, N.; Ward, A.C.; McCulloch, D.R. Biosynthesis and expression of a disintegrin-like and metalloproteinase domain with thrombospondin-1 repeats-15. A novel versican-cleaving proteoglycanase. *J. Biol. Chem.* **2013**, *288*, 37267–37276. [CrossRef] [PubMed]

31. Bukong, T.N.; Maurice, S.B.; Chahal, B.; Schaeffer, D.F.; Winwood, P.J. Versican: A novel modulator of hepatic fibrosis. *Lab. Investig.* **2016**, *96*, 361–374. [CrossRef] [PubMed]

32. McCulloch, D.R.; Nelson, C.M.; Dixon, L.J.; Silver, D.L.; Wylie, J.D.; Lindner, V.; Sasaki, T.; Cooley, M.A.; Argraves, W.S.; Apte, S.S. ADAMTS metalloproteases generate active versican fragments that regulate interdigital web regression. *Dev. Cell* **2009**, *17*, 687–698. [CrossRef] [PubMed]

33. Hope, C.; Foulcer, S.; Jagodinsky, J.; Chen, S.X.; Jensen, J.L.; Patel, S.; Leith, C.; Maroulakou, I.; Callander, N.; Miyamoto, S.; et al. Immunoregulatory roles of versican proteolysis in the myeloma microenvironment. *Blood* **2016**, *128*, 680. [CrossRef] [PubMed]

34. Malik, V.; Rodino-Klapac, L.R.; Mendell, J.R. Emerging drugs for duchenne muscular dystrophy. *Expert Opin. Emerg. Drugs* **2012**, *17*, 261–277. [CrossRef] [PubMed]

35. Sali, A.; Guerron, A.D.; Gordish-Dressman, H.; Spurney, C.F.; Iantorno, M.; Hoffman, E.P.; Nagaraju, K. Glucocorticoid-treated mice are an inappropriate positive control for long-term preclinical studies in the *mdx* mouse. *PLoS ONE* **2012**, *7*, e34204. [CrossRef] [PubMed]

36. Angelini, C.; Peterle, E. Old and new therapeutic developments in steroid treatment in duchenne muscular dystrophy. *Acta Myol.* **2012**, *31*, 9–15. [PubMed]

37. Anderson, J.E.; McIntosh, L.M.; Poettcker, R. Deflazacort but not prednisone improves both muscle repair and fiber growth in diaphragm and limb muscle in vivo in the *mdx* dystrophic mouse. *Muscle Nerve* **1996**, *19*, 1576–1585. [CrossRef]

38. Guiraud, S.; Davies, K.E. Pharmacological advances for treatment in duchenne muscular dystrophy. *Curr. Opin. Pharmacol.* **2017**, *34*, 36–48. [CrossRef] [PubMed]

39. Belanto, J.J.; Diaz-Perez, S.V.; Magyar, C.E.; Maxwell, M.M.; Yilmaz, Y.; Topp, K.; Boso, G.; Jamieson, C.H.; Cacalano, N.A.; Jamieson, C.A. Dexamethasone induces dysferlin in myoblasts and enhances their myogenic differentiation. *Neuromuscul. Disord.* **2010**, *20*, 111–121. [CrossRef] [PubMed]

40. Passaquin, A.C.; Metzinger, L.; Leger, J.J.; Warter, J.M.; Poindron, P. Prednisolone enhances myogenesis and dystrophin-related protein in skeletal muscle cell cultures from *mdx* mouse. *J. Neurosci. Res.* **1993**, *35*, 363–372. [CrossRef] [PubMed]

41. Hunt, L.C.; Gorman, C.; Kintakas, C.; McCulloch, D.R.; Mackie, E.J.; White, J.D. Hyaluronan synthesis and myogenesis: A requirement for hyaluronan synthesis during myogenic differentiation independent of pericellular matrix formation. *J. Biol. Chem.* **2013**, *288*, 13006–13021. [CrossRef] [PubMed]

42. Kuroda, M.; Sasamura, H.; Shimizu-Hirota, R.; Mifune, M.; Nakaya, H.; Kobayashi, E.; Hayashi, M.; Saruta, T. Glucocorticoid regulation of proteoglycan synthesis in mesangial cells. *Kidney Int.* **2002**, *62*, 780–789. [CrossRef] [PubMed]

43. Todorova, L.; Gurcan, E.; Miller-Larsson, A.; Westergren-Thorsson, G. Lung fibroblast proteoglycan production induced by serum is inhibited by budesonide and formoterol. *Am. J. Respir. Cell Mol. Biol.* **2006**, *34*, 92–100. [CrossRef] [PubMed]

44. Gebhardt, C.; Averbeck, M.; Diedenhofen, N.; Willenberg, A.; Anderegg, U.; Sleeman, J.P.; Simon, J.C. Dermal hyaluronan is rapidly reduced by topical treatment with glucocorticoids. *J. Investig. Dermatol.* **2010**, *130*, 141–149. [CrossRef] [PubMed]

45. Zhang, W.; Watson, C.E.; Liu, C.; Williams, K.J.; Werth, V.P. Glucocorticoids induce a near-total suppression of hyaluronan synthase mRNA in dermal fibroblasts and in osteoblasts: A molecular mechanism contributing to organ atrophy. *Biochem. J.* **2000**, *349*, 91–97. [CrossRef] [PubMed]

46. Grounds, M.D.; Radley, H.G.; Lynch, G.S.; Nagaraju, K.; De Luca, A. Towards developing standard operating procedures for pre-clinical testing in the *mdx* mouse model of duchenne muscular dystrophy. *Neurobiol. Dis.* **2008**, *31*, 1–19. [CrossRef] [PubMed]

47. Lynch, G.S.; Hinkle, R.T.; Chamberlain, J.S.; Brooks, S.V.; Faulkner, J.A. Force and power output of fast and slow skeletal muscles from *mdx* mice 6–28 months old. *J. Physiol.* **2001**, *535*, 591–600. [CrossRef] [PubMed]

48. Stupka, N.; Michell, B.J.; Kemp, B.E.; Lynch, G.S. Differential calcineurin signalling activity and regeneration efficacy in diaphragm and limb muscles of dystrophic *mdx* mice. *Neuromuscul. Disord.* **2006**, *16*, 337–346. [CrossRef] [PubMed]

49. Huang, P.; Cheng, G.; Lu, H.; Aronica, M.; Ransohoff, R.M.; Zhou, L. Impaired respiratory function in *mdx* and *mdx/utrn+/−* mice. *Muscle Nerve* **2011**, *43*, 263–267. [CrossRef] [PubMed]

50. O'Connor, R.S.; Steeds, C.M.; Wiseman, R.W.; Pavlath, G.K. Phosphocreatine as an energy source for actin cytoskeletal rearrangements during myoblast fusion. *J. Physiol.* **2008**, *586*, 2841–2853. [CrossRef] [PubMed]

51. Abmayr, S.M.; Pavlath, G.K. Myoblast fusion: Lessons from flies and mice. *Development* **2012**, *139*, 641–656. [CrossRef] [PubMed]

52. Horsley, V.; Pavlath, G.K. Forming a multinucleated cell: Molecules that regulate myoblast fusion. *Cells Tissues Organs* **2004**, *176*, 67–78. [CrossRef] [PubMed]

53. Pavlath, G.K. Spatial and functional restriction of regulatory molecules during mammalian myoblast fusion. *Exp. Cell Res.* **2010**, *316*, 3067–3072. [CrossRef] [PubMed]

54. Horsley, V.; Jansen, K.M.; Mills, S.T.; Pavlath, G.K. IL-4 acts as a myoblast recruitment factor during mammalian muscle growth. *Cell* **2003**, *113*, 483–494. [CrossRef]

55. Pavlath, G.K.; Horsley, V. Cell fusion in skeletal muscle: Central role of NFATC2 in regulating muscle cell size. *Cell Cycle* **2003**, *2*, 420–423. [CrossRef] [PubMed]

56. Seidah, N.G.; Mayer, G.; Zaid, A.; Rousselet, E.; Nassoury, N.; Poirier, S.; Essalmani, R.; Prat, A. The activation and physiological functions of the proprotein convertases. *Int. J. Biochem. Cell Biol.* **2008**, *40*, 1111–1125. [CrossRef] [PubMed]

57. Mylona, E.; Jones, K.A.; Mills, S.T.; Pavlath, G.K. CD44 regulates myoblast migration and differentiation. *J. Cell. Physiol.* **2006**, *209*, 314–321. [CrossRef] [PubMed]

58. Goetsch, K.P.; Myburgh, K.H.; Niesler, C.U. In vitro myoblast motility models: Investigating migration dynamics for the study of skeletal muscle repair. *J. Muscle Res. Cell Motil.* **2013**, *34*, 333–347. [CrossRef] [PubMed]

59. Li, F.; Li, S.; Cheng, T. TGF-β1 promotes osteosarcoma cell migration and invasion through the miR-143-versican pathway. *Cell. Physiol. Biochem.* **2014**, *34*, 2169–2179. [CrossRef] [PubMed]

60. Bu, P.; Yang, P. MicroRNA-203 inhibits malignant melanoma cell migration by targeting versican. *Exp. Ther. Med.* **2014**, *8*, 309–315. [CrossRef] [PubMed]

61. Henderson, D.J.; Ybot-Gonzalez, P.; Copp, A.J. Over-expression of the chondroitin sulphate proteoglycan versican is associated with defective neural crest migration in the *pax3* mutant mouse (splotch). *Mech. Dev.* **1997**, *69*, 39–51. [CrossRef]

62. Wight, T.N. Versican: A versatile extracellular matrix proteoglycan in cell biology. *Curr. Opin. Cell Biol.* **2002**, *14*. [CrossRef]

63. Carthy, J.M.; Abraham, T.; Meredith, A.J.; Boroomand, S.; McManus, B.M. Versican localizes to the nucleus in proliferating mesenchymal cells. *Cardiovasc. Pathol.* **2015**, *24*, 368–374. [CrossRef] [PubMed]

64. Kharraz, Y.; Guerra, J.; Pessina, P.; Serrano, A.L.; Munoz-Canoves, P. Understanding the process of fibrosis in duchenne muscular dystrophy. *Biomed. Res. Int.* **2014**, *2014*, 965631. [CrossRef] [PubMed]

65. Chen, Y.W.; Zhao, P.; Borup, R.; Hoffman, E.P. Expression profiling in the muscular dystrophies: Identification of novel aspects of molecular pathophysiology. *J. Cell Biol.* **2000**, *151*, 1321–1336. [CrossRef] [PubMed]

66. Haslett, J.N.; Sanoudou, D.; Kho, A.T.; Bennett, R.R.; Greenberg, S.A.; Kohane, I.S.; Beggs, A.H.; Kunkel, L.M. Gene expression comparison of biopsies from Duchenne muscular dystrophy (DMD) and normal skeletal muscle. *Proc. Natl. Acad. Sci. USA* **2002**, *99*, 15000–15005. [CrossRef] [PubMed]

67. Dours-Zimmermann, M.T.; Zimmermann, D.R. A novel glycosaminoglycan attachment domain identified in two alternative splice variants of human versican. *J. Biol. Chem.* **1994**, *269*, 32992–32998. [PubMed]

68. Pallafacchina, G.; Francois, S.; Regnault, B.; Czarny, B.; Dive, V.; Cumano, A.; Montarras, D.; Buckingham, M. An adult tissue-specific stem cell in its niche: A gene profiling analysis of in vivo quiescent and activated muscle satellite cells. *Stem Cell Res.* **2010**, *4*, 77–91. [CrossRef] [PubMed]

69. Wight, T.N.; Kang, I.; Merrilees, M.J. Versican and the control of inflammation. *Matrix Biol.* **2014**, *35*, 152–161. [CrossRef] [PubMed]

70. Zimmermann, D.R.; Ruoslahti, E. Multiple domains of the large fibroblast proteoglycan, versican. *EMBO J.* **1989**, *8*, 2975–2981. [PubMed]

71. Carrino, D.A.; Oron, U.; Pechak, D.G.; Caplan, A.I. Reinitiation of chondroitin sulphate proteoglycan synthesis in regenerating skeletal muscle. *Development* **1988**, *103*, 641–656. [PubMed]

72. Stupka, N.; Schertzer, J.D.; Bassel-Duby, R.; Olson, E.N.; Lynch, G.S. Stimulation of calcineurin Aα activity attenuates muscle pathophysiology in *mdx* dystrophic mice. *Am. J. Physiol. Regul. Integr. Comp. Physiol.* **2008**, *294*, R983–R992. [CrossRef] [PubMed]

73. Mikami, T.; Koyama, S.; Yabuta, Y.; Kitagawa, H. Chondroitin sulfate is a crucial determinant for skeletal muscle development/regeneration and improvement of muscular dystrophies. *J. Biol. Chem.* **2012**, *287*, 38531–38542. [CrossRef] [PubMed]

74. Foulcer, S.J.; Nelson, C.M.; Quintero, M.V.; Kuberan, B.; Larkin, J.; Dours-Zimmermann, M.T.; Zimmermann, D.R.; Apte, S.S. Determinants of versican-V1 proteoglycan processing by the metalloproteinase ADAMTS5. *J. Biol. Chem.* **2014**, *289*, 27859–27873. [CrossRef] [PubMed]

75. Heier, C.R.; Damsker, J.M.; Yu, Q.; Dillingham, B.C.; Huynh, T.; Van der Meulen, J.H.; Sali, A.; Miller, B.K.; Phadke, A.; Scheffer, L.; et al. VBP15, a novel anti-inflammatory and membrane-stabilizer, improves muscular dystrophy without side effects. *EMBO Mol. Med.* **2013**, *5*, 1569–1585. [CrossRef] [PubMed]

76. Hoffman, E.P.; Reeves, E.; Damsker, J.; Nagaraju, K.; McCall, J.M.; Connor, E.M.; Bushby, K. Novel approaches to corticosteroid treatment in duchenne muscular dystrophy. *Phys. Med. Rehabil. Clin. N. Am.* **2012**, *23*, 821–828. [CrossRef] [PubMed]

77. Zhou, H.; Sivasankar, M.; Kraus, D.H.; Sandulache, V.C.; Amin, M.; Branski, R.C. Glucocorticoids regulate extracellular matrix metabolism in human vocal fold fibroblasts. *Laryngoscope* **2011**, *121*, 1915–1919. [CrossRef] [PubMed]

78. Hartel, J.V.; Granchelli, J.A.; Hudecki, M.S.; Pollina, C.M.; Gosselin, L.E. Impact of prednisone on TGF-β1 and collagen in diaphragm muscle from *mdx* mice. *Muscle Nerve* **2001**, *24*, 428–432. [CrossRef]

79. Ma, Z.; Zhong, Z.; Zheng, Z.; Shi, X.M.; Zhang, W. Inhibition of glycogen synthase kinase-3β attenuates glucocorticoid-induced suppression of myogenic differentiation in vitro. *PLoS ONE* **2014**, *9*, e105528. [CrossRef] [PubMed]

80. Bolkenius, U.; Hahn, D.; Gressner, A.M.; Breitkopf, K.; Dooley, S.; Wickert, L. Glucocorticoids decrease the bioavailability of TGF-β which leads to a reduced Tgf-β signaling in hepatic stellate cells. *Biochem. Biophys. Res. Commun.* **2004**, *325*, 1264–1270. [CrossRef] [PubMed]

81. Wen, F.Q.; Kohyama, T.; Skold, C.M.; Zhu, Y.K.; Liu, X.; Romberger, D.J.; Stoner, J.; Rennard, S.I. Glucocorticoids modulate TGF-β production by human fetal lung fibroblasts. *Inflammation* **2003**, *27*, 9–19. [CrossRef] [PubMed]

82. Schoepe, S.; Schacke, H.; May, E.; Asadullah, K. Glucocorticoid therapy-induced skin atrophy. *Exp. Dermatol.* **2006**, *15*, 406–420. [CrossRef] [PubMed]

83. Snow, H.E.; Riccio, L.M.; Mjaatvedt, C.H.; Hoffman, S.; Capehart, A.A. Versican expression during skeletal/joint morphogenesis and patterning of muscle and nerve in the embryonic mouse limb. *Anat. Rec. A Discov. Mol. Cell. Evol. Biol.* **2005**, *282*, 95–105. [CrossRef] [PubMed]

84. Kawashima, H.; Hirose, M.; Hirose, J.; Nagakubo, D.; Plaas, A.H.; Miyasaka, M. Binding of a large chondroitin sulfate/dermatan sulfate proteoglycan, versican, to L-selectin, P-selectin, and CD44. *J. Biol. Chem.* **2000**, *275*, 35448–35456. [CrossRef] [PubMed]

85. Aruffo, A.; Stamenkovic, I.; Melnick, M.; Underhill, C.B.; Seed, B. CD44 is the principal cell surface receptor for hyaluronate. *Cell* **1990**, *61*, 1303–1313. [CrossRef]

86. Nikitovic, D.; Zafiropoulos, A.; Katonis, P.; Tsatsakis, A.; Theocharis, A.D.; Karamanos, N.K.; Tzanakakis, G.N. Transforming growth factor-β as a key molecule triggering the expression of versican isoforms V0 and V1, hyaluronan synthase-2 and synthesis of hyaluronan in malignant osteosarcoma cells. *IUBMB Life* **2006**, *58*, 47–53. [CrossRef] [PubMed]

87. Wight, T.N. Arterial remodeling in vascular disease: A key role for hyaluronan and versican. *Front. Biosci.* **2008**, *13*, 4933–4937. [CrossRef] [PubMed]

88. Cross, N.A.; Chandrasekharan, S.; Jokonya, N.; Fowles, A.; Hamdy, F.C.; Buttle, D.J.; Eaton, C.L. The expression and regulation of ADAMTS-1, -4, -5, -9, and -15, and TIMP-3 by TGFβ1 in prostate cells: Relevance to the accumulation of versican. *Prostate* **2005**, *63*, 269–275. [CrossRef] [PubMed]

89. Carthy, J.M.; Meredith, A.J.; Boroomand, S.; Abraham, T.; Luo, Z.; Knight, D.; McManus, B.M. Versican V1 overexpression induces a myofibroblast-like phenotype in cultured fibroblasts. *PLoS ONE* **2015**, *10*, e0133056. [CrossRef] [PubMed]

90. Choocheep, K.; Hatano, S.; Takagi, H.; Watanabe, H.; Kimata, K.; Kongtawelert, P.; Watanabe, H. Versican facilitates chondrocyte differentiation and regulates joint morphogenesis. *J. Biol. Chem.* **2010**, *285*, 21114–21125. [CrossRef] [PubMed]

91. Wight, T.N.; Frevert, C.W.; Debley, J.S.; Reeves, S.R.; Parks, W.C.; Ziegler, S.F. Interplay of extracellular matrix and leukocytes in lung inflammation. *Cell. Immunol.* **2017**, *312*, 1–14. [CrossRef] [PubMed]

92. Kang, I.; Harten, I.A.; Chang, M.Y.; Braun, K.R.; Sheih, A.; Nivison, M.P.; Johnson, P.Y.; Workman, G.; Kaber, G.; Evanko, S.P.; et al. Versican deficiency significantly reduces lung inflammatory response induced by polyinosine-polycytidylic acid stimulation. *J. Biol. Chem.* **2017**, *292*, 51–63. [CrossRef] [PubMed]

93. Helliwell, T.R. Lectin binding and desmin staining during bupivicaine-induced necrosis and regeneration in rat skeletal muscle. *J. Pathol.* **1988**, *155*, 317–326. [CrossRef] [PubMed]

94. Liu, N.; Garry, G.A.; Li, S.; Bezprozvannaya, S.; Sanchez-Ortiz, E.; Chen, B.; Shelton, J.M.; Jaichander, P.; Bassel-Duby, R.; Olson, E.N. A Twist2-dependent progenitor cell contributes to adult skeletal muscle. *Nat. Cell Biol.* **2017**, *19*, 202–213. [CrossRef] [PubMed]

95. Emde, B.; Heinen, A.; Godecke, A.; Bottermann, K. Wheat germ agglutinin staining as a suitable method for detection and quantification of fibrosis in cardiac tissue after myocardial infarction. *Eur. J. Histochem.* **2014**, *58*, 2448. [CrossRef] [PubMed]

96. Blau, H.M.; Pavlath, G.K.; Hardeman, E.C.; Chiu, C.P.; Silberstein, L.; Webster, S.G.; Miller, S.C.; Webster, C. Plasticity of the differentiated state. *Science* **1985**, *230*, 758–766. [CrossRef] [PubMed]
97. Bains, W.; Ponte, P.; Blau, H.; Kedes, L. Cardiac actin is the major actin gene product in skeletal muscle cell differentiation in vitro. *Mol. Cell. Biol.* **1984**, *4*, 1449–1453. [CrossRef] [PubMed]
98. Wang, L.; Chen, X.; Zheng, Y.; Li, F.; Lu, Z.; Chen, C.; Liu, J.; Wang, Y.; Peng, Y.; Shen, Z.; et al. MiR-23a inhibits myogenic differentiation through down regulation of fast myosin heavy chain isoforms. *Exp. Cell Res.* **2012**, *318*, 2324–2334. [CrossRef] [PubMed]

International Journal of
Molecular Sciences

Article

Decellularized Diaphragmatic Muscle Drives a Constructive Angiogenic Response In Vivo

Mario Enrique Alvarèz Fallas [1,2], Martina Piccoli [1], Chiara Franzin [1], Alberto Sgrò [2], Arben Dedja [3], Luca Urbani [4], Enrica Bertin [1], Caterina Trevisan [1,2], Piergiorgio Gamba [2], Alan J. Burns [4,5], Paolo De Coppi [4] and Michela Pozzobon [1,2,*]

1   Stem Cells and Regenerative Medicine Lab, Fondazione Istituto di Ricerca Pediatrica Città della Speranza, Padova 35127 Italy; marioe.alvarezf@gmail.com (M.E.A.F.); m.piccoli@irpcds.org (M.P.); c.franzin@irpcds.org (C.F.); enrica.bertin@gmail.com (E.B.); trevisan-caterina@libero.it (C.T.)
2   Department of Women and Children Health, University of Padova, Padova 35100, Italy; albertosgro@gmail.com (A.S.); piergiorgio.gamba@unipd.it (P.G.)
3   Department of Cardiac, Thoracic and Vascular Sciences, University of Padova, Padova 35100, Italy; arben.dedja.pd@gmail.com
4   Stem Cells & Regenerative Medicine Section, Developmental Biology & Cancer Programme, UCL Great Ormond Street Institute of Child Health, London WC1N 1EH, UK; urbani81@gmail.com (L.U.); a.burns@ac.ucl.uk (A.J.B.); Paolo.DeCoppi@gosh.nhs.uk (P.D.C.)
5   Department of Clinical Genetics, Erasmus Medical Centre, Wytemaweg 80 3015 CN, Rotterdam, The Netherlands
*   Correspondence: m.pozzobon@irpcds.org; Tel.: +39-049-964-0126

Received: 19 March 2018; Accepted: 24 April 2018; Published: 28 April 2018

Abstract: Skeletal muscle tissue engineering (TE) aims to efficiently repair large congenital and acquired defects. Biological acellular scaffolds are considered a good tool for TE, as decellularization allows structural preservation of tissue extracellular matrix (ECM) and conservation of its unique cytokine reservoir and the ability to support angiogenesis, cell viability, and proliferation. This represents a major advantage compared to synthetic scaffolds, which can acquire these features only after modification and show limited biocompatibility. In this work, we describe the ability of a skeletal muscle acellular scaffold to promote vascularization both ex vivo and in vivo. Specifically, chicken chorioallantoic membrane assay and protein array confirmed the presence of pro-angiogenic molecules in the decellularized tissue such as HGF, VEGF, and SDF-1α. The acellular muscle was implanted in BL6/J mice both subcutaneously and ortotopically. In the first condition, the ECM-derived scaffold appeared vascularized 7 days post-implantation. When the decellularized diaphragm was ortotopically applied, newly formed blood vessels containing CD31[+], αSMA[+], and vWF[+] cells were visible inside the scaffold. Systemic injection of Evans Blue proved function and perfusion of the new vessels, underlying a tissue-regenerative activation. On the contrary, the implantation of a synthetic matrix made of polytetrafluoroethylene used as control was only surrounded by vWF[+] cells, with no cell migration inside the scaffold and clear foreign body reaction (giant cells were visible). The molecular profile and the analysis of macrophages confirmed the tendency of the synthetic scaffold to enhance inflammation instead of regeneration. In conclusion, we identified the angiogenic potential of a skeletal muscle-derived acellular scaffold and the pro-regenerative environment activated in vivo, showing clear evidence that the decellularized diaphragm is a suitable candidate for skeletal muscle tissue engineering and regeneration.

Keywords: skeletal muscle; tissue engineering; angiogenesis; microenvironment

---

1. Introduction

In complex organs and tissues the vascular network, besides assuring an appropriate supply of nutrients and oxygen for regular tissue function, provides a specific interphase to allow the balance of tissue homeostasis and immune functions [1,2]. The growth of new blood vessels requires interactions between endothelial cells, soluble growth factors, and a complex network of extracellular matrix (ECM) components. For all these reasons, to repair a damaged tissue it is of paramount importance to unravel strategies that keep, and at the same time, stimulate vessels with the final aim of obtaining a functional regeneration. Thus, for a wide range of clinical applications, the ability to stimulate and control angiogenesis has a significant impact. Specifically, successful tissue engineering strategies rely on the efficiency of vascularization and subsequent tissue integration. This aspect is indeed fundamental to avoid transplant failure and, to date, experimental approaches for organ and tissue replacement have been focused on different approaches: the functionalization of natural or synthetic scaffolds with proteins or cells known to be involved in angiogenesis (in situ vascularization) [3,4], the generation of an efficient vascular network before graft implantation (pre-vascularization) [5,6]. Since angiogenesis in situ relies mainly on the host cells and pre-existing vasculature, limitations on the use of scaffolds become evident when little porous, non elastic materials and large defects are considered. Indeed, for skin wounds and small or mid-size ulcers, scaffold implantation supports spontaneous regeneration [7]. Differently, in cases of transplantation of large portions of tissue or even whole organs, the prolonged time required for vascularization of the construct through correct anastomosis with the host could result in the formation of a necrotic core or in the development of fibrotic tissue, thereby leading to graft failure [8–11]. With the aim of accelerating angiogenesis, a natural scaffold possessing similar biological and structural features to the native damaged tissue could be advantageous, since new vessels can be driven from the host tissue toward the three-dimensional scaffold [12–14]. Moreover, the molecules originally hidden in healthy tissue, called cryptic molecules, are released from the remodeled ECM and are known to be supportive of several biological processes [15–17].

As for the other tissues, in skeletal muscle the vascular network also plays a key role during remodeling and regeneration [12–22]. It is established that the implantation of any foreign device, from synthetic products or acellular biomaterials to even living tissue constructs, always results in its interactions with the host immune system, remodeling ECM and influencing neo-angiogenesis [23]. Immune cells, particularly macrophages, play a critical role in regulating wound healing and tissue remodeling [24]. They can indeed shift the response towards a positive outcome by means of stimulating new vases formation, matrix remodeling, and immunotolerance [25].

Recently, our group developed an acellular matrix derived from mouse diaphragmatic muscle, which was able to positively modulate immune response and led to a local proregenerative environment upon transplantation, both in healthy and diseased mouse models [26].

In this study, we characterized the angiogenic properties of a decellularized diaphragm (DD) scaffold both ex vivo and in vivo. Specifically, in vivo we have chosen two different murine models where we applied DD and a synthetic nonfunctionalized material, the expanded polytetrafluoroethylene (ePTFE) used in clinical practice, as control: the first model expects the implant under the murine back skin and for the second we used the already established orthotopic implant in healthy diaphragm [26]. Our findings not only confirm the angiogenic properties of the ECM obtained by skeletal muscle decellularization, but also underline a highly stable and prolonged angiogenic stimulus, endorsing the use of this scaffold as an alternative support for skeletal muscle defect repair

2. Results

2.1. Diaphragm-Derived Decellularized Matrix Retains Angiogenic Potential

Chorion allantoic membrane (CAM) assay was used to evaluate blood vessel chemo-attractive activity of DD scaffold. Samples along with positive (polyester loaded with VEGF) and negative (polyester only) controls were analyzed daily under a stereomicroscope. Seven days after implantation, neovessels were organized in a network surrounding the tissue samples (Figure 1A). Quantification of vessel growth (i.e., blood converging towards the matrix) displayed a significant increase in DD with respect to controls at the same time-point ($p < 0.05$; Figure 1B).

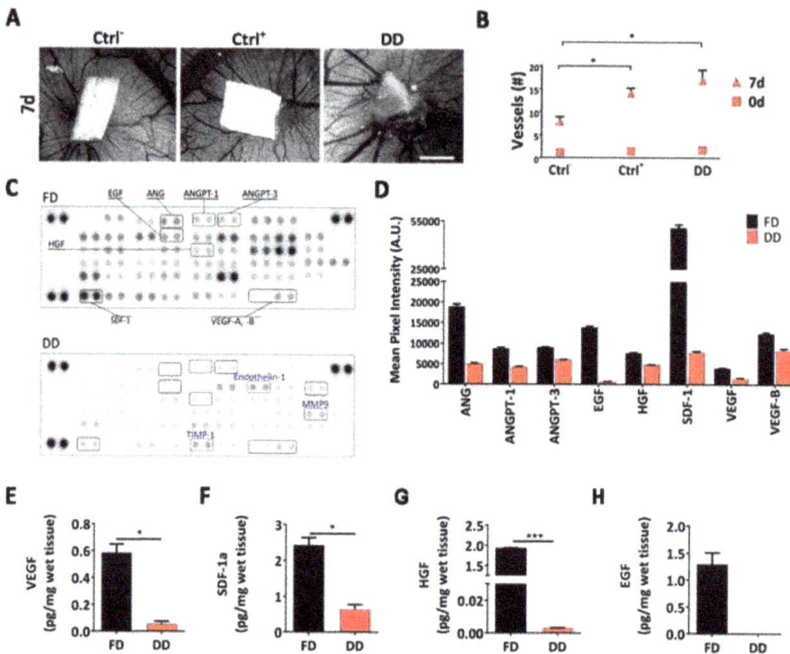

Figure 1. Characterization of diaphragmatic acellular scaffold angiogenic potential. (**A**) Chorion allantoic membrane (CAM) assay appearance after 7 days of incubation with negative control, positive control, and decellularized diaphragm. Scale bar = 1 mm; (**B**) Number of vessels converging towards each sample at baseline (0d) and after 7 days; vessels were counted in a blinded fashion; (**C**) Immunoblotted membrane array of fresh and decellularized tissue homogenate. 53 angiogenic factors were detected; (**D**) Pixel intensity of some of the principal angiogenic factors. Intensity was calculated using UVITEC Cambridge software (mean ± SD). Quantification of (**E**) VEGF; (**F**) SDF-1a; (**G**) HGF; and (**H**) EGF from fresh and decellularized tissue homogenate (mean ± SD). Ctrl$^-$ = Negative Control (inert polyester only); Ctrl$^+$ = Positive Control (polyester with VEGF); DD = Decellularized Diaphragm; FD = Fresh Diaphragm. * $p < 0.01$; *** $p < 0.0001$

To both confirm CAM assay results and verify whether the DD matrix retained pro-angiogenic factors, an immune array directed against 53 proteins specifically involved in mouse angiogenesis was performed. Most of the analyzed proteins were detected in the decellularized tissue (Figure 1C,D), although with lower intensity compared to fresh tissue. In particular, specific angiogenic cytokines (e.g., VEGF and Angiopoietins), matrikines, and enzymes necessary for endothelialization (e.g., endothelin or metalloproteases such as MMP9) were detected. Furthermore, for VEGF, SDF-1α, HGF, and EGF, key factors regulating vasculogenesis and angiogenesis, we have also measured the amount retained in the decellularized tissue via ELISA test. Despite a significant decrease, 3 out of 4 factors were still present in a detectable amount in the tissue after decellularization (VEGF: 0.580 ± 0.06 pg/mg in FD, 0.068 ± 0.02 pg/mg in DD; SDF-1α: 2.432 ± 0.204 pg/mg in FD, 0.642 ± 0.13 pg/mg in DD; HGF: 1.939 ± 0.02 pg/mg in FD, 0.003 ± 0.0001 pg/mg in DD. Figure 1E–H).

2.2. Cell-Scaffold Interaction

Once the attraction of new vessels was confirmed, their origin was investigated to identify whether neovascularization was due to new vessel formation, or if the pre-existing vessels were recellularized by host cells. To this aim, we performed implantation in GFP$^+$ mice for host cell-tracking (Figure 2A).

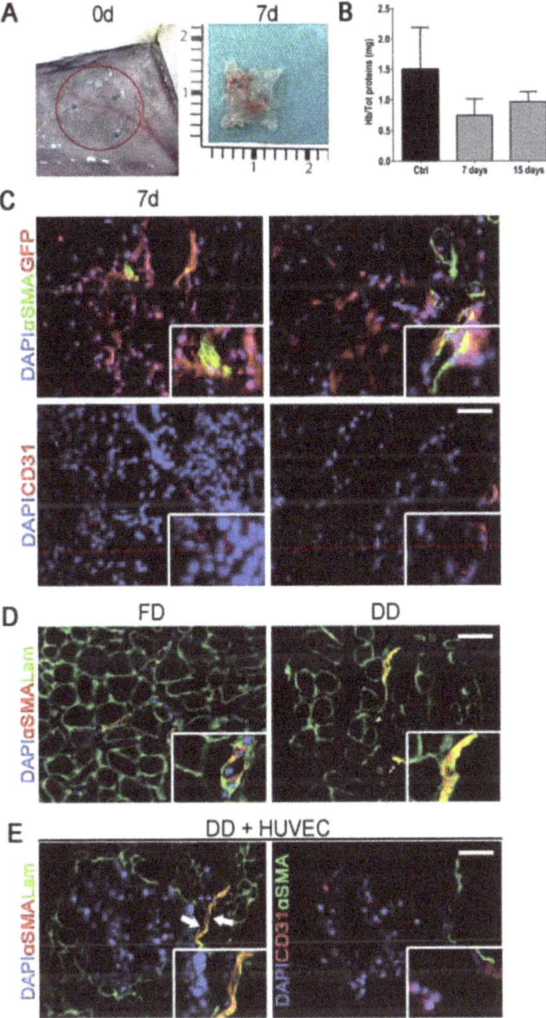

Figure 2. Scaffold subcutaneous implantation and vessel after DET treatment. (**A**) Macroscopic appearance of the implanted DD at day 0 (red circle) and of DD explanted after 7 and 15 days of implantation; (**B**) Quantification of hemoglobin over total protein content in control (skin) and DD after 7 and 15 days of implantation (mean ± SD); (**C**) Immunofluorescence staining against αSMA and GFP performed in explanted DD after 7 and 15 days from implantation (upper panel) and CD31 (lower panel); nuclei were counterstained with DAPI; (**D**) Immunofluorescence staining against αSMA and Laminin performed in fresh diaphragm (FD) and DD; nuclei were counterstained with DAPI; (**E**) Immunofluorescence staining against αSMA and Laminin or CD31 performed in DD after 48 h of incubation with HUVECs; nuclei were counterstained with DAPI; Scale bar = 100 μm.

Transplanted patches appeared vascularized at 7 days (Figure 2B), confirming CAM assay results. The amount of hemoglobin (Hb) quantified on the harvested DD patches, corroborated the previous data with a tendency of Hb to increment with time, suggesting also a functional angiogenesis with perfused vessels inside the applied patch (Figure 2C). Immunofluorescence performed on DD excised 7 days post implant revealed the presence of αSMA$^+$ vessels with uneven appearance and morphology (Figure 2D). Among all, some of them were αSMA$^+$ but did not have any GFP$^+$ cell inside, indicating the presence of empty vessels (Figure 2D, 7-day right inset). Fifteen days after implant, the majority of the analyzed vessels had a uniform shape and dimension as pointed out by the presence of GFP$^+$ cells inside the vessel structure (Figure 2D, 15-day right inset). To confirm the hypothesis that endothelial cells recolonized leftover vascular structures in the transplanted matrix, HUVEC were seeded in vitro on top of the DD and 48 h later their migration was analyzed. The vast majority of pre-existing vascular structures was recognized by HUVEC (CD31$^+$ cells), confirming the influence of old vasculature structure on patch angiogenic features and host endothelial cell behavior (Figure 2D,E)

2.3. Angiogenic Response to Orthotopic Transplantation of DD vs. ePTFE

Having demonstrated formation of functional vasculature, we evaluated DD angiogenic properties by orthotopic implantation (Figure 3). Patches were applied on host diaphragm, in order to challenge the DD with a physiological environment, without any local injury. We compared the angiogenic effect exerted by our scaffold with that of ePTFE, a prosthetic patch already used in clinic [27]. From the histological point of view, the H & E staining showed that the biological DD underwent remodeling coherently to the outcome seen during the previous characterization [26], while ePTFE was encapsulated by a thick cellular layer, suggesting a beginning of foreign body reaction (Figure 4A).

Concerning endothelial cell presence, CD31$^+$ cells migrated from the native diaphragm (ND) in the applied scaffold after 7 and 15 days post implantation (Figure 3A). In both the analyzed time points endothelial cells increased significantly ($p < 0.0001$) in DD samples compared to a fresh untreated diaphragm (Figure 3B). Furthermore, although CD31$^+$ cells could be detected also in the fibrotic capsule surrounding the implanted ePTFE, their number was not significantly different from a basal untreated condition (Figure 3B). These results suggest a stronger angiogenic effect of the applied naturally-derived ECM with respect to prosthetic material.

We co-stained αSMA as a vessels perimetric marker, together with vWF to confirm both vessel function and luminal size, which allowed the measurement of the vessel size (Figure 3C,D). In the DD implants, at 7 days postsurgery the distribution was between small and mid-size vessels (from 2.0 to 1.75, and from 1.75 to 1.55, respectively), then after 15 days it resembled a physiological condition (vessel diameter from 1.55 to lower values) (Figure 3D). On the contrary, at the interface between ePTFE and ND, vessels had a distribution similar to native condition 7 days postsurgery, whereas middle-size vessels appeared at the later time point (Figure 3D).

To confirm the presence of functioning and perfused vessels highlighted previously by the co-expression of αSMA and vWF inside the applied DD, Evans Blue dye was systemically injected. Red color stain was detected in the lumen of αSMA$^+$ vessels at both time points. Moreover, the proportion of nonperfused αSMA$^+$ vessels diminished through the time points, indicating an increase of functional vessels (Figure 4B,C).

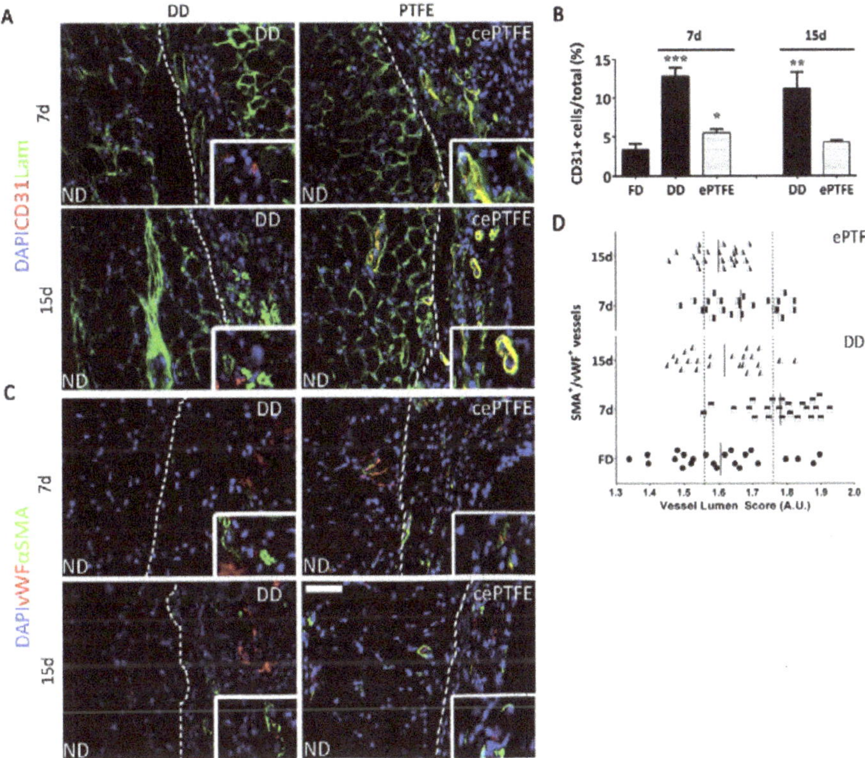

Figure 3. Comparison of angiogenic response after orthotropic implantation of DD vs. ePTFE. White line marks the normal diaphragm (ND) versus the patch (decellularized diaphragm –DD- or plastic –ePTFE- or capsule expanded –cePTFE-) (**A**) Immunofluorescence staining against CD31 and Laminin performed in explanted DD and ePTFE after 7 and 15 days from implantation; nuclei were counterstained with DAPI; (**B**) Quantification of CD31$^+$ cells over total cell content in explanted DD and ePTFE after 7 and 15 days from implantation (mean $\pm$ SD); (**C**) Immunofluorescence staining against vWF and αSMA performed in explanted DD and ePTFE after 7 and 15 days from implantation; nuclei were counterstained with DAPI; (**D**) αSMA$^+$/vWF$^+$ vessels dimension distribution calculated in fresh tissue, DD and ePTFE after 7 and 15 days from implantation. Symbol legend. Black dots: vessel dimension of the fresh diaphragm (control); black and white squares: vessel dimension 7 days post implant; black and white triangles: vessel dimension 15 days post implant. Scale bar = 100 µm. * $p < 0.01$; ** $p < 0.001$; *** $p < 0.0001$.

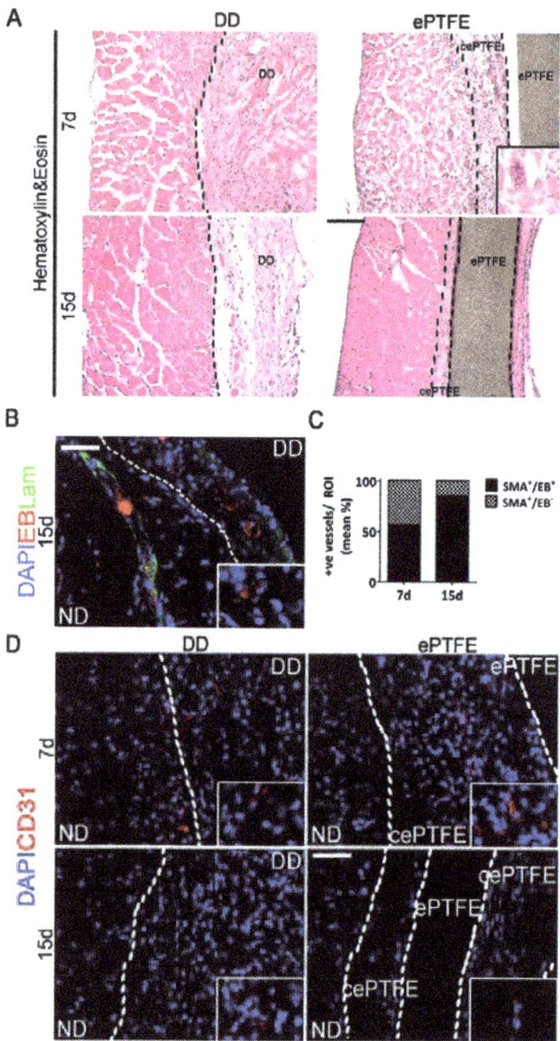

Figure 4. Detection of vases in DD vs. ePTFE after orthotropic implantation. White or black line marks the normal diaphragm (ND) versus the patch (decellularized diaphragm –DD- or plastic –ePTFE- or capsule expanded –cePTFE-); (**A**) Histological appearance of DD and ePTFE-implanted diaphragm. Haematoxylin&Eosin staining performed in DD and ePTFE after 7 and 15 days from implantation. No vessels inside the ePTFE; ePTFE 7 days inlet displays a foreign body giant cell. Scale bar = 100 μm; (**B**) Functional vessels after 15 days of DD implantation. Immunofluorescence staining against Laminin performed on DD-implanted diaphragm after 15 days of implantation; Evans Blue dye is autofluorescent in the red spectrum; nuclei were counterstained with DAPI; (**C**) Quantification of αSMA$^+$/EB$^+$ and αSMA$^+$/EB$^-$ vessels content in DD after 7 and 15 days from implantation (mean %); (**D**) Immunofluorescent staining against CD31$^+$ cells performed n DD- and ePTFE-implanted diaphragms 7 and 15 days post implantation. Scale bar = 100 μm.

2.4. Molecular Profiling of Host-Scaffold Interaction

The general pattern of the array data displayed by ePTFE vs. DD implants reflected the difference between the two materials in terms of properties and exerted tissue response. To aid data interpretation, proteins of interest were plotted under different categories according to the function or involvement in angiogenesis. Seven days after surgery, protein quantification in ePTFE implants displayed an abrupt increase compared to DD, supporting the rapid vessel constitution seen by immunofluorescence (Figure 4D). Specifically, of the three proteins commonly associated with neoangiogenesis, VEGF, FGF2, and PDGF-BB, only FGF2 had a different behavior when comparing the two implanted materials (Figure 5A), increasing in the ePTFE-treated diaphragms after 15 days. Regarding proteins associated with vessel turnover, Angiopoietin-1 (Angpt-1) raised in both cases, CD105 decreased in DD-implanted muscles while increasing in ePTFE implants, and Dll4 decreased towards the baseline, returning to fresh diaphragm (FD) values particularly after 15 days of DD implantation (Figure 5B). The difference among the stimuli driving angiogenesis in the two implants was confirmed by the mRNA expression of *Flt1* (VEGFR1), which appears to be progressively down- or upregulated after DD or ePTFE implantation, respectively.

In both treated samples, inflammation-associated proteins seem to indicate the persistence of an inflammatory state (Figure 5D), since Il-1a and Il-1b lingered overexpressed. Interestingly, the mRNA levels of TNFα remained considerably high after 15 days in the ePTFE-implanted diaphragms (Figure 5C). On the contrary, proteins associated with a profibrotic and proangiogenic response (Il-10, MMP-9, OPN) or with an antifibrotic and antiangiogenic effect (CXC4, MMP-8, PEDF) were expressed differently upon stimulation of the two materials at both time points. Specifically, in the ePTFE-treated samples Il-10, MMP-9, and OPN were upregulated also at 15 days, whereas antifibrotic proteins such as MMP-8 and PEDF were downregulated at the same time point, suggesting a continuous stimulation through a foreign body reaction. On the contrary, in DD-treated diaphragm MMP-9 and OPN decreased during the time points, and at the same time, MMP-8 and CXCL4 were more expressed, with PEDF highly present at the early time point (Figure 5I). In addition, CD26, a protein with a demonstrated pathogenic role in development of fibrosis of various organs, is differently regulated in DD and ePTFE with an important overexpression in the latter after 15 days (Figure 5C).

Lastly, although tissue regeneration is generally associated with M2 type macrophages, it is known that a balance response among M1 and M2 phenotype is essential for a constructive remodelling, while excess M2 can result in fibrosis. Gene expression of *Nos2* (commonly associated with M1), *Arg1*, and *Klf4* (both associated with the shift toward M2) confirmed our previous results with DD being able to stimulate a balanced shift from M1 to M2. ePTFE-implanted muscles expressed the three markers simultaneously, suggesting a copresence of the two different polarization states (Figure 5E,F,H).

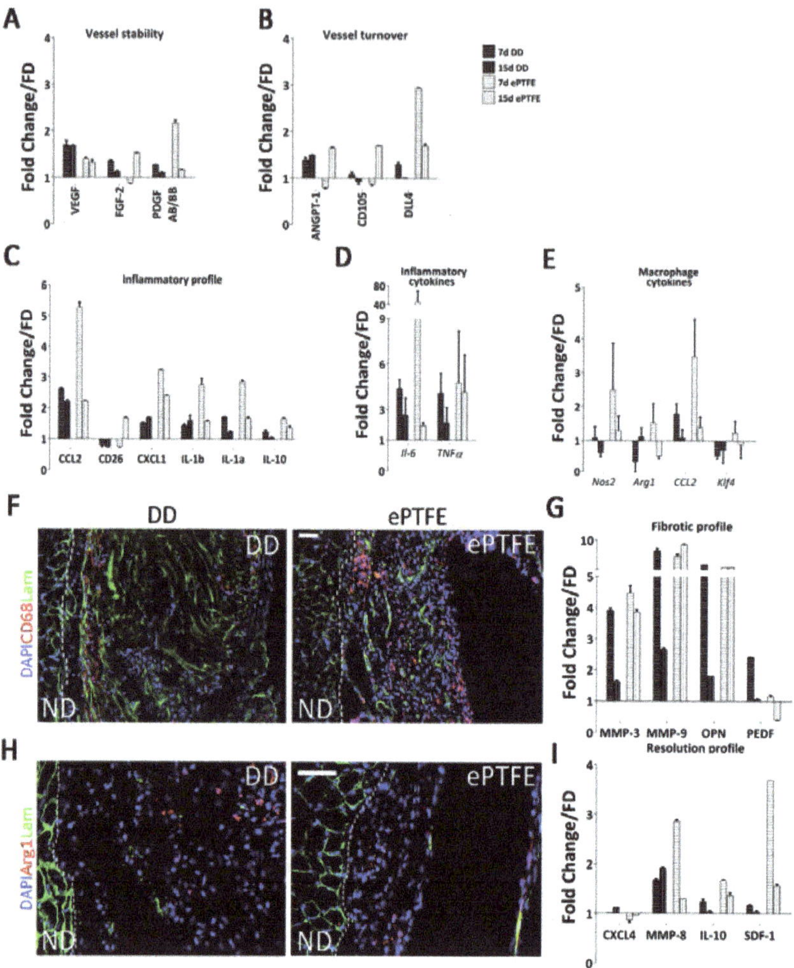

Figure 5. Molecular response and immunoreaction elicited by DD vs. ePTFE. White or black dot line marks the normal diaphragm (ND) versus the patch (decellularized diaphragm –DD- or plastic –ePTFE-) (**A,B**) Quantification of angiogenic factors from DD and ePTFE-implanted diaphragms after 7 and 15 days compared to native tissue (set at 1), grouped by the processes involved (mean ± SD); (**C**) Quantification of *Il-6* and TNFα mRNA extracted from DD and ePTFE after 7 and 15 days from implantation, compared to native tissue (set at 1) (mean ± SD); (**D**) Quantification of factors involved in both angiogenic and immunomodulatory processes from DD and ePTFE-implanted diaphragms after 7 and 15 days, compared to native tissue (set at 1) (mean ± SD); (**E**) Quantification of *Nos2*, *ArgI*, *CCl2* and *Klf4* mRNA extracted from DD and ePTFE after 7 and 15 days from implantation, compared to native tissue (set at 1) (mean ± SD); (**F**) Immunofluorescence staining against pan-macrophages CD68 antigen performed in DD and ePTFE implanted diaphragms after 15 days; (**G**) Quantification of factors involved in both angiogenic and fibrosis processes from DD and ePTFE-implanted diaphragms after 7 and 15 days, compared to native tissue (set at 1) (mean ± SD); (**H**) Immunofluorescence staining against M2 macrophages Arginase I antigen performed in DD and ePTFE-implanted diaphragms after 15 days; (**I**) Quantification of factors involved in both angiogenic and reconstructive processes from DD and ePTFE-implanted diaphragms after 7 and 15 days, compared to native tissue (set at 1) (mean ± SD); Scale bar = 100 μm.

3. Discussion

Angiogenesis remodels and extends the pre-existing vascular tree forming new capillaries [28] and this process in the context of tissue regeneration is crucial to modulate the response towards a proregenerative environment [29,30].

This work was specifically focused on defining the angiogenic properties of DD following previous studies on the murine model that already demonstrated several proregenerative effects after in vivo transplantation [26]. At first, we started by characterizing the angiogenic profile of our DD scaffold. Since it is established that the ECM provides both a mechanical support and a molecular reservoir able to influence cell behavior [31,32], it was hypothesized that the presence of molecules regulating vessel formation could be the trigger accounting for CAM result. This hypothesis was supported by several studies which demonstrated the preservation of the ECM components after decellularization [33,34], also by the means of bioactive molecules [35–37]. Using a specific panel against 53 molecules involved in mouse angiogenesis, we more deeply characterized the reservoir quality of our scaffold. Almost all the spots in the prepatterned mouse angiogenesis array developed a signal, indicating that most of the essayed angiogenesis-involved molecules were still present and detectable after decellularization. Focusing on the quantification of four specific cytokines, we were able to detect VEGF, known to be the master regulatory factor for angiogenesis [38], HGF, involved in both angiogenesis and skeletal muscle growth/repair [39], and SDF-1a, linked not only to angiogenesis but also to other proregenerative processes (e.g., chemotaxis of immune cells) [40]. Considering these data, it was clear that the angiogenic potential persists and is effective despite reduction in the ECM reservoir. The presence of almost all of the array proteins can be advantageous not only because the scaffold does not necessarily require preconditioning [41], but also because a desired effect (angiogenesis, for example) could be further enhanced if a chosen factor is additionally loaded. In this respect, quantification of the cytokines retained by the DD could be the initial step for a tunable control of this process.

Next, subcutaneous implant [42] was used as common method to evaluate angiogenic potential of the biological scaffold. In an allogeneic setting, GFP+ mouse was chosen as recipient to mark and recognize the host cells and vessels. At both analyzed time points, GFP+/αSMA+ vessels indicated a complete integration with the host vasculature. Hemoglobin quantification in the explanted tissue highlighted the functionality of these vessels and blood flow activity. Interestingly, decellularization preserved the innermost (as intima was depleted) connective layer, but also the hollow structure of the media, since αSMA+ vessels could be found in the DD. We decided to use endothelial-like cells such as HUVEC [43] as a tool to investigate whether the vessels preserved after decellularization could be recognized as vascular mold. We demonstrated that HUVEC migrated towards the vessel structures considered as their natural location. This result might be a clue of what happens in vivo when cells get in touch with ECM; however, to really understand the process further, studies are required.

Although the use of synthetic materials is common practice in other areas of medicine, several products, from both xeno and allogeneic origin, have been already commercialized [30], while others, both unseeded or seeded with cells, are currently in clinical use (bladder, urethra, and trachea [44–46]. Organs with a high modular complexity, such as the heart, lung, and liver [47–49], are still undergoing preclinical studies, as well as skeletal muscle [50,51]. With the orthotropic scaffold implantation, our group already proved the ability of the muscle-derived ECM to exert a proregenerative effect. Here, in line with the findings on decellularized tissue advantages over synthetic materials [52], we showed that the regeneration ability was due to the solid angiogenic properties of DD underlying strong differences between the inert ePTFE, one of the gold standard materials in clinical settings [53]. After histological analyses, DD implants attracted vessels from the host while modulating its response towards a constructive environment. On the contrary, the ePTFE seemed to have elicited an abrupt response, which after 15 days appeared to have stabilized. Recipient CD31+ cells could be detected in the ECM of the decellularized scaffold, as well as functional and pre-existing vessels. Presence of functional αSMA+/vWf+ vases, as well as retention of the Evans Blue dye [54], indicated the constitution of both epithelium and blood flow. ePTFE implants on one hand exerted an effect in which CD31$^+$ cells were present only inside functional vessels, while on the other hand the rapidly elicited angiogenesis appeared to be simultaneously regressing, as part of a chronic turnover process. Indeed, while distribution among the size of the vessels found in DD-implanted animals ultimately resembled the natural environment with dis-homogeneous vessel dimensions (from small to big), the synthetic material stabilized on middle size vase dimensions. It is known that the prolonged contact of any device, either synthetic or biological, results in activation of angiogenesis during the cascade of events involved in the foreign body reaction [24]. Nevertheless, differences through the analyzed time points highlighted the properties of the biological scaffold in modulating the host response towards a functional remodeling. The response to the ePTFE indeed confirmed the behavior seen in a classic foreign body reaction [55], in which the host enhances angiogenesis to efficiently deliver immune cells [56] and increases the local angiogenesis (high level of FGF2 and PDGF-BB after 7 days) aiming only at restoring physiological conditions rather than constituting a proregenerative environment (Figure 4, ePTFE bars). Instead, implanted DD was recognized as tissue to be remodeled and reabsorbed, stimulating a regenerative functional angiogenesis, as demonstrated by the comparable expression of FGF2 and PDGF-BB (Figure 4, DD bars) [57]. The increased downregulation of VEGFr1 upon DD implantation, which conversely seems to be trending towards restoration in ePTFE-implanted muscles, could be a further indication of the difference between the angiogenic stimuli. A decrease in VEGFr1 could suggest either an increase in vascular sprouting or a decrease in vessel regression [58].

The two different responses achieved by DD and ePTFE patches could be appreciated also by the gene expression results. In general, the protein pattern displayed by ePTFE implants showed an intense increase after 7 days compared to DD, but in both cases for many proteins it trended similarly towards normal levels (healthy diaphragm line), suggesting that, although different, some extent of overlap had occurred in the responses. The overall mRNA and protein level (i.e., Il-6, TNFα and Il-1β, Il-10, Ccl2) were coherent to results previously seen with these two types of materials [59].

Considering the immunomodulation of the innate immune system [60], it is now established that constructive remodeling requires a fine-tuned balance in immune system cell intervention, not only to avoid rejection of the graft, but also to avoid the establishment of a chronic response [61,62]. In particular, macrophages were found to be of paramount significance, being a heterogeneous population influencing inflammation, angiogenesis, and fibrosis, both directly and indirectly [63,64]. In this respect, DD implants displayed a balanced presence of cytokines and chemokines (e.g., FGF2, CXCL4, Il-10, MMP-3, MMP-8, MMP-9, PEDF) known to be up- or downregulated by cells of the immune system to keep the physiological equilibrium as already shown [26] and involved in the remodeling of ECM, which can in turn be supportive of angiogenesis [65,66]. In this context, Real Time PCR confirmed previous results of a specific time course in macrophage polarization, from M1 to M2, in DD-transplanted animals, while ePTFE triggered an abrupt increase in the mRNA expression of *Il-6*, *TNFα*, *Nos2*, and *Ccl2*, commonly associated with inflammatory cells. The overall profile suggested that differently polarized macrophages were present simultaneously, probably due to the management of the fibrotic capsule surrounding the foreign material [67–72].

With the synthetic material, the lack of a balanced progression ultimately resulted in a nonconstructive response where angiogenesis was not consistent.

Understanding construct–host interactions is essential to avoid any adverse effect by both modulating foreign body reaction (i.e., graft vs. host) and enhancing regeneration (i.e., functional integration). Decellularized scaffolds represent a promising tool for tissue engineering. However, improvement in the development of complex organ decellularization protocols is needed to address issues that are still unanswered, such as complete xenogeneic antigen depletion, sterilization strategies to avoid pathogen transmission, and costs and availability for scale-up and large-scale production. Nevertheless, DD proved to be a promising material, with both pro-angiogenic potential and immunomodulatory properties. Precisely, the ability to direct the immune system was reflected in the foreign body reaction, which was driven towards (1) a constructive remodeling, (2) the induction of the myogenic program, (3) the formation of stable vessels, and (4) the ultimate reabsorption of the scaffold into a restoration of normal condition.

4. Materials and Methods

4.1. Animals

All surgical procedures and animal husbandry were carried out in accordance with the University of Padua's Animal care and Use Committee (CEASA, protocol number 67/2011 approved on 21 September 2011). Eight 12-week-old male and female C57BL/6J mice (B6) were used as both donors (diaphragms for scaffolds generation) and recipients (orthotropic implantation). Twelve-week-old male and female B6 UBC-GFP mice (GFP$^+$) were used as recipients for subcutaneous implantation. For the experiments carried out in the UK, all surgical procedures and animal husbandry were carried out in accordance with UK Home Office guidelines under the Animals (Scientific Procedures) Act 1986 and the Local Ethics Committee.

4.2. Diaphragm Decellularization

Twenty four diaphragms, harvested with the whole rib cage, were washed in sterile phosphate buffered saline (PBS) and either used as fresh diaphragm (FD) controls, or immediately treated with three cycles of a detergent-enzymatic treatment protocol as previously described [26]. Each cycle was composed of deionized water at 4 °C for 24 h, 4% sodium deoxycholate (Sigma, Milan, Italy) at room temperature (RT) for 4 h and 2000 kU DNase-I (Sigma, Milan, Italy) in 1 M NaCl (Sigma, Milan, Italy) at RT for 3 h. At the end of the third cycle, samples (DD) were left in milliQ water for a final washing step of 72 h, after which they were immediately used or preserved at 4 °C in PBS supplemented with 1% Penicillin/Streptomycin (PBS-P/S).

4.3. Chicken Chorioallantoic Membrane Assay

CAM assay was performed as previously described [73]. Fertilized chicken eggs (Henry Stewart and Co. 6 for negative control and 13 for the experimental conditions) were incubated at 37 °C and constant humidity. At 3 days of incubation, an oval window of approximately 3 cm in diameter was cut into the shell with small dissecting scissors to reveal the embryo and CAM vessels. The window was sealed with tape and the eggs were returned to the incubator for a further 5 days. At day 8 of incubation, 1 mm diameter acellular matrices were placed on the CAM between branches of the blood vessels. Polyester sections (ePTFE) soaked overnight either in PBS or in 200 ng/mL VEGF in PBS were used as negative and positive controls, respectively. Eggs with patches (acellular biological matrices and ePTFE) were examined daily until 7 days after placement wherein they were photographed in ovo with a stereomicroscope equipped with a Camera System (Leica, Italy) to quantify the blood vessels surrounding the matrices. The number of blood vessels converging towards the placed tissues was counted blindly by assessors, with the mean of the counts being considered.

4.4. Proteome Profiler Angiogenesis Array

Samples designated for protein array (5 for each group) were snap frozen and stored at −80 °C until the beginning of the assay. Before start, tissue was thawed and lysed by mechanical homogenization, and the resulting protein solution was collected in PBS with protease inhibitors (10 µg/mL Aprotinin, 10 µg/mL Leupeptin, and 10 µg/mL Pepstatin, all Sigma-Aldrich (Milan, Italy). Nonoperated B6 diaphragms served as control (FD). Three tissue homogenates for each sample (FD, DD, implanted diaphragms after 7 and 15 days with DD or ePTFE) were pooled for analysis and concentration of each pool was determined with BCA protein assay kit (Pierce, Thermo Fisher Scientific, Waltham, MA, USA, performed according to manufacturers' instructions). Proteome profiler mouse angiogenesis array ARY015 (R&D Systems, Milan, Italy) was used according to manufacturers' instructions. Luminescence acquisition and quantification of the pixel density of each spot was determined with Alliance (UVITEC, Cambridge, UK). For data analysis, average background signal (negative control spots) was subtracted from protein spots signal, which was subsequently normalized to positive control spots and related to signals detected from FD pool (Fold change/FD).

4.5. ELISA Test

Samples designated for ELISA test (5 for each group) were snap frozen and stored at $-80\ ^\circ$C until the beginning of the assay. Before start, tissue was thawed, lysed by mechanical homogenization, and immediately assayed. Nonoperated B6 diaphragms served as control. Three decellularized diaphragm quarters (from different samples) were pooled and used for each ELISA. Quantikine for mouse VEGF, HGF, EGF, and SDF-1a (R&D systems) were carried out according to manufacturers' instructions. Luminescence acquisition and sample quantification was performed using SpectraMax Plus 384 (Molecular devices, San Jose, CA, USA).

4.6. Subcutaneous Implantation

Mice were gently handled in general anesthesia with O_2 and 1.5–2.0% isoflurane (Forane, Merial, Padova, Italy) inhalation. While in procumbent position, a medial incision was performed on the back of the mouse allowing the skin to be gently detached from the underlying fascia. For each animal, up to three portions (0.7×0.7 cm each) of acellular scaffold (5 pieces from different animals and for each time point) were positioned and fixed to the skin side with a Prolene 7/0 suture. Skin was then closed with a Prolene 6/0 suture and animal left to recover under a heating lamp. Mice were euthanized by cervical dislocation at 7 and 15 days post-implantation.

4.7. Hemoglobin Quantification

For hemoglobin quantification, DD only were carefully dissected from the host skin to prevent host tissue contamination. Drabkin's reagent lyses red blood cells and oxidizes all forms of hemoglobin (Hb), [74]. Samples were homogenized and diluted 1/1000 with Drabkin's reagent (Sigma) so that the final HiCN concentration fell within the range of the calibration curve (0–0Æ8 g/L HiCN) produced using purchased hemoglobin (Sigma). After mixing, samples and standards were incubated at RT for 30 min, protected from light. Absorbance was read at 550 nm with SpectraMax Plus 384 and the Hb concentration of each sample was calculated from the linear equation of the calibration curve.

4.8. Evans Blue Injection

Evans Blue (Sigma) was diluted to 0.5% in 0.9% NaCl solution. Two hundred µL of the prepared solution was injected via the tail vein and left diffusing in the bloodstream for 30 min, before animal euthanasia. To avoid loss of the dye, diaphragm muscles were fixed right after harvesting with 0.25% Glutaraldehyde (Sigma) in PBS. The procedure was performed only in DD-implanted animals and a total of 5 for each time point were essayed.

4.9. Human Umbilical Vein Endothelial Cells Culture and Seeding

Human umbilical vein endothelial cells (HUVEC, PromoCell) were expanded in endothelial medium (PromoCell) and used after passage 2 or 3. For the in vitro experiment of seeding HUVEC on the top of the decellularized ECM, cells were detached from the culture plate using 0.025% Trypsin-EDTA (Invitrogen, Carlsbad, CA, USA) and counted, aiming to obtain a concentration of 5×10^5 cells/15 µL, which was the amount distributed on each scaffold (empty of other cells). Samples were finally left in culture for 48 h and were subsequently fixed and analyzed (5 samples).

4.10. Orthotopic Implantation

Surgical procedure was carried out as previously described [26]. Briefly, while placed in supine position with its caudal part towards the operator, a medial incision was performed in the abdomen of the mouse. To visualize the diaphragm, liver and stomach were then gently moved on the right downward part of the abdomen with the help of a sterile gauze soaked in warm saline solution. DD or ePTFE patches (0.7 × 0.7 cm each) were fixed on the left side of the native diaphragm with 4 Prolene 8/0 stitches. The abdominal wall was closed in two layers with 5/0 running suture and the animals were left to wake up under a heating lamp. Mice were euthanized by cervical dislocation at 7 and 15 days post-implantation (8 mice for each time point).

4.11. Histological Stain and Immunofluorescence

Frozen sections (thickness of 6–8 μm) were stained with HE kit for rapid frozen section (Bio-Optica, Milano, Italy) under the manufacturer's instruction. For immunofluorescence analyses, sections were permeabilized with 0.5% Triton X-100 (10 min), blocked with 5% Horse serum/5% Goat serum (30 min) and incubated with primary antibodies overnight at 4 °C. Slides were then incubated with secondary antibodies Alexa Fluor-conjugated. Antibodies used are listed in Table S1. Nuclei were counterstained with fluorescent mounting medium plus 100 ng/mL 4′,6-diamidino-2-phenylindole (DAPI) (Sigma-Aldrich). For each count performed, $n = 8$ random pictures were collected and analyzed.

4.12. Vessel Size Analysis

Eight images per group (fresh, transplanted after 7 days, and 15 days) were analyzed to examine the difference in vessel dimension. To this aim, for each image the perimeter and the area of alpha smooth muscle actin (αSMA) and von Willebrand Factor (Vwf) double positive vessels were calculated and their proportion was analytically obtained by means of the fractal dimension index (FRAC, http://www.umass.edu/landeco/research/fragstats/documents/Metrics/Shape%20Metrics/Metrics/P9%20-%20FRAC.htm). The formula is commonly used to calculate the proportion of area and perimeter in irregular shape surfaces. Subsequently, the shape data expressed as vessel lumen were collected and the groups compared.

4.13. Real Time PCR

Total RNA was extracted using RNeasy Plus Mini kit (QIAGEN GmbH, Milan Italy) following the supplier's instructions. RNA was quantified with an ND-2000 spectrophotometer and 1 μg total was reverse transcripted with SuperScript II and related products (all from Life Technologies, Milan, Italy) in a 20 μL reaction. Real-time PCR reactions were performed using a LightCycler II (Roche, Milan, Italy). Reactions were carried out in triplicate using 4 μL of FASTSTART SYBR GREEN MASTER (Roche) and 2 μL of primers mix Forward and Reverse (final concentration, 300/300 nM) in a final volume of 20 μL. Serial dilutions of a positive control sample were used to create a standard curve for the relative quantification. The amount of each mRNA was normalized over the expression of β2-microglobulin. Primer sequences used are listed in Table S2.

4.14. Statistical Analysis

Image based counts and measurements were performed with Fiji 30 or alternatively with Imago 1 (Mayachitra, Santa Barbara, CA USA). For each analysis, at least five random pictures were used for data output. All graphs displayed were produced with GraphPad software (GraphPad Software Inc, CA, USA) 5 or 6. Data are expressed as means ± SEM. Statistical significance was determined using an equal-variance Student's t-test for qRT-PCR analyses (t-test was performed between: DD 7d vs. DD 15d, DD 7d vs. ePTFE 7d, DD 15d vs. ePTFE 15d, ePTFE 7d vs. ePTFE 15d). A p value below 0.05 was considered to be statistically significant. See Table S3.

5. Conclusions

Summarizing, to understand and make reproducible tissue regeneration approaches, it is of paramount importance to deepen the investigation of the interactions between scaffold and host, unveiling the specific players from either the cellular, mechanical, or molecular side. In this study we used a tailored scaffold closely resembling the tissue of origin, proving that angiogenic and, consequently, immunomodulatory properties persist after decellularization and may lead to an efficient regeneration process [75,76]. These characteristics make our DD a promising scaffold to be implemented for clinical purposes in tissue engineering-based approaches.

Supplementary Materials: Supplementary materials can be found at http://www.mdpi.com/1422-0067/19/5/1319/s1.

Author Contributions: M.E.A.F.: Conception and design, collection and assembly of data, data analysis and interpretation, manuscript writing; M.P.: Conception and design, collection and assembly of data, data analysis and interpretation, manuscript writing; C.F., L.U.: experiments performance, data analysis and interpretation, manuscript writing; E.B. and C.T.: experiments performance, data analysis; A.D. and A.S.: surgical procedures; P.G.: financial support and final approval of the manuscript; A.J.B.: provision of study material and results; P.D.C.: manuscript writing and final approval of the manuscript; M.P.: project supervision, data interpretation, manuscript writing and final approval of the manuscript.

Acknowledgments: M.P. and C.F. are supported by Fondazione Città della Speranza, M.P. is funded by University of Padova, Grant number GRIC15AIPF, Assegno di Ricerca Senior. P.D.C. is funded by GOSH charity. C.F., M.P., L.U. and M P. are co-inventors of Italian Patent N. 0001422436 entitled "Matrice acellulare per ricostruzione in vivo di muscolo scheletrico".

Conflicts of Interest: The authors declare no conflict of interest.

Abbreviations

DD	Decellularized diaphragm
ePTFE	Expanded polytetrafluoroethylene
cePTFE	Capsule Expanded polytetrafluoroethylene
ECM	Extracellular matrix
CAM	Chorion allantoic membrane
ND	Native diaphragm

References

1. Carmeliet, P.; Jain, R.K. Angiogenesis in cancer and other diseases. *Nature* **2000**, *407*, 249–257. [CrossRef] [PubMed]

2. Rajendran, P.; Rengarajan, T.; Thangavel, J.; Nishigaki, Y.; Sakthisekaran, D.; Sethi, G.; Nishigaki, I. The vascular endothelium and human diseases. *Int. J. Biol. Sci.* **2013**, *9*, 1057–1069. [CrossRef] [PubMed]

3. Conconi, M.T.; Bellini, S.; Teoli, D.; de Coppi, P.; Ribatti, D.; Nico, B.; Simonato, E.; Gamba, P.G.; Nussdorfer, G.G.; Morpurgo, M.; et al. *In vitro* and *in vivo* evaluation of acellular diaphragmatic matrices seeded with muscle precursors cells and coated with VEGF silica gels to repair muscle defect of the diaphragm. *J. Biomed. Mater. Res. Part A* **2009**, *89A*, 304–316. [CrossRef] [PubMed]

4. Lee, T.T.; García, J.R.; Paez, J.I.; Singh, A.; Phelps, E.A.; Weis, S.; Shafiq, Z.; Shekaran, A.; del Campo, A.; García, A.J. Light-triggered in vivo activation of adhesive peptides regulates cell adhesion, inflammation and vascularization of biomaterials. *Nat. Mater.* **2014**, *14*, 352–360. [CrossRef] [PubMed]

5. Gholobova, D.; Decroix, L.; van Muylder, V.; Desender, L.; Gerard, M.; Carpentier, G.; Vandenburgh, H.; Thorrez, L. Endothelial Network Formation Within Human Tissue-Engineered Skeletal Muscle. *Tissue Eng. Part A* **2015**, *21*, 150901071945000. [CrossRef] [PubMed]

6. Miller, J.S.; Stevens, K.R.; Yang, M.T.; Baker, B.M.; Nguyen, D.-H.T.; Cohen, D.M.; Toro, E.; Chen, A.A.; Galie, P.A.; Yu, X.; et al. Rapid casting of patterned vascular networks for perfusable engineered three-dimensional tissues. *Nat. Mater.* **2012**, *11*, 768–774. [CrossRef] [PubMed]

7. Tremblay, P.L.; Hudon, V.; Berthod, F.; Germain, L.; Auger, F.A. Inosculation of tissue-engineered capillaries with the host's vasculature in a reconstructed skin transplanted on mice. *Am. J. Transplant.* **2005**, *5*, 1002–1010. [CrossRef] [PubMed]

8. Atala, A.; Kasper, F.K.; Mikos, A.G. Engineering complex tissues. *Sci. Transl. Med.* **2012**, *4*, 160rv12. [CrossRef] [PubMed]

9. Baranski, J.D.; Chaturvedi, R.R.; Stevens, K.R.; Eyckmans, J.; Carvalho, B.; Solorzano, R.D.; Yang, M.T.; Miller, J.S.; Bhatia, S.N.; Chen, C.S. Geometric control of vascular networks to enhance engineered tissue integration and function. *Proc. Natl. Acad. Sci. USA* **2013**, *110*, 7586–7591. [CrossRef] [PubMed]

10. Folkman, J.; Hochberg, M. Self-regulation of growth in three dimensions. *J. Exp. Med.* **1973**, *138*, 745–753. [CrossRef] [PubMed]

11. Laschke, M.W.; Menger, M.D. Prevascularization in tissue engineering: Current concepts and future directions. *Biotechnol. Adv.* **2015**. [CrossRef] [PubMed]

12. Burns, J.S.; Kristiansen, M.; Kristensen, L.P.; Larsen, K.H.; Nielsen, M.O.; Christiansen, H.; Nehlin, J.; Andersen, J.S.; Kassem, M. Decellularized Matrix from Tumorigenic Human Mesenchymal Stem Cells Promotes Neovascularization with Galectin-1 Dependent Endothelial Interaction. *PLoS ONE* **2011**, *6*, e21888. [CrossRef] [PubMed]

13. Moore, M.C.; Pandolfi, V.; McFetridge, P.S. Novel human-derived extracellular matrix induces in vitro and in vivo vascularization and inhibits fibrosis. *Biomaterials* **2015**, *49*, 37–46. [CrossRef] [PubMed]

14. Totonelli, G.; Maghsoudlou, P.; Georgiades, F.; Garriboli, M.; Koshy, K.; Turmaine, M.; Ashworth, M.; Sebire, N.J.; Pierro, A.; Eaton, S.; et al. Detergent enzymatic treatment for the development of a natural acellular matrix for oesophageal regeneration. *Pediatr. Surg. Int.* **2013**, *29*, 87–95. [CrossRef] [PubMed]

15. Brown, M.D.; Hudlicka, O. Modulation of physiological angiogenesis in skeletal muscle by mechanical forces: Involvement of VEGF and metalloproteinases. *Angiogenesis* **2003**, *6*, 1–14. [CrossRef] [PubMed]

16. Christov, C.; Chretien, F.; Abou-Khalil, R.; Bassez, G.; Vallet, G.; Authier, F.-J.; Bassaglia, Y.; Shinin, V.; Tajbakhsh, S.; Chazaud, B.; et al. Muscle Satellite Cells and Endothelial Cells: Close Neighbors and Privileged Partners. *Mol. Biol. Cell* **2007**, *18*, 1397–1409. [CrossRef] [PubMed]

17. Fuoco, C.; Rizzi, R.; Biondo, A.; Longa, E.; Mascaro, A.; Shapira-, K.; Kossovar, O.; Benedetti, S.; Salvatori, M.L.; Santoleri, S.; et al. In vivo generation of a mature and functional artificial skeletal muscle. *EMBO Mol. Med.* **2015**, 1–13. [CrossRef] [PubMed]

18. Jank, B.J.; Xiong, L.; Moser, P.T.; Guyette, J.P.; Ren, X.; Cetrulo, C.L.; Leonard, D.A.; Fernandez, L.; Fagan, S.P.; Ott, H.C. Engineered composite tissue as a bioartificial limb graft. *Biomaterials* **2015**, *61*, 246–256. [CrossRef] [PubMed]

19. Juhas, M.; Engelmayr, G.C.; Fontanella, A.N.; Palmer, G.M.; Bursac, N. Biomimetic engineered muscle with capacity for vascular integration and functional maturation in vivo. *Proc. Natl. Acad. Sci. USA* **2014**, *111*, 5508–5513. [CrossRef] [PubMed]

20. Levenberg, S.; Rouwkema, J.; Macdonald, M.; Garfein, E.S.; Kohane, D.S.; Darland, D.C.; Marini, R.; van Blitterswijk, C.A.; Mulligan, R.C.; D'Amore, P.A.; et al. Engineering vascularized skeletal muscle tissue. *Nat. Biotechnol.* **2005**, *23*, 879–884. [CrossRef] [PubMed]

21. Shimizu-Motohashi, Y.; Asakura, A. Angiogenesis as a novel therapeutic strategy for Duchenne muscular dystrophy through decreased ischemia and increased satellite cells. *Front. Physiol.* **2014**, *5*, 50. [CrossRef] [PubMed]

22. Uchida, C.; Nwadozi, E.; Hasanee, A.; Olenich, S.; Olfert, I.M.; Haas, T.L. Muscle-derived vascular endothelial growth factor regulates microvascular remodelling in response to increased shear stress in mice. *Acta Physiol.* **2015**, *214*, 349–360. [CrossRef] [PubMed]

23. Van Amerongen, M.J.; Molema, G.; Plantinga, J.; Moorlag, H.; van Luyn, M.J. Neovascularization and vascular markers in a foreign body reaction to subcutaneously implanted degradable biomaterial in mice. *Angiogenesis* **2002**, *5*, 173–180. [CrossRef] [PubMed]

24. Kwee, B.J.; Mooney, D.J. Manipulating the Intersection of Angiogenesis and Inflammation. *Ann. Biomed. Eng.* **2015**, *43*, 628–640. [CrossRef] [PubMed]

25. Anderson, J.M.; Rodriguez, A.; Chang, D.T. Foreign body reaction to biomaterials. *Semin. Immunol.* **2008**, *20*, 86–100. [CrossRef] [PubMed]

26. Piccoli, M.; Urbani, L.; Alvarez-Fallas, M.E.; Franzin, C.; Dedja, A.; Bertin, E.; Zuccolotto, G.; Rosato, A.; Pavan, P.; Elvassore, N.; et al. Improvement of diaphragmatic performance through orthotopic application of decellularized extracellular matrix patch. *Biomaterials* **2016**, *74*, 245–255. [CrossRef] [PubMed]

27. Romao, R.L.P.; Nasr, A.; Chiu, P.P.L.; Langer, J.C. What is the best prosthetic material for patch repair of congenital diaphragmatic hernia? Comparison and meta-analysis of porcine small intestinal submucosa and polytetrafluoroethylene. *J. Pediatr. Surg.* **2012**, *47*, 1496–1500. [CrossRef] [PubMed]

28. Vailhé, B.; Vittet, D.; Feige, J.J. In vitro models of vasculogenesis and angiogenesis. *Lab. Investig.* **2001**, *81*, 439–452. [CrossRef] [PubMed]

29. Bland, E.; Dréau, D.; Burg, K.J.L. Overcoming hypoxia to improve tissue-engineering approaches to regenerative medicine. *J. Tissue Eng. Regen. Med.* **2013**, *7*, 505–514. [CrossRef] [PubMed]

30. Brown, B.N.; Londono, R.; Tottey, S.; Zhang, L.; Kukla, K.A.; Wolf, M.T.; Daly, K.A.; Reing, J.E.; Badylak, S.F. Macrophage phenotype as a predictor of constructive remodeling following the implantation of biologically derived surgical mesh materials. *Acta Biomater.* **2012**, *8*, 978–987. [CrossRef] [PubMed]

31. Kyriakides, T.R.; Bornstein, P. Matricellular proteins as modulators of wound healing and the foreign body response. *Thromb. Haemost.* **2003**, *89*, 986–992. [CrossRef] [PubMed]

32. Ricard-Blum, S.; Salza, R. Matricryptins and matrikines: Biologically active fragments of the extracellular matrix. *Exp. Dermatol.* **2014**, *23*, 457–463. [CrossRef] [PubMed]

33. Hill, R.C.; Calle, E.A.; Dzieciatkowska, M.; Niklason, L.E.; Hansen, K.C. Quantification of Extracellular Matrix Proteins from a Rat Lung Scaffold to Provide a Molecular Readout for Tissue Engineering. *Mol. Cell. Proteom.* **2015**, *1*, 961–973. [CrossRef] [PubMed]

34. Li, Q.; Uygun, B.E.; Geerts, S.; Ozer, S.; Scalf, M.; Gilpin, S.E.; Ott, H.C.; Yarmush, M.L.; Smith, L.M.; Welham, N.V.; et al. Proteomic analysis of naturally-sourced biological scaffolds. *Biomaterials* **2016**, *75*, 37–46. [CrossRef] [PubMed]

35. Caralt, M.; Uzarski, J.S.; Iacob, S.; Obergfell, K.P.; Berg, N.; Bijonowski, B.M.; Kiefer, K.M.; Ward, H.H.; Wandinger-Ness, A.; Miller, W.M.; et al. Optimization and critical evaluation of decellularization strategies to develop renal extracellular matrix scaffolds as biological templates for organ engineering and transplantation. *Am. J. Transplant.* **2015**, *15*, 64–75. [CrossRef] [PubMed]

36. Hoganson, D.M.; O'Doherty, E.M.; Owens, G.E.; Harilal, D.O.; Goldman, S.M.; Bowley, C.M.; Neville, C.M.; Kronengold, R.T.; Vacanti, J.P. The retention of extracellular matrix proteins and angiogenic and mitogenic cytokines in a decellularized porcine dermis. *Biomaterials* **2010**, *31*, 6730–6737. [CrossRef] [PubMed]

37. Reing, J.E.; Zhang, L.; Myers-Irvin, J.; Cordero, K.E.; Freytes, D.O.; Heber-Katz, E.; Bedelbaeva, K.; McIntosh, D.; Dewilde, A.; Braunhut, S.J.; et al. Degradation products of extracellular matrix affect cell migration and proliferation. *Tissue Eng. Part A* **2009**, *15*, 605–614. [CrossRef] [PubMed]

38. Moens, S.; Goveia, J.; Stapor, P.C.; Cantelmo, A.R.; Carmeliet, P. The multifaceted activity of VEGF in angiogenesis—Implications for therapy responses. *Cytokine Growth Factor Rev.* **2014**, *25*, 473–482. [CrossRef] [PubMed]

39. Latroche, C.; Gitiaux, C.; Chrétien, F.; Desguerre, I.; Mounier, R.; Chazaud, B. Skeletal Muscle Microvasculature: A Highly Dynamic Lifeline. *Physiology* **2015**, *30*, 417–427. [CrossRef] [PubMed]

40. Brzoska, E.; Kowalewska, M.; Markowska-Zagrajek, A.; Kowalski, K.; Archacka, K.; Zimowska, M.; Grabowska, I.; Czerwińska, A.M.; Czarnecka-Góra, M.; Stremińska, W.; et al. Sdf-1 (CXCL12) improves skeletal muscle regeneration via the mobilisation of Cxcr4 and CD34 expressing cells. *Biol. Cell* **2012**, *104*, 722–737. [CrossRef] [PubMed]

41. Anderson, E.M.; Kwee, B.J.; Lewin, S.A.; Raimondo, T.; Mehta, M.; Mooney, D.J. Local Delivery of VEGF and SDF Enhances Endothelial Progenitor Cell Recruitment and Resultant Recovery from Ischemia. *Tissue Eng. Part A* **2015**, *21*, 1217–1227. [CrossRef] [PubMed]

42. Simons, M.; Alitalo, K.; Annex, B.H.; Augustin, H.G.; Beam, C.; Berk, B.C.; Byzova, T.; Carmeliet, P.; Chilian, W.; Cooke, J.P.; et al. State-of-the-Art Methods for Evaluation of Angiogenesis and Tissue Vascularization: A Scientific Statement From the American Heart Association. *Circ. Res.* **2015**. [CrossRef] [PubMed]

43. Chen, Z.; Htay, A.; Santos, W.D.; Gillies, G.T.; Fillmore, H.L.; Sholley, M.M.; Broaddus, W.C. In vitro angiogenesis by human umbilical vein endothelial cells (HUVEC) induced by three-dimensional co-culture with glioblastoma cells. *J. Neurooncol.* **2009**, *92*, 121–128. [CrossRef] [PubMed]

44. Elliott, M.J.; de Coppi, P.; Speggiorin, S.; Roebuck, D.; Butler, C.R.; Samuel, E.; Crowley, C.; McLaren, C.; Fierens, A.; Vondrys, D.; et al. Stem-cell-based, tissue engineered tracheal replacement in a child: A 2-year follow-up study. *Lancet* **2012**, *380*, 994–1000. [CrossRef]

45. Atala, A.; Bauer, S.B.; Soker, S.; Yoo, J.J.; Retik, A.B. Tissue-engineered autologous bladders for patients needing cystoplasty. *Lancet* **2006**, *367*, 1241–1246. [CrossRef]

46. Raya-Rivera, A.; Esquiliano, D.R.; Yoo, J.J.; Lopez-Bayghen, E.; Soker, S.; Atala, A. Tissue-engineered autologous urethras for patients who need reconstruction: An observational study. *Lancet* **2011**, *377*, 1175–1182. [CrossRef]

47. Uygun, B.E.; Soto-Gutierrez, A.; Yagi, H.; Izamis, M.-L.; Guzzardi, M.A.; Shulman, C.; Milwid, J.; Kobayashi, N.; Tilles, A.; Berthiaume, F.; et al. Organ reengineering through development of a transplantable recellularized liver graft using decellularized liver matrix. *Nat. Med.* **2010**, *16*, 814–820. [CrossRef] [PubMed]

48. Ott, H.C.; Matthiesen, T.S.; Goh, S.-K.; Black, L.D.; Kren, S.M.; Netoff, T.I.; Taylor, D.A. Perfusion-decellularized matrix: Using nature's platform to engineer a bioartificial heart. *Nat. Med.* **2008**, *14*, 213–221. [CrossRef] [PubMed]

49. Petersen, T.H.; Calle, E.A.; Zhao, L.; Lee, E.J.; Gui, L.; Raredon, M.B.; Gavrilov, K.; Yi, T.; Zhuang, Z.W.; Breuer, C.; et al. Tissue-engineered lungs for in vivo implantation. *Science* **2010**, *329*, 538–541. [CrossRef] [PubMed]

50. Vigodarzere, G.C.; Mantero, S. Skeletal muscle tissue engineering: Strategies for volumetric constructs. *Front. Physiol.* **2014**, *5*, 1–13. [CrossRef]

51. Garg, K.; Ward, C.L.; Corona, B.T. Asynchronous inflammation and myogenic cell migration limit muscle tissue regeneration mediated by acellular scaffolds. *Inflamm. Cell Signal.* **2014**, 6–8. [CrossRef]

52. De Coppi, P.; Deprest, J. Regenerative medicine for congenital diaphragmatic hernia: Regeneration for repair. *Eur. J. Pediatr. Surg.* **2012**, *22*, 393–398. [CrossRef] [PubMed]

53. Grethel, E.J.; Cortes, R.A.; Wagner, A.J.; Clifton, M.S.; Lee, H.; Farmer, D.L.; Harrison, M.R.; Keller, R.L.; Nobuhara, K.K. Prosthetic patches for congenital diaphragmatic hernia repair: Surgisis vs Gore-Tex. *J. Pediatr. Surg.* **2006**, *41*, 29–33. [CrossRef] [PubMed]

54. Wälchli, T.; Mateos, J.M.; Weinman, O.; Babic, D.; Regli, L.; Hoerstrup, S.P.; Gerhardt, H.; Schwab, M.E.; Vogel, J. Quantitative assessment of angiogenesis, perfused blood vessels and endothelial tip cells in the postnatal mouse brain. *Nat. Protoc.* **2014**, *10*, 53–74. [CrossRef] [PubMed]

55. Moore, L.B.; Sawyer, A.J.; Charokopos, A.; Skokos, E.a.; Kyriakides, T.R. Loss of monocyte chemoattractant protein-1 alters macrophage polarization and reduces NFκB activation in the foreign body response. *Acta Biomater.* **2015**, *11*, 37–47. [CrossRef] [PubMed]

56. Pober, J.S.; Sessa, W.C. Inflammation and the blood microvascular system. *Cold Spring Harb. Perspect. Biol.* **2015**, *7*, 1–12. [CrossRef] [PubMed]

57. Cao, Y. Therapeutic angiogenesis for ischemic disorders: What is missing for clinical benefits? *Discov. Med.* **2010**, *9*, 179–184. [PubMed]

58. Korn, C.; Augustin, H.G. Born to Die: Blood vessel regression research coming of age. *Circulation* **2012**, *125*, 3063–3065. [CrossRef] [PubMed]

59. Boersema, G.S.A.; Grotenhuis, N.; Bayon, Y.; Lange, J.F.; Bastiaansen-Jenniskens, Y.M. The Effect of Biomaterials Used for Tissue Regeneration Purposes on Polarization of Macrophages. *BioRes. Open Access* **2016**, *5*, 6–14. [CrossRef] [PubMed]

60. Fishman, J.M.; Lowdell, M.W.; Urbani, L.; Ansari, T.; Burns, A.J.; Turmaine, M.; North, J.; Sibbons, P.; Seifalian, A.M.; Wood, K.J.; et al. Immunomodulatory effect of a decellularized skeletal muscle scaffold in a discordant xenotransplantation model. *Proc. Natl. Acad. Sci. USA* **2013**, *110*, 14360–14365. [CrossRef] [PubMed]

61. Eaton, K.V.; Yang, H.L.; Giachelli, C.M.; Scatena, M. Engineering macrophages to control the inflammatory response and angiogenesis. *Exp. Cell Res.* **2015**, *339*, 300–309. [CrossRef] [PubMed]

62. Kreuger, J.; Phillipson, M. Targeting vascular and leukocyte communication in angiogenesis, inflammation and fibrosis. *Nat. Rev. Drug Discov.* **2015**, *15*. [CrossRef] [PubMed]

63. Kzhyshkowska, J.; Gudima, A.; Riabov, V.; Dollinger, C.; Lavalle, P.; Vrana, N.E. Macrophage responses to implants: Prospects for personalized medicine. *J. Leukoc. Biol.* **2015**, *98*, 953–962. [CrossRef] [PubMed]

64. Lech, M.; Anders, H.J. Macrophages and fibrosis: How resident and infiltrating mononuclear phagocytes orchestrate all phases of tissue injury and repair. *Biochim. Biophys. Acta Mol. Basis Dis.* **2013**, *1832*, 989–997. [CrossRef] [PubMed]

65. Neve, A.; Cantatore, F.P.; Maruotti, N.; Corrado, A.; Ribatti, D. Extracellular matrix modulates angiogenesis in physiological and pathological conditions. *Biomed Res Int.* **2014**, *2014*, 756078. [CrossRef] [PubMed]

66. Stetler-Stevenson, W.G. Tissue inhibitors of metalloproteinases in cell signaling: Metalloproteinase-independent biological activities. *Sci. Signal.* **2008**, *1*, re6. [CrossRef] [PubMed]

67. Barrientos, S.; Stojadinovic, O.; Golinko, M.S.; Brem, H.; Tomic-Canic, M. Growth factors and cytokines in wound healing. *Wound Repair Regen.* **2008**, *16*, 585–601. [CrossRef] [PubMed]
68. Giannandrea, M.; Parks, W.C. Diverse functions of matrix metalloproteinases during fibrosis. *Dis. Model. Mech.* **2014**, *7*, 193–203. [CrossRef] [PubMed]
69. Ho, T.-C.; Chiang, Y.-P.; Chuang, C.-K.; Chen, S.-L.; Hsieh, J.-W.; Lan, Y.-W.; Tsao, Y.-P. PEDF-derived peptide promotes skeletal muscle regeneration through its mitogenic effect on muscle progenitor cells. *Am. J. Physiol. Cell Physiol.* **2015**, *309*, C159–C168. [CrossRef] [PubMed]
70. Sahin, H.; Wasmuth, H.E. Chemokines in tissue fibrosis. *Biochim. Biophys. Acta Mol. Basis Dis.* **2013**, *1832*, 1041–1048. [CrossRef] [PubMed]
71. Van Putten, S.M.; Ploeger, D.T.A.; Popa, E.R.; Bank, R.A. Macrophage phenotypes in the collagen-induced foreign body reaction in rats. *Acta Biomater.* **2013**, *9*, 6502–6510. [CrossRef] [PubMed]
72. Zhang, Y.; Yang, P.; Cui, R.; Zhang, M.; Li, H.; Qian, C.; Sheng, C.; Qu, S.; Bu, L. Eosinophils Reduce Chronic Inflammation in Adipose Tissue by Secreting Th2 Cytokines and Promoting M2 Macrophages Polarization. *Int. J. Endocrinol.* **2015**, *2015*, 2–7. [CrossRef] [PubMed]
73. Ponce, M.L.; Kleinmann, H.K. The chick chorioallantoic membrane as an in vivo angiogenesis model. *Curr. Protoc. Cell Biol.* **2003**, *19*. [CrossRef]
74. Drabkin, D.L. The standardization of hemoglobin measurement. *Am. J. Med. Sci.* **1948**, *215*, 110. [CrossRef] [PubMed]
75. Browne, S.; Monaghan, M.G.; Brauchle, E.; Berrio, D.C.; Chantepie, S.; Papy-Garcia, D.; Schenke-Layland, K.; Pandit, A. Modulation of inflammation and angiogenesis and changes in ECM GAG-activity via dual delivery of nucleic acids. *Biomaterials* **2015**, *69*, 133–147. [CrossRef] [PubMed]
76. Brudno, Y.; Ennett-Shepard, A.B.; Chen, R.R.; Aizenberg, M.; Mooney, D.J. Enhancing microvascular formation and vessel maturation through temporal control over multiple pro-angiogenic and pro-maturation factors. *Biomaterials* **2013**, *34*, 9201–9209. [CrossRef] [PubMed]

International Journal of
Molecular Sciences

Review

Decellularized Tissue for Muscle Regeneration

Anna Urciuolo [1,2] and Paolo De Coppi [2,*]

[1] Department of Industrial Engineering, University of Padova, 35122 Padova, Italy; anna.urciuolo@unipd.it
[2] Great Ormond Street Institute of Child Health, University College of London, London WC1N 1EH, UK
* Correspondence: p.decoppi@ucl.ac.uk; Tel.: +44-020-7905-2641; Fax: +44-020-7404-6181

Received: 14 July 2018; Accepted: 10 August 2018; Published: 14 August 2018

Abstract: Several acquired or congenital pathological conditions can affect skeletal muscle leading to volumetric muscle loss (VML), i.e., an irreversible loss of muscle mass and function. Decellularized tissues are natural scaffolds derived from tissues or organs, in which the cellular and nuclear contents are eliminated, but the tridimensional (3D) structure and composition of the extracellular matrix (ECM) are preserved. Such scaffolds retain biological activity, are biocompatible and do not show immune rejection upon allogeneic or xenogeneic transplantation. An increase number of reports suggest that decellularized tissues/organs are promising candidates for clinical application in patients affected by VML. Here we explore the different strategies used to generate decellularized matrix and their therapeutic outcome when applied to treat VML conditions, both in patients and in animal models. The wide variety of VML models, source of tissue and methods of decellularization have led to discrepant results. Our review study evaluates the biological and clinical significance of reported studies, with the final aim to clarify the main aspects that should be taken into consideration for the future application of decellularized tissues in the treatment of VML conditions.

Keywords: skeletal muscle engineering; tissue engineering; volumetric muscle loss; decellularized tissue; decellularized muscle; acellular tissue; acellular muscle; skeletal muscle regeneration

---

1. Introduction

The crucial role of the ECM environment in the stem cell niche, and in the regulation of stem cell identity and differentiation, organogenesis and tissue homeostasis has been a topic of extensive and intriguing study [1,2]. In the field of regenerative medicine this has allowed for the development of an increasing number of tissue engineering strategies, in which scaffold materials are used to mimic in vivo the biological microenvironment of the ECM, providing the components needed to drive cells toward the regeneration of the tissue of interest. Despite the incredible improvements that have been made in 3D bioprinting technology, the bona-fide reproduction of a scaffold capable to accurately mimic complex tissues, such as skeletal muscle, remains a matter that cannot be technically solved [2]. The 3D interactions existing among different components of the ECM is far for being a simple overlay of proteins organized in a layer-by-layer fashion. Indeed, ECM components not only interact with each other in specific fashions, but each single component, and also defined isoforms of a same component, are tissue-specific and even site-specific inside a defined tissue [2,3]. Such complexity, which likely has a biological meaning for cells, can be preserved in scaffolds only by taking advance of the native tissue themselves, that is achieved by decellularizing tissues or whole organs. Upon the removal of nuclear content and cellular elements, decellularized or acellular scaffolds still retain the architecture and complexity of the native tissues, including vasculature and biofactors present in the ECM. These characteristics make decellularized matrices the ideal bio-scaffold necessary to guide host or donor cells toward the regeneration of new and functional tissues. Several studies have already demonstrated the possibility of successfully obtaining acellular scaffolds from many organs, such as heart, kidney, pancreas, lung, liver, esophagus and intestine. Importantly, some of these decellularization protocols have been adopted to decellularize simple hollow organs such as the trachea, which have then been successfully transplanted in patient after autologous cell seeding, i.e., trachea [4–6]. Importantly, the trachea transplant has been achieved without immunosuppression, a great advantage over conventional transplantation because it avoids potential risks for patients, including frequent infections and cancer. Acellular tissues are biocompatible and the absence of rejection after allogeneic or xenogeneic transplantation make them the ideal scaffold for translational medicine applications and organ replacement [7,8].

Even though skeletal muscle has a remarkable capacity to undergo regeneration, several pathological conditions can lead to extensive and irreversible muscle loss: i.e., congenital defects, traumatic injuries, surgical ablations, and neuromuscular diseases [2]. Failure of normal regeneration results in VML, with loss of muscle function, often associated with scar tissue formation and adipose tissue substitution. Current treatments for such conditions have limited success [7], leading to considerable social and economic burden. Therefore, there is a great need for new regenerative medicine strategies aimed at treating VML conditions. Skeletal muscle is a complex tissue in which myofiber 3D organization and function is intimately linked to other tissue components, such as motor neurons, vasculature, myogenic stem cells (including satellite cells, SCs), interstitial cells, and ECM [9–11]. A tubular network of ECM is intimately connected to all the different cellular components. However, the ECM does not just provide mechanical support for bearing force transmission, but also a dynamic signaling environment that is crucial for muscle development, homeostasis, and regeneration [12,13]. SCs are mitotically quiescent muscle stem cells necessary for muscle regeneration and located between the basal lamina and myofibers [11]. Defined composition and mechanical properties of the ECM in SC niche is required to allow SC self-renewal and efficient muscle regeneration in vivo [11,14–18].

2. Acellular Tissues and Biomaterials for VML Treatment: Types and Methods

The ideal biomaterial for VML repair would need to fill the volumetric loss and sustain SC activity, while guaranteeing access to both vascular and neural cells for a correct revascularization and reinnervation of the tissue [7,8,19]. Based on the above points, it is not surprising that acellular scaffolds derived form a range of different tissues have already been tested in animal models [20–34] and in small cohorts of patients affected by VML [35,36].

Acellular tissues are mainly generated by using physical, enzymatic, and/or chemical mechanisms [9,37–39]. Based on its simplicity, the most commonly used method to obtain acellular tissue is immersion and agitation of the sample, in presence of decellularizing agents. However, this approach does not allow homogeneous decellularization of large samples and thick tissues, such as skeletal muscle. To overcome such limitation, perfusion methods have been developed. Indeed, by using the native vascular tree of the tissue/organ, decellularizing reagents can be homogeneously distributed across the tissue, allowing better access and deep tissue exposure, ultimately improving the removal of cellular components from large tissues [40–42]. Jank and colleagues reported the ability to decellularize rat and primate forearms by perfusion of detergents. The preservation of the composite architecture allowed the repopulation of muscle and vasculature of the construct with cells of appropriate phenotypes in vitro, and also supported blood perfusion following in vivo transplantation [43]. Moreover, a recent report by Gerli and colleagues also successfully generated a large-scale, acellular composite tissue scaffold from a full cadaveric human upper extremity using a perfusion method of decellularization [44]. This construct retained its morphological architecture and perfusable vascular conduits with the preservation of the native ECM components. Such biocompatible constructs could have significant advantages over the currently implanted matrices, in terms of nutrient distribution, size-scalability, and immunological response [44]. These studies demonstrate the possibility of developing more complex and reproducible decellularized organs, which hold higher regenerative potential and translational efficacy for the treatment of VML conditions.

Small intestine submucosa matrix (SIS), urinary bladder matrix (UBM), and skeletal muscle decellularized scaffolds have been the most commonly used materials to repair VML defects. SIS and UBM scaffolds are prepared with standardized protocols, and are commercially available, clinically approved, and have also been used in patients [30]. Conversely, skeletal muscle scaffolds have only been used in animal models, and their preparation has been shown to be a more complex procedure, mainly due to tissue complexity and thickness. The main protocols applied to generate decellularized scaffolds from skeletal muscle are essentially of three types: enzymatic-not detergent [9,24,29], detergent [31–34], and detergent-enzymatic [33,45,46] treatments. In the literature such scaffolds are defined as decellularized; however, some of them would be better defined as "anucleated" rather than "acellular" scaffolds. Indeed, together with the elimination of the nuclear content and the maintenance of the ECM, such treatments can partially preserve cytoplasmic components of the original tissue, in particular those belonging to myofibers [24,32,33,39,45,46].

As mentioned above, different sources of tissues and methods of decellularization have been used to generate acellular scaffolds for the treatment of VML, complicating the understanding of which scaffold is most ideal. The biological properties of the scaffolds depend on the agent, or combined-agents, used for inducing decellularization, as well as on the method applied (i.e., immersion vs. perfusion) and on the characteristics of the primary tissue/organ (density, cellularity, dominant component in tissue, and thickness) [37]. Another important factor to consider is whether using a tissue-matched decellularized scaffold can have positive effects on regeneration. SIS and UBM are scaffolds that can fill the volumetric loss of tissue but do not have any muscular specific components. Based on the complexity of the biological meaning of ECM—discussed above— it is reasonable to speculate that matched acellular tissues should be better at instructing host and/or seeded-donor cells toward the regeneration of the tissue of interest.

While this is not the focus of the present work, it is important to recognize that major work in the field of skeletal muscle tissue engineering has been undertaken using polymers. Polymers have some advantages over decellularized tissue, e.g., their manufacture is more reliable and consistent, allowing a precise design of their geometry as well as mechanical and structural properties. As with natural scaffolds, they can be loaded with bioactive molecules. Scaffolds used for skeletal muscle regeneration include (i) synthetic polymers such as poly(ethylene glycolic) (PEG), polyglycolic acid (PGA), poly(lactic-co-glycolic acid) (PLGA), poly(l-lactic acid) (PLLA), their copolymers (PLLA/PLGA), and polycaprolactone (PCL); (ii) natural polymers such as alginate, collagen, fibrin, or hyaluronic acid, or (iii) a combination of the two [47–60]. Rat myoblasts seeded onto PGA meshes and implanted into the omentum of syngeneic animals generated vascularized constructs in which myoblasts were found to be organized between strands of degrading polymer mesh [48,49]. In comparison to synthetic materials, natural polymers may have the advantage of closely resembling the original ECM, or ECM function of a tissue, and will thus facilitate an efficient regeneration. For example, fibrin itself can efficiently promote regeneration in skeletal muscle. Fibrin microthreads seeded with adult human cells improved regeneration of a large defect in the tibialis anterior muscle in murine model of VML, with new muscle tissue formation and presence of Pax7-positive cells [55]. Efficient muscle regeneration using fibrin was also previously demonstrated using a 3D fibrin matrix seeded with expanded primary rat syngeneic myoblasts [56]. Another chief component of the ECM, hyaluronan, has been shown to promote muscle regeneration when seeded with muscle progenitor cells in a murine model of VML. Notably, SCs embedded in photo-cross-linkable hyaluronan also promoted functional skeletal muscle recovery, with the formation of both neural and vascular networks and the reconstitution of a functional SC niche [57]. Growth factors can also be efficiently associated to synthetic or natural polymers. Mouse or human mesoangioblasts engineered to express placental-derived growth factor and encapsulated inside PEG-fibrinogen hydrogels were able to efficiently replace an ablated tibialis anterior [61]. In another study, an injectable, degradable alginate gel loaded with vascular endothelial growth factor (VEGF) and insulin-like growth factor 1 (IGF1) led to parallel angiogenesis, reinnervation, and myogenesis upon ischemic damage in skeletal muscle, with SCs activation, proliferation, and simultaneous protection from inflammation and apoptosis [51]. VEGF can also be delivered by genetically engineering grafted cells. Rat myoblasts transfected with a plasmid encoding VEGF and suspended in collagen type I showed increased generation of tissue mass after subcutaneous injection into nude mice compared with nonfunctional VEGF-transfected cells [52]. Besides using factors promoting angiogenesis, factors that induce cell activation and migration, such as hepatocyte growth factor (HGF) and fibroblast growth factor (FGF), can also promote scaffold grafting and myogenesis [53]. The role of growth factors and natural polymers was further investigated by a recent paper in which VEGF, IGF-1, FGF, HGF, and other factors were released locally from alginate microbeads. Such constructs induced differentiation of urine-derived stem cells into a myogenic lineage, enhanced revascularization and innervation of the implants and stimulated in vivo resident cell growth [54].

3. VML Models for Testing Acellular Tissues

The translational potential of decellularized tissues has mainly been evaluated in vivo using surgical animal models of VML. Therefore, the use of such models mainly aim to test the ability of decellularized tissues to promote (i) cell homing; (ii) regeneration of myofibers and motor neuron axons; and (iii) angiogenesis. These are all essential steps necessary to have a functional skeletal muscle in the site of transplantation. Moreover, acellular tissues have been used as natural devices in which host cells can drive muscle regeneration, as well as used as scaffolding material to deliver donor cells with the aim to improve cell therapy strategies. In order to evaluate the studies that have been reported so far, here we compare the results obtained among different acellular scaffolds and strategies— i.e., with or without donor cells, used for the treatment of VML in comparable animal models (Table 1).

Mice are commonly used for studying muscle regeneration and testing the regenerative potential of acellular scaffolds in VML. Sicari and colleagues studied the ability of porcine SIS to promote muscle regeneration after xenotransplantation in a VML model in which the quadriceps muscle was ablated. Although no details were reported in terms of functional activity of the implant, the authors concluded that the scaffold within the defect was associated with constructive tissue remodeling, including the formation of site-appropriate skeletal muscle tissue [26]. The same model was used in another study to test the regenerative potential of porcine UBM. In this instance, Fisher and colleagues showed that the scaffolds promoted the formation of functional skeletal muscle cells, with perivascular stem cell mobilization and their accumulation within the site of injury [35]. Yet another study reported the ability of sodium dodecil sulfate (SDS)-derived acellular skeletal muscle to promote the formation of islands of myofibers after implantation in a tibialis anterior VML model [21]. Recently, few studies strongly supported the idea that decellularized muscles can offer a favorable environment to donor or host cells that promotes functional muscle regeneration [31,33,34]. We recently demonstrated that decellularized skeletal muscles derived with three different perfusion methods per se were able to generate functional artificial muscles in a xenogeneic immune-competent model of VML, in which the EDL muscle was surgically resected. In particular, decellularized tissues promoted migration and differentiation of the host myogenic cells, as well as SC homing, the formation of nervous fibers, neuromuscular junction and vascular networks [33]. Quarta and colleagues showed that decellularized muscle seeded with adult muscle stem cells and muscular resident cells were able to generate functional muscle tissue in a VML model. More in detail, SDS-based acellular muscles were used to deliver and promote the maintenance of human muscular cells in an immunocompromised murine model of VML in which the tibialis anterior muscle was ablated. Both innervation and in vivo force production were enhanced when implantation of bioconstructs was followed by an exercise regimen [31]. Another study showed the ability of decellularized skeletal muscle to support functional muscle regeneration by host cells, with less fibrosis and more de novo neuromuscular receptors than either autograft or collagen implants in a rat model of gastrocnemius VML [32].

307

Table 1. Decellularized tissues used in vivo. The table reports the studies in which decellularized tissues have been applied in vivo. The type of decellularized tissues, the method of decellularization, the type of the eventual seeded cells, the species subject to implantation, and the in vivo outcome have been reported for each reference. –, not performed.

Decellularized Tissue	Method of Decellularization	Seeded Cells	Implanted Species	In Vivo Outcome	Ref.
Porcine small intestine submucosa (SIS); Canine skeletal muscle	Immersion	–	Rat	Remodeling and partial skeletal muscle regeneration. Comparable results between SIS and skeletal muscle scaffolds	[21]
Murine skeletal muscle	Immersion	–	Mouse	Remodeling and partial skeletal muscle regeneration	[22]
Rat skeletal muscle	Immersion	Rat SC–derived myoblasts	Rat	Partial skeletal muscle regeneration	[23]
Rat skeletal muscle	Immersion	–	Rat	Partial skeletal muscle regeneration. No force restoration	[24]
Rat skeletal muscle	Immersion	Murine myoblasts	Rat	Improvement of donor cells survival	[25]
Porcine urinary bladder matrix (UBM)	Immersion	Minced muscle	Rat	Fibrosis and scarce skeletal muscle regeneration	[26]
Porcine SIS	Immersion	–	Mouse	Remodeling and partial skeletal muscle regeneration	[27]
Porcine SIS	Immersion	–	Dog	Remodeling and partial skeletal muscle regeneration. No functional recovery	[28]
Porcine SIS; Porcine skeletal muscle	Perfusion	–	Rat	Skeletal muscle regeneration with partial functional recovery. Improved results for skeletal muscle scaffolds	[29]
Porcine SIS; Carbodiimide-crosslinked porcine SIS	Immersion	–	Rat	Skeletal muscle regeneration with partial functional recovery. Improved results for SIS scaffolds	[30]
Porcine SIS; Porcine UBM	Immersion	–	Pig	Remodeling and fibrosis. No functional recovery	[31]
Murine skeletal muscle	Immersion	Co-culture of adult murine or human muscle stem cells and muscle resident cells	Mouse	Functional skeletal muscle regeneration improved after exercise regimen	[32]
Rat skeletal muscle	Patent	–	Rat	Functional skeletal muscle regeneration	[33]
Rat skeletal muscle	Perfusion	–	Mouse	Functional skeletal muscle regeneration	[34]
Rat skeletal muscle	Infusion	Minced muscle vs no cells	Rat	Functional skeletal muscle regeneration improved when cell seeded scaffolds are used	[35]
Porcine UBM	Immersion	–	Mouse	Skeletal muscle regeneration with partial functional recovery	[36]
Rat and primate forearm	Perfusion	Co-culture of C2C12 cells, fibroblasts and HUVEC	Rat	Reperfused vascular tree	[44]

The rat represents the most commonly used animal model for assessing muscle regeneration upon treatment of VML with acellular tissues. More than 10 years ago, we used a full-thickness defect of the abdominal wall to evaluate the syngeneic regenerative potential of cell-matrix constructs composed of SC–derived myoblasts seeded on muscle acellular matrices. Acellular muscles were obtained by a detergent-enzymatic method, and showed the preservation of both FGF and transforming growth factor-beta. The implanted construct promoted the formation of skeletal muscle and allowed the survival of donor cells nine months after surgery. Interestingly, a vesicular acetylcholine transporter was present on the surface of muscle fibers identified in the implant, suggesting the possible integration of the nervous system [22]. A similar approach was performed later in a xenogeneic model. Fishman and colleagues used a VML model of tibialis anterior for the implantation of a construct composed by rat skeletal muscle scaffolds and murine myoblasts. The decellularized muscle was prepared using an enzymatic protocol performed under agitation. Unfortunately, the aim of the study was focused on the immunomodulatory activity of the scaffold and muscle regeneration at the site of implantation was poorly characterized. However, the authors demonstrated that donor cells displayed better survival when delivered thought the scaffold, adding relevant information regarding the use of acellular matrix as scaffolding material for cell therapy approaches [24]. More recently, another study tested the possible application of UBM as scaffolding material for syngeneic cell delivery. Goldman and colleagues combined UBM with minced muscle grafts, promoting de novo muscle fiber regeneration and neuromuscular strength recovery in a VML model in which the tibialis anterior muscle was ablated. However, this functional improvement was associated with a concomitant reduction in graft-mediated regeneration, with coincident fibrous matrix deposition and interspersed islands of nascent muscle fibers. This effect was linked to the inability of UBM to create a favorable environment for efficiently promoting muscle regeneration [25]. Such results indicate that acellular scaffolds can ameliorate the survival of delivered cells. However, we can speculate that acellular muscles are better than UBM scaffolds when it comes to support muscle regeneration and donor cell maintenance.

Studies aimed at using acellular scaffolds with no implementation of cellular therapy have also been reported in rat VML models. Merritt and colleagues derived skeletal muscle decellularized tissues from rat by using an SDS-based immersion protocol. The scaffolds were implanted in a model in which a portion of the lateral gastrocnemius had been removed. After 42 days, growth of blood vessels and myofibers into the ECM was apparent, but no restoration of force occurred [23]. In another study, the histomorphologic characteristics and the physiologic activity of the implants were evaluated in an abdominal wall VML model upon implanting either porcine SIS, carbodiimide-cross-linked porcine SIS, autologous tissue, or polypropylene mesh. The best histomorphologic results were obtained with SIS scaffold, which showed islands and sheets of skeletal muscle. On the other hand, functional recovery was similar between SIS- and autologous tissue-implants. Conversely, implants of carbodiimide-cross-linked SIS and polypropylene mesh were characterized by a chronic inflammatory response and produced little or no measurable tetanic force [29]. Despite the range of final responses, these studies also supported the idea that acellular tissues can per se promote muscle regeneration in VML models.

Muscular or nonmuscular decellularized tissues implanted in the same VML model have been used to directly compare their ability to promote muscle regeneration. Wolf and colleagues used a partial thickness abdominal wall defect model in which they xeno-transplanted muscle-derived scaffold or SIS-derived matrix. In this study, the muscular tissue was derived from dogs and decellularized muscles were obtained with an agitation method and an enzymatic and chemical decellularization process. Even though acellular muscles were shown to have preserved growth factors, glycosaminoglycans, and basement membrane structural proteins, which differed from those present in SIS, the remodeling outcome was comparable between the two implanted scaffolds [20]. On the other hand, a different study reported that upon xenotransplantation decellularized muscles allowed better neovascularization, myogenesis, and functional recellularization than that obtained with porcine SIS implants. Interestingly, the muscular scaffolds used in this particular study, prepared from porcine rectus abdominis, were obtained with a perfusion method and retained the intricate 3D microarchitecture and vasculature networks of the native tissue, many of the bioactive ECM components and the mechanical properties [28]. Altogether these results strongly support the idea that the protocol used to decellularize skeletal muscle is fundamental in determining if and to what extent the obtained scaffold is capable to improve the regenerative response of the hosting tissue.

It is important to underline that the large volume of tissue that needs to be regenerated in patients affected by VML can be a major limiting step for acellular tissue application in clinic. To address this point, some studies attempted to apply decellularized matrix to larger animal models. Turner and colleagues have used a canine VML model, in which the distal third of the vastus lateralis and a portion of the vastus medialis muscles were resected. This defect was replaced with a scaffold composed of SIS. Even though the initial remodeling process followed a pattern similar to that reported in other studies, in the long term the implanted scaffolds showed dense collagenous tissue formation and islands of skeletal muscle with no successful restoration of tissue functionality [27]. In a more recent study, similar results were obtained in a porcine VML model in which peroneous tertius muscle was ablated. By comparing the outcome of implanted SIS, UBM, and hyaluronic acid hydrogel, Greising and colleagues showed that in all three cases ECM implantation could not orchestrate a skeletal muscle regeneration capable to lead to a physiological improvement. Instead, a significant deposition of fibrotic tissue was observed within the defect region three months post-injury [30]. Still, SIS and UBM have been shown to allow functional improvement in patients affected by VML. In particular, Fisher and colleagues showed that functional improvement was observed in three out of the five patients after implantation of SIS scaffolds [35]. More recently, a 13-patient cohort study was performed, using commercially available ECMs—BioDesign (SIS), Matristem (UBM), and Xenatrix (dermis-ECM)—to repair VML defects. The authors reported that in all cases the acellular scaffold facilitated functional tissue remodeling, thanks to the recruitment of myogenic progenitor cells and improved innervation [32]. These clinical results are in contrast with those obtained from studies performed in large animal models [27,30]. More studies will be needed to determine if the reason for such discrepancies is related to species-specific biologic responses to decellularized scaffold implantation. Besides, it is also important to emphasis that so far muscular acellular scaffolds have not been tested in large animal models or patients.

4. Conclusions and Perspectives

Studies performed on murine and rat VML models have conclusively demonstrated the ability of acellular scaffolds to promote myogenesis both as stand-alone devices and when associated with cell therapy strategies. Interestingly, by comparing results obtained between this two animal models one could speculate that acellular scaffolds derived from skeletal muscle are to be the best candidate to promote skeletal muscle regeneration in vivo, when compared to SIS and UBM scaffolds. This conclusion seems to be valid not only when scaffolds are used as devices, but also when they are associated with cell therapy. The possibility of using decellularized scaffolds as devices represent an important aspect for their translational application, as it eliminates the limiting steps specifically related to cell therapy. Perfusion methods of decellularization appear to better preserve instructive cues necessary to promote functional muscle regeneration by host cells. The recently published method of composite tissue decellularization, strongly suggest the feasibility of applying such muscular scaffolds in large animals and patients [43,44].

Overall, we can therefore conclude that despite incredible improvements in the design and development of synthetic or natural materials, decellularized tissues possess the irreplaceable advantage of preserving the biological signals that mimics the normal ECM of an in vivo tissue. Hence, we strongly believe that for the next foreseeable future application of decellularized tissues will represent a valid and powerful strategy to develop new therapeutic approaches for the cure of VML conditions.

Author Contributions: P.D.C. and A.U. have designed the concept, organized the writing, and proofread the review.

Funding: This research was founded by the National Institute of Health Research (RP 2014-04-046) and Great Ormond Street Hospital Children's Charity (13SG30).

Acknowledgments: P.D.C. and A.U. have been supported by the National Institute of Health Research (RP 2014-04-046) and Great Ormond Street Hospital Children's Charity (13SG30). All research at Great Ormond Street Hospital NHS Foundation Trust and UCL Great Ormond Street Institute of Child Health is made possible by the NIHR Great Ormond Street Hospital Biomedical Research Centre. The views expressed are those of the author(s) and not necessarily those of the NHS, the NIHR or the Department of Health.

Conflicts of Interest: The authors declare no conflicts of interest.

Abbreviations

3D	tridimensional
ECM	extracellular matrix
FGF	fibroblast growth factor
HGF	hepatocyte growth factor
IGF1	insulin-like growth factor 1
PCL	polycaprolactone
PEG	poly(ethylene glycolic)
PGA	polyglycolic acid
PLGA	poly(lactic-co-glycolic acid)
PLLA	poly(l-lactic acid)
SC	satellite cells
SDS	sodium dodecil sulfate
SIS	Small intestine submucosa matrix
UBM	urinary bladder matrix
VEGF	vascular endothelial growth factor
VML	volumetric muscle loss

References

1. Gattazzo, F.; Urciuolo, A.; Bonaldo, P. Extracellular matrix: A dynamic microenvironment for stem cell niche. *Biochim. Biophys. Acta Gen. Subj.* **2014**, *1840*, 2506–2519. [CrossRef] [PubMed]

2. Qazi, T.H.; Mooney, D.J.; Pumberger, M.; Geissler, S.; Duda, G.N. Biomaterials based strategies for skeletal muscle tissue engineering: Existing technologies and future trends. *Biomaterials* **2015**, *53*, 502–521. [CrossRef] [PubMed]

3. Muiznieks, L.D.; Keeley, F.W. Molecular assembly and mechanical properties of the extracellular matrix: A fibrous protein perspective. *Biochim. Biophys. Acta Mol. Basis Dis.* **2013**, *1832*, 866–875. [CrossRef] [PubMed]

4. Juhas, M.; Bursac, N. Engineering skeletal muscle repair. *Curr. Opin. Biotechnol.* **2013**, *24*, 880–886. [CrossRef] [PubMed]

5. Moran, E.C.; Dhal, A.; Vyas, D.; Lanas, A.; Soker, S.; Baptista, P.M. Whole-organ bioengineering: Current tales of modern alchemy. *Transl. Res.* **2014**, *163*, 259–267. [CrossRef] [PubMed]

6. Elliott, M.J.; De Coppi, P.; Speggiorin, S.; Roebuck, D.; Butler, C.R.; Samuel, E.; Crowley, C.; McLaren, C.; Fierens, A.; Vondrys, D.; et al. Stem-cell-based, tissue engineered tracheal replacement in a child: A 2-year follow-up study. *Lancet* **2012**, *380*, 994–1000. [CrossRef]

7. Kwee, B.J.; Mooney, D.J. Biomaterials for skeletal muscle tissue engineering. *Curr. Opin. Biotechnol.* **2017**, *47*, 16–22. [CrossRef] [PubMed]

8. Hinds, S.; Bian, W.; Dennis, R.G.; Bursac, N. The role of extracellular matrix composition in structure and function of bioengineered skeletal muscle. *Biomaterials* **2011**, *32*, 3575–3583. [CrossRef] [PubMed]

9. Gillies, A.R.; Smith, L.R.; Lieber, R.L.; Varghese, S. Method for Decellularizing Skeletal Muscle without Detergents or Proteolytic Enzymes. *Tissue Eng. Part C Methods* **2011**, *17*. [CrossRef] [PubMed]

10. Buckingham, M. Skeletal muscle formation in vertebrates. *Curr. Opin. Genet. Dev.* **2001**, *11*, 440–448. [CrossRef]

11. Yin, H.; Price, F.; Rudnicki, M.A. Satellite cells and the muscle stem cell niche. *Physiol. Rev.* **2013**, *93*, 23–67. [CrossRef] [PubMed]

12. Grzelkowska-Kowalczyk, K. *The Importance of Extracellular Matrix in Skeletal Muscle Development and Function*; World's Largest Science, Technology & Medicine Open Access Book Publisher: London, UK, 2016.

13. Rosenblatt, J.D.; Lunt, A.I.; Parry, D.J.; Partridge, T.A. Culturing satellite cells from living single muscle fiber explants. *In Vitro Cell. Dev. Biol. Anim.* **1995**, *31*, 773–779. [CrossRef] [PubMed]

14. Baghdadi, M.B.; Castel, D.; Machado, L.; Fukada, S.-I.; Birk, D.E.; Relaix, F.; Tajbakhsh, S.; Mourikis, P. Reciprocal signalling by Notch-Collagen V-CALCR retains muscle stem cells in their niche. *Nature* **2018**, *557*, 714–718. [CrossRef] [PubMed]

15. Bentzinger, C.F.; Wang, Y.X.; Von Maltzahn, J.; Soleimani, V.D.; Yin, H.; Rudnicki, M.A. Fibronectin regulates Wnt7a signaling and satellite cell expansion. *Cell Stem Cell* **2013**, *12*, 75–87. [CrossRef] [PubMed]

16. Urciuolo, A.; Quarta, M.; Morbidoni, V.; Gattazzo, F.; Molon, S.; Grumati, P.; Montemurro, F.; Tedesco, F.S.; Blaauw, B.; Cossu, G.; et al. Collagen VI regulates satellite cell self-renewal and muscle regeneration. *Nat. Commun.* **2013**, *4*. [CrossRef] [PubMed]

17. Quarta, M.; Brett, J.O.; DiMarco, R.; De Morree, A.; Boutet, S.C.; Chacon, R.; Gibbons, M.C.; Garcia, V.A.; Su, J.; Shrager, J.B.; et al. An artificial niche preserves the quiescence of muscle stem cells and enhances their therapeutic efficacy. *Nat. Biotechnol.* **2016**, *34*, 752–759. [CrossRef] [PubMed]

18. Rayagiri, S.S.; Ranaldi, D.; Raven, A.; Mohamad Azhar, N.I.F.; Lefebvre, O.; Zammit, P.S.; Borycki, A.G. Basal lamina remodeling at the skeletal muscle stem cell niche mediates stem cell self-renewal. *Nat. Commun.* **2018**, *9*. [CrossRef] [PubMed]

19. Ngan, C.G.Y.; Quigley, A.; Kapsa, R.M.I.; Choong, P.F.M. Engineering skeletal muscle—From two to three dimensions. *J. Tissue Eng. Regen. Med.* **2017**, *12*, e1–e6. [CrossRef] [PubMed]

20. Wolf, M.T.; Daly, K.A.; Reing, J.E.; Badylak, S.F. Biologic scaffold composed of skeletal muscle extracellular matrix. *Biomaterials* **2012**, *33*, 2916–2925. [CrossRef] [PubMed]

21. Perniconi, B.; Costa, A.; Aulino, P.; Teodori, L.; Adamo, S.; Coletti, D. The pro-myogenic environment provided by whole organ scale acellular scaffolds from skeletal muscle. *Biomaterials* **2011**, *32*, 7870–7882. [CrossRef] [PubMed]

22. Coppi, P.D.E.; Bellini, S.; Conconi, M.T.; Sabatti, M.; Simonato, E.; Gamba, P.G.; Nussdorfer, G.G.; Parnigotto, P.P. Myoblast-Acellular Skeletal Muscle Matrix Constructs Full-Thickness Abdominal Wall Defects. *Tissue Eng.* **2006**, *12*, 1929–1936. [CrossRef] [PubMed]

23. Merritt, E.K.; Hammers, D.W.; Tierney, M.; Suggs, L.J.; Walters, T.J.; Farrar, R.P. Functional Assessment of Skeletal Muscle Regeneration Utilizing Homologous Extracellular Matrix as Scaffolding. *Tissue Eng. Part A* **2010**, *16*, 1395–1405. [CrossRef] [PubMed]

24. Fishman, J.M.; Lowdell, M.W.; Urbani, L.; Ansari, T.; Burns, A.J.; Turmaine, M. Immunomodulatory effect of a decellularized skeletal muscle scaffold in a discordant xenotransplantation model. *Proc. Natl. Acad. Sci. USA* **2013**, *110*. [CrossRef] [PubMed]

25. Goldman, S.M.; Henderson, B.E.P.; Walters, T.J.; Corona, B.T. Co-delivery of a laminin-111 supplemented hyaluronic acid based hydrogel with minced muscle graft in the treatment of volumetric muscle loss injury. *PLoS ONE* **2018**, 1–15. [CrossRef] [PubMed]

26. Sicari, B.M.; Agrawal, V.; Siu, B.F.; Medberry, C.J.; Dearth, C.L.; Turner, N.J.; Badylak, S.F. A Murine Model of Volumetric Muscle Loss and a Regenerative Medicine Approach for Tissue Replacement. *Tissue Eng. Part A* **2012**, *18*, 1941–1948. [CrossRef] [PubMed]

27. Turner, N.J.; Badylak, J.S.; Weber, D.J.; Badylak, S.F. Biologic scaffold remodeling in a dog model of complex musculoskeletal injury. *J. Surg. Res.* **2012**, *176*, 490–502. [CrossRef] [PubMed]

28. Zhang, J.; Hu, Z.Q.; Turner, N.J.; Teng, S.F.; Cheng, W.Y.; Zhou, H.Y.; Zhang, L.; Hu, H.W.; Wang, Q.; Badylak, S.F. Perfusion-decellularized skeletal muscle as a three-dimensional scaffold with a vascular network template. *Biomaterials* **2016**, *89*, 114–126. [CrossRef] [PubMed]

29. Valentin, J.E.; Turner, N.J.; Gilbert, T.W.; Badylak, S.F. Functional skeletal muscle formation with a biologic scaffold. *Biomaterials* **2010**, *31*, 7475–7484. [CrossRef] [PubMed]

30. Greising, S.M.; Rivera, J.C.; Goldman, S.M.; Watts, A.; Aguilar, C.A.; Corona, B.T. Unwavering Pathobiology of Volumetric Muscle Loss Injury. *Sci. Rep.* **2017**, 1–14. [CrossRef] [PubMed]

31. Quarta, M.; Cromie, M.; Chacon, R.; Blonigan, J.; Garcia, V.; Akimenko, I.; Hamer, M.; Paine, P.; Stok, M.; Shrager, J.B.; et al. Bioengineered constructs combined with exercise enhance stem cell-mediated treatment of volumetric muscle loss. *Nat. Commun.* **2017**. [CrossRef] [PubMed]

32. McClure, M.J.; Cohen, D.J.; Ramey, A.N.; Bivens, C.B.; Mallu, S.; Isaacs, J.E.; Imming, E.; Huang, Y.-C.; Sunwoo, M.; Schwartz, Z.; et al. Decellularized muscle supports new muscle fibers and improves function following volumetric injury. *Tissue Eng. Part A* **2018**. [CrossRef] [PubMed]

33. Urciuolo, A.; Urbani, L.; Perin, S.; Maghsoudlou, P.; Scottoni, F.; Gjinovci, A.; Collins-Hooper, H.; Loukogeorgakis, S.; Tyraskis, A.; Torelli, S.; et al. Decellularised skeletal muscles allow functional muscle regeneration by promoting host cell migration. *Sci. Rep.* **2018**, *8*. [CrossRef] [PubMed]

34. Kasukonis, B.; Kim, J.; Brown, L.; Jones, J.; Ahmadi, S.; Washington, T.; Wolchok, J. Codelivery of Infusion Decellularized Skeletal Muscle with Minced Muscle Autografts Improved Recovery from Volumetric Muscle Loss Injury in a Rat Model. *Tissue Eng. Part A* **2016**, *22*, 1151–1163. [CrossRef] [PubMed]

35. Fisher, L.E.; Sicari, B.M.; Rubin, J.P.; Dearth, C.L.; Wolf, M.T.; Ambrosio, F.; Boninger, M.; Turner, N.J.; Weber, D.J.; Simpson, T.W.; et al. An Acellular Biologic Scaffold Promotes Skeletal Muscle Formation in Mice and Humans with Volumetric Muscle Loss an Acellular Biologic Scaffold Promotes Skeletal Muscle Formation in Mice and Humans with Volumetric Muscle Loss. *Sci. Transl. Med.* **2014**. [CrossRef]

36. Dziki, J.; Badylak, S.; Yabroudi, M.; Sicari, B.; Ambrosio, F.; Stearns, K.; Turner, N.; Wyse, A.; Boninger, M.L.; Brown, E.H.P.; et al. An acellular biologic scaffold treatment for volumetric muscle loss: Results of a 13-patient cohort study. *NPJ Regen. Med.* **2016**, *1*, 16008. [CrossRef] [PubMed]

37. Hussein, K.H.; Park, K.M.; Kang, K.S.; Woo, H.M. *Biocompatibility Evaluation of Tissue-Engineered Decellularized Scaffolds for Biomedical Application*; Elsevier B.V.: Amsterdam, The Netherlands, 2016; Volume 67, ISBN 8233244236.

38. Yu, Y.; Alkhawaji, A.; Ding, Y.; Mei, J. Decellularized scaffolds in regenerative medicine. *Oncotarget* **2016**, *7*, 58671–58683. [CrossRef] [PubMed]

39. Rana, D.; Zreiqat, H.; Benkirane-Jessel, N.; Ramakrishna, S.; Ramalingam, M. Development of decellularized scaffolds for stem cell-driven tissue engineering. *J. Tissue Eng. Regen. Med.* **2017**, *11*, 942–965. [CrossRef] [PubMed]

40. Ott, H.C.; Matthiesen, T.S.; Goh, S.K.; Black, L.D.; Kren, S.M.; Netoff, T.I.; Taylor, D.A. Perfusion-decellularized matrix: Using nature's platform to engineer a bioartificial heart. *Nat. Med.* **2008**, *14*, 213–221. [CrossRef] [PubMed]

41. Guyette, J.P.; Gilpin, S.E.; Charest, J.M.; Tapias, L.F.; Ren, X.; Ott, H.C. Perfusion decellularization of whole organs. *Nat. Protoc.* **2014**, *9*, 1451–1468. [CrossRef] [PubMed]

42. Gupta, S.K.; Mishra, N.C. Decellularization Methods for Scaffold Fabrication. *Methods Mol. Biol.* **2017**. [CrossRef]

43. Jank, B.J.; Xiong, L.; Moser, P.T.; Guyette, J.P.; Ren, X.; Cetrulo, C.L.; Leonard, D.A.; Fernandez, L.; Fagan, S.P.; Ott, H.C. Engineered composite tissue as a bioartificial limb graft. *Biomaterials* **2015**, *61*, 246–256. [CrossRef] [PubMed]

44. Gerli, M.F.M.; Guyette, J.P.; Evangelista-Leite, D.; Ghoshhajra, B.B.; Ott, H.C. Perfusion decellularization of a human limb: A novel platform for composite tissue engineering and reconstructive surgery. *PLoS ONE* **2018**, *13*, e0191497. [CrossRef] [PubMed]

45. Porzionato, A.; Sfriso, M.M.; Pontini, A.; Macchi, V.; Petrelli, L.; Pavan, P.G.; Natali, A.N.; Bassetto, F.; Vindigni, V. Decellularized Human Skeletal Muscle as Biologic Scaffold for Reconstructive Surgery. *Int. J. Mol. Sci.* **2015**, 14808–14831. [CrossRef] [PubMed]

46. Piccoli, M.; Trevisan, C.; Maghin, E.; Franzin, C.; Pozzobon, M. Mouse Skeletal Muscle Decellularization. *Methods Mol. Biol.* **2017**. [CrossRef]

47. Fuoco, C.; Petrilli, L.L.; Cannata, S.; Gargioli, C. Matrix scaffolding for stem cell guidance toward skeletal muscle tissue engineering. *J. Orthop. Surg. Res.* **2016**, *11*, 86. [CrossRef] [PubMed]

48. Saxena, A.K.; Willital, G.H.; Vacanti, J.P. Vascularized three-dimensional skeletal muscle tissue-engineering. *Biomed. Mater. Eng.* **2001**, *11*, 275–281. [PubMed]

49. Saxena, A.K.; Marler, J.; Benvenuto, M.; Willital, G.H.; Vacanti, J.P. Skeletal muscle tissue engineering using isolated myoblasts on synthetic biodegradable polymers: Preliminary studies. *Tissue Eng.* **1999**, *5*, 525–532. [CrossRef] [PubMed]

50. Boldrin, L.; Malerba, A.; Vitiello, L.; Cimetta, E.; Piccoli, M.; Messina, C.; Gamba, P.G.; Elvassore, N.; De Coppi, P. Efficient delivery of human single fiber-derived muscle precursor cells via biocompatible scaffold. *Cell Transplant.* **2008**, *17*, 576–584. [CrossRef]

51. Borselli, C.; Storrie, H.; Benesch-Lee, F.; Shvartsman, D.; Cezar, C.; Lichtman, J.W.; Vandenburgh, H.H.; Mooney, D.J. Functional muscle regeneration with combined delivery of angiogenesis and myogenesis factors. *Proc. Natl. Acad. Sci. USA* **2010**, *107*, 3287–3292. [CrossRef] [PubMed]

52. De Coppi, P.; Delo, D.; Farrugia, L.; Udompanyanan, K.; Yoo, J.J.; Nomi, M.; Atala, A.; Soker, S. Angiogenic gene-modified muscle cells for enhancement of tissue formation. *Tissue Eng.* **2005**, *11*, 1034–1044. [CrossRef] [PubMed]

53. Hill, E.; Boontheekul, T.; Mooney, D.J. Regulating activation of transplanted cells controls tissue regeneration. *Proc. Natl. Acad. Sci. USA* **2006**, *103*, 2494–2499. [CrossRef] [PubMed]

54. Liu, G.; Pareta, R.A.; Wu, R.; Shi, Y.; Zhou, X.; Liu, H.; Deng, C.; Sun, X.; Atala, A.; Opara, E.C.; et al. Skeletal myogenic differentiation of urine-derived stem cells and angiogenesis using microbeads loaded with growth factors. *Biomaterials* **2013**, *34*, 1311–1326. [CrossRef] [PubMed]

55. Page, R.L.; Malcuit, C.; Vilner, L.; Vojtic, I.; Shaw, S.; Hedblom, E.; Hu, J.; Pins, G.D.; Rolle, M.W.; Dominko, T. Restoration of Skeletal Muscle Defects with Adult Human Cells Delivered on Fibrin Microthreads. *Tissue Eng. Part A* **2011**, *17*, 2629–2640. [CrossRef] [PubMed]

56. Beier, J.P.; Stern-Straeter, J.; Foerster, V.T.; Kneser, U.; Stark, G.B.; Bach, A.D. Tissue engineering of injectable muscle: Three-dimensional myoblast-fibrin injection in the syngeneic rat animal model. *Plast. Reconstr. Surg.* **2006**, *118*, 1113–1121. [CrossRef] [PubMed]

57. Rossi, C.A.; Flaibani, M.; Blaauw, B.; Pozzobon, M.; Figallo, E.; Reggiani, C.; Vitiello, L.; Elvassore, N.; De Coppi, P. In vivo tissue engineering of functional skeletal muscle by freshly isolated satellite cells embedded in a photopolymerizable hydrogel. *FASEB J.* **2011**, *25*, 2296–2304. [CrossRef] [PubMed]

58. Lesman, A.; Koffler, J.; Atlas, R.; Blinder, Y.J.; Kam, Z.; Levenberg, S. Engineering vessel-like networks within multicellular fibrin-based constructs. *Biomaterials* **2011**, *32*, 7856–7869. [CrossRef] [PubMed]

59. Cronin, E.M.; Thurmond, F.A.; Bassel-Duby, R.; Williams, R.S.; Wright, W.E.; Nelson, K.D.; Garner, H.R. Protein-coated poly(L-lactic acid) fibers provide a substrate for differentiation of human skeletal muscle cells. *J. Biomed. Mater. Res.* **2004**, *69A*, 373–381. [CrossRef] [PubMed]

60. Choi, J.S.; Lee, S.J.; Christ, G.J.; Atala, A.; Yoo, J.J. The influence of electrospun aligned poly(ε-caprolactone)/ collagen nanofiber meshes on the formation of self-aligned skeletal muscle myotubes. *Biomaterials* **2008**, *29*, 2899–2906. [CrossRef] [PubMed]

61. Fuoco, C.; Rizzi, R.; Biondo, A.; Longa, E.; Mascaro, A.; Shapira-Schweitzer, K.; Kossovar, O.; Benedetti, S.; Salvatori, M.L.; Santoleri, S.; et al. In vivo generation of a mature and functional artificial skeletal muscle. *EMBO Mol. Med.* **2015**, *7*, 411–422. [CrossRef] [PubMed]

Review

Beyond the Matrix: The Many Non-ECM Ligands for Integrins

Bryce LaFoya [1], Jordan A. Munroe [2], Alison Miyamoto [3], Michael A. Detweiler [2], Jacob J. Crow [1], Tana Gazdik [2] and Allan R. Albig [1,2,*]

[1] Biomolecular Sciences PhD Program, Boise State University, Boise, ID 83725, USA; brycelafoya@u.boisestate.edu (B.L.); jacobcrow@u.boisestate.edu (J.J.C.)
[2] Department of Biological Sciences, Boise State University, Boise, ID 83725, USA; jordanmunroe@u.boisestate.edu (J.A.M.); mikedetweiler@u.boisestate.edu (M.A.D.); tanagazdik@u.boisestate.edu (T.G.)
[3] Department of Biological Science, California State University, Fullerton, CA 92831, USA; almiyamoto@fullerton.edu
* Correspondence: AllanAlbig@boisestate.edu; Tel.: +1-208-426-1541

Received: 31 December 2017; Accepted: 30 January 2018; Published: 2 February 2018

Abstract: The traditional view of integrins portrays these highly conserved cell surface receptors as mediators of cellular attachment to the extracellular matrix (ECM), and to a lesser degree, as coordinators of leukocyte adhesion to the endothelium. These canonical activities are indispensable; however, there is also a wide variety of integrin functions mediated by non-ECM ligands that transcend the traditional roles of integrins. Some of these unorthodox roles involve cell-cell interactions and are engaged to support immune functions such as leukocyte transmigration, recognition of opsonization factors, and stimulation of neutrophil extracellular traps. Other cell-cell interactions mediated by integrins include hematopoietic stem cell and tumor cell homing to target tissues. Integrins also serve as cell-surface receptors for various growth factors, hormones, and small molecules. Interestingly, integrins have also been exploited by a wide variety of organisms including viruses and bacteria to support infectious activities such as cellular adhesion and/or cellular internalization. Additionally, the disruption of integrin function through the use of soluble integrin ligands is a common strategy adopted by several parasites in order to inhibit blood clotting during hematophagy, or by venomous snakes to kill prey. In this review, we strive to go beyond the matrix and summarize non-ECM ligands that interact with integrins in order to highlight these non-traditional functions of integrins.

Keywords: integrin; extracellular matrix; counterreceptor; disintegrin; immune system; stem cell; pathogen; virus; bacteria; venom; growth factor; hormone

---

1. Introduction

The adhesion of cells to extracellular matrices is a fundamental requirement for multicellular organisms, and animals employ many mechanisms to fulfill this demand. Amongst these mechanisms of adhesion, integrins are perhaps the most ubiquitous. Integrins are heterodimeric transmembrane proteins, made up of non-covalently paired α and β subunits, which serve as adhesion and signaling hubs at the cell surface. In mammals, there are 18 α-integrin subunits and eight β-integrin subunits that can combine to form as many as 24 unique heterodimeric receptor complexes [1]. Typically, ligand binding is carried out through integrin receptor recognition of small peptide sequences. Target sequences for integrins can be as simple as the RGD or LDV tripeptides, or more complex as in the case of the GFOGER peptide [1]. Many classical extracellular matrix (ECM) proteins contain these short integrin recognition motifs. RGD sequences are found in both vitronectin and fibronectin, an LDV motif is present in fibronectin, GFOGER is found within collagen, and the target sequence within laminin has not yet been defined [1]. These sequences are not globally recognized by all integrins; therefore, integrin heterodimers are often grouped by the target sequences they specialize in recognizing (Figure 1). Once bound to its ligand, an integrin not only provides adhesion, but also initiates signaling mechanisms which allow cells to respond to the mechanical and chemical properties of the cellular microenvironment. The primary signaling mediators working downstream of integrins include focal adhesion kinase (FAK), Src-family protein tyrosine kinases, and integrin-linked kinase (ILK) [2]. Upon adhesion, cytoskeletal proteins are recruited to the cytoplasmic tails of integrins, forming a linkage between the ECM and cytoskeleton [2].

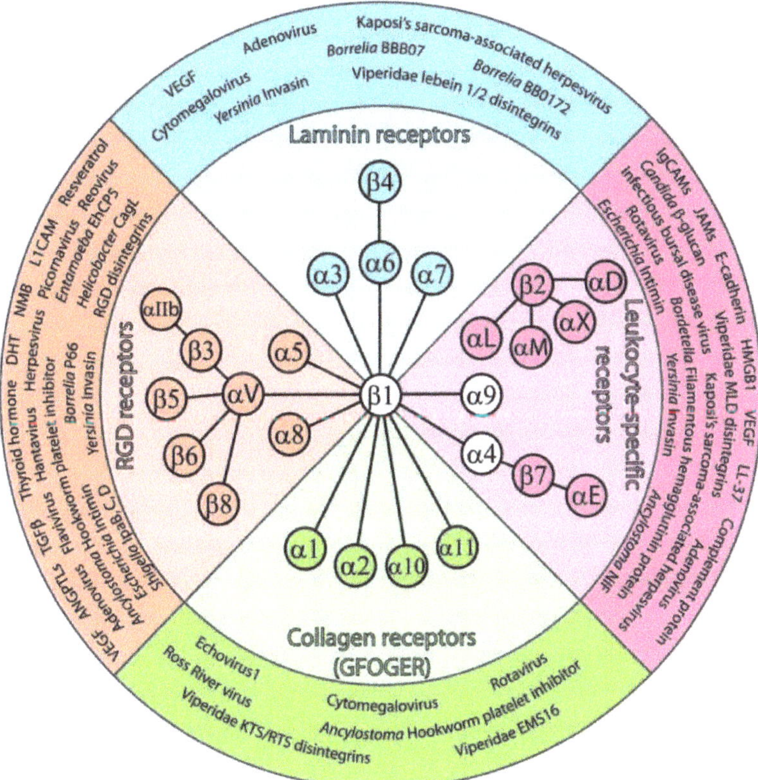

Figure 1. Integrin heterodimers and their ligands. Integrins are heterodimeric cell surface receptors that bind extracellular matrix (ECM) molecules. In addition to this role, integrins also bind many non-ECM ligands. Integrin subunits connected by a ray represent heterodimeric α/β binding partners. The inner ring depicts integrin heterodimers grouped into families based upon their classical binding profile. These families include RGD receptors, collagen (GFOGER) receptors, laminin receptors, or leukocyte-specific receptors. Within the outer ring, the non-ECM ligands of these families are listed. Non-ECM ligands include growth factors, hormones, venomous compounds, disintegrins, bacterial proteins, fungal polysaccharides, viruses, polyphenols, and counterreceptors.

As a family of proteins, integrins and many of their downstream signaling intermediates have a long evolutionary history. Beginning at the root of the metazoan lineage, sponges have been shown to express α- and β-integrin subunits [3,4] that bind to peptides in a fashion similar to mammalian integrins [5]. Interestingly, integrin-encoding genes have been found in the single-celled eukaryotic relatives of metazoans, thus the origin of integrins predates the emergence of metazoans [6]. Moreover, components of integrin signaling machinery such as FAK, Src, and ILK, and integrin-interacting cytoskeletal proteins such as α-actinin, talin, vinculin, and paxillin, have pre-metazoan origins [6]. This suggests that integrins and their aforementioned signaling machinery may have played an important role in the evolution of multicellularity.

Beyond their traditional role as mediators of ECM attachment, a vast literature has developed that describes interactions between integrins and ligands that are not located in the classical extracellular matrix. For example, integrins have been shown to interact with various proteins on the surfaces of eukaryotic, prokaryotic, and fungal cells, as well as a range of viruses. Within eukaryotes specifically, integrin-mediated cell-cell adhesion has been shown to coordinate a range of interactions and processes including leukocyte extravasation, stem cell homing, tumor cell migration, erythrocyte development, and interactions in the immune system. For infectious prokaryotes, integrins are exploited as cell surface adhesion receptors that mediate colonization and/or the bypassing of epithelial or endothelial barriers. Beyond mediating cellular interactions, integrins can also serve as cell surface receptors for hormones, growth factors, and polyphenols. Finally, integrins are also common targets for a class of small molecules called disintegrins, which are components of various snake venoms, and are also employed by hematophagous parasites. Collectively, the range of non-ECM molecules that interact with integrins is vast, making integrins indispensable mediators of cell biology at large. The goal of this review is to highlight some of the best understood non-ECM ligands of integrins and discuss the diverse biological roles for these interactions.

2. Integrin-Mediated Cell-Cell Interactions

The first integrins discovered were isolated based on their ability to bind to fibronectin, which had itself just recently been identified (reviewed in [7]). However, in the early days of integrin research, several groups studying cell-cell adhesion in the immune system were also on the forefront of integrin identification (reviewed in [8,9]). In fact, integrins that mediate cell-cell adhesion in the immune system were among the first integrins to be characterized [8]. As more integrins were discovered, it became apparent that the majority of integrins established cell-ECM connections rather than cell-cell connections. Nonetheless, it is important to understand that integrins are important mediators of cell-cell adhesion. The term counterreceptor has often been used to describe membrane-bound, non-matrix integrin ligands which facilitate cell-cell contact and will be used to differentiate them from the other non-matrix ligands in this review. While there are many types of counterreceptors, the best-known examples include the immunoglobulin superfamily cell adhesion molecules (IgCAMs) and junctional adhesion molecules (JAMs). Collectively, interactions between integrins and these counterreceptors mediate a range of immune cell functions including leukocyte extravasation from the blood stream, immunological surveillance in the gut, and hematopoietic stem cell homing and mobilization. Additionally, non-ECM ligands enhance the interaction between pathogens and phagocytic immune cells, acting as phagocytic primers and inducers of neutrophil extracellular traps. Beyond the immune system, non-ECM-based integrin interactions are important during the transmigration and metastasis of tumor cells, and during erythrocyte development. Integrins and the non-ECM ligands that mediate these cell-cell interactions are listed in Table 1.

Table 1. Selected non-ECM ligands which mediate cell-cell interactions.

Integrin Dimers	Common Name		Non-ECM Ligand	Function of Interaction [Key Refs]
α4β1	VLA-4	Very late antigen-4	MAdCAM1 VCAM1 AM-B	Leukocyte adhesion [10–12] Leukocyte adhesion [10–12] Erythrocyte differentiation [13–15] Cancer cell metastasis [16] Leukocyte transmigration [11]
α4β7	LPAM	Lymphocyte Peyer's patch adhesion molecule	MAdCAM1	T-lymphocyte homing [17] HSC homing to bone marrow [18]
α5β1	Fibronectin receptor	Fibronectin receptor	Glycoprotein NMB	Cancer cell growth, metastasis [19]
αEβ7			E-cadherin	Cytotoxic T cell targeting of tumor cells [20]
αLβ2	LFA-1	Lymphocyte function associated antigen-1	ICAM1, 2, 3 JAM-A	Leukocyte adhesion [10,11] Leukocyte transmigration [11]
αMβ2	Mac-1/CR3	Macrophage antigen-1/Complement receptor-3	ICAM1 β-glucan Complement C3 LL-37 JAM-C HMGB1	Leukocyte adhesion [10,11] NETosis [21,22] Phagocytosis [23,24] Bacterial opsonization [25–29] Leukocyte transmigration [11] NETosis [30]
αVβ3	Vitronectin receptor	Vitronectin receptor	L1CAM	Cancer cell metastasis [31,32]
αXβ2	CR4/CD11c/CD18	Complement receptor-4	Complement C3	Phagocytosis [23,24]

mucosal addressin cell adhesion molecule (MAdCAM), vascular cell adhesion molecule (VCAM), intercellular adhesion molecule (ICAM), junctional adhesion molecule (JAM), glycoprotein non-metastatic gene B (Glycoprotein NMB), high mobility group box protein (HMGB), L1 cell adhesion molecule (L1CAM), hematopoietic stem cell (HSC), neutrophil extracellular trap (NET).

2.1. Integrin-Counterreceptor Interactions in Leukocyte Extravasation

Integrin-counterreceptor interactions play multiple roles during extravasation, a process in which white blood cells are recruited from the blood stream to a site of inflammation (depicted in Figure 2). Extravasation begins when glycoproteins on the leukocyte cell surface, such as P-selectin glycoprotein ligand-1 (PSGL-1), bind endothelial selectins, which allows the leukocyte to slow down as it rolls along the vessel wall [33]. Next, local chemokines stimulate leukocyte integrins to adopt a high-affinity state, causing them to bind specific immunoglobulin superfamily cell adhesion molecules (IgCAMs) on endothelial cells [34]. There are many integrin-IgCAM pairs involved in this process: $\alpha L\beta 2$ (LFA-1) integrin binds to ICAM1, 2, or 3, $\alpha M\beta 2$ (Mac-1) integrin binds to ICAM1, and $\alpha 4\beta 1$ (VLA-4) integrin binds to VCAM1 or MAdCAM1 (reviewed in [10]). Additionally, leukocyte integrins can bind a family of proteins known as junctional adhesion molecules (JAMs) found on endothelial cells. Similar to integrin-IgCAM interactions, integrins display specificity for particular JAM proteins: JAM-A binds to $\alpha L\beta 2$, JAM-B binds to $\alpha 4\beta 1$, and JAM-C binds to $\alpha M\beta 2$ [11]. All of these integrin-counterreceptor binding events serve to tightly adhere the leukocyte to the endothelium, enabling the white blood cell to cross the endothelial layer (a process known as transendothelial migration) in order to reach the inflamed tissue.

2.2. Non-ECM Integrin Ligands as Primers for Phagocytosis

One of the best-characterized examples of non-ECM integrin-binding ligands in the immune system involves the interplay of integrins with the complement system. Complement proteins aid in the immune system's clearance of pathogens by attaching to invaders and tagging them for destruction. Integrin $\beta 2$ is essential for complement recognition by the complement receptors $\alpha M\beta 2$ (Mac-1, CR3) and $\alpha X\beta 2$ (CR4) integrins [23]. $\alpha M\beta 2$ and $\alpha X\beta 2$ ligation with the iC3b component of complement induces the phagocytosis of complement opsonized pathogens and particles by phagocytic immune cells (depicted in Figure 2) [24]. Despite high homology between both integrins, they bind the iC3b fragment of complement via distinctive receptor sites, which may afford a greater diversity of leukocytes in opsonized target recognition modes [35]. This leads to the intriguing possibility of cooperativity between two integrins binding the same complement molecule [35].

Phagocytosis mediated by integrins is not strictly complement dependent. Human cathelicidin peptide LL-37, an antimicrobial peptide that binds to the prokaryotic cell wall, inserts itself into the membrane, and enhances phagocytosis by interacting with $\alpha M\beta 2$ integrin present on neutrophils and macrophages [26,27]. As an important part of innate defenses, LL-37 is expressed in various mammalian tissues and released upon contact with bacterial invaders [29]. For example, upon infection by *Helicobacter pylori*, gastric epithelial cells express and secrete LL-37, thus tagging the bacterial invaders for destruction by phagocytic immune cells (depicted in Figure 2) [28]. Interestingly, LL-37 binds $\alpha M\beta 2$ with a comparable strength to complement C3d, a ligand with one of the strongest known affinities for $\alpha M\beta 2$ [25].

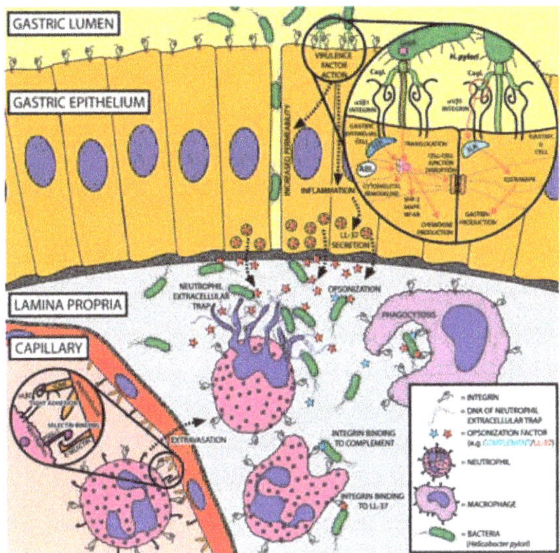

Figure 2. Integrins act as "double agents" during *Helicobacter pylori* infection in the stomach, serving to potentiate bacterial pathogenicity while also aiding in the immune response. *H. pylori* bacteria in the gastric lumen bind integrins on gastric epithelial cells in order to inject the virulence factor CagA. As shown in the magnified view of this process, docking of α5β1 integrin is achieved through integrin affinity for the RGD motif of the CagL protein component of the type IV secretion system (T4SS). Integrin α5β1-mediated stabilization of the T4SS facilitates the translocation of CagA while activating intracellular kinases. Once in the cytosol, CagA is phosphorylated by Src family kinases (SFKs) and Abelson (ABL) kinases, which potentiates its virulence. Phospho-CagA activates Src homology 2 domain-containing phosphatase-2 (SHP-2) and mitogen-activated protein kinase (MAPK) signaling, triggering cytoskeletal remodeling. CagA disrupts cell-cell junctions, activates the nuclear factor-κB (NF-κB) pathway, and stimulates cytokine production. Alternatively, CagL docking with αVβ5 integrin on gastric G cells activates integrin-linked kinase (ILK), which stimulates epidermal growth factor receptor (EGFR) and MAPK activation, inducing gastrin production. These mechanisms increase the permeability of the gastric epithelium, which aids *H. pylori* dissemination into the underlying lamina propria. This stimulates an inflammatory response causing the release of the antimicrobial peptide LL-37 from gastric epithelial cells and recruitment of immune cells from the blood stream. As shown in the magnified view of the recruitment process, leukocytes first stick to inflamed endothelium through selectin binding, which facilitates integrin-mediated tight adhesion. This leads to leukocyte extravasation into the lamina propria, where neutrophils and macrophages phagocytize bacteria. Phagocytosis is mediated through integrin recognition of the opsonization factors LL-37 and complement. Neutrophil extracellular traps (NETs) are stimulated through integrin interaction with pathogens.

2.3. Non-ECM Integrin Ligands as Triggers for NETosis

Another example of non-ECM integrin ligation at work in the innate immune system is the neutrophil extracellular trap (NET). In the process of NETosis, chromatin is ejected from neutrophils upon interaction with pathogens, thus entangling foreign invaders in a web of DNA and histones (depicted in Figure 2) [36]. This process is mediated through pathogen recognition by neutrophil integrins. For example, the pathogen-associated molecular pattern, β-glucan, found on *Candida albicans* is recognized by αMβ2 at a unique lectin-like domain, and its binding stimulates NETosis [21]. Once stimulated, anti-microbial peptides are integrated into NETs. These include defensins and the αMβ2 ligand LL-37 [22]. NETosis is not exclusively used to trap foreign invaders, as it is also involved in wound healing and sterile inflammation [37]. For instance, during cell necrosis the chromatin protein high-mobility group box 1 (HMGB1 aka amphoterin) is released extracellularly and recruits neutrophils by binding integrin β2 [38]. HMGB1 has been demonstrated to be an inducer of NETosis when presented on platelets during thrombosis [30]. This evidence suggests that HMGB1 serves as a molecule that is capable of signaling to white blood cells the presence of tissue damage through leukocyte integrins. Although αMβ2 plays a starring role in the literature connecting NETosis and integrins, other integrins may be involved. Bacterial invasin proteins from *Yersinia pseudotuberculosis* interact with neutrophil integrin β1, stimulating phagocytosis while also causing the release of NETs [39]. In addition to trapping cells within a tangle of DNA and histones, fibronectin has been identified in NETs, which ligates to αVβ3 and α5β1 integrins found on neutrophils and cancer cells, thus potentially enhancing cancer cell-leukocyte interaction [40]. Collectively, this information demonstrates rich evidence for the importance of integrin engagement during NETosis.

2.4. Non-ECM Integrin Ligands in Immune Surveillance

The intestinal immune system must display tolerance towards commensal microbiota and food antigens while still maintaining immunogenicity against pathogens. In the gut mucosa, resident antigen-presenting cells (APCs) have the job of sampling foreign antigens. APCs then transport these antigens to specialized gut-associated lymphoid tissue where they can interact with naïve T cells to promote their maturation. Additionally, the APCs imprint intestinal homing properties on the T cells through inducing the expression of α4β7 integrin and C-C chemokine receptor type 9, a receptor for the gut-associated C-C motif chemokine 25 (CCL25) [41]. Mature T effector cells then reenter the circulation and can be recruited back to the gastrointestinal tract during times of inflammation through gut endothelial expression of CCL25 and the α4β7 counterreceptor, MAdCAM1 [17]. There is also a role for α4β1-VCAM1 interactions in the gut; this pair mediates the binding of effector T cells to inflamed gut epithelium [12].

Integrins in the gut also bind to cadherins to modulate the immune response. For instance, cadherin 26 binding to integrins αE and α4 can lead to a T cell immunosuppression phenotype [42]. Moreover, this study found that a similar phenotype is provoked through the treatment of T cells with a soluble form of cadherin 26. So, unlike the integrin-mediated IgCAM interactions in the gut, cadherin binding appears to moderate the immune response. It has been suggested that this interaction may therefore be involved in resolving inflammation [42]. Cadherin-integrin interactions in the lungs have been shown to mediate the engagement of cytotoxic T lymphocytes (CTLs) with cancer cells. Here, CTLs employ αEβ7 integrin to engage E-cadherin on cancer cell surfaces in order to facilitate accurate targeting and release of cytotoxic granules [20].

2.5. Integrin-Mediated Stem Cell Homing

The homing and mobilization of hematopoietic stem cells (HSCs) to and from the bone marrow is also regulated by integrins (reviewed in [13]). After the treatment of hematologic malignancies with large doses of radiation and/or chemotherapy, transplantation of HSCs is commonly performed. Success of the HSC engraftment within the bone marrow is dependent upon proper HSC homing to a bone marrow niche where they can regenerate hematopoietic lineages. New evidence is revealing that integrin engagement of counterreceptors plays a critical role in this homing process. For example, Murakami et al. determined that a subpopulation of murine HSCs expressing integrin β7 have enhanced homing capabilities to bone marrow niches compared to their counterparts which do not express β7 [18]. Mechanistic insight was provided when it was revealed that α4β7 integrins on HSCs were binding MAdCAM1 present on endothelial cells within the bone marrow niche, and β7 knockout HSCs showed decreased CXCR4 homing receptor expression [18].

In addition to α4β7-MAdCAM1 interactions, α4β1-VCAM1 binding also mediates HSC retention in bone marrow niches. The importance of α4 integrin to this interaction is supported by the phenotypes of multiple α4 knockout mouse models that show elevated numbers of HSCs in the bloodstream relative to wild-type littermates (reviewed in [13]). The treatment of mice with Bortezomib, which inhibits the expression of VCAM1, also increases HSC mobilization [14]. Together, these results support a role for integrins in holding HSCs within the bone marrow and have raised great clinical interest in using Bortezomib-induced mobilization for the harvesting of HSCs from the peripheral blood of healthy individuals for use in transplantation.

Another integrin-targeting small molecule antagonist is the drug Firategrast; it inhibits $\alpha4\beta1$ and $\alpha4\beta7$ activity and can also be used to mobilize HSCs from the bone marrow to the circulation, making HSC harvesting much less invasive. There is particular interest in using Firategrast for in utero hematopoietic cell transplants (IUHCT). These transplants can be especially useful for diseases where a more mature immune system can thwart the therapeutic benefit of the transplanted cells (reviewed in [43]). Firategrast was tested in a mouse model of IUHCT and found to increase long-term engraftment of HSCs; there was 15% engraftment at six months with Firategrast, compared to 3% with vehicle alone [44]. The current thinking is that the mobilization of endogenous HSCs through the disruption of integrin adhesion by Firategrast makes room in the bone marrow for transplanted HSCs to compete with endogenous cells for niche binding. Although still in preclinical studies, Firategrast is well-tolerated by adults but has not yet been tested in children (reviewed in [43]).

Some interesting new data on mesenchymal stem cell (MSC) homing demonstrates that the role of integrin αL (CD11a) in MSC transmigration across vessel endothelium differs from that of leukocyte extravasation [45]. Using zebrafish whose endothelium was labeled with green fluorescent protein as a model system, mammalian leukocytes, cardiac stem cells, and MSCs were transplanted to determine their transmigration properties. As expected, leukocyte extravasation proceeded in an αL-dependent fashion, as αL-blocking antibodies inhibited leukocyte extravasation. However, the blocking antibodies did not inhibit the transmigration of cardiac stem cells or MSCs, indicating that these cells were traversing the endothelium in an αL-independent fashion that was found to rely on the remodeling of the endothelium for vascular expulsion of these types of stem cells. Based on this evidence and additional phenotypic differences in the transmigration of cardiac stem cells and MSCs, the authors have named this alternate process angiopellosis [45].

2.6. Integrin-Counterreceptor Interactions in Tumor Cell Migration

Integrin binding to IgCAMs also mediates tumor cell binding to endothelial cells, influencing metastasis. Many of these interactions involve L1CAM (reviewed in [46]); this protein contains an RGD motif that binds to $\alpha V\beta3$ integrin [47,48]. The expression of L1CAM by various types of cancer cells is utilized to engage $\alpha V\beta3$ on endothelial cells. It has been demonstrated that L1CAM expression in glioma tumor cells serves to promote the motility of both cancer cells [31] and endothelial cells [32], thus having important implications for both metastasis and angiogenesis, respectively. Other non-ECM integrin ligands have been implicated in tumor cell migration. For example, when expressed on cancer cells, VCAM1 has been identified as a driver of metastasis due to its ability to bind $\alpha4\beta1$ integrin expressed on lymph node endothelium (reviewed in [16]). Additionally, metastatic breast cancer cells express the transmembrane glycoprotein NMB that contains an RGD motif and can bind to $\alpha5\beta1$ integrin on adjacent tumor cells. This interaction activates Src and FAK signaling within the tumor and leads to increased growth and metastasis [19].

2.7. Integrin-Counterreceptor Interactions in Erythrocyte Development

Since integrins can bind to both ECM and other cells, it is perhaps not surprising that there are modulators that can push integrins towards either a cell-ECM or a cell-cell interaction. During erythrocyte differentiation in the bone marrow, immature erythroblasts cluster around a central macrophage, forming what is known as erythroblastic islands. This cell-cell interaction is mediated by $\alpha4\beta1$ on erythroblasts and VCAM1 on macrophages and is an essential part of the maturation process [15]. The same $\alpha4\beta1$ integrin can bind to fibronectin in the ECM, and the modulation of $\alpha4\beta1$ binding to either macrophages or ECM is in part due to the activity of erythrocyte tetraspanin proteins CD81, CD82, and CD151 [49]. These tetraspanins are co-expressed with α4β1 on human proerythroblasts, where they increase the affinity and/or clustering of integrins to favor α4β1-VCAM1 interactions over α4β1-fibronectin interactions [49].

3. Non-ECM Integrin Ligands of Viruses

Although there is debate as to when viruses first emerged in the evolution of life, it is likely that viruses (in one form or another) have co-existed with cells for nearly as long as cells have existed [50]. It is also safe to assume that viruses have a long history of exploiting cell surface receptors to facilitate their infectious cycles. As already discussed, integrins are first present in evolutionary history at the root of the metazoan lineage, and perhaps predate metazoans [3,4,6]. Therefore, it is not surprising that many species of viruses have exploited (and continue to exploit) integrins as a major point of cell attachment, entry, and eventually infection of target cells. A common theme among many of the viruses discussed here is the display of RGD motifs on viral capsids to bind to integrins that are commonly found on either epithelial or endothelial surfaces [51,52]. Presumably, the RGD motif serving as a minimal integrin-binding unit accommodates the viral quest for genomic minimization. Additionally, RGD-recognizing integrins are common in tissues targeted by invading viruses. However, RGD-based mechanisms are not the only means of integrin engagement by viruses, as some viruses employ other integrin targeting motifs. The virus-integrin interactions we have chosen to highlight are in no way an exhaustive list (for a more comprehensive review of the subject, refer to [53,54]). Integrins that participate in viral interactions that we discuss are listed in Table 2 and depicted in Figure 3.

Table 2. Selected integrin binding by viruses.

Integrin	Virus Name [Key Refs]
α1β1	Ross River virus [55]
α2β1	Echovirus 1 [56,57] Cytomegalovirus [58] Rotavirus [59,60]
α3β1	Kaposi's sarcoma-associated herpesvirus [61] Adenovirus [62]
α4β1	Infectious bursal disease virus [63] Rotatvirus [60]
α5β1	Foot-and-mouth disease virus [64] Epstein-Barr virus [65] Adenovirus [66]
α6β1	Cytomegalovirus [58]
α9β1	Kaposi's sarcoma-associated herpesvirus [67]
αMβ2	Adenovirus [68]
αVβ1	Echovirus 22 [69,70] Adenovirus [71]
αVβ3	Echovirus 9 [72] Coxsackievirus A9 [73] Foot-and-mouth disease virus [74] Japanese encephalitis virus [75] Kaposi's sarcoma-associated herpesvirus [76] Cytomegalovirus [58] Andes virus [77] Adenovirus [78] Rotavirus [79,80] Sin Nombre virus [81]
αVβ5	Kaposi's sarcoma-associated herpesvirus [82] Adenovirus [78] Epstein-Barr virus [83]
αVβ6	Coxsackievirus A9 [73] Foot-and-mouth disease virus [84,85] Epstein-Barr virus [83] Herpes simplex virus [86]
αVβ8	Epstein-Barr virus [83] Herpes simplex virus [86]
αXβ2	Rotavirus [60]
αIIbβ3	Sin Nombre virus [81]

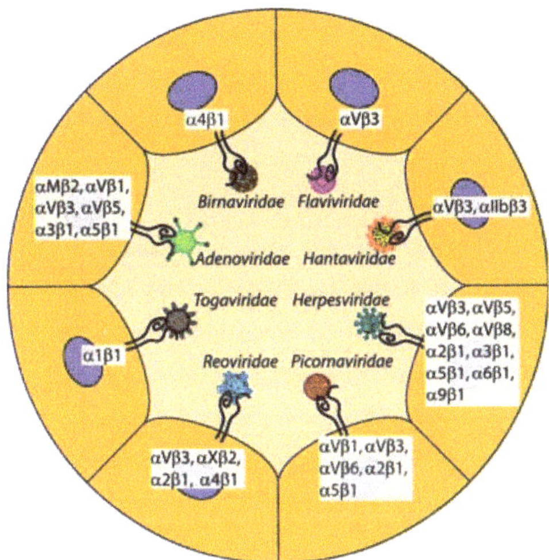

Figure 3. Viruses hijack integrins for adhesion and infectivity. Virus families use specific integrins in order to adhere to target cells for the purposes of internalization and infectivity. Members of the family *Adenoviridae* are non-enveloped viruses with icosahedral capsids that have penton base structures which facilitate RGD-dependent docking with αVβ1, αVβ3, αVβ5, and α5β1 integrins as well as the RGD-independent engagement of α3β1. Adenoviruses also target αMβ2 integrin through an undetermined mechanism. *Birnaviridae* contains members who employ a fibronectin-mimicking IDA peptide to bind α4β1 integrin. Members of the *Flaviviridae* family have an RGD-containing E-protein which binds αVβ3 integrin. Viruses in the family *Hantaviridae* target the plexin-semaphorin-integrin (PSI) domain of αVβ3 and αIIbβ3 integrins. *Herpesviridae* has members that employ a few different mechanisms of integrin engagement for the purposes of viral entry. The envelope protein BMRF-2 contains an RGD sequence that docks α5β1 integrin. The envelope proteins gH and gL dock with αVβ5, αVβ6, and αVβ8. Another envelope protein, known as gB, contains both an RGD motif and disintegrin-like domain, which affords viral targeting of αVβ3, αVβ5, α2β1, α3β1, α6β1, and α9β1 integrins. Members of the *Picornaviridae* family use capsid proteins to target integrins. The targeting of α2β1 integrin proceeds in an RGD-independent manner, while αVβ1, αVβ3, αVβ6, and α5β1 integrins are bound in an RGD-dependent fashion. *Reoviridae* contains members which employ a DGE sequence within a VP4 capsid protein to engage α2β1. Additionally, the reovirus VP7 capsid protein has a GPR tripeptide which recognizes αXβ2, an LDV tripeptide that ligates α4β1, and a novel NEWLCNPDM amino acid sequence that targets αVβ3. *Togaviridae* has members which have a collagen-mimicking spike protein that docks α1β1 integrin.

3.1. Non-ECM Integrin Ligands of Picornaviridae

Viruses of the *Picornaviridae* family cause a variety of human diseases including aseptic (viral) meningitis, paralysis, hepatitis, and poliomyelitis [87], and there are currently no approved treatments to minimize picornavirus infection. Picornaviruses are non-enveloped viruses with icosahedral capsids, with each face of the 20-sided capsid consisting of three capsid proteins (VP1-3) to form a protomer with 60 subunits. The VP4 protein is contained within the capsid and is thought to help package the single-stranded RNA genome (reviewed in [88]). Several picornaviruses have been shown to exploit integrins as cell surface receptors to facilitate cell invasion. In most cases, the non-ECM ligands that enable picornavirus binding to integrins are located on the VP1-3 capsid proteins.

Members of *Picornaviridae* include the enteric cytopathic human orphan (echo) viruses. Echovirus 1 (EV1) utilizes the α2I functional domain of α2β1 integrin as a docking receptor on the surface of a target cell [56,57]. Although the precise peptide sequence of EV1 that binds to α2β1 integrin has not been discovered, it is known that EV1 binds to the α2I domain of α2β1 integrin

10 times more tightly than collagen [56,89,90]. The structure of the EV1 capsid provides a pentameric arrangement of binding sites for α2β1, which induces the clustering of α2β1 integrins, and is thought to promote the entry of the virus [56]. During infection, EV1 along with α2β1 integrin are taken into the host cell via caveolar endocytosis and moved to a caveosome, where it is thought that the virus ejects its genome into the cytosol [91–94]. While EV1 utilizes a non-RGD signal to bind its integrin receptor, echovirus 9 (EV9) docks to integrin αVβ3 via an RGD domain located on the EV9 VP1 capsid protein [72]. RGD motifs are also thought to be critical in host cell attachment for echovirus 22 (EV22) to αVβ1 integrins [69,70]. Another member of the *Picornaviridae* family, coxsackievirus A9, utilizes the coxsackievirus and adenovirus receptor (CAR) together with an RGD motif situated in the C-terminal of its VP1 to bind αVβ3 and αVβ6 integrins and gain cellular entry [73,95]. Yet another member of the *Picornaviridae* family, foot-and-mouth disease virus (FMDV), is a major scourge of animal husbandry. The VP1 protein of FMDV has an exposed flexible loop, termed the GH loop, which contains an RGD motif and mediates binding to host α5β1, αVβ3, and αVβ6 integrins [64,74,84,85].

3.2. Non-ECM Integrin Ligands of Flaviviridae

Flaviviridae is a family of single-stranded, positive sense RNA viruses that are commonly transmitted to human hosts from arthropods such as ticks and mosquitos [96]. Japanese encephalitis virus (JEV), a mosquito-borne member of the genus *Flavivirus*, is a leading cause of viral encephalitis in humans and animals [97,98]. JEV has an envelope protein, called E protein, which contains an RGD motif [99]. Data suggest that JEV utilizes this RGD motif to bind αVβ3 integrin to aid in cellular infection. Specifically, JEV infectivity is reduced by shRNA knockdown of integrin αV and β3 subunits, pretreatment of cells with soluble RGD peptides, or αV/β3 blocking antibodies. Conversely, the expression of β3 integrin promotes infectivity in otherwise resistant cell lines [75]. Finally, the utilization of integrin receptors appears to be a common infection strategy for the *Flaviviridae* family, since other members such as West Nile virus [100–102], Murray Valley encephalitis virus [103], dengue virus [104], and yellow fever virus [105] have all been connected with integrin-mediated infection or have at least been demonstrated to possess RGD-containing E proteins.

3.3. Non-ECM Integrin Ligands of Herpesviridae

Members of the *Herpesviridae* family of viruses also use integrins for cellular attachment and entry. Epstein-Barr virus (EBV) utilizes α5β1 integrin for infectivity in tongue and nasopharyngeal epithelium by binding host cell integrins with its RGD-containing envelope glycoprotein, BMRF-2 [65]. In addition, the engagement of EBV envelope glycoproteins gH and gL with αVβ5, αVβ6, and αVβ8 integrins induces a conformation in these glycoproteins which facilitates fusion with the target cell membrane [83]. More mechanistic insight is provided by herpes simplex virus (HSV), which also uses gH and gL envelope glycoproteins to dock αVβ6 and αVβ8 integrins, and this engagement routes HSV to acidic endosomes, thus promoting viral entry [86]. Another herpes virus, Kaposi's sarcoma-associated herpesvirus (KSHV), uses αVβ3 [76], αVβ5 [82], and α3β1 [61] integrins as entry receptors. The expression of the envelope protein, known as glycoprotein B (gB), which is highly conserved across *Herpesviridae* and contains an RGD sequence near its N-terminus, affords KSHV its integrin-binding capacity. However, RGD-mediated binding is not the only mechanism of KSHV-integrin interaction. KSHV glycoprotein B also contains a disintegrin-like domain (DLD) which is capable of binding integrin β1 in an RGD-independent fashion [58]. Walker et al. discovered that α9β1 is the integrin target of the glycoprotein B DLD and plays a critical role in KSHV infection [67]. This mechanism is not unusual among *Herpesviridae* family members, as human cytomegalovirus (HMCV) also uses gB to bind αVβ3, α2β1, and α6β1 through its disintegrin-like domain [58,106].

3.4. Non-ECM Integrin Ligands of Togaviridae

Ross River fever is a mosquito-borne disease caused by the Ross River virus (RRV), a member of the *Togaviridae* family. This disease induces arthritis by the viral infection of macrophages within synovial joints [107]. It is believed that the RRV spike protein, known as E2, contains two conserved domains which fold in a manner that mimics collagen IV [55]. This allows for the infection of mammalian cells by docking the collagen receptor, $\alpha 1\beta 1$ integrin, in matrix-binding adherent cell types [55].

3.5. Non-ECM Integrin Ligands of Adenoviridae

Human adenoviruses, known for causing respiratory, gastrointestinal, and ocular infections, are non-enveloped viruses with icosahedral capsids. At each capsid vertex, a penton base supports a fiber protein [108]. Many adenoviruses require two receptors for efficient infection of cells. The coxsackievirus and adenovirus receptor (CAR) is required for the initial adhesion of adenoviral particles to target cells, while subsequent integrin engagement is required for the internalization of the viral particle [109]. It is the penton base structure that affords adenoviruses a diverse array of integrin targets. RGD peptide sequences are located atop each monomer of the penton base, forming an RGD ring around the fiber protein [78]. The RGD peptides mediate docking to $\alpha V\beta 1$, $\alpha V\beta 3$, $\alpha V\beta 5$, and $\alpha 5\beta 1$ integrins for the purpose of internalization [66,71,110–112]. Mechanistically, it is thought that the pentameric structure of the base stimulates integrin clustering and downstream integrin signaling, which further facilitates viral internalization [113–115]. Adenoviruses also interact with the laminin binding integrin, $\alpha 3\beta 1$, via its penton base, but in a manner that is RGD-independent [62]. Additionally, $\alpha M\beta 2$ integrin on myeloid cells can be targeted by adenoviruses, but this interaction is dictated through an as yet undetermined sequence within the penton base [68].

3.6. Non-ECM Integrin Ligands of Hantaviridae

As a member of the *Hantaviridae* family, the rodent-targeting Andes virus can spread to humans through the inhalation of aerosolized excreted virus, targeting human endothelial cells and resulting in several fatal diseases such as hantavirus hemorrhagic fever with renal syndrome and hantavirus pulmonary syndrome [116]. The infection of $\alpha V\beta 3$ integrin-expressing endothelial cells occurs through the viral targeting of the PSI domain within the $\beta 3$ subunit [77]. Interestingly, a human polymorphism that has a leucine to proline substitution at position 33 of the integrin $\beta 3$ PSI domain was experimentally shown to abolish Andes virus infectivity [77]. Sin Nombre virus also utilizes $\beta 3$-containing integrins, such as $\alpha IIb\beta 3$ and $\alpha V\beta 3$, for viral attachment [81]. Using atomic force microscopy (AFM) to study membrane dynamics upon Sin Nombre virus interaction, more mechanistic insight was provided for integrin-dependent hantavirus infectivity. Bondu et al. used AFM data to propose a model in which viral docking to the $\beta 3$ PSI domain of $\alpha IIb\beta 3$, when the integrin is in a low affinity state, enhances integrin *cis* interaction with an RGD-containing G-protein coupled receptor known as P2Y$_2$R [117]. This *cis* interaction is thought to induce a switchblade-like conformational change within the integrin that ultimately leads to the endocytosis of the viral bound integrin [117]. Other pathogenic hantaviruses also bind and cause the dysregulation of $\beta 3$ integrins, resulting in the blockade of endothelial cell migration [118], and the enhancement of vascular endothelial growth factor (VEGF)-mediated vascular permeability [119].

3.7. Non-ECM Integrin Ligands of Birnaviridae

Infectious bursal disease virus (IBDV) is an immunosuppressive avian pathogen in the *Birnaviridae* family that attacks the bursa of Fabricius (the site of hematopoiesis in birds) of young chickens, having a major negative impact on the poultry industry. The IBDV capsid is built by 260 trimers of the VP2 polypeptide arranged in an icosahedral lattice [120]. VP2 is the only component of the virus capsid, and contains a conserved, fibronectin-mimicking IDA peptide sequence that binds to $\alpha 4\beta 1$ integrins present on target cell membranes [63]. IBDV binding to $\alpha 4\beta 1$ integrin triggers c-Src tyrosine phosphorylation and actin rearrangement, which creates membrane protrusions that internalize the virus [121].

3.8. Non-ECM Integrin Ligands of Reoviridae

The family *Reoviridae* includes the gastrointestinal pathogens, known as the rotaviruses, which are the leading etiological factor of diarrheal disease in young children worldwide [122]. The outer layer of the rotavirus capsid consists of 60 VP4 spike proteins protruding from a VP7 protein shell [123]. It is these outermost structures which mediate host cell binding and infectivity. The VP4 spike protein subunit, VP5, contains a DGE tripeptide sequence that serves to recognize α2β1 integrin on target cells [59,60]. Rotavirus VP7 contains an αXβ2-recognizing GPR tripeptide, as well as an α4β1 ligating LDV motif, embedded in a disintegrin-like domain of the protein [60]. Additionally, rotaviruses can target αVβ3 integrin for the purpose of cellular entry; however, this binding does not occur within the RGD pocket [80]. Rather, it is a novel αVβ3-targeting NEWLCNPDM amino acid sequence within the VP7 protein that is thought to mediate rotavirus-αVβ3 interaction [79]. It has been proposed that reoviruses employ a sequential binding mechanism to multiple receptors for the purpose of internalization. Initial binding to the counterreceptor JAM-A is thought to position the virus for subsequent binding to β1 containing integrins that facilitate internalization [124].

4. Non-ECM Integrin Ligands in Venoms

Selectively blocking integrins is a major therapeutic goal when combatting a number of pathologies, and a wide variety of approaches have been initiated. One rich source for anti-integrin compounds are venoms from various snake species [125,126], and the study of venom-derived integrin antagonists remains an active area of research. A venom is defined as a secreted toxin, produced by various types of animals, which is injected into another animal for the purpose of defense or predation. The Viperidae family of snakes (collectively known as the vipers) produce a venom which causes local necrosis and blood coagulation within their prey. The discovery of small integrin-targeting peptides found in the venom of these snakes initiated the study of disintegrins. These small molecular weight (40–100 amino acids in length), non-enzymatic proteins were originally characterized by their platelet-disrupting properties through antagonistic targeting of αIIbβ3 integrin [127]. Since the identification of the first disintegrins, the field has grown with the discovery of many more examples. As discussed below, major families of venom-derived disintegrins include the RGD, MLD, PIII, and KTS/RTS disintegrins. On the other hand, C-type lectin-like proteins are an example of non-disintegrin toxins, which also disrupt integrin activity. Integrin-targeting venomous compounds are summarized in Table 3.

Table 3. Integrin binding by small molecules, hormones, growth factors, and venoms.

Integrin	Non-ECM Ligand	Function [Key Refs]
α1β1	KTS/RTS disintegrins	Block cell adhesion [128,129]
α2β1	EMS16 CLP	Block adhesion to collagen [130,131]
α3β1	VEGF	Cell adhesion [132]
	Disintegrin Lebein 1/2	Block cell adhesion [133]
α4β1	MLD disintegrins	Block cell adhesion [128]
α4β7	MLD disintegrins	Block cell adhesion [128]
α5β1	ANGPTL2	Cancer cell migration/proliferation [134]
		Macrophage pro-inflammatory response [135]
α6β1	Disintegrin Lebein 1/2	Block cell adhesion [133]
α7β1	Disintegrin Lebein 1/2	Block cell adhesion [133]
α9β1	VEGF-A, -C, -D	Endothelial adhesion & lymphangiogenesis [136]
	MLD disintegrins	Block cell adhesion [128]
αVβ3	Resveratrol	Anti-angiogenesis [137–139]
	Thyroid hormones (T3/T4)	Cell proliferation/angiogenesis [140–142]
	DHT	Cancer cell proliferation [143,144]
	ANGPTL3	Podocyte motility [145]
	ANGPTL4	Enhanced endothelial junctions [146]
	VEGF	Endothelial cell adhesion [132]
αVβ5	ANGPTL4	Reduce proteinuria [147]
αVβ6	Pro-TGFβ	TGFβ activation [148,149]

Abbreviations: lysine-threonine-serine (KTS), arginine-threonine-serine (RTS), C-type lectin-like protein (CLP), vascular endothelial growth factor (VEGF), methionine-leucine-aspartic acid (MLD), angiopoietin-like protein (ANGPTL), dihydrotestosterone (DHT), transforming growth factor β (TGFβ)

The RGD family of disintegrins is the largest family, although RGD amino acid sequences are not strictly required to be members in this family. Instead, disintegrins containing RGD or similar amino acid motifs, such as KGD, MGD, VGD, and WGD, are all capable of targeting RGD-binding integrins, serving to disrupt their physiological functions. Moreover, not all RGD disintegrins target RGD-binding integrins exclusively. For example, lebein1 and lebein2 are two RGD-containing disintegrins found in the venom of *Macrovipera lebetina*, which have the unusual property of targeting the laminin-binding integrins α3β1, α6β1, and α7β1 in an RGD-independent fashion [133]. They are thought to mimic the integrin-binding motif of laminin, thus allowing these molecules to disrupt cellular attachment to the laminin-rich basement membrane [133].

Other disintegrin families include the MLD-, PIII-, and KTS/RTS-containing disintegrins. Whereas the RGD family of disintegrins possesses an RGD tripeptide (or similar motif) within the integrin-binding loop of the protein, the MLD motif is found at this same position in MLD-containing disintegrins [128]. These MLD disintegrins appear in heterodimeric complexes and are highly dependent on adjacent sequences to target the α4β1, α4β7, and α9β1 leukocyte specific receptor family of integrins [128]. PIII class disintegrins are large multi-domain toxins (60–100 kDa) which use an ECD integrin-targeting tripeptide and contain a metalloprotease domain which is a close homolog to the ADAM (a disintegrin and metalloprotease) family of metalloproteases [150]. The disintegrin known as alternagin uses an ECD tripeptide motif to target α2β1 integrin and disrupt matrix binding [151]. Once bound, alternagin uses its protease domain to cleave β1, causing integrin shedding and further disruption of collagen-induced platelet aggregation [152]. Finally, the KTS/RTS group of disintegrins, found in Viperidae venom, are monomeric proteins which bind the collagen receptor α1β1 integrin [129]. This high level of specificity is not matched by RGD and MLD disintegrins, as KTS/RTS disintegrins only target α1β1 integrin [128].

Another class of toxin found in Viperidae venom is the C-type lectin-like proteins (CLPs). These proteins do not exhibit the sugar-binding capabilities of C-lectin proteins, but instead target collagen-binding integrins [153]. The viper species *Echis carinatus multisquamatus* produces EMS16, a potent and selective inhibitor of α2β1 integrin [130]. X-ray crystallography reveals that EMS16 spatially blocks collagen-integrin ligation through docking with the α2I domain of α2β1 integrin and stabilizing a low matrix affinity integrin conformation [131]. Several studies have shown that many viper-derived CLPs target endothelium and block angiogenesis [130,154,155], while applying CLPs to cancer cells can inhibit cell-collagen binding [153] and metastasis [156]. Integrins that interact with CLPs are summarized in Table 3.

5. Bacterial Use of Non-ECM Integrin Ligands

For many bacterial cells, successful adhesion to host cell surfaces is a prerequisite for successful colonization and/or infection. Many bacteria take advantage of the binding capabilities of integrins on cell membranes for infectious purposes. Some bacteria utilize specific integrin dimers for cellular binding, while others exploit extracellular fibrous proteins that naturally bind to integrins for the purpose of translocating virulence factors. For this review, we will highlight three of the most commonly studied interactions between bacteria and integrins. There are other notable examples of bacterial cells using integrins as host cell receptors that we will not discuss: the intimin protein of *Escherichia coli* that binds to α4β1 and α5β1 integrins [157], the IpaB, C, and D proteins of *Shigella flexneri* that bind to α5β1 integrin [158], and the filamentous hemagglutinin protein of *Bordetella pertussis* that binds to αMβ2 integrin [159]. Integrins that participate in bacterial interactions are listed in Table 4.

5.1. Non-ECM Integrin Ligands of Borrelia burgdorferi

The spirochete *Borrelia burgdorferi* is the causative agent of Lyme disease, a devastating disease of the nervous system. The natural reservoir for *B. burgdorferi* includes mice, birds, and lizards [160]. These spirochetes are transmitted to humans via tick vectors of the *Ixodes* genus [160]. Once injected into the blood stream, *B. burgdorferi* spirochetes adhere to the microvasculature, transmigrate through the endothelium, and disseminate into various tissues [161]. Characterizing the proteins that enable this pathological mechanism illustrates several interesting examples of how microbes take advantage of host integrins.

A variety of screening techniques have identified at least 19 *B. burgdorferi* proteins that mediate or enhance adhesion to target cells [162]. The majority of these proteins mediate indirect adhesion to mammalian cells via interactions with various ECM molecules. Three proteins however, P66, BBB07, and BB0172, have been shown to interact with integrins on platelets and a variety of cells such as endothelial cells. Prior to the discovery of the P66 protein, it had been known for some time that *B. burgdorferi* cells could adhere to β3 chain-containing integrins [163,164]. The P66 protein was later identified by phage display and shown to bind αVβ3 and αIIbβ3 integrins [165]. P66 displays no typical integrin-binding sites [165], although the adhesion of P66 to integrins can be blocked by soluble RGD peptides, suggesting that P66 may bind into the RGD pocket of β3 integrins [166]. Moreover, a minimal seven-amino acid sequence (QENDKDT) from P66 was found to bind integrins, and the deletion of the aspartic acid residues from this peptide eliminated P66 integrin binding [167]. Despite the integrin-binding activity of P66, the deletion of P66 does not appear to affect *B. burgdorferi* adhesion to microvasculature, a key step proceeding tissue invasion [168]. Instead, the P66 protein (presumably via its integrin-binding activity) appears to be essential for the endothelial transmigration and dissemination of *B. burgdorferi* spirochetes into host tissues [167,168]. Although P66 deletion did not affect microvascular adhesion, *B. burgdorferi* binding to various cells can be blocked by soluble RGD peptides [163], suggesting the presence of other integrin-binding proteins. In support of this, two additional integrin-binding outer membrane surface proteins, BBB07 and BB0172, have been detected on *B. burgdorferi* [169,170]. Although both BBB07 and BB0172 have been shown to interact with α3β1 integrins, only BBB07 contains an RGD motif [170]. Currently, there is little known about the function of α3β1 integrins in endothelial biology, although it has been proposed that α3β1 binding to Laminin 511 in the basal lamina may be linked to endothelial barrier function [171], which could provide a link to the transendothelial migration of *B. burgdorferi* during infection.

5.2. Non-ECM Integrin Ligands of Helicobacter pylori

Helicobacter pylori infects roughly half of the world's human population and shares responsibility for gastric complications including stomach ulcers and gastric adenocarcinoma through its infection of gastric epithelial cells [172–174]. *H. pylori* utilizes a type IV protein secretion system (T4SS) involving the cytotoxin-associated gene L (CagL) adhesion tip protein to infect target cells with the virulence factor, cytotoxin-associated gene A (CagA) [175,176]. The efficiency of CagA injection is enhanced by an RGD domain present on the CagL protein [177]. CagL interacts primarily with α5β1 integrin; however, αVβ3, αVβ5, and αVβ6 have also been implicated [178–181]. Interestingly, while the CagL RGD domain is necessary for CagA injection, additional CagL sequences have been identified that enhance integrin binding. For example, an RGD helper sequence, FEANE, is located in close proximity to the RGD domain of CagL and reinforces integrin engagement [177]. Additional domains on CagL that enhance RGD binding include a TSPSA sequence [182], an LXXL sequence that is directly adjacent to the RGD domain [181], and a TASLI sequence located opposite the RGD domain in the CagL integrin-binding domain [182]. CagL-α5β1 interaction leads to the activation of the kinases Src and FAK [179], followed by subsequent tyrosine phosphorylation of the CagA EPIYA amino acid motifs by Src and ABL kinases [183]. These phosphorylation events potentiate CagA pathogenicity (reviewed in [184]). Phospho-CagA interacts with Shp-2 while initiating mitogen-activated protein kinase (MAPK) signaling, and inducing cytoskeletal rearrangements which serve to cause an elongation of epithelial cells and enhance their mobility. CagA also disrupts cell-cell junctions while triggering an inflammatory response, including nuclear factor-κB (NF-κB) activation and chemokine production. Additionally, in a negative feedback loop phospho-CagA downregulates Src activity, ensuring that a reservoir of nonphospho-CagA remain in the cell, which is necessary for a prolonged infection. As mentioned previously, CagL is capable of interacting with other integrins. Interestingly, a novel mechanism of CagL-αVβ5-induced production of gastrin has been uncovered. It was found that CagL ligation to αVβ5 on gastric epithelial cells activates ILK, which in turn activates the epidermal growth factor receptor (EGFR) and subsequently MAPK pathways, serving to induce gastrin expression [178]. This mechanism may explain *H. pylori* induced hypergastrinemia, which is a major risk factor for gastric adenocarcinoma. The integrin-dependent mechanisms of *H. pylori* infection discussed here are depicted in Figure 2.

5.3. Non-ECM Integrin Ligands of Yersinia

The Gram-negative bacteria *Yersinia enterocolitica* and *Yersinia pseudotuberculosis* commonly cause foodborne illnesses. These *Yersinia* species express two adhesion proteins that facilitate cellular attachment and invasion of target cells in the small intestine. The *Yersinia* adhesion A (YadA) protein indirectly binds to integrins via interaction with various molecules of the ECM, but is dispensable for cellular invasion [185,186]. However, the *Yersinia* invasin protein directly binds to a variety of β1 subunit-containing integrins (α3, α4, α5, α6, αV) and is crucial for cellular adhesion and invasion [187,188]. *Yersinia* species invading through the small intestine target the apical membrane of Peyer's patch M-cells, which express integrin β1 [189,190]. Invasins lack the typical RGD domain used to bind integrins, although RGD peptides prevent invasin binding to β1 integrins [191]. This suggests that invasin proteins interact with the RGD binding domain of β1-containing integrin heterodimers. In support of this, the structural analysis of the invasin protein, and comparison to fibronectin, reveals similar structures with key conserved integrin-binding residues, suggesting the convergent evolution of invasins to match fibronectin [192].

Table 4. Integrin binding by bacteria and parasitic organisms.

Integrin	Species	Binding Protein [Key Refs]
α2β1	*Ancylostoma caninum*	Hookworm platelet inhibitor (HPI) [193,194]
αIIbβ3	*Ancylostoma caninum* *Macrobdella decora* *Tabanus yao* *Ornithodoros moubata* *Ixodes pacificus* *Dermacentor variabilis*	Hookworm platelet inhibitor (HPI) [193,194] Decorsin [195] Vasotab TY [196] Tablysin-15 [197] Disagregin [198] YY-39 [199] Variabilin [200]
α3β1	*Borrelia burgdorfori* *Yersinia*	BBB07, BB0172 [170] Invasin [187,188]
α4β1	*Escherichia coli* *Yersinia*	Intimin [157] Invasin [187,188]
α5β1	*Helicobacter pylori* *Escherichia coli* *Shigella flexneri* *Entamoeba histolytica* *Yersinia*	CagL [177,179] Intimin [157] Ipa B, C, D [158] EhCP5 [201] Invasin [187]
α6β1	*Yersinia*	Invasin [187,188]
αMβ2	*Bordetella pertussis* *Ancylostoma caninum*	Filamentous hemagglutinin protein [159] Neutrophil inhibitor factor (NIF) [202]
αVβ1	*Yersinia*	Invasin [187]
αVβ3	*Borrelia burgdorfori* *Helicobacter pylori* *Entamoeba histolytica*	P66 [165] CagL [177] EhCP5 [203,204]
αVβ5	*Helicobacter pylori*	CagL [178,180]
αVβ6	*Helicobacter pylori*	CagL [181]

6. Protists and Multicellular Parasites That Use Non-ECM Integrin Ligands

A broad array of examples of non-ECM ligands for integrins are employed by many parasitic organisms. Here we discuss just a few examples, including non-ECM ligands produced by the amoebozoa *Entamoeba histolytica* and a range of hematophagic (blood-sucking) organisms. These examples illustrate the importance of non-ECM ligands to parasitic infections. Although compared to bacteria and viruses, there is far less literature on the subject of non-ECM ligands as components of pathogenicity in protozoan and multicellular parasites. Non-ECM integrin ligands derived from parasitic organisms are summarized in Table 4.

6.1. Non-ECM Integrin Ligands of Entamoeba histolytica

Entamoeba histolytica (Eh) causes amoebic dysentery and liver abscess [205] and is responsible for ~100,000 deaths/year [206]. Eh invasion into host tissues involves multiple integrin-mediated steps. The best-characterized of these integrin-mediated steps involves the Eh cysteine protease 5 (EhCP5) binding to αVβ3 integrins [203,204]. The binding of EhCP5 to αVβ3 integrins on colonic epithelial cells via an RGD domain triggers NF-κB-mediated inflammation [203] and mucin exocytosis [204]. The EhCP5 protein has also been shown to interact with α5β1 integrins to mediate local inflammation, which is crucial to Eh invasion into host tissues [201]. Additional involvement of integrins in Eh invasion has been linked to β2 integrin activation and release of reactive oxygen species [207,208] as well as an integrin β1-like receptor present on Eh trophocytes that mediates adhesion to host fibronectin [209].

6.2. Non-ECM Integrin Ligands of Hookworms

The hookworm platelet inhibitor (HPI) protein illustrates another fascinating example of non-ECM integrin ligands. Hookworms are blood-feeding intestinal parasites and a leading cause of iron deficiency in humans. HPI was isolated from the hookworm *Ancylostoma caninum* based on its ability to inhibit the function of integrins αIIbβ3 and α2β1 [193,194]. HPI appears to block platelet aggregation and blood clothing, thus enabling continued feeding. Interestingly, sequence and structural analysis has failed to identify any integrin-binding domains in the HPI protein [210]. In addition to the HPI protein, *Ancylostoma caninum* also expresses the neutrophil inhibitor factor (NIF) that interacts with αMβ2 integrins present on neutrophils [202,211]. NIF disrupts αMβ2 interaction with ICAM1 [202], which is necessary for stable neutrophil adhesion to the endothelium and transendothelial migration, thus suppressing local inflammation. Collectively, the combined actions of HPI and NIF help ensure that hookworms are able to feed from their host for a prolonged period of time.

6.3. Non-ECM Integrin Ligands of Blood-Sucking Parasites

In addition to *Entamobea histolytica* and *Ancylostoma caninum*, several other examples of integrin inhibition by hematophagic (blood-sucking) animals have been described in the literature (reviewed in [212]). Many of these strategies involve non-ECM integrin ligands that interfere with various integrin-mediated steps that are essential for blood coagulation. The majority of these non-matrix ligands block platelet αIIbβ3 integrin interactions with fibrin, von Willebrand factor, and vitronectin, which are collectively essential for blood coagulation. Many of these integrin disrupting molecules are found in the saliva of hematophagic organisms and not only inhibit platelet aggregation, but also disrupt neutrophil function and angiogenesis [212]. Examples of these integrin disrupting proteins include the decorsin protein from the leech *Macrobdella decora* [195], the vasotab TY and tablysin-15 proteins from the horsefly *Tabanus yao* [196,197], and the disagregin (*Ornithodoros moubata*), YY-39 (*Ixodes pacificus* and *Ixodes scapularis*), and variabilin (*Dermacentor variabilis*) proteins from ticks [198–200]. Many of these proteins contain RGD or similar integrin-binding amino acid motifs (KGD, VGD, MLD, KTS, RTS, WGD, or RED) which bind to and interfere with αIIbβ3 integrin function on platelets. Additional RGD or RGD-like integrin antagonists have been identified in silico from other blood-sucking arthropods such as mosquitos and sand flies [212], but have yet to be explored.

7. Hormones, Small Molecules, and Growth Factors That Mimic Integrin Ligands

To this point, we have focused on the non-ECM integrin ligands utilized by various organisms to mediate adhesion to target cell membranes. However, as it turns out, a wide variety of small molecules (including hormones and growth factors) can also interact with integrins, thus broadening the role for integrins in non-ECM interactions. As described in the examples below, integrins binding to small molecules serve a number of cellular functions ranging from cell surface receptor-signaling roles, as in the case of thyroid hormone, dihydrotestosterone, angiopoietin-like proteins (ANGPTLs), and VEGF, to the activation of growth factors, as in the case of TGFβ. Integrins that interact with hormones, small molecules, or growth factors are summarized in Table 3 and depicted in Figure 4.

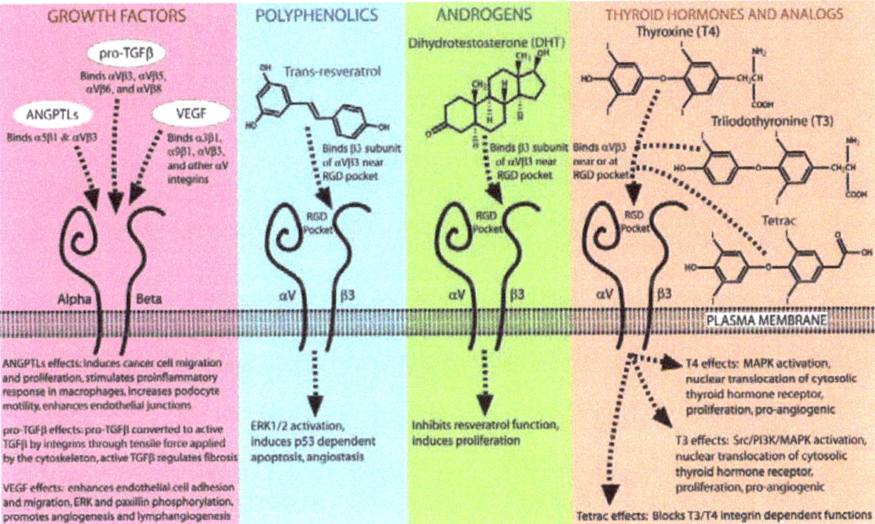

Figure 4. Integrins serve as cell surface receptors for growth factors, hormones, and small molecules. Various growth factors use integrins as cell surface receptors. Angiopoietin-like proteins (ANGPTLs) bind α5β1 and αVβ3 integrins to facilitate a host of cellular effects. Pro-TGFβ is activated by αVβ3, αVβ5, αVβ6, and αVβ8 through the integrin-dependent dissociation of an RGD-containing latency-associated peptide (LAP), thus converting it to its active form. Activated TGFβ acts as a master regulator of fibrosis, among other roles. Vascular endothelial growth factor (VEGF) ligates α3β1, α9β1, αVβ3, and other αV-containing integrins, resulting in cellular effects that promote angiogenesis and lymphangiogenesis. The polyphenol trans-resveratrol, which is derived from grapevines, binds the β3 subunit of αVβ3 integrin near the RGD recognition pocket. This binding event induces extracellular signal-regulated kinase (ERK) activation and p53-dependent apoptosis, while promoting angiostasis. Like trans-resveratrol, the active form of testosterone (DHT) also binds the β3 subunit of αVβ3 integrin near the RGD pocket. DHT-αVβ3 interaction inhibits trans-resveratrol-induced effects and stimulates cellular proliferation. The thyroid hormones, T3 and T4, utilize αVβ3 integrin as a cell surface receptor to activate a range of signaling molecules which induce angiogenesis. When binding to αVβ3 integrin, the thyroid hormone analog tetrac blocks T3/T4 integrin-induced effects.

7.1. Small Molecules and Hormones That Bind Integrins (Resveratrol, Thyroid Hormone, DHT)

Trans-resveratrol is a stilbenoid produced in plants such as grapevines that is well-known for its anti-inflammatory activity [213], anti-angiogenic function [214], and anticancer properties [215–217].

Resveratrol binds the extracellular portion of the β3 monomer of αVβ3 integrin near the RGD pocket [137]. This binding inhibits αVβ3 integrin-dependent endothelial cell adhesion to vitronectin-coated plates, while also exhibiting angiostatic function and inhibiting tumor growth [139]. Resveratrol binding to αVβ3 integrin induces extracellular signal-regulated kinase (ERK1/2) activation, which leads to p53-induced apoptosis in various cancer cell lines [137,138]. This evidence implicates resveratrol binding αVβ3 integrin as being at least in part responsible for resveratrol's ability to mitigate angiogenesis and tumorigenesis.

Integrin αVβ3 bears a receptor site for the thyroid hormones T3 and T4 as well as thyroid hormone analogs (reviewed in [140]). Perhaps the first evidence of this interaction was uncovered when Hoffman et al. [218] used an αVβ3 inhibitor (SB-273005) to block T4-induced bone resorption in rats. The binding of T3 and T4 to αVβ3 integrin induces cell proliferation and angiogenesis through MAPK activation, and this effect is negated by a T4 derivative tetraiodothyroacetic acid (tetrac), RGD peptide, and αVβ3 integrin-blocking antibodies, suggesting that the thyroid hormone receptor site is at or near the RGD binding pocket [141–143]. Through radioligand binding experiments, it was shown that purified αVβ3 integrin binds T4 preferentially over T3, and binds T4 with high affinity, having a dissociation constant (K_d) of 333 pM and an EC_{50} of 371 pM [142]. Lin et al. proposed a model for the thyroid hormone receptor activity of αVβ3 integrin that describes two distinct thyroid hormone binding sites on αVβ3 [219]. The site known as "site 1" appears to bind T3 but not T4, while another site called "site 2" binds both T3 and T4 [219]. T3 binding at site 1 leads to Src and phosphatidylinositol 3-kinase (PI3K) activation, which induces nuclear translocation of thyroid hormone receptor (TR) α1, and these effects can be disrupted through the addition of RGD peptide [219]. Meanwhile, T3/T4 binding at site 2 induces ERK activation, which causes nuclear translocation of TRβ1, and only T4-induced effects at this site are disrupted by RGD peptides [219]. This suggests that αVβ3-dependent thyroid hormone signaling acts as a complex, hierarchical system capable of mediating distinct site-specific activities. Since some of these activities are disrupted through RGD binding, and this leads to the possibility that cells embedded in an RGD-rich matrix may respond differentially to thyroid hormone compared to those embedded in an RGD-deficient matrix. Perhaps this is a mechanism by which a ubiquitous receptor, such as αVβ3, can provide tissue-specific responses to thyroid hormone.

In addition to thyroid hormones, αVβ3 integrin also interacts with the biologically active form of testosterone, dihydrotestosterone (DHT). Whether or not this interaction is involved in the normal physiological roles of DHT is unknown; however, DHT binding to αVβ3 has been implicated in cancer cell growth. For example, DHT binding to αVβ3 stimulates MDA-MB-231 breast cancer cell proliferation [143]. Additionally, DHT binding to αVβ3 integrin inhibits resveratrol-induced, p53-dependent apoptosis effects in MDA-MB-231 cells [144], thus highlighting the complexity of hormone signaling through αVβ3 integrin. Through these examples, it is clear that αVβ3 integrin has diverse receptor activity which affords hormones additional non-canonical signaling capacity.

7.2. Growth Factors That Bind Integrins (ANGPTLs, TGFβ, VEGF)

Many growth factors are capable of binding integrins. An interesting example is the angiopoietin-like proteins (ANGPTLs), also known as angiopoietin-related proteins (ARPs), which consist of a family of proteins that display structural similarity to the growth factor angiopoietin, although they do not bind classical angiopoietin receptors [220]. Instead, ANGPTLs have been demonstrated to bind various integrins through a C-terminal fibronectin-like domain containing a conserved RGD sequence [221]. In human prostate cancer (LNCaP) cells, ANGPTL2 binds α5β1 integrin, inducing migration and proliferation, and this effect can be negated by use of integrin-blocking antibodies [134]. Furthermore, ANGPTL2 binding α5β1 integrin on macrophages mediates pro-inflammatory responses in mice, and ANGPTL2 knockout mice have muted immune responses, leaving them more susceptible to infections [135]. In the kidney, glomerular podocyte motility is enhanced through cytoskeletal rearrangement induced by ANGPTL3 binding podocyte αVβ3 integrin [145]. The deletion of ANGPTL3 can reduce proteinuria in mouse models of nephropathy, and ANGPTL3 activation of integrin β3 has been identified in patients with nephrotic syndrome [222]. The ANGPTL family also affects vascular integrity. In response to decreased albumin levels during peak proteinuria, podocytes and extrarenal tissues secrete ANGPTL4 into the blood, which binds glomerular endothelial αVβ5 integrin and serves to reduce proteinuria [147]. This effect may be explained by another study where surface plasmon resonance and proximity ligation assays were used to discover that ANGPTL4 also binds another endothelial integrin, αVβ3, which serves to recruit Src kinase and enhance endothelial junction stability, thereby reducing vascular permeability [146]. Taken collectively, these studies suggest that ANGPTL3 binding podocyte integrins enhances proteinuria, whereas ANGPTL4 binding glomerular endothelial integrins decreases proteinuria. The ANGPTLs are a good example of a protein family that mimics a classical extracellular matrix protein in order to bind integrins and implement their cellular effects.

Integrins also play a critical role in the activation of TGFβ (reviewed in [148]). An inactive form of TGFβ (pro-TGFβ) is secreted from cells with an RGD containing latency-associated peptide (LAP) non-covalently bound to TGFβ, which must be removed before TGFβ is biologically active. While the RGD binding αVβ6 integrin plays a key role in separating LAP from TGFβ, other αV-containing integrins, including αVβ3, αVβ5, and αVβ8, have been implicated in this process. Mutation of the LAP integrin-binding site in mice yields normal levels of pro-TGFβ, but results in a lethal phenotype which appears identical to TGFβ deletion [223]. LAP separation is mediated by a tensile force generated by a cell's cytoskeleton that is transmitted via αVβ6 integrin in order to reshape and activate the pro-TGFβ [149]. The dependence of pro-TGFβ on αVβ6 for activation, and the fact that TGFβ is a well-known master regulator of fibrosis [224], has led to the suggestion that the inhibition of αVβ6 integrin binding may represent a clinical strategy to treat diseases characterized by fibrosis, such as scleroderma [225]. This idea is supported by observations showing that αVβ6 knockout mice [226] or treatment with αVβ6 blocking antibodies [227,228] substantially decrease fibrosis in mouse models of lung fibrosis.

The vascular endothelial growth factors (VEGFs) comprise a group of cytokines which are important mediators of angiogenesis and lymphangiogenesis. VEGF signaling functions through VEGF binding to a group of receptor tyrosine kinases, known as VEGF receptors (VEGFRs). Since this pathway is an inducer of angiogenesis, it has been the target of many anticancer therapies with the hope of inhibiting tumor vascularization. One therapeutic strategy involves inhibiting VEGF-VEGR binding through the targeting of VEGFRs with monoclonal antibodies [229]. However, this approach has not proven as effective as drug developers and clinicians envisioned [229,230]. One reason for this failure may be that VEGFRs are not the only membrane-bound receptor of VEGFs, as these growth factors are also known to bind integrins. Some VEGF isoforms are integrated into the extracellular matrix, where they bind α3β1, αVβ3, and other αV integrins to promote endothelial cell adhesion [132]. Interestingly, the solubility of VEGF ligands greatly effects the integrin response. Vlahakis et al. found that when α9β1 integrin binds immobilized VEGF-A, it induces the recruitment of VEGFR2 into macromolecular structures at the cell membrane [136]. This serves to permit endothelial cell adherence and migration on VEGF-A functionalized Petri dishes, and stimulates the phosphorylation of the downstream effectors paxillin and ERK [136]. In contrast, when soluble VEGF binds α9β1 integrin, paxillin is phosphorylated, but neither the phosphorylation of ERK nor the formation of VEGFR2 macromolecular complexes are induced [136]. Moreover, VEGF-A is not the only VEGF member to have these functions. VEGF-C and VEGF-D also bind α9β1 integrin, stimulating the phosphorylation of paxillin and ERK, while contributing to lymphangiogenesis [231]. Taken together, these findings suggest a VEGF-induced synergy between VEGFR and integrins. Therefore, it may be beneficial to co-target integrins when employing an anti-VEGF therapeutic strategy during cancer treatment.

8. Conclusions

Throughout this review, we have sought to venture beyond the matrix and highlight biological examples of integrin ligands that do not fit the classical model of ECM-mediated integrin function.

Given the strong conservation of integrins across much of the biological world, it is no surprise that there exists an extremely diverse array of these non-ECM integrin ligands. Consequently, interactions between integrins and non-ECM ligands are actively being exploited for a number of applications in the biotechnology realm. RGD peptides are being used to target liposomes and small molecules to specific tissues for various purposes, including the improvement of chemotherapeutic delivery to cancer cells [232–234]. Similarly, RGD peptides are also being used to target viral particles to various tissues. For instance, the new field of "chemical virology" seeks to load viral capsids with chemotherapeutics that, in some instances, utilize RGD functionalization to deliver these nanoparticles to specific tissues [235]. In a related example, a plant virus known as the cowpea mosaic virus, which does not normally target mammalian cells, was functionalized with RGD peptides to successfully target cancer cell lines [236]. Demonstrating another example of applied integrin biotechnology, various artificial "extracellular matrices" are now being created and designed with incorporated RGD peptides to enable cell seeding and growth [237]. Two exciting examples include the development of graphene that has been functionalized with RGD peptides, which is being used to detect nitric oxide release from living cells [238], and DNA origami tubes that have been tagged with RGD peptides and shown to bind neural stem cells and promote their differentiation [239]. These instances and many others provide fascinating examples of how the unique binding properties of integrins continue to be uncovered and utilized.

Acknowledgments: This work was supported by funding from the National Institute of General Medical Sciences to Allan R. Albig (2R15GM102852-02) and it was also supported by Institutional Development Awards (IDeA) from the National Institute of General Medical Sciences of the National Institutes of Health under Grants #P20GM103408 and P20GM109095.

Author Contributions: Bryce LaFoya, Jordan A. Munroe, Alison Miyamoto, Michael A. Detweiler, Jacob J. Crow, Tana Gazdik, and Allan R. Albig contributed to the literature research of the topic, writing individual sections, and proofreading the final manuscript. Bryce LaFoya designed and constructed all figures. Bryce LaFoya, Jordan A. Munroe, and Allan R. Albig prepared the final manuscript.

Conflicts of Interest: The authors declare no conflicts of interest.

References

1. Humphries, J.D.; Byron, A.; Humphries, M.J. Integrin ligands at a glance. *J. Cell Sci.* **2006**, *119*, 3901–3903. [CrossRef] [PubMed]
2. Harburger, D.S.; Calderwood, D.A. Integrin signalling at a glance. *J. Cell Sci.* **2009**, *122*, 159–163. [CrossRef] [PubMed]
3. Pancer, Z.; Kruse, M.; Muller, I.; Muller, W.E. On the origin of Metazoan adhesion receptors: Cloning of integrin α subunit from the sponge Geodia cydonium. *Mol. Biol. Evol.* **1997**, *14*, 391–398. [CrossRef] [PubMed]
4. Brower, D.L.; Brower, S.M.; Hayward, D.C.; Ball, E.E. Molecular evolution of integrins: Genes encoding integrin β subunits from a coral and a sponge. *Proc. Natl. Acad. Sci. USA* **1997**, *94*, 9182–9187. [CrossRef] [PubMed]
5. Wimmer, W.; Blumbach, B.; Diehl-Seifert, B.; Koziol, C.; Batel, R.; Steffen, R.; Muller, I.M.; Muller, W.E. Increased expression of integrin and receptor tyrosine kinase genes during autograft fusion in the sponge Geodia cydonium. *Cell Adhes. Commun.* **1999**, *7*, 111–124. [CrossRef] [PubMed]
6. Sebe-Pedros, A.; Roger, A.J.; Lang, F.B.; King, N.; Ruiz-Trillo, I. Ancient origin of the integrin-mediated adhesion and signaling machinery. *Proc. Natl. Acad. Sci. USA* **2010**, *107*, 10142–10147. [CrossRef] [PubMed]
7. Hynes, R.O. The emergence of integrins: A personal and historical perspective. *Matrix Biol.* **2004**, *23*, 333–340. [CrossRef] [PubMed]
8. Hynes, R.O. Integrins: A family of cell surface receptors. *Cell* **1987**, *48*, 549–554. [CrossRef]
9. Hynes, R.O. Integrins: Versatility, modulation, and signaling in cell adhesion. *Cell* **1992**, *69*, 11–25. [CrossRef]
10. Sullivan, D.P.; Muller, W.A. Neutrophil and monocyte recruitment by PECAM, CD99, and other molecules via the LBRC. *Semin. Immunopathol.* **2014**, *36*, 193–209. [CrossRef] [PubMed]

11. Imhof, B.A.; Aurrand-Lions, M. Adhesion mechanisms regulating the migration of monocytes. *Nat. Rev. Immunol.* **2004**, *4*, 432–444. [CrossRef] [PubMed]

12. Zundler, S.; Fischer, A.; Schillinger, D.; Binder, M.T.; Atreya, R.; Rath, T.; Lopez-Posadas, R.; Voskens, C.J.; Watson, A.; Atreya, I.; et al. The α4β1 Homing Pathway Is Essential for Ileal Homing of Crohn's Disease Effector T Cells In Vivo. *Inflamm. Bowel Dis.* **2017**, *23*, 379–391. [CrossRef] [PubMed]

13. Rettig, M.P.; Ansstas, G.; DiPersio, J.F. Mobilization of hematopoietic stem and progenitor cells using inhibitors of CXCR4 and VLA-4. *Leukemia* **2012**, *26*, 34–53. [CrossRef] [PubMed]

14. Ghobadi, A.; Rettig, M.P.; Cooper, M.L.; Holt, M.S.; Ritchey, J.K.; Eissenberg, L.; DiPersio, J.F. Bortezomib is a rapid mobilizer of hematopoietic stem cells in mice via modulation of the VCAM-1/VLA-4 axis. *Blood* **2014**, *124*, 2752–2754. [CrossRef] [PubMed]

15. Bungartz, G.; Stiller, S.; Bauer, M.; Muller, W.; Schippers, A.; Wagner, N.; Fassler, R.; Brakebusch, C. Adult murine hematopoiesis can proceed without β1 and β7 integrins. *Blood* **2006**, *108*, 1857–1864. [CrossRef] [PubMed]

16. Seguin, L.; Desgrosellier, J.S.; Weis, S.M.; Cheresh, D.A. Integrins and cancer: Regulators of cancer stemness, metastasis, and drug resistance. *Trends Cell Biol.* **2015**, *25*, 234–250. [CrossRef] [PubMed]

17. Sun, H.; Liu, J.; Zheng, Y.; Pan, Y.; Zhang, K.; Chen, J. Distinct chemokine signaling regulates integrin ligand specificity to dictate tissue-specific lymphocyte homing. *Dev. Cell* **2014**, *30*, 61–70. [CrossRef] [PubMed]

18. Murakami, J.L.; Xu, B.; Franco, C.B.; Hu, X.; Galli, S.J.; Weissman, I.L.; Chen, C.C. Evidence that β7 Integrin Regulates Hematopoietic Stem Cell Homing and Engraftment Through Interaction with MAdCAM-1. *Stem Cells Dev.* **2016**, *25*, 18–26. [CrossRef] [PubMed]

19. Maric, G.; Annis, M.G.; Dong, Z.; Rose, A.A.; Ng, S.; Perkins, D.; MacDonald, P.A.; Ouellet, V.; Russo, C.; Siegel, P.M. GPNMB cooperates with neuropilin-1 to promote mammary tumor growth and engages integrin α5β1 for efficient breast cancer metastasis. *Oncogene* **2015**, *34*, 5494–5504. [CrossRef] [PubMed]

20. Le Floc'h, A.; Jalil, A.; Vergnon, I.; Le Maux Chansac, B.; Lazar, V.; Bismuth, G.; Chouaib, S.; Mami-Chouaib, F. αEβ7 integrin interaction with E-cadherin promotes antitumor CTL activity by triggering lytic granule polarization and exocytosis. *J. Exp. Med.* **2007**, *204*, 559–570. [CrossRef] [PubMed]

21. O'Brien, X.M.; Reichner, J.S. Neutrophil integrins and matrix ligands and NET release. *Front. Immunol.* **2016**, *7*, 1–7. [CrossRef] [PubMed]

22. Doke, M.; Fukamachi, H.; Morisaki, H.; Arimoto, T.; Kataoka, H.; Kuwata, H. Nucleases from Prevotella intermedia can degrade neutrophil extracellular traps. *Mol. Oral Microbiol.* **2017**, *32*, 288–300. [CrossRef] [PubMed]

23. Wu, C.Y.; Liang, M.X.; Chen, Q. Production and stabilization of an integrin-binding moiety of complement component 3. *Mol. Biol.* **2015**, *49*, 723–727. [CrossRef]

24. Lukacsi, S.; Nagy-Balo, Z.; Erdei, A.; Sandor, N.; Bajtay, Z. The role of CR3 (CD11b/CD18) and CR4 (CD11c/CD18) in complement-mediated phagocytosis and podosome formation by human phagocytes. *Immunol. Lett.* **2017**, *189*, 64–72. [CrossRef] [PubMed]

25. Zhang, X.; Bajic, G.; Andersen, G.R.; Christiansen, S.H.; Vorup-Jensen, T. The cationic peptide LL-37 binds Mac-1 (CD11b/CD18) with a low dissociation rate and promotes phagocytosis. *Biochim. Biophys. Acta Proteins Proteom.* **2016**, *1864*, 471–478. [CrossRef] [PubMed]

26. Podolnikova, N.P.; Podolnikov, A.V.; Haas, T.A.; Lishko, V.K.; Ugarova, T.P. Ligand recognition specificity of leukocyte integrin αMβ2 (Mac-1, CD11b/CD18) and its functional consequences. *Biochemistry* **2015**, *54*, 1408–1420. [CrossRef] [PubMed]

27. Lishko, V.K.; Moreno, B.; Podolnikova, N.P.; Ugarova, T.P. Identification of Human Cathelicidin Peptide LL-37 as a Ligand for Macrophage Integrin αMβ2 (Mac-1, CD11b/CD18) that Promotes Phagocytosis by Opsonizing Bacteria. *Res. Rep. Biochem.* **2016**, *2016*, 39–55. [PubMed]

28. Hase, K.; Murakami, M.; Iimura, M.; Cole, S.P.; Horibe, Y.; Ohtake, T.; Obonyo, M.; Gallo, R.L.; Eckmann, L.; Kagnoff, M.F. Expression of LL-37 by Human Gastric Epithelial Cells as a Potential Host Defense Mechanism Against Helicobacter pylori. *Gastroenterology* **2003**, *125*, 1613–1625. [CrossRef] [PubMed]

29. Dürr, U.H.N.; Sudheendra, U.S.; Ramamoorthy, A. LL-37, the only human member of the cathelicidin family of antimicrobial peptides. *Biochim. Biophys. Acta Biomembr.* **2006**, *1758*, 1408–1425. [CrossRef] [PubMed]

30. Maugeri, N.; Campana, L.; Gavina, M.; Covino, C.; de Metrio, M.; Panciroli, C.; Maiuri, L.; Maseri, A.; D'Angelo, A.; Bianchi, M.E.; et al. Activated platelets present high mobility group box 1 to neutrophils, inducing autophagy and promoting the extrusion of neutrophil extracellular traps. *J. Thromb. Haemost.* **2014**, *12*, 2074–2088. [CrossRef] [PubMed]

31. Yang, M.; Li, Y.; Chilukuri, K.; Brady, O.A.; Boulos, M.I.; Kappes, J.C.; Galileo, D.S. L1 stimulation of human glioma cell motility correlates with FAK activation. *J. Neurooncol.* **2011**, *105*, 27–44. [CrossRef] [PubMed]

32. Burgett, M.E.; Lathia, J.D.; Roth, P.; Nowacki, A.S.; Galileo, D.S.; Pugacheva, E.; Huang, P.; Vasanji, A.; Li, M.; Byzova, T.; et al. Direct contact with perivascular tumor cells enhances integrin $\alpha v \beta 3$ signaling and migration of endothelial cells. *Oncotarget* **2016**, *7*, 43852–43867. [CrossRef] [PubMed]

33. Sundd, P.; Pospieszalska, M.K.; Ley, K. Neutrophil rolling at high shear: Flattening, catch bond behavior, tethers and slings. *Mol. Immunol.* **2013**, *55*, 59–69. [CrossRef] [PubMed]

34. Montresor, A.; Toffali, L.; Constantin, G.; Laudanna, C. Chemokines and the signaling modules regulating integrin affinity. *Front. Immunol.* **2012**, *3*, 127. [CrossRef] [PubMed]

35. Xu, S.; Wang, J.; Wang, J.-H.; Springer, T.A. Distinct recognition of complement iC3b by integrins $\alpha X \beta 2$ and $\alpha M \beta 2$. *Proc. Natl. Acad. Sci. USA* **2017**, *114*, 3403–3408. [CrossRef] [PubMed]

36. Kazzaz, N.M.; Sule, G.; Knight, J.S. Intercellular Interactions as Regulators of NETosis. *Front. Immunol.* **2016**, *7*, 453. [CrossRef] [PubMed]

37. Delgado-Rizo, V.; Martínez-Guzmán, M.A.; Iñiguez-Gutierrez, L.; García-Orozco, A.; Alvarado-Navarro, A.; Fafutis-Morris, M. Neutrophil Extracellular Traps and Its Implications in Inflammation: An Overview. *Front. Immunol.* **2017**, *8*, 1–20. [CrossRef] [PubMed]

38. Orlova, V.V.; Choi, E.Y.; Xie, C.; Chavakis, E.; Bierhaus, A.; Ihanus, E.; Ballantyne, C.M.; Gahmberg, C.G.; Bianchi, M.E.; Nawroth, P.P.; et al. A novel pathway of HMGB1-mediated inflammatory cell recruitment that requires Mac-1-integrin. *EMBO J.* **2007**, *26*, 1129–1139. [CrossRef] [PubMed]

39. Gillenius, E.; Urban, C.F. The adhesive protein invasin of Yersinia pseudotuberculosis induces neutrophil extracellular traps via $\beta 1$ integrins. *Microbes Infect.* **2015**, *17*, 327–336. [CrossRef] [PubMed]

40. Monti, M.; Iommelli, F.; de Rosa, V.; Carriero, M.V.; Miceli, R.; Camerlingo, R.; di Minno, G.; del Vecchio, S. Integrin-dependent cell adhesion to neutrophil extracellular traps through engagement of fibronectin in neutrophil-like cells. *PLoS ONE* **2017**, *12*, 1–15. [CrossRef] [PubMed]

41. Mora, J.R.; Bono, M.R.; Manjunath, N.; Weninger, W.; Cavanagh, L.L.; Rosemblatt, M.; Von Andrian, U.H. Selective imprinting of gut-homing T cells by Peyer's patch dendritic cells. *Nature* **2003**, *424*, 88–93. [CrossRef] [PubMed]

42. Caldwell, J.M.; Collins, M.H.; Kemme, K.A.; Sherrill, J.D.; Wen, T.; Rochman, M.; Stucke, E.M.; Amin, L.; Tai, H.; Putnam, P.E.; et al. Cadherin 26 is an α integrin-binding epithelial receptor regulated during allergic inflammation. *Mucosal Immunol.* **2017**, *10*, 1190–1201. [CrossRef] [PubMed]

43. Dvorak, C.C. Musical chairs: In utero HCT via mobilization. *Blood* **2016**, *128*, 2378–2380. [CrossRef] [PubMed]

44. Kim, A.G.; Vrecenak, J.D.; Boelig, M.M.; Eissenberg, L.; Rettig, M.P.; Riley, J.S.; Holt, M.S.; Conner, M.A.; Loukogeorgakis, S.P.; Li, H.; et al. Enhanced in utero allogeneic engraftment in mice after mobilizing fetal HSCs by $\alpha 4 \beta 1/7$ inhibition. *Blood* **2016**, *128*, 2457–2461. [CrossRef] [PubMed]

45. Allen, T.A.; Gracieux, D.; Talib, M.; Tokarz, D.A.; Hensley, M.T.; Cores, J.; Vandergriff, A.; Tang, J.; de Andrade, J.B.; Dinh, P.U.; et al. Angiopellosis as an Alternative Mechanism of Cell Extravasation. *Stem Cells* **2017**, *35*, 170–180. [CrossRef] [PubMed]

46. Kiefel, H.; Pfeifer, M.; Bondong, S.; Hazin, J.; Altevogt, P. Linking L1CAM-mediated signaling to NF-κB activation. *Trends Mol. Med.* **2011**, *17*, 178–187. [CrossRef] [PubMed]

47. Voura, E.B.; Ramjeesingh, R.A.; Montgomery, A.M.; Siu, C.H. Involvement of integrin $\alpha v \beta 3$ and cell adhesion molecule L1 in transendothelial migration of melanoma cells. *Mol. Biol. Cell* **2001**, *12*, 2699–2710. [CrossRef] [PubMed]

48. Montgomery, A.M.; Becker, J.C.; Siu, C.H.; Lemmon, V.P.; Cheresh, D.A.; Pancook, J.D.; Zhao, X.; Reisfeld, R.A. Human neural cell adhesion molecule L1 and rat homologue NILE are ligands for integrin $\alpha v \beta 3$. *J. Cell Biol.* **1996**, *132*, 475–485. [CrossRef] [PubMed]

49. Spring, F.A.; Griffiths, R.E.; Mankelow, T.J.; Agnew, C.; Parsons, S.F.; Chasis, J.A.; Anstee, D.J. Tetraspanins CD81 and CD82 facilitate $\alpha 4 \beta 1$-mediated adhesion of human erythroblasts to vascular cell adhesion molecule-1. *PLoS ONE* **2013**, *8*, e62654. [CrossRef] [PubMed]

50. Holmes, E.C. What does virus evolution tell us about virus origins? *J. Virol.* **2011**, *85*, 5247–5251. [CrossRef] [PubMed]

51. Stupack, D.G.; Cheresh, D.A. ECM remodeling regulates angiogenesis: Endothelial integrins look for new ligands. *Sci. STKE* **2002**, *2002*, PE7. [CrossRef] [PubMed]

52. Sheppard, D. Airway epithelial integrins: Why so many? *Am. J. Respir. Cell Mol. Biol.* **1998**, *19*, 349–351. [CrossRef] [PubMed]

53. Stewart, P.L.; Nemerow, G.R. Cell integrins: Commonly used receptors for diverse viral pathogens. *Trends Microbiol.* **2007**, *15*, 500–507. [CrossRef] [PubMed]

54. Hussein, H.A.; Walker, L.R.; Abdel-Raouf, U.M.; Desouky, S.A.; Montasser, A.K.; Akula, S.M. Beyond RGD: Virus interactions with integrins. *Arch. Virol.* **2015**, *160*, 2669–2681. [CrossRef] [PubMed]

55. La Linn, M.; Eble, J.A.; Lübken, C.; Slade, R.W.; Heino, J.; Davies, J.; Suhrbier, A. An arthritogenic αvirus uses the α1β1 integrin collagen receptor. *Virology* **2005**, *336*, 229–239. [CrossRef] [PubMed]

56. Xing, L.; Huhtala, M.; Pietiainen, V.; Kapyla, J.; Vuorinen, K.; Marjomaki, V.; Heino, J.; Johnson, M.S.; Hyypia, T.; Cheng, R.H. Structural and functional analysis of integrin α2I domain interaction with echovirus 1. *J. Biol. Chem.* **2004**, *279*, 11632–11638. [CrossRef] [PubMed]

57. Marjomaki, V.; Turkki, P.; Huttunen, M. Infectious Entry Pathway of Enterovirus B Species. *Viruses* **2015**, *7*, 6387–6399. [CrossRef] [PubMed]

58. Feire, A.L.; Roy, R.M.; Manley, K.; Compton, T. The glycoprotein B disintegrin-like domain binds β 1 integrin to mediate cytomegalovirus entry. *J. Virol.* **2010**, *84*, 10026–10037. [CrossRef] [PubMed]

59. Graham, K.L.; Takada, Y.; Coulson, B.S. Rotavirus spike protein VP5* binds α2β1 integrin on the cell surface and competes with virus for cell binding and infectivity. *J. Gen. Virol.* **2006**, *87*, 1275–1283. [CrossRef] [PubMed]

60. Coulson, B.S.; Londrigan, S.L.; Lee, D.J. Rotavirus contains integrin ligand sequences and a disintegrin-like domain that are implicated in virus entry into cells. *Proc. Natl. Acad. Sci. USA* **1997**, *94*, 5389–5394. [CrossRef] [PubMed]

61. Akula, S.M.; Pramod, N.P.; Wang, F.Z.; Chandran, B. Integrin α3β1 (CD 49c/29) is a cellular receptor for Kaposi's sarcoma-associated herpesvirus (KSHV/HHV-8) entry into the target cells. *Cell* **2002**, *108*, 407–419. [CrossRef]

62. Salone, B.; Martina, Y.; Piersanti, S.; Cundari, E.; Cherubini, G.; Franqueville, L.; Failla, C.M.; Boulanger, P.; Saggio, I. Integrin α3β1 is an alternative cellular receptor for adenovirus serotype 5. *J. Virol.* **2003**, *77*, 13448–13454. [CrossRef] [PubMed]

63. Delgui, L.; Ona, A.; Gutierrez, S.; Luque, D.; Navarro, A.; Caston, J.R.; Rodriguez, J.F. The capsid protein of infectious bursal disease virus contains a functional α4β 1 integrin ligand motif. *Virology* **2009**, *386*, 360–372. [CrossRef] [PubMed]

64. Jackson, T.; Blakemore, W.; Newman, J.W.; Knowles, N.J.; Mould, A.P.; Humphries, M.J.; King, A.M. Foot-and-mouth disease virus is a ligand for the high-affinity binding conformation of integrin α5β1: Influence of the leucine residue within the RGDL motif on selectivity of integrin binding. *J. Gen. Virol.* **2000**, *81*, 1383–1391. [CrossRef] [PubMed]

65. Tugizov, S.M.; Berline, J.W.; Palefsky, J.M. Epstein-Barr virus infection of polarized tongue and nasopharyngeal epithelial cells. *Nat. Med.* **2003**, *9*, 307–314. [CrossRef] [PubMed]

66. Davison, E.; Diaz, R.M.; Hart, I.R.; Santis, G.; Marshall, J.F. Integrin α5β1-mediated adenovirus infection is enhanced by the integrin-activating antibody TS2/16. *J. Virol.* **1997**, *71*, 6204–6207. [PubMed]

67. Walker, L.R.; Hussein, H.A.M.; Akula, S.M. Disintegrin-like domain of glycoprotein B regulates Kaposi's sarcoma-associated herpesvirus infection of cells. *J. Gen. Virol.* **2014**, *95*, 1770–1782. [CrossRef] [PubMed]

68. Huang, S.; Kamata, T.; Takada, Y.; Ruggeri, Z.M.; Nemerow, G.R. Adenovirus interaction with distinct integrins mediates separate events in cell entry and gene delivery to hematopoietic cells. *J. Virol.* **1996**, *70*, 4502–4508. [PubMed]

69. Stanway, G.; Kalkkinen, N.; Roivainen, M.; Ghazi, F.; Khan, M.; Smyth, M.; Meurman, O.; Hyypia, T. Molecular and biological characteristics of echovirus 22, a representative of a new picornavirus group. *J. Virol.* **1994**, *68*, 8232–8238. [PubMed]

70. Pulli, T.; Koivunen, E.; Hyypia, T. Cell-surface interactions of echovirus 22. *J. Biol. Chem.* **1997**, *272*, 21176–21180. [CrossRef] [PubMed]

71. Li, E.; Brown, S.L.; Stupack, D.G.; Puente, X.S.; Cheresh, D.A.; Nemerow, G.R. Integrin $\alpha v \beta 1$ is an adenovirus coreceptor. *J. Virol.* **2001**, *75*, 5405–5409. [CrossRef] [PubMed]

72. Nelsen-Salz, B.; Eggers, H.J.; Zimmermann, H. Integrin $\alpha v \beta 3$ (vitronectin receptor) is a candidate receptor for the virulent echovirus 9 strain Barty. *J. Gen. Virol.* **1999**, *80*, 2311–2313. [CrossRef] [PubMed]

73. Roivainen, M.; Piirainen, L.; Hovi, T.; Virtanen, I.; Riikonen, T.; Heino, J.; Hyypia, T. Entry of coxsackievirus A9 into host cells: Specific interactions with $\alpha v \beta 3$ integrin, the vitronectin receptor. *Virology* **1994**, *203*, 357–365. [CrossRef] [PubMed]

74. Neff, S.; Sa-Carvalho, D.; Rieder, E.; Mason, P.W.; Blystone, S.D.; Brown, E.J.; Baxt, B. Foot-and-mouth disease virus virulent for cattle utilizes the integrin $\alpha v \beta 3$ as its receptor. *J. Virol.* **1998**, *72*, 3587–3594. [PubMed]

75. Fan, W.; Qian, P.; Wang, D.; Zhi, X.; Wei, Y.; Chen, H.; Li, X. Integrin $\alpha v \beta 3$ promotes infection by Japanese encephalitis virus. *Res. Vet. Sci.* **2017**, *111*, 67–74. [CrossRef] [PubMed]

76. Garrigues, H.J.; Rubinchikova, Y.E.; Dipersio, C.M.; Rose, T.M. Integrin $\alpha V \beta 3$ Binds to the RGD motif of glycoprotein B of Kaposi's sarcoma-associated herpesvirus and functions as an RGD-dependent entry receptor. *J. Virol.* **2008**, *82*, 1570–1580. [CrossRef] [PubMed]

77. Matthys, V.S.; Gorbunova, E.E.; Gavrilovskaya, I.N.; Mackow, E.R. Andes virus recognition of human and Syrian hamster $\beta 3$ integrins is determined by an L33P substitution in the PSI domain. *J. Virol.* **2010**, *84*, 352–360. [CrossRef] [PubMed]

78. Wickham, T.J.; Mathias, P.; Cheresh, D.A.; Nemerow, G.R. Integrins $\alpha v \beta 3$ and $\alpha v \beta 5$ promote adenovirus internalization but not virus attachment. *Cell* **1993**, *73*, 309–319. [CrossRef]

79. Zarate, S.; Romero, P.; Espinosa, R.; Arias, C.F.; Lopez, S. VP7 mediates the interaction of rotaviruses with integrin $\alpha v \beta 3$ through a novel integrin-binding site. *J. Virol.* **2004**, *78*, 10839–10847. [CrossRef] [PubMed]

80. Guerrero, C.A.; Mendez, E.; Zarate, S.; Isa, P.; Lopez, S.; Arias, C.F. Integrin $\alpha v \beta 3$ mediates rotavirus cell entry. *Proc. Natl. Acad. Sci. USA* **2000**, *97*, 14644–14649. [CrossRef] [PubMed]

81. Gavrilovskaya, I.N.; Shepley, M.; Shaw, R.; Ginsberg, M.H.; Mackow, E.R. $\beta 3$ Integrins mediate the cellular entry of hantaviruses that cause respiratory failure. *Proc. Natl. Acad. Sci. USA* **1998**, *95*, 7074–7079. [CrossRef] [PubMed]

82. Veettil, M.V.; Sadagopan, S.; Sharma-Walia, N.; Wang, F.Z.; Raghu, H.; Varga, L.; Chandran, B. Kaposi's Sarcoma-Associated Herpesvirus Forms a Multimolecular Complex of Integrins (V5, V3, and 31) and CD98-xCT during Infection of Human Dermal Microvascular Endothelial Cells, and CD98-xCT Is Essential for the Postentry Stage of Infection. *J. Virol.* **2008**, *82*, 12126–12144. [CrossRef] [PubMed]

83. Chesnokova, L.S.; Hutt-Fletcher, L.M. Fusion of Epstein-Barr virus with epithelial cells can be triggered by $\alpha v \beta 5$ in addition to $\alpha v \beta 6$ and $\alpha v \beta 8$, and integrin binding triggers a conformational change in glycoproteins gHgL. *J. Virol.* **2011**, *85*, 13214–13223. [CrossRef] [PubMed]

84. Burman, A.; Clark, S.; Abrescia, N.G.; Fry, E.E.; Stuart, D.I.; Jackson, T. Specificity of the VP1 GH loop of Foot-and-Mouth Disease virus for αv integrins. *J. Virol.* **2006**, *80*, 9798–9810. [CrossRef] [PubMed]

85. Berryman, S.; Clark, S.; Monaghan, P.; Jackson, T. Early events in integrin $\alpha v \beta 6$-mediated cell entry of foot-and-mouth disease virus. *J. Virol.* **2005**, *79*, 8519–8534. [CrossRef] [PubMed]

86. Gianni, T.; Salvioli, S.; Chesnokova, L.S.; Hutt-Fletcher, L.M.; Campadelli-Fiume, G. $\alpha v \beta 6$- and $\alpha v \beta 8$-integrins serve as interchangeable receptors for HSV gH/gL to promote endocytosis and activation of membrane fusion. *PLoS Pathog.* **2013**, *9*, e1003806. [CrossRef] [PubMed]

87. Zell, R. Picornaviridae-the ever-growing virus family. *Arch. Virol.* **2017**. [CrossRef] [PubMed]

88. Tuthill, T.J.; Groppelli, E.; Hogle, J.M.; Rowlands, D.J. Picornaviruses. *Curr. Top. Microbiol. Immunol.* **2010**, *343*, 43–89. [PubMed]

89. Johnson, M.S.; Lu, N.; Denessiouk, K.; Heino, J.; Gullberg, D. Integrins during evolution: Evolutionary trees and model organisms. *Biochim. Biophys. Acta* **2009**, *1788*, 779–789. [CrossRef] [PubMed]

90. Emsley, J.; Knight, C.G.; Farndale, R.W.; Barnes, M.J.; Liddington, R.C. Structural basis of collagen recognition by integrin $\alpha 2 \beta 1$. *Cell* **2000**, *101*, 47–56. [CrossRef]

91. Wary, K.K.; Mariotti, A.; Zurzolo, C.; Giancotti, F.G. A requirement for caveolin-1 and associated kinase Fyn in integrin signaling and anchorage-dependent cell growth. *Cell* **1998**, *94*, 625–634. [CrossRef]

92. Pietiainen, V.; Marjomaki, V.; Upla, P.; Pelkmans, L.; Helenius, A.; Hyypia, T. Echovirus 1 endocytosis into caveosomes requires lipid rafts, dynamin II, and signaling events. *Mol. Biol. Cell* **2004**, *15*, 4911–4925. [CrossRef] [PubMed]

93. Parton, R.G. Caveolae and caveolins. *Curr. Opin. Cell Biol.* **1996**, *8*, 542–548. [CrossRef]

94. Marjomaki, V.; Pietiainen, V.; Matilainen, H.; Upla, P.; Ivaska, J.; Nissinen, L.; Reunanen, H.; Huttunen, P.; Hyypia, T.; Heino, J. Internalization of echovirus 1 in caveolae. *J. Virol.* **2002**, *76*, 1856–1865. [CrossRef] [PubMed]

95. Bergelson, J.M.; Cunningham, J.A.; Droguett, G.; Kurt-Jones, E.A.; Krithivas, A.; Hong, J.S.; Horwitz, M.S.; Crowell, R.L.; Finberg, R.W. Isolation of a common receptor for Coxsackie B viruses and adenoviruses 2 and 5. *Science* **1997**, *275*, 1320–1323. [CrossRef] [PubMed]

96. Huang, Y.J.S.; Higgs, S.; Horne, K.M.E.; Vanlandingham, D.L. Flavivirus-Mosquito interactions. *Viruses* **2014**, *6*, 4703–4730. [CrossRef] [PubMed]

97. Unni, S.K.; Růžek, D.; Chhatbar, C.; Mishra, R.; Johri, M.K.; Singh, S.K. Japanese encephalitis virus: From genome to infectome. *Microbes Infect.* **2011**, *13*, 312–321. [CrossRef] [PubMed]

98. Luca, V.C.; AbiMansour, J.; Nelson, C.A.; Fremont, D.H. Crystal Structure of the Japanese Encephalitis Virus Envelope Protein. *J. Virol.* **2012**, *86*, 2337–2346. [CrossRef] [PubMed]

99. Das, S.; Laxminarayana, S.V.; Chandra, N.; Ravi, V.; Desai, A. Heat shock protein 70 on Neuro2a cells is a putative receptor for Japanese encephalitis virus. *Virology* **2009**, *385*, 47–57. [CrossRef] [PubMed]

100. Chu, J.J.; Rajamanonmani, R.; Li, J.; Bhuvanakantham, R.; Lescar, J.; Ng, M.L. Inhibition of West Nile virus entry by using a recombinant domain III from the envelope glycoprotein. *J. Gen. Virol.* **2005**, *86*, 405–412. [CrossRef] [PubMed]

101. Chu, J.J.H.; Ng, M.L. Interaction of West Nile virus with αVβ3 integrin mediates virus entry into cells. *J. Biol. Chem.* **2004**, *279*, 54533–54541. [CrossRef] [PubMed]

102. Bogachek, M.V.; Zaitsev, B.N.; Sekatskii, S.K.; Protopopova, E.V.; Ternovoi, V.A.; Ivanova, A.V.; Kachko, A.V.; Ivanisenko, V.A.; Dietler, G.; Loktev, V.B. Characterization of glycoprotein E C-end of West Nile virus and evaluation of its interaction force with αVβ3 integrin as putative cellular receptor. *Biochemistry* **2010**, *75*, 472–480. [CrossRef] [PubMed]

103. Lee, E.; Lobigs, M. Substitutions at the putative receptor-binding site of an encephalitic flavivirus alter virulence and host cell tropism and reveal a role for glycosaminoglycans in entry. *J. Virol.* **2000**, *74*, 8867–8875. [CrossRef] [PubMed]

104. Wan, S.-W.; Lin, C.-F.; Lu, Y.-T.; Lei, H.-Y.; Anderson, R.; Lin, Y.-S. Endothelial cell surface expression of protein disulfide isomerase activates β1 and β3 integrins and facilitates dengue virus infection. *J. Cell. Biochem.* **2012**, *113*, 1681–1691. [CrossRef] [PubMed]

105. van der Most, R.G.; Corver, J.; Strauss, J.H. Mutagenesis of the RGD motif in the yellow fever virus 17D envelope protein. *Virology* **1999**, *265*, 83–95. [CrossRef] [PubMed]

106. Feire, A.L.; Koss, H.; Compton, T. Cellular integrins function as entry receptors for human cytomegalovirus via a highly conserved disintegrin-like domain. *Proc. Natl. Acad. Sci. USA* **2004**, *101*, 15470–15475. [CrossRef] [PubMed]

107. Assuncao-Miranda, I.; Cruz-Oliveira, C.; Da Poian, A.T. Molecular mechanisms involved in the pathogenesis of αvirus-induced arthritis. *BioMed Res. Int.* **2013**, *2013*, 973516. [CrossRef] [PubMed]

108. Mangel, W.F.; San Martin, C. Structure, function and dynamics in adenovirus maturation. *Viruses* **2014**, *6*, 4536–4570. [CrossRef] [PubMed]

109. Nemerow, G.R.; Stewart, P.L. Role of αv integrins in adenovirus cell entry and gene delivery. *Microbiol. Mol. Biol. Rev. MMBR* **1999**, *63*, 725–734. [PubMed]

110. Wickham, T.J.; Filardo, E.J.; Cheresh, D.A.; Nemerow, G.R. Integrin αvβ5 selectively promotes adenovirus mediated cell membrane permeabilization. *J. Cell Biol.* **1994**, *127*, 257–264. [CrossRef] [PubMed]

111. Wickham, T.J. Targeting adenovirus. *Gene Ther.* **2000**, *7*, 110–114. [CrossRef] [PubMed]

112. Bai, M.; Harfe, B.; Freimuth, P. Mutations that alter an Arg-Gly-Asp (RGD) sequence in the adenovirus type 2 penton base protein abolish its cell-rounding activity and delay virus reproduction in flat cells. *J. Virol.* **1993**, *67*, 5198–5205. [PubMed]

113. Miyamoto, S.; Akiyama, S.K.; Yamada, K.M. Synergistic roles for receptor occupancy and aggregation in integrin transmembrane function. *Science* **1995**, *267*, 883–885. [CrossRef] [PubMed]

114. Li, E.; Stupack, D.; Klemke, R.; Cheresh, D.A.; Nemerow, G.R. Adenovirus endocytosis via αv integrins requires phosphoinositide-3-OH kinase. *J. Virol.* **1998**, *72*, 2055–2061. [PubMed]

115. Li, E.; Stupack, D.; Bokoch, G.M.; Nemerow, G.R. Adenovirus endocytosis requires actin cytoskeleton reorganization mediated by Rho family GTPases. *J. Virol.* **1998**, *72*, 8806–8812. [PubMed]

116. Schmaljohn, C.; Hjelle, B. Hantaviruses: A global disease problem. *Emerg. Infect. Dis.* **1997**, *3*, 95–104. [CrossRef] [PubMed]

117. Bondu, V.; Wu, C.; Cao, W.; Simons, P.C.; Gillette, J.; Zhu, J.; Erb, L.; Zhang, X.F.; Buranda, T. Low-affinity binding in cis to P2Y2R mediates force-dependent integrin activation during hantavirus infection. *Mol. Biol. Cell* **2017**, *28*, 2887–2903. [CrossRef] [PubMed]

118. Gavrilovskaya, I.N.; Peresleni, T.; Geimonen, E.; Mackow, E.R. Pathogenic hantaviruses selectively inhibit β3 integrin directed endothelial cell migration. *Arch. Virol.* **2002**, *147*, 1913–1931. [CrossRef] [PubMed]

119. Gavrilovskaya, I.N.; Gorbunova, E.E.; Mackow, N.A.; Mackow, E.R. Hantaviruses direct endothelial cell permeability by sensitizing cells to the vascular permeability factor VEGF, while angiopoietin 1 and sphingosine 1-phosphate inhibit hantavirus-directed permeability. *J. Virol.* **2008**, *82*, 5797–5806. [CrossRef] [PubMed]

120. Bottcher, B.; Kiselev, N.A.; Stel'Mashchuk, V.Y.; Perevozchikova, N.A.; Borisov, A.V.; Crowther, R.A. Three-dimensional structure of infectious bursal disease virus determined by electron cryomicroscopy. *J. Virol.* **1997**, *71*, 325–330. [PubMed]

121. Ye, C.; Han, X.; Yu, Z.; Zhang, E.; Wang, L.; Liu, H. Infectious Bursal Disease Virus Activates c-Src To Promote α4β1 Integrin-Dependent Viral Entry by Modulating the Downstream Akt-RhoA GTPase-Actin Rearrangement Cascade. *J. Virol.* **2017**, *91*. [CrossRef] [PubMed]

122. Tate, J.E.; Burton, A.H.; Boschi-Pinto, C.; Parashar, U.D. Global, Regional, and National Estimates of Rotavirus Mortality in Children <5 Years of Age, 2000–2013. *Clin. Infect. Dis. Off. Publ. Infect. Dis. Soc. Am.* **2016**, *62*, S96–S105. [CrossRef] [PubMed]

123. Yeager, M.; Dryden, K.A.; Olson, N.H.; Greenberg, H.B.; Baker, T.S. Three-dimensional structure of rhesus rotavirus by cryoelectron microscopy and image reconstruction. *J. Cell Biol.* **1990**, *110*, 2133–2144. [CrossRef] [PubMed]

124. Maginnis, M.S.; Forrest, J.C.; Kopecky-Bromberg, S.A.; Dickeson, S.K.; Santoro, S.A.; Zutter, M.M.; Nemerow, G.R.; Bergelson, J.M.; Dermody, T.S. B1 integrin mediates internalization of mammalian reovirus. *J. Virol.* **2006**, *80*, 2760–2770. [CrossRef] [PubMed]

125. Marcinkiewicz, C. Applications of snake venom components to modulate integrin activities in cell-matrix interactions. *Int. J. Biochem. Cell Biol.* **2013**, *45*, 1974–1986. [CrossRef] [PubMed]

126. Huang, T.-F.; Hsu, C.-C.; Kuo, Y.-J. Anti-thrombotic agents derived from snake venom proteins. *Thromb. J.* **2016**, *14*, 18. [CrossRef] [PubMed]

127. Musial, J.; Niewiarowski, S.; Rucinski, B.; Stewart, G.J.; Cook, J.J.; Williams, J.A.; Edmunds, L.H., Jr. Inhibition of platelet adhesion to surfaces of extracorporeal circuits by disintegrins. RGD-containing peptides from viper venoms. *Circulation* **1990**, *82*, 261–273. [CrossRef] [PubMed]

128. Walsh, E.M.; Marcinkiewicz, C. Non-RGD-containing snake venom disintegrins, functional and structural relations. *Toxicon* **2011**, *58*, 355–362. [CrossRef] [PubMed]

129. Calvete, J.J. The continuing saga of snake venom disintegrins. *Toxicon* **2013**, *62*, 40–49. [CrossRef] [PubMed]

130. Marcinkiewicz, C.; Lobb, R.R.; Marcinkiewicz, M.M.; Daniel, J.L.; Smith, J.B.; Dangelmaier, C.; Weinreb, P.H.; Beacham, D.A.; Niewiarowski, S. Isolation and characterization of EMS16, a C-lectin type protein from Echis multisquamatus venom, a potent and selective inhibitor of the α2β1 integrin. *Biochemistry* **2000**, *39*, 9859–9867. [CrossRef] [PubMed]

131. Horii, K.; Okuda, D.; Morita, T.; Mizuno, H. Crystal structure of EMS16 in complex with the integrin α2-I domain. *J. Mol. Biol.* **2004**, *341*, 519–527. [CrossRef] [PubMed]

132. Hutchings, H.; Ortega, N.; Plouet, J. Extracellular matrix-bound vascular endothelial growth factor promotes endothelial cell adhesion, migration, and survival through integrin ligation. *FASEB J.* **2003**, *17*, 1520–1522. [CrossRef] [PubMed]

133. Eble, J.A.; Bruckner, P.; Mayer, U. Vipera lebetina venom contains two disintegrins inhibiting laminin-binding β1 integrins. *J. Biol. Chem.* **2003**, *278*, 26488–26496. [CrossRef] [PubMed]

134. Sato, R.; Yamasaki, M.; Hirai, K.; Matsubara, T.; Nomura, T.; Sato, F.; Mimata, H. Angiopoietin-like protein 2 induces androgen-independent and malignant behavior in human prostate cancer cells. *Oncol. Rep.* **2015**, *33*, 58–66. [CrossRef] [PubMed]

135. Yugami, M.; Odagiri, H.; Endo, M.; Tsutsuki, H.; Fujii, S.; Kadomatsu, T.; Masuda, T.; Miyata, K.; Terada, K.; Tanoue, H.; et al. Mice Deficient in Angiopoietin-like Protein 2 (Angptl2) Gene Show Increased Susceptibility to Bacterial Infection Due to Attenuated Macrophage Activity. *J. Biol. Chem.* **2016**, *291*, 18843–18852. [CrossRef] [PubMed]

136. Vlahakis, N.E.; Young, B.A.; Atakilit, A.; Hawkridge, A.E.; Issaka, R.B.; Boudreau, N.; Sheppard, D. Integrin $\alpha9\beta1$ directly binds to vascular endothelial growth factor (VEGF)-A and contributes to VEGF-A-induced angiogenesis. *J. Biol. Chem.* **2007**, *282*, 15187–15196. [CrossRef] [PubMed]

137. Lin, H.Y.; Lansing, L.; Merillon, J.M.; Davis, F.B.; Tang, H.Y.; Shih, A.; Vitrac, X.; Krisa, S.; Keating, T.; Cao, H.J.; et al. Integrin $\alpha V\beta3$ contains a receptor site for resveratrol. *FASEB J.* **2006**, *20*, 1742–1744. [CrossRef] [PubMed]

138. Lin, H.-Y.; Tang, H.-Y.; Keating, T.; Wu, Y.-H.; Shih, A.; Hammond, D.; Sun, M.; Hercbergs, A.; Davis, F.B.; Davis, P.J. Resveratrol is pro-apoptotic and thyroid hormone is anti-apoptotic in glioma cells: Both actions are integrin and ERK mediated. *Carcinogenesis* **2008**, *29*, 62–69. [CrossRef] [PubMed]

139. Belleri, M.; Ribatti, D.; Savio, M.; Stivala, L.A.; Forti, L.; Tanghetti, E.; Alessi, P.; Coltrini, D.; Bugatti, A.; Mitola, S.; et al. $\alpha v\beta3$ Integrin-dependent antiangiogenic activity of resveratrol stereoisomers. *Mol. Cancer Ther.* **2008**, *7*, 3761–3770. [CrossRef] [PubMed]

140. Lin, H.Y.; Cody, V.; Davis, F.B.; Hercbergs, A.A.; Luidens, M.K.; Mousa, S.A.; Davis, P.J. Identification and functions of the plasma membrane receptor for thyroid hormone analogues. *Discov. Med.* **2011**, *11*, 337–347. [PubMed]

141. Cayrol, F.; Flaqué, M.C.D.; Fernando, T.; Yang, S.N.; Sterle, H.A.; Bolontrade, M.; Amorós, M.; Isse, B.; Farías, R.N.; Ahn, H.; et al. Integrin $\alpha v\beta3$ acting as membrane receptor for thyroid hormones mediates angiogenesis in malignant T cells. *Blood* **2015**, *125*, 841–851. [CrossRef] [PubMed]

142. Bergh, J.J.; Lin, H.Y.; Lansing, L.; Mohamed, S.N.; Davis, F.B.; Mousa, S.; Davis, P.J. Integrin $\alpha V\beta3$ contains a cell surface receptor site for thyroid hormone that is linked to activation of mitogen-activated protein kinase and induction of angiogenesis. *Endocrinology* **2005**, *146*, 2864–2871. [CrossRef] [PubMed]

143. Lin, H.Y.; Sun, M.; Lin, C.; Tang, H.Y.; London, D.; Shih, A.; Davis, F.B.; Davis, P.J. Androgen-induced human breast cancer cell proliferation is mediated by discrete mechanisms in estrogen receptor-α-positive and -negative breast cancer cells. *J. Steroid Biochem. Mol. Biol.* **2009**, *113*, 182–188. [CrossRef] [PubMed]

144. Chin, Y.-T.; Yang, S.-H.; Chang, T.-C.; Changou, C.A.; Lai, H.-Y.; Fu, E.; Huangfu, W.-C.; Davis, P.J.; Lin, H.-Y.; Liu, L.F. Mechanisms of dihydrotestosterone action on resveratrol- induced anti-proliferation in breast cancer cells with different ERα status. *Oncotarget* **2015**, *6*, 35866–35879. [CrossRef] [PubMed]

145. Lin, Y.; Rao, J.; Zha, X.L.; Xu, H. Angiopoietin-like 3 induces podocyte f-actin rearrangement through integrin $\alpha v \beta 3$ /FAK/PI3K pathway-mediated rac1 Activation. *BioMed Res. Int.* **2013**. [CrossRef] [PubMed]

146. Gomez Perdiguero, E.; Liabotis-Fontugne, A.; Durand, M.l.; Faye, C.M.; Ricard-Blum, S.; Simonutti, M.; Augustin, S.B.; Robb, B.M.; Paques, M.; Valenzuela, D.M.; et al. ANGPTL4-$\alpha v\beta3$ interaction counteracts hypoxia-induced vascular permeability by modulating Src signalling downstream of vascular endothelial growth factor receptor 2. *J. Pathol.* **2016**, *240*, 461–471. [CrossRef] [PubMed]

147. Clement, L.C.; Macé, C.; Avila-Casado, C.; Joles, J.A.; Kersten, S.; Chugh, S.S. Circulating angiopoietin-like 4 links proteinuria with hypertriglyceridemia in nephrotic syndrome. *Nat. Med.* **2014**, *20*, 37–46. [CrossRef] [PubMed]

148. Worthington, J.J.; Klementowicz, J.E.; Travis, M.A. TGFβ: A slooping giant awoken by integrins. *Trends Biochem. Sci.* **2011**, *36*, 47–54. [CrossRef] [PubMed]

149. Dong, X.; Zhao, B.; Iacob, R.E.; Zhu, J.; Koksal, A.C.; Lu, C.; Engen, J.R.; Springer, T.A. Force interacts with macromolecular structure in activation of TGF-β. *Nature* **2017**, *542*, 55–59. [CrossRef] [PubMed]

150. Juárez, P.; Comas, I.; González-Candelas, F.; Calvete, J.J. Evolution of snake venom disintegrins by positive Darwinian selection. *Mol. Biol. Evol.* **2008**, *25*, 2391–2407. [CrossRef] [PubMed]

151. Souza, D.H.F.; Iemma, M.R.C.; Ferreira, L.L.; Faria, J.P.; Oliva, M.L.V.; Zingali, R.B.; Niewiarowski, S.; Selistre-de-Araujo, H.S. The Disintegrin-like Domain of the Snake Venom Metalloprotease Alternagin Inhibits $\alpha2\beta1$ Integrin-Mediated Cell Adhesion. *Arch. Biochem. Biophys.* **2000**, *384*, 341–350. [CrossRef] [PubMed]

152. Kamiguti, A.S.; Hay, C.R.; Zuzel, M. Inhibition of collagen-induced platelet aggregation as the result of cleavage of $\alpha 2 \beta 1$-integrin by the snake venom metalloproteinase jararhagin. *Biochem. J.* **1996**, *320*, 635–641. [CrossRef] [PubMed]

153. Jakubowski, P.; Calvete, J.J.; Eble, J.A.; Lazarovici, P.; Marcinkiewicz, C. Identification of inhibitors of $\alpha2\beta1$ integrin, members of C-lectin type proteins, in Echis sochureki venom. *Toxicol. Appl. Pharmacol.* **2013**, *269*, 34–42. [CrossRef] [PubMed]

154. Pilorget, A.; Conesa, M.; Sarray, S.; Michaud-Levesque, J.; Daoud, S.; Kim, K.S.; Demeule, M.; Marvaldi, J.; El Ayeb, M.; Marrakchi, N.; et al. Lebectin, a Macrovipera lebetina venom-derived C-type lectin, inhibits angiogenesis both in vitro and in vivo. *J. Cell. Physiol.* **2007**, *211*, 307–315. [CrossRef] [PubMed]

155. Momic, T.; Cohen, G.; Reich, R.; Arlinghaus, F.T.; Eble, J.A.; Marcinkiewicz, C.; Lazarovici, P. Vixapatin (VP12), a C-type lectin-protein from Vipera xantina palestinae venom: Characterization as a novel anti-angiogenic compound. *Toxins* **2012**, *4*, 862–877. [CrossRef] [PubMed]

156. Rosenow, F.; Ossig, R.; Thormeyer, D.; Gasmann, P.; Schlüter, K.; Brunner, G.; Haier, J.; Eble, J.A. Antimetastatic Integrin as Inhibitors of Snake Venoms. *Neoplasia* **2008**, *10*, 168–176. [CrossRef] [PubMed]

157. Frankel, G.; Lider, O.; Hershkoviz, R.; Mould, A.P.; Kachalsky, S.G.; Candy, D.C.; Cahalon, L.; Humphries, M.J.; Dougan, G. The cell-binding domain of intimin from enteropathogenic Escherichia coli binds to $\beta1$ integrins. *J. Biol. Chem.* **1996**, *271*, 20359–20364. [CrossRef] [PubMed]

158. Watarai, M.; Funato, S.; Sasakawa, C. Interaction of Ipa proteins of Shigella flexneri with $\alpha5\beta1$ integrin promotes entry of the bacteria into mammalian cells. *J. Exp. Med.* **1996**, *183*, 991–999. [CrossRef] [PubMed]

159. Ishibashi, Y.; Claus, S.; Relman, D.A. Bordetella pertussis filamentous hemagglutinin interacts with a leukocyte signal transduction complex and stimulates bacterial adherence to monocyte CR3 (CD11b/CD18). *J. Exp. Med.* **1994**, *180*, 1225–1233. [CrossRef] [PubMed]

160. Tilly, K.; Rosa, P.A.; Stewart, P.E. Biology of Infection with Borrelia burgdorferi. *Infect. Dis. Clin. N. Am.* **2008**, *22*, 217–234. [CrossRef] [PubMed]

161. Hyde, J.A. Borrelia burgdorferi Keeps Moving and Carries on: A Review of Borrelial Dissemination and Invasion. *Front. Immunol.* **2017**, *8*, 114. [CrossRef] [PubMed]

162. Caine, J.A.; Coburn, J. Multifunctional and Redundant Roles of Borrelia burgdorferi Outer Surface Proteins in Tissue Adhesion, Colonization, and Complement Evasion. *Front. Immunol.* **2016**, *7*, 442. [CrossRef] [PubMed]

163. Coburn, J.; Magoun, L.; Bodary, S.C.; Leong, J.M. Integrins $\alpha v\beta3$ and $\alpha5\beta1$ mediate attachment of lyme disease spirochetes to human cells. *Infect. Immun.* **1998**, *66*, 1946–1952. [PubMed]

164. Coburn, J.; Leong, J.M.; Erban, J.K. Integrin α IIb β 3 mediates binding of the Lyme disease agent Borrelia burgdorferi to human platelets. *Proc. Natl. Acad. Sci. USA* **1993**, *90*, 7059–7063. [CrossRef] [PubMed]

165. Coburn, J.; Chege, W.; Magoun, L.; Bodary, S.C.; Leong, J.M. Characterization of a candidate Borrelia burgdorferi $\beta3$-chain integrin ligand identified using a phage display library. *Mol. Microbiol.* **1999**, *34*, 926–940. [CrossRef] [PubMed]

166. Defoe, G.; Coburn, J. Delineation of Borrelia burgdorferi p66 sequences required for integrin αIIb$\beta3$ recognition. *Infect. Immun.* **2001**, *69*, 3455–3459. [CrossRef] [PubMed]

167. Ristow, L.C.; Bonde, M.; Lin, Y.P.; Sato, H.; Curtis, M.; Wesley, E.; Hahn, B.L.; Fang, J.; Wilcox, D.A.; Leong, J.M.; et al. Integrin binding by Borrelia burgdorferi P66 facilitates dissemination but is not required for infectivity. *Cell. Microbiol.* **2015**, *17*, 1021–1036. [CrossRef] [PubMed]

168. Kumar, D.; Ristow, L.C.; Shi, M.; Mukherjee, P.; Caine, J.A.; Lee, W.Y.; Kubes, P.; Coburn, J.; Chaconas, G. Intravital Imaging of Vascular Transmigration by the Lyme Spirochete: Requirement for the Integrin Binding Residues of the B. burgdorferi P66 Protein. *PLoS Pathog.* **2015**, *11*, 1–20. [CrossRef] [PubMed]

169. Wood, E.; Tamborero, S.; Mingarro, I.; Esteve-Gassent, M.D. BB0172, a Borrelia burgdorferi outer membrane protein that binds integrin $\alpha3\beta1$. *J. Bacteriol.* **2013**, *195*, 3320–3330. [CrossRef] [PubMed]

170. Behera, A.K.; Durand, E.; Cugini, C.; Antonara, S.; Bourassa, L.; Hildebrand, E.; Hu, L.T.; Coburn, J. Borrelia burgdorferi BBB07 interaction with integrin $\alpha3\beta1$ stimulates production of pro-inflammatory mediators in primary human chondrocytes. *Cell. Microbiol.* **2008**, *10*, 320–331. [PubMed]

171. Song, J.; Zhang, X.; Buscher, K.; Wang, Y.; Wang, H.; di Russo, J.; Li, L.; Lutke-Enking, S.; Zarbock, A.; Stadtmann, A.; et al. Endothelial Basement Membrane Laminin 511 Contributes to Endothelial Junctional Tightness and Thereby Inhibits Leukocyte Transmigration. *Cell Rep.* **2017**, *18*, 1256–1269. [CrossRef] [PubMed]

172. Wroblewski, L.E.; Peek, R.M.; Wilson, K.T. Helicobacter pylori and gastric cancer: Factors that modulate disease risk. *Clin. Microbiol. Rev.* **2010**, *23*, 713–739. [CrossRef] [PubMed]

173. Saber, T.; Ghonaim, M.M.; Yousef, A.R.; Khalifa, A.; Al Qurashi, H.; Shaqhan, M.; Samaha, M. Association of Helicobacter pylori cagA Gene with Gastric Cancer and Peptic Ulcer in Saudi Patients. *J. Microbiol. Biotechnol.* **2015**, *25*, 1146–1153. [CrossRef] [PubMed]

174. Parsonnet, J.; Friedman, G.D.; Orentreich, N.; Vogelman, H. Risk for gastric cancer in people with CagA positive or CagA negative Helicobacterpylori infection. *Gut* **1997**, *40*, 297–301. [CrossRef] [PubMed]

175. Wallden, K.; Rivera-Calzada, A.; Waksman, G. Type IV secretion systems: Versatility and diversity in function. *Cell. Microbiol.* **2010**, *12*, 1203–1212. [CrossRef] [PubMed]

176. Terradot, L.; Waksman, G. Architecture of the Helicobacter pylori Cag-type IV secretion system. *FEBS J.* **2011**, *278*, 1213–1222. [CrossRef] [PubMed]

177. Conradi, J.; Tegtmeyer, N.; Woźna, M.; Wissbrock, M.; Michalek, C.; Gagell, C.; Cover, T.L.; Frank, R.; Sewald, N.; Backert, S. An RGD helper sequence in CagL of Helicobacter pylori assists in interactions with integrins and injection of CagA. *Front. Cell. Infect. Microbiol.* **2012**, *2*, 70. [CrossRef] [PubMed]

178. Wiedemann, T.; Hofbaur, S.; Tegtmeyer, N.; Huber, S.; Sewald, N.; Wessler, S.; Backert, S.; Rieder, G. Helicobacter pylori CagL dependent induction of gastrin expression via a novel v 5-integrin-integrin linked kinase signalling complex. *Gut* **2012**, *61*, 986–996. [CrossRef] [PubMed]

179. Kwok, T.; Zabler, D.; Urman, S.; Rohde, M.; Hartig, R.; Wessler, S.; Misselwitz, R.; Berger, J.; Sewald, N.; König, W.; et al. Helicobacter exploits integrin for type IV secretion and kinase activation. *Nature* **2007**, *449*, 862–866. [CrossRef] [PubMed]

180. Conradi, J.; Huber, S.; Gaus, K.; Mertink, F.; Royo Gracia, S.; Strijowski, U.; Backert, S.; Sewald, N. Cyclic RGD peptides interfere with binding of the Helicobacter pylori protein CagL to integrins αvβ3 and α5β1. *Amino Acids* **2012**, *43*, 219–232. [CrossRef] [PubMed]

181. Barden, S.; Niemann, H.H. Adhesion of several cell lines to helicobacter pylori CagL Is mediated by integrin αvβ6 via an rgdlxxl motif. *J. Mol. Biol.* **2015**, *427*, 1304–1315. [CrossRef] [PubMed]

182. Bönig, T.; Olbermann, P.; Bats, S.H.; Fischer, W.; Josenhans, C.; Blaser, M.J.; Atherton, J.C.; Kusters, J.G.; Vliet, A.H.V.; Kuipers, E.J.; et al. Systematic site-directed mutagenesis of the Helicobacter pylori CagL protein of the Cag type IV secretion system identifies novel functional domains. *Sci. Rep.* **2016**, *6*, 38101. [CrossRef] [PubMed]

183. Mueller, D.; Tegtmeyer, N.; Brandt, S.; Yamaoka, Y.; de Poire, E.; Sgouras, D.; Wessler, S.; Torres, J.; Smolka, A.; Backert, S. c-Src and c-Abl kinases control hierarchic phosphorylation and function of the CagA effector protein in Western and East Asian Helicobacter pylori strains. *J. Clin. Investig.* **2012**, *122*, 1553–1566. [CrossRef] [PubMed]

184. Tohidpour, A. CagA-mediated pathogenesis of Helicobacter pylori. *Microb. Pathogen.* **2016**, *93*, 44–55. [CrossRef] [PubMed]

185. Tertti, R.; Skurnik, M.; Vartio, T.; Kuusela, P. Adhesion protein YadA of Yersinia species mediates binding of bacteria to fibronectin. *Infect. Immun.* **1992**, *60*, 3021–3024. [PubMed]

186. El Tahir, Y.; Skurnik, M. YadA, the multifaceted Yersinia adhesin. *Int. J. Med. Microbiol. IJMM* **2001**, *291*, 209–218. [CrossRef] [PubMed]

187. Isberg, R.R.; Leong, J.M. Multiple β1 chain integrins are receptors for invasin, a protein that promotes bacterial penetration into mammalian cells. *Cell* **1990**, *60*, 861–871. [CrossRef]

188. Hamzaoui, N.; Kerneis, S.; Caliot, E.; Pringault, E. Expression and distribution of β1 integrins in in vitro-induced M cells: Implications for Yersinia adhesion to Peyer's patch epithelium. *Cell. Microbiol.* **2004**, *6*, 817–828. [CrossRef] [PubMed]

189. Schulte, R.; Kerneis, S.; Klinke, S.; Bartels, H.; Preger, S.; Kraehenbuhl, J.P.; Pringault, E.; Autenrieth, I.B. Translocation of Yersinia entrocolitica across reconstituted intestinal epithelial monolayers is triggered by Yersinia invasin binding to β1 integrins apically expressed on M-like cells. *Cell. Microbiol.* **2000**, *2*, 173–185. [CrossRef] [PubMed]

190. Clark, M.A.; Hirst, B.H.; Jepson, M.A. M-cell surface β1 integrin expression and invasin-mediated targeting of Yersinia pseudotuberculosis to mouse Peyer's patch M cells. *Infect. Immun.* **1998**, *66*, 1237–1243. [PubMed]

191. Leong, J.M.; Morrissey, P.E.; Marra, A.; Isberg, R.R. An aspartate residue of the Yersinia pseudotuberculosis invasin protein that is critical for integrin binding. *EMBO J.* **1995**, *14*, 422–431. [PubMed]

192. Hamburger, Z.A.; Brown, M.S.; Isberg, R.R.; Bjorkman, P.J. Integrin-Binding Protein Crystal Structure of Invasin: A Bacterial Crystal Structure of Invasin: A Bacterial Integrin-Binding Protein. *Science* **1999**, *286*, 291–295. [CrossRef] [PubMed]

193. Del Valle, A.; Jones, B.F.; Harrison, L.M.; Chadderdon, R.C.; Cappello, M. Isolation and molecular cloning of a secreted hookworm platelet inhibitor from adult Ancylostoma caninum. *Mol. Biochem. Parasitol.* **2003**, *129*, 167–177. [CrossRef]

194. Chadderdon, R.C.; Cappello, M. The hookworm platelet inhibitor: Functional blockade of integrins GPIIb/IIIa (αIIbβ3) and GPIa/IIa (α2β1) inhibits platelet aggregation and adhesion in vitro. *J. Infect. Dis.* **1999**, *179*, 1235–1241. [CrossRef] [PubMed]

195. Seymour, J.L.; Henzel, W.J.; Nevins, B.; Stults, J.T.; Lazarus, R.A. Decorsin. A potent glycoprotein IIb-IIIa antagonist and platelet aggregation inhibitor from the leech Macrobdella decora. *J. Biol. Chem.* **1990**, *265*, 10143–10147. [PubMed]

196. Zhang, Z.; Gao, L.; Shen, C.; Rong, M.; Yan, X.; Lai, R. A potent anti-thrombosis peptide (vasotab TY) from horsefly salivary glands. *Int. J. Biochem. Cell. Biol.* **2014**, *54*, 83–88. [CrossRef] [PubMed]

197. Ma, D.; Xu, X.; An, S.; Liu, H.; Yang, X.; Andersen, J.F.; Wang, Y.; Tokumasu, F.; Ribeiro, J.M.; Francischetti, I.M.; et al. A novel family of RGD-containing disintegrins (Tablysin-15) from the salivary gland of the horsefly Tabanus yao targets αIIbβ3 or αVβ3 and inhibits platelet aggregation and angiogenesis. *Thromb. Haemost.* **2011**, *105*, 1032–1045. [CrossRef] [PubMed]

198. Karczewski, J.; Endris, R.; Connolly, T.M. Disagregin is a fibrinogen receptor antagonist lacking the Arg-Gly-Asp sequence from the tick, Ornithodoros moubata. *J. Biol. Chem.* **1994**, *269*, 6702–6708. [PubMed]

199. Tang, J.; Fang, Y.; Han, Y.; Bai, X.; Yan, X.; Zhang, Y.; Lai, R.; Zhang, Z. YY-39, a tick anti-thrombosis peptide containing RGD domain. *Peptides* **2015**, *68*, 99–104. [CrossRef] [PubMed]

200. Wang, X.; Coons, L.B.; Taylor, D.B.; Stevens, S.E., Jr.; Gartner, T.K. Variabilin, a novel RGD-containing antagonist of glycoprotein IIb-IIIa and platelet aggregation inhibitor from the hard tick Dermacentor variabilis. *J. Biol. Chem.* **1996**, *271*, 17785–17790. [CrossRef] [PubMed]

201. Mortimer, L.; Moreau, F.; Cornick, S.; Chadee, K. The NLRP3 Inflammasome Is a Pathogen Sensor for Invasive Entamoeba histolytica via Activation of α5β1 Integrin at the Macrophage-Amebae Intercellular Junction. *PLoS Pathog.* **2015**, *11*, e1004887. [CrossRef] [PubMed]

202. Muchowski, P.J.; Zhang, L.; Chang, E.R.; Soule, H.R.; Plow, E.F.; Moyle, M. Functional interaction between the integrin antagonist neutrophil inhibitory factor and the I domain of CD11b/CD18. *J. Biol. Chem.* **1994**, *269*, 26419–26423. [CrossRef] [PubMed]

203. Hou, Y.; Mortimer, L.; Chadee, K. Entamoeba histolytica cysteine proteinase 5 binds integrin on colonic cells and stimulates NFkappaB-mediated pro-inflammatory responses. *J. Biol. Chem.* **2010**, *285*, 35497–35504. [CrossRef] [PubMed]

204. Cornick, S.; Moreau, F.; Chadee, K. Entamoeba histolytica Cysteine Proteinase 5 Evokes Mucin Exocytosis from Colonic Goblet Cells via αvβ3 Integrin. *PLoS Pathog.* **2016**, *12*, e1005579. [CrossRef] [PubMed]

205. Kucik, C.J.; Martin, G.L.; Sortor, B.V. Common intestinal parasites. *Am. Fam. Physician* **2004**, *69*, 1161–1168. [PubMed]

206. Gunther, J.; Shafir, S.; Bristow, B.; Sorvillo, F. Short report: Amebiasis-related mortality among United States residents, 1990–2007. *Am. J. Trop. Med. Hyg.* **2011**, *85*, 1038–1040. [CrossRef] [PubMed]

207. Sim, S.; Park, S.J.; Yong, T.S.; Im, K.I.; Shin, M.H. Involvement of β 2-integrin in ROS-mediated neutrophil apoptosis induced by Entamoeba histolytica. *Microbes Infect.* **2007**, *9*, 1368–1375. [CrossRef] [PubMed]

208. Pillai, D.R.; Kain, K.C. Entamoeba histolytica: Identification of a distinct β2 integrin-like molecule with a potential role in cellular adherence. *Exp. Parasitol.* **2005**, *109*, 135–142. [CrossRef] [PubMed]

209. Sengupta, K.; Hernandez-Ramirez, V.I.; Rosales-Encina, J.L.; Mondragon, R.; Garibay-Cerdenares, O.L.; Flores-Robles, D.; Javier-Reyna, R.; Pertuz, S.; Talamas-Rohana, P. Physical, structural, and functional properties of the β1 integrin-like fibronectin receptor (β1EhFNR) in Entamoeba histolytica. *Infect. Genet. Evol. J. Mol. Epidemiol. Evol. Genet. Infect. Dis.* **2009**, *9*, 962–970. [CrossRef] [PubMed]

210. Ma, D.; Francischetti, I.M.; Ribeiro, J.M.; Andersen, J.F. The structure of hookworm platelet inhibitor (HPI), a CAP superfamily member from Ancylostoma caninum. *Acta Crystallogr. Sect. F Struct. Biol. Commun.* **2015**, *71*, 643–649. [CrossRef] [PubMed]

211. Moyle, M.; Foster, D.L.; McGrath, D.E.; Brown, S.M.; Laroche, Y.; de Meutter, J.; Stanssens, P.; Bogowitz, C.A.; Fried, V.A.; Ely, J.A.; et al. A hookworm glycoprotein that inhibits neutrophil function is a ligand of the integrin CD11b/CD18. *J. Biol. Chem.* **1994**, *269*, 10008–10015. [PubMed]

212. Assumpcao, T.C.F.; Ribeiro, J.M.C.; Francischetti, I.M.B. Disintegrins from hematophagous sources. *Toxins* **2012**, *4*, 296–322. [CrossRef] [PubMed]

213. Fremont, L. Biological effects of resveratrol. *Life Sci.* **2000**, *66*, 663–673. [CrossRef]

214. Tseng, S.H.; Lin, S.M.; Chen, J.C.; Su, Y.H.; Huang, H.Y.; Chen, C.K.; Lin, P.Y.; Chen, Y. Resveratrol suppresses the angiogenesis and tumor growth of gliomas in rats. *Clin. Cancer Res.* **2004**, *10*, 2190–2202. [CrossRef] [PubMed]

215. Varoni, E.M.; Lo Faro, A.F.; Sharifi-Rad, J.; Iriti, M. Anticancer Molecular Mechanisms of Resveratrol. *Front. Nutr.* **2016**, *3*, 82–83. [CrossRef] [PubMed]

216. Chin, Y.T.; Hsieh, M.T.; Yang, S.H.; Tsai, P.W.; Wang, S.H.; Wang, C.C.; Lee, Y.S.; Cheng, G.Y.; HuangFu, W.C.; London, D.; et al. Anti-proliferative and gene expression actions of resveratrol in breast cancer cells in vitro. *Oncotarget* **2014**, *5*, 12891–12907. [CrossRef] [PubMed]

217. Buhrmann, C.; Shayan, P.; Goel, A.; Shakibaei, M. Resveratrol Regulates Colorectal Cancer Cell Invasion by Modulation of Focal Adhesion Molecules. *Nutrients* **2017**, *9*, 1073. [CrossRef] [PubMed]

218. Hoffman, S.J.; Vasko-Moser, J.; Miller, W.H.; Lark, M.W.; Gowen, M.; Stroup, G. Rapid inhibition of thyroxine-induced bone resorption in the rat by an orally active vitronectin receptor antagonist. *J. Pharmacol. Exp. Therap.* **2002**, *302*, 205–211. [CrossRef]

219. Lin, H.Y.; Sun, M.; Tang, H.Y.; Lin, C.; Luidens, M.K.; Mousa, S.A.; Incerpi, S.; Drusano, G.L.; Davis, F.B.; Davis, P.J. L-Thyroxine vs. 3,5,3′-triiodo-L-thyronine and cell proliferation: Activation of mitogen-activated protein kinase and phosphatidylinositol 3-kinase. *Am. J. Physiol. Cell Physiol.* **2009**, *296*, C980–C991. [CrossRef] [PubMed]

220. Santulli, G. Angiopoietin-like proteins: A comprehensive look. *Front. Endocrinol.* **2014**, *5*, 1–6. [CrossRef] [PubMed]

221. Zhang, Y.; Hu, X.; Tian, R.; Wei, W.; Hu, W.; Chen, X.; Han, W.; Chen, H.; Gong, Y. Angiopoietin-related growth factor (AGF) supports adhesion, spreading, and migration of keratinocytes, fibroblasts, and endothelial cells through interaction with RGD-binding integrins. *Biochem. Biophys. Res. Commun.* **2006**, *347*, 100–108. [CrossRef] [PubMed]

222. Liu, J.; Gao, X.; Zhai, Y.; Shen, Q.; Sun, L.; Feng, C.; Rao, J.; Liu, H.; Zha, X.; Guo, M.; et al. A novel role of angiopoietin-like-3 associated with podocyte injury. *Pediatr. Res.* **2015**, *77*, 732–739. [CrossRef] [PubMed]

223. Yang, Z.; Mu, Z.; Dabovic, B.; Jurukovski, V.; Yu, D.; Sung, J.; Xiong, X.; Munger, J.S. Absence of integrin-mediated TGFβ1 activation in vivo recapitulates the phenotype of TGFβ1-null mice. *J. Cell Biol.* **2007**, *176*, 787–793. [CrossRef] [PubMed]

224. Wynn, T.A. Common and unique mechanisms regulate fibrosis in various fibroproliferative diseases. *J. Clin. Investig.* **2007**, *117*, 524–529. [CrossRef] [PubMed]

225. Katsumoto, T.R.; Violette, S.M.; Sheppard, D. Blocking TGFβ via Inhibition of the αvβ6 Integrin: A Possible Therapy for Systemic Sclerosis Interstitial Lung Disease. *Int. J. Rheumatol.* **2011**, *2011*, 208219. [CrossRef] [PubMed]

226. Munger, J.S.; Huang, X.; Kawakatsu, H.; Griffiths, M.J.; Dalton, S.L.; Wu, J.; Pittet, J.F.; Kaminski, N.; Garat, C.; Matthay, M.A.; et al. The integrin αvβ6 binds and activates latent TGF β 1: A mechanism for regulating pulmonary inflammation and fibrosis. *Cell* **1999**, *96*, 319–328. [CrossRef]

227. Puthawala, K.; Hadjiangelis, N.; Jacoby, S.C.; Bayongan, E.; Zhao, Z.; Yang, Z.; Devitt, M.L.; Horan, G.S.; Weinreb, P.H.; Lukashev, M.E.; et al. Inhibition of integrin αvβ6, an activator of latent transforming growth factor-β, prevents radiation-induced lung fibrosis. *Am. J. Respir. Crit. Care Med.* **2008**, *177*, 82–90. [CrossRef] [PubMed]

228. Horan, G.S.; Wood, S.; Ona, V.; Li, D.J.; Lukashev, M.E.; Weinreb, P.H.; Simon, K.J.; Hahm, K.; Allaire, N.E.; Rinaldi, N.J.; et al. Partial inhibition of integrin αvβ6 prevents pulmonary fibrosis without exacerbating inflammation. *Am. J. Respir. Crit. Care Med.* **2008**, *177*, 56–65. [CrossRef] [PubMed]

229. Sullivan, L.A.; Brekken, R.A. The VEGF family in cancer and antibody-based strategies for their inhibition. *mAbs* **2010**, *2*, 165–175. [CrossRef] [PubMed]

230. Jayson, G.C.; Hicklin, D.J.; Ellis, L.M. Antiangiogenic therapy—evolving view based on clinical trial results. *Nat. Rev. Clin. Oncol.* **2012**, *9*, 297–303. [CrossRef] [PubMed]

231. Vlahakis, N.E.; Young, B.A.; Atakilit, A.; Sheppard, D. The lymphangiogenic vascular endothelial growth factors VEGF-C and -D are ligands for the integrin α9β1. *J. Biol. Chem.* **2005**, *280*, 4544–4552. [CrossRef] [PubMed]

232. Temming, K.; Schiffelers, R.M.; Molema, G.; Kok, R.J. RGD-based strategies for selective delivery of therapeutics and imaging agents to the tumour vasculature. *Drug Resist. Updates Rev. Comment. Antimicrob. Anticancer Chemother.* **2005**, *8*, 381–402. [CrossRef] [PubMed]

233. Marelli, U.K.; Rechenmacher, F.; Sobahi, T.R.; Mas-Moruno, C.; Kessler, H. Tumor Targeting via Integrin Ligands. *Front. Oncol.* **2013**, *3*, 222. [CrossRef] [PubMed]

234. Garanger, E.; Boturyn, D.; Dumy, P. Tumor targeting with RGD peptide ligands-design of new molecular conjugates for imaging and therapy of cancers. *Anti-Cancer Agents Med. Chem.* **2007**, *7*, 552–558. [CrossRef]

235. Wen, A.M.; Steinmetz, N.F. Design of virus-based nanomaterials for medicine, biotechnology, and energy. *Chem. Soc. Rev.* **2016**, *45*, 4074–4126. [CrossRef] [PubMed]

236. Hovlid, M.L.; Steinmetz, N.F.; Laufer, B.; Lau, J.L.; Kuzelka, J.; Wang, Q.; Hyypia, T.; Nemerow, G.R.; Kessler, H.; Manchester, M.; et al. Guiding plant virus particles to integrin-displaying cells. *Nanoscale* **2012**, *4*, 3698–3705. [CrossRef] [PubMed]

237. Anderson, E.H.; Ruegsegger, M.A.; Murugesan, G.; Kottke-Marchant, K.; Marchant, R.E. Extracellular matrix-like surfactant polymers containing arginine-glycine-aspartic acid (RGD) peptides. *Macromol. Biosci.* **2004**, *4*, 766–775. [CrossRef] [PubMed]

238. Guo, C.X.; Ng, S.R.; Khoo, S.Y.; Zheng, X.; Chen, P.; Li, C.M. RGD-peptide functionalized graphene biomimetic live-cell sensor for real-time detection of nitric oxide molecules. *ACS Nano* **2012**, *6*, 6944–6951. [CrossRef] [PubMed]

239. Stephanopoulos, N.; Freeman, R.; North, H.A.; Sur, S.; Jeong, S.J.; Tantakitti, F.; Kessler, J.A.; Stupp, S.I. Bioactive DNA-peptide nanotubes enhance the differentiation of neural stem cells into neurons. *Nano Lett.* **2015**, *15*, 603–609. [CrossRef] [PubMed]

MDPI
St. Alban-Anlage 66
4052 Basel
Switzerland
Tel. +41 61 683 77 34
Fax +41 61 302 89 18
www.mdpi.com

International Journal of Molecular Sciences Editorial Office
E-mail: ijms@mdpi.com
www.mdpi.com/journal/ijms